Ministero BB.CC.AA.
Soprintendenza per i Beni Ambientali e
Architettonici di Napoli e Provincia
Palazzo Reale - Napoli

Centro Materiali Compositi
AMME-ASMECCANICA

Università degli Studi di Napoli
"Federico II"

The
Institute of
Materials

EUROPEAN ASSOCIATION
FOR COMPOSITE MATERIALS

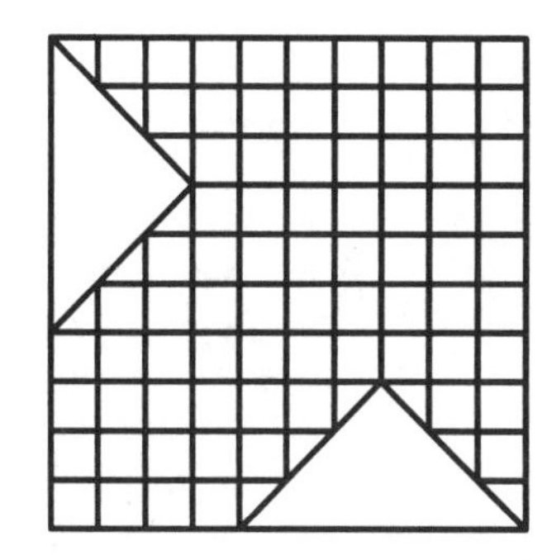

ECCM - 8

EUROPEAN CONFERENCE ON COMPOSITE MATERIALS

SCIENCE, TECHNOLOGIES and APPLICATIONS

3 - 6 JUNE 1998 NAPLES - ITALY

VOLUME 2

EDITOR: I. CRIVELLI VISCONTI

WOODHEAD PUBLISHING LIMITED

Published by Woodhead Publishing Limited,
Abington Hall, Abington, Cambridge CB1 6AH, England

First published 1998, Woodhead Publishing Limited

British Library Cataloguing in Publication Data
A catalogue record for this book is available from the British Library.

ISBN 1 85573 408 7 (Vol 2)
ISBN 1 85573 377 3 (Four volume set)

ECCM-8

EUROPEAN CONFERENCE ON COMPOSITE MATERIALS
SCIENCE, TECHNOLOGIES AND APPLICATIONS

EXECUTIVE COMMITTEE

ECCM-8
EUROPEAN CONFERENCE ON COMPOSITE MATERIALS
SCIENCE, TECHNOLOGIES AND APPLICATIONS

CONTENTS

SYMPOSIUM 3

SYMPOSIUM 4

SYMPOSIUM 3

Bridge Strengthening by Carbon Fibres

J. Luyckx* - R. Lacroix - J.P Fuzier****

* Soficar, 3 avenue du Chemin de Presles, 94410 Saint Maurice, France
** Freyssinet, 10 rue Paul Dautier, 78140 Vélizy, France

Abstract

An ever expanding world demand for highways means a global stock of bridges not only increasing in number, but also in age. At the same time, traffic and bridge loading are also increasing, leading to associated expansion of bridge strengthening programs. These maintenance and rehabilitation programs reflect not only the desire to build better quality structures but also an economical need.

New materials and technology are presently being developed to strengthen concrete structures. This paper deals with the recent developments achieved with carbon fibre fabrics. These fabrics are glued to the structure. Several tests are presented together with operating procedures.

1. Introduction

In the civil engineering field, the mass construction of the 60s and 70s was followed by maintenance works which often highlighted the need to strengthen the structures.

One of the most widely used has been the L'Hermitte process, which consists of gluing external steel plates by epoxyde resin. This process, although very efficient showed some limits :

- the steel is subject to corrosion, it requires a costly protection and maintenance,
- due to the flexure of the stiff steel plates, the mechanical properties can not be fully taken into account,
- the gluing requires a specific surface treatment, and the application of a constant pressure until the complete bonding of the resin,
- the steel plates are heavy, very stiff and cumbersome, they cannot be applied on curved surfaces,
- althrough some improvements were brought by opening hloes through the plates, it is very difficult to avoid the presence of air bubbles between the plate and the support.

In order to go beyond the L'Hermitte process, a French working group was been active, for more than three years now, to develop a new method based on carbon fibre properties.

2. Carbon fibre properties

Compared to steel, carbon fibres present many advantages :

- longitudinal high strength and high tensile modulus,
- low density,
- high fatigue strength,
- high wear resistance,
- vibration absorption,
- high dimensional stability,
- high thermal stability,
- high chemical and corrosion resistance against acids, alkalis, salts, organic solvents.

In order to benefit from the carbon properties, a specific epoxy resin has been developped, allowing the application by temperatures in a broad range (+7° to +35° C or more) dissegarding the condition of the support : dry or wet.

Properties of Composites based on Carbon Fibres as compared with Steel Reinforcement

Reinforcement	Units	Steel (E 24-3)	TFC bidirectional*	Concrete
Density	kg / m³	7,8	1,8	2,3
Thickness	mm	3	0,22	--
Weight	g / m²	23 400	400	
Longitudinal Tensile Strength	MPa	240	2 700	3-5
Longitudinal tensile modulus	GPa	200	160	--
Longitudinal failure load	kN / cm	7,2	5,9	4,5 (safety factor : 3)
Transverse failure load	kN / cm	7,2	2,5	

*All properties reported for dry fibres.

3. Research works

The aim of the research was to develop the strengthening of structures made of three basic materials : concrete, steel and timber.

Tests have been carried out in the LCPC (French official laboratory of Public Works Department) as well as in the laboratories of the industrial group participants. After a preliminary testing phase of two years, the most convenient components were selected : epoxyde resin and uni or bi-directional carbon fibre dry fabrics, named **TFC**[(R)].

Test results

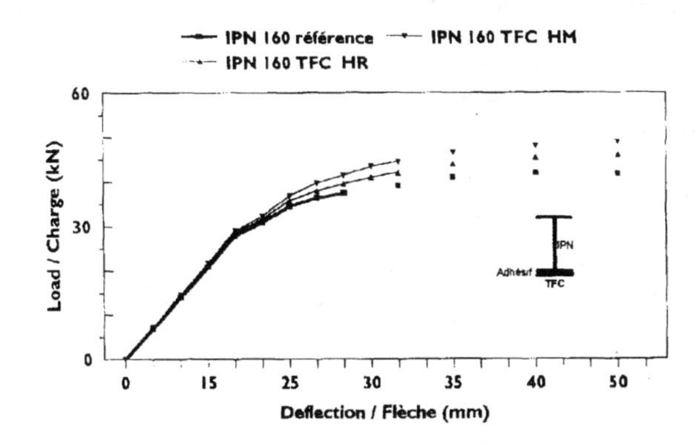

LCPC tested the strengthening of steel profiles in four points flexure configuration ; after the TFC[(R)] reinforcement, the mechanical properties have been improved by 10 to 15%.

The increasing load capacity of the girders was limited, due to the girder torsion.

These first laboratory test results were encouraging for this kind of reinforcement.

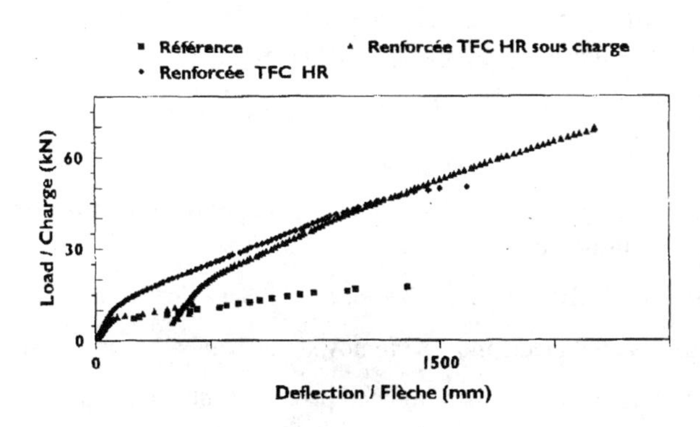

Similary, LCPC obtained very good test results with the concrete beams reinforced by TFC[(R)] and tested in four points flexure configuration. The load resistance capacity has been multiplied by 5 or more.

For the reinforced concrete beam the failures always occured inside the concrete (concrete delamination or shear cracking), and not by debonding or by rupture of the carbon composite. All tests gave similar strength results.

The fabric composite (TFC(R)) reinforcement is more flexible than the steel plates and also the pultruded laminates which have been used for structural strengthening so far.

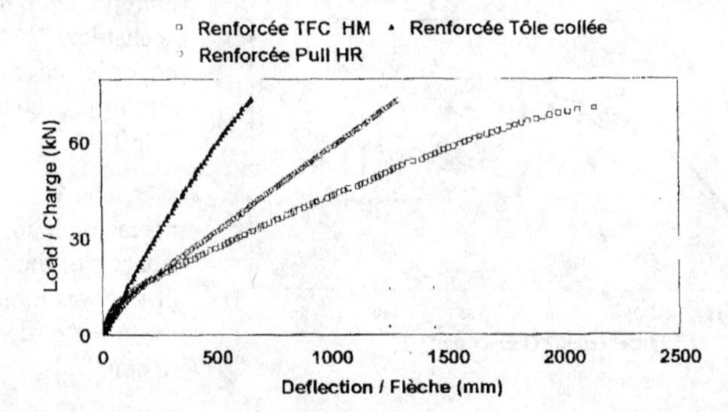

The standart thickness of the TFC(R) is 0.4 mm only, to be compared with the standard values of 3 to 5 mm for steel plates and up to 1.0 mm for carbon laminates.

As compared with steel plates or pultruded laminates, the TFC(R) shows definitive advantages :

- easier to transport and handling,
- easier cutting and shaping, including on non-developable surfaces,
- easier application, with no need of applying pressure during hours or days,
- possibilty of applying a multi-layer reinforcement,
- easier control and achievement of the right resin thickness.

4. Procedure

As for any other bonding process, the surface condition of the support material and adhesiv resin are very important factors.

Steel : the oxidized particules must be removed mechanically.

Concrete : the carbonated layers must be removed mechanically (sandblasting).

Wood : the exterior paints or contaminated parts must be removed.

The TFC(R) procedure application can be summarized as follows :

- cleaning surfaces, degreasing,
- bonding the surface - first resin layer -,
- TFC(R) application
- bonding the fabrics - second resin layer -.

Although the TFC(R) reinforcement procedure is quite independant of the weather conditions, it should be applied at a minimum of + 7°C.

The curing time is a few hours, depending on the external temperature.

5. Industrial application

The first application took place in 1996, October. With the owner's agreement, the group undertook to strengthen the first French bridge by using the TFC(R) process. This is an overpass located on the motorway Paris-Chartres (A10).

After steel passivation and filling holes by additional concrete and sandblasting, the TFC(R) reinforcement has been layed on end of October and early November 1996. Weather conditions were quite unfavourable : temperatures reached from +3°C to +12°C, under constant rain, with a relative humidity near 100%. The wind blowed up to 80 km/h.

Despite these bad weather conditions and with a limited number of handworkers, the work went quite well, without interruption of the motorway traffic.

Futher tests were satisfactorily carried out. All gauges, put on the beams and on the TFC reinforcements, showed the right process has been applied.

In October, 1997, this first work was selected innovant and received the IVOR Label (Innovation Validée sur Ouvrage de Référence).

In 1997, first semester, up to 20 applications appeared, to strenghten bridges (Motorway A6) or walls and floors slabs in buildings. Several hundreds square meters of TFC(R) have been applied successfully by Freyssinet. Due to this success, TFC(R) became a Freyssinet trade mark.

In 1997, during the second half, the contractor of the first bridge reinforced (A10), committed Freyssinet for a new 3 bridge reinforcement, using the TFC(R) process.

We are now expecting for some reinforcement works where several thousands square meters of TFC(R) will be applied.

Further research works

The group is now working on fatigue tests and durability, included the fire resistance of the TFC(R) reinforcement.

With EDF and CEA collaboration, the group is also working on structural seismic reinforcement.

6. Conclusion

The experimental work and the first in situ applications have fully confirmed the advantages of the TFC$^{(R)}$ process for repair and strengthening of all kinds of structures.

Compared to the steel or other carbon composite reinforcements, the TFC$^{(R)}$ has both technical and economic advantages.

It is undoubted that this technology is very promising and the application fields are larger than the current field of steel reinforcement.

This development also is a good example of a fruitful cooperation between suppliers, civil engineering contractors and official laboratories. "L'esprit d'équipe" was a necessary condition of success.

COMPOSITE MATERIALS IN CONSTRUC THE FUTURE

Ph. Martineau*, M. Olivares Santiago**

* Institut des Matériaux Composites – Site Montesquieu – 33650 Martill
** Ets Arquitectura – Avda Reina Mercedes – 41012 Seville – Espagne

Abstract

Today, Composite materials have numerous applications in every indus can't imagine living without them in some fields, such as Aeronautic Nautism, for example.
Their development in construction essentially in Europe is more rec reasons. In this paper, we'll present the main advantages offered by t for applications in construction but also the principal obst development.
Some applications will be presented and through the experience des Matériaux Composites (IMC), we'll try to give some orientati and develop them in this sector.

1. Introduction

The choice of materials used in Construction responds to several economical criteria. Composite materials – based on a resin reinforced among the solutions offered for constructors.
Several studies are presented in literature in which the abilities of tho demonstrated [1, 2, 3]. We also find experiences of applications, mor [4, 5].

2. Advantages

The numerous advantages of composites have been well known for some time.
In this industry, where transport and handling of large pieces are the daily program, a reduction of the weight is considered of great interest.
The comparison of two pieces with the same volume, one in composite materials and the other in traditional material (stone, concrete, ...). give a positive note for composite solution (Table 1).

	Composites	**PVC**	**Wood**	**Reinforced concrete**	**Aluminium**	**Steel**
Density	1.5–2	1.35–1.45	0.6-1.2	2.2-2.5	2.7	7.8

Table 1 : Densities of principal materials used in construction

Yet, while weighing less than concrete or steel, they are among the toughest materials available on the market. When we compare the mechanical specific characteristics of the principal known materials we obtain very high values with composites, due to the use of fibers with high mechanical properties (table 2).
One other advantage for engineer and design offices is in the possibility to produce materials adapted to particular specifications, by using ratio of adapted fibers with specific orientations. The thickness of composite panels can also be varied, there by increasing resistance and rigidity at certain points.

Materials Specific properties	**Polyester 40% short glass fibers**	**Polyester 80% Unid. Glass fibers**	**PVC**	**Wood**	**Reinforced concrete**	**Steel**
Specific strength	93	700	41	80-92	1	38-77
Specific modulus	6500	24000	2100	10000 I 12000	17000	27000

Table 2 : specific mechanical properties of principal materials used in construction

One other advantage presented by composite materials, is their excellent behaviour facing corrosion. Those materials present resistance to corrosion attack from chemical environments or bad weather. Numerous applications exist in chemistry plants.
Composite materials present many other advantages which explain the interest of architects for their use. In particular, they offer aesthetic and decorative opportunities. Those materials allow a wide freedom in design sometimes impossible to realise with traditional materials (complex shapes, reproductibility, surface aspect (smooth or rough), transparency or opaque.

We also quote their ideal thermal characteristics (good insulation, low heat conductivity and good electrical insulating properties (apart with carbon fibers). Composite materials can be also transparent to electromagnetic waves and used for realisation of special buildings where interferences are prohibited.

3. The constraints to development

Obstacles can be of technical or economical aspects, but the principal ones are essentially in relation with the youth of composites. They are little known among professional construction people.
The information and training are insufficient and incomplete, based in most cases on the experiences, in the field of aeronautic or transportation industries. We have to develop special informations adapted to this profession.
Other obstacles are in the image, construction people have of composite materials. Glass/polyester often have a bad reputation, because of colour changes, moistures or fire behaviour. If it was the case in the past, today we have found solutions and, for example, we realise parts with good fire behaviour (M1, F0).
In the economic study, most of the time, it is realised a comparison of the cost, but on the basis of the price per kilogram. With composite materials, or, in general with high performance light materials, we have to change those criteria of comparison. For example, to reach the same properties we have to realise different thicknesses of panels if we choose composites or concrete. In fact, there is an ignorance of real comparing costs between composite and traditional materials and an absence of method for calculating the global cost with the integration of reduction in transport and handling, the absence of maintenance, possibility of integrated functions (e.g. sandwich panels) ...
In the other problems encountered, we quote the fact that it doesn't exist material (parts) in stocks. When we explain to the architects that we realise the piece and the material simultaneously, they don't follow us.
To this ignorance, we can add the lack in reglementation, technical advice and warranty

4. Applications

Most applications are in Japan, USA, Canada or in Arabian countries [4,5]. In Europe we find some realisation but not with the same success and development.
We find building totally raised with only composite materials. In Europe, in most cases, they stay at prototype or small series state. But in general composite materials are used to fabricate parts of construction. We find them in structures (framework), pultruded beams, on the roof synthetic tiles, slates or translucent panels.
Cladding panels, in general sandwich panels, are used for walls [6]. Many examples of use in interior or exterior decorations exist.
Window frames or doors are increasing. The bathroom, the kitchen and the sanitary apparatus are largely penetrated by composites.

5. Conclusions

Architects, designers, manufacturers and transformers are, day by day, becoming more interested in the architectural applications of composite.
Since its creation in 1984, IMC has participated in this development with informations special training or feasibility studies in collaboration with building professionals. In 1995-1997 IMC realised, in the frame of an European project KOTEMM (Adapt) important actions of informations, demonstration and training to develop composites in this field.
To exchange experiences, IMC takes part in an world wide network of specialists and organises congresses like ARQUIMACOM'98, in Bordeaux in October 98.
In conclusion, we think that tomorrow, composites will take a large place among construction materials.
We have to consider them not only as a new generation of materials but as materials of a new generation.

References

[1] Ph. Martineau, Les Matériaux Composites dans le Bâtiment
Série Matériaux Composites et Industrie – 1987
Les éditions techniques de l'Institut des Matériaux Composites

[2] Ph. Martineau, Les matériaux composites dans le Bâtiment
Journées Européennes "Composites et Bâtiment"
IMC ed. Octobre 1994

[3] R. Blancon, Les fils de verre de renforcement dans le Bâtiment
Composites. May/June 1992. N°3, pp 272-277

[4] N. Sprecher, Composites and Building : new applications
Composites. May/June 1992. N°3, pp 250-263

[5] G. Gerin – Lajoie, les composites – la solution a plusieurs problèmes que posent les matériaux vulnérables a plusieurs formes de pollution
Composites. May/June 1992. N°3, pp 277-283

[6] M. Tasnon, M. Missihoun; G. Gerin – Lajoie, restoration of bulding facades with composite materials
Advanced composites materials in bridges and structures
Montreal – Quebec – 1996, pp. 589 - 593

COMPOSITES IN CONSTRUCTION

Dott. Ing. E. Piccioli
Eptec Pty. Ltd. 245 Victoria Road Gladesville NSW 2111 Australia

INTRODUCTION

Composites are notorious for their light weight, corrosion resistance, high strength and great design flexibility.

This Paper investigates the use of Composites in large scale civil projects where the well known advantages of composites have to be combined with a competitive cost basis to be used extensively in the Construction Industry.

The Case Study deals with the logistic problems of design, construction and installation of some 55,000 sq.m. of Composite large diameter Tank Covers, and 12,000 lineal metres of complicated foul air duct extraction systems for a large Sewage Treatment Plant in a major capital city in Asia.

The advantages of production "insitu" for large structures, which are typical of Reinforced Plastic Composites, are utilised to their maximum benefit to ensure that composites are the only choice in many difficult civil construction projects such as this Case Study.

BACKGROUND

The major capital cities around the world have sprawled in area over the years, and more particularly in the last 30 years. With industrialisation, the move from the country to the city has meant cities have grown to accommodate more inhabitants and very often the infrastructure centres such as Waste Water Treatment Plants and Power Stations, which were once outside the city, have now become "part of" the city.

One classic example of such development in Asia has been Singapore, which is a tiny country of 600 sq. kms. with some 4 million inhabitants. The emergence of Singapore as a major banking and financial centre in South East Asia has brought wealth and spiralling land prices.

The City is served by 5 Sewage Treatment Plants which were once well distanced from any dwellings and are today very much part of the City. By law, and for health reasons, no dwellings or offices were allowed to be built within a radius of less than 1,000m. from these Sewage Treatment Plants. With the average cost of land in the City precinct being US$4,000-US$6,000/sq.m., it did not take long for the Singaporean Government to work out that the "wasted, uninhabitable land" around the Sewage Treatment Plants had a value of approximately US$6 billion. The Government looked at two options:

(i) Relocating the old plants away from the City and build new plants in remote areas. This meant massive infrastructure costs to extend the collection system besides building new plants.

(ii) Cover the existing plants with gas tight covers and collect the malodorous gases and treat them. At a total cost of approximately US$600 million, the Government of Singapore lost no time in opting for the second alternative, which was the quickest and gave a 10:1 return on investment.

The engineering task was now to set up a viable, long-term, maintenance free "cover and gas collection system" and suitable process treatment of the collected gases.

Given the very corrosive environment with gases containing high proportions of hydrogen sulphide, methane and other toxic gases which, with a particular combination of oxygen can be explosive, the design of this collection system posed many challenges. Various systems were trialled using FRP, aluminium, mild steel (coated) and stainless steel.

In the end, THE MOST TECHNICALLY EFFICIENT AND ECONOMICALLY COMPETITIVE SYSTEM proved to be an ALL FRP SOLUTION of large diameter covers, duct and odour control scrubbers.

THE PROJECT

This Paper relates to just one of these Sewage Treatment Plants which required some 55,000 sq. m. of covered tanks, 12,000 lineal metres of ducting, and a massive odour treatment system for handling all the waste gas.

The problems facing the industry were many. Firstly, the variety of FRP equipment needed and, secondly, the size of this equipment and delivery requirements. The equipment consisted mainly of:

FRP Tank Covers — FRP Wet Chemical Scrubbers
FRP Duct Work and accessories — FRP Activated Carbon Absorption Units
FRP Exhaust Fans — FRP Chemical Storage Tanks
FRP Components such as grates, dampers, and miscellaneous equipment.

The specification was rigorous and detailed, and requested only fabricators with substantial experience in this type of work. The work was quoted as supplied from all over the world, and components came from Europe, America, China, Australia and Singapore.

Given the large dimensions of the FRP products involved, we decided that to be successful, we had to take on the complete responsibility of design, supply, and installation, and that all the work needed to be manufactured close to the Plant in Singapore.

The logistics of setting up a new industry in a Foreign Country, transferring materials and equipment, training machine operators and laminators, monitoring quality, maintaining safety standards and delivery of the finished product on time and within budget to the complete satisfaction of the Client, proved to be our greatest challenge for some years, and it is this experience which this Paper presents to you.

DESIGN & PRODUCTION

The scope of supply included:

i) **Covers - Rectangular and Circular**

Covers to rectangular aeration tanks and channels of varying spans 20m. wide, 9m. wide and 2-3m. wide. The length of these tanks was some 50m. long and the channels often 100m. long.

Furthermore, there were circular tanks such as sedimentation tanks and saturation tanks which needed covers of 35m. diameter.

The design criteria was that all covers had to be sealed, self-supporting and required to withstand severe corrosion conditions under the following operational loads:

Aeration Tank Central Walkway Cover-live load	4.0kN/sq. m.
Flat, barrel arch and corrugated covers	
Point load	1.10kN
Live load	1.0kN/sq. m.
Wind load	100km./h.
Max. vacuum inside cover (for all FRP covers)	0.5kN/sq. m. above atmospheric pressure
Max. pressure inside cover (for all FRP covers)	0.5kN/sq. m. below atmospheric pressure
Maximum cover deflection	Span/230
Safety factor, compensation of ageing, environmental effects and non-homogeneity	5.0

THIS CREATED THREE DESIGN CONDITIONS:

Case One

- 100 km./h. wind creates an estimated wind suction load of 2.1kN/sq. m.
- This is in combination with a maximum pressure inside the cover of 0.5kN/sq. m. overpressure
- Acting in opposite direction is the cover's self weight.

Case Two

- Point load of 1.10kN at mid span plus the cover's own weight.

Case Three

- Live load of 1.0kN/sq. m. at mid span.
- In combination of 0.5kN/sq. m. suction due to vacuum inside cover.
- In combination with cover's own weight.

ii) Ducting

Having sealed the tanks, the foul air needed to be collected and converged to the central odour control treatment building. This was done by the design and supply of 12,000 lineal metres of ducting and some 4,000 fittings, including elbows, tees, reducers and flanges. The ducting system also required FRP flow dampers, expansion joints and stainless steel supports and brackets.

Design conditions for the ducting were as follows:

Water column (negative pressure)	100mm.
Water column (positive internal pressure)	200mm.

MANUFACTURING

The manufacture was therefore undertaken on the island of Batam, Indonesia. An average workforce of 300 Indonesians had to be trained in the safety aspects of FRP, together with the basic technical understanding of the material. A special training programme was set up which, with the assistance of six ex-patriates specialised in FRP manufacture, gave the basic training to the 300 workers over several weeks. In the meantime, production was being ramped up to achieve total laminate lay-ups of 5,000 kg. per day.

From the initial supply of drawings by the Client, through the design and preparation of detailed drawings, the set up of the production line, and completion of manufacture, all the work was completed in 20 months.

TRANSPORT, STORAGE AND HANDLING

Using the light weight of FRP, both the tank covers and the ducts were manufactured to the largest possible managable dimension to minimise the work on site. With this optimisation of size, the handling in the factory, during storage, during transport and on site, proved to be a major cost factor.

As an example, the selection of factory space and external storage area required the manufacture and storage of covers up to 20m. long x 2.5m. wide and ducting up to 2.3m. diameter and 12 metres long. Furthermore, we needed space to manufacture and store fittings such as long radius elbows and tees which were 2.3m. in diameter x 4m.

wide. This implied a minimum factory floor area of 2,500 sq. m. and a minimum external storage area of 10,000 sq.m.

To find a suitably sized factory both for manufacture and storage, it was necessary to move offshore from Singapore onto the nearby island of Batam, Indonesia which is 50 kms. away by barge. Each barge load consisted of approximately 1500 cu. m. of FRP by volume, or a maximum 84,000 kg. by weight. At the receiving site in Singapore, each barge required an average of 100-120 truck loads for delivery to site. A lot of "air" was transported.

The total contract consisted of over 400,000 cu. m., and this is where the logistics of transport and handling became of major importance. This is also the main reason why production for this type of work needs to be relatively close to the installation site, and cannot be contemplated in the "home base" factory because of the large components to be moved.

QUALITY

All work was undertaken to an ISO9002 Quality System which was set up in Batam for this contract. The quality of the finished product can be said to have been excellent, even though all ex-patriates and local supervisors were often exerted to their limit to ensure that everything to be manufactured was right the first time.

SAFETY

Safety was also a major consideration with the newly trained workers. Fundamentally, we had to set up a course for safety training of staff, supervisors and workers, to ensure that, albeit in a very fact manufacturing schedule, safety was never jeopardised.

An unblemished safety record with no LTI (lost time incident) for 1,500,000 hours production is the best we could expect to achieve.

INSTALLATION

Once on site, these large quantities of bulky components made installation the next challenge. Cranage and access on site, as in all operating plants was limited, and a large quantity of installation work needed to be done by hand using purpose built equipment.

For the rectangular tanks this involved the use of tracks and trolleys running underneath the covers on the concrete wall to move the covers from their cranage lift on point to join 50m. down a tank at the furthermost location point away from the crane position. The covers then needed to be removed from the trolleys and lowered onto a prepared concrete or stainless steel support base where a sealing neoprene gasket had already been placed.

The covers had to be butted against each other and bolted to each other by a "joint strap" system which would take into account the high Poisson's ratio and the low Young's modulus of FRP. This was an interesting task on its own.

The support flanges at the ends of the covers were then drilled and fixed into the concrete with the use of chemical anchors into the concrete walls.

The ducts of various sizes from 100mm. diameter up to 2,300mm. diameter presented other difficulties. Much of the ductwork is in the air and needed to be "launched" along the support system from one end for several hundred metres at a time. The joining between ducts was to be done mostly in the air, 3 metres above ground, and in a tropical rain climate where heavy downfalls were most frequent. A truly difficult environment for FRP layup!

The installation crew needed to be trained and closely monitored for good performance of safety and quality, and the crew varied between 30 and 50 men with 4-6 ex patriates over a period of 18 months for the complete installation contract.

Housing, accommodation, Government Legislation for foreign workers, and language problems made the challenge all the more exciting. Nonetheless, the job was completed with excellent quality and safety records, and within the Client's programme.

The odour control building which was not in our scope also involved large quantities of FRP, requiring large diameter FRP wet chemical scrubbers, chemical storage tanks, FRP fans, FRP dampers and ducting. This too was manufactured off site and completely installed with similar ingenuity as the covers and ducts.

CONCLUSION

55,000 sq. m. of tank covers, 12,000 metres of ducts and 4,000 fittings is an impressive specification to design, manufacture, deliver and install in less than 2 years in a foreign country with inexperienced labour under difficult climatic conditions.

The challenge was to complete the project on time, maintaining quality, safety, and of course make a profit. We managed to achieve all of this, but not without a lot of effort.

Having overcome many difficulties on this project, and achieved our target of completing the project on time and within budget, is not only reassuring to us, but hopefully to all future Clients with this type of FRP work in Europe, America, Asia or Africa.

No doubt, we will see many more billions of dollars in capital spent for Water and Water treatment facilities, and we now know that properly engineered FRP has some major advantages to boast for this industry - most importantly: LOWER INITIAL COST AND EVEN LOWER THROUGH LIFE MAINTENANCE COST.

22 8 97

22 8 '97

Structural Characterization of Fiber-Reinforced Composite Short- and Medium-Span Bridge Systems

V.M. Karbhari, F. Seible, R. Burgueño, A. Davol, M. Wernli and L. Zhao

Division of Structural Engineering
University of California, San Diego
La Jolla, CA 92093-0085, USA

Abstract

The paper describes the development of a new structural system for short and medium span bridges wherein use is made of both advanced composites and conventional materials such as concrete. The concept uses prefabricated composite tubes as girders which are then filled with concrete, after which a conventional precast or cast-in-place, or advanced composite, deck system is integrated to form the bridge superstructure. The paper presents experimental results of large-scale tests aimed towards the structural characterization of the girders, anchorages, and girder-deck assemblies for both serviceability and ultimate limit states.

1. Introduction

Fiber reinforced polymer matrix composites (PMCs), originally developed for aerospace and defense applications, are extremely attractive for use in civil infrastructure in applications such as bridge systems because of their high stiffness-to-weight and strength-to-weight ratios, light weight, corrosion resistance and environmental durability, and tailorable mechanical characteristics. To date these materials have been used effectively in the repair and strengthening of existing concrete bridges and in the seismic retrofit of bridge columns, superstructure and shear walls. Attempts to design and manufacture complete advanced composite replacement members or complete bridge systems have shown that, except for niche applications, although technically feasible and structurally sound, these systems will have a difficult time competing with conventional systems on a first cost basis as long as no provisions are made in the comparison for increased durability and reduced maintenance, and with it for overall reduced life-cycle costs. Recent developments at the University of California, San Diego (UCSD) have focused on hybrid systems which combine PMCs with conventional materials such as concrete. A previously investigated application using this concept is that of carbon or carbon/glass hybrid shells for bridge columns, wherein the prefabricated filament wound tubes serve the dual functions of formwork for concrete and as the reinforcement [1]. The concrete filled composite shell system replaces conventional reinforcing steel and formwork while providing enhanced confinement to the concrete core, increased durability and greatly enhanced ease of handling and erection speeds. Initial tests on these systems demonstrated that the design objectives of ductile structural response or elastic strength can be achieved and that the system compare very favorably with the corresponding reinforced concrete columns response [1,2].

These successful investigations present the possibility of expanding the concept to the development of complete structural systems. It is envisaged that complete bridge systems can be composed of linear segments of carbon shell tubes assembled together by means of longitudinal and off-angle joints with appropriate mechanisms depending on the response requirements for the structural component. The light weight PMC shell system provides significant advantages in the construction process of structural members since no heavy lifting equipment is required on site to place the shells, and reinforcement cage construction and placement, as well as formwork removal are no longer necessary.

2. Prototype Carbon Shell Bridge System

The prototype bridge system consists of a two span beam-and-slab bridge superstructure with an intermediate pier. The design concept proposes the use of the carbon shell system technology for both the pier and superstructure systems. The beam-and-slab bridge system is composed of longitudinal carbon shell girders connected across their tops with a continuous road surface, which can be either a reinforced concrete or an advanced composite deck system. The light weight composite shell system provides significant advantages in the construction process since no heavy lifting equipment is required on site to place the shells, and reinforcement cage construction and placement, as well as formwork removal are no longer necessary.

Structural performance requirements are currently being defined for advanced composite shell bridge systems by means of Functionality Limit States and Safety Limit States and in compliance with both AASHTO (American Association of State and Highway Transportation Officials) and CALTRANS (California Department of Transportation) design codes. Characterization studies are in progress with the objective of establishing a proper set of design criteria for carbon shell beam-and-slab bridge systems.

In order to implement the carbon shell technology to complete bridge systems and develop appropriate performance-based design guidelines experimental evaluation of the critical components and connections is required. The following describes an experimental program currently in progress which uses a building block approach to characterize the behavior of key components as steps towards characterization and ultimate field demonstration of the prototype bridge system. The test units consist of columns, grouted and ungrouted carbon shell beams, PMC deck systems, carbon shell beam-and-slab assemblies, and FRP reinforcement anchorages.

3. Carbon Shell Beam Components

Building on the previous successful testing of carbon shell column systems, the experimental characterization of the concrete filled Carbon Shell System (CSS) for girders on short and medium span bridges is being pursued by a series of four-point bending tests as shown in Fig. 1. The dimensions for the test units were determined from a design study performed on the prototype bridge. These bending tests represent the portion of the girder with positive bending moment under maximum live load conditions, that is the region between the abutment and the point of inflection of the bending moment diagram. The simple support of the test units is achieved by means of a low-friction pin connection detail through cast-in-place concrete end blocks supported by steel fixtures. The carbon shell geometry is that of a cylindrical tube with an inside diameter of 343 mm and a wall thickness of 10 mm. The carbon/epoxy laminated shell has a lay-up architecture with approximately 80% longitudinal ($\pm 10^{\circ}$ helical) and 20% transverse (90°) fiber reinforcement with an average fiber volume ratio of 55%. The carbon epoxy shell has equivalent longitudinal and transverse moduli of 97.8 GPa and 25.7 GPa respectively, and a longitudinal-to-transverse Poisson's ratio of 0.18.

The first test consisted of the non-destructive testing of a hollow shell (see Figure 1a) with the purpose of characterizing stiffness. The second bending test consisted of a carbon shell filled with a 20.7 MPa lightweight concrete pump mix, schematically shown in Figure 1b. This test served to verify the analytical and design models developed to predict the stress and strain state in the shell due to bending loads including the expansion of the enclosed concrete. Failure occurred on the compression side of the shell, in the constant moment area, at a longitudinal compression strain of approximately 0.55%. The mechanical response of the girders is compared to analytical models that were developed to predict the confined system behavior. Figure 2a shows the load-displacement history at mid-span, while Figures 2b through 2d compare the longitudinal, hoop and shear strains respectively.

4. PMC Deck Systems

The use of light-weight, corrosion resistant materials such as PMCs for decks has considerable attraction and advantages over conventional concrete bridge decks. Potentially PMC decks could be used as replacement decks in damaged or deteriorated bridges, or in new construction. A typical PMC deck designed to provide stiffness between the cracked and uncracked stiffness of an equivalent concrete structural deck weighs 100-150 kg/m^2, or between 1/4th - 1/5th the weight of reinforced concrete decks. This weight saving translates to an increase in load-carrying capacity and/or reduced load demand for the substructure.

The results of a building block approach that considered a large number of deck systems and manufacturing methods are reported in [3] and a further optimized deck system was fabricated using trapezoidal pultruded cores and hand layup face sheets [4] was selected for application in the current investigation. Tests have been conducted successfully to show the feasibility of using a conventional concrete barrier system attached to the composite deck using polymer concrete to embed the steel reinforcement bars of the barrier locally into the pultruded cores.

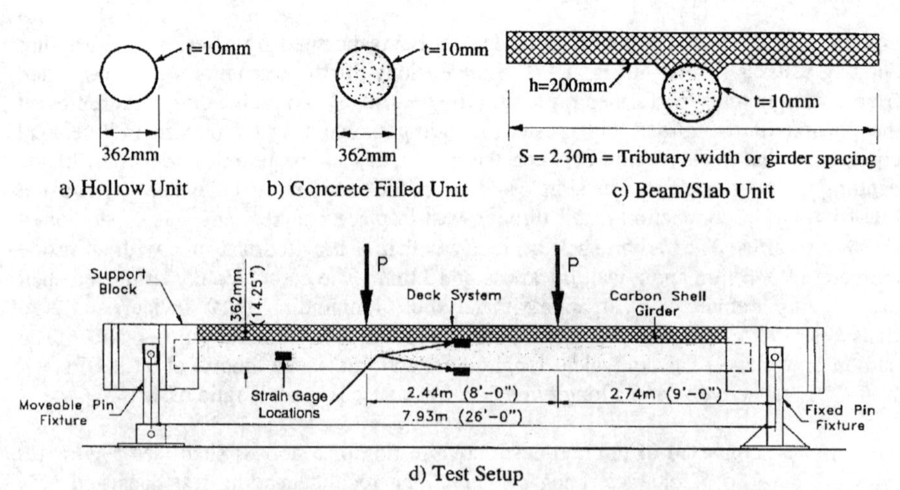

Figure 1. Carbon Shell System Bridge Girder Characterization

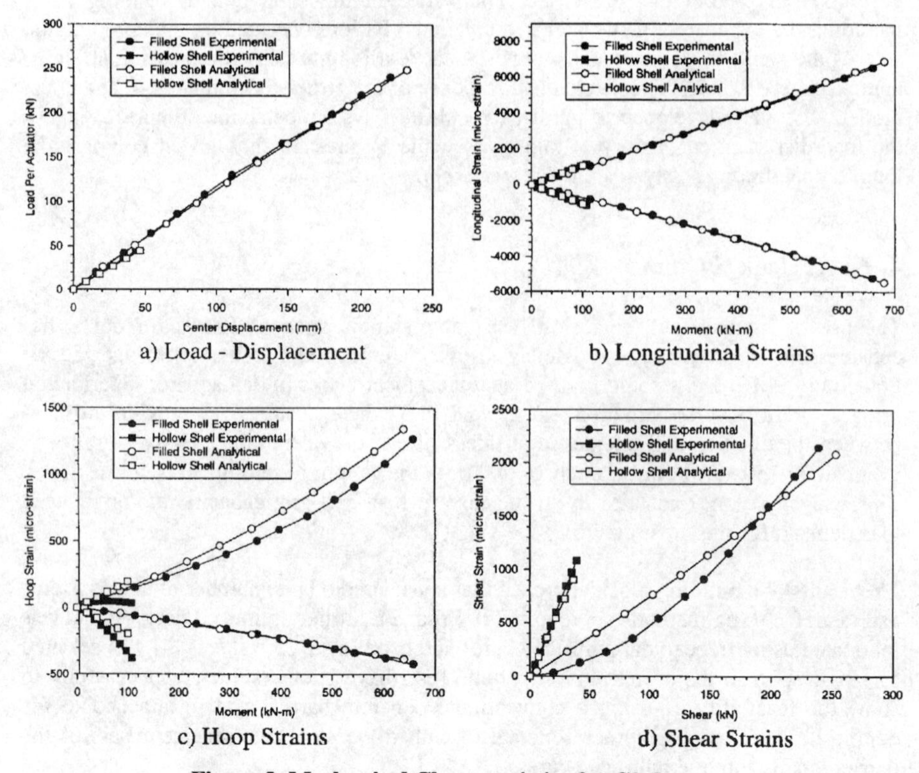

Figure 2. Mechanical Characteristics for Girders

5. PMC Reinforcement Anchorage Systems

As an alternative to steel bars used in anchorages to the abutments or to the column footings, rods made of advanced composite materials were considered as connection reinforcement for the carbon shells. To take advantage of the high strength of these rods, the force transfer between concrete and the advanced composite has to be guaranteed. Typically, the highest force transfer is demanded in regions where the concrete is under tension. Thus the concrete might be cracked and stress concentration might occur in the cracked zone. Hence not only high bond stresses between composite and concrete are required, but also a ductile behavior of the interface to minimize stress concentrations and avoid delamination. This requirement has to be fulfilled not only by the rod-concrete interface, but also by the carbon shell-concrete interface, since typically the force is introduced from the shell into the concrete and from the concrete into the connection reinforcement and vice versa.

Pullout tests have been conducted on rectangular pultruded bars containing carbon fiber in a polyester matrix [5]. Three different types of rod surface treatments and end geometry were chosen to improve the bond behavior at the end of the rods (Fig. 3). In the first type the rod was sand blasted, then painted with epoxy and dipped in sand. The second type was identical to the first, except that two slots were cut along the sides at the end of the rods, and the sides were widened so that the end had a lens shape. The third was also identical to the first, but with seven longitudinal cuts being made at the end and splayed apart. Tests show that the bond behavior can be tailored by choosing a certain geometry and surface treatment of the anchorage (Fig. 4). By combining the three anchorage types described above, a nearly elastic plastic resistance during pullout can be achieved, which comes close to a linear plastic behavior with strain hardening of steel during yielding. A splice test was also conducted using the third type of anchorage to join two shells that were then pumped full of concrete and tested in a fashion similar to the girder tests described in section 3.

Figure 3 Types of Anchorages

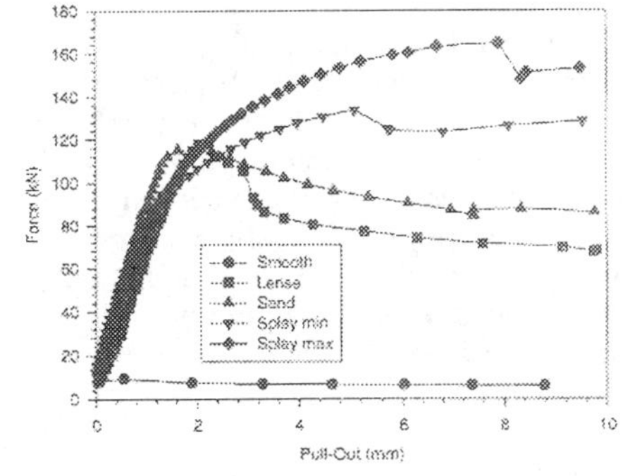

Figure 4 Typical Pull Out P-δ Curves for Different Anchorages

6. Carbon Shell Beam-and-Slab Assemblies

The final effort towards the development of the composite shell concept for bridge super-structures consists in the characterization of beam-and-slab components. Two full-scale experimental test units have been considered. The first consists of a single beam/slab assembly with the composite shell affixed to a normal weight concrete slab, while the second unit is composed of three girders with an integral advanced composite deck system. Again, the beam-and-slab assemblies are tested in a four-point bending type test with simple spans representing the portion of the bridge superstructure under positive bending moments. The cross sections are modeled after the preliminary design of the previously discussed prototype modular carbon shell bridge. The experimental investigation will allow the following issues to be addressed: (1) stiffness and strength characterization of carbon shell beam/slab assemblies, (2) characterization of the shear transfer mechanism between the carbon tube and the deck system, (3) characterization of the system load distribution, (4) assessment of system durability in terms of structural integrity and damage tolerance under fatigue loading and, (5) development of design recommendations for concrete filled carbon shell bridge systems.
The single beam/slab assembly shown in Figure 1c represents the final test of the four-point bending test series. The test unit is composed of a carbon shell girder filled with lightweight concrete affixed to a normal weight concrete slab through steel shear dowels. The shear connectors consisted of two #6 G60 bars spaced at every 610 mm. This test allowed the verification of the bending stiffness of the beam-and-slab system as well as demonstrated the efficacy of the shear connectors. The unit was cyclically loaded for 1000 cycles, at a frequency of 1 Hz, at a load level chosen to match the shear demand on the connection dowels at peak service conditions (67 kN per actuator). Subsequently, the unit was cycled up to three times the service load with no significant stiffness degradation. Degradation of the system was gradually observed, starting at 3.3 times the service load, as the beam-slab interface slipped. The test was stopped as the load-deformation response plateaued due to excessive deformations within the end blocks.

As a final effort towards the characterization of beam-and-slab assemblies, the testing of a full-scale superstructure component of three longitudinal girders with an integral advanced composite deck system is in progress. The assembly consists of three concrete filled carbon shell girders spaced 2.3m on center connected along their tops with a 181 mm deep PMC E-glass deck system. Again, the shear connectors consist of two #6 G60 bars spaced at every 610 mm. The girders are connected to a continuous reinforced concrete end diaphragm simply supported on top of 6 load cells, one under each girder end. Loading is applied vertically to the top of the deck system by means of 4 servo-controlled hydraulic actuators attached to a load frame. Views of the overall test setup geometry are shown in Figure 5. Only initial stiffness characterization tests at service load levels have been performed yet. In this case the load level was chosen to match the shear demand on the connection dowels of the center girder at peak service loads (56 kN per actuator).

A summary of the preliminary comparison of experimental and analytical response of the carbon shell system for bridge beam-and-slab assemblies is shown in Figure 6. The comparison is based on the response of the CSS girder center displacements and strains.

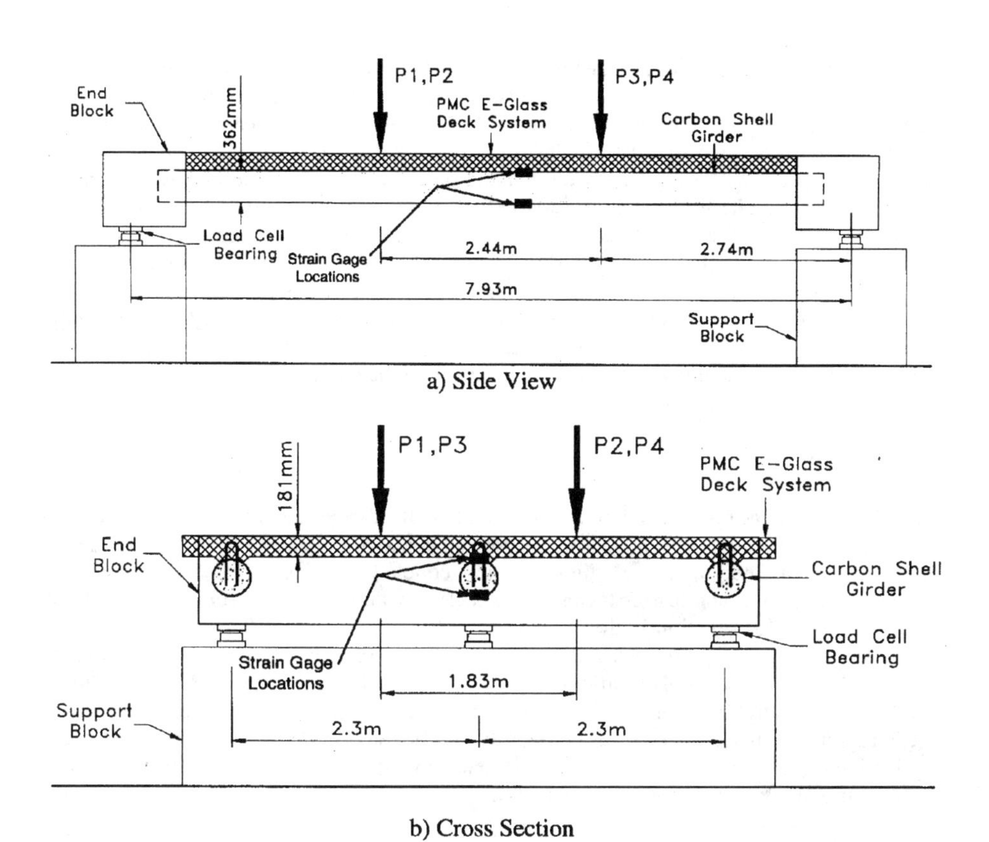

Figure 5. 3-Girder Beam-and-Slab Assembly Test Setup Geometry

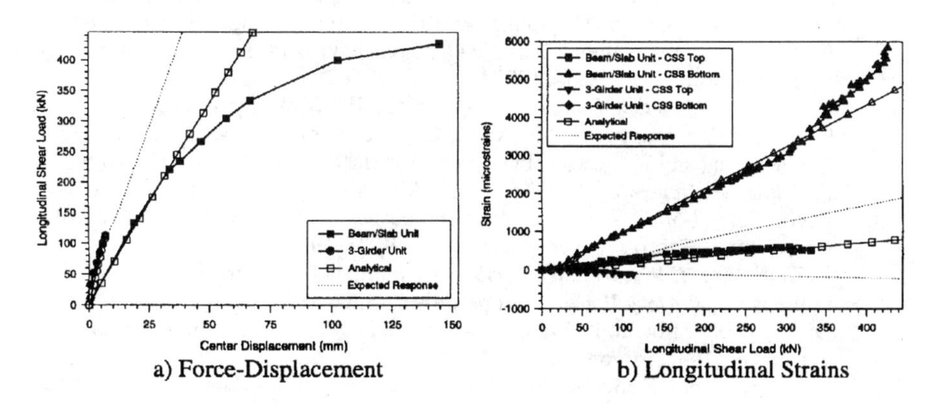

Figure 6. Beam-and-Slab Assemblies Characteristics

The results shown for the 3-girder unit are based on the response of the center CSS girder of the assembly. The load-displacement profiles in Figure 6a indicate the increased stiffness and moment capacity of the beam/slab assemblies in comparison to the single girders (see Figure 2a). The strains along the extreme fibers of the girder shown in Figure 6b show the reduced demand on the compression strains at the top of the carbon shell. The limiting compression failure mode of the individual filled carbon shell unit is therefore not expected on beam-and-slab assemblies unless the system becomes uncoupled. Figure 6 also shows the analytical predictions together with the experimental traces. The analytical evaluations are based on models developed for the characterization of the concrete filled carbon shell system acting integrally with the slab assuming full shear interaction. Furthermore, the predicted response shown for the 3-girder unit assumes a uniform curvature profile along the complete cross section.

7. Summary

Based on the demonstrated technical advantages of advanced composite materials in civil engineering applications of high directional and tailored strength and stiffness, high chemical inertness, and light weight coupled with simplified construction procedures, future applications can be expected which use these characteristics to integrate form and function in the develop new structural systems. The extent of actual acceptance of these systems will depend on (1) the resolution of outstanding technical issues such as repeatability in the low cost fabrication of large PMC structural components, durability of the materials in a civil engineering environment, and repairability in the field, (2) the availability of validated codes, standards and design guidelines, and (3) the education of engineers in aspects related to both composite materials and civil structural analysis and design.

References

[1] *F. Seible, F., R. Burgueño, M.G. Abdallah, and R. Nuismer*, "Advanced Composite Carbon Shell Systems for Bridge Columns under Seismic Loads", Proceedings of the National Seismic Conference on Bridges and Highways, San Diego, California, December 1995.

[2] *F. Seible*, “Advanced Composites for Bridge Infrastructure Rehabilitation and Renewal,” Proceedings of the IABSE Conference on Composite Construction, Conventional and Innovative, Innsbruck, Austria, Sept. 1997, pp. 741-746.

[3] *F. Seible, G. Hegemier, V. Karbhari, A. Davol, R. Burgueño, M. Wernli and L. Zhao*, “The I-5/Gilman Advanced Composite Cable Stayed Bridge Study”, University of California, San Diego, SSRP-96/05, 1996

[4] *L. Zhao, V.M. Karbhari and F. Seible*, “ Experimental Results for a Large-Scale Pultruded Core Deck Panel,” Test report submitted to Martin Marietta Materials, Division of Structural Engineering, University of California, San Diego, September, 1997, Report No. TR-97/14

[5] *M. Wernli, and F. Seible*, “Ductile Connections of Modular Bridge Girders Made of Concrete Filled FRP-Tubes”, Second International Conference on Composites in Infrastructure, Tucson, Arizona, January 1998.

Shear Strengthening of Prestressed Concrete Bridge Girders Using Bonded CFRP Sheets

R. Hutchinson*, D. Donald*, A. Abdelrahman*, S. Rizkalla*

* ISIS Canada, University of Manitoba, Winnipeg, Manitoba, Canada R3T 5V6

Abstract

The use of heavier trucks demands upgrading of a twenty-nine year old prestressed concrete bridge in Manitoba, Canada. The use of Carbon Fibre Reinforced Polymer (CFRP) sheets for shear strengthening could provide a low-cost solution due to a reduction of construction time and traffic interruption.

An experimental program has been undertaken to investigate the use of bonded CFRP sheets for the shear strengthening of I-shaped concrete AASHTO girders. Seven 1:3.5 scale model precast prestressed concrete girders are tested to failure at both ends to determine the most efficient shear strengthening scheme. The bond properties of the CFRP systems are evaluated using eighteen specially designed concrete specimens strengthened with CFRP. Test results suggests that shear strengthening using CFRP sheets is an efficient rehabilitation solution for the bridge. This paper presents test results and design recommendations for the use of this technique on this particular girder shape.

1. Introduction

The city of Winnipeg, Manitoba, Canada, is considering upgrading a twenty-nine year old bridge in response to the demand for using heavier truck loads. The twin five-span continuous prestressed concrete structures were designed in 1969 according to the American Association of State Highway and Transportation Officials (AASHTO) Code. Analysis using the current AASHTO Code indicates that the shear strength of the girders is not sufficient to withstand the new truck load. Carbon Fibre Reinforced Polymer (CFRP) sheets provide an excellent solution for shear strengthening since they are light-weight, corrosion-free, and have a high tensile strength. When compared with conventional methods, this technique provides a low-cost solution due to significant reduction of construction time without traffic interruption.

An experimental program has been undertaken at the University of Manitoba, Canada, to test scale models of the I-shaped bridge girders strengthened with CFRP sheets. Seven prestressed concrete beams were tested to failure at each end to determine the most efficient strengthening scheme. The contribution of the CFRP sheets to the enhanced shear capacity of the girder has been examined with emphasis on the effect of this particular girder shape. Since the bond between the sheets and the concrete is a critical component of this strengthening method, a series of bond specimens have been tested in order to determine the bond characteristics. This paper summarizes test results available to date and recommendations for the use of this strengthening technique on the I-shaped girders.

2. Experimental Program

Seven prestressed concrete girders were tested at the University of Manitoba, Canada. The ten meter long beams are 1:3.5 scale models of the I-shaped bridge girders. All of the beams had a depth of 415 mm with a top slab of 480 mm wide and 60 mm deep as shown in Figure 1. The compressive strength of the concrete at the time of testing ranged from 44 to 55 MPa. The beams were pretensioned with 7-wire steel strands with a diameter of 13 mm and an ultimate tensile strength of 1860 MPa. To increase the flexural capacity and avoid premature failure due to flexure, non-prestressed strands were also provided.

The beams were designed to carry the same shear stress at ultimate as the girders of the bridge. The stirrup shape used for the first series of four beams is shown in Figure 1 and is identical to those used in the bridge girders. An alternate straight-legged stirrup shape, shown in Figure 2, was used for the second series of beams. The spacing of the stirrups was identical in all of the test beams.

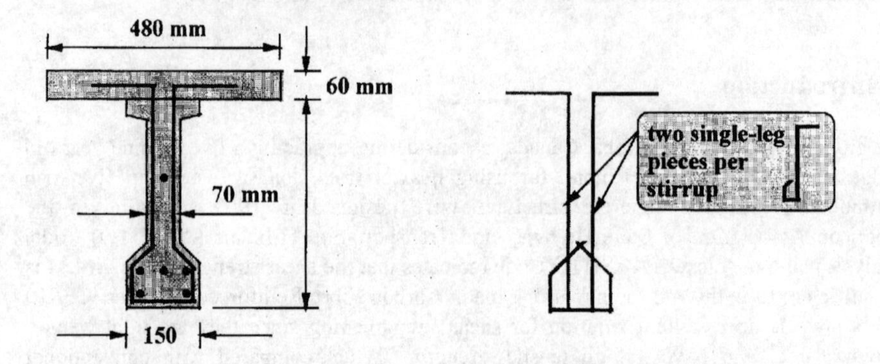

Figure 1. Test Beam Dimensions *Figure 2. Alternate Straight-legged Stirrups*

One beam from each series, Series I with bent stirrup legs and Series II with straight-legged stirrups, were tested as control beams. The remaining beams were strengthened using the three different types of CFRP sheets indicated in Table 1. The beams were tested to failure at each end to determine the most efficient strengthening scheme.

Table 1. Properties of the CFRP Sheets

Property:	Type A*	Type B*	Type C+
Design Thickness (mm):	0.11	0.11	0.79
Fibre Areal Weight (g/m^2):	200	200	660
Tensile Strength (MPa):	3350	3400	760
Tensile Strength (N/mm width):	390	375	600
Tensile Modulus (GPa):	235	230	76
Strain at Rupture:	0.0151	0.0148	0.01

*All properties reported for dry fiber sheets

+All properties reported for composite fiber and resin sheets

During testing of the first end of the control beam with bent stirrup legs, premature failure occurred due to the shape of the stirrups. An outward force was observed by spalling of the concrete cover as shown in Figure 3. This force is the resultant of the tensile forces in the vertical and diagonal legs of the stirrups and causes the stirrup to straighten. To control this outward force, the second end of the control beam in Series I was strengthened by a clamping scheme as shown in Figure 4.

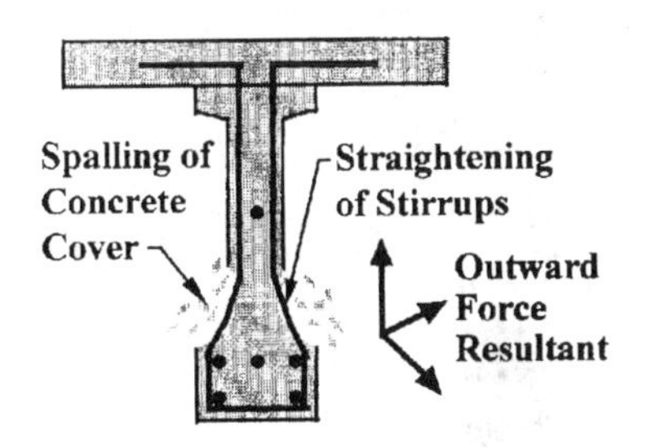

Figure 3. Premature Stirrup Failure

Figure 4. Web Clamping Scheme

Prior to application of the CFRP sheets, the concrete surface of one beam was prepared using a grinder, wire brush and high pressure air for cleaning the surface after grinding. The surfaces of the remaining beams were prepared using a high pressure water-blasting technique. Any sharp corners on the beams were rounded using a grinder.

The CFRP sheets were applied to the beams using well defined procedures and proprietary products supplied by each manufacturer. An epoxy primer was first applied to the surface of the beam to seal the concrete. After setting of the primer, any significant surface irregularities were filled using an epoxy putty or an epoxy resin with additional filler material. For the majority of the beams, the epoxy resin was applied to the surface of the beam followed by the CFRP sheets and a second layer of epoxy resin was applied as a top coat. In the case of multiple layers of sheets, the top coat served as a base coat for the next layer of CFRP sheets. For one beam, the Type C CFRP sheets were impregnated with epoxy resin in a separate resin bath prior to application on the beam and subsequent layers were applied using the same technique.

The test beams were strengthened using one layer of 250 mm wide CFRP sheets with the fibres oriented either vertically or diagonally at 45 °. A gap of either 20 or 100 mm was provided between each sheet to allow drainage of any moisture accumulation. The vertical or diagonal CFRP sheets were applied on each side of the cross-section from the top of the beam immediately below the slab to the underside of the beam where they were overlapped for a minimum length of 100 mm. For some beams, a 220 mm wide horizontal CFRP sheet was applied on the top of the vertical or diagonal sheets. The effect of using two layers of diagonal sheets was also examined.

The simply supported beams were subjected to two equivalent non-symmetric point loads. The shear span of 1940 mm was kept constant for all of the tests while the overall span was varied in order to test both ends of each beam. The beams were tested using monotonic load and stroke control.

Bond Specimens

The bond specimens consisted of 100 x 275 x 900 mm reinforced concrete prisms strengthened on opposite faces with 200 mm wide CFRP sheets and subjected to uniaxial tension as shown in Figure 5. Due to the arrangement of the internal reinforcing, cracking of the concrete was initated at mid-height of the specimen as shown in Figure 6. For some specimens, steel reinforcement crossing the crack was used to simulate the effect of stirrups in order to evaluate load sharing between the CFRP sheets and the stirrups. In all specimens, the sheets were applied with the continuous longitudinal fibres parallel to the applied load while the orientation of the crack and the steel reinforcement crossing the crack was varied as shown in Figure 6. The effect of multiple layers of CFRP sheets on the bond characteristics was also examined.

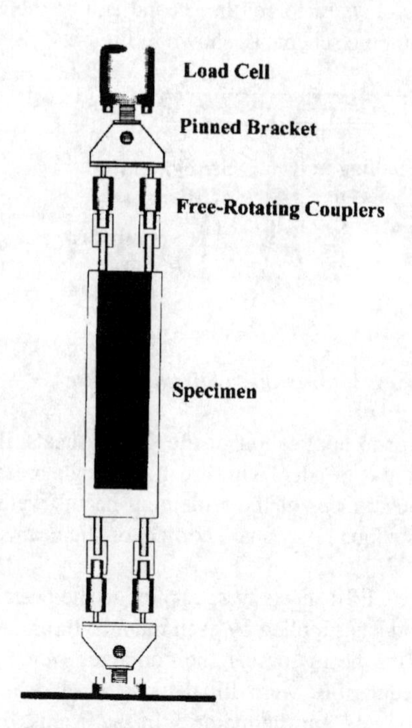

Figure 5. Bond Test Set-up

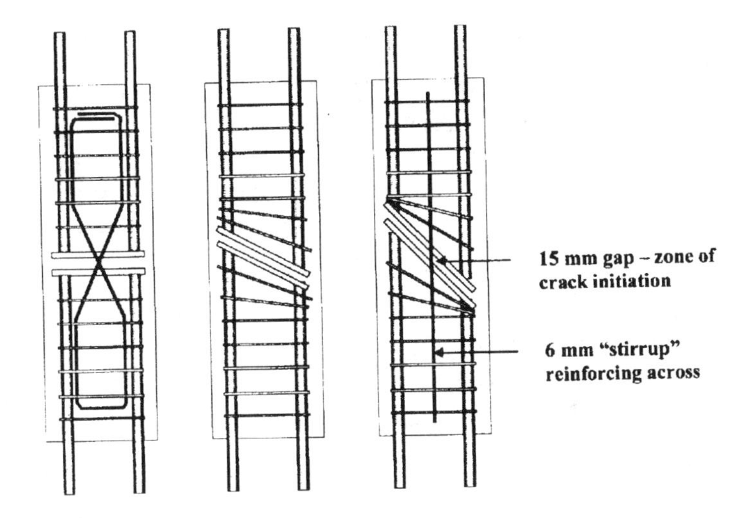

Figure 6. Section Through Typical Bond Specimens

The CFRP sheets were applied to the bond specimens using the same procedures and two surface preparation techniques as described previously for the test beams. The CFRP sheets, steel reinforcing crossing the crack, and the concrete used for the bond specimens were all selected to have similar properties as the materials used for the test beams.

The bond specimens were tested under monotonic loading and stroke control. Instrumentation was applied on both sides of the specimens to measure the ultimate strain in the CFRP sheets at failure, the length of sheet over which stress is effectively transferred to the concrete, and the distribution of axial strain both along the length and across the width of the sheets.

3. Test Results and Discussion: Beam Series I

The first series of four beams have been tested to failure at each end for a total of eight tests. The beams in this series were reinforced using stirrups with a shape that is identical to those used in the existing bridge. The strengthening schemes used for these beams and the experimental parameters examined in each test are described in Table 2. Also provided in Table 2, is a comparison of the ultimate shear failure load, V_u, and the initial shear cracking load, V_c. Typically, V_c is determined by the load corresponding to the initiation of the first diagonal cracks. However, since the presence of CFRP sheets made the observation of cracks difficult, V_c was determined as the load corresponding to first measurements of strain in the stirrups.

Table 2. Comparison of Specimen Parameters and Test Results: Beam Series I

Strengthening Scheme	Gap Size (mm)	CFRP Thickness (mm)	Surface Preparation	f_c' (MPa)	V_c (kN)	V_u (kN)	$\frac{(V_u)test}{(V_u)control}$
Control (none)	n.a.	n.a.	n.a.	46	66	137	1.00
Clamped Stirrups	n.a.	n.a.	n.a.	46	66	174	1.27
Vertical CFRP	100	0.11 A	grinding	53	74	151	1.10
Vertical CFRP	20	0.11 B	hydro-blast	44	76	161	1.17
Diagonal CFRP	100	0.79 C	hydro-blast	55	72	177	1.29
Diagonal CFRP	20	0.11 B	hydro-blast	44	76	173	1.26
Horiz. & Vert. CFRP	100	0.11 A	grinding	53	74	185	1.34
Horiz. & Diag. CFRP	100	0.79 C	hydro-blast	55	72	186	1.36

In all tests, flexural shear cracks were observed within the shear span and extended toward the top flange at ultimate. Figure 7 shows the control beam at failure. Just prior to failure, spalling of the concrete cover was observed due to the outward tensile force resultant causing straightening of the bent corner of the stirrups

Due to the shape of the stirrups, only one of the stirrups in the control beam reached yield before failure occured due to spalling of the concrete cover and straightening of the adjacent stirrups.

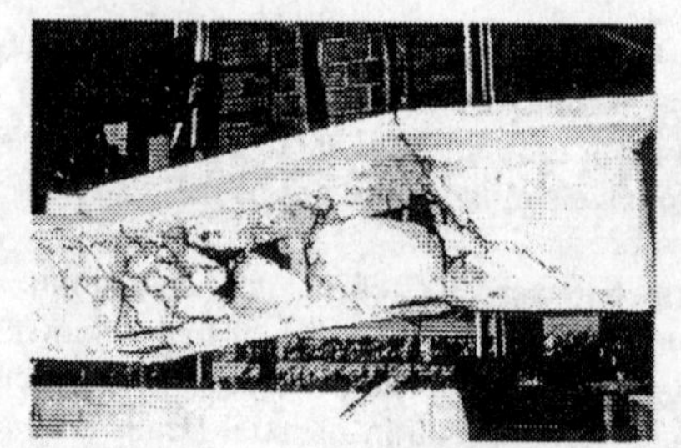

Figure 7. Series I Control Beam at Failure

The clamping scheme applied to the second end of the control beam was effective in preventing failure due to straightening of the stirrups. All of the measured stirrup strains were significantly higher than those observed for the the control beam. The distribution of forces between the clamped stirrups was improved, contributing to a 27 % increase in the ultimate shear capacity, V_u, when compared to the control beam.

For the beams strengthened with vertical sheets, two different types of CFRP sheets with similar thicknesses and material properties were used. Two different surface preparation techniques and two gap sizes were used, as shown in Table 2. The ultimate shear capacity, V_u, of the beam with the smaller gap size increased by 17 %, while only a 10 % increase was observed for the beam with the larger gap size. In both beams, failure was initiated by straightening of the stirrups and debonding of the CFRP sheets above and below the diagonal cracks. Debonding of the CFRP sheets was less extensive on the beam prepared using hydro-blasting when compared with the beam using the grinding technique.

The thickness and material properties of the two types of CFRP sheets used for the diagonal configurations vary significantly as shown in Tables 1 and 2. The gap size was also varied for these two tests while the surface preparation technique was not. For both beams, similar 26 % and 29 % increases in ultimate shear capacity, V_u, were achieved. Due to the shape of the girder, debonding and straightening of the CFRP sheets was observed on both beams at the lower part of the thin web. After the test, the CFRP sheets were removed to examine the stirrups. Due to the efficiency of the diagonal CFRP contribution and the resulting reduced stress in the stirrups at ultimate, the stirrups did not straighten significantly.

The efficiency of the diagonal CFRP sheets is illustrated in Figure 8. The stirrup strain at any level of applied shear is lower for the beams with diagonal sheets. Although the beam with horizontal and vertical sheets reached a higher ultimate shear load, the stirrup strain was greater. Figure 8 also shows that in spite of the larger gap size, the beam with thicker sheets exhibited lower stirrup strains and therefore a greater contribution from the CFRP sheets at the same level of applied shear load.

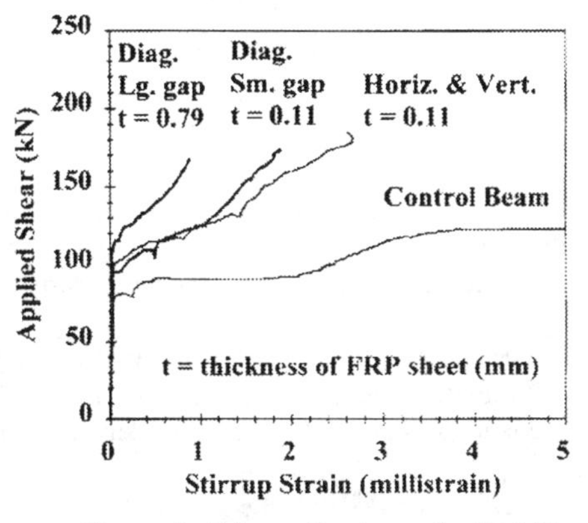

Figure 8. Stirrup Strain vs Applied Shear

Both beams with a horizontal CFRP sheet applied on top of the vertical or diagonal sheets, achieved similar 34 % and 36 % increases in ultimate shear capacity, V_u. Figure 9 shows the beam with horizontal and vertical sheets at failure. Similar to the beams with vertical CFRP sheets only, the beam with both horizontal and vertical sheets, demonstrated spalling on the lower part of the thin web due to the outward force in the stirrups at this location. Due to the

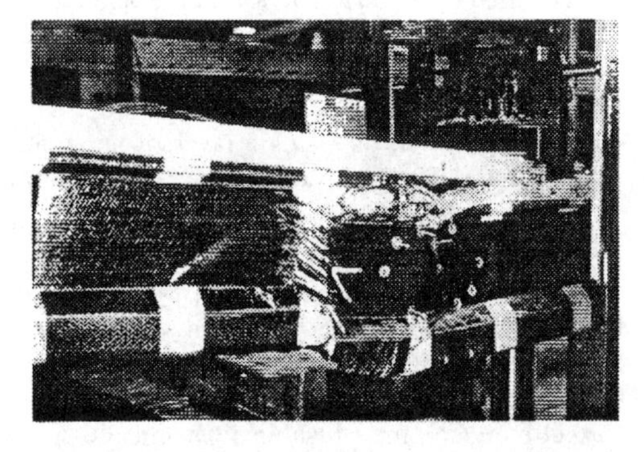

Figure 9. Beam with Horizontal & Vertical CFRP Sheets at Failure

presence of both horizontal and vertical sheets, however, spalling was observed at a higher load level than in the control beam or the beam with vertical sheets only. For the beam with both horizontal and diagonal sheets, debonding and straightening of the diagonal sheets was observed at the lower part of the thin web similar to the beams with diagonal sheets only, however, the observed debonding was less extensive.

Conclusions

Seven ten meter long prestressed concrete beams and eighteen smaller bond specimens were tested to determine the most efficient shear strengthening scheme using CFRP sheets for I-shaped girders. The beams are scale models of existing bridge girders which require shear capacity upgrading in order to carry increased truck loads. Test results available as of January 1998 suggest that shear strengthening using CFRP sheets is an efficient solution for the bridge. The following conclusions are based on this preliminary analysis of test results available at the time of submission of this paper:

1. CFRP sheets are effective in reducing the tensile force in the stirrups under the same applied shear load, delaying or preventing a premature failure due to straightening of the bent stirrup legs. The clamping scheme was also effective in preventing straightening of the stirrups and premature failure.

2. The reduced gap size for beams with vertical CFRP sheets increased the improvement in ultimate shear capacity from 10 % to 17 %. The use of hydro-blasting for surface preparation reduced the extent of debonding of the sheets.

3. Diagonal CFRP sheets are the most efficient configuration in reducing the tensile force in the stirrups at the same level of applied shear load. The ultimate shear capacity was increased by 26 % to 29 % with the application of diagonal sheets. Due to the shape of the girder, debonding and straightening of the CFRP sheets was observed on the lower part of the web in beams with diagonal sheets.

4. The application of a single layer of horizontal CFRP sheets on top of the vertical or diagonal sheets was shown to increase the ultimate shear capacity by 34 % to 36 %.

Acknowledgements

Funds for this program were provided by the ISIS Canada Network of Centres of Excellence on Intelligent Sensing for Innovative Structures, the City of Winnipeg, Dillon Consulting Engineers, Vector Construction Ltd., Concrete Restoration Services Ltd., the Hexcel Fyfe Co., the Mitsubishi Chemical Corp., and the Tonen Corp. Special assistance was provided by Mr. Moray McVey of the ISIS Canada Network.

REPAIR AND REHABILITATION OF AN EXISTING RC STRUCTURE USING CFRP SHEETS

F. Focacci*, A. Nanni, F. Farina***, P. Serra****, C. Canneti*******

* Dipartimento di Scienza dei Materiali, Università degli Studi di Lecce, Lecce, Italy
** University of Missouri, Rolla, USA
*** Studio Farina, Bologna, Italy
**** DISTART, Università degli Studi di Bologna, Bologna, Italy
***** Comune della Spezia, La Spezia, Italy

Abstract

The application of Carbon FRP sheet material for repair and strengthening of an RC floor system is presented in this paper. The floor is formed by four parallel RC beams, transversally connected by a one-way RC slab. Both the beams and the slab present some damage due to reinforcement corrosion. The floor system needed strengthening due to the increasing of dead and live service loads.
The solution of using CFRP sheets was based on low cost, speed, and simplicity of the installation. Material characterization was conducted in order to evaluate the actual conditions of the system. Based on the experimental data, an analytical model was used to compute the moment - curvature and the load - deflection diagrams.
Flexural and shear CFRP reinforcement was applied to the beams; only flexural CFRP reinforcement was applied to the slab.

1. Introduction

Carbon FRP sheets externally bonded to concrete can be successfully used for repair and strengthening of RC beams and columns. The main advantages of using CFRP sheets are: light weight, high tensile strength, high Young's modulus, low profile, flexibility and speed and simplicity of the installation.
The application of Carbon FRP sheet material for the repair and strengthening of an RC floor system is presented in this paper. The floor is formed by four parallel RC beams, transversally connected by a one-way RC slab. The spans of the beams are between 9 m and 10 m, and their spacing is about 2.50 m. The beams are simply supported over two masonry walls. Both the beams and the slab present some damage caused by the corrosion of the steel reinforcing bars.
This floor system is the flat roof of an old building in La Spezia, Italy. As the building becomes part an art gallery (Museo Civico Amedeo Lia), it was decided to open its roof to visitors and make it a terrace. The aim of the rehabilitation was: 1) to repair the existing damage, 2) to improve the flexural capacity of beams and one-way slab for the new live load as prescribed by the building code for public areas, 3) to improve the shear capacity of the beams, and 4) to limit deflections of the beams under the new service load.

Concrete material characterization was conducted in order to evaluate the actual conditions of the system. Based on the experimental data, an analytical model was used to compute the moment - curvature and the load - deflection diagrams for different combinations of CFRP reinforcement. The model is based on the classical flexural theory and takes into account the nonlinear behavior of concrete. The model can also account for the fact that CFRP sheets are installed while the permanent load is acting. For the beams, the adopted solution consisted of the application of three plies of CFRP sheets (0 deg.) adhered to their soffit, and two plies of CFRP sheets (0 and 90 deg.) on their lateral surfaces in the maximum shear zone. The 90 deg. plies were extended around the beam to prevent peeling of the flexural plies adhered to the soffit. For the one-way slab, one 250 mm single ply strip was applied every meter.

2. Situation before renovation

The cross section and the longitudinal shape of the beams are reported in figure 1.

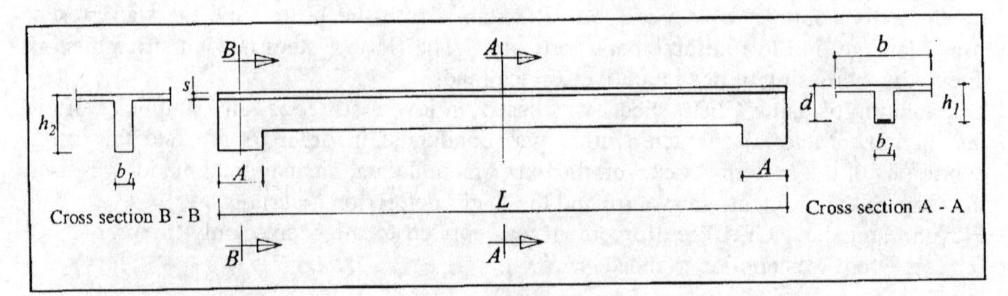

Figure 1 The structure before renovation

The following dimensions were measured:

$h_1 = 600\ mm$ $b = 2400\ mm$ $s = 100\ mm$ $A_s = 4241\ mm^2$ ($6\phi30\ mm$)

$h_2 = 900\ mm$ $b_1 = 300\ mm$ $A = 700\ mm$ $d = 530\ mm$

The four beams have spans: $L = 8.93\ m$, $L = 9.20\ m$, $L = 9.72\ m$, $L = 10.1\ m$.

Microsismic sclerometric and compression tests on cores were carried out in order to evaluate the concrete properties. Tests results are reported in tables 1 and 2.

Spec.	f_c *(MPa)*
1	*24.2*
2	*14.6*
3	*20.3*
4	*14.7*

Table 1 Results of compression tests

Test	E_d *(MPa)*	f_c *(MPa)*
1	*30995*	*19.4*
2	*26640*	*11.6*
3	*32282*	*17.5*
4	*33544*	*23.1*
5	*35447*	*19.8*
6	*32367*	*17.2*

Table 2 Results of microsismic and sclerometric tests

The following lower-bound properties were considered for the evaluation of the existing structure:

compressive strength: $f_c = 12\ MPa$; Young's modulus: $E_c = 24000\ MPa$.

Beams and the slab present spalling of the concrete cover, caused by the corrosion of the steel reinforcing bars, as shown in figures 2a and 2b.

Figure 2a

Figure 2b

Referring to the beam having span $L = 9.20\ m$ the following loads act before renovation:

Dead load:	$W_{DLE} = 13.83\ kN/m$	Service load:	$W_{WLE} = 17.97\ kN/m$
Live load:	$W_{LLE} = 4.84\ kN/m$	Load at renov.	$W_{ren} = 13.83\ kN/m$

At the time of renovation, only the dead load was acting, $W_{ren} = 13.83\ kN/m$.

3. Description

The renovation operation included the following steps:

1. Removal of deteriorated concrete, which was done mechanically by means of a small pneumatic hammer.
2. Cleaning of existing steel reinforcement, which was done mechanically by means of a rotating metal brush. The existing steel bars were cleaned to white metal.
3. Steel reinforcement protection, steel bars were treated with a corrosion-protective coating.
4. Surface restoration, all the surfaces interested by FRP adhesion were restored by means of proper no-shrinkage mortar.
5. Primer application, a first layer of low-viscosity epoxy resin was applied on the concrete surface for better adhesion between concrete and FRP sheet.
6. First Coat, epoxy resin is applied after the primer has cured and its purpose is to impregnate the fiber sheet and to adhere it to the concrete.
7. Sheet adhesion.
8. Finish coat application, the epoxy coat is meant to utterly impregnate the fibers and protect them from the environment. When another FRP ply is to be adhered, this coat has the function of the first coat.

Details of CFRP reinforcement are shown in figures 3a, 3b and 4.

Due to jobsite management necessities and to air humidity, the different steps of the application were carried out over a long period of time between November 1997 and January 1998, but it was estimated that the whole renovation took approximately 24 man-work days.

Figure 3a

Figure 3b

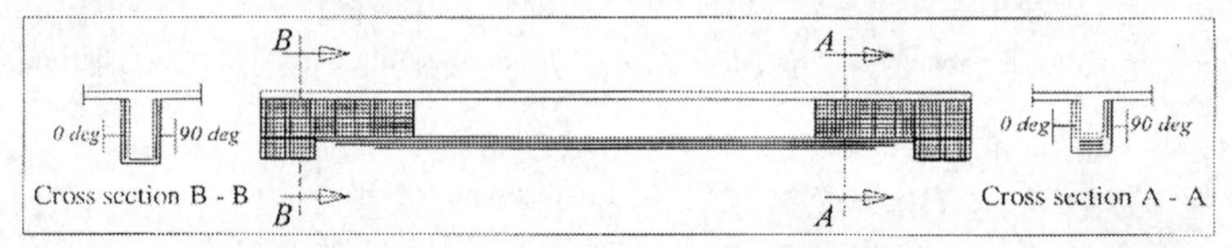

Figure 4 Configuration of CFRP sheets.

4. Flexural design

An analytical model based on the classical flexural theory is used for design and verification. The model takes into account for the fact that the CFRP sheets are applied when the load W_{ren} acts on the beam and the beam is cracked as a consequence of the load history before the renovation, W_{WLE}. Thus the strain on CFRP is determined as the difference between the actual concrete strain and the concrete strain at the time of renovation. The latter strain is determined on the cracked moment-curvature diagram if the bending moment due to W_{WLE} exceeds the cracking moment or on the uncracked moment - curvature diagram otherwise.

4.1 Material properties

CFRP sheets having the following characteristics were used:

Single ply thickness: $t_d = 0.16\ mm$ Ply width: $b_d = 250\ mm$
Tensile strength: $f_F = 2600\ MPa$ Young's modulus: $E_F = 230000\ MPa$

Linear elastic constitutive law is considered for CFRP sheet.

Considered concrete constitutive law is:

Compression:

$$\sigma = f_c \frac{k\frac{\varepsilon}{\varepsilon_{c0}} - \left(\frac{\varepsilon}{\varepsilon_{c0}}\right)^2}{1+(k-2)\frac{\varepsilon}{\varepsilon_{c0}}}$$

Tension:

$$\sigma = f_{ct}\left[k_t\frac{\varepsilon}{\varepsilon_{ct}} - (2k_t - 3)\left(\frac{\varepsilon}{\varepsilon_{ct}}\right)^2 + (k_t - 2)\left(\frac{\varepsilon}{\varepsilon_{ct}}\right)^3\right]$$

$f_c = 12\ MPa$ $\quad f_{ct} = 1.2\ MPa$

$\varepsilon_{c0} = 2‰$ $\quad \varepsilon_{cu} = 3.5‰$

$\varepsilon_{ct} = 0.15‰$ $\quad E_c = 24\ GPa$

$k = E_c\varepsilon_{c0}/f_c$ $\quad k_t = E_c\varepsilon_{ct}/f_{ct}$

Concrete constitutive law is represented in figure 5.

Prandtl constitutive law is considered for existing steel rebars, being:

$f_y = 165\ MPa$

$E_s = 205\ GPa$

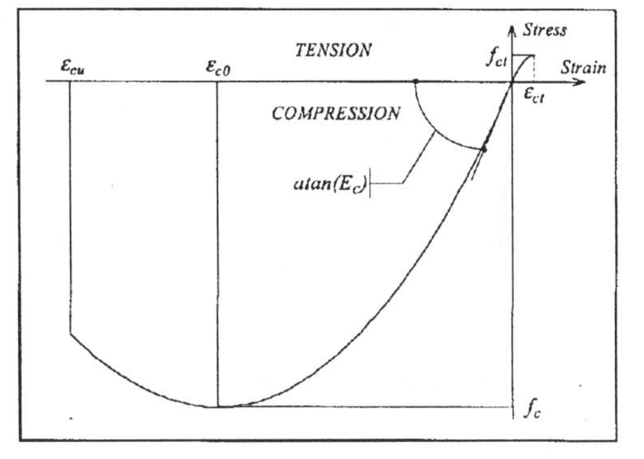

Figure 5 Concrete constitutive law.

4.2 Design at ultimate

The following loads act after renovation:

Dead load: $W_{DLP} = 16.09\ kN/m$ $\quad$ Working load: $W_{WLP} = 30.31\ kN/m$

Live load: $W_{LLP} = 14.92\ kN/m$ $\quad$ Ultimate load: $W_{ULP} = 43.86\ kN/m$

The strains along the soffit of the beam due to W_{ren} are represented in figure 6. For the determination of these strains it is assumed that the beam is cracked due to the service load before renovation, that is $W_{WLE} = 17.97\ kN/m$. Since the cracking moment of the beam is $M_{cr} = 99.84\ kNm$, the distance from the supports of the first cracked cross section, x_{cr}, can be determined as:

$$x_{cr} = \frac{W_{WLE}L - \sqrt{(W_{WLE}L)^2 - 8W_{WLE}M_{cr}}}{2W_{WLE}} = 1.4m$$

The strain represented in figure 6 are determined by considering the uncracked cross section in the range $0 \le x \le x_{cr}$ and the cracked cross section in the range $x_{cr} \le x \le L - x_{cr}$. For the midspan cross section, the strain in concrete at the level of FRP sheet at time of renovation is $\varepsilon_{ren} = 0.377‰$.

By considering the described assumptions and materials constitutive laws, the moment curvature of each cross section can be determined. Figure 7 reports the moment - curvature diagram of the midspan cross section when 1, 2 and 3 FRP plies are applied.

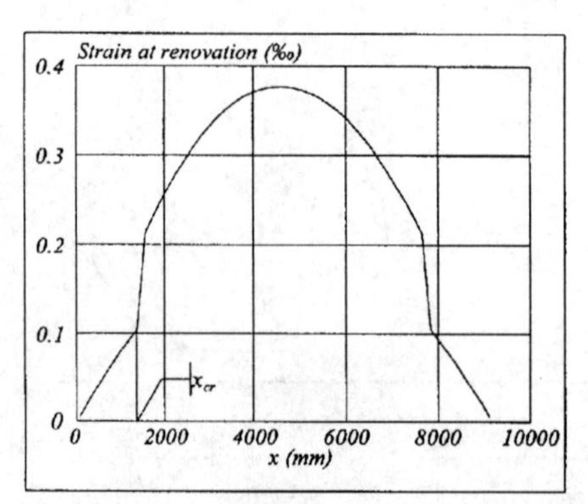

Figure 6 Strain at the FRP level at renovation

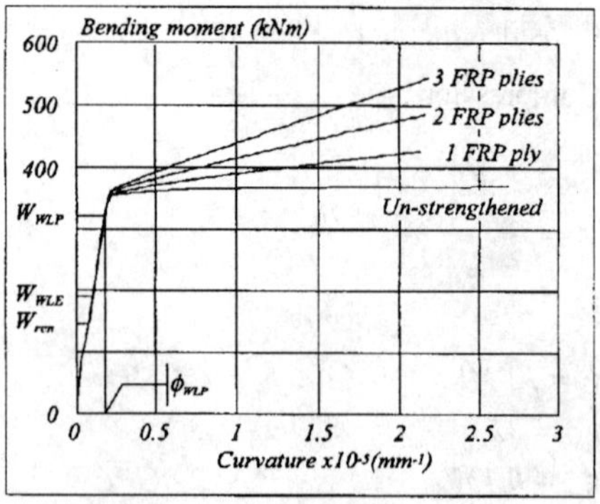

Figure 7 Moment - Curvature diagrams.

Since the moment at ultimate is $M_u = 464\ kNm$, the configuration with three CFRP plies was adopted (figure 7) for which CFRP failure is reached corresponding to the bending moment $M_n = 543\ kNm$.

4.3 Stress at working loads

From the moment - curvature diagram of the midspan cross section (figure 7), the curvature $\phi_{WLP} = 1.77 \cdot 10^{-6}\ mm^{-1}$ can be obtained as corresponding to the moment $M_{WLP} = 320.7\ kNm$. Equilibrium of axial forces gives the depth to neutral axis $y_{WLP} = 139.1\,mm$. Corresponding maximum stresses are:

a) concrete: $\sigma_c = 4.6\ MPa$

b) steel: $\sigma_s = 154\ MPa$

c) CFRP: $\sigma_F = \left[\phi_{WLP}\left(h_l - y_{WLP}\right) - \varepsilon_{ren}\right]E_F = 101\ MPa$

4.4 Deflection under transitory loads

By computing the moment - curvature diagrams for a certain number of cross sections (note that they are different even if the geometrical characteristics are constants along the beam, due to the changing of ε_{ren} along the beam), it is possible to find the load – midspan deflection diagram reported in figure 8.

The deflection due to the working loads acting after renovation is:

$\Delta_{WLP} = 15.5\ mm$

The deflection at the time of renovation is:

$\Delta_{ren} = 7.0\ mm$

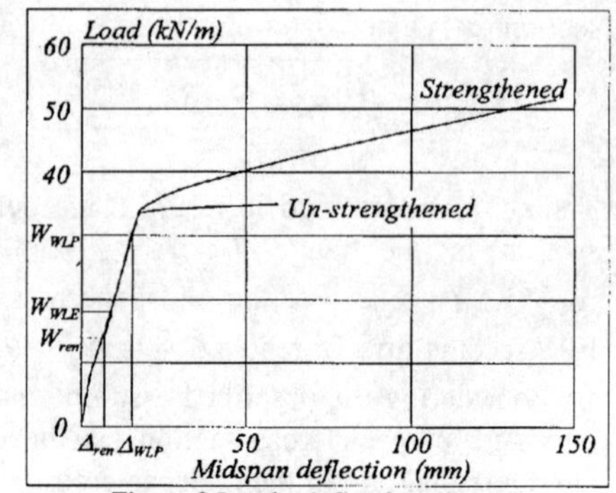

Figure 8 Load - deflection diagram

Therefore, the increasing of deflection due to loads acting after renovation is: $\Delta_{r} = \Delta_{WLP} - \Delta_{ren} = 8.5\ mm$

5. Shear design

Two CFRP sheets (0 - 90 deg) were applied on the lateral surface of the beam to provide shear reinforcement, as shown in figure 5. Shear design is made by following the EC2 code substituting mechanical characteristics of steel with those of CFRP. The presence of existing steel stirrups is neglected where CFRP reinforcement is applied. CFRP effective depth is reduced by the amount $l_a = 50\ mm$ to take into account development length.

i) Cross section at $x = 0$, shear force at ultimate: $V_{Sd} = 202\ kN$
Shear CFRP area is:

$$A_{sF} = 2\frac{(2b_d - l_a)t_d}{A} = 0.206\ mm^2/mm$$

$$V_{cd} = 0.25kf_{ct}(1.2 + 40\rho_l)b_w d = 80.6\ kN$$

being:

$k = 1$ $\quad$ $b_w = 280\ mm$

$\rho_l = \frac{A_{sl}}{b_w d} = 0$ $\quad$ $d = 800\ mm$

$$V_{Wd} = 0.9 A_{sF} d f_{Fu} = 385.6\ kN$$

ii) Cross section at $x = A = 700\,mm$, shear force at ultimate: $V_{Sd} = 171\ kN$
Shear CFRP area is:

$$A_{sF} = 2\frac{(4b_d - 2l_a)t_d}{1000} = 0.288\ mm^2/mm$$

$$V_{cd} = 0.25kf_{ct}(1.2 + 40\rho_l)b_w d = 53.4\ kN$$

being:

$k = 1$ $\quad$ $b_w = 280\ mm$

$\rho_l = \frac{A_{sl}}{b_w d} = 0$ $\quad$ $d = 530\ mm$

$$V_{Wd} = 0.9 A_{sF} d f_{Fu} = 357.2\ kN$$

6. Conclusions

The application of CFRP sheets externally bonded to concrete for repair and rehabilitation of an RC floor system is described in this paper. Description of the installation and flexural / shear design is included. This method proved advantageous with respect to traditional solutions for: low cost, speed and simplicity of the installation, no change in structure dimension or weight, and no demolition.
From an analysis point of view, stress in concrete and steel reinforcement under working loads may govern design. This is because CFRP sheets are applied when the

dead load acts on the structure. The magnitude of these stresses could be reduced by jacking up the structure while applying the CFRP sheet.

Acknowledgments

The authors are thankful to Prof. A. Di Tommaso, Prof. A. La Tegola and Dr. Marco Arduini for their advice, Mr. F. Palumbo and the contractor EDIL ATELLANA for their positive attitude in using this new technology, Mr. D. Betti for his practical suggestions and Messers R. Carli, R. Roffi and F. Maccaferri for the concrete testing.

References

[1]*Nanni A., Focacci F., Cobb C.*, PROPOSED PROCEDURE FOR THE DESIGN OF RC FLEXURAL MEMBERS STRENGTHENED WITH FRP SHEETS - Proceedings, ICCI-98, Tucson, AZ, Jan. 5-7, 1998, Vol. I, pp.187-201.

[2]*Arduini M., Di Tommaso A., and Nanni A.*, BRITTLE FAILURE IN FRP PLATE AND SHEET BONDED BEAMS - ACI Structural Journal, Vol. 94, No. 4, July-Aug. 1997, pp. 363-370.

[3]*Arduini M., Nanni A.*, PARAMETRIC STUDY OF BEAMS WITH EXTERNALLY BONDED FRP REINFORCEMENT - ACI Structural Journal, Vol. 94, No. 5, Sept.-Oct. 1997, pp. 493-501.

[4]*Greenfield, T. K.*, CARBON FIBER LAMINATES FOR REPAIR OF CONCRETE STRUCTURES - Materials Performance, Vol. 34, No. 3, March 1995, pp. 36-38.

[5]*Meier, U. and Kaiser, H. P.*, STRENGTHENING OF STRUCTURES WITH CFRP LAMINATES - Proceedings of a Conference on Advanced Composite Materials in Civil Engineering Structures, Las Vegas, Nevada, 1991, ASCE, pp. 224-232.

[6]*Triantafillou, T. C. and Plevris, N.*, STRENGTHENING OF RC BEAMS WITH EPOXY-BONDED FIBRE-COMPOSITE MATERIALS - Materials and Structures, Vol. 25, 1992, pp. 201-211.

[7]*Chajes, M., Januszka, T., Mertz, D., Thomson, T., Jr., and Finch, W., Jr.* SHEAR STRENGTHENING OF REINFORCED CONCRETE BEAMS USING EXTERNALLY APPLIED COMPOSITE FABRICS - ACI Structural Journal, Vol. 92, No. 3, May-June 1995, pp. 295-303.

[8]*Arduini, M. and Nanni, A.*, BEHAVIOR OF PRE-CRACKED RC BEAMS STRENGTHENED WITH CARBON FRP SHEETS - ASCE, Journal of Composites in Construction, Vol. 1, No. 2, May 1997, pp. 63-70.

[9]*Oehlers, D. J.*, REINFORCED CONCRETE BEAMS WITH PLATES GLUED TO THEIR SOFFITS - Journal of Structural Engineering, Vol. 118, No. 8, August 1992, pp. 2023-2038.

[10]*Sharif, A., Al-Sulaimani, G., Basunbul, A., Baluch, M., and Ghaleb, B.*, STRENGTHENING OF INITIALLY LOADED REINFORCED CONCRETE BEAMS USING FRP PLATES - ACI Structural Journal, Vol. 91, No. 2, March - April 1994 pp. 60-168.

Fibreglass/Epoxy Composites for the Seismic Upgrading of Reinforced Concrete Beams with Shear and Bar Curtailment Deficiencies

J.I. Restrepo[1], Y.C. Wang[1], R.W. Irwin[2] and B. DeVino[3]

1 Department of Civil Engineering, University of Canterbury, Christchurch, New Zealand
2 Construction Techniques Group, Auckland, New Zealand
3 Composite Retrofit International, Montreal, Canada

Abstract

Many reinforced concrete buildings and bridges built prior to the mid 1970's in seismically prone regions of the world may need to be retrofitted if they are to survive without collapse during a strong seismic event. Advanced composite materials (ACM) offer the versatility and ease of application at a reasonable cost to carry out seismic retrofitting in those regions of critical elements that are deficient. This paper deal with the seismic retrofitting of flanged reinforced concrete beams of older frame buildings. Two full-scale tests were built and tested under reversed cyclic load conditions simulating seismic loading. A test unit was tested in its original condition, then repaired using fibreglass/epoxy strips and re-tested. The other unit was retrofitted prior to the test. Details of the test, the retrofitting technique and a discussion of the test results are given in this paper.

1. Introduction

For many years structural engineers have designed building structures in seismically prone regions of the world for lateral loads that are significantly less than those required to ensure elastic response. As a result, critical regions of the lateral load resisting elements are expected to undergo inelastic excursions and dissipate energy. If these regions are to perform adequately they must be detailed for ductility. While the concept of inelastic response was known since the late 1950's, reinforced concrete building codes began to incorporate ductility requirements for seismic design in the 1970's only. The American Concrete Institute building code, ACI 318-71 [1], following the San Fernando Earthquake of 1971, introduced seismic design recommendations by requiring the detailing for ductility of those regions in the earthquake resisting structural system assumed to be critical. The main recommendations to ensure ductile response were the use of closely spaced transverse reinforcement at the beam and column ends. In the 1970's the ACI 318-71 building code became accepted as a model code for seismic design in many parts of the world.

It is known now that buildings designed according to ACI 318-71 may not perform adequately during strong earthquakes. Recent earthquakes and experimental work have shown that shear failures, premature longitudinal bar buckling, and lap splices are some of the most common deficiencies found in structures designed with this code. ACMs have been shown to be able to successfully correct most of these deficiencies [6].

An aspect ignored in the ACI 318-71 building code was the participation of the slab longitudinal reinforcement toward the negative flexural strength of flanged beams of moment resisting frames designed to provide the earthquake resistance. It can be speculated that this practice was considered conservative as it was believed that the slab contribution increased the negative flexural strength of the beam and increased the lateral load resistance of the building and, as a result, decreased the ductility demand in the critical regions. The main problem when ignoring the participation of the slab reinforcement is that the curtailment of the beam reinforcement and the transverse reinforcement provided for shear resistance may be inadequate. Nowadays, modern codes require that the contribution of the reinforcement in cast-in-place concrete solid slabs be assessed and be accounted for in design [7, 9]. It can be shown that, especially in the upper floors of buildings where the slab contribution is more significant in relative terms, the critical region for the development of a negative plastic hinge may not occur at the beam ends but at a distance away towards midspan. The relocation of the critical region implies that plasticity will develop where no special detailing for ductility has been provided. Such beams may fail in a rather brittle manner and at a load less than that required to attain the flexural strength when accounting for the longitudinal reinforcement of the beam alone.

The detailing deficiency induced by the presence of the slab reinforcement can be corrected to ensure that a negative plastic hinge will develop and be maintained in the beam at the face of the column as initially anticipated in design. The retrofit of the beam is achieved by increasing the flexural strength at the top of the beam in and around the deficient regions. ACMs can be particularly effective in providing the additional flexural strength through the use of bonded strips. This paper summarizes an experimental programme to assess the effectiveness of ACM strips in retrofitting and repairing reinforced concrete flanged beams with bar curtailment deficiencies leading to flexure-shear failures.

2. Experimental Programme

Description of the Test Units

A research work, currently underway in the University of Canterbury, is looking at ways to assess the seismic behaviour of flanged beams designed according to older codes and to find ways to correct the potential deficiencies using ACMs. As part of the research work an eight storey building was designed following the seismic design recommendations contained ACI 318-71 building code. The primary earthquake resisting system of the building was formed by a grid of moment resisting frames spaced at 5.5 m and 5.0 m in two orthogonal directions. The slab was designed to transfer gravity loading in two-way action. The lateral force coefficient for the design of the building for seismic induced forces was 0.09. A refined assessment of the structure in this floor indicated that negative plastic hinges would not form at the beam ends due to the reinforcement arrangement in the slab and the curtailment of the longitudinal reinforcement in the beam.

Two identical full-scale beam subassemblages were built and tested under reversed cyclic loading conditions simulating seismic loading. Figure 1 shows reinforcing details of the units tested. Three tests were performed in the two units:

Test Unit T1 - Stage 1 Test and observe the seismic performance of the "as-built" unit
Test Unit T1 - Stage 2 Repair the beam in the damaged region and resume testing.
Test Unit T2 Retrofit the beam before testing. Test and observe the seismic performance of the retrofitted unit.

The units were cast using ready-mixed concrete. The concrete compressive strength at the day of testing, measured using 100 mm diameter by 200 mm high cylinders, was 24 and 18 MPa for Units T1 and T2, respectively. This relatively low strength concrete was deliberately preferred as it would provide a worst case scenario. The measured yield strength of the deformed (D) 28, 24 and 10 mm diameter bars was 316, 320 and 316 MPa, respectively. The measured yield strength of the 10 mm diameter plain round stirrups was

354 MPa. The units were repaired and retrofitted using TYFO S fibrwrap fibreglass/epoxy strips with 1.27 mm ply thickness, a measured tensile strength of 375 MPa and 20 GPa elastic modulus in the principal direction of the fibres. TYFO S fibrwrap is a proprietary ACM primarily reinforced in one direction.

Test Arrangement

Figure 2 shows the test set-up. The position of the hydraulic actuator at the beam end emulated reasonably well the position of the point of inflection in the beam when plastic hinges have formed in the bay of the frame. Figure 3 shows the test regime. Load controlled cycles were initially imposed to the units to find the secant stiffness and vertical displacement at 75% of the estimated capacity of the unit, see Figure 3. Displacement controlled cycles were applied to the units when loaded beyond the elastic range. The cycles were controlled in terms of the displacement ductility, μ_Δ, which is defined as the ratio between the applied vertical displacement, Δ, and the vertical displacement at first yield, Δ_y. The vertical displacement at first yield is defined here as 4/3 times the vertical displacement observed in the load controlled cycles to 75% of the capacity of the unit [7]. Vertical displacements were measured at the point of application of loading with any component of vertical displacement due to the rigid body rotation of the column stub being removed.

Test Results

Figure 4 shows the hysteretic response of the test units. The test units behaved as expected when loaded upwards. A positive plastic hinge formed in the beam at the column face and extensive yielding of the beam bottom longitudinal reinforcement was observed in this region. The measured and predicted capacity agreed very well, as anticipated.

Under downward loading, cracking in the beam extended from the column face to near the point of application of loading. When the beam was pushed into the inelastic range, a large diagonal crack inclined at 42° to the horizontal opened up at 1.3 m from the column face. The beam top longitudinal reinforcement remained elastic at the column face but yielded in the region at and adjacent to the large diagonal crack. Note that there are not closely spaced stirrups in this part of the beam. Further downward cycles resulted in extensive yielding of the stirrups and crushing of the concrete in this region of the beam. The extent of cracking at the end of the test is shown in Figure 5 (a). It is apparent in Fig. 4 (a) that the negative flexural strength, calculated at the face of the column and considering the slab reinforcement, was not attained. At -1% drift angle the main diagonal crack was 4.8 mm wide and the capacity of the beam had began to drop due to an imminent flexure-shear failure. This drift angle and the associated damage were considered to be the limit at which a satisfactory repair scheme could be carried out in a structure. Consequently, the test was halted and the repair work conducted.

The damaged unit T1 was repaired by epoxy injecting the main cracks, by applying two 1.38 m wide layers of TYFO S fibrwrap on the slab and by applying TYFO S fibrwrap on the sides of the beam at the damaged region. Figure 1 illustrates details of the repair work with ACM strips. The concrete surface where the ACM strips were to be applied was carefully ground to remove any laitance and then cleaned with oil-free air pressure. A thick layer of epoxy adhesive was applied to the concrete surface before the ACM. Since the interface bond between the ACM and the concrete surface was considered critical, care was taken to ensure that the steel plate at the point of application of downward loading would transfer the load directly onto the beam without clamping the ACM. Unit T2 was retrofitted in the same manner as unit T1.

The aim of the repair and retrofit work carried out with the 1.38 m wide fibreglass/epoxy strips was to force a negative plastic hinge to form at the column face, where closely spaced stirrups had been provided. Forcing the plastic hinges at the beam ends also reduces the plastic rotation demand in both the positive and negative plastic hinges in the bay of the frame. The repair/retrofit work was done by providing additional passive flexural resistance to the beam throughout its length, except at beam ends. The strength deficit was found by modelling the beam as variable angle truss [3]. Consideration was given during the design of the repair/retrofit scheme to limiting the longitudinal tensile strain in the ACM strip bonded to the slab to 0.4%, and also to limiting the bond stress between the ACM strip and the concrete to be less than $0.17\sqrt{f'_c}$ [MPa]. The tensile strain limit was chosen to avoid premature deterioration of the shear strength mechanism in the beam where the longitudinal reinforcement had been cut-off. The bond stress limit was selected to avoid delamination of the ACM strip. The three-sided layers of the fibreglass/epoxy were designed to control the width of the diagonal cracks and, hence, to delay the loss of the shear transfer mechanism through aggregate interlock.

The hysteretic response of the repaired unit T1 and the retrofitted unit T2 can be compared with the response of the prototype unit T1 in Fig. 4. It is evident that the ACM strip bonded to the slab was very effective in forcing a negative plastic hinge to develop in the beam at the column face. The significant strength increase under negative loading is due to the contribution of the slab reinforcement. The slab reinforcement was observed to yield across the full flange width in a yield line passing through the column face. This was also confirmed as the predicted capacity P_n accounting for all the slab longitudinal reinforcement agreed very well with the measured load.

Cracking due to negative loading in these two tests was more distributed than in the first test. The major cracks occurred in the beams at the column face. In these two tests delamination of the three-sided fibreglass/epoxy strips commenced from the top in the region where the beam top longitudinal reinforcement had been cut-off and slowly propagated downwards and sideways. This suggest that side strips bonded to the web of reinforced concrete beams may not be a reliable way to enhance the shear strength. Both units reached two cycles to -2% drift angle at a displacement ductility μ_Δ= -2 with minimum strength degradation. The hysteresis loops show some pinching due to the formation of diagonal cracks that developed in the beam at the top longitudinal bar cut-off points, see Fig. 5. The end of the test in these two units occurred in the cycles to a displacement ductility μ_Δ= -3 at -3% drift angle when reaching the peak displacement the top ACM strip suddenly peeled off.

3. Discussion

The effectiveness of ACM strips to provide additional flexural strength in beams with bar curtailment deficiencies was demonstrated in the experimental programme. The assessment and design of the repair/retrofit scheme can be carried out using a variable angle truss model [3], ensuring that the limit imposed on the ACM strip used to enhance the flexural strength at the bar cut-off points is always satisfied. Figure 6 depicts the strain profile of the ACM strip bonded to the slab of test unit T2 to correct the flexural strength deficiency at the beam longitudinal bar cut-off points. It is apparent that the observed behaviour of the ACM strip compares very well with the limiting design strain of 0.4% chosen for the retrofit. The maximum strain gradient occurred in the ACM strips between 1.7 to 2.5 m from the column face, see Fig. 6. The average bond stress in this region was $0.07\sqrt{f'_c}$ [MPa] and is the value associated with the bond failure observed in the tests. This value is much lower than values observed in tests loaded under monotonic conditions [2, 4, 8]. Several factors may be combined that result in such a low bond strength. For example the surface preparation and the effect of reversed cyclic loading may be variables that could influence the bond strength. It has also been observed that in ACM and steel strips bonded to concrete members large bond stresses tend to concentrate at the end of the strips ad lead to bond failures [4, 5, 8]. This stress concentration was not discernible from the strain readings taken during the experimental programme. It is worth noting that in a real structure ACM strips will be applied between column ends and the bond failure observed in the tests is unlikely to occur.

4. Conclusions

1. The experimental programme carried out in this research is conclusive that under some circumstances the critical region in beams of moment resisting frames designed for earthquake resistance and designed with older code provisions may form negative plastic hinges in apparently unexpected regions. This is due to the effect that the slab reinforcement has in the overall seismic response of the frame. This deficiency may result in an unexpected flexure-shear failure of the beam at the shifted critical location.

2. Tests showed that ACM strips bonded to the top of the slab can be used to increase the flexural strength of the beam in those deficient regions and force a plastic hinge to form at the beam ends where is desirable and where closely spaced stirrups are likely to be present.

3. It appears that, under reversed cyclic loading conditions, bond failures at the interface between the concrete surface and the ACM can occur at bond stresses as low as 0.07 $\sqrt{f'_c}$ [MPa], which is a value significantly lower than the limits commonly found from monotonic tests.

5. References

[1] *American Concrete Institute*, BUILDING CODE REQUIREMENTS FOR REINFORCED CONCRETE ACI 318-71, Detroit, Michigan, 1971.

[2] Arduini, M., Tommaso, A.D., and Nanni, A., BRITTLE FAILURE IN FRP PLATE AND SHEET BONDED BEAMS, ACI Structural Journal, Vol.94, No.4, 1997, pp.363-370.

[3] Collins, M. and Mitchell, D., PRESTRESSED CONCRETE STRUCTURES, Prentice Hall, New Jersey, 1991.

[4] *Mohamed Ali, M.S., and Oehlers, D.J.*, PREVENTING DEBONDING THROUGH SHEAR OF TENSION FACE PLATED CONCRETE BEAMS, Conference on the Mechanics of Structures and Materials, Grzebieta, Al-Mahaidi & Wilson, Australia, 1997, pp.123-127.

[5] *Norris, T., Saadatmanesh, H., and Ehsani, M.R.*, SHEAR AND FLEXURAL STRENGTHENING OF R/C BEAMS WITH CARBON FIBER SHEETS, Journal of Structural Engineering, ASCE, Vol.123, No.7, 1997, pp.903-911.

[6] *Priestley, M.J.N., Seible, F., and Calvi, G.M.*, SEISMIC DESIGN AND RETROFIT OF BRIDGES, John Wiley & Sons, New York, 1996.

[7] *Paulay, T. and Priestley, M.J.N.*, SEISMIC DESIGN OF REINFORCED CONCRETE AND MASONRY BUILDINGS, John Wiley & Sons, New York, 1992.

[8] *Roberts, T.M.*, APPROXIMATE ANALYSIS OF SHEAR AND NORMAL STRESS CONCENTRATIONS IN THE ADHESIVE LAYER OF PLATED RC BEAMS, The Structural Engineer, Vol.67, No.12, 1989, pp.229-233.

[9] *Standards Association of New Zealand*, NZS 3101: CONCRETE STRUCTURES STANDARD PART 1 AND PART 2 - THE DESIGN OF CONCRETE STRUCTURES, 1995, Wellington, New Zealand.

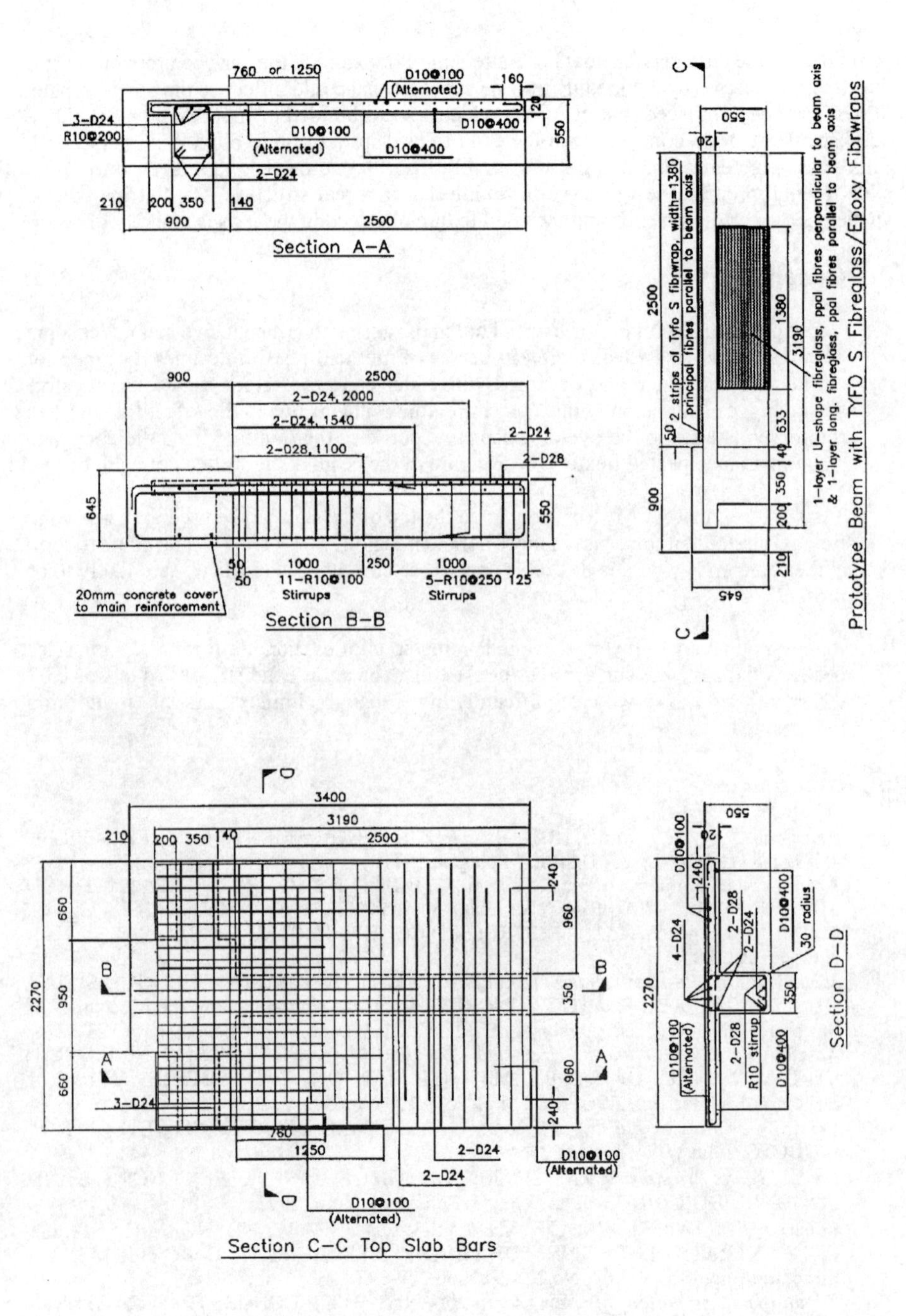

Figure 1 - Main reinforcing details of the units tested.

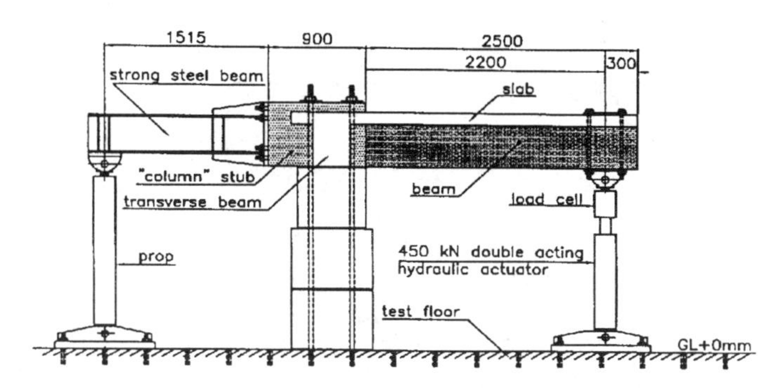

Figure 2 - Test set-up.

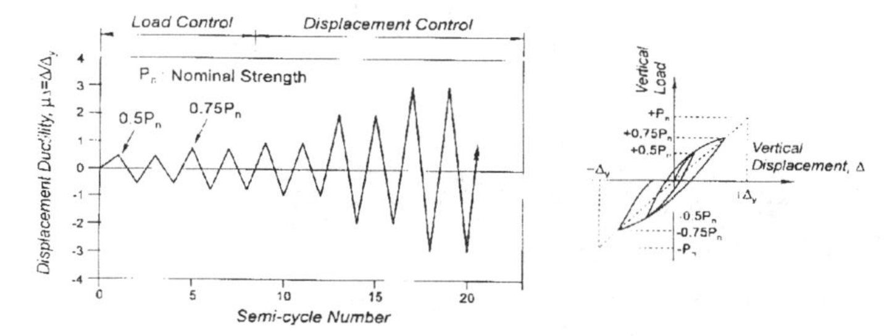

Figure 3 - Test regime and definition of first yield displacement.

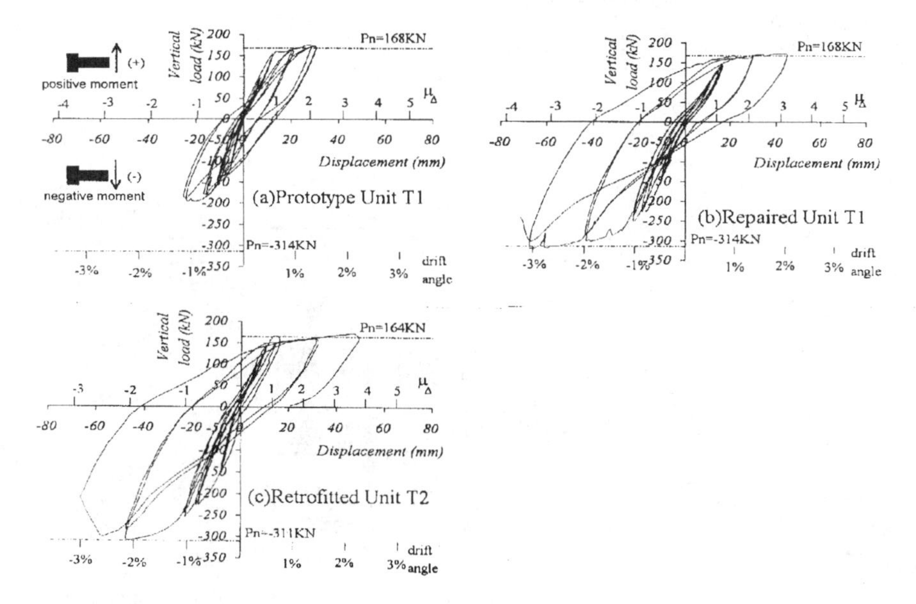

Figure 4 - Measured hysteretic response.

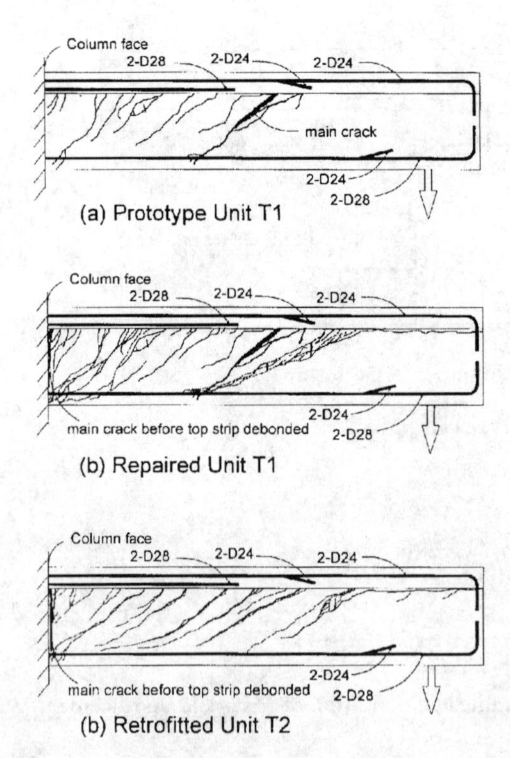

Figure 5 - Crack pattern developed in the test units when loaded downwards.

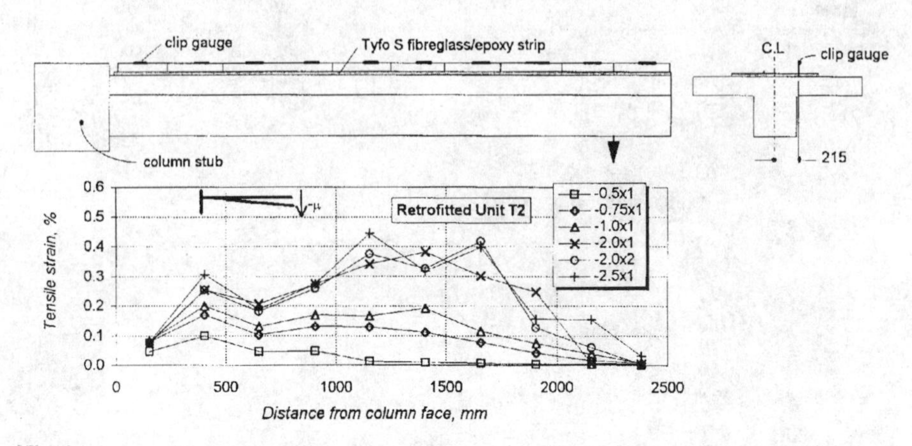

Figure 6 - Measured longitudinal strains in the top ACM strip of Unit T2.

On Carbon Fibers Strengthening of Heat Damaged Prestressed Concrete Elements

A. Benedetti (1), A. Nanni (2)

(1) DISTART, University of Bologna, Viale Risorgimento 2, I-40136 Bologna, ITALY
(2) Engineering Faculty, University of Missouri at Rolla, MO, USA

Abstract

In the paper we examine a recent rehabilitation work completed in Italy using C-FRP strips; in particular, due to the heat radiating from a flue, a double curvature prestressed concrete shell included in the roof of an industrial building experienced a severe localised damage, requiring so the definition of a proper repair technique.
The selection of this last involved the screening of different methodologies with relation to the constraints imposed by the requirements of the owner: small disturbances to the industrial process, safety of the occupants, low cost, durability of the repair; in addition, the position of the concrete element in the centre of the building did not allow the use of a crane for the lifting off of the damaged shell element.
The C-FRP repair showed to be capable to overcome all the requirements, and the final construction process demonstrated the high reliability of these new techniques; in fact new materials are highly successful in restoration, as far as we decide to open our mind and to consider not only customary solutions.

1. Introduction

The evolution of the construction industry has led in the past to the development of a variety of precast prestessed roof elements capable to cover the very large spans requested for free plan industrial buildings.
All these elements show very thin sections which are possible only mixing a pretensioning system and a by form resisting shape; however in this case we arrive generally to a very low fire resistance of the roof.
A very common solution in the European countries is the hypar HPV Silberkhul P/C shell (fig. 1), which covers a 20·2,5 m^2 area with only 75 to 100 mm thickness; tanks to the geometrical form of the surface, the pretensioning system is arranged in the form of two intersecting bundles of straight slanted high strength wires laying on the mean surface of the shell.
In the case history presented below, one of these element positioned in the roof of a major industrial building site near Perugia (Italy), has been damaged by the thermal shock due to the heat radiating from a flue where the firebrick insulation was fallen down (fig. 2).
Two types of problems appeared immediately, in order to set a suitable restoration technique: firstly, due to the large area of the building, it was not possible to find a crane capable to remove the damaged element from outside but, as a consequence of the machinery, it was almost impossible to replace the element from inside too. Secondly, due to the doubly curved surface of the element, a candidate solution would have to accomplish with a large bend.
After the first recognition, only two types of techniques presented the required features: a system of external steel wires and a system of carbon fiber reinforced plastic sheets (C-FRP). However, the close spacing of the prestressing wires in fact did not allow for the execution of the holes needed for the anchorage bolts fixing the steel external rods; so, ultimately the only strengthening technique compelling with the requirements was identified in the C-FRP [1].

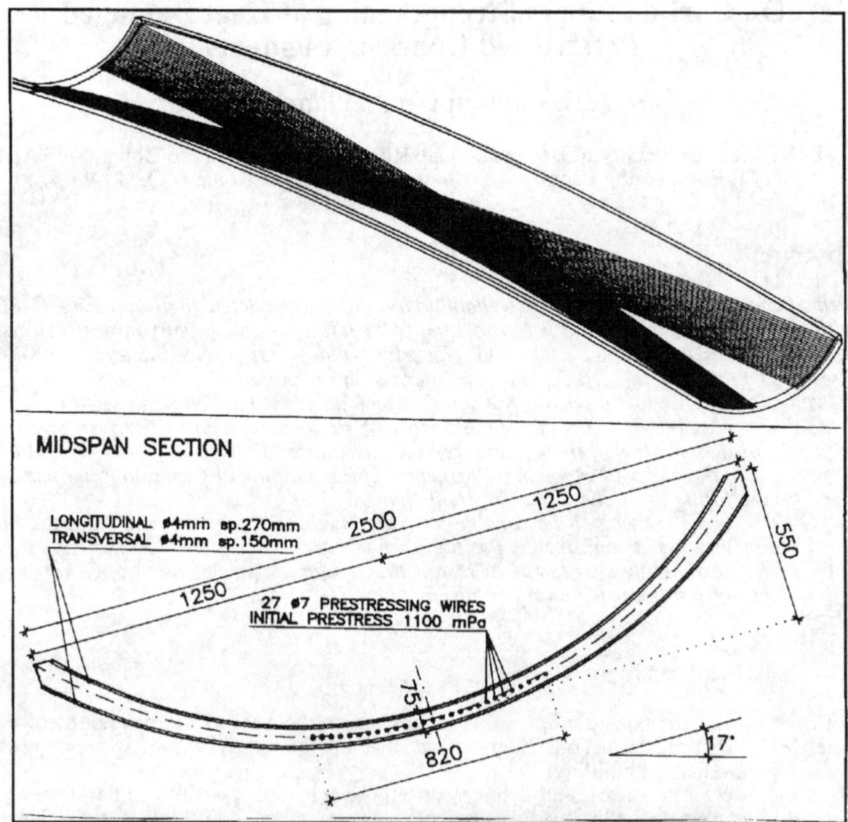

Fig. 1 : *Schematic view of HPV Silberkuhl roof shell*

In what follows we describe briefly the adopted execution technique and the features of the designed carbon fiber reinforcement net; after, we discuss the identification tests made before and after the strengthening of the P/C shell, showing that in this case the reduction of shear stiffness in the damaged zone plays a major role. Finally we conclude presenting the relevant data of the completed work.

2. Roof Shell design Data

The shells have been designed firstly in the sixties, and their evolution has signed several improvements leading finally to a well optimised structural element; all the Italian factories make use of accelerated wet-thermal ageing, arriving finally to a production of one shell on each casting form per day.
In fact the design includes a main loop assuming a beam like behaviour of the roof shell, and several local checks, where the appropriate 3-D behaviour is modelled; more precisely the checks concern the shear and torsional stresses at the supports, the effect of transversal bending moments, the anchoring splitting stresses and, if required, the seismic behaviour.

According to the Italian Regulations the R/C shell is dimensioned following the data reported in table I; it is worth noting that actually the Italian Regulations fit the Eurocode 2 Norms, assuming a safety factor for ultimate loads of 1.5.

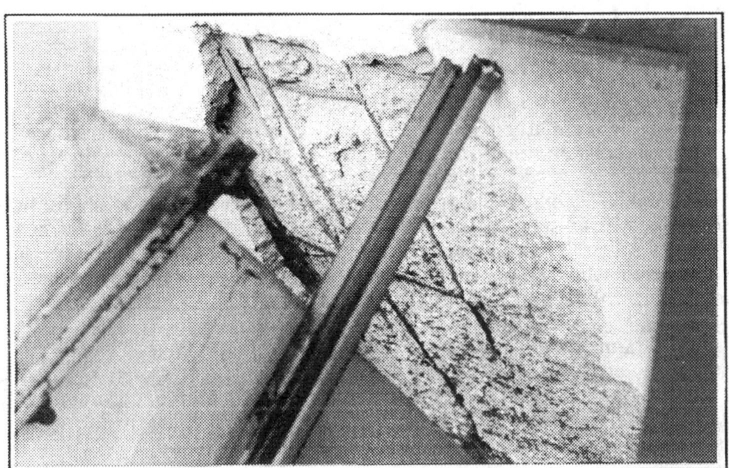

Fig. 2 : *View of the damaged zone in contact with the heat radiating flue*

Table I

Action	*Characteristic Value*	*Service Condition*	*Ultimate Snow*	*Ultimate Wind*	
Self Weight	2.4	2.4	3.36	3.36	kN/m^2
Windows	0.8	0.8	1.12	1.12	kN/m
Snow	1.2	1.2	1.8	1.2	kN/m^2
Wind	1.0	1.0	1.0	1.5	kN/m^2

The stress range allowed for P/C elements are as follows:

Table II

Component	*Service Initial*	*Service Long Term*	*Ultimate*
Concrete-Compression	0.60 f_{ck}	0.50 f_{ck}	0.55 f_{ck}
Concrete-Tension	0.10 f_{ck}	0.07 f_{ck}	-
Prestressing Wires	0.90 f_{pyk}	0.60 f_{ptk}	f_{yk}

The data assumed for the design are typical of medium technology Italian precasting companies:

f_{ck} (at the ageing end) = 25 MPa, f_{ck} (long term) = 42 MPa,

f_{pyk} = 1400 MPa, f_{ptk} = 1650 MPa.

The initial prestress, as a consequence of the technology, cannot be too high; in practice values around 1100 MPa are quite general for this type of shells. As a consequence, the prestress loss including the elastic one, is approximately 25 % of the initial one. On the other hand the collapse of this type of shells is mainly due to a change of form, showing almost no variation in a wide range of prestress forces.

3. Repair Technique

As can be seen in fig. 2 presenting a zoom of the damaged zone, the high thermal gradient in the concrete cover led to the crushing instability of the cover itself partly as a consequence of the thermal expansion and partly due to the pore pressure generated by vaporisation and moisture migration.
As a concern, the high temperature level raised by prestressing wires undoubtedly lowered to an unknown extent the prestressing force at least in the neighbourhood of the collapsed cover; we can suppose that the tension loss in the part of the wires heated above 500 °C has destroyed the steel concrete adherence for a certain length in the undamaged sections near the heating point, leading finally to a continuously varying tension from a minimum (unknown) value, to the initial value at a sufficient distance from the minimum one.
Before the Edilninno Company began with the commissioned work, a load test assessment was carried out from SGM Company; in this test, driven by a concentrated force in the middle of the shell, we measured the deflection of five points located under the load, at the quarter sections and in the supported ends. It is worth to mention that we had to stop the test due to the evident low stiffness of the element.

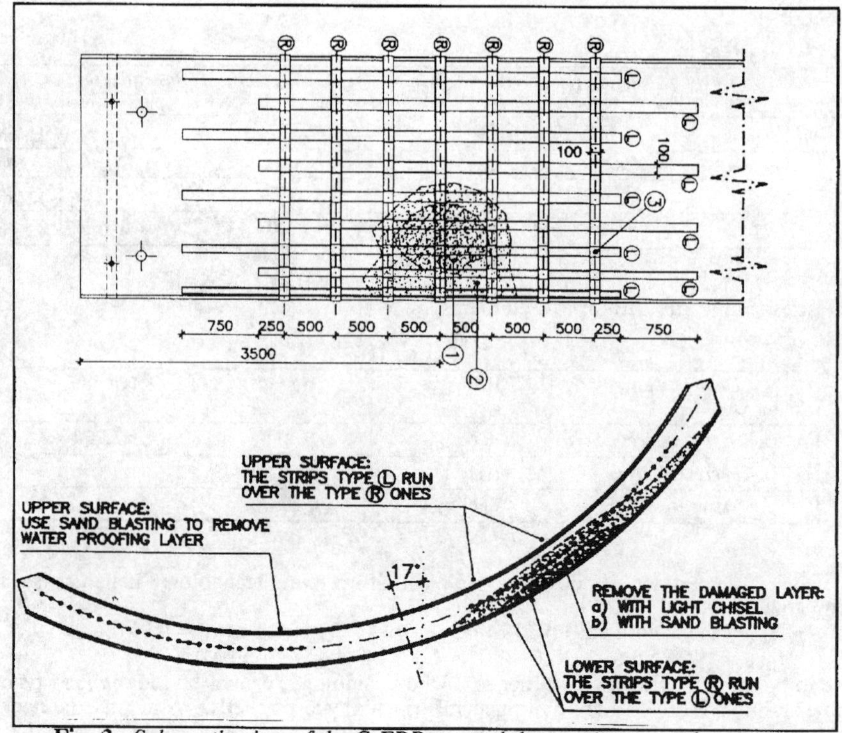

Fig. 3 : *Schematic view of the C-FRP net and the repair protocol activities*

3.1 Execution of the repair

As we ascertained that the problem was a mixed flexure-shear deformability, and taking into account the torsional behaviour of such type of non symmetrical shell placement, we decided to define a reinforcement net able to withstand both flexural and shear stresses.
Since the strips selected are unidirectional, no shear force can be resisted by the C-FRP in its transverse direction; thus the transmission of shear force requires the introduction of a mesh system able to generate an ideal Mörsch truss inside the shell surface. This can be practically accomplished defining a C-FRP grid composed of longitudinal and transverse strips with arrangements of +45°/-45° or 0°/90° acting as stringers [2-5].
The 0°/90° layout was finally selected for ease of installation resulting also in a cost minimisation of the total repair work (fig. 3).
Due to the negative Gaussian curvature of the roof shell, the bond strength of the C-FRP strips can be severely affected by the peeling effect of the normal stresses in the resin. The problem generated by the curvature of reinforcing strips can be effectively circumvented resorting to the "ordered" or "layered" layout concept; in fact, stating in the design phase not only the pattern of the reinforcing mesh, but also the working sequence, we can considerably increase the performance of the repair.
In the presented case we adopted the following sequence: in the lower surface of the shell the longitudinal concave strips were adhered first, whereas the convex transverse ones come after to restrain the others against peeling; in the upper surface the situation is reversed and the longitudinal strips were adhered last.
We defined finally the following repair protocol:

a) demolition of the damaged cover and sand blasting of the surface;
b) protection of the steel bars with a corrosion inhibitor paint;
c) replacement of fallen concrete with an high strength no shrinkage sand - cement mixture; the hand execution suggests the use of a tixotropic compound reinforced with polypropylene microfibers;
d) smoothing of the sticking areas with dry sand polishing and levelling of the concrete with an impregnation primer (fig. 4);
e) painting of the first epoxy resin base layer;
f) gluing of the C-FRP sheets in the desired direction and order (fig. 5);
g) fiber impregnation of the strengthening mesh;
h) execution of the protection layer in the form of a brush painted film.

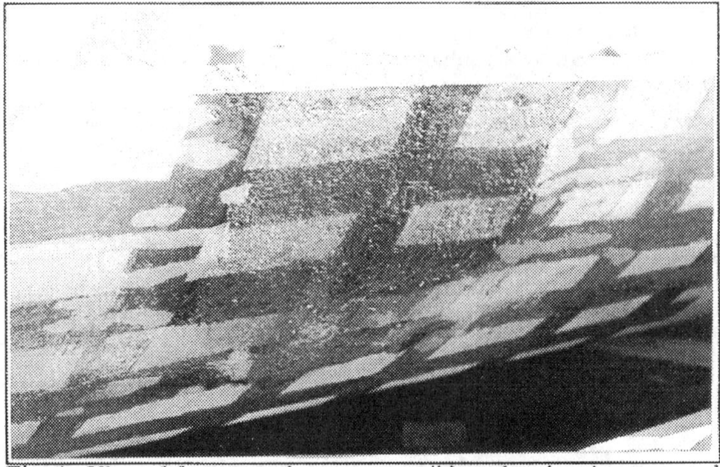

Fig. 4 : *View of the restored concrete sandblasted and preimpregnated*

Fig. 5 : *View of the C-FRP bonding operations*

3.2. Dimensioning of the repair

The system of 27+27 shell prestressing wires, passing through the damaged zone positioned at 3.5 m from the support, shows no superposition and an almost constant spacing of 50 mm among the wires (fig. 3); we can thus assume a total prestress loss of those crossing the 400·400 mm^2 central area where the cover is crushed, and a partial one in the eight neighbouring wires. It was computed finally an average total prestress loss of 25 %.

We define the repair extent needed as the quantity of longitudinal fibers that renders the final safety factor as high as the initial one; performing a limit state design we must equate the bending moment decrease due to the prestress loss with the increase sustained by the fibers. It is to recall that the wide elastic range of C-FRP imply that the ultimate stress of the fibers has to be defined according to the ultimate strain of the steel, although this condition is not a true collapse one; so defining as ρ, f_{py} and ε_{pu} the prestress loss ratio, the design stress and the ultimate strain of the steel wires, we have the condition:

$$\rho \, A_{ps} \, f_{py} \, h_p = A_{frp} \, E_{frp} \, \varepsilon_{pu} \, h_{frp} \, ,$$

where for generality we assumed a different bending arm for the steel and the FRP repair. Solving for the needed carbon fiber area we can define the geometry of longitudinal sheets. With the same scheme we can analyse the shell for shear and transversal bending moments, defining the entity of tie strips necessary in order to compose the mesh.

Finally, making use of Sumitomo Corp. Replark 30 sheets divided in three strips of 100 mm breadth each [6], we put in work approximately 110 m of repair; more precisely, the final design included nine longitudinal strips four meters long onto the lower surface, the same on the upper surface, and seven tie strips, six meters long each.

3. Identification of the Structural Behaviour

According to the uncertain characterisation of the prestress loss and with the motivation of checking the efficiency of the C-FRP repair, we prepared an experimental program based on the comparison of a load test before and one after the rehabilitation works. More precisely, we thought useful to carry out a load test with a concentrated force at the midspan of the shell, because by this way we can impose both bending moment and shear force to each section.
Main attention was focused on the midspan and quarter length vertical displacements; for sake of safety in the damaged state the load raised up only to 12 kN, while during the second test, aiming to reproduce the service condition of the roof, we fixed a target load of 36 kN. In table III the relevant data of the tests are shown.

Table III

Parameter	Test I	Test II	
Load	12	36	kN
Hysteresis Area Ratio	6,93	1,3	%
Midspan Deflection	10.0	23.9	mm
Quarter Length Deflection	8.4	16.8	mm

As a first consideration we note that, although the load is far higher in the second test, is the first that shows the larger hysteresis; in fact the anelastic deformation of the damaged zone can be considered the dissipation mechanism involved in the shell.
A very interesting information can be extracted from the midspan to quarter length displacement ratio; as a matter of fact, for a simply supported beam under concentrated load we have:

$$\eta = \frac{w_{L/4}}{w_{L/2}} = 0{,}6875 \,.$$

As far as the ratio runs toward this limit, we can infer a linear elastic behaviour.
Examining thus the ratio showed by the two tests we have a 0.84 value for the first while the second stops to 0.70; we conclude that in the damaged shell the bending and shear distortion caused by the rigidity variation along the axis line can deeply change the behaviour of the shell.
For sake of clarity we can cite that, numerical experiments with shell finite element meshes carried out in order to analyse the behaviour of the damaged and restored situations, let us conclude that the deformations after the repair are rather similar to the linear elastic hypothesis.
On the other hand, the modification of the elastic modulus and the inertia of the heated zone in order to fit the experimental data of the damaged shell, resulted in a factor ten reduction of the concrete section local rigidity.

4. Conclusion

The presented case history demonstrates that, for a strengthening to be done under the convergence of several limitation and needs, only FRP repair techniques have the sufficient ductility to fit all the requirements. Moreover, the tests carried out confirmed the expected behaviour of the repaired system.
If the today trend in growing information and reduction in price will continue in the future, we can forecast a sharp widening of FRP repair applications. In general the low

architectural impact, the negligible weight, the ability to fit to non planar surfaces, and the possibility of on site impregnation render the C-FRP sheets superior to steel plating reinforcement.
In the case of a double curvature shell weighting approximately 100 kN, the use of C-FRP could dramatically reduce the production downtime caused by a more traditional demolition/reconstruction technique.
The load tests completed, despite limited to assess the behaviour in a quasi linear range, suggest that the heat damage exerted to the structure played a major role in terms of load deflection performance of the member, and that the adopted strengthening technique successfully restored the original stiffness.

Acknowledgements
The restoration works have been completed by TEC.INN Company in partnership to Edilninno Srl of Fabriano (Italy); the experimental tests have been executed under the supervision of the authors by SGM company of Perugia (Italy).

References

[1] NANNI A.,Ed., *"Fiber Reinforced Plastics (FRP) Reinforcement for Concrete Strutures: Properties and Applications."*, Developments in Civil Engineering n. 42, Elsevier, Amsterdam 1993.
[2] CHAJES M., JANUSZKA T., MERTZ D., THOMSON T., FINCH W., "Shear strengthening of reinforced concrete beams using externally applied composite fabrics", *ACI Jour.*, **92**-3, May-June 1995, 295-303.
[3] AN W., SAADTMANESH H., ESHANI M., "R/C Beams Strengthened with GFRP Plates: Analysis and parametric Study", *ASCE Jour. of Struct. Engnrg*, **117**-11, Nov. 1994, 3434-3455.
[4] SHARIF A., AL-SULAIMANI G.J., BASUNBUL I.A., BALUCH M.H., GHALEB B.N., "Strengthening Of Initially Loaded Reinforced Concrete Beams Using FRP Plates", *ACI Jour.*, **91**-3, Feb. 94, 160-168.
[5] ACI COMMITTE 440 "State-of-the-Art Reporton FRP Concrete Structures", *ACI 440R-96,* ACI Farmington Hills, MI, 68 pp.
[6] MITSUBISHI CHEMICAL *"Replark Manufacturer's Brochure"*, Tokio, Japan 1996, 12p.

FRP COMPOSITES FOR MODULAR CONSTRUCTION OF HOUSING PROJECTS

Simone Starnini[1], Marco Arduini[2], Luisella Gelsomino[3], Antonio Nanni[4]

Keywords: panel system, floor system, FRP, connections, low-cost, construction.

Abstract

This paper is the result of a study about the use of FRP in the field of housing. Several structural applications take profit of the great worthies of composite materials, but rarely these benefits have been applied to the construction of residential units. The building technique described in this paper allows an easy and quick construction of two story houses through the assembly of FRP panels and connectors.

Introduction

The development of new materials, especially FRP, is bringing innovation with regard to building techniques and structural performances. As many applications have been successfully tested and adopted, the use of FRP in the construction industry has basically been limited to reinforcing bars for RC elements, or specifically shaped structural parts, especially in chemically aggressive environments.

The proposed low-cost construction solutions described in this paper show a way to employ the great worthies of FRP in the field of housing. The FRP panels building system presented in this paper, is meant for easy and quick construction, able to provide two-story houses in a week. Since all the pieces have modular dimensions, the system offers a significant versatility. After being installed, the hollow panels are filled with cast-in-place concrete. A further hollow space in the external walls may be filled with insulating foam. Several panel systems for load bearing walls and partitions are being produced, but connections between vertical and horizontal diaphragms present a certain degree of complexity. Thus, particular attention is paid in this study to the achievement of an appropriate FRP flooring system to be adopted in the construction of two story residential units. A general analysis of costs is also provided to show how the proposed system may be competitive with the traditional building techniques. Prices may be affected by several variations, depending on economic, social and geographical factors, which are not addressed. Nevertheless, the expected mass production of FRP has to be accounted as a significant cause in the decrease of prices.

[1] Professional engineer, University of Bologna, Italy
[2] Professional engineer, University of Bologna, Italy
[3] Professor, University of Bologna, Italy
[4] Professor, University of Missouri, Rolla

I. FLOOR SYSTEM

Construction of slab on grade for the first floor and suspended floor system is described in this section.

First, the necessary soil is excavated to accommodate the perimeter ring beam and the interior walls footings. Formwork is constructed, and when required, the insulation panels can be placed together with the vapor barrier (see Figure 1). All the mechanical and electrical services are arranged in the correct position, and they are capped and extend above the surface of the slab. The granular material is placed for the specified depth and it is compacted. The foundation dowel is arranged to anchor the wall assembly components to the concrete footing. The steel bar is cast into the concrete footing or grouted with a cementitious grout into holes drilled into the concrete footing. The foundation slab is poured and the final surface is smoothed and leveled. Before processing with the assembly of the building, all the formworks must be removed from the slab.

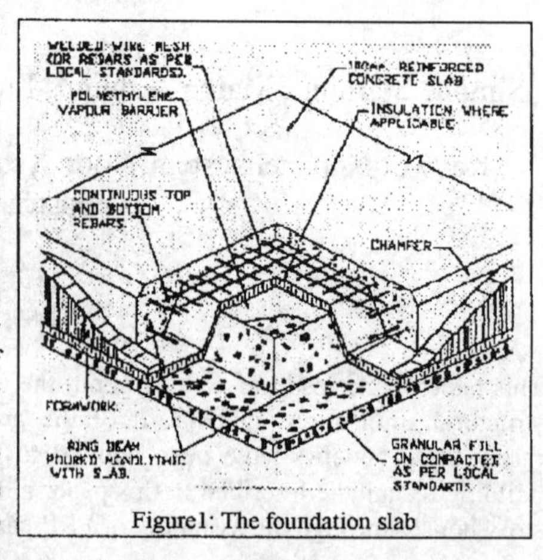

Figure1: The foundation slab

In the construction of a two-story house using the proposed FRP panel system, special attention is devoted to construction and detailing of the suspended floor. Several different floor may be adapted to the proposed modular technique, either clay tiles floors or prestressed concrete ones. Specific solutions have already been proposed but they are not discussed in this paper. For the aim of this study, an FRP panel floor system is presented, meant to show the performances of composite materials in terms of strength, design flexibility, and light-weight. A pultruded FRP panel is the part of the floor that provides tensile reinforcement for the complete system and permanent formwork (see Figure2).

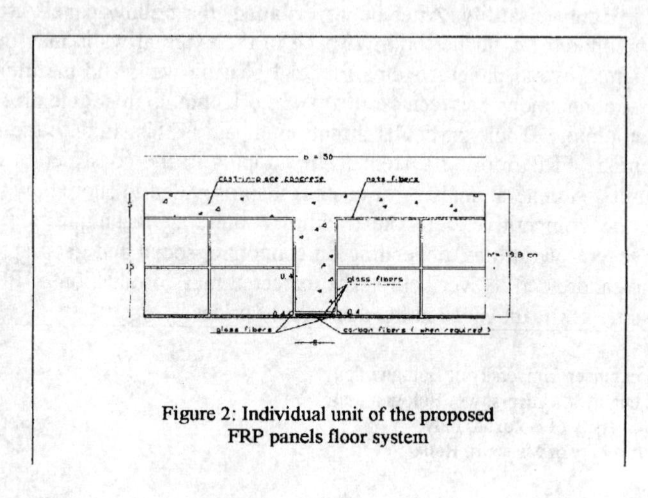

Figure 2: Individual unit of the proposed FRP panels floor system

The cross-section is 4 mm thick and it is composed of 420 mm adjacent hollow blocks, at 80 mm spacing. When the concrete is poured this space is filled, while the hollow blocks allow the floor to be light-weight. Beside this function, the void supplies a space where water, heating and electrical arrangements may be located. If a failure should occur, the ducts are easy to reach and repair from the lower ceiling, where appropriate access is provided. A rough surface or a layer of adhesive guarantees the bond of concrete and FRP. The cast-in-place concrete fills the spaces between the blocks, providing continuity and shear resistance. The top slab could be 50 mm or any required thickness. The FRP section is reinforced with continuous glass fibers except for the top surface which is made of glass fiber mats. The continuous fibers provide strength where it is needed, and less expensive mats provide a support for the fresh concrete. A 4 mm thick carbon FRP sheet may be equipped in the lower zone of the section subjected to tensile stresses. When required by the need of higher strength, it is enclosed in a 4 mm thick glass fiber reinforcing element. This is usually not necessary for a housing floor with span smaller than 6 meters (see Calculation in Appendix A). Preliminary cost analysis should account for the fact that carbon fibers are significantly more expensive than glass fibers. Two additional glass fiber reinforced corners are provided, shaped as a quarter of a circle and running all through the span. For low loads or reasonable spans these corners are sufficient to carry the tensile force (see the calculation section). The adjacent panels are connected through a simple hooked key which is meant to help and ease the operations of spanning of the floor (see Figure 3). Once the panel is locked to the adjacent one, it is slid till the end (position "a") and then released (position "b"). As a result the speed of construction is significantly improved. The panel system is designed to be self-supporting during concrete placement, so there is no need of shoring (see Calculations in Appendix A). Time savings are even more important in consideration of the possibility of serial casting of several attached units at a time.

II. WALL SYSTEM

The vertical diaphragms are constructed with modular panels and box connectors. Bearing walls and their connectors are made of glass FRP, while partitions have no reinforcing fibers. Wall panels have hollow cross-section, 200 mm high and 600 mm wide, with 4 mm thick walls. Wall pieces may be cut straight across the top or cut at an angle. Angled pieces form the walls that follow the slope of the roof. The box connectors have hollow section too, the same thickness and height, but they are 200 mm wide. These elements are provided with a set of connecting legs on two, or three, or all four sides, depending on their position and on the number of vertical panels they are supposed to connect (see Figure 4). Assembly procedures are quick and easy, and no special support is required during construction. Starting from a corner, a box connector is placed over the appropriate bar coming out from the foundation slab (or from the lower panel for the second floor walls), then the panels slid into each side of the corner box connector (see Figure 5). The assembling of appropriate panels, box connectors, door and window frames, and other pieces, continues in both directions away from the corner until all the pieces are in place. As mentioned, door and window frames and windows are provided. These elements slide on the appropriate panel and then are reinforced on top with steel bars.

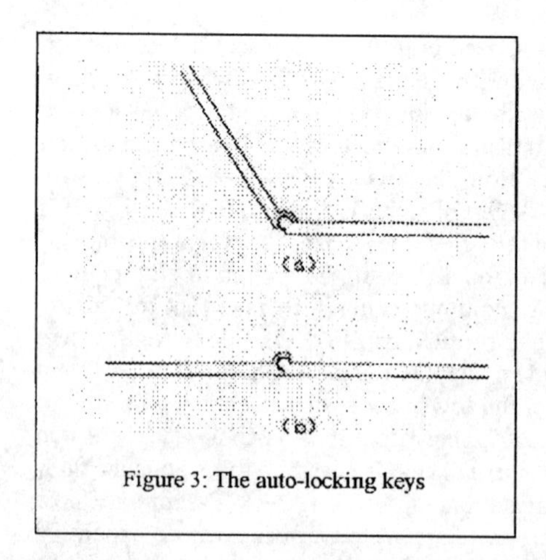

Figure 3: The auto-locking keys

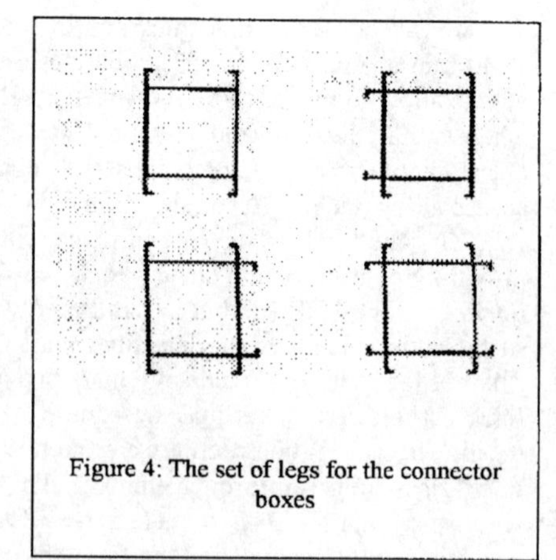

Figure 4: The set of legs for the connector boxes

Reinforcing bars must extend 90 mm minimum into each lateral box connector and lap 450 mm minimum in the header (see Figure 6).

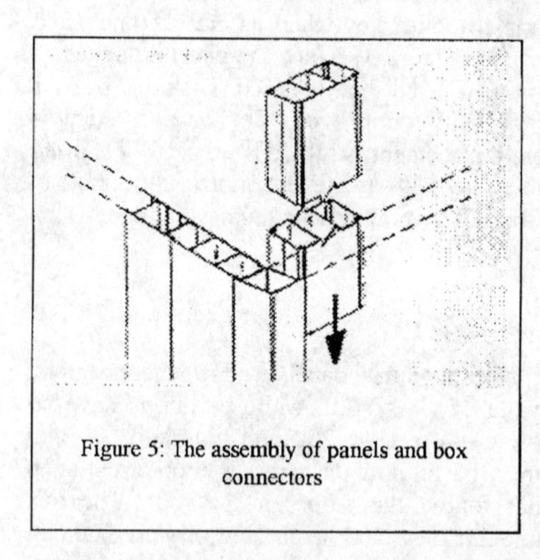

Figure 5: The assembly of panels and box connectors

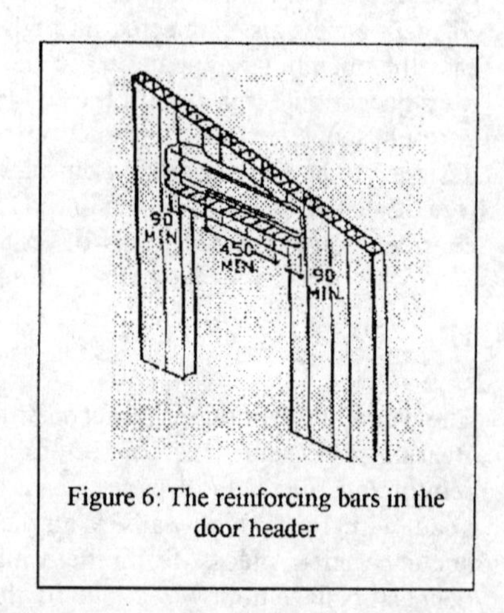

Figure 6: The reinforcing bars in the door header

When these operations are completed, the roof beam may be installed. At each box connector that supports a beam, a reinforcing bar is inserted, extending from the slab to the underside of the beam itself. An anchor bolt tied with tie wire to the bar and locked with a nut to the beam provide the required connection. Each end of the beam is lowered over the anchor bolts extending through the hole in the bottom and top of the beam. The screws are now to be installed so that the beam is held in place until the concrete is poured. Before pouring the concrete inside the panels, some operations must be accomplished. First of all, all the electrical channels are to be installed at each location where required. The electrical channel slides over the small legs, from the top of the panel until it reaches the slab. The function of this channel is to keep the electrical wires apart from concrete and at a safe distance from the surface of the wall, according to the requirements of the building codes. Where the electrical fixtures are to be located, a hole is cut in the wall. After ensuring that all the walls are plumb, square, and correctly positioned, concrete may be placed. Concrete must have a minimum compressive strength of 20 MPa at 28 days and the concrete mix must have a maximum aggregate size at 10 mm and be properly consolidated. A water reducing additive and an air-entraining agent is recommended. Concrete needs at least two days to reach a proper strength, and when the first floor panels concrete has set, the floor may be installed. Once the floor system is installed and the connections ready, the second floor panels may be erected following the same procedures of the first floor walls. Immediately after the concrete is poured in these panels, the roof anchors are to be properly installed (see Figure 7). Before installing the roof it is essential to let the concrete set for at least three days. In this study, a typical gable roof with tiles is considered, but different types of roof may be accomplished. On a roof without tiles, the roof deck is assembled from a series of panels and box connectors so that they form a smooth surface. On a roof with tiles, the box connectors in the roof deck are installed so that each has a set of legs facing upwards to accommodate the tiles adapters. The gable roof is assembled beginning from one angled wall, sliding the appropriate roof gable box connector into place. Moving from one side, this operation is repeated for each box connector (see Figure 8).

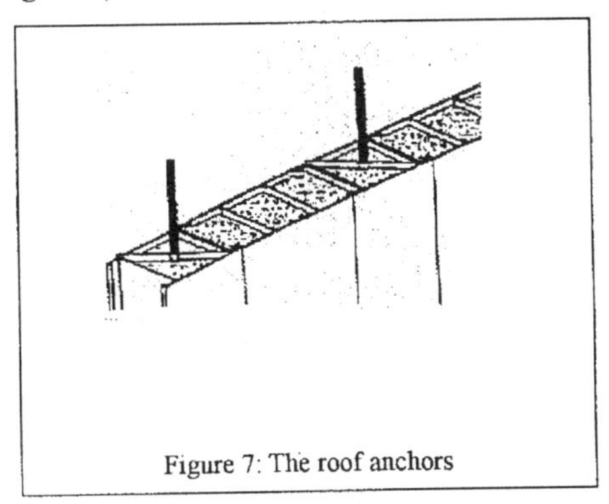

Figure 7: The roof anchors

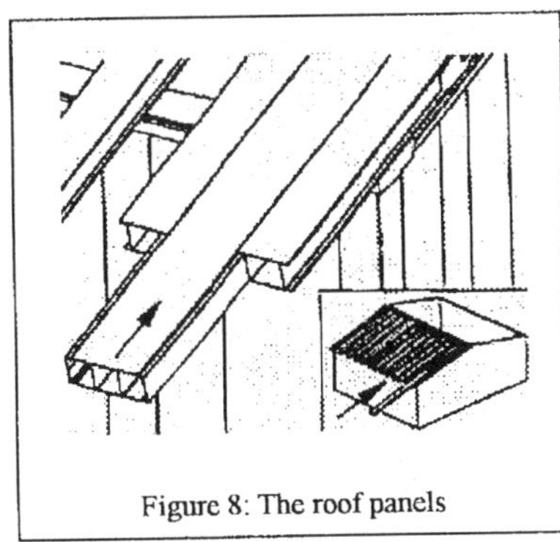

Figure 8: The roof panels

Once all the connectors are in place, the roof panels are to be slid in between and then fixed driving a screw through both sides of the panel, through the legs of the box connector and into the ridge beam. At the top of the panel, two additional screws are driven at the center webs into the ridge beam. The same procedures are followed for the other side of the roof, then all the panels and the connectors are fixed either to the angled walls and to the flat walls through metal wedges. Several varieties of roof tiles may be installed, essentially in the same manner. For the aims of a housing project, a clay tile roof is considered. The roof panels contain grooves to accommodate the snap legs on the roof tiles. The first row of tiles is also fixed to the roof deck with screws, then all the other rows are assembled behind the first. The roof tiles rows are added until the edge of the roof deck is reached for both sides, and a roof peak is then installed. A number of different solutions may be chosen regarding the roof tiles, the external wall finishes, and the architectural settlements. Many different vinyl casings may be easily installed driving screws, and they all perfectly replicate the original material: brick, stone, stucco or vinyl siding.

III. ANALYSIS OF COSTS

The costs for the alternative proposal are inspected in this section (see Table 1). Bearing walls, partitions and the roof are constructed with the panel system seen before. The innovative floor system is made of FRP panels. These interlocking elements are shaped in such a way to accomplish a light-weight floor, easy and fast to construct. Moreover, the strength of the fibers in the direction of the span allows the panel to support the weight of the cast in place concrete beside its own weight. The price of these floor system is determined through square foot costs[1], so it may be affected by significant variations depending on the production technique, the amount of production, the span and the type of required fibers (Glass fibers panels are considered in the following analysis; Carbon fibers panel are highly more expensive). Clay tiles could be used for the floor, even if carpet is suggested to achieve a better connection with the vertical diaphragms.

Table 1: Costs evaluation for the experimental proposal

	Product-Work	Cost per Sq. Foot of living area [US $]	Extension of living area [Square Foot]	Total [US $]
1	Site Work	0.52	1,165	606
2	Foundation	4.30	1,165	5,009.5
3	Walls and Partitions	14.00	1,165	16,310
4	FRP Floor System	4.80	1,165	5,592
5	Floor Finishing	2.65	1,165	2,530.75
6	Roofing	11.0	1,165	12,815
7	Stairs			2,205
8	Electrical and Piping	3.25	1,165	3,786.25
9	Cost of the Land			40,010
10	Overhead	3.60	1,165	4,185
	TOTAL			**93,632**

Comparing these data to the average cost for an attached unit in the Italian housing market (around US $ 90,000) the gap is irrelevant. Not only it is just the 3.8% of the final price, but the absence of maintenance costs must be accounted. Furthermore, the constituent materials of the panels don't degrade during the years, so no further expenses are required after the construction is completed. Even considering that costs may be affected by variations, and that an exact determination is hardly predictable, this analysis shows a very competitive product, which is likely to be found attractive by several groups of customers.

[1] Source: Means "Residential Cost Data", 15th Annual Edition, 1996

Appendix A: Calculations

The following calculations are made to check the resistance to loads of the experimental FRP floor system; further calculations are intended to prove the effective self-supporting function of the system itself.
The assumptions made in this section are:
i) Material is linear elastic
ii) Concrete resists compression only
iii) Perfect bond between concrete and FRP

The position of the neutral axis is found by imposing the first moment of area of the compressive zone equals that of the tensile zone. The area of the quarter of a circle reinforcement is chosen equal to *2ϕ 16* bars.
The depth of the neutral axis results equal to: x = 5.53 cm
The moment of inertia, supposing that the concrete below the neutral axis is all cracked, is: $I = 10466\ cm^4$

Loads analysis

With reference to ACI values for loads:
Live load........................2 kN/m²
Partitions.....................0.4 kN/m²
Floor tiles....................0.3 kN/m²
TOTAL........................2.7 kN/m²

Under this assumptions, the midspan deflection results equal to:

$$\delta = \frac{5}{384} \cdot \frac{ql^4}{EI} = 0.72\ cm$$

This value is widely lower than $\frac{1}{360}$ of the span which is 1.67 cm.
If the carbon fiber sheet is provided, a further amount of tensile strength is given to the section; assuming its thickness equal to 0.4 cm, the values seen before change into:

$$x = 6.53cm$$

$$I_x = 18258\ cm^4$$

$$\delta = 0.416\ cm$$

This value shows that the reduction in the midspan deflection (= 42 %) reached inserting the carbon fiber FRP layer is significant but, being both very small values, the sheet of carbon fiber FRP may not be convenient with regard to the cost of it. A self supporting panel system has to be able to resist to the weight of the concrete that is going to be cast on top. Supposing to use light-weight concrete, its weight is:

$$P = 18\ \frac{kN}{m^3}$$

and we can assume as a load : $q = 0.66\ {kN}/{m}$
With this load the bending moment is: $M = 3.015 kN \cdot m$
In this case the depth of the neutral axis is x = 9.05 cm

and the moment of inertia is equal to: $I_x = 2147.4 \ cm^4$

Under these assumptions the stress on the fibers is : $\sigma = 126 \ kg/cm^2 = 12.6 \ MPa$

Taking as a value for the Young's modulus $E = 3 \cdot 10^6 \ psi \cong 210921 \ kg/cm^2$, we find

a midspan deflection equal to : $\delta = \frac{5}{384}\frac{q \cdot l^4}{EI} = 1.23 cm \ \left(\leq \frac{l}{360} = 1.67 cm \right)$

These are reasonable and acceptable values, but if a smaller stress or deflection are required, a significant decrease of their value may be reached by a lightly bigger thickness of the superior fibers. Considering in fact a thickness equal to 0.635 mm (= 1/4 in), the stress is: $\sigma = 82 \ kg/cm^2 = 8.2 \ MPa$

and the midspan deflection is equal to: $\delta = 0.95 \ cm \quad \left(\leq \frac{l}{360} = 1.67 cm \right)$

The self supporting panels must also resist to shear stresses: considering that the fibers resisting to the shear forces are the vertical supports of the blocks, the shear tension results to be: $V = \frac{q \cdot l}{2} = 198 \ kg$

and $\tau = 8.25 \ kg/cm^2 = 0.82 \ MPa$

which is a value much smaller than the maximum shear stress allowable in the section, which may be found as:

$$\tau_{max} = \left(\frac{3}{2}\right)\left(\frac{V_u}{2td}\right) = 125 \ kg/cm^2 = 12.5 \ MPa$$

At the same time there are no problems of buckling, being the shear stress so far from the value of shear buckling which may be determined theoretically as follows (Timoshenko and Gere 1961; Holmes and Just 1983):

$$\tau_b^* = \frac{4K\sqrt[4]{D_L D_L^3}}{td^2} \quad for \ \vartheta \succ 1 \qquad \text{or}$$

$$\tau_b^* = \frac{4K\sqrt{D_T H}}{td^2} \quad for \ \vartheta \prec 1 \qquad \text{where}$$

$$D_L = \frac{E \cdot t}{12(1-\upsilon_L \cdot \upsilon_T)} \qquad D_T = \frac{E \cdot t}{12(1-\upsilon_L \cdot \upsilon_T)}$$

$$H = \frac{1}{2}(\upsilon_L D_T + \upsilon_T D_L) + \frac{G \cdot t^3}{6(1-\upsilon_L \upsilon_T)}$$

$$\vartheta = \frac{\sqrt{D_L D_T}}{H}$$

This value of the shear stress is :

$$\tau_b = 125 \ kg/cm^2 = 12.5 \ MPa$$

Because the shear stress in the considered section is 0.82 MPa, there are no risks of buckling.

Strengthening of PC Slabs with Carbon FRP Composites

A. Nanni* and W. Gold*

*The University of Missouri – Rolla, Department of Civil Engineering, Rolla, Missouri 65409-0300 USA

Abstract

Projects involving the strengthening of existing prestressed concrete (PC) slabs with composite materials are presented. Each project presents a different aspect of strengthening and a different structural configuration. The slabs are one-way and two-way; and require strengthening based on a change in use, a degradation problem, or a design/construction defect. The strengthening method used in each project involved externally bonding dry carbon fiber reinforced polymer (CFRP) sheets to the concrete surface as a means of providing additional flexural strength. This installation technique is known as wet lay up. The presented projects demonstrate the adaptability of this technology to solve many problems in the repair and rehabilitation of concrete structures.

1. Introduction

The need for strengthening reinforced concrete (RC) and prestressed concrete (PC) structures is becoming more apparent, particularly when there is an increase in load requirements, a change in use, a degradation problem, or some design/construction defects. Potential solutions range from replacement of the structure, to strengthening with steel plates, to the use of composites such as fiber reinforced plastic (FRP) materials [1,2]. Among several possibilities, FRP composites can be used for this purpose in the form of flexible carbon FRP (CFRP) sheets externally bonded to the concrete surfaces. The flexible sheets considered in this paper are made of unidirectional dry fibers and are installed with a technique known as manual lay-up.

In the United States, owners and contractors are realizing that the high material costs associated with CFRP are offset by time saving benefits associated with installation. This coupled with the adaptability of this technology is leading to increased use of this strengthening method. This paper shows the adaptability of CFRP sheet bonding through investigations of three projects in which a variety of structural repair problems are solved using this technology.

2. Correcting design/construction defects

The post-tensioned slab of a parking garage was originally strengthened with gunite RC beams shortly after construction in order to correct a deficiency in the number of steel tendons along the east-west alignment of the building (see Figure 2.1). These beams were 75 mm deep, 1 m wide and reinforced with 6 Number 9 (28.5 mm diameter) bars. The beams were 5.2 m long and ran along the column line, connecting the column drop panels (3 by 3 m). The integrity of the composite action between gunite beam and slab was to be based solely on the strength of the interfacial bond between the two. Since delamination had occurred over time, such action was compromised and epoxy injection was required.

Figure 2.1 – Original gunite beam strengthening solution

In order to find a more permanent solution to the problem, it was suggested that the gunite beams be demolished and replaced with externally bonded CFRP sheets [3]. As shown in the summary calculations given below, two double-ply strips of CFRP were sufficient to strengthen the slab. The resistance and the effect of design loads on the column-strip of the slab were computed using the assumptions listed below:

Uniform Dead Load (w_{DL}) = 4.63 kPa
Uniform Live Load (w_{LL}) = 2.39 kPa
Clear span (l_n) = 7.72 m
Transverse length (l_2) = 9.14 m
Effective column strip width (b) = 4.11 m
Moment due to loads = $0.35\ 0.60\ (w\ l_2\ l_n^2\ /8)$
where 35% is for moment distribution and 60% reduction is middle-strip effect

The results of this calculation are given in Table 2.1. M_u is a factored moment of $1.4M_{DL}+1.7M_{LL}$ and ϕM_n is the design moment capacity.

Table 2.1 – Summary of design values

Condition	M_u or ϕM_n (N-m)	Failure Mode
$0.35\ 0.60\ (w_u\ l_2\ l_n^2\ /8)$	151.1	N/A
Slab only	113.1	Strand yielding
Slab with RC Beam (before de-lam)	319.7	Strand and rebar yielding
Slab with 4 CFRP sheets	295.2	Strand yielding and FRP rupture

Figure 2.2 shows the application of the second ply for one of the strips. The CFRP strips were located on either side of the demolished gunite beam so that adhesion would take place on a smoother concrete surface.

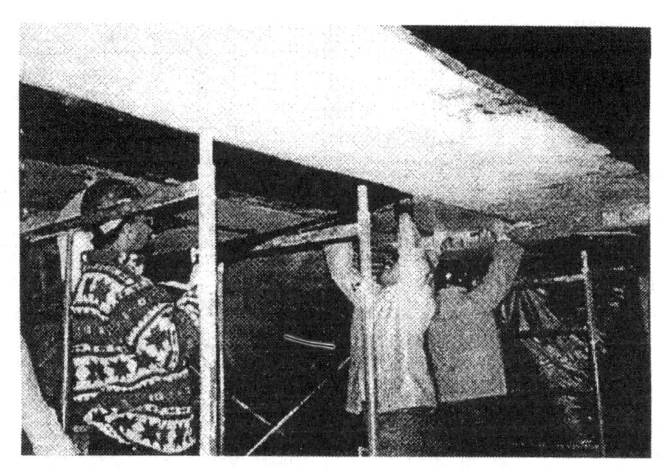

Figure 2.2 – Application of Second CFRP Ply

3. Restoring structural capacity after a degradation problem

The one-way post-tensioned deck slab of an industrial facility had suffered a loss of tendons due to stress corrosion. The slab is 28 cm thick and continuously spans over several wide-flange steel beams. Typical spans are in the range of 2.3-m to 2.6-m. The slab required both positive and negative moment strengthening. Due to accessibility issues, the only options for rehabilitation were the use of bonded CFRP sheets or demolition and replacement of the slab. The CFRP option allowed the facility to remain in service during the repair, and was therefore the method of choice.

The strengthening system consisted of evenly spaced CFRP strips installed on the top of the slab and on the slab soffit. Due to the degradation problem, the existing concrete surface had become moderately cracked. These cracks were repaired by epoxy injection prior to installation of the CFRP.

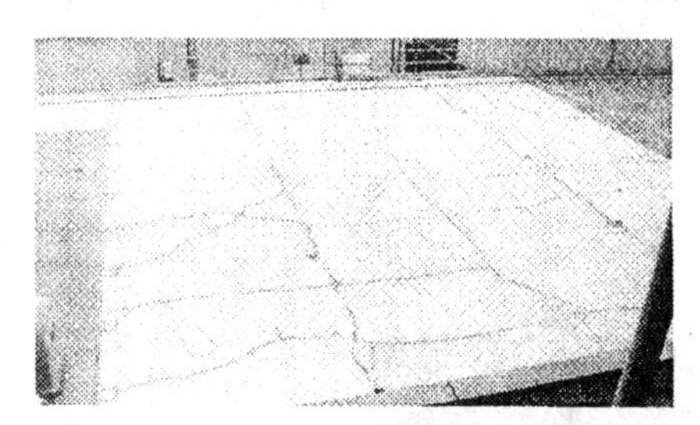

Figure 3.1 – Slab before strengthening

Figure 3.2 – Slab after strengthening

The loads on the slab consisted of dead load, live load, and concentrated loads from a crane that traversed the deck surface. The moments induced by these loads were computed based on one-way slab action. The loads for the 2.6-m span are summarized in Table 3.1.

Table 3.1 – Service loads and bending moments for a typical span

Condition	Uniform Load (kPa)	Positive Moment (kN-m/m)	Negative Moment (k-in/ft)
Dead Load	6.6	2.7	3.7
Live Load	14.4	6.0	8.0
Crane Load	N/A	13.3	17.5

Negative moment strengthening consisted of 25-cm strips of CFRP installed at 1.5-m intervals on the top of the slab. Figure 3.1 shows the top of the slab prior to strengthening and Figure 3.2 shows the slab after strengthening. Table 3.2 summarizes the design moments for the 2.6-m span.

Table 3.2 – Summary of design values for negative moment strengthening

Condition	M_u or ϕM_n (kN-m/m)	Failure Mode
Factored Load Condition	32.8	N/A
Existing Condition	17.5	Strand yielding
Strengthened Condition	33.3	Concrete Crushing

Positive moment strengthening consisted of 25-cm strips of CFRP installed at 1.2-m intervals on the slab soffit as shown in Figure 3.3. (The wood planks in the figure were supporting measuring equipment for a load test.) This additional reinforcement was adequate to return the structure to an ultimate capacity within safe limits. Table 3.3 summarizes the positive design moments for the 2.6-m span.

Table 3.3 –Summary of design values for positive moment strengthening

Condition	M_u or ϕM_n (kN-m/m)	Failure Mode
Factored Load Condition	26.3	N/A
Existing Condition	15.9	Strand yielding
Strengthened Condition	30.3	Concrete Crushing

Figure 3.3 – CFRP strips installed on the slab soffit

4. Upgrading structural capacity to accommodate a change in use

A structural floor constructed as a two-way, post-tensioned, flat slab was strengthened in order to increase its service load capacity. The structure was originally used as a parking garage, but was to be converted to office space. This change in use caused an increase in the expected floor load levels. An analysis of the floor revealed deficiencies in the floor's flexural capacity to resist the increased loads. To compensate for the deficiencies, a strengthening system was installed consisting of externally bonded CFRP sheets installed on the slab's soffit along column lines between drop panels and on top of the slab at the column locations.

In the design of the CFRP strengthening system, it was assumed that the existing structure's performance was adequate for the current use/loading condition. Strengthening is only needed as a result of a change in use of the structure. The strengthening system was, therefore, designed to increase the capacity of the existing system by a percentage. In this way, the safety factor of the original design is maintained.

The percentage was obtained by conservatively using the percentage increase in total loads for the structure (some load increase is due to increased dead load, conservatively all increases are considered live load). It was determined that the structure's capacity needs to be increased by at least 18%.

The slab was considered to be a wide beam spanning between columns in each direction. The width of the slab used was equal to the transverse span for the particular direction. The capacity of the existing structure was computed using standard ACI procedures. The strengthened system was then analyzed using a strain compatibility analysis using a bond reduction factor on the strain in the tendons to account for unbonded installation. The CFRP quantity was designed to increase the capacity by 18%.

The results of the calculations are given in the Table 4.1.

Table 4.1 – Strength increase afforded by CFRP

Condition	CFRP Req'd (cm)	Existing Capacity, $\phi M_{n,e}$ (kN-m)	Strengthened Capacity, $\phi M_{n,s}$ (kN-m)	Capacity Increase
E-W Mid-span	200	886.3	1083.3	18.2%
N-S Mid-span	230	918.1	1136.8	19.3%
N-S Support	250	1620.2	2038.4	20.5%

The CFRP quantities given indicate the total width of CFRP to be used. The CFRP extends between drop panels in the positive regions and 6 feet on either side of the columns in the negative regions.

Double ply, 100-cm wide sheets were, therefore, installed on the slab's soffit between drop panels in the east-west direction to increase the positive moment resistance of the slab (see Figure 4.1).

Figure 4.1 – CFRP strip installed on the slab soffit

The strengthening system allowed the load level on the floor to be increased for use as office space.

5. Conclusions

The strengthening technology consisting of externally-bonded CFRP sheets is easy to perform and results in significant improvements in ultimate load capacity. Several recently completed strengthening projects were presented to demonstrate that CFRP is becoming an acceptable rehabilitation method for buildings. The presented projects show the adaptability of CFRP technology to upgrade an existing structure, provide strengthening after a degradation problem, and correct design/construction defects.

References

[1] ACI Committee 440, 1996, "State-of-the-Art Report on FRP for Concrete Structures," ACI440R-96, Manual of Concrete Practice, ACI, Farmington Hills, MI, 68 pp.

[2] El-Badry, M. (editor). "Advanced Composite Materials in Bridges and Structures." Proceedings ACMBS-II, Montreal, Canada, August 1996, pp. 1027

[3] Nanni, A.; Gold, W.; Thomas, J.; Vossoughi, H. "FRP Strengthening and On-site Evaluation of a PC Slab." Proceedings of the Second International Conference on Composites in Infrastructure, Volume 1, Tucson, Arizona, January 5-7, 1998, pp. 202-212

[4] Barboni, M.; Benedetti, A.; and Nanni, A. "Carbon FRP Strengthening of Doubly Curved Precast PC Shell," ASCE-Journal of Composites for Construction, Vol. 1, No. 4, Nov. 1997, pp. 168-174

DUCTILITY OF HYBRID FIBER COMPOSITE REINFORCEMENT FRP FOR CONCRETE

R.Apinis*, J.Modniks*, V.Tamuzs*, R.Tepfers**

*Institute of Polymer Mechanics - Aizkraukles 23, Riga, LV - 1006, Latvia
**Division for Building Technology - Chalmers University of Technology - S-412 96 Göteborg, Sweden.

Abstract: Reinforcing units of fiber composites, FRP, for concrete have brittle elastic behavior up to tensile failure. For safety reasons a certain stress hardening and an elongation > 3% at maximum load is usually required for steel reinforcement. The same is desirable for FRP rods. Two ways have been investigated to enlarge the ductility of FRP rods. At first ductile behavior can be obtained with unidirectional hybrid fiber rods, if carbon fibers are mixed in certain proportion with aramid or glass fibers. Under sustained load carbon fibers are responsible for most of the resistance. At some critical strain level the carbon fibers break and successively transfer the load to the more ductile aramide or glass fibers. The ductile rod behavior can be obtained with braided fiber strands, too. Braids increase the ductility, when resin breaks up and enables the braids to rotate and compress into an open core. The disadvantage of this method consists in undesirable high creep of braided rods.

Introduction

Corrosion of concrete reinforcement is a serious problem for offshore structures, bridges, industrial buildings with aggressive working environment, and similar structures. Among the ways to solve this problem is to use non-corroding fiber composites instead of steel. Glass, aramid, and carbon fibers moulded with either epoxy or polyester resin have usually been used. However these materials, being well suited for concrete reinforcement, have also some drawbacks. Unlike the cold worked steel, they stay elastic until failure, have small ultimate strain, and fail in a noticeably brittle way. Therefore, the concrete structure can fail momentarily without any foreboding. In order to change that, the properties of the non-metallic reinforcement have to be modified toward those of the cold worked steel, i.e. those materials should have elastic-plastic behaviour and the ultimate strain over 3% [1].

This task has been dealt with in the Polymer Mechanics Institute (Riga, Latvia) in two directions [3-6]. The first is to make the reinforcing rod out of different materials mixture. One of the materials is carbon fiber with a large Young modulus and small ultimate strain, the other is either aramid or glass fiber with a smaller modulus and greater (over 3%) ultimate strain. Theoretical studies show that the elastic-plastic zone should evolve in such a hybrid material before failure [2].

The second direction is creating "special" reinforcing rod. Such a rod is made as a single-material fibers braid with a cylindrical open core shell. The rod fails with plastic zone developing, too, and its ultimate deformation exceeds 3%.

However, it is seen from [3] that the required effect was not achieved with the hybrid composites. One of the reasons why not, is believed to be small cross-section of the

specimens tested, resulting in fairly non-homogeneous mixing of the different kinds of fibers. Research of glass and aramid braids [3, 4, 5] showed developing of plateau and the ultimate strain over 3%, however the Young modulus was small and creep deformation too large.
The purpose of this work is to continue research in the both above-mentioned directions using bigger rods, so as to develop a material capable of replacing steel reinforcement in concrete, in the end.

1. Mechanical properties of the used materials.

The raw materials used for specimens were "Toray" carbon fiber, T400HB-6K-40D produced in Japan; aramid fiber ZSVM-300, produced in Russia; E-glass fiber produced in Valmiera, Latvia; and polyester resin NORPOL 440M850, made in Norway.
In order to determine mechanical properties of the fibers, unidirectional composite specimens where made. Volume fraction of the fibers for those specimens was known beforehand, the polyester resin NORPOL 440M850 was used as the matrix.
The specimens were tested for tension in the fiber direction with a 500 kN MTS testing machine to obtain σ - ε graphs for the studied materials. σ - ε graphs were also obtained for polymerised resin NORPOL 440M850 specimens in tension experiments.
The obtained data, using the linear rule of mixture, allowed to find the mechanical properties of the fibers. The results are shown in Table 1. For comparison there are shown also E-glass and aramid fibers properties referred from literature and carbon fibers properties from the "Toray" prospectuses.

Table 1.

Type of fiber	E, GPa	E_{ref}, GPa	σ^*, GPa	σ^*_{ref}, GPa	ε^*, %	ε^*_{ref}, %
E-glass	75.0	72-75	1.84	1.25-3.8	2.64	3-4.8
Aramid ZSVM	106	115-135	2.70	3.8-4.2	3.29	3-4
Carbon fiber T400HB	244	250	2.93	4.49	1.21	1.8

It can be seen from the table that the elastic moduli determined from the experimental data are close to those found in literature, but respective strengths are noticeably lower. One of the reasons for this difference is the stress concentration in specimens near the grips of the testing machine. It is in this zone, where due to uneven distribution actual stresses are higher than average, specimen usually starts to fail. Another reason is that the fiber strength can not be achieved fully in a composite [7].
In tests determined mechanical properties of the fibers given in Table 1. served as the basis for calculations of the concrete's hybrid composite reinforcing rods.

2. Unidirectional hybrid composite rods.

The exclusive properties of the hybrid composites are achieved by mixing fibers of different materials in an appropriate proportion. The hybrid fiber rods should be made so that aramid or glass fibers alone are able to carry at rupture of rod 1.1 times the load, which was taken by the carbon fibers together with aramid or glass fibers at rupture of carbon fibers. The effect of resin is disregarded because the resin may micro crack from the shocks at rupture of carbon fiber filaments. Further it is known that the contribution from uncracked resin in taking tensile force is small. The shocks from the failing carbon fibers will also rupture probably 5% of the aramid or glass fibers. Consequently the following formula can be set up for determination of the relation between the area of carbon and area of aramid or glass fibers:

$$1.1 \cdot 1.05\,(\varepsilon_c \cdot E_c \cdot A_c + \varepsilon_c \cdot E_a \cdot A_a) = \varepsilon_a \cdot E_a \cdot A_a \qquad (1)$$

where A_a – area of aramid or glass fibers, A_c – area of carbon fibers, E_a – modulus of aramid or glass fibers, E_c – modulus of carbon fibers, ε_a – strain at failure of aramid or glass fibers, ε_c – strain at failure of carbon fibers.

The hybrid composite test specimens were made as unidirectional rods. To facilitate fixing the rods in the testing station, cones were attached to their ends, made of the polyester filled with short chaotically oriented glass fiber cuts. To prevent the rods from being torn out from the cones, the rod-making fibers were knotted at the bases of the cones and the knots moulded with polyester resin (Figure 1). The diameter of the rods is 16 mm and the length of the working segment 100 mm. The length of each cone is 200 mm, the diameter of the cones' bases is 56 mm.

Figure 1. The specimen.

The rods were loaded with the MTS testing machine, at a constant loading rate. Rather low loading rate (20.8 N/s) was chosen in order to be able to study deformation of the rods just before failure in details. The applied force and rod deformation were measured constantly during the loading, force with the machine's own dynamometer and deformation with an attachable electronic extensometer with the measuring base of 20 mm. Measuring results were input into a computer memory during experiment and printed out after the end of it. Beside that, the testing machine's analogue two-co-ordinate plotter would plot the loading curve.
During this work, two different sorts of hybrid composites were made and tested, carbon-aramid and carbon-glass.

2.1. Carbon-aramid hybrid composite.

Carbon-aramid rods were made of T400HB carbon fiber and Russia-made ZSVM 300 aramid fiber. The required proportion of the fiber kinds was found using (1). However, since $\sigma-\varepsilon$ curve of the aramid fiber used is not linear and its modulus just before failure differs from the initial modulus E_a, in the right-hand side of (1) term $\varepsilon_a \cdot E_a$ changes to σ^*, the critical stress. With the data taken from Table 1, the required proportion was 74% aramid to 26% carbon. Due to some practical reasons actual fibers contents in specimens was a little different: 76% aramid and 24% carbon. Total volume fraction of fibers in the rod was $\mu_{tot} = 0.36$.

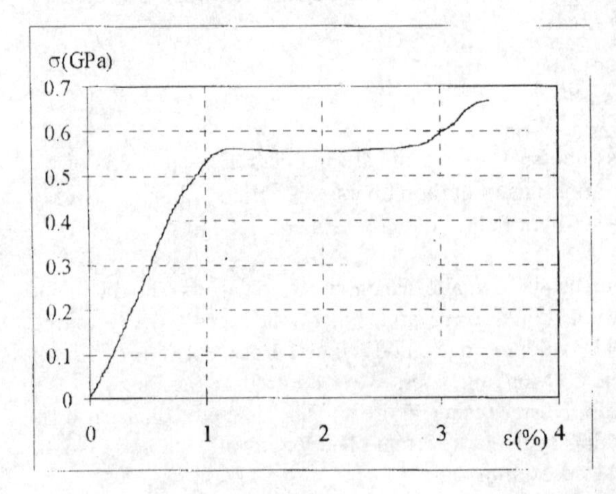

Figure 2. Carbon-aramid hybrid composite σ-ε diagram.

Figure 2 shows σ-ε curve obtained on one of such specimens. Three distinct parts are clearly seen on the curve. In the initial stage, the rod deforms as a monolith material with the modulus determined by the constituting materials amount and moduli. Having reached the critical strain, the carbon fibers rupture rather quickly and the aramid fibers takes all the load. Now (the second stage) the strain grows quickly and reaches the value appropriate for the aramid fibers at this load. In the third stage, the specimen deforms, up to the rupture, with the modulus determined by the modulus and amount of the aramid fibers left. According to calculations, the initial modulus of the material is E = 51.7 GPa. Experiments give E = 52 GPa. The calculated value of the modulus in the final stage of loading is 22.4 GPa (the modulus of the aramid fiber used decreases as strain grows so that just before rupture it is about 75% of the initial value; this effect is taken into account here). Experimental value is ≈ 20 GPa. Therefore, the experiment confirms that the wanted hybrid effect was attained. This effect was not attained before [3] with hybrid unidirectional specimens of cross-sections only 10% of these used here. Increasing the specimens cross-section decreases ratio of the fiber bundle cross-section to the total specimen cross-section and improves homogeneity of fibers mixing. Obviously, there exists some critical value of the ratio of the fiber bundle cross-section to the total specimen cross-section, such that hybrid effect vanishes when the value is exceeded.

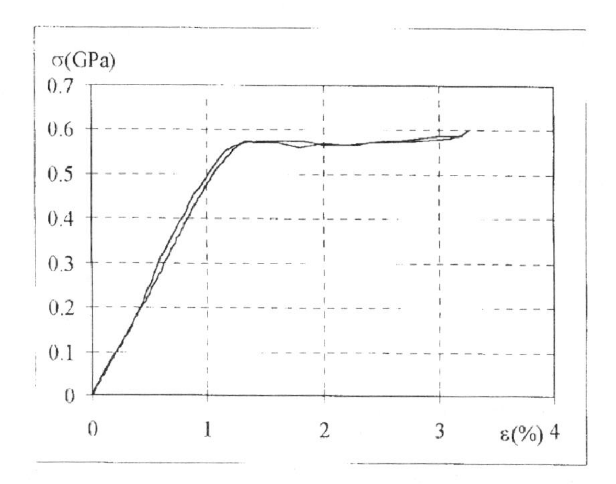

Figure 3. Carbon-aramid hybrid composite σ-ε diagrams.

In additional experiments with rods of the same material, two more σ-ε curves were obtained, shown in Figure 3. One of the specimens was loaded at rate 20.8 N/s, the other – 46 N/s. Making the loading rate twice as high, as can be seen, does not influence process of deformation. The curves shown in Figure 2 and Figure 3 are similar, however the stress growth in the final loading stage in the latter is not more than 5%. It is possible that a small change in the fibers proportion can make the effect more pronounced.

2.2. Carbon-glass hybrid composite

Glass fiber is much cheaper than aramid fiber, which makes prospects of using it for concrete reinforcing very attractive and motivates research in hybrid rods with E-glass fibers instead of aramid. Calculations using (1) show that such a material must contain 81% glass fiber and 19% carbon fiber. Two hybrid composite specimens containing 81% E-glass fiber and 19 % T400HB carbon fiber were made and tested.Total volume fraction of the fibers was $\mu_{tot} = 0.47$. The tension tests were carried out at a constant loading rate of 20.8 N/s. Figure 4 shows the results of these experiments.
It is seen that also this hybrid composite has a plateau. 3% strain was almost reached in these first experiments and the stress growth before the rupture was about 3 to 5%. This composite will, hopefully, be improved during subsequent work. It can be added that the initial modulus and strength of the rod are close to the respective values of the carbon-aramid hybrid composite (Figure 3). The explanation to this is that although E-glass has lower values of both the modulus and strength, comparing to the aramid fiber properties, the carbon-glass rods can have higher total volume fraction of the fibers.

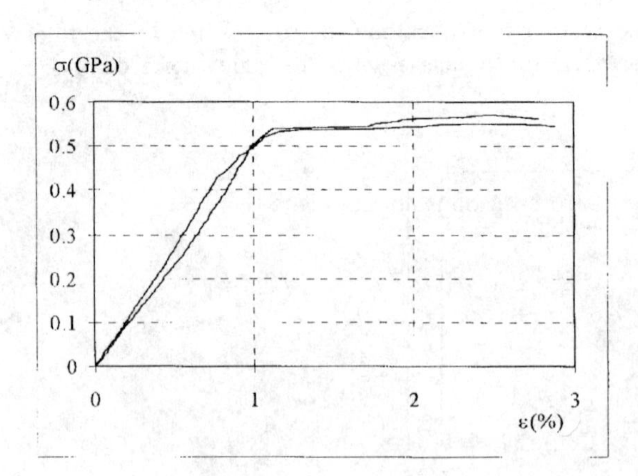

Figure 4. Carbon-glass hybrid composite σ-ε diagram.

3. Braids

In our previous research we have attain required properties for concrete reinforcement (strain level, plateau in σ-ε curve) with "special" rods [3-6]. They were made as a single material fiber braid with a cylindrical core shell in the middle. Ready braids were resin-moulded and polymerized (Figure 5).

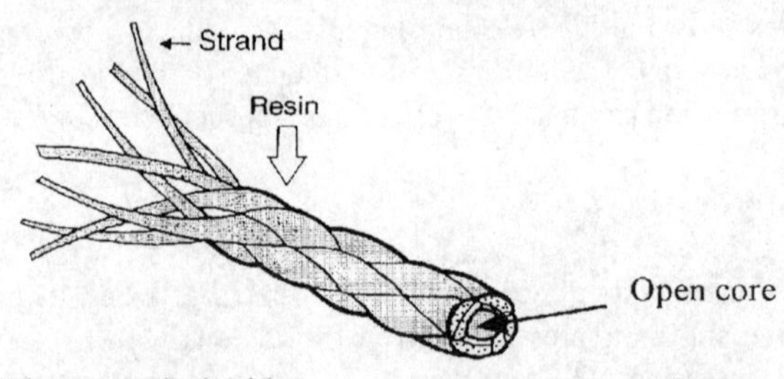

Figure 5. The braid.

When a tensile load is applied to such a rod, the core shell is subject to growing inward pressure and fails at a certain load level. This allows the braid to constrict so that its elongation sharply increases. Influence of the geometry of the structure and of the moduli of its parts on the shell failure was studied theoretically and experimentally and parameters were found allowing the rod elongation to reach 3% at the highest load [5,6]. Small cross-section braids were used in this research (outer diameter ≈ 10 mm, inner core opening diameter 6 mm) with thin wall.
To continue work in this direction, braided rods were made with thick wall and empty core inside. The rod's outer diameter was ≈ 16 mm and the diameter of the empty core diameter was 6 mm. No core shell was used in this case because it was believed that it cannot essentially raise elasic modulus in the initial stage when the wall is thick and the braiding angle small. Braiding technique was identical to that for smaller specimens.

A plaited rod was made of aramid fiber, with plaiting angle of 20°. In the same way as for the straight-reinforced rods, to fix the specimen in the testing machine it was provided with the cones at its ends. Tension experiments were carried out at the constant deformation rate of 2 mm/min. The total fibers cross-section in the aramid braid was $A_{fib} = 56.5 mm^2$. Determining reinforcement coefficient of the braid or estimating stress is possible only approximately, since its cross-secion is not constant. This is why the load-strain curve is shown in Figure 6, not the σ-ε curve. Estimated values of the elastic modulus E, ultimate stress σ*, and reinforcement coefficient μ are given in the caption. If the test had been performed with constant loading rate, the measured curve would have been the upper envelop of the saw-tooth curve. This envelop is followed in an overload situation of a structure.

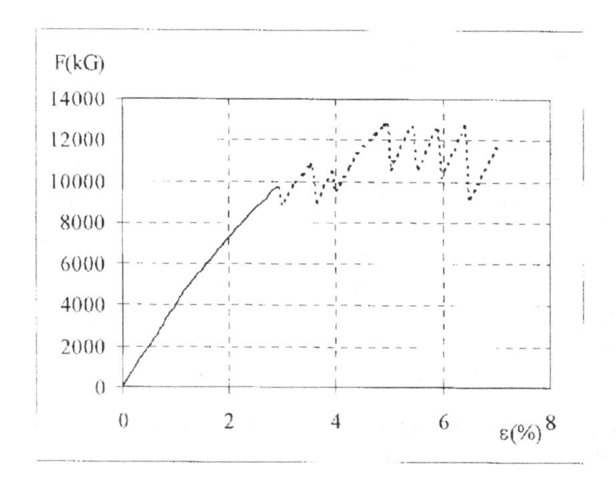

Figure 6. Loading curve of the aramid braid.
$E \approx 26$ GPa, $\sigma^* \approx 0.89$ GPa, $\mu \approx 0.38$.

The peaks seen in Figure 6 are believed to correspond to failures of the matrix at the fiber bundles crosspoints. Due to effect of strong shocks appearing when the matrix gets fractured have on the strain meter, the ultimate strain shown, 7%, may be incorrect. However, it obviously exceeded 3%. Homogenous polyester starts to fracture at a strain of about 1.3% and in the braid obviously at about 2.5%.

It must be emphasized that braided rods have some drawbacks, essential to their usage as reinforcement for concrete. One of them is that mechanical properties of fibers are used to a rather low degree. The elastic modulus of the rod in this experiment is by 30% lower than that of a straight-reinforcement composite. Also, due to the braid geometry, such rods are more subjected to creep.

4. Conclusions.

1 A ductile carbon-aramid hybrid composite suited to usage as reinforcement for concrete has been created.
2 Properties of carbon-glass hybrid composite are close to those of the ductile carbon-aramid material, and, hopefully, can be improved as the work continues.
3 A braid aramid fiber rod, having been tested, shows big deformability just before tensile rupture, but its initial elastic modulus is noticeably reduced.

5. Acknowledgement

The project was supported by SBUF, FoU-Väst - The Development Fund of the Swedish Construction Industry and Chalmers University of Technology in Göteborg.

6. References.

1. Editor R. Tepfers, FIBER COMPOSITES AS NONMETALLIC REINFORCEMENT IN CONCRETE. Seminar , March 15-16, 1993 at Chalmers University of Technology, Göteborg. Department of Building Materials, Chalmers University of Technology, P-93:3, Work No 539, Göteborg 1993. p. 199.
2. V.Tamuzs, STIFFNESS, STRENGTH, DAMAGE AND DUCTILITY OF HYBRID COMPOSITES. Seminar paper. Fiber composites as nonmetallic reinforcement in concrete. Seminar March 15-16, 1993 at Chalmers University of Technology, Göteborg. Department of Building Technology, Chalmers University of Technology, P-93:3, Work No 539, Göteborg 1993, pp. 52-61.
3. V.Tamuzs, R. Apinis, J.Modniks, U. Vilks: HYBRID FIBROUS COMPOSITES AND NONMETALLIC REINFORCEMENT OF CONCRETE. Part I. Report on the project. Institute of Polymer Mechanics, Latvian Academy of Sciences, Riga, June 1994. P. 35.
4. V. Tamuzs, R. Tepfers, DUCTILITY OF NONMETALLIC HYBRID FIBER COMPOSITE FOR CONCRETE. Chalmers University of Technology, Department of Building Technology, Publication No 95:2, Work No 3. Göteborg, 1995-02-10. P. 8. Contribution to the Second International RILEM Symposium FRPRCS-2 on "Non-metallic (FRP) Reinforcement for Concrete Structures", Proceedings 29, 23-25 August 1995, Ghent, Belgium. pp. 18-25.
5. V.Tamuzs, R.Tepfers, R.Apinis, U.Vilks, J.Modniks. DUCTILITY OF NONMETALLIC HYBRID FIBER COMPOSITE REINFORCEMENT FOR CONCRETE. Chalmers University of Technology, Division of Building Technology, Publication No 95:3, Work No 5. Göteborg, 1995-09-26. Contribution to the First International Conference on Composites in Infrastructure, ICCI'96, 15-17, January 1996, Tucson Arizona, USA, Department of Civil Engineering and Engineering Mechanics, University of Arizona, Tucson, Arizona 85721, USA. Tucson 1996. pp. 109-122.
6. R.Tepfers, V.Tamuzs, R.Apinis, U.Vilks and J.Modniks. DUCTILITY OF NONMETALLIC HYBRID FIBRE COMPOSITE REINFORCEMENT FOR CONCRETE. Mechanics of Composite Materials. - 1996. - Vol. 32, Nr. 2. - pp. 167-179.
7. Ju.G. Korabelnikov, V.P. Tamuzh, O.F. Silujanov, V.M. Bondarenko, M.T. Azarova, SCALE EFFECT OF FIBRE STRENGTH AND THE PROPERTIES OF UNIDIRECTIONAL COMPOSITES ON THEIR BASIS. Mechanics of Composite Materials, 1984. Nr. 2, pp. 195-200 (in Russian).

Influence of steel microfibers and microsilica on behavior of high performance cement based composites

L. Biolzi*, G. L. Guerrini^ and G. Rosati*

* Dipartimento di Ingegneria Strutturale, Politecnico di Milano, Piazza L. da Vinci 32, 20133 Milano, Italy.
^ CTG Italcementi Group Direzione Laboratori, Bergamo, Italy.

Abstract

The effects of steel microfibers and microsilica content on the mechanical properties of a very high strength cement-based material in tension and compression are examined. Tests were conducted on specimens of high performance concrete under a strain controlled system through a closed-loop testing machine. Moduli of elasticity and fracture energies are evaluated and compared. A notable outcome of the present investigation is that a dosage of around a 10% of microsilica is favorable to replacement of an equal amount of cement.

1. Introduction

Cement based materials, characterized by very high mechanical properties, are becoming more and more common in the field of construction of high-rise buildings and large span bridges [1]. Numerous structures now exist, where strengths of 100 MPa (and above) have been prescribed and achieved consistently on a routine basis [2]. Beside traditional high strength concretes, particular cement based composites (microconcretes) with very interesting mechanical properties have been developed. Essentially they are so called DSP (Densified Systems containing homogeneously arranged, ultrafine Particles) [3] and the similar materials [4] that contain, in some measure, additions (such as microsilica, a by-product of the silicon or ferrosilicon industry, and/or fine mineral powder), reinforcing fibers and processing that includes heat treatment and pressure. Due to a dense microstructure and a reduced porosity these materials show an enhanced mechanical, physical and chemical durability [5]. Therefore they are truly high performance materials according to the definition given by Aitcin and Neville [6].

In this paper one will examine some factors that can affect the performance of very high strength cement based composites. After a description of such materials, one will discuss the influence on the mechanical properties of reinforcing steel microfibers volume and fiber length. In the next section are illustrated experimental data on the behavior of specimens with a microsilica content of 0, 10 and 20 percent by mass of the portland cement are described. Conclusions, that apply to the specific materials and mixtures proportions used in this investigation, complete the paper.

2. Materials and test setup

The materials considered in this work are similar to the so-called densified system particles (DSP). The cement-based materials were obtained with:

a) Portland cement CEM I 52.5 R, according ENV 197/1 European Standard with a Blaine fineness of 4590 cm^2/g;

b) Gray microsilica in the form of a dry, uncompacted powder having a surface area of 20 m^2/g (B.E.T. method);

c) acrylic copolymer superplasticizer, 30% solid content;

d) natural crystalline quartz of high purity (99% SiO_2);

e) carbon steel microfibers, diameter 0.15 mm, length 6 mm.

The concrete had a maximum aggregate size of 3 mm, an aggregate/binder ratio of two and a water/binder ratio of 0.22. The superplasticizer dosage, that is the ratio between the dry mass of superplasticizer solids and the mass of cement, was 0.02. Specimens were obtained with three different microsilica content (0, 10, 20 % by weight of binder) and three different quantities of randomly dispersed steel microfibers (0, 2, 4 % by volume). Table I shows the seven combinations of microsilica and microfibers considered in this investigation.

As is well known, the main functions of the microsilica (characterized by a particle size distribution of two orders of magnitude finer that normal portland cement), are to fill the voids (thus reduces the amount of freezable water), to develop a lubrication effect and to produce secondary hydrates by pozzolanic reaction. It results a material with reduced porosity, improved mechanical properties and durability.

The acrylic copolymer superplasticizer was found to be the most efficient because melaminic and naftalene-sulfonate superplasticizers yielded materials with poorer mechanical characteristics [7].

Mix proportions for the materials are given in Table I. The results are compared with a control material that is a material without microsilica and reinforcing fibers.

Table I. Mix proportions.

microsilica/cement ratio	0, 10, 20%
superplasticizer/binder ratio	0.02
aggregate/cement ratio	2
water/binder ratio	0.225
% fibers in volume	0, 2, 4, 6

The fracture tests were carried out at the concrete age of about four weeks and at least three specimens were tested for each geometry and material considered.

The testing system consisted of a closed-loop electromechanical Instron testing machine with a maximum capacity of 100 kN. The main characteristic are:

a) electromechanical controls (actuator-motoreducing), with a minimum speed of 2 μm/hour;

b) three control channels, one of which can be external (giving the possibility to choose the feedback signal that allows a stable test control);

c) closed control loop with integral and derivative gain (in order to remove the effect of the finite axial stiffness of the machine).

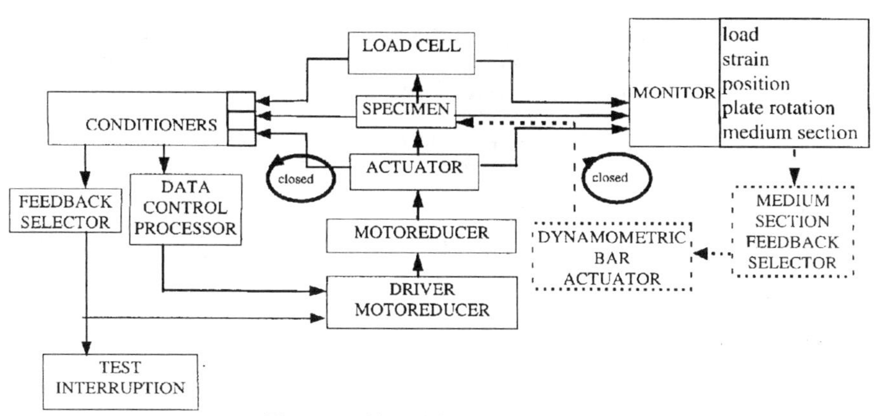

Figure 1. Closed-loop testing system.

It is well know that strain softening structures can be studied in the laboratory only within a closed-loop testing system. In fact, to avoid unstable failure and to have a material characterization, with the necessity to quantify the fundamental behaviors that involve non linear mechanisms (such as microcrack, fracture and localization), strain control tests were performed with an appropriate choice of the feedback signal to the servocontroller. Fig. 1 shows the line diagram of the test setup.

3. Reinforcing steel microfibers

a) Fiber volume

Tensile tests with cylindrical specimens and bending tests with the classical three point bending scheme were performed in a closed loop testing system. The direct tension tests were performed on cylindrical specimens with an external circumferential notch, 7 mm deep. In bending, beams 40x40x200 mm in size were tested over a simple supported span of 140 mm. In a tensile test, the response of a specimen (Fig. 2)

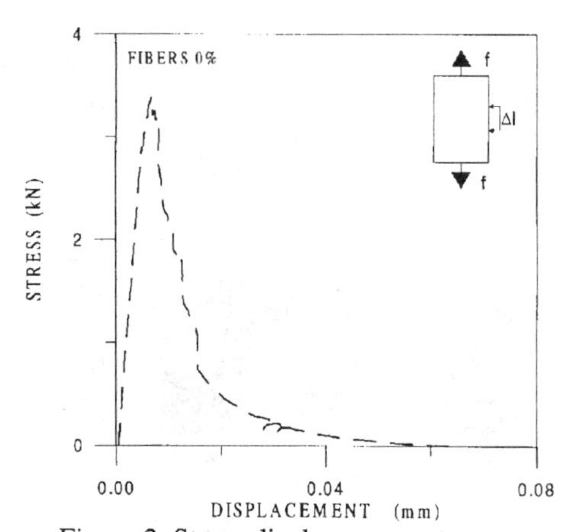

Figure 2. Stress-displacement response

composed of material without dispersed fibers was quite brittle. The fibers (Fig. 3) produced an increment in the peak load and a more ductile softening behavior. In the first elastic branch of the load-displacement curve, the specimens of identical size

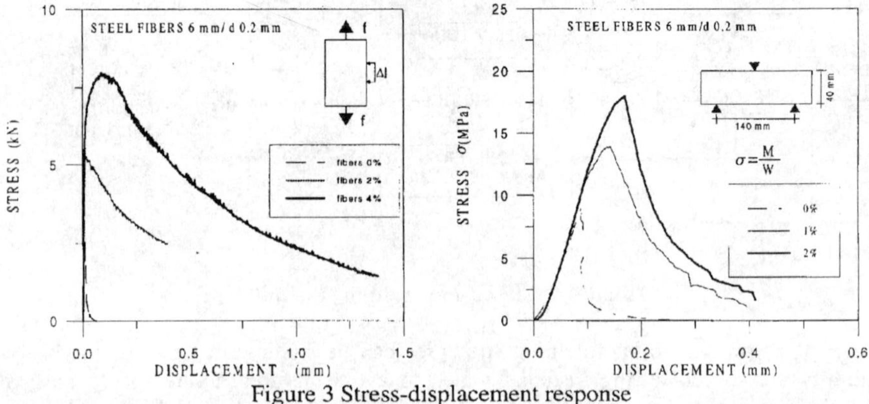

Figure 3 Stress-displacement response

showed the same stiffness. Therefore it may be inferred that, in the considered range of fibers volume, Young's modulus was not particularly influenced by fibers content. The material in flexure may be characterized by the apparent or nominal strength σ_n, that is the maximum allowable stress evaluated according to the elementary methods of classical beam theory, defined by

$$\sigma_n = M_{max}/W,$$

where M_{max} is the maximum bending moment and W is the elastic section modulus. The nominal strength as a function of the fibers volume is displayed in Fig. 4, together with the tensile strength evaluated by the direct tension tests. As demonstrated in the Fig. 4,

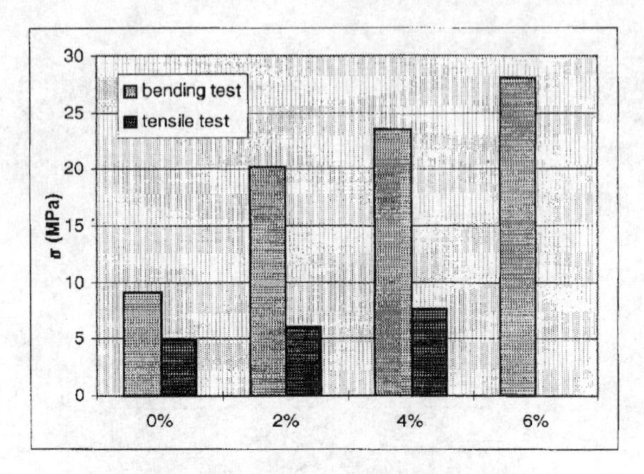

Figure 4. Tensile and bending strength.

the contribution of fiber reinforcement is usually more important for the flexural strength. This fact is caused by the fibers bridging the crack so that relevant tensile stresses are transmitted in the stretched region. In addition, the increase of the maximum moment (at the peak load) for the fiber reinforced materials also produced higher compressive stresses and consequently a better exploitation of the material. It is usually defined a critical volume fraction of fibers V_{cr} , that is the fibers amount that allows in the cracked phase a load carrying capacity equal to cracking strength of the matrix. It can be easily shown (see for instance [8]) that V_{cr} is equal to

$$V_{cr} = \sigma_{mu} / [\sigma_{mu}(\sigma_{fu} - \sigma'_{fu})],$$

where σ_{mu} is the matrix tensile strength, σ_{fu} is the fiber tensile strength and σ'_{fu} is the stress in the fiber at the incipient matrix cracking strain. For aligned fiber V_{cr} is lower than 1% but for randomly oriented fibers it is significantly higher.
A notable advantage given by the fiber-reinforced materials is the improvement in fracture energy. Fig. 5 is a plot of the specific fracture energy represented by the area

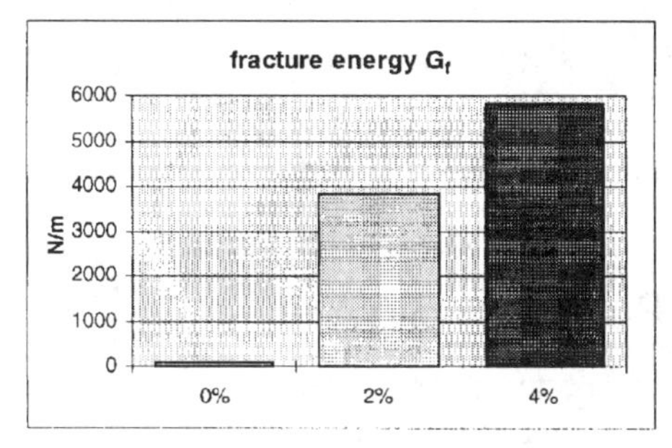

Figure 5. Specific fracture energy.

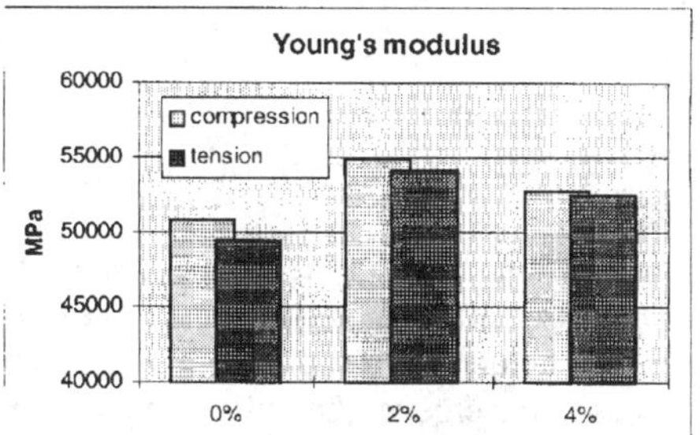

Figure 6.Young's modulus

under the load-displacement curve divided by the cross sectional area and evaluated from the tensile tests and bending tests. In all cases the area calculation was obtained with a complete load relaxation (the load was zero although in Fig. 3 this is not shown). The failure primarily due to pull-out or debonding in the fiber reinforced materials produced high values of fracture energy compared to those exhibited from the plain materials (65.81 N/m and 71.1 N/m in tension and flexure, respectively). It is interesting to observe that the values of fracture energy measured in the tensile tests were smaller than those evaluated in flexure. Nevertheless, in all cases, they were comparable. Some specimens of cylindrical shape were used to evaluate Young's modulus in compression and in tension. The deformation was measured with four electrical resistance strain gages. Fig 6 shows the test results for the modulus versus the volume fraction of fibres. First, it can be observed that the modulus practically does not increases with an increase in the volume fraction of fibres. Second, there is no difference between the values of the modulus in tension and compression. Therefore stress-strain curves in tension or in compression have, at the beginning, the same slope.

b) Fiber length

An important parameter is the fiber length; indeed a better exploitation of the reinforcing fibers can be obtained trying to optimize their length. The influence of the fiber length in shown in Fig. 7. The longer fibers allowed a improved behavior due to

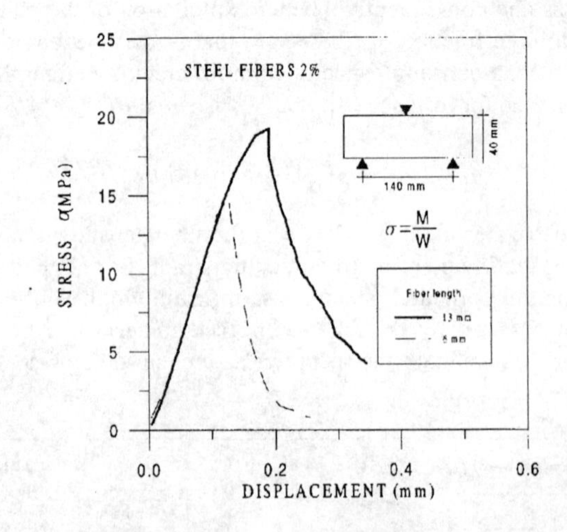

Figure 7. Influence of the fiber length on the stress-displacement curves.

increased stresses in the fibers interested by the crack. However, also in this case, the fiber strength was not achieved because the failure was governed by debonding or matrix failure. A 13 mm length is lower than the critical length, that is of the length that allows to achieve the full strength capacity of the fibers. In any case, the fibers considered in these experiments did not show a significant bond strength.

4. Microsilica content

Microsilica and/or fine powder are considered essential mineral admixtures to produce very high strength concrete [9-10]. The use of microsilica in cement based materials has become widespread in Europe and North America in the areas where high performance materials is of prime concern. Table II presents the strengths of the material as

Table II. Compressive strength and tensile strength.

microsilica dosage	Compressive strength (MPa)	tensile strength (MPa)
0%	113.25	7.08
10%	146.47	7.26
20%	151.00	7.35

a function of the microsilica dosage. It is interesting to observe that the materials containing microsilica showed comparable strength at the beginning and only after two week curing did the materials show an increase of mechanical strengths The greatest improvement in strength was obtained with 10% of microsilica. Considering the typical situation of microsilica costing ten times as much as a portland cement [6], it seems that from an economical point of view the larger microsilica content is not justified. In addition experimental results reported in [10] reveal that, at 1 year curing age concrete incorporating microsilica between 5-15% by mass as a cement replacement, does not exhibit significant differences in compressive strength. In any case, the strengths of the materials with microsilica are higher than the values exhibited from the control material. It is interesting to observe that the long term strength gain of materials with microsilica is sully very low in contrast with the portland cement materials that continued to gain

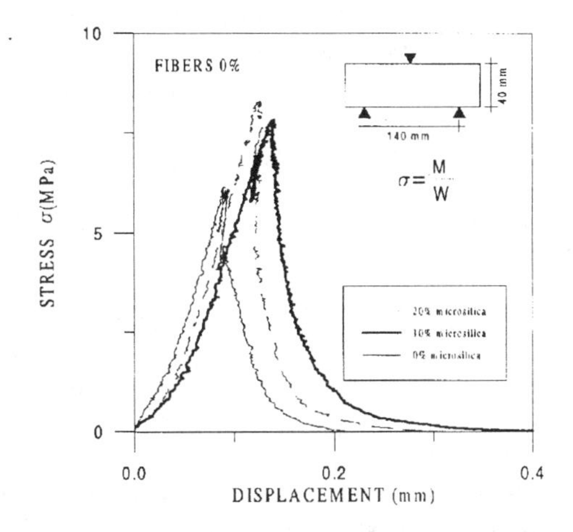

Figure 8. Typical load-displacement curves for different microsilica dosage

strength at later ages [11]. Typical load-displacement curves obtained are shown in Fig. 8. The results suggest than an optimum value of microsilica dosage as a cement replacement is close to a 10 %.

5. Conclusions

An increase in the microfibers content leads to significant enhancement in strength and fracture energy.

In compression, the strength is not particularly influenced by the fiber content whereas a contribution to ductility is observed.

Young's modulus shows practically no change with an increase in the volume fraction of fibers; the recorded values in tension and compression were more or less the same.

The results suggest than an optimum value of microsilica acting as an efficient pozzolan and as a cement replacement is close to a 10 % dosage.

ACKNOWLEDGEMENTS. This investigation is a part of a research program on very-high performance cement-based materials financed by CTG Italcementi Group.

References

[1] *J. Webb*, HIGH-STRENGTH CONCRETE: ECONOMICS, DESIGN AND DUCTILITY - Concrete International, Vol. 25, No. 1, 1993, pp. 75-80.
[2] *J. Walraven*, HIGH PERFORMANCE CONCRETE: EXPLORING A NEW MATERIAL - Structural Engineering International, Vol. 12, No. 3, 1995, pp. 50-58.
[3] *H. H. Bache*, DENSIFIED CEMENT ULTRAFINE PARTICLE-BASED MATERIALS - 2nd Int. Conf. on Superplasticizers in Concrete, Ottawa, 1981, p. 33.
[4] *P. Richard*, THE FUTURE OF HS/HPC - Proc. 4th Int. Symp. On Utilization of High Strength/High Performance Concrete, ENPC, Paris 1996, pp. 101-106.
[5] *P.C. Aitcin*, DEVELOPMENTS IN THE APPLICATION OF HIGH PERFORMANCE CONCRETE - Constructions and Building Materials, Vol. 9, No.1, 1995, pp. 13-17.
[6] *P.C. Aitcin, A. Neville*, HIGH PERFORMANCE CONCRETE DEMYSTIFIED - Concrete International, Vol. 15, No. 1, 1983, pp. 21-26.
[7] *L. Biolzi, G. L. Guerrini and G. Rosati*, OVERALL STRUCTURAL BEHAVIOR OF HIGH STRENGTH CONCRETE SPECIMENS - Construction and Building Materials, Vol. 12, No. 1, 1997, pp. 50-58.
[8] *P.N. Balaguru, S.P. Shah*, FIBER REINFORCED CEMENT COMPOSITES - Mc Graw-Hill, 1992.
[9] *S.P. Shah*, Special HPCs I: FIBER REINFORCED HPC, ULTRA HIGH STRENGTH CONCRETE - Proc. 4th Int. Symp. On Utilization of High Strength/High Performance Concrete, ENPC, Paris 1996, pp. 75-82.
[10] *H.A. El-Sayed, A.H. Ali, B.A. El-Sabbagh*, SOME ASPECT OF MICROSILICA-MODIFIED CEMENT AND CONCRETE - Proc. Int. Exhibition and Conference ConChem, Les Pyramides Brussels 28th-30th Nov. 1995, pp. 283-293.
[11] *R.D. Hooton*, INFLUENCE OF SILICA FUME REPLACEMENT OF CEMENT ON PHYSICAL PROPERTIES AND RESISTANCE TO SULFATE ATTACK, FREEZING AND THAWING, AND ALKALI SILICA REACTIVITY - ACI Material J., Vol. 90, No. 2, 1993, pp. 143-151.

Structure-property relations of annual fibres and their influence on composite properties

H.L. Bos, M.J.A. van den Oever, O.C.J.J. Peters
ATO-DLO, P.O.Box 17, 6700 AA Wageningen, the Netherlands

Abstract

In this paper the specific influence of annual fibre characteristics on composite properties is presented. Due to the composite-like nature of annual fibres their properties strongly differ from those of man-made fibres. The tensile and lateral fibre strengths depend on the degree of fibre opening. Furthermore, tensile strength depends on clamping length. The resulting composite properties are found to be strongly dependent on the specific fibre structure and the degree of fibre opening.

1. Introduction

Composites of thermoplastic and thermoset polymers with annual fibres (f.i. flax, hemp or jute) have in recent years received increasing attention in light of the growing environmental awareness. Especially with respect to the specific stiffness and specific tensile strength, annual fibres can in some areas compete with glass fibres. The structure of the fibres, however, is incomparable to the structure of man-made fibres. If the fibre structure is not taken into account, resulting composite properties can be lower than is expected purely on basis of tensile data.
This paper addresses the influence of the fibre structure on the fibre properties and the way in which these aspects influence the properties of the resulting composites.

2. Fibre Characteristics

Fibre structure
An annual fibre is in fact a composite in itself. The general structure of jute, hemp and flax is comparable. Figure 1 shows the structure of a flax fibre. The 1m long flax fibres, which are isolated from the plant by the breaking and scutching processes, are usually called fibre bundles. These fibre bundles are still quite coarse and contain many weak lateral bonds. The fibre bundles can be further refined by hackling, into so called technical fibres. In the traditional textile industry these technical fibres are spun into yarns for the production of linen. The technical fibres are composed of elementary fibres, generally with diameters around 15 μm and lengths between 2 and 5 mm. The elementary fibres are bound together by a pectin interphase. Depending on the extent of the hackling process, the technical fibres contain 10 to 30 elementary fibres in diameter. The elementary fibres are single plant cells, they consist of a primary cell wall, a secondary cell wall and a lumen, which is a small open channel in the centre. The primary cell wall is relatively thin, about 0.2 μm [1] and consists of pectin, some lignin and cellulose [2]. The secondary cell wall makes up most of the fibre

diameter. It consists mainly of highly crystalline cellulose fibrils, oriented approximately in fibre direction, and amorphous hemicellulose. The secondary cell wall gives the fibre its high tensile strength. However, due to the fibrillar, highly crystalline nature of the secondary cell wall, the fibres are sensitive to kink band formation under compression.

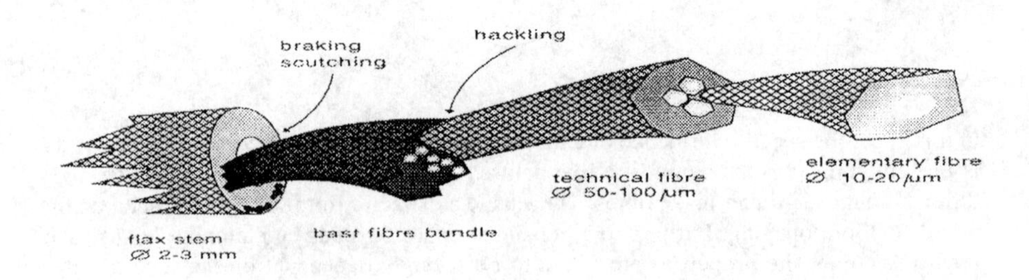

Figure 1. Schematic drawing of the structure of a flax fibre.

Fibre properties
The composite nature of the technical flax fibres leads to a strong dependence of tensile strength upon clamping length (see figure 2.) Down to a clamping length of approximately 25 mm, the fibre tensile strength remains relatively constant, around 500 MPa. Below 25 mm the fibre strength begins to increase towards a value of 810 MPa at a clamping length of 3 mm. At large clamping lengths the technical fibre fails through the relatively weak pectin interphase between the elementary fibres. The rise in tensile strength at shorter clamping lengths is caused by a change in failure mechanism. At clamping lengths below the elementary fibre length the crack must run through the stronger secondary cell wall of the elementary fibres. If one single elementary fibre is clamped at 3 mm length, a tensile strength of ca 1500 MPa is found. The large difference in strength at this clamping length between a technical and an elementary fibre is most likely caused by the bundle effect.

The lateral strength of flax fibres is especially low in unhackled fibre bundles. In these coarse fibres the pectin interphase is at some places along the fibres virtually non-existent. Hackling the fibres generally splits the fibres along these weak interphases into the thinner and stronger technical fibres. However, also in the technical fibres the pectin interphase still provides possible paths for lateral fibre failure. The lateral strength of technical fibres consequently is expected to be far lower than of elementary fibres.

Also the behaviour of the elementary fibre is governed by its build. The fibrillar highly crystalline structure of the secondary cell wall makes the fibre sensitive to kink band formation under compressive loading. The compressive strength of hand isolated elementary fibres is lower than the tensile strength, approximately 1200 MPa [3]. However, the presently available decortication processes, and especially the hackling process, induce substantial damage in the elementary fibres, which leads to an additional decrease in compressive strength. The tensile strength is less affected by these processes.

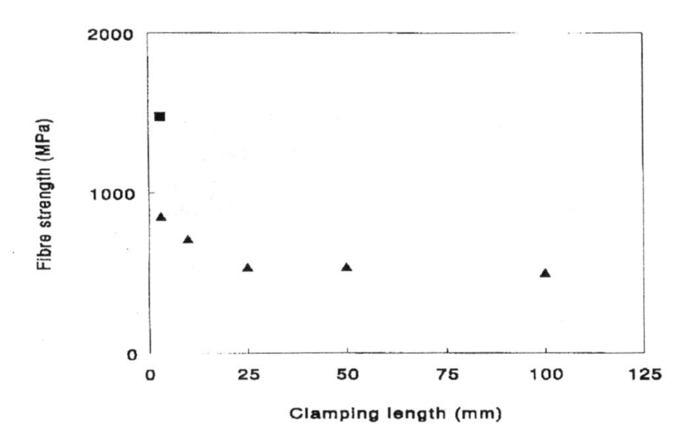

Figure 2. Tensile strength of elementary (■) and technical (▲) flax fibres versus clamping length.

3. Composite Properties

Technical agrofibres, as mentioned above, can be processed into non-wovens, which can form the basis for natural fibre thermoplastic composites (NMT), the agrofibre counterpart of glass mat reinforced thermoplastics (GMT). The agrofibre non-wovens can be used as a basis for thermoset composites as well. The composite-like structure of the fibres, however, introduces a considerable influence of fibre characteristic, e.g. scutched or hackled, on the effective use of the fibre properties in the composite, additional to the general influence of fibre-matrix adhesion.

In NMT composites the agrofibres are 2D-randomly oriented, which implies that in the composite not only the axial fibre properties are important, but also the lateral fibre properties. The influence of fibre characteristics, i.e. scutched versus hackled fibres, on composite strength is presented in figure 3. The scutched flax fibre composites show a lower composite strength than the hackled fibre reinforced composites, which is due to the very low lateral strength of the scutched fibre bundles. The weak lateral bonds in the scutched fibre bundles undermine the effect of enhanced fibre-matrix adhesion by maleic anhydride modified polypropylene (MAPP), and as a result the critical link in this case is not only the fibre-matrix interface. In the hackled fibres the weakest lateral bonds have been removed by the combing process. Consequently, the composites with hackled fibres show higher strength values than those with scutched fibres. However, also the lateral strength of these hackled fibres is still far lower than their tensile strength, due to the pectin interphases. These remaining weak interphases cause the hackled fibre composite strength to be lower than the Kelly-Tyson prediction [4], which is based on a fibre tensile strength of 810 MPa (see figure 3). The NMT strength can be increased strongly if well separated elementary fibres become available for the production of fibre mats, and comes close to the strength of GMT material. In this case also the Kelly-Tyson prediction will increase, since now the high tensile strength of elementary fibres will form the basis for this prediction. The predicted specific elementary flax fibre reinforced composite strength is even higher than the predicted specific GMT strength. It is expected that the actual properties of elementary fibre NMT will come close to this new Kelly-Tyson prediction and also come much closer to the properties of commercial GMT materials (figure 3).

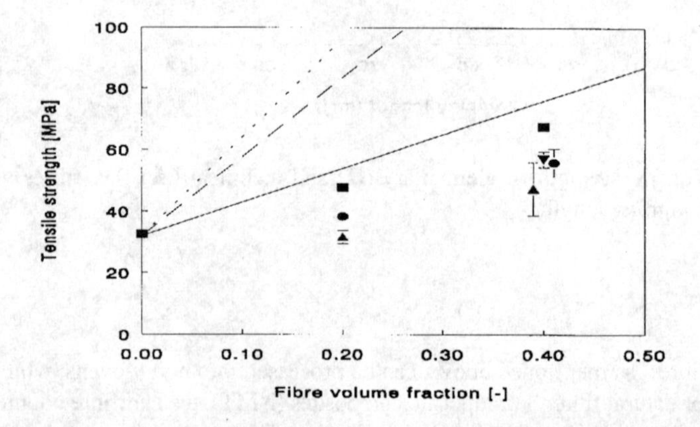

Figure 3. Tensile strength of flax/PP composites versus fibre volume fraction;
▲ = scutched flax/PP; ▼ = scutched flax/PP/MAPP;
● = hackled flax/PP; ■ = hackled flax/PP/MAPP;
——— = Kelly-Tyson prediction for technical fibres ($\sigma = 810$ MPa);
- - - - - = Kelly-Tyson prediction for elementary fibres ($\sigma = 1500$ MPa);
. . . . = Experimental data for commercial GMT [5].

The tensile modulus of the composites apparently is not influenced by the weak lateral bonds, however, the flexural modulus is. The effect of fibre characteristics on the flexural modulus is similar to the effect on tensile strength (figure 4).

Another interesting application for annual fibres is in injection moulding compounds. These compounds are normally produced via extrusion, starting from scutched fibres. During the extrusion process the fibres are split into elementary fibres and at the same time broken up to lengths of about 1 mm. Since in these compounds the weak lateral bonds are virtually absent, the composite strength of these compounds consequently is relatively high, over 70% of the strength of commercial short fibre glass thermoplastic compounds (figure 5). The injection moulding compounds show a large influence of improved fibre-matrix adhesion on composite strength. Clearly in this case the fibre-matrix interface is the weakest link. However, since the fibres are broken down to lengths around the critical fibre length [6], the composite strength can be further enhanced if one succeeds to choose processing parameters that split the scutched fibres into elementary fibres without decreasing the fibre length too much. These processing parameters will be different from the optimal parameters for glass compounds, since splitting the scutched fibres into elementary fibres requires higher shear forces than dispersing glass fibres.

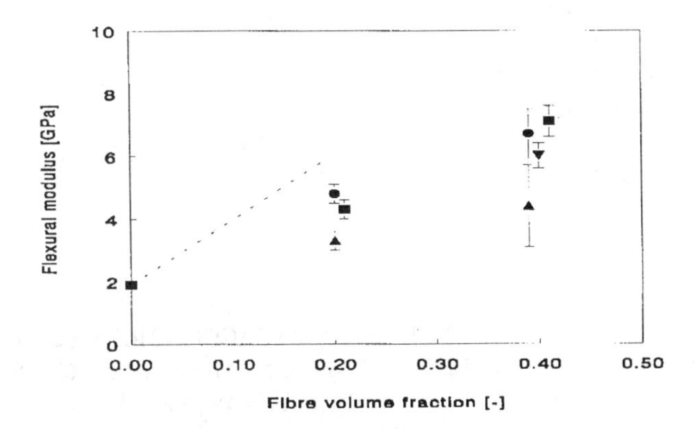

Figure 4. Flexural modulus of flax/PP composites versus fibre volume fraction;
▲ = scutched flax/PP; ▼ = scutched flax/PP/MAPP;
● = hackled flax/PP; ■ = hackled flax/PP/MAPP.
. . . . . . = Experimental data for commercial GMT [5].

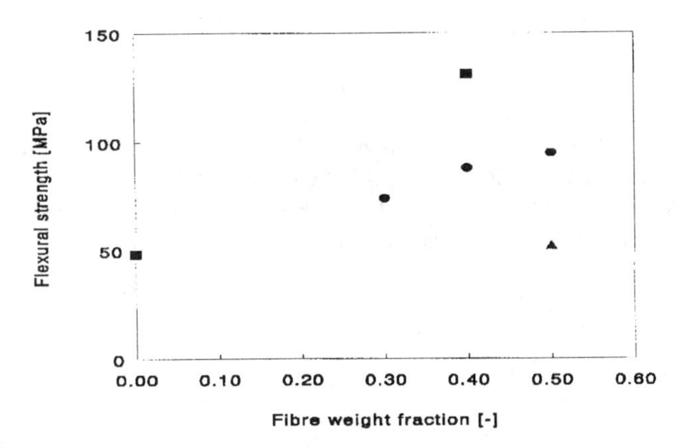

Figure 5. Flexural strength of extruded and subsequently injection moulded PP compounds versus fibre weight fraction;
▲ = flax/PP; ● = flax/PP/MAPP; ■ = glass/PP/MAPP [7].

4. Conclusions

Annual fibres have a composite-like structure which influences their mechanical properties. The strength of technical flax fibres depends on the clamping length, failure either takes place through the weak pectin interphase or through the strong secondary cell wall. The technical fibre strength is lower than the elementary fibre strength due to the bundle effect. The strength of a flax fibre reinforced composite is determined both by the tensile strength of the fibre and by the presence of weak lateral fibre bonds. Consequently elementary fibres give stronger composites than hackled fibres, which give better results than scutched fibres. Therefore, in order to further improve the strength of annual fibre reinforced composites, one should aim for further elementarisation of the fibres.

References

[1] *Thiery I.P.*, MISE EN EVIDENCE DES POLYSACCHARIDES SUR COUPES FINES EN MICROSCOPIE ELECTRONIQUE, J. Microsc., **6**, 1967, 987

[2] *Sal'nikov V.V., Ageeva M.V., Yumashev V.N.*, ULTRASTRUCTURAL ANALYSIS OF BAST FIBRES, Russian Plant Phys., **40**, 1993, 416

[3] *Bos H.L., Oever M.J.A. van den, Peters O.C.J.J.*, THE INFLUENCE OF FIBRE STRUCTURE AND DEFORMATION ON THE FRACTURE BEHAVIOUR OF FLAX FIBRE REINFORCED COMPOSITES, Proc. 4th Int. Conf. Def. Fract. Comp., Umist, Manchester, 24-26 march, 1997, 499

[4] *Thomason J.L., Vlug M.A., Schipper G., Krikort H.G.L.T.*, INFLUENCE OF FIBRE LENGTH AND CONCENTRATION ON THE PROPERTIES OF GLASS FIBRE-REINFORCED POLYPROPYLENE: PART 3. STRENGTH AND STRAIN AT FAILURE, Comp Part A, **27A**, 1996, 1075

[5] *Berglund L.A. and Ericson M.L. in* POLYPROPYLENE - STRUCTURE, BLENDS AND COMPOSITES - volume 3, ed. Karger-Kocsis J., Chapman & Hall, 1995

[6] *Oever M.J.A. van den, Bos H.L.*, submitted to Adv. Comp. Lett.

[7] *Sanadi A.R., Caulfield D.F., Jacobson R.E., Rowell R.M.*, RENEWABLE AGRICULTURAL FIBRES AS REINFORCING FILLERS IN PLASTICS: MECHANICAL PROPERTIES OF KENAF FIBER-POLYPROPYLENE COMPOSITES, Ind. Eng. Chem. Res., **34**, 1995, 1889

Possibilities to Improve the Properties of Natural Fiber Reinforced Plastics by Fiber Modification

J. Gassan, A.K. Bledzki

Institut für Werkstofftechnik - Kunststoff- und Recyclingtechnik - University of Kassel, Moenchebergstrasse 3, 34109 Kassel - Germany, Tel.: (+49) 561-804-3673, Fax: (+49) 561-804-3692, e-mail: kutech@hrz.uni-kassel.de

Abstract

Improvements in the characteristic properties of jute-polypropylene composites were obtained with the application of MAH grafted PP copolymers as coupling agent to the fiber. Flexural, tensile, and dynamic strength were increased distinctly (up to 50%), impact energy was reduced due to the lower energy absorption in the interphase.
SEM investigations demonstrated that fiber pull out is reduced after the modification with the coupling agent. This improved fiber-matrix adhesion further leads to a lower creep strain in the outer fibers. This was demonstrated for composites with two different fiber contents.

1. Introduction

Natural fiber composites combine good mechanical properties with a low specific mass. But, their high level of moisture absorption, poor wettability, and insufficient adhesion between untreated fibers and the polymer matrix, leads to debonding with age.
The strongly polar cellulose fibers are inherently incompatible with hydrophobic polymers. There are several mechanisms of coupling in materials [1]:

- Weak boundary layers - coupling agents eliminate weak boundary layers,
- Deformable layers - coupling agents produce a tough, flexible layer,
- Restrained layers - coupling agents develop a highly crosslinked interphase region, with a modulus intermediate between that of substrat and of the polymer,
- Wettability - coupling agents improve the wetting between polymer and substrat (critical surface tension factor),
- Chemical bonding - coupling agents form covalent bonds with both materials,
- Acid - base effect - coupling agents alter acidity of substrat surface.

2. Modification Methods for Natural Fibers

2.1 Physical Treatment Methods

Electric discharge (corona, cold plasma) is the classical way for physical treatment of fibers. Corona treatment is one of the most interesting techniques for surface oxidation activation. This process changes the surface energy of the cellulose fibers and in case of wood surface activation increases the amount of aldehyde groups.
The same effects are reached by cold plasma treatment. Depending on type and nature of the used gases, a variety of surface modification could be achieved. Surface crosslinkings could be introduced, surface energy could be increased or decreased, reactive free radicals and groups could be produced.
Electric discharge methods are known to be very effective for "nonactive" polymer substrats as polystyrene, polyethylene, polypropylene etc. [2].

2.2 Chemical Treatment Methods

The most important type of chemical modification is the chemical coupling method. This method improves the interfacial adhesion. The fiber surface is treated with a compound, that forms a bridge of chemical bonds between fiber and matrix. Possible methods are [2]:

- Graft Copolymerization

An effective chemical modification method for natural fibres is graft copolymerization. This reaction is initiated by free radicals of the cellulose molecule. The cellulose is treated with an aqueous solution with selected ions and is exposed to a high energy radiation. Then the cellulose molecule cracks and radicals are formed. Afterwards the radical sites of the cellulose are treated with a suitable solution (compatible with the polymer matrix), for example vinyl monomer, acrylonitrile, methyl methacrylate, and polystyrene. The resulting co-polymer posesses properties characteristic of both, fibrous cellulose and grafted polymer. For example, the treatment of cellulose fibers with hot polypropylene - maleic anhydride (MAH-PP) copolymers, provides covalent bonds across the interface. The mechanism of reaction can be devided in two steps, the activation of the copolymer by heating ($t = 170°$ C) (before fiber treatment) and the esterification of cellulose. After this treatment the surface energy of the fibers is changed to a level much closer to the surface energy of the matrix. Thus, a better wettability and a higher interfacial adhesion is obtained. The polypropylene (PP) chain permits segmental crystallization and cohesive coupling between the modified fiber and the PP matrix.

- Treatment with Compounds which Contain Methylol Groups

Chemical compounds which contain methylol groups ($-CH_2$ OH) form stable, covalent bonds with cellulose fibers. Those compounds are well known and widely used in textile chemistry. Hydrogen bonds with cellulose, can be formed in this reaction as well.
The treatment of cellulose with methylolmelamine compounds before forming cellulose unsaturated polyesters (UP) composites, decreases the moisture pick up, and increases the wet strength of reinforced plastic.

- Treatment with Isocyanates

The mechanical properties of composites reinforced with wood-fibers and PVC or PS as resin can be improved by an isocyanate treatment of those cellulose fibers or the polymer matrix. Polymethylene-polyphenyl-isocyanate (PMPPIC) in pure state or solution in plasticizer can be used. PMPPIC is chemically linked to the cellulose matrix through strong covalent bonds. Both PMPPIC and PS contain benzene rings, and their delocalized π- electrons provide strong interactions, so that there is an adhesion between PMPPIC and PS. Comparing both methods, treatment with silanes or treatment with isocyanats, it is obvious, that the isocyanatic treatment is more effective than the treatment with silane. Equal results are obtained, when PMPPIC is used for the modification of the fibers or polymer matrix.

- Triazine Coupling Agents

Triazine derivatives form covalent bonds with cellulose fibers. The reduction of the moisture absorption of cellulose-fibers and their composites, treated with triazine derivates is explained by reducing the number of cellulose hydroxyl groups, which are available for moisture pick-up, reducing the hydrophility of the fiber`s surface, and restraint of the swelling of the fiber, by creating a crosslinked network due to covalent bonding, between matrix and fiber.

- **Organosilanes as Coupling Agents**

Organosilanes are the main group of coupling agents for glass-fiber reinforced polymers. They have been developed to couple virtually any polymer to the minerals, which are used in reinforced composites. The organofunctional group in the silane coupling agent causes the reaction with the polymer. This could be a co-polymerization, and/or the formation of an interpenetrating network. The curing reaction of a silane treated substrate enhances the wetting by the resin. Analog to glass-fibers, silanes are used as coupling agents for natural fiber polymer composites. For example, the treatment of wood-fibers with methacryl silane (A-174 from Union Carbide) improves the dimensional stability of wood. Contrastingly, a decrease of mechanical properties was observed for coir - unsaturated polyester composites after a fiber modification with dichlormethylvinyl silane. The treatment of mercerized sisal-fiber with aminosilane (A-1100 from Union Carbide) before forming sisal-epoxy composites, markedly improves moisture repellency of the composite. These examples show, that theories used for the silane treatment of natural fibers are contradictory, therefore further studies are necessary.

3. Materials and Experimental Technique

3.1 Materials

The composites tested were made of tossa jute-fibers ($n_g = 2$ - woven by J. Schilgen GmbH & Co.), 280 tex, embedded in a PP matrix from Vestolen GmbH Germany (Vestolen® P 6000F-Table 1) using a film stacking technique.

Table 1: Technical specification of Vestolen® P 6000F - homopolymer

Characteristic Properties of Vestolen® P 6000 F	Value
Viscosity	240 cm³/g
Melting Flow Index - MVR/230/2.16	7.4 cm³/10 min.
Melting Point	164 - 168 °C
Young's Modulus	1500 N/mm²
Shear Modulus	800 N/mm²
Vicat Softening Temperature	90°C

To obtain a higher fiber matrix adhesion, fibers were modifed with MAH-PP (®Hostaprime HC 5 from Hoechst, Germany). First, the fibres had to be dewaxed in an alcohol solution for 24 hours, removing the weaving size (potato starch and waxes) and then washed with distilled water. The MAH-PP treatment was carried out in a toluene solution with 0.1 wt.-% MAH-PP content for 5 minutes at 100°C. The fiber treatment was completed by 2-hours drying in a vacuum oven at 75°C.

3.2 Experimental Technique

A DIN EN 61 tensile test was used to measure the Young`s moduls (test speed = 1 mm/min) and the tensile strength (test speed = 2 mm/min), a DIN EN 63 flexural test (test speed = 2 mm/min) for flexural modulus and strength of the composites. Ten samples

with a geometry of (160*25*4)mm³ and (80*15*4)mm³, respectively, were investigateted in each case.
The creep behaviour was investigated according to DIN EN 20899 in flexural mode for 2 samples in each case with a geometry of (140 x 10 x 4) mm³. The support span was 100 mm.
The fatigue tests were carried out in accordance to DIN 50 100 in load increasing mode, by measuring the dynamic strength. The load increasing tests with 10^4 cycles per stress level were carried out as repeated tensile stress tests, with 4 samples (geometry: 130*25*4 mm³) for each treatment. Testing frequency "f_{test}" and stress ratio "R" were chosen to 10Hz and 0.1. The maximum heating of the samples did not exeed 7°C.
Charpy impact tests were carried out for 10 samples according to DIN 53 453 for specimens with a geometry of (50*6*4) mm³.
All tests were carried out at 50 % rel. humidity and 23°C

4. Results and Dicussion

The effectiveness of the MAH-PP coupling agent depends on concentration and treatment time, as was shown for combinations of flax and PP by Mieck et al. [3] and jute and PP by Gassan et al. [4]. As with glass-fiber composites, a monomolecular layer of the coupling agent on the fiber surface is ideal [5].

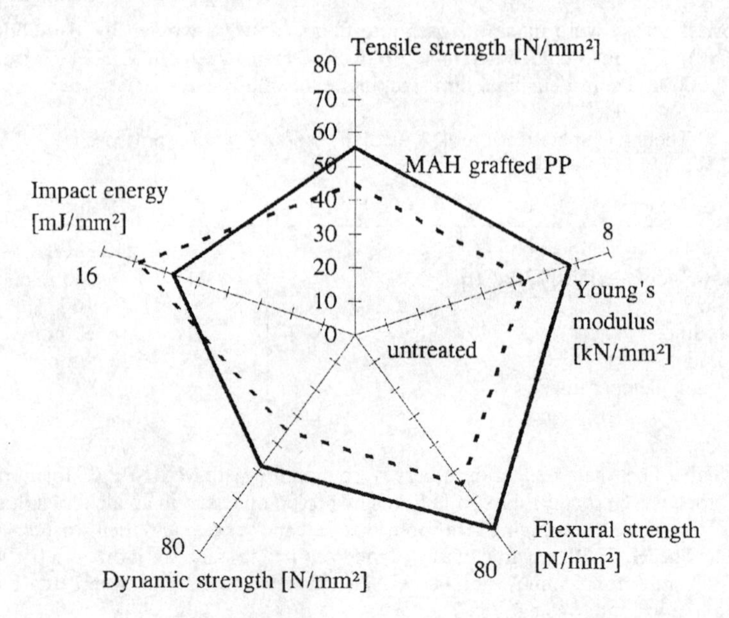

Figure 1: Influence of coupling agent (MAH grafted PP) on the mechanical properties of jute polypropylene composites (fiber content = 37 vol.-%)

Figure 1 demonstrates, that tensile, flexural and dynamic strength as well as Young's modulus could be achieved by a treatment with a toluene MAH-PP solution (0.1 wt.-%) for 5 minutes. As anticipated, impact energy decreases with improved fiber-matrix adhesion. A lot of other papers [3,6-8] are concerned with the effectiveness of maleic anhydride-polypropylene copolymers as a coupling agent. Mieck et al. [3] determined similar to our observations an increased in tensile strength of about 25% for flax-polypropylene composites, when the coupling agent was applied to the flax fibers before the composite was processed. His values were dependent on the grafting rate and on the average molar mass of the graft copolymer. Similarly increased values as in figure 1 could be obtained with a PP matrix material modified with MAH-PP. The acidic anhydride groups of the MAH coupling agent lead to hydrogen as well as chemical bonds with the hydroxyl groups of the flax fiber, tight by anchoring the coupling agent onto the fiber surface. Furthermore, the long PP-chains of the MAH-PP coupling agent lead to an adaptation of the very different surface energies of matrix and reinforcement fiber, which allows a good wetting of the fiber by the viscous polymer. Again an improved wetting can increase adhesion strength by an increased work of adhesion.

The increased strength and modulus properties due to the MAH-PP coupling agent are mainly based on a reduction of fiber pull-out and less fiber-matrix debonding. This should lead to less micropores at the interface [9]. Fiber pull-out is illustrated by the SEM photographs in figure 2. The improved fiber-matrix interface leads further to a lowering of the critical fiber length for effective stress transfer [7].

Scanning electron microscopy investigations on MAH-PP modified cellulose fibers (filter paper) [7] showed similar results as in our case.

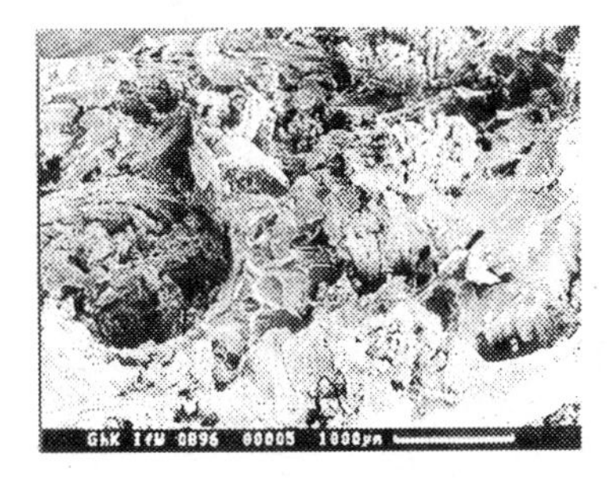

a) untreated

b) MAH-PP treated

Figure 2: SEM studies of the influence of coupling agent on the fracture behaviour of jute polypropylene composites (fiber content = 30 vol.-%)

To investigate the influence of the used MAH- PP coupling agent as well as to observe the effect of the fiber content, creep tests were carried out up to a maximum time of 300 min. After this time, all samples reached the so-called secondary or steady creep state [10] with a nearly constant creep rate.

The results of the creep investigations for composites without coupling agent (figure 3a) show a strain of the outer fiber of around 5 % after 300 min. test period for a fiber content of 23 vol.-%. The experimental data were fitted by using the creep law according to Abbott (eq.1).

$$\epsilon(t) = a + \ln t^{b} \tag{1}$$

with b as the creep kinetic coefficient.

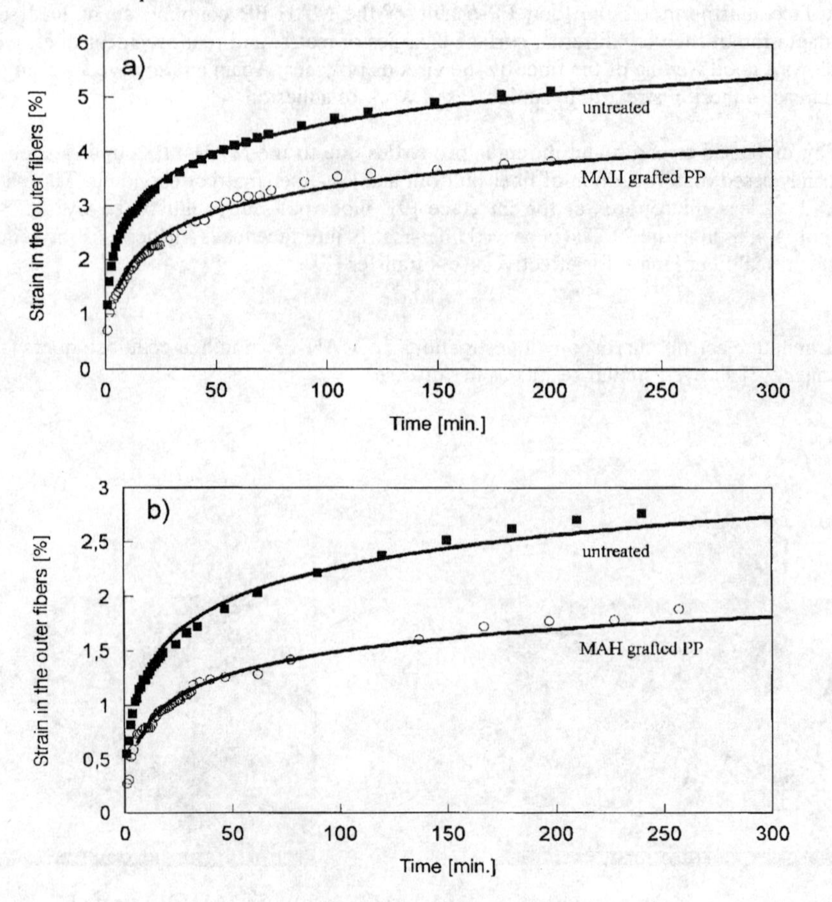

Figure 3: Influence of the fiber treatment on the creep behaviour for different fiber content (applied flexural stress = 5.1 N/mm²)
a) fiber content = 23 vol.-%; b) fiber content = 36 vol.-%

An increase of fiber content up to 36 vol.-% leads to half of the strain at 300 min. as well as to a lower creep kinetics.
As known from the creep behaviour of glass-fiber-reinforced plastics [11], the improved fiber-matrix interphase reduces the creep strain distinctly (figure 3) because of the better matrix/fiber load transfer.
In comparison to composites with untreated fibers, the strain in the outer fiber was reduced about 20 - 25 % by the use of coupling agent. The coupling agent is necessary to get an effective stress absorption through the fibers [12], which is the assumption for the observed materials' behaviour.

5. Conclusion

Improvements in the characteristic strength and stiffness properties of jute polypropylene composites were obtained with the application of MAH-PP copolymers to the fiber. The most noticeable increase in performance was obtained with a 5-minute application of the coupling agent in a toluene solution of 0.1 wt.-%. Impact energy decreases with improved fiber-matrix adhesion.
SEM investigations demonstrated that fiber pull out is reduced after the modification with the coupling agent. The improved fiber-matrix adhesion is due to the chemical bonds between fiber and matrix provided by the coupling agent, and the stress transfer from matrix to fiber is improved leading to improved reinforcing effect.
This improved fiber-matrix adhesion further leads to a lower creep strain in the outer fibers. This was demonstrated for composites with two different fiber contents. For an applied stresses of 5.1 N, the creep behaviour for the composites with unmodified, as well as for those with MAH-PP modified, fibers obeys the creep law according to Abbott.

6. References

[1] K.L. Mittal, SILANES AND OTHER COUPLING AGENTS, VSP BV Netherlands, 1992

[2] A.K. Bledzki and J. Gassan, NATURAL FIBER REINFORCED PLASTICS, N.P. Cheremisinoff, Ed., Handbook of Engineering Polymeric Materials, Marcel Dekker, Inc., New York 1997

[3] K.-P. Miek, A. Nechwatal, C. Knobelsdorf, Die Angewandte Makromolekulare Chemie, 1995, Vol. 225, pp. 37

[4] J. Gassan and A.K. Bledzki, Composites, 1997, Vol. 28 A, pp. 1001

[5] G. Wacker, EXPERIMENTELL GESTÜTZTE IDENTIFIKATION AUSGEWÄHLTER EIGENSCHAFTEN GLASFASERVERSTÄRKTER EPOXIDHARZE UNTER BERÜCKSICHTIGUNG DER GRENZSCHICHT, Ph.D. Thesis at the University of Kassel, Kassel 1996

[6] M. Avella, C. Bozzi, R. dell' Ebra, B. Focher, A. Marzetti and E. Martuscelli, Die Angewandte Makromolekulare Chemie, 1995, Vol. 233, pp. 149

[7] J. Felix and P. Gatenholm, Journal of Applied Polymer Science, 1991, Vol. 42, pp. 609

[8] A. Karmaker and J. Schneider, Journal of Materials Science Letters, 1996, Vol. 15 pp. 201

[9] S. Dong, S. Sapieha and H.P. Schreiber, Polymer Engineering and Science, 1993, Vol. 33, pp. 343

[10] J.T. Boyle and J. Spence, STRESS ANALYSIS FOR CREEP, Butterworth Southampton 1983

[11] J. Hugo, M. Sova and J. Cizinsky, Composite Structures, 1993, Vol. 24, pp. 233
[12] R. Yosomiya, K. Morimoto, A. Nakajima, Y. Ikada and T. Suzuki, ADHESION AND BONDING IN COMPOSITES, Marcel Dekker, Inc., New York 1990

Acknowledgement

The authors wish to express their gratitude to company J. Schilgen GmbH & Co. and Vestolen GmbH for providing the jute fibres and the polypropylene films, respectively.

Natural-fibre-mat-reinforced thermoplastics based on upgraded flax fibres for improved moisture resistance

T. Peijs[a], H.G.H. van Melick[a], S.K. Garkhail[a], G.T. Pott[b] & C.A. Baillie[c]

[a] Eindhoven University of Technology, Centre for Polymers and Composites, PO Box 513, 5600 MB Eindhoven, The Netherlands
[b] CERES B.V., Wildekamp 1b, 6704 AT Wageningen, The Netherlands
[c] Imperial College of Science, Technology and Medicine, Department of Materials, Prince Consort Rd., SW72BP London, UK

Abstract

In this paper the environmental properties of thermoplastic composites based on upgraded flax fibres and a polypropylene (PP) matrix are presented. The flax used in this study was upgraded using a novel treatment process for improved moisture and rot sensitivity. The treatment consists of heating the straw-flax in an aqueous environment in an autoclave at temperatures above 160°C, followed by drying- and curing at temperatures above 150°C. In this study the effect of this flax fibre treatment on the environmental properties of natural-fibre-mat-reinforced thermoplastics (NMTs) is investigated. This is done by monitoring the moisture absorption and swelling, and measuring the residual mechanical properties of the samples at different moisture levels. Results showed that the moisture absorption and swelling of these upgraded flax fibre composites is approximately 30% lower than that of composites based on green flax fibres.

1. Introduction

Because of their excellent price-performance ratio, E-glass fibres are by far the most important reinforcing fibres for polymer composites. However, these fibres do have some disadvantages. Glass fibres are non-renewable and give problems with respect to ultimate disposal at the end of a materials lifetime since they can not be thermally recycled by incineration. They are also very abrasive which leads to increased wear of processing equipment such as extruders and moulds. Next to some ecological disadvantages, glass fibres can cause problems with respect to health and safety. For example, they give skin irritations during handling of fibre products, and processing and cutting of fibre-reinforced parts. Nowadays, ecological concern has resulted in a renewed interest in natural materials and issues such as recyclability and environmental safety are becoming increasingly important for the introduction of new materials and products. In view of all this an interesting environmentally friendly alternative for the use of glass fibres as reinforcement in engineering composites are natural fibres based on ligno-cellulose such as flax, hemp, sisal and jute [1]. These vegetable fibres are renewable, nonabrasive and can be incinerated for energy recovery since they posses a good calorific value. Moreover, they give less concern with safety and health during handling of fibre products. In addition, they exhibit excellent mechanical properties, especially when their low price and density (1.4 $g.cm^{-3}$) in comparison to E-glass fibres (2.5 $g.cm^{-3}$) is taken into account. Despite these attractive features, ligno-cellulose fibres have found so far few applications as a reinforcement material in engineering composites. A major restriction in the successful use of natural fibres in durable composite applications is their high moisture absorption and poor dimensional stability (swelling), as well as their susceptibility to rotting. Swelling of fibres can lead to microcracking of the composite and degradation of mechanical properties. The poor environmental- and dimensional stability of natural fibres can however be significantly improved by a recently developed upgrading process for ligno-cellulose based materials [2]. This upgrading process has also proven its applicability to natural fibres and has let to the development of upgraded flax, the so-called Duralin flax. The availability of this upgrading process for natural fibres could remove one of the main restrictions for the successful application of natural fibres in high-quality engineering composites. In this research the environmental properties of natural-fibre-mat-reinforced thermoplastics (NMTs), being glass-mat-reinforced thermoplastic (GMT) -like materials [3] based on treated flax fibres and a polypropylene (PP) matrix [4-9], are investigated. The effect of the upgrading treatment is evaluated by comparing the moisture absorption, swelling and residual mechanical properties of NMTs based on treated Duralin-flax-fibre mats and green-flax-fibre mats. Furthermore, the influence of the fibre/matrix interface on these properties is investigated by using both isotactic-PP and maleic-anhydride grafted PP (MA-PP) as an adhesion promotor [5,6,10,11].

2. Experimental

2.1. Flax fibre treatment

The flax-upgrading process used here uses full rippled (deseeded) straw-flax. The use of straw-flax turned out to be beneficial for both strength and reproducibility (no dew-retting required) of the treated fibres. In addition, a valuable by-product, the treated flax-shives, is produced from which e.g. water resistant chip-boards can be made. The treatment consist of a steam or water heating step of the sheaves of rippled straw-flax at temperatures above 160°C during approximately 30 minutes in an autoclave. This is followed by a drying step and a heating (curing) step above 150°C during approximately two hours. During this treatment, the hemi-cellulose and lignin is depolymerized into low molecular aldehyde- and phenolic functionalities, which are combined by the subsequent curing reaction into a water resistant resin. After the treatment the fibres can easily be separated from the stem by a simple breaking and scutching operation. These treated flax fibres where subsequently converted into a non-woven mat by Eco Fibre Products B.V. (The Netherlands) via a conventional fleeze making and needling process. The treated Duralin fibres have mechanical properties comparable to green or dew-retted flax, with strength values in the order of 500-700 MPa and a stiffness of about 50 GPa. The equilibrium moisture absorption of the upgraded fibres at 90% R.H. and 20°C is about 11%, and 19% for green flax. Up to now most of the work on upgrading of ligno-cellulose fibres has been done on flax but since other fibres like hemp, kenaf and sisal are also based on ligno-cellulose, and only differ in the relative amounts of cellulose, hemi-cellulose and lignin, it is expected that the technology initially developed for flax, can be directly transferred to other vegetable fibres.

2.2. *Composite manufacturing*

In this study random non-woven flax fibre mats in combination with an isotactic-polypropylene (PP) matrix of Montell (XS6500S) with a melt flow index of 38 were used. As an adhesion promotor maleic-anhydride modified polypropylene (MA-PP) was used [10,11]. The blending of PP with a commercially available MA-PP (Polybond® 3002, BP Chemical Ltd.) was performed on a Haake extruder. In this study 5 wt.% of MA-PP is added to the PP. For convenience, this blend will simply be designated as MA-PP. Next, the PP and MA-PP pellets were compression moulded into 0.1 mm thick sheets using a hot-press. NMT composites were made using the film stacking method. First, flax fibre mats (250x250 mm) were cut and dried in an oven at 60°C for 2 hours. Alternating layers of non-woven flax mats and PP-sheets were stacked and impregnation was achieved by applying heat (200°C) and pressure for about 15 minutes. The composites obtained after cooling had a thickness of approximately 3 mm and a fibre volume fraction of about 38%. As a reference, similar composite plates based on non-woven green flax mats were made. The obtained composite plates were pre-dried in an oven at 60°C for one day to reach the initial moisture level. Next, these plates were cut into specimens of 200x200 mm, which were immersed into a tank filled with tap water. Weight and thickness were measured as a function of time. At certain moisture levels tensile tests were performed on test specimens cut from these plates to investigate the effect of moisture on strength and stiffness. Three different types of NMTs were evaluated: (i) treated Duralin flax/PP, (ii) treated Duralin flax/MA-PP and (iii) green flax/PP as a reference.

3. Moisture absorption

In order to study the moisture absorption of materials, different models have been developed [12]. A problem in which the temperature and the moisture distribution inside the material are to be determined, is often referred to as the 'moisture problem'. Such problems can be solved analytically when the following conditions are met:

- Heat transfer is by conduction only and can be described by Fourier's law.
- Moisture diffusion can be described by a concentration dependent form of Fick's law
- The temperature inside the material approaches equilibrium much faster than the concentration, hence the energy (Fourier) and the mass transfer (Fick) equations are decoupled
- The thermal conductivity and the mass diffusivity depend only on the temperature and are independent of moisture concentration or of the stress levels inside the material

When these assumptions are met the diffusion problem is said to be 'Fickian'. Calculations can be made requiring knowledge of the following parameters:

- Geometry (material thickness h in case of a 1-dimensional problem).
- Boundary conditions: ambient temperature and relative humidity (100% in case of immersion).
- Initial conditions: temperature and moisture concentration M_i inside the material.
- Material properties: density ρ, specific heat C, thermal conductivity K, mass diffusivity

D, maximum moisture content M_m and a relationship between the maximum moisture content and the ambient conditions.

The moisture content M_t as function of the square root of time for a typical Fickian process is schematically given in Figure 1.

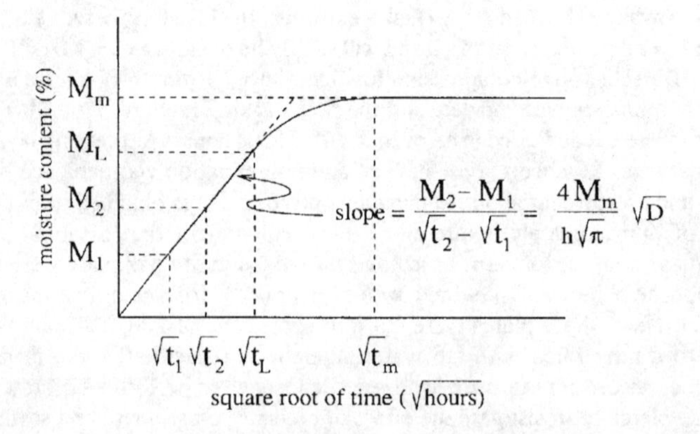

Figure 1. Moisture content as a function of time for a typical Fickian process [12].

The relative moisture absorption can be described by the following equation:

$$\frac{M_t}{M_m} = 1 - \frac{8}{\pi^2} \sum_{j=0}^{\infty} \frac{1}{(2j+1)^2} e^{-\frac{D(2j+1)^2 \pi^2 t}{h^2}} \tag{1}$$

This equation will be used to characterize the diffusivity (D) and maximum moisture content (M_m) of the flax composites.

4. Results

The weight and thickness of the NMT plates is measured to characterize the moisture absorption of the different composites. Figure 2 shows the moisture absorption as a function of time. All samples were immersed in water for about 60 days at room temperature. By fitting Equation 1 on the experimental data, the maximum moisture content (M_m) and the diffusivity (D) of the different types of composites can be determined (see Table 1). Green flax fibre based composites are clearly more sensitive to moisture than the other two types of composites based on treated flax. The maximum moisture content in treated flax fibre composites is some 30% lower than that of green flax fibre systems. Also the diffusivity of the upgraded flax fibre composites, as calculated from the initial slope of the moisture uptake curves, is much lower than that for the NMTs based on green-flax-fibre mats.

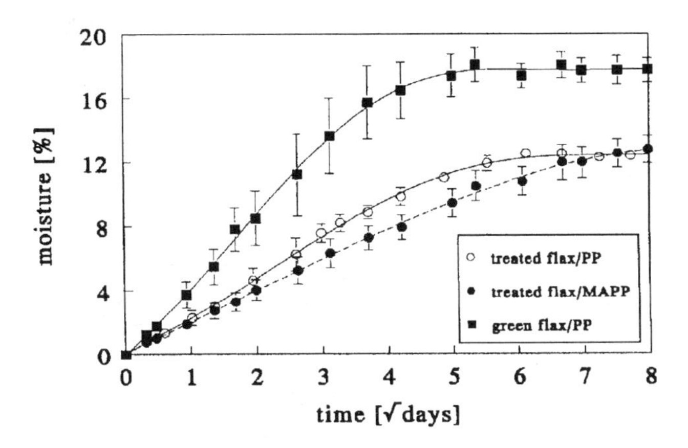

Figure 2. Moisture content (M_t) as a function of the square root of time

It is interesting to note that the use of MA-PP as a compatibiliser further lowers the diffusivity. Clearly, the initial moisture uptake in this composite system is taken place at a lower rate than for the PP system without compatibiliser. The maximum moisture content level is, however, similar for both PP and MA-PP based treated flax systems. The higher diffusivity for the PP system without compatibiliser, indicates that initially a fair amount of moisture uptake takes place along the fibre/matrix interface. For the treated flax/PP, as well as for the green flax/PP, a plateau value for moisture uptake is reached after about one month immersion in water.

Table 1. Maximum moisture content and diffusivity

Composite	Max. Moisture content M_m (%)	Diffusivity D ($m^2.s^{-1}$)
Green flax/PP	18.0	$1.3*10^{-6}$
Treated flax/PP	12.8	$7.8*10^{-7}$
Treated flax/MA-PP	13.5	$5.0*10^{-7}$

It can be expected that the absorption of moisture has its effect on the dimensional stability of the composite plates. Figure 3 shows the thickness-swell of the samples immersed in water for 60 days at room temperature. Again the performance of the treated flax fibre composites is much better than that of the green flax based system. The thickness of the green flax based composite increases by about 13%, whereas the treated Duralin flax composites swell about 9 to 10%.

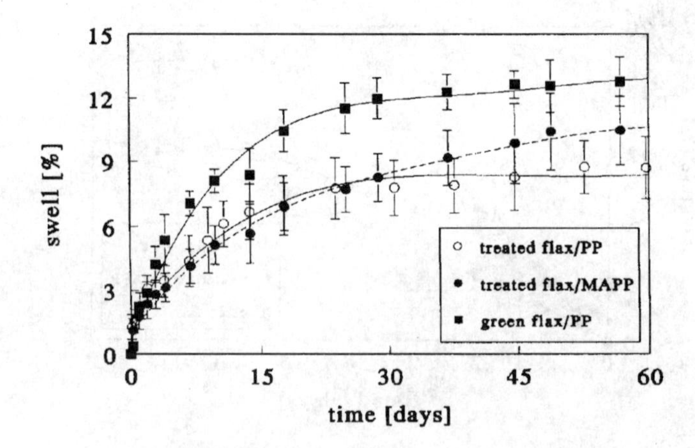

Figure 3. Thickness-swell of composite plates as a function of time

With respect to the durability of natural fibre composites, the most serious effect of moisture absorption is the degradation of mechanical properties. Therefore, the stiffness and strength of the NMTs based on upgraded Duralin flax was measured as a function of the moisture content. The initial stiffness and strength values for these type of natural-fibre-reinforced composites is about 8 GPa and 50 MPa, respectively. In Figure 4 the normalized stiffness is plotted as a function of moisture content, showing a clear drop in composite stiffness. The modulus of the saturated samples is about 30% lower than that of dry samples. Figure 5 shows the tensile strength as a function of moisture content. Compared to the drop in composite stiffness, the tensile strength is not very significantly affected by the water uptake.

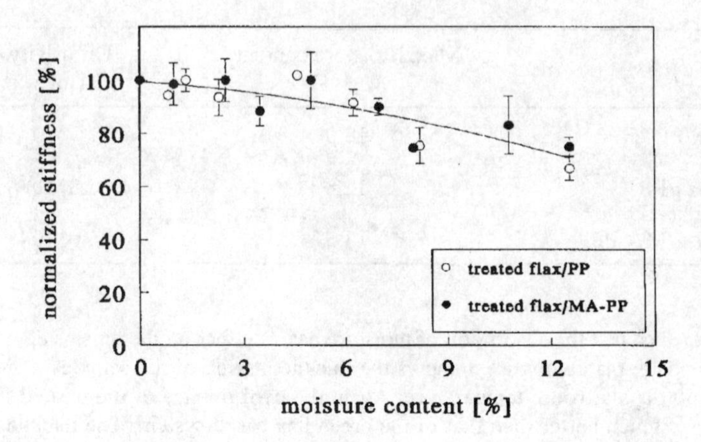

Figure 4. Effect of moisture content on modulus of treated flax composites

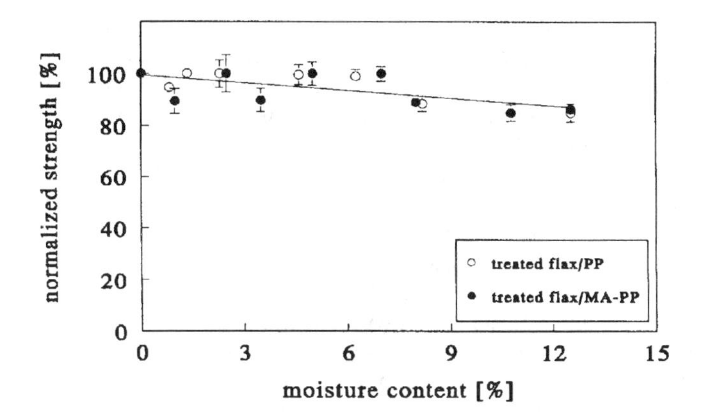

Figure 5. Effect of moisture content on tensile strength of treated flax composites

5. Conclusions

The moisture resistance of natural-fibre-mat-reinforced thermoplastics (NMTs) based on flax fibres and a polypropylene matrix can be improved by the use of upgraded Duralin flax fibres. These upgraded fibres were treated using a novel process for improved moisture and rot sensitivity of ligno-cellulose based materials. The treatment consists of heating the straw-flax in an aqueous environment in an autoclave, followed by a drying- and curing step at elevated temperatures. The effect of this upgrading treatment on the environmental performance of NMTs was investigated by monitoring the moisture absorption and swelling, and measuring the residual mechanical properties of the samples at different moisture levels. Results showed that the moisture absorption of upgraded flax/PP composites is about 30% lower than that of green flax based composites, whereas also the dimensional stability (swelling) is significantly improved. The absorption of moisture by random flax-fibre-reinforced composites can adequately be described by Fick's law. The moisture content is linear to the square root of time until about two weeks of immersion in water. After that, the moisture content levels of to a maximum moisture content. The use of maleic-anhydride modified PP as a compatibiliser for improved interfacial bonding lowers the diffusivity, being the water uptake rate, significantly. The maximum moisture content is however not affected by the interfacial properties. The modulus of treated flax based composites decreases with increasing moisture content. At the saturation level the stiffness reduction is about 30%. The tensile strength, however, is not significantly affected by the water uptake.

References

1. Morton, W.E. and Hearle, J.W.S., *Physical Properties of Textile Fibres,* 2nd ed., John Wiley & Sons, Inc., New York (1975).
2. Ruyter, H.P. and Hortulanus, A., *European patent application* EP 373 726, 1993.

3. Berglund, L.A. and Ericson, M.L. in: *Polypropylene: Structure, blends and composites, Vol. 3*, ed. J. Karger-Kocsis, Chapman & Hall, London (1995) 202.
4. Heijenrath, R. and Peijs, T., *Advanced Composite Letters*, 5(3) (1996) 81
5. Garkhail, S.K., Heijenrath, R., Peijs, T., Van den Oever, M. and Bos, H., *Composites, Part A,* (1998) submitted.
6. Garkhail, S., Heijenrath, R., Van den Oever, M., Bos, H. and Peijs, T., in: *Proc. of the 11th International Conference on Composite Materials (ICCM/11)*, Vol. II, ed. M.L. Scott, Gold Coast 14-18 July 1997, Woodhead Publ. Ltd. (1997) 794.
7. Mieck K.-P., Lutzkendorf, R. and Reussmann, T., *Polymer Comp.,* 17 (6) 1996, 873
8. Mieck, K.-P., Reussmann, T., Knobelsdorf, C., *Kunststoffberater*, 5, 1997, 29.
9. Pott, G.T., Pilot, R.J. and van Hazendonk, J.M., in: *Proc. of the 5th European Conference on Advanced Materials and Processes and Applications (EUROMAT 97)*, Vol.2 Polymers and Ceramics, Maastricht, 21-23 April 1997,107.
10. Felix, J.M. and Gatenholm, P., *J. Applied Polymer Science.*, 42 (1991) 609.
11. Mieck K.-P., Nechwatal, A. and Knobelsdorf, C., *Die Angewandte Makromolekulare Chemie*, 225, 1995, 37.
12. Springer, G.S., *Environmental effects on composite materials*, Westport, 1981.

Cork/Thermoplastic Particleboards

L. Gil*

*Unidade de Tecnologia da Madeira e da Cortiça - Instituto Nacional de Engenharia e Tecnologia Industrial - Estrada das Palmeiras, Queluz de Baixo, 2745 Barcarena - Portugal

Abstract

In this work the manufacture of cork powder particleboards is described, where the agglomeration of the particles and in some cases the gluing of surface layers, is achieved by using different kinds of thermoplastics such as polyethylene and polypropylene. Boards obtained by this process have physico-mechanical characteristics wich allow for a wider spectrum of applications than those of current corkboards.

1. Introduction

Cork from *Quercus suber* L. has several uses. For many applications, cork must be reduced to particles. Cork grinding gives rise to large quantities of cork powder, which is mainly burned in boilers for steam and energy production.
The aim of this work was to find out other industrial uses of cork powder, produced at the different steps of cork processing and which constitutes about 25% (w/w) of the raw material i.e. 17000 to 45000 ton/year.
In this work a process was developed for the manufacture of particleboards from cork and thermoplastic resins, where these polymers work not only as binding agents, but also give some special characterisics to the agglomerates, allowing applications different from those of the common corkboards.

2. Experimental

The cork waste powder used is called "CALDEIRA" and is produced at the initial step of the manufacturing process in the granulation zone. This cork waste powder has a lot of impurities (interior and exterior parts of cork and dirt). The thermoplastic resins, were polyethylene and polypropylene.
The main physico-chemical characteristics of the materials - density, moisture content, waxy fraction, ash content - were determined.
The following step was the study of the agglomeration conditions (P. pressure, T. temperature; t. time) and the physico-mechanical characterisation of the most promising particleboards.
The cork particles were mixed with a certain quantity of a thermoplastic polymer, in powder form. The mixing was done in a mixing device for a period of time needed for good homogeneity. The mixture was introduced into a metallic mould in such a quantity that corresponds to the desired initial thickness of the material (e.g. 2 cm), this thickness being a function of the pressure to be used, which by itself depends on the material used, and on the density and type of application required for the agglomerate.

The mould had removables top and bottom, in order to get a good release at the end of the manufacturing operations. The mould also had orifices on the side wall for the introduction of transverse bars above the cover after compression.
The distance of these orifices to the bottom of the mould depends on the final thickness desired for the board (e.g. 1 cm). At the bottom and top surfaces of the deposited material, release sheets were used (e.g. paper) to avoid the gluing of the agglomerate to the cover and bottom of the mould.
When polyethylene was used, the gluing of the paper sheet to the board occurred. Wood veneers also glued to the board, improving the aesthetic aspect and the mechanical resistance of the board.
Concerning the boards made of cork powder and polypropylene, it was impossible to use the wood veneer as covering, even with high temperature (190°C).
The mould with the mixture and cover was pressed at 15-30 Kgf/cm^2 reducing the material's thickness by half.
Before releasing the pressure, the transverse bars were introduced into the side orifices. The mould containing the compressed material was then placed in a heating system (e.g. oven), to melt thermoplastics. The heating temperature was higher than the melting point of the thermoplastics, usually 40°C more (e.g. 130° C for the polyethylene and 190° C for the polypropylene) to allow heat to reach the inner part of the board. However, the temperature should not exceed 200°C, at which carbonisation of the cork particles begins. Due to the low thermal conductivity of the cork, the heating time used was 1h - 1h30m, for final thicknesses lower than 1 cm when using releasing/covering sheets.
After the heating period, the moulds were cooled in air, with the material compressed and the transverse bars in position, generally for 24 hours, for dimensional stabilisation.
Samples of the boards produced were tested for density, moisture content, dimensional change, bending and tensile strength, and others according to standards.

3. Results and discussion

Cork powder and thermoplastics data are summarised in Tables 1 and 2, respectively.
Particle size analysis data are given in Tables 3 and 4.

Table 1. Physico-chemical characteristics of the cork powder.

Density (Kg/m^3)	Moisture content (% w/w)	Waxy fraction (% w/w)	Ash content (% w/w)
287	14	4,2	4,6

Table 2. Physico-chemical characteristics of the thermoplastics.

Thermoplastics	Density (Kg/m^3)	Melting temperature vicat (°C)	Melting index (g/10min)
Polyethylene (PE)	272	78	16-24
Polypropylene (PP)	604	152	16-20

Table 3. Particle size analysis of the cork powder ("caldeira").

Mesh	% (w/w)
1.25mm	11,20
800μm	8,30
400μm	18,45
250μm	11,73
100μm	20,05
Under	30,23

Table 4. Particle size analysis of the thermoplastics.

Thermoplastics	Mesh	% (w/w)
Polyethylene (<2mm)	1,00 mm	43,08
	500μm	33,99
	250μm	17,08
	Under	5,49
Polypropylene (<2mm)	500μm	71,66
	250μm	21,51
	Under	6,10

The choice of low melting index thermoplastic materials as agglutinants was made on the basis of their advantages over conventional glues, namely absence of solvents and non toxicity. Agglomeration conditions are given in Table 5.
In the case of some thermoplastics (e.g. polyethylene) with convenient characteristics, if the release sheets (e.g. paper sheet, wood venner) allowed the exit of the water (moisture) and the penetration of melted plastic, the gluing of these sheets was achieved, and the manufacture of covered boards, in a single operation was possible.
To reach the same final thickness having the same initial volume of material, with the *polypropylene:cork* mixture more pressure is needed. To obtain resistant boards the heating time must be higher than 1 hour. Increasing the temperature (to certain limits), can reduced time of operation.
Results of the physico-mechanical tests of the selected particleboards with promissing characteristics are shown in Table 6.

Table 5. Examples of types and operational conditions of the productions of the boards.

Board N°	Material	Vol. ratio	Initial thickness - Final thickness (cm)	Pressure (Kgf/cm^2)	Oven Temp. (°C)	Heating time (min)	Comments
1	Cork:PE	4:1	2,0-0,80	24.4	130	90	Covered with a veneer of Eucalyptus wood on both sides of the board

2	Cork:PE	4:1	2,0-0,81	14.4	130	90	Covered with a sheet of paper on both sides of the board.
3	Cork:PE	4:1	2,0-0,85	14.4	160	90	Covered with a sheet of paper on both sides of the board
4	Cork:PP	5:1	2,0-0,83	14.4-17.3	170	90	Covered with a sheet of paper on both sides of the board.
5	Cork:PP	5:1	2,0-0,91	28.9	190	90	Covered with a sheet of paper on both sides of the board. This kind of board seems very resistant.
6	Cork:PP	5:1	2,0-0,88	28.9	190	60	Covered with a sheet of paper on both sides of the board.

Table 6. Physico-mechanical characteristics of the best particleboards.

Board No.	Material	Thermal conductivity (Kcal/m.h °C)	Oxygen index (%)	Density (Kg/m^3)	Thickness swelling (%)	Flexure test Ef=mod. of elasticity nf=bending strength (da N/cm^2)	Tensile strength (daN/cm^2)
1	Cork:PE	0,041-0,059	24,4	561	11,2	Ef= 58501 nf= 26,2	0,46
2	Cork:PE	0,042-0,049	24,5	475	7,5	Ef= 60020 nf= 19,7	0,36
3	Cork:PE	0,042	24,0	460	9,4	Ef= 8615 nf= 26,5	0,51
4	Cork:PP	0,044-0,049	23,0	574	1,6	Ef= 19787 nf= 36,9	3,55
5	Cork:PP	0,042-0,046	22,5	580	2,7	Ef= 29855 nf= 55,1	6,16
6	Cork:PP	---	---	549	2,5	Ef= 19803 nf= 29,8	2,96

4. Conclusions

Present manufacturing processes, using compression in a mould and heating, or hot pressing, cause problems when using small dimension cork particles, wich require the use of larger amounts of glues containinf solvents and toxic components, and may turn the process non profitable. The agglomerates obtained are flexible and compressible, characteristics which may limit their applications, namely in structural panels.

The cork board manuacturing process based on the present work produces a rigid agglomerate that is covered or not by an adequate sheet. It can be used as decorative board (e.g. false ceiling and wall coverings) and as divise panels and door panel providing a better thermal and acoustic insulation and lower water absorption than those of wood particleboards.
The main advantages of this type of corkboard are the following:
- The process of mixing the binding agent in a powder form with the cork particles, is simple.
-It is possible to manufacture agglomerates with particles having smaller dimensions than the usual, without toxic and technical problems.
For some thermoplastics, it is possible to produce the agglomerate, by gluing the surface layer only in a single operation, avoiding the steps of lamination, sanding and sheet gluing. It is also possible to use thermoplastics wastes, and mixtures of several thermoplastics.
- The main manufacturing operation and the equipment to be used are not very much different from those of the usual method used for some composition corkboards, and so the cork industry does not need a lot of modifications and investments to adopt and to manufactire this new kind of corkboard.
The excellent results led to a patent application.

References

Gil, 1993: New cork powder particleboards with thermoplastic binding agents, Wood Sci. Technol., 27: 173-182.

Thermoplastic Composites Based on Natural Fibres

Kristiina Oksman and Peter Nilsson
Swedish Institute of Composites (SICOMP)
Box 271, SE-941 26, Piteå, Sweden

Abstract

The use of natural fibres instead of man made fibres, as reinforcements in thermoplastics, gives interesting alternatives for production of low cost and ecologically friendly composites. In this work different commercially available semi-finished natural fibre mat reinforced thermoplastics (NMT) composites have been studied. Mechanical properties and microstructure of different NMT composites were investigated and compared to conventional GMT (glass fibre mat reinforced thermoplastic) composites and pure polypropylene (PP). The results showed that NMT composites had high stiffness compared to pure polymer and the NMT with a high fibre content (50 % by weight) showed even better stiffness than the GMT. The GMT composites had superior strength and impact properties compared to the NMT which might depend on the relatively low strength of the natural fibres but also on poor adhesion to the PP matrix.

1. Introduction

During the last few years there has been an increasing environmental consciousness which has increased the interest to use natural fibres instead of man made fibres in composite materials. Therefore several manufacturers have started to produce semi-finished NMT materials. The advantages using natural fibres instead of glass fibres are: low density, lower price, low abrasive wear and they are virtually available everywhere. Further the natural fibres are recyclable, biodegradable and carbon dioxide neutral and can therefore be energy recovered in an environmentally acceptable way [1-3]. The natural fibres normally involved in these kinds of materials can be for example flax, and the matrix is usually polypropylene (PP). The commercially available NMT materials can be nonwoven fibre mats impregnated by polymers in the form of sheets but also like a nonwoven fibre mat where the flax and polymer fibres are mixed together in the form of fleece. During processing the mats or sheets need to be heated at first to the melting temperature of the polymer matrix, and then compression moulded to products.

Several investigations have been made to study the potential of natural fibres as reinforcement in thermoplastics [1-6]. The results have shown that natural fibres have potential to be used as reinforcement for plastics but they do not attain the strength level of glass fibre reinforced plastics [2-4]. Mieck et al. [2, 3] showed that when flax fibres were used as reinforcements in thermoplastics (PP) the modulus was increased to the level of glass fibre reinforced thermoplastics. Heijenrath and Peijs [4] reported that NMT composites made by film stacking method resulted in composites with comparable stiffness to conventional GMT composites while the strength of NMTs was lower. It is well known that the stress transfer efficiency between natural fibres and synthetic polymers is poor due to incompatibility between the polar and hydrophilic fibre and nonpolar and hydrophobic polymer [6-8]. Several investigations have been made to study the adhesion between natural fibres and synthetic polymers and the results showed that the composite strength and toughness are significantly improved when coupling agents are used [3, 5-8].

The objective of this work was to investigate the mechanical properties and morphology of semi-finished NMT composites and compare those with conventional GMT.

2. Experimental

2.1 Materials

Materials used in this study are semi-finished NMT from different manufacturers where two different types of processing technique have been used:

1. PP melt-impregnated fibre mats, available as semi-finished product in the form of sheets

2. Nonwoven of PP fibres and flax fibres, available as semi-finished product in the form of fleece

The composition of the studied materials is shown in Table I.

Table I. Composition (% by weight) and sample codes of the various NMT composites.

Sample code	Type	Matrix, PP	Fibre content
PP	-	100	-
S	melt impregnated sheet	75	25
I_1	melt impregnated sheet	63	37
I_2	melt impregnated sheet	60	40
M	fleece	50	50
GMT	melt impregnated sheet	60	40

Typically GMT had a glass fibre content of 40 wt% and was melt impregnated. According to supplier data, the NMT materials are reinforced by flax fibres, fibre content varying between 25-50 wt%.

2.2 Manufacturing composites

The processing of the semi-finished NMT materials is made through three steps:

1. cutting blanks of the sheets (or nonwoven fibre mats)
2. heating the blanks above PP melting temperature
3. moulding the heated materials by compression moulding

Rectangular blanks were heated in a combined hot air and infrared (IR) oven using a temperature range of 200-220°C and afterwards quickly moved to the conventional hydraulic compression moulding press (Fjellman press AB, Mariestad, Sweden, capacity of 3100 kN).

The fleeces of PP and flax fibres were difficult to heat without burning the material and were therefore contact heated between thick aluminium plates to the temperature of 195°C. The temperature was controlled by a thermocouple placed between the nonwoven mats during heating. Materials were then compression moulded.

The pressure was about 10 MPa. Moulded composites had a size of 350 x 450 mm and thickness between 2.6 to 3.8 mm. All materials had a weight between 700-800 g.

2.3 Mechanical testing

Tensile testing of the specimens was performed on an Instron test machine (model 8510) using a crosshead speed of 2 mm/min. Samples for the tensile testing were cut to the width of 30 mm and length of 250 mm. The thickness varied between 2.6 to 3.8 mm. Izod impact testing was performed according to ASTM D 256 (Izod impact) on a Karl Frank GmbH impact tester. Test samples for impact testing were machined with a milling cutter. At least 5 specimens of every composition were tested.

2.4 Microscopy study

Fractured surfaces of Izod impact specimens were sputter coated with platinum and examined using a Jeol JSM-5200 scanning electron microscope (SEM) at an acceleration voltage of 20 kV.

3. Composite properties

The mechanical properties of the composites obtained from tensile and impact tests are summarised in Table II.

Table II. Mechanical properties of different commercially available NMT composites, GMT and pure PP. [a]

Composite	Tensile properties			Izod impact properties	
	Strength	Modulus	Elongation at break	Notched	Unnotched
	(MPa)	(GPa)	(%)	-------- (J/m)	--------
PP	28.5 (±0.6)	1.5 (±0.1)	-	24 (±1)	553 (±81)
S	21.2 (±0.9)	2.8 (±0.3)	1.3 (±0.1)	38 (±0)	66 (±5)
I_1	25.4 (±2.0)	3.6 (±0.6)	1.3 (±0.4)	49 (±8)	100 (±13)
I_2	33.3 (±1.4)	4.4 (±0.4)	1.4 (±0.1)	88 (±10)	141 (±19)
M	56.6 (±2.1)	6.2 (±0.6)	1.8 (±0.2)	150 (±10)	309 (±30)
GMT	74.9 (±7.5)	5.1 (±0.4)	1.8 (±0.1)	410 (±75)	717 (±91)

[a] Standard deviations in parentheses

Mechanical tests show that maximum tensile strength of NMT composites is increased with increased fibre content. The I_2 and M did not reach the level of GMT, which was expected due to higher strength of glass fibres, 3400 MPa, compared to flax fibres about 1000 MPa. It should be mentioned that the NMT with the fibre content of 50 wt% has a tensile strength of 57 MPa which is twice of pure PP, 29 MPa.

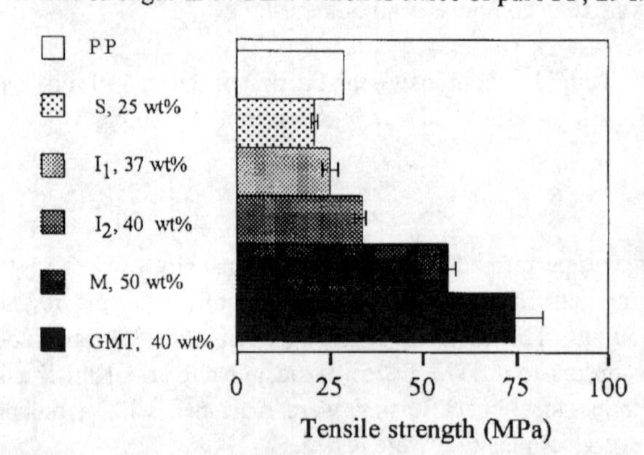

Figure 1. Tensile strength of different NMT composites compared to conventional GMT and pure PP.

Izod impact tests showed that NMT composites have inferior impact properties compared to GMT. In notched samples both NMT and GMT had better impact strength than pure PP. Generally, the impact strength of NMT composites was improved with increased fibre content except compared to unnotched pure PP samples. Composites impact properties are shown in Figure 2.

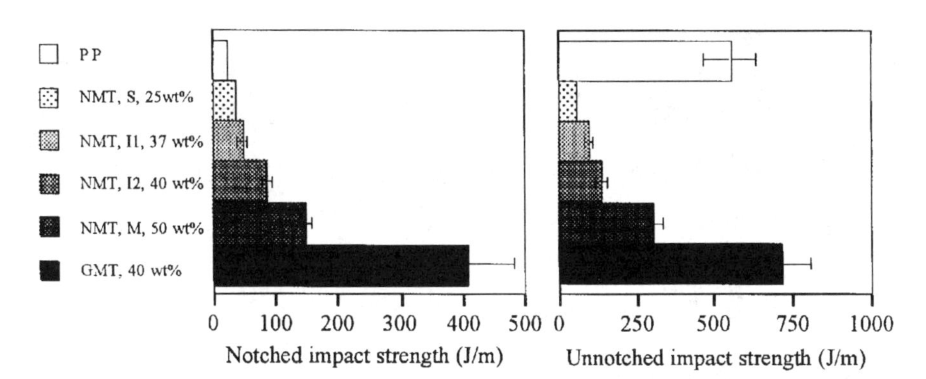

Figure 2. Izod impact properties of notched and unnotched composite

Figure 3 shows that stiffness of the NMT composites with higher or at least the same fibre content by weight as GMT are comparable with the glass fibre composites, it is even higher for NMT with 50 wt% flax fibre.

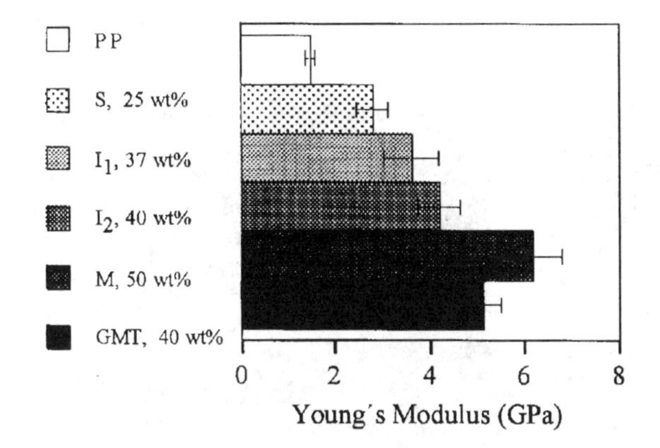

Figure 3. Stiffness of different NMT composites compared to GMT and pure PP.

4. Composite morphology

Scanning electron microscopy of fractured surfaces of different NMT and GMT composites were made to study the morphology and adhesion between the fibres and the matrix. Generally, the fractured surfaces of the NMT composites were not as sharp as the GMT fractured surfaces.

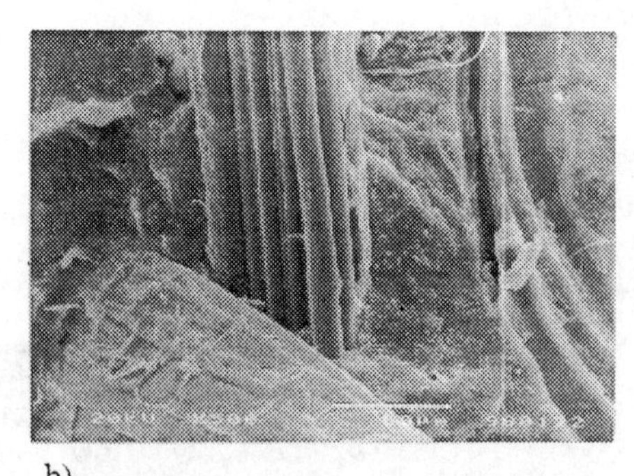

a) b)

Figure 4. Scanning electron micrographs of the fracture surface of a NMT composite with 25 wt% flax fibre a) overview b) detail.

Figure 4 a) shows a typical fracture surface of NMT composites with fibre content of 25 wt%. It can be seen that there is a lot of fibre pull-outs and that the fibres are mostly as fibre bundles. Figure 4 b) shows a more detailed micrograph of the composite structure. It can be seen that there appears to be poor adhesion between the fibre bundles and the PP matrix. The surface of the fibre bundle is clean and it is also possible to see where the fibres have been located before fracture.

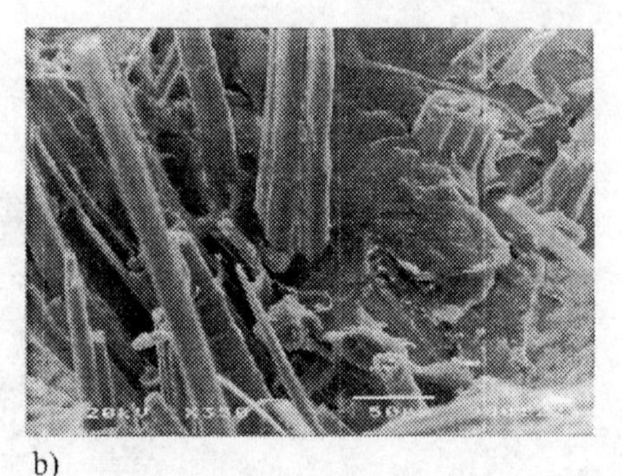

a) b)

Figure 5. Scanning electron micrographs of the fracture surface of a NMT composites with 37 wt% flax fibre a) overview b) detail.

Figure 5 a) shows an NMT composite with 37 wt% fibre content and compared to the Figure 4 it has higher fibre content. It is again possible to observe many fibre pull-outs indicating poor adhesion. Compared to the micrographs in Figure 4 there are more individual fibres in these composites. In b) a more detailed micrograph of the interface shows that the fibre surfaces are clean. In larger magnification there were voids visible between the fibre and matrix.

For the NMT composite with fibre content of 50 wt% it was impossible to study the interface region between the matrix and the fibres. The fibres pull-outs are so long and so many that it was difficult to see any matrix.

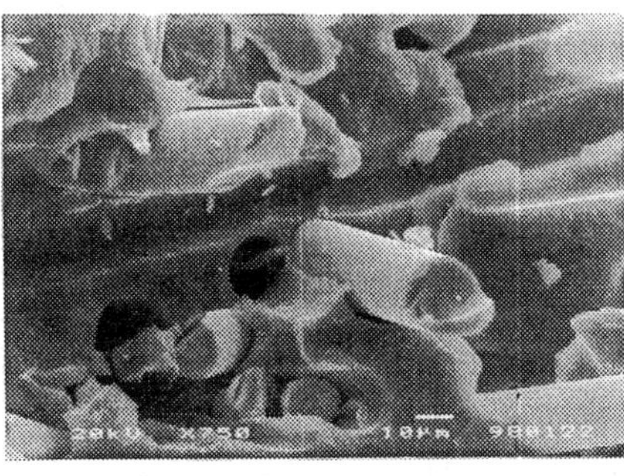

a) b)

Figure 6. SEM micrographs of the fracture surface of a GMT composite with 40 wt% glass fibre a) overview b) detail.

Figure 6 a) shows that the fibre content (by volume) of the GMT is low compared to the NMT in figure 5 even if the composite has almost the same fibre content by weight. The difference is explained by the higher glass fibre density, which is about 2.6 g/cm^3 compared to natural fibres 1.5 g/cm^3. It is also possible to see that the adhesion between glass fibres and PP matrix is poor, there is a lot of long pull-outs, the interfacial region shows voids and the fibre surfaces are clean.

5. Conclusions

The objective of this study was to investigate the mechanical properties and morphology of four different commercially available semi-finished NMT composites and compare these with conventional GMT and the pure polymer matrix.

The results of mechanical tests show that the stiffness of NMT composites is comparable with conventional GMTs when the fibre content is higher or at least as high as the glass fibre content by weight.

The tensile strength of NMT materials is lower than the GMT, but it is increased with increased fibre content and it is expected to improve further if coupling agents are used. The NMT composites impact properties showed much lower Izod impact strength compared to GMT composites in both notched and unnotched samples. Impact-, tensile properties and the morphology study indicates that there is poor fibre/matrix adhesion.

These results show that NMT composites have a potential to be used instead of conventional GMT in engineering applications where low weight, easily recyclable and environmental friendly materials are desirable.

Acknowledgements

The authors would like to thank Lear Corporation Sweden, Interior Systems AB, Tanum and Symalit AG, Switzerland for supplied materials.

References

[1] *T. Fölster and W. Michaeli*, FLAX - A RENEWABLE SOURCE OF REINFORCING FIBRE FOR PLASTICS, Kunststoffe German Plastics, Vol. 83, No. 9, 1993, p. 687-691

[2] *K-P, Mieck, A, Nechwatal and C, Knobelsdorf*, POTENTIAL APPLICATIONS OF NATURAL FIBRE IN COMPOSITE MATERIALS, International Textile Reports, Vol. 75, No. 11, 1994, p. 892-898

[3] *K-P. Mieck, R. Lützkendorf and T. Reussmann*, NEEDLE-PUNCHED HYBRID NONWOVENS OF FLAX AND PP FIBRES-TEXTILE SEMIPRODUCTS FOR MANUFACTURING OF FIBRE COMPOSITES, Polymer Composites, Vol. 17, No. 6, 1996, p. 873-878

[4] *R. Heijenrath and T. Peijs*, NATURAL-FIBRE-MAT-REINFORCED THERMOPLASTIC COMPOSITES BASED ON FLAX FIBRES AND POLYPROPYLENE, Advanced Composites Letter, Vol. 5, No 3, 1996, p. 81-85

[5] *P. R. Hornsby, E. Hinrichen and K. Taverdi*, PREPARATION AND PROPERTIES OF POLYPROPYLENE COMPOSITES REINFORCED WITH WHEAT AND FLAX STRAW FIBRES, Journal of Material Science, Vol. 32, No. 4, 1997, p. 1009-1015

[6] *A. R. Sanadi, D. F. Cauldfield and R. M, Rowell*, REINFORCING POLYPROPYLENE WITH NATURAL FIBRERS, Plastic Engineering, No. 4, 1994, p. 27-28

[7] *K. Oksman and H. Lindberg*, INTERACTION BETWEEN WOOD AND SYNTHETIC POLYMERS, Holzforschung, Vol. 49, 1995, p. 249-254

[8] *K. Oksman*, IMPROVED INTERACTION BETWEEN WOOD AND SYNTHETIC POLYMERS IN WOOD/POLYMER COMPOITES, Wood Science and Technology, Vol. 30, 1996, p. 197-205

APPLICATION OF RENEWABLE FIBRE REINFORCED PLASTIC PRODUCTS

J.C.M. de Bruijn (*, #)

(*) Polynorm Plastics B.V., P.O. Box 1267, 4700 BG Roosendaal, The Netherlands
tel.: +31 165 575 454; fax: +31 165 561 606
(#) University of Technology Delft, faculty of Industrial Design Engineering,
Jaffalaan 9, 2628 BX Delft, The Netherlands

Abstract

Natural fibres have significant advantages, as an alternative fibre reinforcement material over glass. Natural fibres are more environmentally friendly, healthier and safer, and cause less abrasive wear of processing equipment. On the other hand, their mechanical properties show a large scatter, and are at best equivalent to glass (natural fibres, however, have a lower density). Further disadvantages of the current natural fibre reinforced materials are their moisture sensitivity - which makes them prone to swelling and rotting -, their smell and their current cost level.
Our own tests with the application of Natural Mat Thermoplastics (NMT) on current automotive products proved the disadvantages. On the other hand it yielded several new research themes concerning property limits and gave insight in the area's where to optimize in order to get a broad application of natural fibre reinforced plastic product.
Looking towards the long term, other alternatives, like eco-composites or all-PP composites should be further explored.

1. Introduction

Polynorm Plastics - subsidiary of the Polynorm Group with over 2000 employees and a turnover of approximately 280 million ECU - is a major player in the field of design, engineering and manufacturing of Sheet Moulding Compounds (SMC) and Glass Mat Thermoplastic (GMT) parts, mainly for the automotive industry.

As an innovative partner towards the automotive industry several 'new' technologies are continuously scanned and evaluated, i.e. new processes, for example extrusion compression molding (ECM) and new materials, for example carbon/SMC, carbon/PA as well as natural mat reinforced thermoplastics (NMT).

2. Advantages of natural fibres over glass

Natural fibres, like for example flax, hemp and sisal are generally thought to be more environmentally friendly, healthier, safer and cause less abrasive wear than glass fibres. These advantages are elaborated below.

Environmentally friendly

Natural fibres propagate by nature and can be defined as renewable. The amount of CO2-uptake during its growth is matched - besides the efforts necessary to grow and harvest the fibres -, with the CO2 which is released during the rotting or burning process. Natural fibres are therefore favoured for the overall CO_2 - balance.
The amount of energy necessary for its "production" is lower than that of glass. More interestingly, however, is the fact that natural fibres have generally a significantly lower density (>40%) compared to glass, which gives an interesting improvement of the fuel efficiency (or maximum load capacity) when these parts are applied in transportation applications.
Another interesting environmental aspect is the fact that there is almost no residue with thermal recycling, exactly the opposite to glass!

Healthier and safer

Due to the different nature of the fibres, natural fibres do not cause skin irritation during handling and natural fibres are (although still not scientifically proven) less suspected to affect the lungs both during production and use.
When the natural fibres are applied in thermoplastic products, no sharp fibres or sharp splintering occurs during collision.

Less abrasive wear

Natural fibres bring less mechanical damage to processing equipment during processing (e.g. mould shear edges), recycling (e.g. extruders), etc. compared to glass, thus lowering overall maintenance costs.

3. Disadvantages of natural fibres over glass

Natural fibres are moisture sensitive; i.e. prone to rotting and swelling. When the fibres are incorporated in a plastic matrix, the latter can lead to:

- poor dimensional stability
- cracking
- degradation of mechanical properties

Even without the influence of moisture, natural fibre reinforced plastics suffer from lower mechanical properties compared to glass reinforced plastics. The E-modulus is affected somewhat less than the strength, with the impact strength being affected most drastically, yielding values of 20% of the values which can be obtained with glass reinforcement. Even when the lower density of the natural fibres is taken into account, the values of the specific impact strength still cannot match those of glass reinforcement. This is especially important to note, because in the case of products of long glass reinforced plastics like GMT, the impact strength often determined the material and processing choice! Figure 1 shows a comparison between the mechanical data taken from datasheets, for both NMT and GMT.

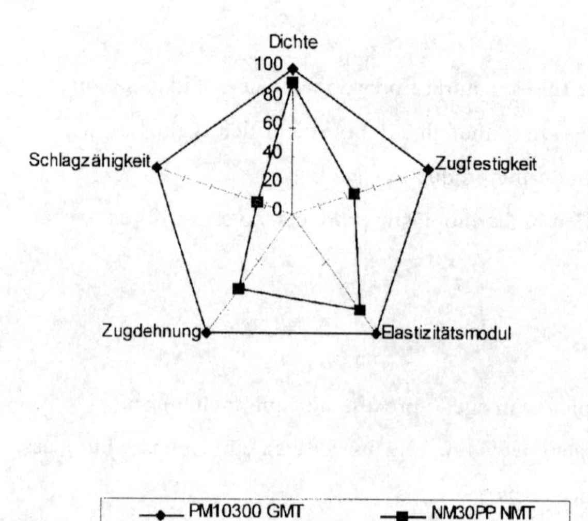

Figure 1 Comparison of some key mechanical data for NMT [1] and GMT [2], both at 30wt% fibers

Apart from the value, the (mechanical) properties show a larger scatter in properties, i.e. lower consistency of quality, compared to glass. Evidently this can be traced back to the natural source of the fibres.

Although natures scent may be very pleasing, the smell of natural fibres, processed at a high temperature is not comfortable for very long, which can be especially annoying in a production facility, in use however, the smell has the habit of disappearing within a few weeks!

The raw material costs of natural fibres - suitable for reinforcement -, may well be somewhat below that of glass fibres. However, the activities to achieve a NMT lead currently - and probably also in the future - to a price level around 0 to 15% above that of GMT. Moreover, one has to take into account that the cycle time increases with the use of NMT.

4. Threats towards the application of natural fibres

For a wide acceptance of natural fibres, one has to overcome a considerable part of the disadvantages listed above, like for example the moisture sensivity, low impact strength, etc. Lighter parts that do not pass on all functional tests will not be acceptable in the market.

On the other hand, natural fibres have an environmentally friendly image, mainly because they are renewable and leave nearly no residual waste after incineration. To elaborate on the latter, one can ask the question: Is thermal recycling the future for (long fibre reinforced) plastics (e.g. GMT)? From an environmental standpoint the mechanical recycling route, which is functioning well, should be favoured! However, for natural fibres the heating time should be short and temperature should be as low as possible, in order to maintain acceptable residual mechanical values. This is in contradiction with the wish to reprocess (recycle) these kind of products, even in production!

The second question related to this: Doesn' t NMT frustrate mechanical recycling of GMT? The colour and appearance are often similar, thus hampering the selection. The presence of some NMT parts amongst GMT parts may well lower the quality of the reprocessed finished product! This would mean that one has to at least label these kind of products.

When one reads these considerations one may ask if these natural fibre reinforced plastics are the way to go.
Looking at the future, an all-PP composite, may well be the better alternative, even when the logistic problems currently facing mechanical recycling are overcome.

The current price of NMT is 10 to 15% above that of GMT. Even when economy of scale and process optimization throughout the supply chain is reached, a price level equivalent to GMT could be expected. But what about the availability of natural fibres when agricultural subsidairies dissapear!?

5. Research topics

Looking at the number of disadvantages, threats and questions concerning the application of natural fibres in products, it is very easy to define several research topics.

Due to the lack of data on problems encountered in real products some tests with application of NMT on running automotive (GMT) products were carried out.

Experimental results product application

For this experiment a mass of NMT similar to that of GMT was heated in more or less similar way as the GMT, however a significantly longer oven time was needed. The mould and press conditions were also kept constant. The resulting product, was consequently tested for impact strength using a pendulum with a span of 40 mm, see figure 2. Besides experiments at room temperature conditions, the NMT samples were also kept in water for 68 hours (see NMT wet in figure). Both the GMT and the NMT (dry as well as wet) were further kept at -15 °C and consequently tested.

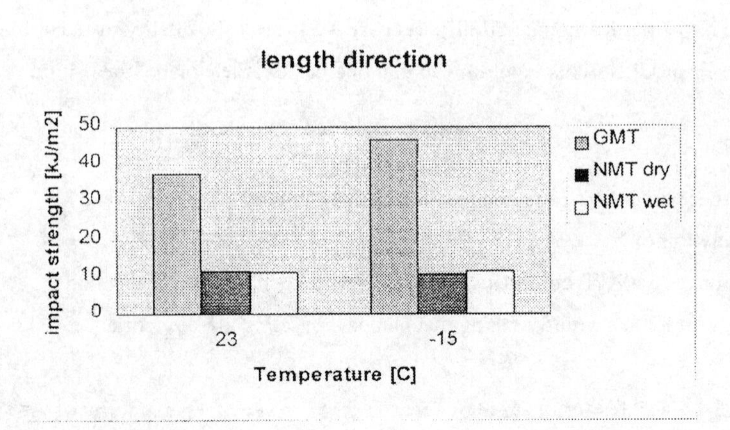

Figure 2 Impact strength for GMT versus NMT samples taken from a product, see text for description

It can be concluded from the samples taken in the flow direction (see figure 2) as well as in the perpendicular direction, that the impact strength of NMT is 20-25% of that of GMT. The water did not affect the properties drastically within the 68 hours and the lower temperature also did not cause drastic changes in properties

Future research themes

From a OEM and processors point of view it is important to gain more insight in the product performance; i.e. answering questions concerning:

- property limits (e.g., acoustics, smell, recycling, fire resistance)
- limits and possibilities in secondary operations (e.g. coating and adhesion)
- processing optimisation
- improvement of mechanical (especially impact) behavior and reduction of moisture sensitivity

From a more general view it is important to get more information to decide upon the optimal concept for recycling and the place of eco-composites (fully renewable materials) and all-PP composites in the future.

6. Possible applications

One could easily conclude from the information given previously that the possible applications should be in area's of low demands such as:

- moisture 'attack'
- (mechanical), especially impact properties
- smell
- surface quality

On the other hand the weight savings and general environmental advantages should be exploited to the full.

Within the automotive arena one would, besides potentially the underbody cladding, look at interior parts such as:

- parcel shelfs
- door trim
- instrument panels
- arm- and headrests
- seat shells

Besides automotive, applications in for example consumer goods, furniture, etc. may also be interesting, for example computer or television housings and seat shells.

7. Conclusions

Natural fibres offer several advantages compared over glass fibres, however, several (research) steps have to be taken to get natural fibre reinforced plastics widely accepted. Polynorm Plastics therefore closely cooperates with several universities, research institutes, suppliers and OEM's in order to evaluate and exploit its potential to the full.

References

[1] Symalit GMT Information, Kennwerte Entwicklungsmaterial, NM30PP, 28.02.1997

[2] GE Plastics Europe, AZDEL Product information PM10300, 13.09.94

Flexural Fatigue of Laminated Wood-Epoxy Beams

S.D. Clark*, R.A. Shenoi*

*Department of Ship Science, University of Southampton, Southampton, SO17 1BJ, U.K.

Abstract

The paper presents findings on the fatigue performance of wooden laminated beams used in the construction of the largest wooden tall ship being built in Britain. The materials used are Siberian larch and epoxy resins. The results show that their is much scatter in both static and fatigue performance of the structures which is mainly influenced by the distribution of knots. The work shows that much improved fatigue performance can be gained if such defects are selectively placed in regions of the structure not susceptible to high tensile/compressive stresses.

1. Introduction

Wood has been used in the construction of ship hull forms for centuries. Currently the use of wood in the marine industry is more limited, being replaced by higher strength and lower weight materials such as steel, aluminium and fibre reinforced plastic (FRP) composites. The Jubilee Sailing Trust (JST) has chosen wood as the construction material of a new traditional 175ft sail training vessel. This vessel is Britain's largest wooden tall ship currently under construction. The aim has been to build a traditional hull form using wood as the main construction material but using more modern building techniques. The vessel is currently being built using the 'strip-planking method' as opposed to the more traditional 'carvel' construction which offers advantages of increased strength and stiffness. The vessel is being built to Lloyds Registry of Shipping (LRS) regulations. However, the chosen type of construction method, though commonly used for sailing dinghies and small yachts, is not specifically covered by the rules for this size of vessel. Thus in many cases the strength of particular components in the structure had to be verified by experimental tests. The long term performance is of concern and the performance when subjected to long term in-service loads is not well known. This is primarily because the mechanical properties of wood can show large variations in strength depending on factors such as the type of wood chosen and the wood quality (i.e. number of knots/defects). Coupled with the fact that this type of structure is novel for a vessel of this size, it was thought prudent to try to characterise the fatigue endurance of the structure.

The paper outlines a fatigue testing programme for wooden laminated structures. The aim is to understand the fatigue endurance of laminated Siberian Larch beams and to investigate the effect of factors such as bond line quality, location of knots and variations in wood grain structure which can influence scatter in the results.

2. Vessel Construction

The overall length, beam, depth and design draught of the vessel are 52m, 10.5m, 10m and 4.5m respectively. The vessel is being constructed by the strip planking technique using epoxy resins to form a wooden laminated structure. 69 transverse frames are mounted along the vessels length at a spacing of 700mm. The total planking thickness in the shell is about 75mm and the transverse frames depth is approximately 290mm. A combination of steel bolts, FRP and wooden dowels are used to add further strength to connections in the vessel structure. The shell and transverse frames in regions where curvature is most pronounced are constructed one layer at a time and bonded using epoxy resin which is then cured before proceeding to the next layer. Typically, smaller laminae thickness and greater numbers of layers are used in regions with more curvature. The orientation of the wood grain in the frames all lies in the frame longitudinal direction. The shell of the vessel has been constructed from a number of layers of wooden planks, the middle layers orientated at 90 degrees to the inner and outer layers. The transverse frames of the vessel are shown in Figure 1.

Figure 1: Wooden Laminated Transverse Frames of JST Vessel

3. Test Programme

3.1 Materials and Specimen Geometry

The wood material chosen for the construction of the Jubilee sailing vessel is Siberian Larch. The wood was chosen because of its abundance, price, durability and workability. Because the material properties are dependent on many other factors such as growth conditions, seasoning conditions and number and type of defects [1-4], it was thought necessary to carry out separate static tests to determine the woods mechanical properties. These were determined after a series of tests on coupon specimens of dimensions 250mm×825mm×5mm for tensile tests and bending tests and 20mm×20mm×70mm for compression tests. The resulting properties are given in Table 1. An epoxy resin (WEST System Epoxy Resin 105/205 and 105/206 [5]) has been used to bond the laminated structure together.

Property	Value
Density (kg/m^3)	638
Flexural Modulus (GPa)	13.96
Flexural Modulus at Rupture (MPa)	97
Tensile Modulus (Gpa) *	16.23
Ultimate Tensile Strength (MPa) *	127
Ultimate Tensile Strength (MPa) #	3.0
Maximum Compressive Crushing Strength (MPa) *	46.5
Fibre Stress at Proportionality limit (MPa) *	47

** parallel to grain; # perpendicular to grain*

Table 1: Material Properties of Siberian Larch

Beam specimens have been chosen to represent a section of the transverse frames of the Jubilee Sailing Trust vessel. 12 beam specimens were constructed each with an overall length of 1600mm. The width used was 85mm. The overall depth was 105mm with the layers orientated through the thickness (i.e. 7 layers; each thickness=15mm). All the layers have the grain running in the longitudinal direction of the beam.

3.2 Test Rig

In the fatigue rig used, the beam was loaded using compressed air at eight equally spaced points along its length in order to simulate a uniform load; see Figure 2. The ends of the beam were simply supported with a span between supports of 1300mm. Generally, the nature of the pneumatic system means that the loading rate [(maximum load)/(time to reach maximum load)] and the unloading rate are constant for a given beam. Thus a symmetrical trapezoidal wave profile was achieved. The load was applied to the bottom of the beam therefore the top face was in tension. Measurements of the applied load and central deflection were taken. The testing frequency for fatigue tests was 0.5Hz with an applied stress ratio (min/max load), R=0. All tests were carried out at 23±5°C. Further details of the apparatus are discussed in detail elsewhere [6] and hence not elaborated here.

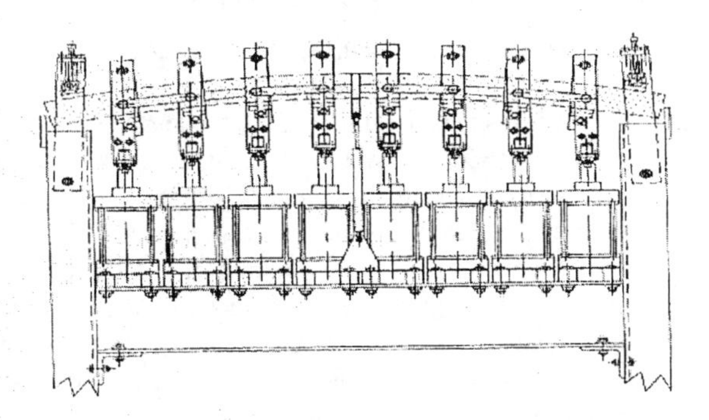

Figure 2: Fatigue Test Rig

3.3 Static Tests

3 beam specimens were tested on the test rig at a loading rate of approximately 0.15N/mm per second. The load versus deflection graphs are shown in Figure 3 and the failure loads and calculated tensile stress and flexural modulus in Table 2.

Beam No.	Failure Load (N/mm)	Tensile Failure Stress (MPa)	Flexural Modulus (GPa)
JB-01	44.94	60.78	8.091
JB-02	16.54	22.37	5.529
JB-03	35.84	48.47	8.390

Table 2: Static Failure Values for Wooden Laminated Beams

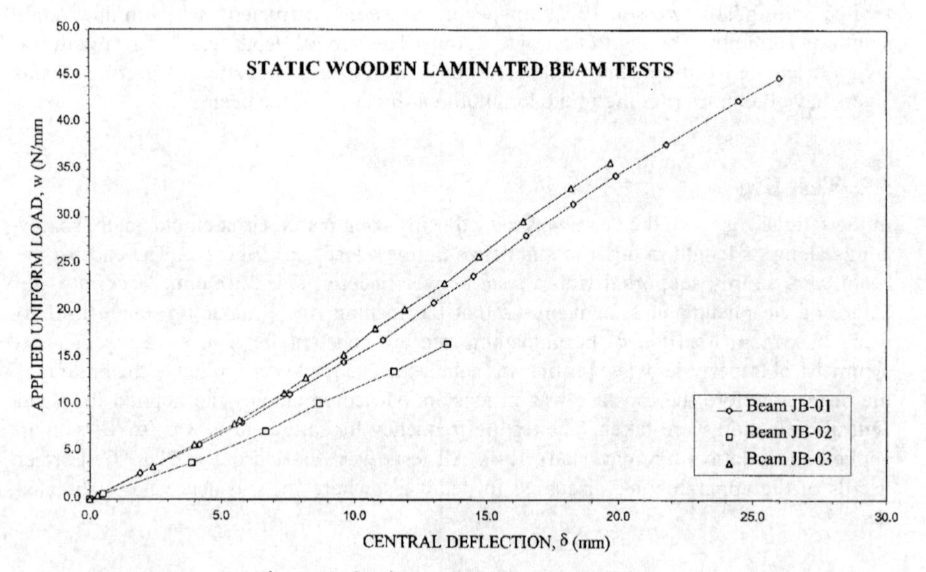

Figure 3: Static Load-Deflection Graphs

The failure mode in all cases was tensile failure on the top surface of the beam at or near midspan. The results show that there is scatter in the ultimate failure stress of the beams. This was found to be primarily due to the distribution of knots in the specimens which influenced the exact failure site. Beam 2 failed prematurely due to a large defect on the top ply at midspan which contributed to the reduction of the initial bending stiffness. Whilst Beam 1 had no major observable knots or defects, Beam 3, had a small knot on the top laminate at midspan which accounted for the premature failure without greatly compromising the beam's stiffness. A typical failure of a beam without major knots is shown in Figure 4 (beam JB-01) in contrast to a premature failure initiated by large knots at midspan which is shown in Figure 5 (Beam JB-02).

4. Fatigue Test Results

4.1 Failure Modes

The main failure mode is the same as that for static tests which is by tensile failure of the upper laminate layers. Again the failure site and fatigue endurance was primarily influenced by the location and size of knots and other defects in the specimens. Figure 6 (JB-04) shows a beam with only small knots at midspan whilst Figure 7 (JB-07) shows a beam with a large knot at midspan which gave a premature failure.

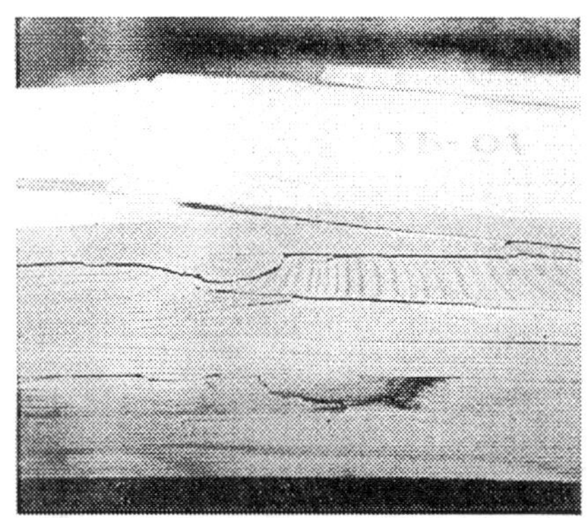

Figure 4: Static Failure Mode: JB-01

Figure 5: Static Failure Mode: JB-02

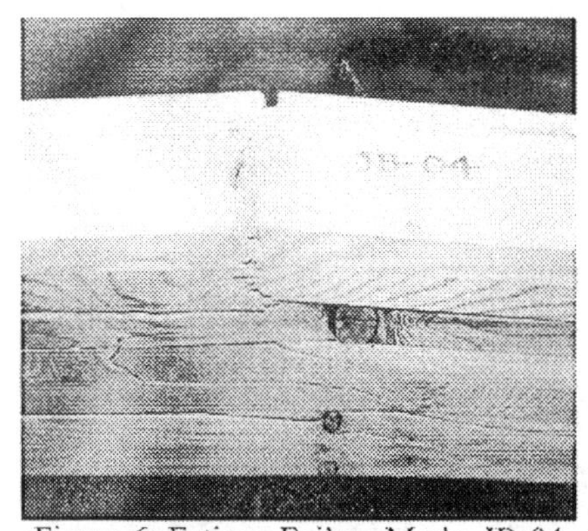

Figure 6: Fatigue Failure Mode, JB-04

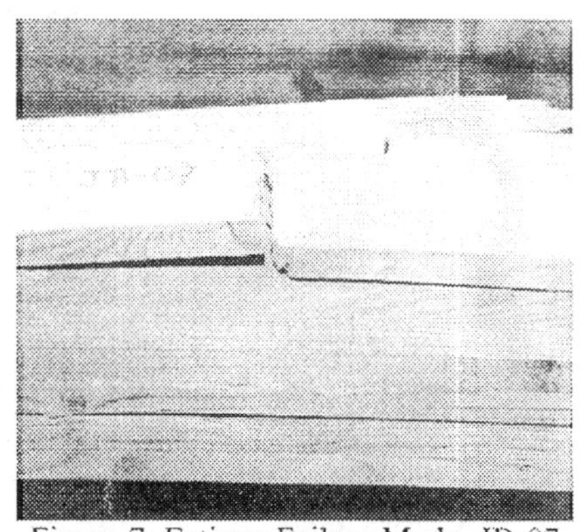

Figure 7: Fatigue Failure Mode, JB-07

4.2 Deflection-Number of Cycles Response

The deflection versus number of cycles response is shown in Figure 8. It can be observed that the increase in deflection is relatively slow and remains constant until failure occurs rather suddenly.

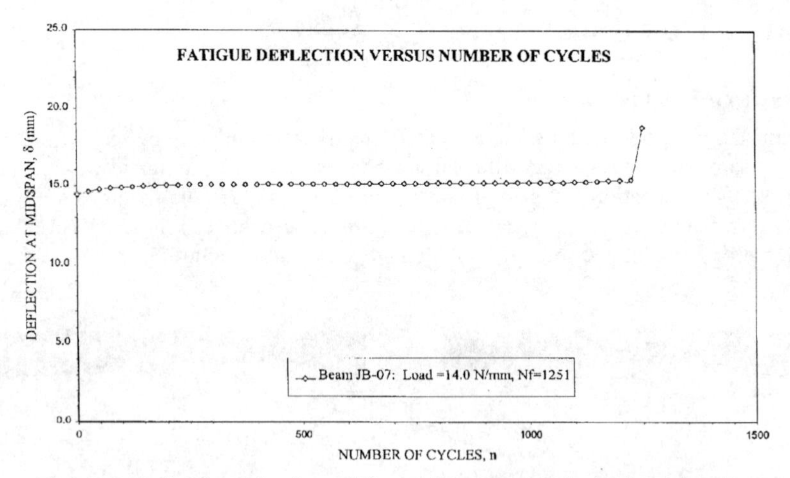

Figure 8: Deflection-Number of Cycles Response under Fatigue Loading

4.3 S-N Curve

The S-N curve for the materials is shown in Figure 9. The scatter in the results is rather noticeable and primarily reflects the differences between the specimens in terms of knot location. It can be observed that the weaker specimens all had knots located at or adjacent to the midspan section in the top or close to the surface laminate layers. All the beams with defects away from the central region showed greater fatigue resistance.

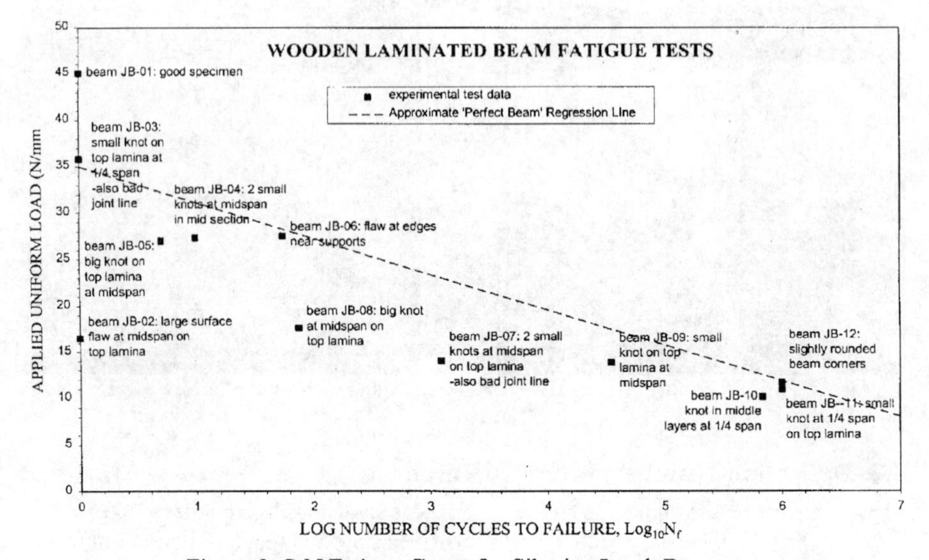

Figure 9: S-N Fatigue Curve for Siberian Larch Beams

5. Discussion

5.1 Comparison Between Coupon Tests and Beam Tests

Wood is a material that has different material properties in different directions (i.e. it is an anisotropic material). Typically wood is stronger in the direction of the grain than perpendicular to it. The results from the coupon tests revealed that the compressive stress was noticeably worse than the tension stress by almost a factor of 3. However, the failure mode for beam specimens was primarily tensile failure. Calculations of tensile failure stress and the flexural modulus both gave values much less than the values obtained in coupon tests, see Table 2. The main difference is thought to be due to the presence of defects in the laminated beams and the fact that though the grain direction was nominally longitudinal, in many layers there was grain run out as the grain ran to the beam surface, thus the wood fibres were not always continuous. The coupon specimens on the other hand were constructed from wood without defects and did not suffer from grain irregularities as the specimens were relatively small.

5.2 Effect of Knots and Grain Structure

A knot is typically much harder and more brittle than the parent wood material and therefore tends to carry the load. The grain structure in many cases can carry much of the induced stresses in the material around the knot but in cases where the grain makes a sharp angle around the knot severe stress concentrations can arise. In most cases, the grain structure did not flow smoothly around the knot which could have led to the many cases of observed premature failure. For both static and fatigue test samples there was an observable amount of scatter in the results which was primarily influenced by the location of knots in the structure. Because the wood appeared to be susceptible to tensile failure, the largest induced tensile stresses occur on the top surface of the laminate at midspan. In cases where knots were located in this vicinity, premature failure was likely to occur. The likelihood of premature failure due to knots at midspan decreases from the tensile outer surface through the laminate layers towards the neutral axis and increases again on the compression side. Though the samples were primarily tensile sensitive, a compression failure was observed for one specimen where a large knot was located at midspan on the compressive face. Where knots were located away from the midspan region, they did not appear to have much influence on the ultimate failure load.

5.3 Effect of Bonding.

A secondary effect affecting failure was the bonding of the individual laminae to each other. During, the testing programme a few dry joints were observed in the specimens, probably due to the laminates being compressed too much during manufacture. Though the areas of dry joints tended to be reasonably localised over the beams length, there were indications that poorly bonded joints tended to lead to premature failure, especially near the supports where the inter-laminar shear stresses are greatest.

5.4 Characterisation of Fatigue Behaviour

In all cases where scatter in the static and fatigue results was observed it was possible to account for variations in fatigue strength by identifying the location of defects in the laminate structure. From the results it is possible to approximate an S-N curve for the specimens not showing defects in the critical region at midspan. It appears as if a log-linear S-N mean regression line could be suitable to approximate the fatigue behaviour of good quality wooden beam specimens not having defects in the same manner show in other fatigue studies on composite sandwich beams [7,8]; see Figure 8. To obtain optimum fatigue performance, attention during construction can be made by selectively placing any wood with defects in non-critical regions of the structure where the tensile/compressive stresses are least.

6. Closure

The paper has presented work concerning fatigue tests on wooden laminated beams. The work has shown that the scatter in the results is large as can be expected from wooden structures. The scatter is mainly a function of the wood quality chosen. Attempts should be made if using knotted timber, to lay them away from the midspan panel area when under construction so as to utilise the full strength of the wood.

Acknowledgements

The authors would like to acknowledge the help and advice of Mr Howard Mackenzie-Wilson and his staff of the Jubilee Sailing Trust on technical aspects concerning construction of the vessel and for provision of the materials for the structural tests.

References

1 '*How Wooden Ships are Built*', H Cole Estep, W.W Norton & Co, New York, 1918.

2 '*Practical Design of Structural Elements in Timber*', 2nd edition, H. Parker, John Wiley & Sons Inc, 1963.

3 '*Timber Construction Manuel*', 3rd edition, American Institute of Timber Construction, John Wiley & Sons Inc, 1985.

4 '*Design of Wood Structures*', 2nd edition, D.E. Breyer, McGraw-Hill Book Co, 1988

5 '*The Gougeon Brothers on Boat Construction-Wood and WEST System Materials*', new revised edition, Gougeon Bros Inc, USA, 1985.

6 Allen, H.G., Shenoi, R.A., 1992, '*Flexural Fatigue Tests on Sandwich Structures*', Sandwich Constructions 2, Proceedings of the 2nd International Conference on Sandwich Construction, editors- D., Weissman-Berman, K. A., Olsson, Vol II, Engineering Materials Advisory Services, pp 499-517.

7 Shenoi, R.A., Clark, S.D. and Allen, H.G., 1995, '*Fatigue Behaviour of Polymer Composite Sandwich Beams*', Journal of Composite Materials, Vol 29, No. 18, pp 2423-2445.

8 Clark, S.D., 1997, '*Long Term Behaviour of FRP Structural Foam Cored Sandwich Beams*', Phd Thesis, University of Southampton, U.K.

Creation of particleboard with predetermined properties

Pavlo Bekhta

Department of Hydrothermal Processing of Timber and Glued Materials - Ukrainian State University of Forestry & Wood Tech. - gen.Chuprynky Str.,103, Lviv, 290057 - Ukraine

Abstract

The particleboards (chaotic structure) strength is approximately equal 30% of natural wood strength, in spite of addition the expensive synthetic glue. Approximately 70% of wood strength is lost during particleboards manufacturing process. This paper shows some possible causes of such loss of wood strength and outlines the main directions of the next investigations which realization can approach scientific for the solution of the problem - reception of particleboards with determined properties, strength among their number. The reserves of increase of the particleboards strength are more compact packing of wood particles in the mattress, improvement their orientation in the mattress and strengthening interaction between particles.

1. Introduction

As it is know [1], the theoretical strength of a solid body considerably exceeds its real strength. And it is not only the proof of imperfection of theoretical calculation, but also a direct evidence of a large not used reserves of strength, inherent to the material.
Today approximately 1.6m^3 of wood raw material is used up on manufacturing of 1m^3 of particleboard. Besides that for security of strengthen properties of particleboard is added glue (8...14% to weight of absolutely dry wood), which cost is approximately 30% of final particleboard cost. However, the particleboard does not achieve theoretically its possible parameters of strength in spite of addition of expensive glue and increasing of their density. The strength of wood particles in a particleboard in none of cases are not used completely according to their strength. If strength is not really great, then potential possibility of a particleboard is not used completely or are causes of reduction of adhesive connection.
Which causes of losses of strength and what are necessary for making to reduce their formation to a minimum? Whether it is possible from wood particles to receive particleboard, strength of their was equalled strength of natural wood? How to use high strength of wood in particleboard? Below are given some of the causes of decrease of particleboard strength and outlines the main directions of investigations which can give the answers on the directed questions and to approach of scientists to the solution of the problem - reception of particleboard with predetermined properties, strength among their number.

2. Influence of wood and glue on the process of forming of particleboards strength

An characteristic peculiarity of particleboards is the fact that their properties should be considered as properties of natural (wood) and synthetic (glue) polymers. That is why it is possible to assume that the components' properties (type of particles and their arrangement in a board; kind of glue and its distribution in chip-glue mass; interaction of glue with particles (the formation of pores and contacts)) will be determined, to a considerable extent, particleboards properties.

Type of particles and their arrangement in a board

The influence of type of wood particles on properties of particleboards is determined, first of all, by the fact that particles are the strongest structural elements of boards and in most cases take upon them the main part of external load causing particleboard deformation.

Particles influence on particleboards properties can also be connected with the fact that, firstly, particles properties effect adhesion and the magnitude of internal stresses in glue, and, secondly, besides adhesive bonds between particles there can be other bonds preconditioned by the particles interaction. Nevertheless the process of chipping is not simply a mechanical one but it is of a physical-chemical nature and it is accompained by the considerable changes not only of wood particles demensions but also by the physical-chemical changes of their surface.

Evidently that for reception of particleboard strength which was identical to strength of natural wood it is necessary to use principles of natural connection between fibres, to use bonded forces of wood fibres. This complicated that we constrained to be used up from already present and often changed by structurally and chemically substance.

During technological process of particleboard production the wood is experience great changes which to cause the degree of it activity to bonds formation. Just that causes are explained that from one wood are received the particleboard which have the different properties. Therefore, the process of formation of particleboard properties it is necessary to consider from noting these changes which are experiened of wood during the process of particleboard production.

Changes begin already on stage of wood making on the chip and then on the wood particles and to continue in the process of wood particles drying and wood particleboard pressing.

In the process of manufacturing of wood particles, the decrease of cohesive strength of wood takes place. The received particles have the surface, which eliminates mutual interaction of them in the process of pressing and gluing. Wilson J.B. and Krahmer R.L. [2] to mark that the surfaces of wood furnish have extensive damage and coarse texture. The depth of this damage could be sufficient to cause weak zones in the particle after bonding.

The results of special studies [3,4] have shown that wood particles strength is much lower than wood strength. So, if tensile strength of wood plankwise as equal 120 MPa then strength of wood particle only 48.3 MPa, that is 40% of wood strength. Tensile strength of wood particle across-the grain partically was equalled zero. Other scientists [5] think that strength of wood particles practically is equal of strength of wood.

Among the causes, which bring to the loss of strength, are stressed condition of wood in first half process of drying; change of chemical composition on the surface of particles (which causes complicacy at gluing); impregnating of resin and chemical substance on to the wood surface, when the temperature becomes too high, that brings to the difficulties at gluing.
Low using of wood strength in the particleboards also is explained by residual influence of elastic resistance in boards during hot pressing.
Structure of a particleboard is very important for the mechanism of transference of the properties of particles to the properties of a particleboard. Efficiency of the transference of the properties depends on the whole from bending of the particles in a board and character of junction of the particles. The quantity of the contacts, pores depend on the character of the mutual situation. Therefore, the character of the situation of particles in the mattress, depending on the mechanical structure of the power framework does not plays less role than the quality of the wood particles.
Every particle should withstand the force which is transfered to it by glue from a neighbouring particle in a board which is under a working load.
The cross-arrangement of most particles and some unparallelity of the patricle areas to the area of movement is an characteristic peculiarity for a chaotic stacking of particles in a mattress. In this case during the mattress compression and pressing particles are subjected not only to the compression deformation but also to the deformation of bending and twisting. During such stacking due to a small particle dispersion in a mattress cavities and areas are formed in places where particles are compressed very much.
In connection with a random wood particle arrangement in a board particles are subjected to stresses under different angles to the direction of fibres. But it is known that wood strength depends on the direction of the force which is applied in relation to the fiber length.
At oriented particle arrangement in a sandwich their longitudinal axes are stacked mainly parallel to each other. They are chiefly subjected to the deformation of compression in a normal direction to a particles area. In this case cavities between particles are minimum and the particles themselves are compressed less and more evenly than in an ordinary board.
When particles are oriented in a packet the strength of their glue joint is 15...24% greater than at a chaotic stacking [6]. Besides, particle orientation permits to eliminate substantially residual stresses in magnitude, to decrease the load of glue joints of the particles which results in improvement and streamlining the board properties.
Stresses are distributed between the particles through the links formed by the interaction of glue simultaneously with many particles. In an ideal case, the strength of particleboard should be equal to the strength of the particles if the stresses between the particles are evenly distributed. However, actual strength of a board is much lower than coordinated and simultaneous resistance to the stresses of all particles has not been observed. Particle propertie unevenness in a board, for example, the presence of imperfect ("weak") particles, plays here a decisive role. But the imperfection of most known methods of board production which do not ensure necessary link distribution among the particles resulting in particleboard structure imperfection is the most decisive.

Kind of glue and its distribution in chip-glue mass

The efficiency of glue action is chiefly determined by its capability to more fully interaction with wood in the process of hot board pressing during which glue is to set.
One of the ways of solution this problem is to improve glue distribution in a board, to increase the area of its contact with wood particles. The strength of glueing together glue and wood during surfactant application is increased due to the increase of moistened area.
Properties of particleboard depend do not only on the type of glue but also on its proportion in a board because it influences the character of glue distribution in a chip mass, the quantity of links between glue and particles, the magnitude of internal stresses in glue and a whole number of other particleboard properties. The low degree of glue using is left. 30% of applied glue physically and 20% functionally are not in the composition of a finished product [7]. This negatively affects board quality.
Small content of glue does not ensure link continuity, and results in increased porosity. If there is not enough glue the link between particle weakens, the conditions of load redistribution between them are getting worse, and the unevenness of their work increases. Besides, particle strength is not used copmletely resulting in an overall decrease of a board strength characteristics.
The volume between particles is filled not only with glue but also with air and gases which are in cavities and pores. This volume amounts to 10...35% of the particle volume [8]. If the number of pores is small and almost all the gaps between particles are filled with glue then the joint work of particles will be greater then in the case of increased pore capacity. In the first case board strength properties will be greater then in the second one though the particle content can be the same.
It is quite clear that in the boards from the same initial components and with the same content of particles but with different glue content greater strength will be in a material in which particles work more evenly and simultaneously, i.e., in which they are better linked between themselves by means of glue.
It is also clear that in a material with a great content of porosity, i.e., where there is almost no link continuity, load redistribution from more loaded particles to less loaded particles will be very much hampered and irregular. Such a board has a reduced average-used strength of particles, as well as limit of strength and modulus of elasticity due to unevenness and irregularity of the work of loaded particles.
When the porosity is reduced the bond between particles improves resulting in their better work when the board is loaded. Therefore, the porous structure plays a considerable role for the quality of a board. No doubt, it gives certain properties to the board as material. However, some knowledge about the porous structure of a particleboard and its influence on quality of a board is to be found in fragmentary condition.

Interaction of glue with particles

The influence of the interaction of glue with particles can be explained by the fact that, first of all, it is conducive to particle glueing together ensuring particleboard integrated structure, and secondly it ensures simultaneous work of most particles during board

deformations because due to the bonds between particles stresses can be distribute from some of them to others.
Mechanism of adhesive bonds creation in particleboard is inseparably connected with the nature of bonds in a contact zone and their number. According to modern notions, the nature of bonds in the zone of adhesive contact formation (on the border of phase division) is determined by the chemical structure of adhesive and substrate which stipulates a corresponding character of interaction: chemical, intermolecular, etc. However, irrespective of the nature of bonds, their number is determined by the actual area of the contact of surfaces which touch each other.
It is necessary to remark only the wood has most unhomogeneous surface among the composite materials. This explains that every small length of surface has very variable physical and chemical characteristics. Therefore, the area of contacts can be different beginning from pointlike and finishing by comparatively large areas. Rough rigid surface of particles provides small area of contacts. Good interconnection between particles can be achieved only if it is possible to obtain during crushing the largest inner surface of particles increasing by it the area of particle touching surface and to render to this surface activity due to functional groups prosessing reactivity.
Maximum contact area is achieved by pressure. However, the advantage of wood as strengthenen component practically are not used in the process of particleboard production thanks to their formation is realized under considerable pressure. The density and strength of wood are increased during it compression if herewith is not happen the destruction of it capillary-porous structure [9].
Thus, the particleboard creation is necessary to be done with pressure which should not to be terminated into considerable compression (deformation) of wood particles for maximum realization the specific characteristics of wood in the process of it using as structural component of particleboard.
So, with consideration for aforesaid the general losses of particleboard strength are equal

$$\sigma_l = \sigma_d + \sigma_i + \sigma_o,$$

where σ_d - losses as a result of decrease of particleboard density, and presence of water and glue; σ_i - losses as result of slackening of interaction of particles; σ_o - losses as result of orientation of wood particles.
The investigation of author is determined [10] that summary of loss of particleboard strength as result of all structural factors is equal of difference between wood particles strength (σ_w) and particleboard strength (σ_b)

$$\sigma_l = \sigma_w - \sigma_b.$$

If to presume that wood particles strength is equal of wood strength, for example 87 MPa (it is ultimate static bending strength of wood of pine), and bending strength of particleboard (with unorientation of wood particles) is equal approximately 27 MPa, then losses of strength are equal approximately 60 MPa. However, ensuring good strength of the elements is not a sufficient condition for obtaining high-quality products. A separate testing of the initial elements can be quite satisfactory but at unsatisfactory joining of elements of a structure and properties of its input materials a product during exploitation does not ensure necessary indices of quality.

3. Development further research

All components of wood have the great number of reactive groups, which secure, on the one side, strong interconnection of all structural elements, and on other side - responsibility for the definite properties of wood [11]. Chemical composition of wood is decisive not only for the whole set of properties but also for technical use of wood. Elementary composition (carbon 50%, oxygen 43%, hydrogen 6% and nitrogen 1%) for all wood species is approximately the same; that's why the reactionary capacity and the structure of wood are determined by distribution and chemical macrocomposition. Therefore, only exact knowltdge about the reactive groups of all components of wood and possibilities the purposeful influence over its mechanical, thermal, thermodynamic or chemical processes will be subserve the further successes in the particleboard creation with predetermined properties.

Besides it is necessary to carry out the following research:

1) in detail to study the data about the chemically-physical activity of the surface of wood particles (availability of chemially active groups on the surface of wood, distribution of active groups on the surface of wood, topographical characteristic of the particles etc.) in order to get the maximum use of the binding power, inherent of wood in the process of manufacturing of the particleboard;

2) to determine, which quantitative characteristic the glue and the surface of wood particles must have for reaching high strength of gluing;

3) to investigates the question about as far as wood particles are impregnated under particleboard manufacturing and how part of impregnation used under particleboard pressing is unintelligible and also how this processes are characterize quantitatively;

4) to investigate the porous structure of particleboards.

References

[1]. MECHANICS OF FRACTURE AND STRENGTH OF MATERIALS: Reference book.Vol.1.Bases of the mechanics of fracture /*Panasyuk V.V., Andreykiv A.E., Parton V.Z.*-Kyiv, 1988.-488p.

[2]. *Wilson I.B., Krahmer R.L.* PARTICLEBORD: MICROSCOPIC OBSERVATIONS OF RESIN DISTRIBUTION AND WOOD FRACTURE //Forest Products Journal.-1976.-Vol.26.-№11.-P.42-45.

[3]. *Bazhenov V.A., Bolotov S.H.* STRENGTH OF WOOD PARTICLES AND INFLUENCE ON ITS OF THE VARIOUS FACTORS /Technology of wood-based panels and composite boards.-Moscow,1984.-Vol.159.-P.109-112.

[4]. *Bazhenov V.A., Bolotov S.H.* STRENGTH OF WOOD PARTICLES /Technology of wood-based panels.-Moscow,1985.-Vol.171.-P.107-111.

[5]. *Potashev O.E., Lapshin Yu.H.* MECHANICS OF WOOD-BASED PANELS.-Moscow,1982.-112p.

[6]. *Karasev E.I., Piltser M.Sh.* PROSPECTS OF MANUFACTURE OF PARTICLEBOARDS WITH THE INCREASED MECHANICAL PROPERTIES /Technology and equipment of Woodworking manufactures.-Leningrad,1985.-P.65-67.
[7]. *Salah E.O.* UTJECAJ RESPONDJELE LJEPILA PO IVERJU NA IZRADU I KVALITETU IVERICA //Drvna industrija.-1981,32,№9-10,P.243-258.
[8]. *Kühne G.,Niemz P.* EIN BEITRAG ZUR THEORIE DER BILDUNG UND FORMUNG VON HOLZPARTIKELWERKSTOFFEN //Wissenschaftliche Zeitschrift,1984,33, №5,S.211-218.
[9]. *Byely V.A., Vrublevskaya V.I., Kupchinov B.I.* WOOD-POLYMERIC MATERIALS AND PRODUCTS.-Minsk,1980.-280p.
[10]. *Bekhta P.A.* SCIENTIFIC-TECHNICAL FUNDAMENTALS OF PRODUCTION OF PARTICLEBOARD WITH THE GIVEN STRUCTURE AND PROPERTIES. Dissertation thesis for the Degree of Doctor of technical science.-Lviv,1996.-42p.
[11]. *Pecina H.* ZUR INTERPRETATION DES MECHANISMUS LIGNOCELLULOSER FASER-FASER-BINDUNGEN //Holztechnologie.-1983.-24.-№1.-S.26-33.

TENSILE TESTING PROBLEMS IN GLASS FIBER REINFORCED PLASTIC RODS MADE BY PULTRUSION.

G. Porco* R. Zinno**

* Department of Structural Engineering University of Calabria - Cosenza- ITALY.
* * CNR - ICITE - via Tiburtina 770 - Rome, ITALY.

Abstract

In the present work some experimental procedures to make tensile tests on specimens made by composite materials, undamaged or thermically degraded are presented. By the proposed methodology, it is possible to define the representative mechanical parameters, such as the longitudinal Young modulus, the ultimate load and the ultimate deformation. The set up of the experimental program was preceded by an analysis of the most important aspects relative to the particular type of material under consideration. The problems connected with the continuum modelling of the composites are evidenced to better define a correct tensile test.

The experimental analysis was conduct on specimens of cylindrical shape, defining suitable grip tools and an opportune measurement system.

The specimens were made by glass fibers and polyester or vinylester resin matrix.

1. Introduction

The composite materials are widely used in structural engineering field in substitution or in conjuction with traditional materials, such as concrete or steel.

The most important reasons of this diffusion are related to their potentialities such as: corrosion resistance, electromagnetic transparence, low weight connected to an high resistance.

Moreover, they can offer the advantage of a tailorability of the material mechanical properties in function of the future applications.

The presence of fibrous elements, oriented in a predefined direction, make the composite materials more or less anisotropic.

It is clear that the anisotropic degree depends, for a fixed angle of fiber orientation, on the elastic characteristics of the fibres and on the reinforcement percentage adopted.

Thus, it is necessary to set up suitable experimental procedures which permit to determine the mechanical parameters of the material, such as: elastic moduli, ultimate loads and ultimate elongations[1,2].

To make the passage from the reality to the continuum model, it is needed to outline some important concepts already well defined for the isotropic materials.

Moreover the consideration of a possible utilization of the pultruded rods to substitute steel elements in civil engineering structures, suggested to analyze the effects on the mechanical parameters produced by an exposure to high temperatures of the pultruded composite specimens.

2. Description levels and decay length

The qualification methods for the definition of the mechanical parameters in composite materials are different and their selection depends on the particular material under consideration.

For example, composites with continuum parallel fibers subjected to a tensile action in the reinforcement direction, will follow the Hooke law very well; instead, if the loading direction is inclined with respect to the fiber direction, the stress-strain law will be nonlinear and characterized by a difficult analytical description.

In the experimental procedure, the structural response will easily change also for factors generally considered of secondary importance, such as: loading speed, power and type of the tensile equipments. For all these reasons, it is necessary to normalize the experimental procedures more strictly than that for isotropic traditional materials usually utilized in civil engineering [3,4]. When we have metals, the passage from a not homogeneous medium to an homogeneous one involves only to neglect the actual molecular structure of the material.
Also for composite materials it is possible the passage from a not homogeneous medium to an homogeneous one, but it is needed to identificate the level where this passage will be correct.

In the case of materials reinforced by continuous fibres, it is generally possible the individuation of a regular and repetitive structure, thus becomes possible to create an homogeneous continuum medium by using the so called "Energetic level method". To utilize this method it is needed to individuate three levels of description, depending on the dimensions of the material components.

The lower level is called "level h" and here it is possible to evidenziate the inhomogeneity of the material. This dimension is practically equal to the fiber dimensions.
After that, we can find the "level H", where it is possible to substitute the not homogeneous material with a locally homogeneous one. The macroscopic characteristics of strength and stiffness are obtained through laboratory tests, also macroscopic. As last, it is possible to individuate the "level λ". At this stage, we can distiguish the changes in the stress and strain fields and it is possible to utilize all the well known continuum mechanics analytical expressions.

All these considerations are valid at a certain distance from the load application zone, where the actual distribution of the loads becomes not important (St. Venant principle). This zone it is already well known for the homogeneous and isotropic materials. It present the following decay law [6]:

$$e^{-2.1062\, x/r}$$

where x is the specimen axis and r is the half of the maximum transversal dimension.

It is easy to verify that the percentage of the residual stresses on the original one is on the order of 1% when x is the maximum transversal direction.
For anisotropic materials, the disturbance will interest the "level λ", will be more evident and the decay length will become [7,8,9,10,11,12]:

$$e^{\frac{-2\pi x}{D}\sqrt{\frac{G_{xy}}{E_x}}}$$

where $D=2r$, E_x e G_{xy} are the elastic longitudinal and shear moduli.
As an example, for graphite-epoxy, we observe 1% of residual stresses at a distance four times bigger than that for isotropic materials.

3. Experimental program

After the identification of the specimen shape and of the zone where it is correct make the measurements, in function of the decay length law, an experimental program to obtain the mechanical parameters of the pultruded circular rods was designed.
The choice of the strain gage was done by taking into account all the considerations described in previous works [13,14], while for the effects produced by the temperature, the testing time and the loading speed, the indications listed in [15] were followed.
The proposed system permitted the evaluation of the mechanical characteristics of undamaged and thermically conditioned specimens of pultruded composite materials.

3.1 Grip system

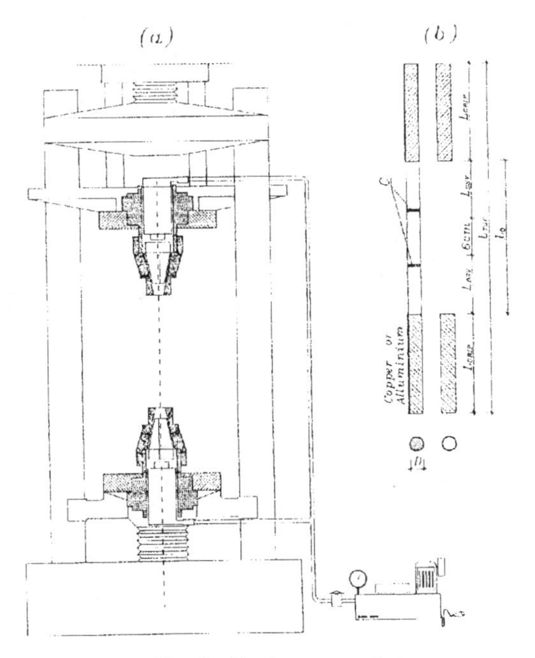

Fig. 1 - Testing apparatus

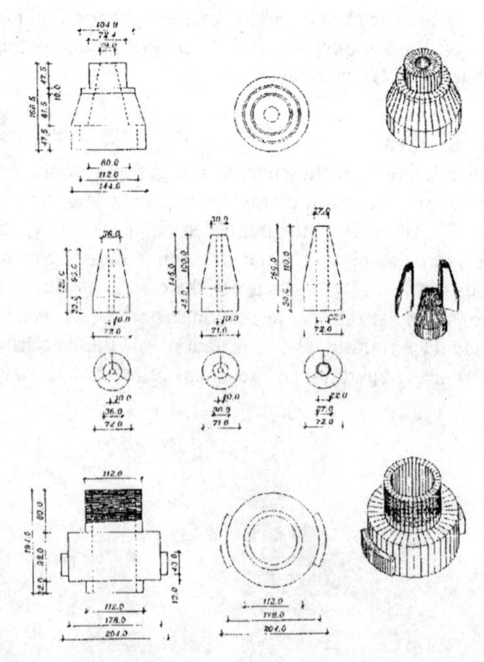

Fig 2 - Grip system

The system must consent the grip, but must avoid failure or delamination in the grip zone. The proposed system [15,16], reported in Figs. 1 and 2, is composed by some wedges which permit the connection to an universal machine UPD 120 of the MFL Systeme.
Moreover, the presence of an hydraulic jack inserted in the two cross-heads of the machine, warranting the development of an opportune pressure on the specimen surface, will avoid possible slips.
In the present experimental program three different types of wedges were adopted, depending on the specimen diameters. The grip length, which permits to avoid damage on the specimes, was identified through the following relation [15]:

$$L = \frac{\chi D \sigma_n^{tn}}{4 \sigma_z^{cn}}$$

where:

$\chi = 1/f$ is the friction coefficient *(f=0.26÷0.40)*
D is the specimen diameter
σ_n^{tn} is the ultimate tensile stress
σ_z^{cn} is the ultimate compression stress

3.2 Load and measurement system.

The load and measurement system utilized in the present work are composed by the following devices:

- Universal electrical-hydraulic machine of the MFL Systeme
- Electronic "ER" device monitoring of the MFL Systeme
- Resistance strain-gages of thermically autocompensated type (350 Ω)
- Load cell of C1 class
- Electronic remote control system UPM 100 of the HBM
- HBM clamp-on strain transducer

In all the tests, by the use of the electronic "ER" monitoring system, the cross-heads displacement speed was fixed to 2 mm/min, both in loading and in unloading phases.
The pressure in the hydraulic jack was fixed to 500 Kg/cmq in all the tests.

3.3 Geometric characteristics of the specimens.

The specimens utilized in the present experimental program are manufactured by the pultrusion technique.

They are composed of glass fibers embedded in a resin which gives the cylindrical shape. The resin is of polyester type and is combined with suitable fillers in a way as to increase the chemical stability, eliminate the surface roughness and reduce the transparence, obtaining the classical white color.
Tables 1 and 2 resume the geometrical data relative to the three different diameters for the specimes used in the experimental program for the determination of the elastic moduli.
In particular, the sigles "L" and "H" are used to distinguish two different reinforcement percentage (L = Low volume fraction (46,60%) and H = High volume fraction (59,11%)).
Moreover, by L_{grip} is marked the length of the copper cylinder used to avoid damage on the specimens produced from the grip pressure, while by L_{DSV} is marked the zone of the specimens were no measurements were taken because it is left free to reach the uniform distribution of the stresses.

Tab. 1 - Geometrical data $L_{DSV}/D = 4$, (five speciments per group).

Group		L Totale	L_{Grip}	L_{DSV}	Diameter
9.5 mm -H	Valori medi	356.00	110.00	38.00	9.30
12 mm -L	//	451.00	145.00	50.50	12.50
22 mm -L	//	548.00	155.00	89.00	22.10

Tab. 2 - Geometrical data $L_{DSV}/D = 10$ (five speciments per group).

Group		L Totale	L_{Grip}	L_{DSV}	Diameter
9.5 mm -L	Valori medi	470.00	110.00	95.00	9.45
9.5 mm -H	//	470.00	110.00	95.00	9.25
12 mm -L	//	602.00	145.00	126.00	12.60

In Tables 3 and 4, instead, are reported the geometrical data of the specimens exposed to the heat and organized in five groups (A, B, C, D, E) depending on the maximum temperature reached during the heat conditioning.
This temperature was mainteined to this value for one hour and was reached by constant increments of 5° C per minute.

Tab. 3 - Geometrical data $L_{DSV}/D = 4$ (D=12 mm, five spec.per group, V_f= 46.60%).

Temp.		L Totale	L_{Grip}	L_{DSV}	Diameter
200 °C - A	Valori medi	451.00	145.00	50.50	12.50
250 °C - B	//	451.00	145.00	50.50	12.70
300 °C - C	//	451.00	145.00	50.50	12.60
325 °C - D	//	451.00	145.00	50.50	12.50

Tab. 4 - Geometrical data $L_{DSV}/D = 4$ (D=22 mm, five spec.per group, V_f= 46.60%).

Temp.		L Totale	L_{Grip}	L_{DSV}	Diameter
200 °C - A	Valori medi	550.00	156.00	89.00	22.30
250 °C - B	//	550.00	156.00	89.00	22.38
300 °C - C	//	550.00	156.00	89.00	22.22
325 °C - D	//	550.00	156.00	89.00	22.28
350 °C - E	//	550.00	156.00	89.00	22.30

4. Experimental results and conclusions

After the individuation of the geometrical data of the specimens, the ultimate load and the elastic modulus on the unconditioned specimens (Tab. 1 and 2) were defined, through two consecutive phases. The test for the determination of the ultimate load has been conducted starting from an initial preload of 3 KN, needed to assure the safety grip of the specimen in the universal machine and to eliminate all the parasite effects produced by the movable parts of the testing apparatus.

Tab. 5 - Experimental results average values.

L_{DSV}/D	Group	Pu (kN)	$\varepsilon_{ul}x10^6$	σ_{ul} (MPa)	E (MPa)	E_{rm} (MPa)	E/E_m (%)
4	9.5 mm-H	55.20	22681.25	814.45	36107.50	44195.00	81.70
	12 mm-L	58.00	14649.67	463.40	31030.36	34445.00	90.08
	22 mm-L	174.00	16109.00	450.30	28542.25	33754.00	84.56
10	9.5 mm-L	31.30	19182.67	448.42	25475.46	35550.00	71.66
	9.5 mm-H	58.60	23704.75	866.58	36507.29	44195.00	82.61
	12 mm-L	57.30	15089.50	461.16	30493.50	34445.00	88.52
	E_f= 72450 MPa		E_m= 3350 Mpa		$E_{rm} = E_m V_m + V_f E_f$		

The complete failure of the specimens was obtained by the use of the "ER" electronic system of monitoring fixing a cross-heads displacement speed to 2 mm/min. The specimen

elongation percentage close to the ultimate load was obtained by a clamp-on mechanical strain transducer. By all the informations collected in this first phase, was possible to program the test to determine the elastic modulus. To identificate this parameter two strain gages were bonded on the specimen on symmetrically opposite positions and connected to a dummy strain gages positioned on an unloaded specimen, to make the thermal compensation.
In Table 5 were reported the Young moduli, the ultimate stresses and strains experimentally obtained.
In particular, the value of the Young modulus for each group is compared to the analogous one obtained by the use of the rule of mixture.

The obtained results show the independence of the measurement from the proposed testing system. Infact, for the two different percentage of reinforcement, by varying the ratio LDSV/D, the elastic moduli, the ultimate loads and the ultimate strains the obtained results are very close one to each others. By utilizing the same methodology for unconditioned specimens, the ultimate strength was experimentally defined for specimens exposed to heat (Tables 3 and 4).
In Figs. 3 and 4, for two different diameters, the effects of the temperatures on the Young modulus and on the ultimate stress are reported.
The characteristic values are referred to the 12 mm(L) and 22 mm(L) with LDSV/D = 4 groups previous analysed.

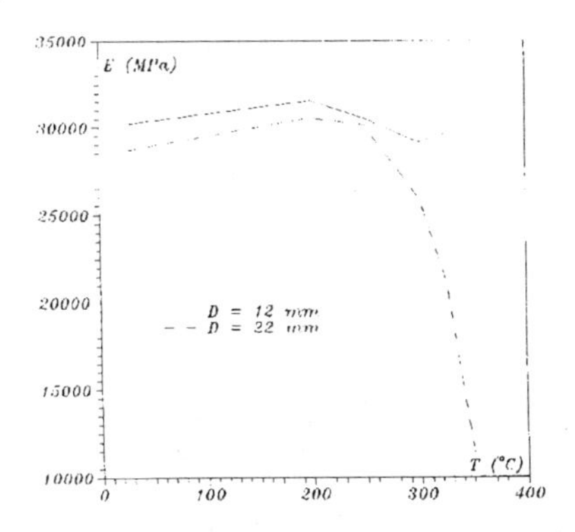

Fig 3 - Effect of the temperature change on the Young modulus

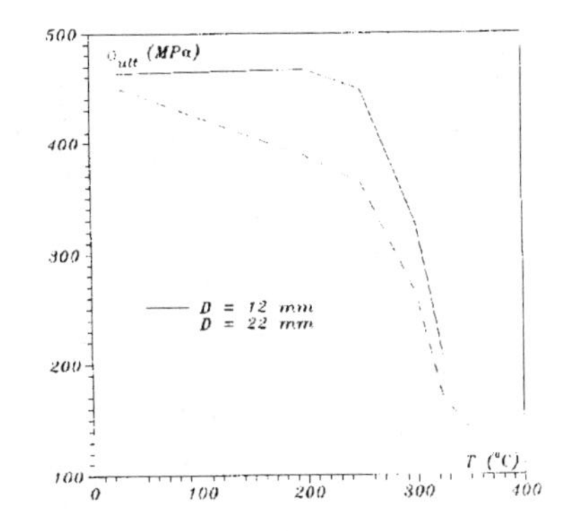

Fig 4 - Effect of the temperature change on the ultimate stresses

From the pictures, clearly appears that the ultimate strength decrease with the increase of the temperature, while for the Young modulus, a small increase in its value it is possible to identificate in correspondence of 200°C.
This can be probably due to the formation of new molecular chains inside the resin matrices used in the composite manufacturing.

All these informations can be used in the definition of the behaviour of a structure where pultruded composites are used and where can be possible heat exposure.

References

1) Carlsson L. A., Pipes R. B. - Experimental Characterization of Advanced Composite Material - Prentice Hall. Inc., Englewood Cliffs, N. J., (1987).
2) Tarnolposky Y. M., Kincis T - Static Test Methods for Composites (tradotto da George Lubin), Van Nostrand Reinhold Company, New York, (1981).
3) ASTM D 3171-73., Fiber Content of Reinforced Resin Composites, (1973).
4) ASTM D 3039-74. - Standard Method of Test for Tensile Properties of Oriented Fiber Composites - (1974).
5) UNI EN 61 - Materie plastiche rinforzate con fibre di vetro: Determinazione delle caratteristiche a trazione, (1978).
6) Timoshenko S. P. - Goodier J. N.,Theory of Elasticity - 3rd ed., Mc Graw Hill, New York, (1970).
7) Choi I., Horgan C.O. - Saint Venant Principle and end effects in anisotropic elasticity - Journal of Applied Mechanics,vol. 44, pp. 424-430, (1977).
8) Horgan C.O., - The Axisymmetric end problem for transversely isotropic circular cylinders - International Journal Solids & Structures, vol. 10, pp. 837-852, (1974).
9) Lekhnitskii S. G. - Theory of elasticity of an anisotropic body - (tradotto da P. Fern), Holden-Day, (1963).
10) Folkes M. J. e Arridge R. G. C. - The measurement of shear Modulus in Higly Anisotropic Materials: the validity of Saint-Venant's Principleù - Journal of Physics D: Applied Physics, vol. 8, pp. 1053-1064, (1975).
11) Arridge R. G. C. et al. - The importance of End Effects in the Measurement of Moduli of Higly Anisotropic Materials - Journal of Materials Science, vol. 11, pp. 788-790, (1976).
12) Arridge R. G. C. and Folkes M. J. - Effect of Sample Geometry on the Measurements of Mechanical Properties of Anisotropic Materials, Polymer, vol. 17, pp. 495-500, (1976).
13) Tuttle M. E., Brinson H. F. - Resistance foil Strain-gage technology as applied to Composite Materials - Experimental Mechanics, vol. 24, n. 3, pp. 54-65, (1984).
14) Strain Gage Selection: Criteria, Procedures, Recommendations, M-M Tech Note TN-132-2, Micro-Measurements Division, Measurements Group Inc., Raleigh, NC, (1976).
15) Pantuso A., Porco G., Zinno R. - Caratterizzazione sperimentale di materiali compositi pultrusi" - Dipartimento di strutture, Università della Calabria, Report n° 162, settembre 1994.
16) Crea F., Porco G., Zinno R. - Experimental evaluation of thermal effect on the tensile mechanical parameters of FRP rods" - In pubblicazione su Applied Composite Materials.

Vibration properties of coated cement concrete: the effect of bitumen content

M. Mayama*, H. Tanaka, Y. Nakazawa*, Y. Hatanaka***

* Department of Civil Engineering - Hokkaido Institute of Technology
7-15 Maeda Teine-ku Sapporo-Japan
** Under Waterworks Bureau - Asahikawa City Office
1 Kamitokiwa Asahikawa-Japan

Abstract

This paper describes the results of the vibration tests on a new composite material, known as coated cement concrete, consisting of cement and aggregates, which were coated with bitumen. The effect of bitumen content on the absorption of vibration by the material as well as the elastic modulus were evaluated. The absorption of vibration of the coated cement concrete increased as temperature increased and as the bitumen content increased. On average, the bitumen film thickness on the gravel was greater than that on the sand, for the same weight of aggregate, and hence the coated gravel had a more effective damping action than the coated sand. Coated cement concrete had a greater ability to absorb vibration than conventional cement concrete, however, there is a slight difference in elastic modulus between the two materials with the coated cement concrete being lower.

1. Introduction

Vibration, caused by the recent marked increase in the heavy traffic, is a serious problem in Japan and road engineers are required to protect the environment from this undesirable effect [1, 2, 3, 4, 5]. One of the reasons that cement concrete has not been a popular road construction material in Japan is the noise and vibration problems which are induced by traffic on cement concrete roads made to current specifications. Nevertheless, these surfacings have been more durable than asphalt.

At present, the most widely used, and economic, aggregates for cement concrete are sand and gravel, however, the cement concrete made with these aggregates has little damping effect so that there is no decrease the vibration caused by traffic. Author's group has developed a new composite material, named coated cement concrete, which consists of cement and aggregate which has been coated with bitumen. This composite material has a damping effect on the vibration induced in the road [6, 7, 8]. This report presents the effect of bitumen content on the logarithmic decrement, which represents the damping effect of coated cement concrete of vibration, and the relationship between logarithmic decrement and dynamic modulus.

2. Experimental Materials and Method

Materials and sample

Samples of cement concrete were prepared following the mix design in Table 1 using Portland cement (specific surface: 3,260 cm^2/g), gravel (maximum size: 13 mm, specific gravity: 2.67), sand (specific gravity: 2.64) and bitumen with a water/cement ratio of 0.5. The bitumen used to coat the surfaces of the aggregates was a penetration-grade bitumen with the physical properties shown in Table 2.
The surfaces of sand and gravel were coated by bitumen in a forced action mixer at a temperature which produced a viscosity of 0.3 Pa s, and these materials are called 'coated sand' and 'coated gravel' respectively. These coated aggregates were mixed with cement and the mixtures were filled in moulds 300 mm long, 300 mm wide and 50 mm thick. The coated cement concretes were cured in a thermostatically controlled water-bath at 20°C and were cut into samples 300 mm long, 30 mm wide and 50 mm thick using a diamond saw the day before testing.

Table.1. Concrete mix design

Sample number	Water	Cement	Coated sand		Coated gravel	
			Sand	Bitumen	Gravel	Bitumen
50-0.0-0.0	190	380.0	842.300	0.000	922.900	0.000
50-0.0-0.6	190	380.0	842.300	0.000	917.363	5.537
50-0.0-1.0	190	380.0	842.300	0.000	913.671	9.229
50-0.8-0.0	190	380.0	835.562	6.738	922.900	0.000
50-0.8-0.6	190	380.0	835.562	6.738	917.363	5.537
50-0.8-1.0	190	380.0	835.562	6.738	913.671	9.229
50-1.0-0.0	190	380.0	833.877	8.423	922.900	0.000
50-1.0-0.6	190	380.0	833.877	8.423	917.363	5.537
50-1.0-1.0	190	380.0	833.877	8.423	913.671	9.229
50-2.0-0.0	190	380.0	825.454	16.85	922.900	0.000
50-2.0-0.6	190	380.0	825.454	16.85	917.363	5.537
50-2.0-1.0	190	380.0	825.454	16.85	913.671	9.229

Table.2. Properties of straight bitumen

Penetration* 1/100 cm	Softening point, R&B °C	Penetration index	Specific gravity
168	39.5	-1.01	1.020

*25 °C, 100 g, 5 s

Test method

The test method used for the free damped oscillation was the fundamental bending vibration for sample hung at the modal points [9]. A basic system for measuring the vibration properties consists of impulse hammer, vibration accelerometer and FFT analyzer. A transient load is applied to the sample and the resulting signals are processed using a two channel FFT analyzer. The logarithmic decrements δ and dynamic modulus E_f were obtained from the damped oscillation curve and resonance frequency curve respectively [3]. Vibration tests were conducted for curing periods of 3, 7, 14, 28 and 91 days using three samples of each composition of coated cement concrete. Because the mechanical properties of composite materials, and particularly bituminous stiffness, are temperature dependent [10, 11], vibration tests were carried out at a wide range of temperatures from 0°C to 60°C in steps of 20°C. Temperature control was maintained in a forced-air convection chamber.

3.Experimental Results

(a) Relationship between logarithmic decrement and bitumen content

Figure 1 shows the relationship between logarithmic decrement and bitumen content at 0°C for cement concretes water-cured at 20°C for four weeks. The logarithmic decrements, calculated from the damped oscillation curve, were similar for all the coated cement concretes and were a little larger than those for the non-coated cement concretes. For example, the logarithmic decrement of current cement concrete (sample number: 50-0.0-0.0) was about 0.054 and that of coated cement concrete with the highest bitumen content (sample number: 50-2.0-1.0) was about 0.08. This is because the coating bitumen between cement and aggregates showed elastic behaviour at low temperature, resulting in no increase damping.

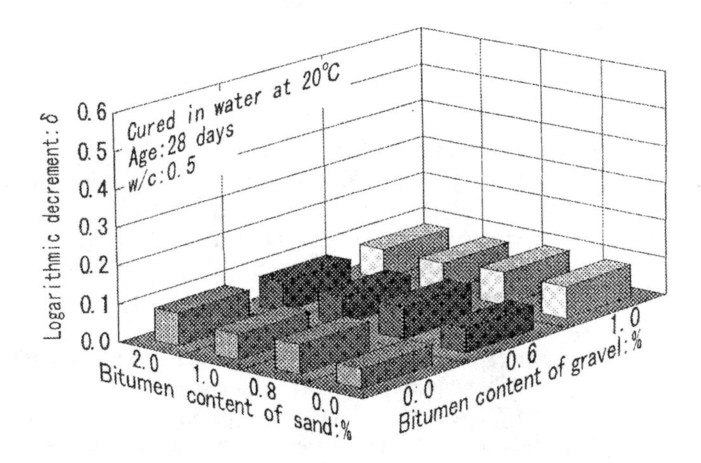

Figure.1. The relationship between logarithmic decrement and coating bitumen content at 0°C for the coated cement concrete cured in water at 20°C for four weeks

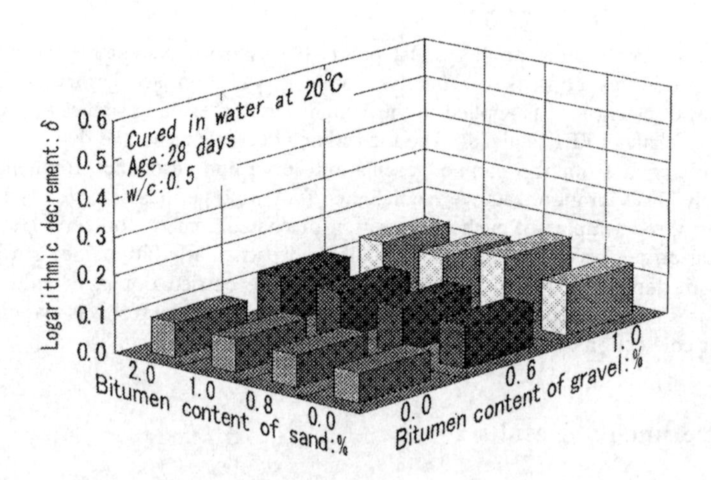

Figure.2. The relationship between logarithmic decrement and coating bitumen content at 20°C for the coated cement concrete cured in water at 20°C for four weeks

Figure 2 shows the relationship between logarithmic decrement and coating bitumen content at 20°C for cement concrete water-cured at 20°C for four weeks. The coated cement concretes show a larger logarithmic decrement than the non-coated cement concrete and this property increased as the bitumen content increased. The coated gravels contribute more than the coated sands. For example, the logarithmic decrement values for the coated cement concrete made of uncoated sand and gravel coated with 1 per cent bitumen (sample number: 50-0.0-1.0) was about 1.7 times that of uncoated cement concrete, whereas the value for the coated cement concrete made of sand coated with 1 per cent bitumen and uncoated gravel (sample number: 50-1.0-0.0) was about 1.1 times that of the uncoated cement concrete.

The apparent surface area of the gravel was smaller than that of the sand so, for the same total weight of aggregates and bitumen, the average bitumen film on the gravel was thicker than that on the sand; this tendency was particularly obvious at the higher temperatures of 40°C and 60°C, as shown in Figure 3 and Figure 4, respectively. For example, the logarithmic decrement values for sample 50-0.0-1.0 at 40°C and 60°C were about 2.7 times and about 3.1 times those of uncoated cement concrete, whereas, the values for sample 50-1.0-0.0 were about 1.2 times and about 1.3 times greater than those of uncoated cement concrete, at 40°C and 60°C respectively.

An aggregate and a cement, with a high rigidity and low damping, respond to a given stress or vibration very quickly whereas a bitumen with a low rigidity and high damping responds to it more slowly. These different rates of response to the stimulation of each component in the composite materials are typical of 'miss-matching', and this 'miss-matching' changed the vibration energy into thermal energy, which resulted in the absorption of vibration. Therefore, the damping effect became larger as the degree of the 'miss-matching' of the aggregate increased.

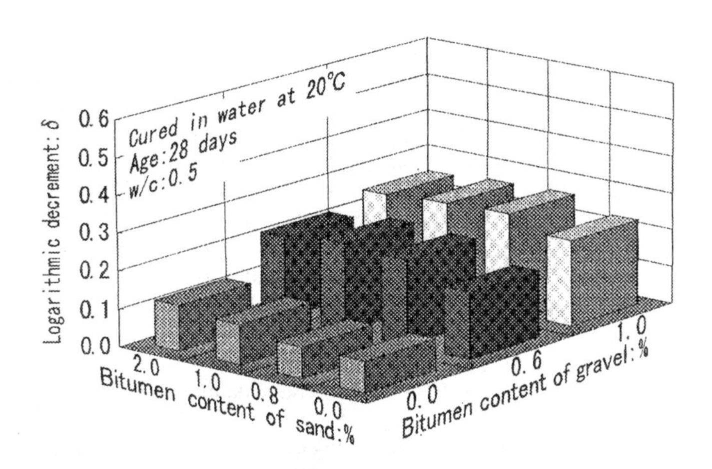

Figure.3. The relationship between logarithmic decrement and coating bitumen content at 40°C for the coated cement concrete cured in water at 20°C for four weeks

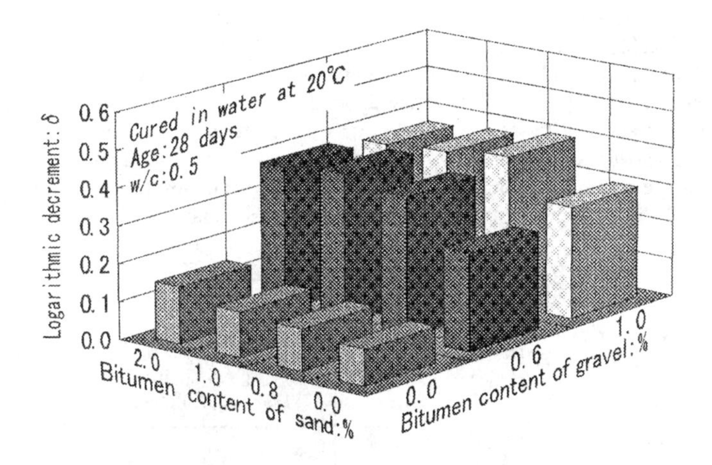

Figure.4. The relationship between logarithmic decrement and coating bitumen content at 60°C for the coated cement concrete cured in water at 20°C for four weeks

(b) Relationship between logarithmic decrement and temperature

In Figures 1 to 4, the test temperatures range from 0°C to 60°C and it was found that the higher the temperature, the larger the logarithmic decrements. These results indicate that the logarithmic decrement is less susceptible to temperature at lower temperatures, because, as the temperature decreases, the behaviour of the coated cement concretes becomes less visco-elastic, approaching that of an elastic solid. The rate of increase of logarithmic decrement with temperature is greater, as the bitumen content of the aggregates, especially that of the gravel, is increased. For example, the values of logarithmic decrement at 20°, 40° and 60°C for the sample number 50-0.0-1.0 were about 1.6, 2.7 and 3.5 times that at 0°C, respectively. Moreover, the values of logarithmic decrement of the sample number 50-0.0-1.0 were 1.3, 1.6, 2.1 and 2.5 times those of the sample number 50-1.0-0.0 at 0°, 20°, 40° and 60°C respectively. Considering that sample number 50-0.0-1.0 represents uncoated sand and coated gravel coated with 1 per cent bitumen and sample number 50-1.0-0.0 represents sand coated with 1 per cent bitumen and uncoated gravel, it is clear that the effect of coated gravel is very much greater than that of coated sand.

(c) Relationship between logarithmic decrement and dynamic modulus

Figure 5 shows the relationship between the logarithmic decrements and the calculated dynamic modulus, or modulus at resonance over the temperature ranges of 0°C to 60°C for water-cured materials for four weeks. Each data point in this figure presents the average of the measured values from the bending vibration test. The logarithmic decrement increased as the dynamic modulus decreased because, as the temperature decreased and the bitumen content decreased, the behaviour of the coated cement concrete becomes more elastic, and less viscoelastic. Damping ability with decreased modulus was achieved at the higher bitumen content of gravel.

On the whole, the coated cement concrete evaluated showed superior behaviour, a high damping effect, with little decrease in the dynamic modulus; for example, the logarithmic decrement was about 4 times that of the uncoated material while the dynamic modulus was about two-thirds, compared with current cement concrete. This means that it is possible for the bitumen-coating technology to increase the vibration absorption without markedly effecting the dynamic modulus of the composite materials.

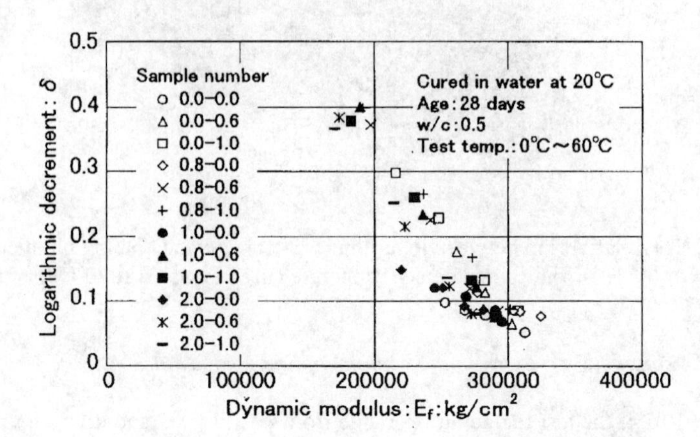

Figure.5. The relationship between logarithmic decrement and dynamic modulus for the coated cement concretes cured in water at 20°C for four weeks

4. Conclusions

The following conclusions are drawn from this investigation.

1) At the lower temperatures, the vibration absorption of the coated and uncoated cement concretes were similar.
2) At higher temperatures the vibration absorption of the coated cement concrete increased as the bitumen content increased.
3) The average bitumen film on the gravel was thicker than that on the sand, for the same weight of aggregate and bitumen. This resulted in greater vibration absorption for coated gravels than for coated sands.
4) The dynamic modulus decreased as the logarithmic decrement increased because the coated cement concretes became less like an elastic solid.
5) Coated cement concretes have greater vibration absorption properties than the uncoated cement concrete but there is some decrement in dynamic modulus of these materials compared to conventional cement concrete.

Acknowledgements

The authors wish to thank Dr. W. D. Powell, Programme Director of Transport Research Laboratory, UK, for valuable comments on the manuscript.

References

[1] *Bureau of construction, Tokyo Metropolitan Government,* REPORT ON THE EXPERIMENTAL RESULTS ON VIBRATION PROPERTIES OF VARIOUS ROAD STRUCTURES, 1975.

[2] *Bureau of construction, Tokyo Metropolitan Government,* REPORTS ON THE PREVENTIVE MEASURES FOR TRAFFIC VIBRATION OF ROAD, 1977.

[3] *M. Mayama, et al.,* VIBRATING AND MECHANICAL PROPERTIES OF FERRITE-RESIN COMPOSITE MATERIALS - International Conference on Composite Materials, pp.559-566, Milan - Italy, 10-12 May 1988.

[4] M. Mayama, M. Mori, VIBRATING AND MECHANICAL PROPERTIES OF FERRITE CONCRETE - Brittle Matrix Composites III, pp.488-497, Elsevier Applied Science, 1991.

[5] *A. Moriyoshi, et al.,* CHARACTERISTICS OF NEW INSULATING MATERIAL USING BITUMINOUS EMULSION - Deuxieme Congres Mondial De L'Emulsion (CME), pp.003/01-003/06 (4-4), Bordeaux - France, 23-26 September, 1997.

[6] *M. Mayama, Y. Hatanaka,* FUNDAMENTAL VIBRATION PROPERTIES OF COATED CEMENT CONCRETE - Proceedings of 36th Annual Meeting on Material Research in Japan Academics Conference, Tokyo - Japan, pp.207-208, 1992.

[7] *Y. Hatanaka, M. Mayama,* EFFECT OF COATING BITUMEN CONTENT ON THE LOGARITHMIC DECREMENT OF CEMENT CONCRETE - Proceedings of Annual Meeting in 1994 of Japan Society for Composite Materials, pp.31-32, Tokyo -

Japan, 1994.
[8] *M. Mayama, Y. Nakazawa,* EFFECT OF COATING BITUMEN CONTENT ON THE MODULUS AT RESONANCE OF CEMENT CONCRETE - Proceedings of 49th Annual Meeting of Japan Society of Civil Engineers, Vol.5-129, pp.258-259, Tokyo - Japan, 1994.
[9] *M. Mayama, H. Tanaka, Y. Nakazawa,* VIBRATION PROPERTIES OF CEMENT CONCRETE USING COATED AGGREGATES - EFFECT OF COATING BITUMEN CONTENT - Proceedings of 51st Annual Meeting of Japan Society of Civil Engineers, Vol.5, 219, pp.438-439, Tokyo - Japan, 1996.
[10] *M. Mayama, M. Yoshino, K. Hasegawa,* AN EVALUATION OF HEAVY DUTY BINDERS IN THE LABORATORY - ASTM STP 1108, pp.61-76, ASTM, Philadelphia - USA, 1992.
[11] *M. Mayama,* THE EVALUATION OF HEAVY DUTY BINDERS IN BITUMINOUS ROAD MATERIALS - Proceedings of Institution of Civil Engineers, Transport, Vol. 123, February, pp.39-52, UK, 1997.

A NEW APPROACH FOR EVALUATING COMPETING OPTIONS USING COMPOSITES FOR RETROFIT PROJECTS

Makarand Hastak[1]

and

Daniel W. Halpin[2]

ABSTRACT

The construction industry is very competitive. Composite materials because of their light weight, ease of handling, and formability offer potential for improved productivity during installation of components. In rehabilitation and retrofit projects, FRP wraps and reinforcement strips are much easier to install and offer productivity improvements over the traditional reinforcement methods. This paper presents a decision making model and decision support system (DSS) to assist construction managers in systematically evaluating whether to opt for a conventional process or a composite based retrofit approach. Factors that affect selection of traditional versus composite based methodologies are ranked and their weights of importance are determined by means of the Analytic Hierarchy Process (AHP). Methods using both composites and traditional technologies can be evaluated to determine advantages to be obtained using advanced repair methods.

INTRODUCTION

With the availability of composite materials for various construction operations the decision to choose between conventional or composite based retrofit systems can play a significant role in construction project planning. As every construction project is

[1]Assistant Professor, Department of Civil & Environmental Engineering, Polytechnic University, Six Metrotech Center, Brooklyn, NY 11201. (718)260-3989; (718)260-3433(FAX); mhastak@duke.poly.edu (E-Mail).

[2]Professor and Head, Division of Construction Engineering and Management, PurdueUniversity, 1294 Civil Engineering Building, West Lafayette, IN 47907-1294. (765)494-2244; (765)494-0644(FAX); halpin@ce.ecn.purue.edu (E-Mail).

unique, it is necessary to evaluate the feasibility of replacing a conventional construction process with a composite based rehabilitation strategy on a project to project basis.

This paper presents a decision support model for evaluation of construction processes and its implementation in a decision support system called AUTOCOP. The decision support model and AUTOCOP have been designed to assist construction managers in systematically evaluating the two options (i.e., conventional versus composite system) with respect to five groups of criteria. The various aspects that should be considered under each group of criteria have been explored and a decision model is proposed to determine the suitability of an automated system for a particular construction project. The decision algorithm behind AUTOCOP has been explained with examples.

DECISION CRITERIA

As shown in Fig. 1, process characteristics define the need for adopting an advanced technology over a conventional method, particularly if a process is labor intensive such as column and beam reinforcement. Also if high skill, dexterity, precision, as well as high productivity are key requirements, process automation is desirable over conventional process to reduce the possibility of human error. Additional ergonomic and process related factors, such as whether the activity is tedious and boring, repetitive, and/or unpleasant and dirty, also govern the decision to select advanced technology over conventional method. The hierarchy of Fig. 1 shows a linkage between all the subcriteria and the available alternatives to illustrate that the relevance of each subcriterion in the hierarchy should be evaluated with respect to each alternative (i.e., automation and conventional) for a given situation.

DECISION FRAMEWORK

AUTOCOP, the proposed decision support system for automation option evaluation for construction processes, includes two models: an analytical model and a group decision model. The objective of the analytical model is to assist the user in evaluating the decision problem through hierarchical assessment of the various criteria and subcriteria. The input required for the assessment is obtained from the user. It might sometimes be important, in a decision problem of this nature, to collect and synthesize the opinion of other team members and experts who are familiar with the various aspects of the project and are in a position to evaluate the benefits and drawbacks of replacing a conventional construction process by an automated process. The group decision model was developed to assist the primary decision maker (PDM) in collecting and evaluating these opinions. This model assists the user in synthesizing the information obtained from other team members into a group decision that is then used in the analytical model to process the various criteria and subcriteria. Both models

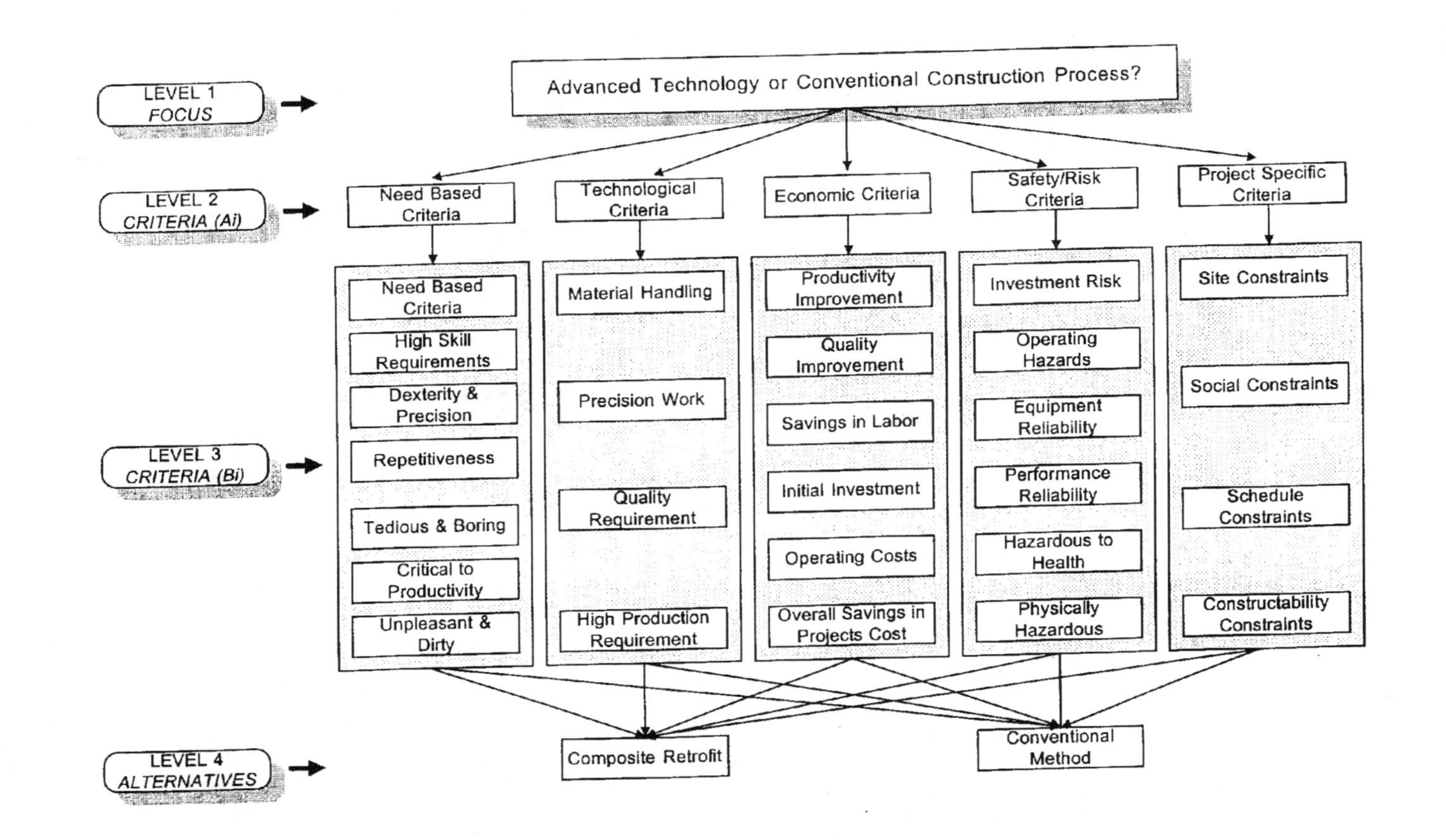

Fig.1 - Sample Hierarchy for Analyzing the Decision Problem

in this decision framework utilize Analytical Hierarchy Process (AHP) to determine the preference among various criteria, subcriteria, and alternatives [1].

The group decision aspect is particularly important because of the subjectivity involved in analyzing the intangible as well as the tangible criteria that are more accentuated due to the lack of adequate historical information for a new technology with respect to performance and effectiveness. Even though the tangible criteria such as cost, productivity, and savings in labor cost can be evaluated for each option separately prior to comparing them in the decision framework, the amount of information available would vary from older to newer technology. Therefore it is important to have a team consensus in comparing the two options and also to perform a more responsive analysis of the decision problem.

EXAMPLE SITUATION

Consider a project which involves retrofitting and strengthening 10 columns supporting an overpass in Chicago. State highway engineers are considering two possible methods of reinforcement. One involves the conventional use of steel jackets which will be specially fabricated and installed at the site. The second option involves use of an automated system which will wrap the columns with a carbon fiber composite material.

Use of the steel jackets requires special fabrication and use of heavy equipment to handle the jackets. Dimensional errors in the fabrication of the jackets can lead to problems in installation or possibly the need for modification of the fabricated elements. The steel jackets will also be subject to corrosion which will impact their overall service life. Installation at best will be cumbersome and time consuming.

The alternate method involves using an automated wrapping machine which precisely wraps the columns with a carbon fiber wrapping. The composite filament is cured by a controlled elevated temperature process and coated to match the existing structure. This approach offers the following advantages:

(1) Safe efficient installation process with 2-3 man crew.
(2) Automated wrapping and monitored cure process.
(3) No fabrication delays.
(4) Precise fit and interface with existing columns.
(5) High strength to weight ratio.
(6) High corrosion resistance.
(7) No seams or welds.
(8) No increase in column stiffness.

The engineers are interested in developing an evaluation or rating of the level of advantage of one alternative over the other. The project situation described is analyzed with respect to the hierarchy shown in Fig. 1 using both the group decision model and the analytical model. Reasonable values for the input variables have been assumed to demonstrate the use of the AUTOCOP system.

GROUP DECISION MODEL

The objective of the group decision model (GDM) is to assist the PDM in collecting and evaluating the opinion of other team members leading to a group decision that establishes the relative preference among criteria, subcriteria, and alternatives. The GDM also allows the primary decision maker to evaluate each member of the team to establish the importance (or weight) of individual input for decision making. The evaluation of the team members is then used to weigh the input provided by each team member and to arrive at a group decision.

In the group decision model (GDM), the primary decision maker (PDM) evaluates each team member with respect to four criteria (i) their technical knowledge, (ii) experience, (iii) current project knowledge, and (iv) knowledge about the firm. This evaluation is performed in two stages. In the first stage, the PDM performs a pairwise comparison of the four criteria to determine his or her value based preference among the four criteria (refer to Fig. 2). The comparison matrix is evaluated based on the AHP to determine the priority vector as explained earlier. Each criteria has four associated intensities or subcriteria namely: extensive, significant, moderate, and low. Pairwise comparison of the subcriteria with respect to each criterion establishes the distinction between the four intensities. The priority vector established for each comparison matrix determines the weight for the subcriteria associated with a particular criterion, e.g.,with respect to technical knowledge one may not consider much difference between significant and moderate whereas in the case of experience one might distinguish significant experience from moderate experience. This analysis has been illustrated in Fig. 2 for the hypothetical project situation described earlier.

In evaluating the group members based on the criteria mentioned above an absolute scale is used as opposed to a relative scale. An absolute scale is favored in situations where the user's preference among criteria and subcriteria are independent of the available alternatives and each alternative is measured on its own merit. In absolute scale measurements the user establishes the preference among various criteria and subcriteria by using the same procedure of pairwise comparison as used in the relative scale. However, the alternatives are not ranked by pairwise comparison but only on their individual merit. The academic grading of A, B, D, C, and F is an example of absolute scale measurement. The procedure for group member evaluation has been illustrated in Fig. 3.

LEVEL A: CRITERIA						
		A1	**A2**	**A3**	**A4**	Weight Factor
A1 = Technical Knowledge	**A1**	1.00	2.00	2.00	4.00	0.44
A2 = Experience	**A2**	0.50	1.00	1.00	2.00	0.22
A3 = Current Project Knowledge	**A3**	0.50	1.00	1.00	2.00	0.22
A4 = Knowledge about the Firm	**A4**	0.25	0.50	0.50	1.00	0.11

INTENSITIES FOR GROUP MEMBER EVALUATION

LEVEL B: INTENSITIES	Analysis with respect to: A1 = Technical Knowledge					
		B1.1	**B1.2**	**B1.3**	**B1.4**	Weight Factor
B1.1 = Extensive	**B1.1**	1.00	2.00	3.00	4.00	0.47
B1.2 = Significant	**B1.2**	0.50	1.00	2.00	3.00	0.28
B1.3 = Moderate	**B1.3**	0.33	0.50	1.00	2.00	0.16
B1.4 = Low	**B1.4**	0.25	0.33	0.50	1.00	0.10

Fig. 2 - Establishing the Basis for Group Member Evaluation Using AHP

Criteria	**Criteria Weight**	**Intensity**	**Intensity Weight**	**Group Members**				
				1	**2**	**3**	**4**	**5**
Technical Knowledge	0.44	Extensive	0.47					
		Significant	0.28		1.00			1.00
		Moderate	0.16			1.00		
		Low	0.10	1.00			1.00	
Experience	0.22	Extensive	0.53	1.00		1.00		
		Significant	0.19					1.00
		Moderate	0.19				1.00	
		Low	0.10		1.00			
Current Project Knowledge	0.22	Extensive	0.47					
		Significant	0.25	1.00	1.00			
		Moderate	0.17			1.00	1.00	
		Low	0.11					1.00
Knowledge About the Firm	0.11	Extensive	0.44					1.00
		Significant	0.22	1.00		1.00		
		Moderate	0.22		1.00		1.00	
		Low	0.11					
Total Score for the Group Member (Structural Rescaling of the Priorities)				0.24	0.23	0.25	0.13	0.24
Normalized Group Member Scores Sum = 1.00				0.22	0.21	0.23	0.12	0.22

NOTE: Total Score for Group Member = SUM (Criteria Weight * Intensity Weight)

Fig. 3 - Sample Group Member Evaluation and Weight Determination

In the second stage of the group evaluation, the PDM evaluates (or grades) each group member's technical knowledge, experience, current project knowledge, and knowledge about the firm as extensive, significant, moderate, or low. The score obtained by each group member is the sum of the weighted score obtained under each category.

ANALYTICAL MODEL

The objective of the analytical model is to evaluate the input provided by the group members and establish the group's preference among the various criteria, subcriteria, and alternatives. As explained earlier, the decision process starts by identifying the various criteria, subcriteria, and alternatives relevant to the decision task. These criteria and subcriteria are then arranged in a hierarchy as shown in Fig. 1. The Analytical Hierarchy process is used for establishing the comparison (or weight) matrix for each level of the hierarchy and for computing the priority vectors as explained earlier.

Fig. 4 shows sample comparison matrices and the corresponding weight vectors established for the hypothetical project situation evaluating the option between the Automated Composite Column Wrapping and conventional steel jacket reinforcement. The comparison matrices of Fig. 4 utilize the five decision criteria as well as the associated subcriteria as illustrated in Fig. 1.

CONCLUSION

The AUTOCOP system incorporates knowledge based preference of team members in arriving at a group decision and establishing priorities among the criteria, subcriteria, and alternatives. The user interface has been designed to facilitate sensitivity analysis for what-if scenarios. There are numerous benefits of the proposed system. These are systematic compilation and analysis of various criteria involved in the decision process. This ensures that all important criteria and their relevance in this decision process have been considered resulting into a more reliable decision. Additionally, model framework and use of AHP allow analysis of both quantifiable and intangible decision criteria that is particularly important in a decision process where sufficient past performance data for automation options are not available. The model analysis leads to the identification of the most suitable option on the basis of priority.

REFERENCES

1) Thomas L. Saaty, The Analytic Hierarchy Process, (New York, NY: McGraw-Hill, 1980).

LEVEL A: CRITERIA

		A1	A2	A3	A4	A5	Weight Vector [WA]
Need Based Criteria	A1	1.00	1.00	2.00	1.00	2.00	0.2476
Technological Criteria	A2	1.00	1.00	2.00	1.00	2.00	0.2476
Economic Criteria	A3	0.50	0.50	1.00	1.00	3.00	0.1882
Safety/Risk Criteria	A4	1.00	1.00	1.00	1.00	1.00	0.1938
Project Specific Criteria	A5	0.50	0.50	0.33	1.00	1.00	0.1227

LEVEL B: SUBCRITERIA

Analysis with respecto to: A1 = Need Based Criteria
WEIGHT (From level -A): 0.2476 (refer to [WA])

		B1.1	B1.2	B1.3	B1.4	B1.5	B1.6	B1.7	Priority Vector [WB1]	Weighted Vector [WB1 * Weight]
Labor Intensiveness	B1.1	1.00	1.00	1.00	0.33	1.00	0.50	1.00	0.11	0.0265
High Skill Requirements	B1.2	1.00	1.00	2.00	1.00	1.00	1.00	2.00	0.17	0.0415
Dexterity & Precision	B1.3	1.00	0.50	1.00	1.00	1.00	1.00	1.00	0.13	0.0312
Repetitiveness	B1.4	3.00	1.00	1.00	1.00	2.00	1.00	3.00	0.21	0.0515
Tedious & Boring	B1.5	1.00	1.00	1.00	0.50	1.00	1.00	1.00	0.12	0.0307
Critical to Productivity	B1.6	2.00	1.00	1.00	1.00	1.00	1.00	3.00	0.18	0.0436
Unpleasant & Dirty	B1.7	1.00	0.50	1.00	0.33	1.00	0.33	1.00	0.09	0.0226

LEVEL C: ALTERNATIVES

Analysis with respecto to: B1.1 = Labor Intensiveness
WEIGHT (From level - B): 0.0265 (refer to level B, Weighted Vector)

		C1	C2	Priority Vector [WCB1.1]	Weighted Vector [WCB1.1 * Weight]
Composite Retrofit	C1	1.00	4.00	0.8	0.0212
Conventional Method	C2	0.25	1.00	0.2	0.0053

CRITERIA: A1	Analysis with respect to:							Aggregate
Weighted Vector	B1.1	B1.2	B1.3	B1.4	B1.5	B1.6	B1.7	Vector
C1	0.0212	0.0208	0.0156	0.0258	0.0154	0.0290	0.0113	0.1390
C2	0.0053	0.0208	0.0156	0.0258	0.0154	0.0145	0.0113	0.1086

AGGREGATE MATRIX

	C1	C2
A1	0.14	0.11
A2	0.14	0.11
A3	0.09	0.10
A4	0.12	0.08
A5	0.07	0.05
SUM	0.55	0.45

FINAL PRIORITY VECTOR
(APLE)
C1 = 0.55
C2 = 0.45

ALTERNATIVES:
C1 = Composite Retrofit
C2 = Conventional Method

Fig. 4 - Sample Comparison Matrices for Criteria, Subcriteria, and Alternatives

On the statical behaviour of reinforced concrete beams wrapped with FRP plates: an experimental investigation

L. Feo*, G. D'Agostino*, D. Tartaglione*

*Department of Civil Engineering, Università degli Studi di Salerno, 84084 Fisciano (Sa), Italy

Abstract

In this paper some preliminary results of an experimental investigation on the statical behaviour of reinforced concrete beams wrapped with CFRP (Carbon Fiber-Reinforced Plastics) plates are presented.
The experimental results are compared with the theoretical predictions furnished by a finite element model presented by the authors in previous papers.
In particular the actual distribution of the interlaminar stresses on the interface between concrete core and CFRP plates is investigated, taking into account its relevant technical interest.

1. Introduction

The examples of applications of reinforced concrete beams wrapped with FRP (Fiber- Reinforced Plastics) sheets or plates are becoming more and more frequent in the Civil Engineering field.
Hermite's traditional technique, which uses steel plates externally bonded to concrete structures, presents many shortcomings such as long term corrosion, high installation cost and weight, and visual intrusion.
In order to eliminate the above mentioned disadvantages, steel plates have been replaced by FRP plates or sheets; in fact, this kind of plates do not corrode and can be formed, fabricated, and bonded more easily than the steel ones.
In technical applications both flexural and shear strengthening can be required: in the first case the FRP wrapping is bonded on the bottom of the beam, while in the second one it is applied to the two-web faces.
Most of the papers available in literature deal with the first kind of wrapping and provide numerical or closed form solutions [1-4]. Only few papers, instead, analyze the mechanical modeling of the shear strengthening [5,6].
The authors have recently analyzed from a numerical point of view the statical behaviour of concrete beams wrapped with FRP plates or sheets, arranged in different ways on their lateral surface [7-10].
The numerical results allow us to predict an accurate interlaminar stress distribution at the interface between concrete core and FRP plates as well as the splitting up of the total shear force and bending moment among the different components of the strengthened beam.

In particular the predicted interaction is characterized by non-constants distributions of interlaminar stresses, which may have values, both negative and positive, three-four times greater than those obtainable within a classical beam theory.
It follows the relevance of developing an experimental analysis in order to study the behaviour of reinforced concrete beams and, specifically, to understand the nature of the actual interactions between the concrete core and FRP plates.
The experimental investigation has been carried out at the Testing Laboratory of Material and Structures of the Department of Civil Engineering of the University of Salerno.

2. Research objectives

The main objectives of the experimental research are:

- to evaluate the actual distribution of the interlaminar stresses at the plate-core interface, comparing experimental and numerical results;
- to study the influence of the strengthening technique on the ultimate limit state;
- to analyze the failure modes exhibited by the CFRP - concrete beams.

In this paper we can only present some preliminary results concerning the first topic of the above outlined program.

3. Materials

Ten beams were manufactured for the experiments. All beams had a cross-section of 150 mm × 250 mm and a total length of 2300 mm (Fig. 1).

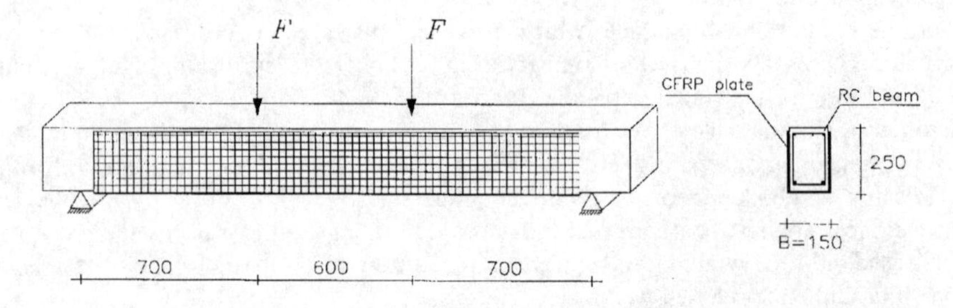

Figure 1. Specimen geometry, CFRP reinforcement type and load configuration (dimensions in mm)

Concrete. Portland cement (32.5R) was used to cast concrete samples. The cement-sand-gravel ratios in the concrete mix were 1:2:3 by weight. The water-cement ratio was 0.67. The maximum size of the aggregate was 13.0 mm. Twenty cubes of side 150 mm were cast and tested to determine the mechanical properties of the hardened concrete. The average compressive strength of the concrete after 28 days was 28.5 MPa, while the elastic modulus was 28 GPa. The Poisson's ratio was assumed to be equal to 0.2.

Reinforcing steel. The longitudinal reinforcement and the shear reinforcement of each beam were two φ 14 rebars and φ 6 stirrups at 150 mm, respectively. Steel had an elastic average yield stress of 435 MPa and an elastic modulus of 210 GPa.

Composite plates. The fiber-composite material consisted of bidirectional (*0°-90° ; warp = weft*) carbon woven textiles (SikaWrap VP®) bonded together with an epoxy matrix (Sikadur 32®). Longitudinal tensile tests (according to ASTM D3039/D3039M) were conducted on the plate to determine its mechanical properties (Fig. 2): the material exhibited a linear-elastic behaviour up to failure (Fig. 3). A summary of the plate properties used in the experimental analysis is given in Table 1.

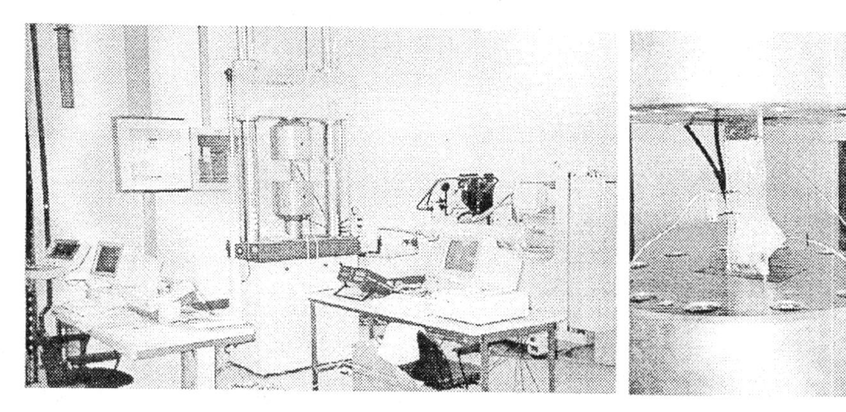

Figure 2. Tensile testing equipment

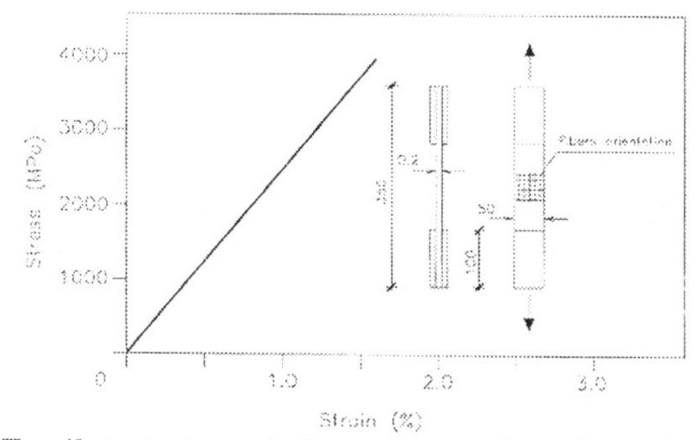

Figure 3. Tensile test: stress-strain response and specimen drawing (in mm)

Table 1. Mechanical properties of the CFRP plates

Material	Nominal thickness *(mm)*	Young's modulus E_{pl} *(GPa)*	Poisson's ratio ν_{pl}	Shear modulus $G_{pl} = E_{pl}/2(1+\nu_{pl})$ *(GPa)*	Ultimate tensile strength *(MPa)*
SikaWrap VP®	0.2	248	0.255	99	3980

4. Experimental Methods

One rectangular reinforced concrete beam was experimentally studied this far. The concrete beam was strengthened by wrapping it with CFRP plates 30 days after casting. Before applying the epoxy adhesive the beam was blasted at the bottom and at the lateral faces by a grinding wheel, and was cleaned using an airjet to insure a good bond between the epoxy glue and the concrete surface.

The epoxy was hand-mixed and hand-applied on the beam lateral surface with a brush and a spatula. Then, 3 layers of carbon woven textiles were used and bonded to the beam by in situ polymerization. The total thickness (t) of the wrapping was about 0.6 mm.

The specimen was instrumented with electrical strain gages. The strain gages were positioned on the bottom CFRP plate (Fig. 4), 300 mm from the support, in order to evaluate the shear stresses $\bar{\tau}$ at three different points (i.e. P_1, P_2, P_3) .

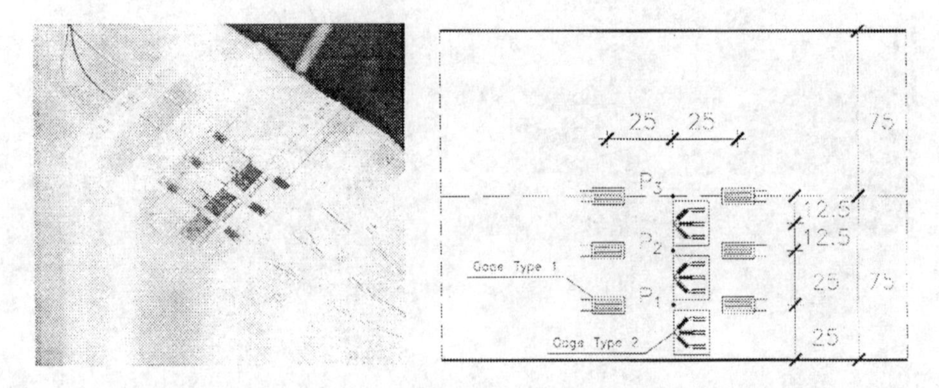

Figure 4. Location of the strain gages on the bottom CFRP plate (dimensions in mm)

Specifically, the above stresses were calculated according to the following expression, obtained from simple equilibrium relationships (Fig. 5):

$$\bar{\tau} = E_{pl} \Delta\varepsilon_{P_i} \frac{t}{\Delta z} + G_{pl} \Delta\gamma_{P_i} \frac{t}{\Delta x} \qquad (1)$$

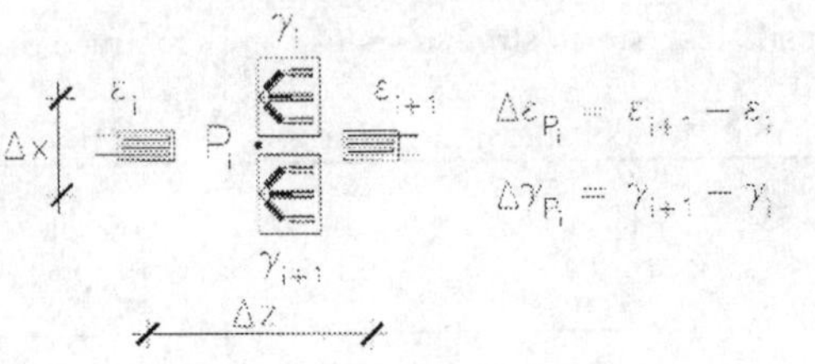

γ_i ε_i P_i ε_{i+1} Δx γ_{i+1} Δz

$\Delta\varepsilon_{P_i} = \varepsilon_{i+1} - \varepsilon_i$

$\Delta\gamma_{P_i} = \gamma_{i+1} - \gamma_i$

Figure 5. Location of the strain gages to evaluate the shear stress at the generic point P_i

In particular only one strain gage type 2 was positioned for P_3 because of the geometry of the system.
The main characteristics of the strain gages used in the experiments are summarized in Table 2.

Table 2. Characteristics of the strain gages

Gage	Gage Code	Gage Length *(mm)*	Overall Length *(mm)*	Grid Width *(mm)*	Overall Width *(mm)*	Gage Resistance *(Ohms at 24°C)*	Gage Factor *(at 24°C)*
Type 1	MM 'C' Feature 250 UW	6.35	11.43	4.57	4.57	120.0 ± 0.3 %	2.070 ± 0.5 %
Type 2	MM 'C' Feature 250 UR	6.35	12.70	3.05	19.30	120.0 ± 0.4 %	2.080 ± 1.0 %

Load was applied by an hydraulic jack and measured by a load cell. The experimental set-up consisted of a simply-supported beam carrying two points forces (F), symmetrically placed from the mid span (Fig. 6).
During the test, load and strains were recorded by an automatic data acquisition system.

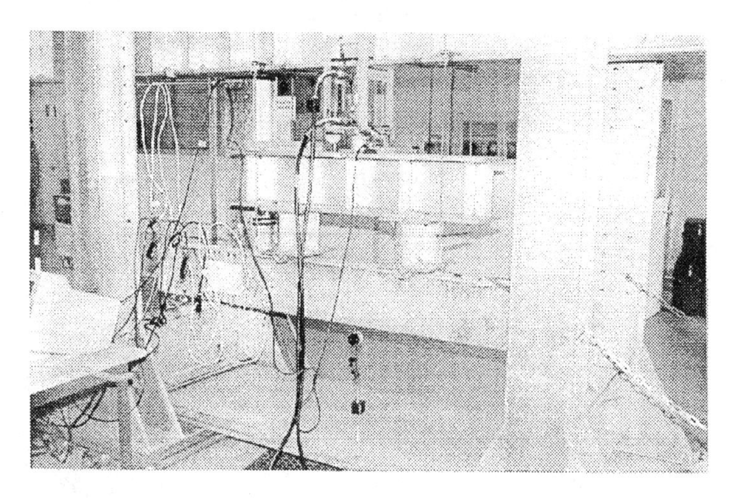

Figure 6. Experimental set-up

5. Analytical model

The mechanical model used to get numerical predictions has been presented by one of the authors in a previous paper [7].
This model assumes that the strengthened beam is composed by a concrete core and a given number of reinforcing plates, bonded to the latter by a continuous distribution of

bilateral elastic springs. The materials constituting core and plates are supposed to be linearly elastic and orthotropic with a possible different behaviour in tension and in compression (bimodular behaviour).
The kinematical model of the beam components (core and plates) derives from a suitable power expansion of the displacement field components with respect to the cross-section coordinates. The non-rigid displacement terms, taken into account, model both in-plane deformations and an out-of-plane warping of the cross-section.
Furthermore, a finite element approximation has been carried out by employing Lagrangian isoparametric element [8]. The discrete problem, non-linear because of the assumed bimodular behaviour, has been solved by the Newton-Raphson method.
We refer the reader to the works [9,10] for an in-depth description of the mechanical and analytical details.

6. Results of experiment and discussion

Figure 7 shows the distribution of the interlaminar shear stresses at the interface between the bottom of the concrete core and the CFRP plates, as obtained from the experimental and the numerical investigations: the predicted values match reasonably well the experimental data.
The maximum load level reached during the experiments (F = 10 kN) was such that no plastic strains occurred both in concrete and in the steel reinforcement.

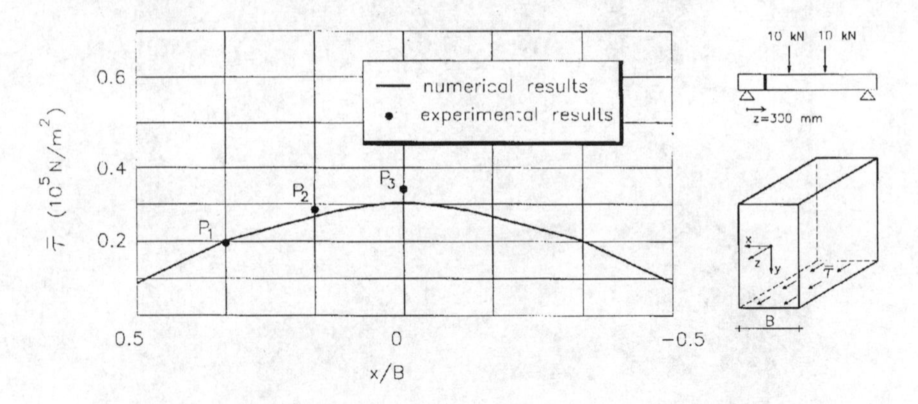

Figure 7. Comparison between numerical and experimental results

The non-constant distribution of the interlaminar stresses, with maximum values in the middle of the contact surface might lead to possible failure mechanisms.
In fact the maximum interlaminar stresses can produce brittle failure of the concrete followed by debonding of the CFRP plates, mainly in presence of thicker strengthening plates, responsible of more relevant interactions.

7. Concluding remarks

Some experimental and numerical results relative to the statical behaviour of reinforced concrete beams wrapped with CFRP plates were presented in this paper.
The preliminary experimental results validate the theoretical predictions and confirm the existence of non-constant interlaminar stresses with maximum values three - four times greater than the mean value.
This behaviour is illuminating in order to predict a possible brittle fracture of the concrete supporting the wrapping followed by the failure of the strengthened beam.
The experimental investigation will be extended to a larger number of specimens to analyze the actual distribution of the interlaminar stresses emerging on the reinforced boundaries, and to study the influence of the geometry and mechanical percentage of the wrapping on the ultimate limit state.

Acknowledgments

This work was carried out with the financial support of the Italian National Research Council (Project 95.C4014.CT11). The authors also gratefully acknowledge Sika Italia S.p.A., and particularly Mr. Francesco Cervaso and Dr. Alberto Grandi, for supplying CFRP and adhesives.

References

[1] *H. Varastehpour, P. Hamelin*, EXPERIMENTAL STUDY OF RC BEAMS STRENGTHENED WITH CFRP PLATE, Second International Conference on Advanced Composite Materials in Bridges and Structures, Montréal, Québec, Canada, 555-563, 1996.
[2] *A. A. Malek, H. Saadatmanesh, M. R. Ehsani*, SHEAR AND NORMAL STRESS CONCENTRATIONS IN RC BEAMS STRENGTHENED WITH FRP PLATES, Second International Conference on Advanced Composite Materials in Bridges and Structures, Montréal, Québec, Canada, 629-638, 1996.
[3] *S. V. Hoa, M. Xie, X. R. Xiao*, REPAIR OF STEEL REINFORCED CONCRETE WITH CARBON/EPOXY COMPOSITES, Second International Conference on Advanced Composite Materials in Bridges and Structures, Montréal, Québec, Canada, 573-580, 1996.
[4] *C. Djelel, E. David, F. Buyle-Bodin*, UTILISATION DE PLAQUES EN COMPOSITE POUR LA REPARATION DE POUTRES EN BETON ARME ENDOMMAGEES, Second International Conference on Advanced Composite Materials in Bridges and Structures, Montréal, Québec, Canada, 581-588, 1996.
[5] *M. Arduini, A. Di Tommaso, O. Manfroni*, FRACTURE MECHANISMS OF CONCRETE BEAMS BONDED WITH COMPOSITE PLATES, Second International Symposium on: Non-Metallic Reinforcement for Concrete Structures FRPRCS-2, Ghent, 483-531, 1995.

[6] *M. Arduini, A. Di Tommaso, O. Manfroni, A. Nanni*, FAILURE MECHANISMS OF CONCRETE BEAMS REINFORCED WITH FRP FLEXIBLE SHEETS, Second International Conference on Advanced Composite Materials in Bridges and Structures, Montréal, Québec, Canada, 253-260, 1996.
[7] *L. Ascione , F. Fraternali, L. Feo*, THE WRAPPING OF REINFORCED CONCRETE BEAMS WITH FRP PLATES: A MECHANICAL MODEL - Proc. Int. Meeting "Advancing with Composites '97", Milan, Italy, 155-170, 1997.
[8] *G. D'Agostino, D. Tartaglione*, THE MECHANICAL BEHAVIOUR OF REINFORCED CONCRETE BEAMS STRENGTHENED BY WRAPPING WITH FRP PLATES: A FINITE ELEMENT APPROACH - Proc. Int. Meeting "Advancing with Composites '97", Milan, Italy, 87-98, 1997.
[9] *G. D'Agostino, L. Feo, D. Tartaglione*, SUL COMPORTAMENTO MECCANICO DI TRAVI RINFORZATE CON PLACCHE IN FRP: UNO STUDIO NUMERICO, Proc. XXVI Naz. Meeting AIAS'97, Catania, Italy, 511-518, 1997.
[10] *G. D'Agostino, L. Feo, D. Tartaglione,* LO STUDIO DELLE INTERAZIONI NEL RINFORZO DI TRAVI IN C.A. MEDIANTE PLACCHE IN FRP, Proc. Meeting on "Materiali e Tecniche per il Restauro", Cassino (FR), Italy, 301-312, 1997.

Stress analysis of reinforced concrete beams wrapped with FRP plates

L. Ascione*, L. Feo*, F. Fraternali*

*Department of Civil Engineering, Università degli Studi di Salerno, 84084 Fisciano (Sa), Italy

Abstract

The present paper deals with the formulation of a mechanical model of a reinforced concrete beam strengthened by externally bonded FRP sheets. The proposed model assumes an elastic and orthotropic behaviour of the FRP wrapping, an elastoplastic behaviour of the concrete core and an adhesive contact between such elements. Particular attention is focused on the shear deformation of the FRP wrapping, which is described as a thin-walled beam, and interfacial FRP-concrete stress distribution.

1. Introduction

The use of Fiber Reinforced Plastics (FRP) to strengthen concrete structures is today diffused in the field of civil engineering. In particular, the wrapping of concrete beams or columns with FRP sheets or plates is come to the attention of engineers as a practical and convenient reinforcing technique, which can be employed, for example, because of change of use of an existing structure, inadequate initial design, deterioration or new earthquake resistance requirements. Besides the improvement of the mechanical properties, it realizes a protection of the concrete structure and steel re-bars against long term corrosion.

The present paper deals with the formulation of a mechanical model of a reinforced concrete beam wrapped with FRP sheets.

The given model allows us to accurately predict the actual state of stress at the interface between the concrete core and the FRP wrapping, which plays a fundamental role in the mechanics of the structures under examination, being able to produce a brittle concrete failure and FRP detach [1-7].

In particular, the FRP wrapping behaviour is described by means of a one-dimensional theory which generalizes the well-known thin-walled beam theory of Vlasov [8], in order to take into account the effects of shear deformation and anisotropic constitutive equations.

2. Mechanical model of the FRP reinforced beam

Let us consider a concrete beam wrapped with FRP sheets over a given portion of its lateral surface (Fig. 1). In the following paragraphs we present a one-dimensional model which is useful to describe the mechanical behaviour of such a beam and, in particular, the interactions between concrete core and FRP wrapping. In the first place, two independent models of the beam components (FRP wrapping and concrete core) are formulated. Then, such models are assembled supposing that they are in adhesive contact.

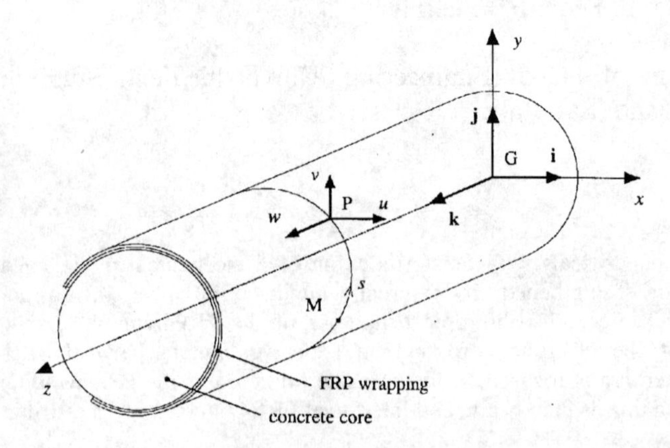

Figure 1. Concrete beam wrapped with FRP sheets

2.1 FRP wrapping

Following the above approach, we firstly consider the FRP wrapping separated from the concrete core. Such a structural element can be considered as a thin-walled beam (Figs. 1,2).

2.1.1 Kinematics

The kinematics of the FRP wrapping can be modelled by assuming that the generic cross-section (Fig. 2) undergoes a rigid transformation in its own plane and an out-of-plane warping during an arbitrary deformation of the beam.
Consequently, an admissible displacement field can be expressed into the following form (Figs. 1,2):

$$u = u_o(z) - \varphi_o(z)(y - y_o), \quad v = v_o(z) + \varphi_o(z)(x - x_o), \quad w = w(x, y, z) \qquad \text{(1a,b,c)}$$

where:
u and v are the displacement components along the axes x and y, respectively;
u_o and v_o are the displacement components of a given point O with coordinates x_o, y_o;
φ_o is the twisting rotation of the current cross-section;
w is the displacement component along the beam axis z.

The infinitesimal strain tensor ε associated with the displacement field (1) has the following component matrix with respect to the basis $\{\mathbf{i}, \mathbf{j}, \mathbf{k}\}$ (Fig. 1):

$$\left[\varepsilon_{ij}\right]=\begin{bmatrix} 0 & 0 & \varepsilon_{xz} \\ 0 & 0 & \varepsilon_{xz} \\ \varepsilon_{xz} & \varepsilon_{xz} & \varepsilon_{z} \end{bmatrix} \tag{2}$$

where:

$$\varepsilon_{xz} = 1/2(\partial u/\partial z + \partial w/\partial x), \quad \varepsilon_{yz} = 1/2(\partial v/\partial z + \partial w/\partial y), \quad \varepsilon_{z} = \partial w/\partial z \tag{3}$$

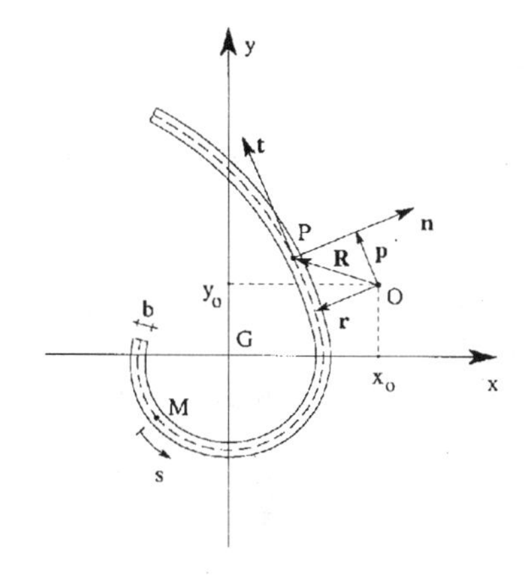

Figure 2. Cross-section of the FRP wrapping

By employing transformation formulae of tensor components under a change of basis, we obtain the component matrix of ε with respect to the local basis {**n**, **t**, **k**} attached to the generic point of the cross-section centerline (Fig. 2):

$$\left[\bar{\varepsilon}_{ij}\right]=\begin{bmatrix} 0 & 0 & \varepsilon_{nz} \\ 0 & 0 & \varepsilon_{tz} \\ \varepsilon_{nz} & \varepsilon_{tz} & \varepsilon_{z} \end{bmatrix} \tag{4}$$

with:

$$\varepsilon_{nz} = \varepsilon_{xz} dy/ds - \varepsilon_{yz} dx/ds, \quad \varepsilon_{tz} = \varepsilon_{xz} dx/ds + \varepsilon_{yz} dy/ds \tag{5}$$

s being the curvilinear coordinate measured along the cross-section centerline (Fig. 2). Due to the FRP wrapping thinness, we are now allowed to introduce the following assumptions:

$$w \cong w\big|_{n=0} = w(s,z) \tag{6}$$

$$\varepsilon_{nz} \cong 0 \tag{7}$$

$$\varepsilon_{tz} \cong \varepsilon_{tz}\big|_{n=0} + \left.\frac{\partial \varepsilon_{tz}}{\partial n}\right|_{n=0} n \tag{8}$$

n being the coordinate measured along the local axis **n**, i.e. along the wrapping thickness (Fig. 2).
On the other hand, it is a simple task to deduce the relation:

$$\frac{\partial \varepsilon_{tz}}{\partial s} - \frac{\partial \varepsilon_{nz}}{\partial n} = \varphi_o' \tag{9}$$

where $\varphi_o' = d\varphi_o / dz$. Such a relation allows us to write:

$$\varepsilon_{tz}(n,s,z) \cong \varepsilon_{tz}\big|_{n=0} + \varphi_o' n = \varepsilon_{xz}\big|_{n=0}\frac{dx}{ds} + \varepsilon_{yz}\big|_{n=0}\frac{dy}{ds} + \varphi_o' n \tag{10}$$

We approximate $\varepsilon_{xz}\big|_{n=0}$ and $\varepsilon_{yz}\big|_{n=0}$ as power series of the variable s truncated at a given order m, that is:

$$\varepsilon_{xz}\big|_{n=0} \cong \frac{1}{2}\left(\gamma_x^{(0)}(z) + \gamma_x^{(k)}(z)\, s^k\right) \qquad (k = 1,2,...m;\ \text{sum on } k) \tag{11a}$$

$$\varepsilon_{yz}\big|_{n=0} \cong \frac{1}{2}\left(\gamma_y^{(0)}(z) + \gamma_y^{(k)}(z)\, s^k\right) \qquad (k = 1,2,...m;\ \text{sum on } k) \tag{11b}$$

where:

$$\gamma_x^{(0)}(z) = 2\varepsilon_{xz}\big|_{s=n=0}, \quad \gamma_x^{(k)}(z) = \frac{2}{k!}\frac{\partial^k \varepsilon_{xz}}{\partial s^k}\bigg|_{s=n=0} \tag{12a}$$

$$\gamma_y^{(0)}(z) = 2\varepsilon_{yz}\big|_{s=n=0}, \quad \gamma_y^{(k)}(z) = \frac{2}{k!}\frac{\partial^k \varepsilon_{yz}}{\partial s^k}\bigg|_{s=n=0} \tag{12b}$$

We hence obtain for ε_{tz}:

$$\varepsilon_{tz}(n,s,z) \cong \frac{1}{2}\left(\gamma_x^{(0)}(z) + \gamma_x^{(k)}(z)\, s^k\right)\frac{dx}{ds} + \frac{1}{2}\left(\gamma_y^{(0)}(z) + \gamma_y^{(k)}(z)\, s^k\right)\frac{dy}{ds} + \varphi_o' n \tag{13}$$

Since, by definition, it results:

$$\varepsilon_{tz}\big|_{n=0} = \frac{1}{2}\left(\frac{\partial u}{\partial z}\frac{dx}{ds} + \frac{\partial v}{\partial z}\frac{dy}{ds} + \frac{\partial w}{\partial s}\right) = \frac{1}{2}\left(u'_o\frac{dx}{ds} + v'_o\frac{dy}{ds} + \varphi'_o r + \frac{\partial w}{\partial s}\right) \tag{14}$$

with (Fig. 2):

$$r = -(y - y_o)\frac{dx}{ds} + (x - x_o)\frac{dy}{ds} \tag{15}$$

upon substituting eqn. (13) into eqn. (14) and integrating with respect to s, we finally obtain:

$$w(s,z)=w_o(z) - \varphi_y(z)x(s) + \varphi_x(z)y(s) - \varphi_o'(z)\omega(s) +$$
$$+ \gamma_x^{(k)}(z)\omega_x^{(k)}(s) + \gamma_y^{(k)}(z)\omega_y^{(k)}(s) \qquad (k=1,2,...m;\ \text{sum on } k) \tag{16}$$

where (Fig. 2):

$$\varphi_x = -v_o' + \gamma_y^{(0)} \tag{17a}$$

$$\varphi_y = u_o' - \gamma_x^{(0)} \tag{17b}$$

$$w_o = w_M + \varphi_y x_M - \varphi_x y_M \tag{17c}$$

$$\omega = \int_M^P r\, ds \tag{17d}$$

$$\omega_x^{(k)} = \int_M^P s^k \frac{dx}{ds} ds \tag{17e}$$

$$\omega_y^{(k)} = \int_M^P s^k \frac{dy}{ds} ds \tag{17f}$$

The quantity defined by eqn. (17d) is known as *sectorial area* in the technical literature. In conclusion, the present model assumes the second members of eqns. (1a,b), (16) as the components of an arbitrary admissible displacement field of the FRP wrapping. The strain field associated with such a displacement field presents only two non-zero components, ε_{tz} and ε_z, whose expressions are respectively given by eqn. (13) and the following equation:

$$\varepsilon_z = w_o' - \varphi_y' x + \varphi_x' y - \varphi_o''\omega + \gamma_x^{(k)}{}'\omega_x^{(k)} + \gamma_y^{(k)}{}'\omega_y^{(k)} \tag{18}$$

It is interesting to notice that the present theory reduces to that of Vlasov if one truncates the power series (11a,b) at the first order and in addition places $\gamma_x^{(0)} = \gamma_y^{(0)} = 0$ into the above equations. The assumption $\gamma_x^{(0)} \neq 0, \gamma_y^{(0)} \neq 0$ instead leads one to obtain a first refinement of Vlasov's theory, which accounts for first-order shear deformation. Finally, higher order shear deformation theories are obtained within the actual model by setting $m \geq 1$.

2.1.2 Principle of virtual work

In view of a numerical approach to the problem here examined, founded upon the finite element method, it is useful to refer to a weak form of the equilibrium equations of the FRP wrapping, given by the Principle of Virtual Work (PVW).
Let us assume the FRP wrapping be loaded only by surface forces acting on its lateral surface. We denote the cartesian components of such surface tractions by p_x, p_y, p_z, the current cross-section of the FRP wrapping (Fig. 2) by Σ, the cross-section centerline by ρ, the FRP thickness by b (Fig. 2), the tangential and normal components of the Cauchy stress over the cross-section by τ_{tz} and σ_z, respectively, the length of the FRP wrapping (i.e. its dimension along the z-axis) by L, and the variational symbol by δ. The insertion of the displacement field (1a,b), (16) into the 3D

expression of the PVW leads us to obtain the following one-dimensional equation:

$$\int_0^L (T_x\delta\gamma_x^{(0)} + T_x\delta\gamma_x^{(0)} + N\delta w_o' + M_x\delta\varphi_x' + M_y\delta\varphi_y' + M_z^{(T)}\delta\varphi_o' - W_\omega\delta\varphi_o'' +$$

$$\Lambda_x^{(k)}\delta\gamma_x^{(k)} + \Lambda_y^{(k)}\delta\gamma_y^{(k)} + \Pi_x^{(k)}\delta\gamma_x^{(k)\prime} + \Pi_y^{(k)}\delta\gamma_y^{(k)\prime})dz =$$

$$= \int_0^L (q_x\delta u_0 + q_y\delta v_0 + q_z\delta w_0 + m_x\delta\varphi_x + m_y\delta\varphi_y + m_z\delta\varphi_0 - m_\omega\delta\varphi_o' +$$

$$\mu_x^{(k)}\delta\gamma_x^{(k)} + \mu_y^{(k)}\delta\gamma_y^{(k)})\,dz \qquad (k = 1,2,...m;\ \text{sum on } k) \tag{19}$$

which must hold for each virtual displacement $\left(\delta u_0, \delta v_0, \delta w_0, \delta\varphi_x, \delta\varphi_y, \delta\varphi_0, \delta\gamma_x^{(k)}, \delta\gamma_y^{(k)}\right)$.

In eqn. (19) we set:

$$T_x = \int_\rho \tau_{tz}\frac{dx}{ds}bds, \qquad T_y = \int_\rho \tau_{tz}\frac{dy}{ds}bds, \qquad N = \int_\rho \sigma_z\, bds \tag{20a,b,c}$$

$$M_x = \int_\rho \sigma_z y\, bds, \qquad M_y = -\int_\rho \sigma_z x\, bds, \qquad M_z^{(T)} = 2\int_\Sigma \tau_{tz} n\, dn\, ds \tag{20d,e,f}$$

$$W_\omega = \int_\rho \sigma_z\omega\, bds, \qquad \Lambda_x^{(k)} = \int_\rho \tau_{tz}\, s^k\frac{dx}{ds}bds, \quad \Lambda_y^{(k)} = \int_\rho \tau_{tz}\, s^k\frac{dy}{ds}bds \tag{20g,h,i}$$

$$\Pi_x^{(k)} = \int_\rho \sigma_z\,\omega_x^{(k)}\, bds, \qquad \Pi_y^{(k)} = \int_\rho \sigma_z\,\omega_y^{(k)}\, bds \tag{20l,m}$$

$$q_x = \int_\rho p_x\, ds, \qquad q_y = \int_\rho p_y\, ds, \qquad q_z = \int_\rho p_z\, ds \tag{21a,b,c}$$

$$m_x = \int_\rho p_z y\, ds, \quad m_y = -\int_\rho p_z x\, ds, \quad m_z = \int_\rho\left[-p_x(y - y_o) + p_y(x - x_o)\right]ds \tag{21d,e,f}$$

$$m_\omega = \int_\rho p_z\omega\, ds, \quad \mu_x^{(k)} = \int_\rho p_z\omega_x^{(k)}\, ds, \quad \mu_y^{(k)} = \int_\rho p_z\omega_y^{(k)}\, ds \tag{21g,h,i}$$

We refer to the quantities (20) as the *generalized stresses* of the present one-dimensional mechanical model. In particular, the quantities (20a)-(20f) represent the classical stress resultants and moments of beam's theory (shear forces, axial force, bending and twisting moments). The quantity W_ω, defined by eqn. (20g), instead coincides with the *bi-moment* of Vlasov's theory. Finally, the quantities $\Lambda_x^{(k)}, \Lambda_y^{(k)}, \Pi_x^{(k)}, \Pi_y^{(k)}$ (k=1,2,...,m) defined by eqns. (20h)-(20m) represent peculiar stress moments of the one-dimensional theory presented in this work.

Similarly, we refer to the quantities (21) as the *generalized forces* of the present model.

2.1.3 Constitutive equations

We assume that the FRP wrapping is formed by the assembly of a given number of sheets whose reinforcing fibres are arranged along arbitrary directions. Moreover, we assume that each FRP sheet (or lamina) behaves as a linearly elastic and orthotropic material. Hence, we write the local stress-strain relations as:

$$\sigma_z = Q_{33}\varepsilon_z + 2Q_{34}\varepsilon_{tz} \tag{22a}$$
$$\tau_{tz} = Q_{34}\varepsilon_z + 2Q_{44}\varepsilon_{tz} \tag{22b}$$

Q_{33}, Q_{34} and Q_{44} being the elastic constants of the lamina under consideration with respect to the structural frame $\{\mathbf{n}, \mathbf{t}, \mathbf{k}\}$.
By substituting eqns (22), (13) and (18) into eqns. (20), we obtain the relations between the generalized stresses (20) and the generalized strains ($\gamma_x^{(0)}, \gamma_y^{(0)}, w_x^{(0)}{}', \varphi_x{}', \varphi_y{}', \varphi_o{}'$, $\varphi_o{}'', \gamma_x^{(k)}, \gamma_y^{(k)}, \gamma_x^{(k)}{}', \gamma_y^{(k)}{}'$) of the FRP wrapping model.

2.2 Concrete core

In this work, we model the mechanical behaviour of the concrete core by means of the one-dimensional model described in the recent paper quoted as [6].
Such a model assumes that an arbitrary admissible displacement of the concrete core can be represented as a power series expansion of the displacement components with respect to the cartesian coordinates *x, y, z* (Fig. 1). Suitably choosing the orders of these power series we can model whether rigid transformations of concrete core cross-sections or in-plane and out-of plane deformations.
Concerning the material behaviour, we adopt an elastoplastic constitutive law.
Due to lack of space, we refer the reader to the paper [6] for an in-depth description of the concrete core model.

2.3 Assembly of concrete core and FRP wrapping

We assume that the concrete core and the FRP wrapping are connected by continuous distributions of elastic springs arranged along the directions of the cartesian reference axes *x, y, z* [6,7].
If $1/\eta$ denotes the stiffness of the interfacial springs, we approximate the condition of adhesive contact between the concrete core and the FRP wrapping by letting $1/\eta$ converge to zero (*penalty technique*). At the same time, we obtain an approximation of the interlaminar stresses, which can be identified with the elastic reactions of the interfacial springs:

$$t_x = \frac{1}{\eta}\left(u^{(core)} - u^{(FRP)}\right),\quad t_y = \frac{1}{\eta}\left(v^{(core)} - v^{(FRP)}\right),\quad t_z = \frac{1}{\eta}\left(w^{(core)} - w^{(FRP)}\right) \tag{23}$$

$u^{(core)}, v^{(core)}, w^{(core)}$ and $u^{(FRP)}, v^{(FRP)}, w^{(FRP)}$ respectively being the displacement components exhibited by the concrete core and by the FRP wrapping along the contact surface.

3. Concluding remarks

In this paper we have presented a mechanical model of a concrete beam strengthened with externally bonded FRP plates. The proposed model generalizes a similar previous one [6] and in particular describes the FRP reinforcing element as a thin-walled beam of arbitrary shape connected to the concrete core by continuous distributions of interfacial elastic springs. As a matter of fact, such a structural element is commonly thin and may be applied both on the bottom and on the web faces of the concrete beam. Due to the weak shear stiffness of Fiber Reinforced Plastics, the given model accurately describes the shear deformation of the reinforced beam, which is instead ignored by classical thin-walled beam theories. Moreover, it allows us to effectively evaluate the FRP-concrete interactions.

In future works, we intend to formulate a finite element approximation of the present mechanical theory, in order to study a large survey of concrete structures which need to be reinforced with Fiber Reinforced Plastics. Particular attention will be focused on the evaluation of the state of stress at the interface between the concrete core and the FRP wrapping.

References

[1] *S. S. Faza, H. V. S. Rhao , E. J. Barbero*, FIBER COMPOSITE WRAP FOR REHABILITATION OF CONCRETE STRUCTURES - Repair and Rehabilitation of the Infrastructure of the Americas, H. A. Toutanji Editor, The National Science Foundation, 181-192 1994.

[2] *A. M. Erki, P. J. Heffernan*, REINFORCED CONCRETE SLABS EXTERNALLY STRENGTHENED WITH FIBRE-REINFORCED PLASTIC MATERIALS - Second Int. Symposium on Non-Metallic (FRP) Reinforcements for Concrete Structures, L. Taerwe Editor, Proc. RILEM 29, Ghent, Belgium, 509-516 1995.

[3] *M. Arduini, A. Di Tommaso, O. Manfroni, A. Nanni*, FAILURE MECHANISMS OF CONCRETE BEAMS REINFORCED WITH FRP FLEXIBLE SHEETS - Second Int. Conference on Advanced Composite Materials in Bridges and Structures, Montréal, Québec, Canada, 253-260 1996.

[4] *R. N. Swamy, C. J. Lynsdale, P. Mukhopadhyaya*, EFFECTIVE STRENGTHENING WITH DUCTILITY: USE OF EXTERNALLY BONDED PLATES OF NON-METALLIC COMPOSITE MATERIALS - Second Int. Conference on Advanced Composite Materials in Bridges and Structures, Montréal, Canada, 481-488, 1996.

[5] *A. M. Malek, H. Saadatmanesh, M. R. Ehsani*, SHEAR AND NORMAL STRESS CONCENTRATIONS IN RC BEAMS STRENGTHENED WITH FRP PLATES - Second Int. Conference on Advanced Composite Materials in Bridges and Structures, Montréal, Canada, 629-638, 1996.

[6] *L. Ascione , F. Fraternali, L. Feo*, THE WRAPPING OF REINFORCED CONCRETE BEAMS WITH FRP PLATES: A MECHANICAL MODEL - Proc. Int. Meeting "Advancing with Composites '97", Milan, Italy, 155-170 1997.

[7] *G. D'Agostino, D. Tartaglione*, THE MECHANICAL BEHAVIOUR OF REINFORCED CONCRETE BEAMS STRENGTHENED BY WRAPPING WITH FRP PLATES: A FINITE ELEMENT APPROACH - Proc. Int. Meeting "Advancing with Composites '97", Milan, Italy, 87-98 1997.

[8] *V. Z. Vlasov*, THIN-WALLED ELASTIC BEAMS - Pergamon Press, New York 1961.

Shear Behaviour of Concrete Beams Reinforced with FRP Wraps

Antonio La Tegola*, Giovanni Noviello*

* Department of Materials Science, Faculty of Engineering, University of Lecce, 73100 Lecce, Italy

Abstract

In recent years, due to the development of advanced composite materials, the wrapping technique with *Fibre Reinforced Polymers* sheets has assumed an important role in the rehabilitation of damaged or under-reinforced concrete structures.
The evaluation of ultimate shear resistance of concrete elements reinforced with external wraps cannot be based on the same provisions and limit analysis equations valid for conventionally reinforced concrete beams. In fact, the particular mechanical behaviour of reinforcing fibres alters the contributions of the conventional shear resistant mechanisms at the u.l.s. as shown by some experimental investigations.
The aim of this paper is to define a formulation in order to understand the shear behaviour of concrete beams reinforced with FRP wraps.

1. Introduction

The shear strength of concrete elements can be increased, such as flexural one, using FRP wraps epoxy-bonded to concrete surface. Experimental investigations on externally wrapped beams evidenced a significant increase in ultimate load-carrying capacity, within the range from 60 to 150 percent respect to the same beams non reinforced in shear [2].
To this scope, reinforcement materials in the form of sheets or plates bonded with epoxy resins are adopted. In Tab. 1 the mechanical characteristics of some commercial fibres are presented. Useful guidelines for the design and installation of composite materials for the strengthening of concrete structures are presented in [1].
It seems necessary to vary the theoretical modelling of shear resistant mechanisms that are mobilised at the u.l.s. in order to consider the particular behaviour and shape of composite materials. The fibres follow a linear stress-strain law until failure and are diffused in the longitudinal and transverse direction over the surface of the element.
It should be noted that, due to the high tensile capacity of the fibres, it is possible to find two more failure mechanisms:

- Adhesive failure at the concrete-composite interface;
- Shear failure in concrete when the shear stress τ reaches τ_u.

It is very difficult to define an analytical model to predict the shear failure of a beam because it is not possible to separate shear failure from the flexural one; on the other hand, it is particularly important to define, with an high degree of accuracy, the shear strength of a beam reinforced with FRP sheets. In fact the shear failure, is particularly dangerous because is brittle and reduces the ductility of structural elements.

Fibre	Average Diameter (μ)	Unit Weight	Young's Modulus GPa	Ultimate Stress GPa	Ultimate Strain a (%)	Coefficient of Thermal Expansion 10^{-6}/°C	Poisson's Ratio
Kevlar™49	11.9	1.45	131	3.62	2.8	-2.0 (long.) +59 (radial)	0.35
Twaron™ 1055	12.0	1.45	127	3.60	2.5	-2.0 (long.) +59 (radial)	0.35
Technora	12.0	1.39	74	3.50	4.6	-3.0 (long.) N/A (radial)	0.35
PAN-Carbon							
T-300	7	1.76	231	3.65	1.4	-0.1+ -0.4 (long.) +7+ 12 (radial)	-0.2
As	7	1.77	220	5.65	1.2	-0.5+ -0.4 (long.) +7+ 12 (radial)	
t-40	6	1.81	276	5.65	2		
HSB	7	1.80	344.5	2.34	0.58		
Fortafil 3™	7	1.80	227	3.80	1.7	-0.1	
Fortafil 5™	7	1.80	345	2.76	0.8		
PITCH-Carbon							
P-555	10	2.00	380	1.90	0.5	-0.9 (long.)	
P-100	10	2.16	758	2.41	0.32	-1.6 (long.)	

Tab. 1. Mechanical characteristics of some commercial fibres

2. Analytical Formulation

Let us consider a concrete beam strengthened with a two laminae externally bonded composite laminate, with fibres oriented at 0/90° respect to the longitudinal direction of the beam.
These following assumptions are adopted:

- the longitudinal reinforcement carries only the flexural tension;
- the tension stresses due to shear act only on the laminate.

Defining θ the average angle of compression concrete struts, let us consider a longitudinal cross section of a beam element of length Δz:
The resultant force of shear stresses is:

$$S_1 = \tau\, b\, \Delta z$$

and being:

$$\tau b = \frac{V}{h^*}$$

where h* is the internal moment lever arm, we obtain:

$$S_1 = \frac{V\, \Delta z}{h^*}$$

This force is balanced by a compression strut in the concrete and a tension force in the vertical fibres of the laminate:

$$S_v = S_1 \ tan\,\theta = \frac{V \ \Delta z}{h^*} \ tan\,\theta$$

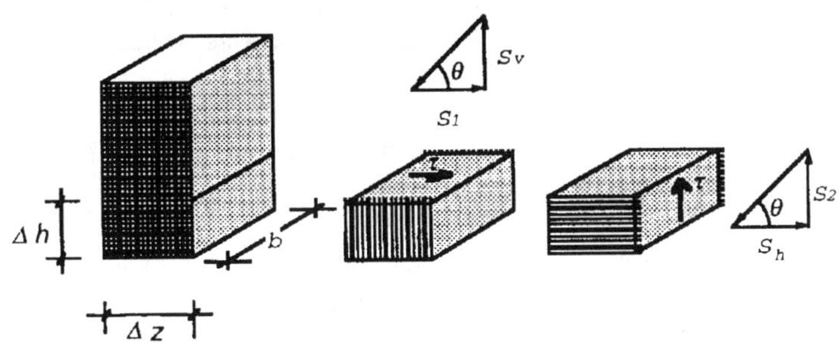

Figure 1. Beam elements considered for equilibrium

The tension force per unit length in the vertical fibres is then:

$$s_v = \frac{S_v}{\Delta z} = \frac{V}{h^*} \ tan\,\theta$$

This tension force per unit length is equal to the product between tension stress and area per unit length of vertical fibres:

$$s_v = \sigma_v \ \omega_v$$

where ωV is the sectional area per unit length of vertical fibres, thus we obtain:

$$\sigma_v = \frac{V}{h^* \ \omega_v} \ tan\,\theta \qquad (1)$$

Let us consider a transverse cross section of a beam element of height Δh.
The resultant force of shear stresses is:

$$S_2 = \tau \, b \, \Delta h = \frac{V \ \Delta h}{h^*}$$

This force is balanced by a compression strut in the concrete and a tension force in the horizontal fibres of the laminate:

$$S_h = S_2 \ cot\,\theta = \frac{V \ \Delta h}{h^*} \ cot\,\theta$$

The tension force per unit length in the horizontal fibres is then:

$$s_h = \frac{S_h}{\Delta h} = \frac{V}{h^*} \ cot\,\theta$$

and being:

$$s_h = \sigma_h \ \omega_h$$

where ω_h is the sectional area per unit length of horizontal fibres, we obtain:

$$\sigma_h = \frac{V}{h^* \ \omega_h} \ cot\,\theta \qquad (2)$$

From equations (1) and (2) it is possible to obtain the ultimate shear related the failure of horizontal and vertical fibres in the laminate:

$$V_{1r} = \sigma_{vu} h^* \omega_v \, cot\,\theta \qquad (3)$$

$$V_{2r} = \sigma_{hu} h^* \omega_h \, tan\,\theta \qquad (4)$$

The collapse of the beam can be caused also by the two above mentioned mechanisms: 1) Adhesive failure at the concrete-composite interface; 2) Shear failure in concrete when the shear stress τ reaches τ_u. It is then possible to calculate two allowable shear stresses which define the failure of the beam. In this case the ultimate shear strength can be written in the following form:

$$V_{int} = \tau_u h^* s_r \qquad (5)$$

where τ_u is the smaller of the two allowable shear stresses, and s_r is the perimeter of the cross section of the beam wrapped by the laminate.
From equations (3),(4) and (5) we have three values for the ultimate shear strength of the beam. The lower value has to be assumed in the design and verification of the beam.
It should be noted that, in the case of equal distribution of fibres both in longitudinal and transverse directions, the choice of an angle $\theta = 45°$ carries to the optimum degree of shear reinforcement. The θ angle of the concrete compression struts can be assumed as a design parameter to find the best distribution of fibres on the web of the beam.
The ultimate value of shear strength of the beam calculated with this model is based on the assumption that the shear stresses are carried only by the fibres without any contribution of the conventional concrete mechanisms. This assumption is considered on the safe side because the different mechanical behaviour of composites compared to steel, such as perfectly elastic-brittle constitutive law or low value of ultimate strain, does not allow the conventional concrete mechanisms to be mobilised.

3. Comparisons

With reference to the experimental tests reported in [2] on simply supported concrete beams strengthened in shear only with FRP fabrics bonded to the concrete surface, in Tab. 2 are presented the results of the tests compared with the theoretical values V_r and V_{int}.

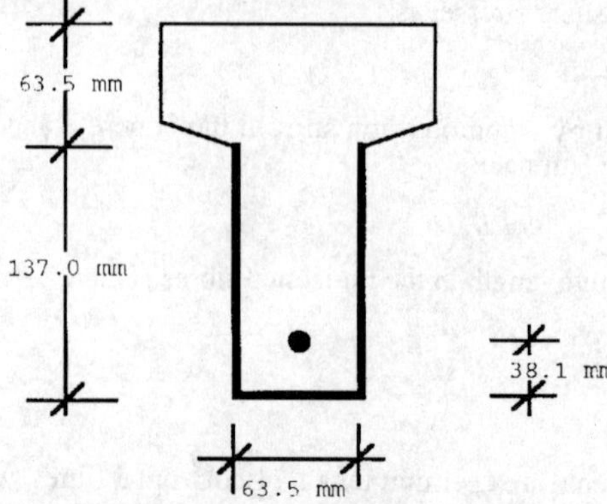

Figure 2. Typical cross section of beams tested in [2]

V_{int} has been calculated assuming τ_u equal to the ultimate shear stress of the concrete adopting the value given in EuroCode 2:

$$\tau_u = 0.25 f_{ctk}$$

where f_{ctk} is the characteristic tensile strength of the concrete.
The value of shear stress adopted is lower than the one which causes the adhesive failure at the concrete-composite interface as given by other authors.

	ω_h (mm²/mm)	ω_v (mm²/mm)	σ_u (N/mm²)	V_{exp} (N)	V_r (N)	V_{int} (N)
Aramid	2x1.04	2x1.04	223.7	34387	67400	31988
E-Glass	2x0.45	2x0.45	171.6	35366	22400	31161
Graphite	2x0.58	2x0.58	185.8	35778	31440	31344

Tab. 2. Theoretical and experimental shear strength.

It is important to note that for graphite fabrics reinforcement, the difference between theoretical and experimental data is in the order of 12 percent, thus justifying that, due to the brittle failure of the graphite fibres, the conventional shear resistant mechanisms were not mobilised.
For E-Glass reinforcement the difference is in the order of 37 percent because the ultimate strain is higher than graphite, so some shear resistant mechanisms have been activated.
As regards the design value of shear strength, according to EuroCode 2, it is necessary to introduce the partial factors γ on the material resistances. Thus the design values of shear strength are:

$$V_{1rd} = \frac{\sigma_{vu} h^* \omega_v \cot\theta}{\gamma_r} \quad (6)$$

$$V_{2rd} = \frac{\sigma_{hu} h^* \omega_h \tan\theta}{\gamma_r} \quad (7)$$

$$V_{intd} = \frac{\tau_u h^* s_r}{\gamma_c} \quad (8)$$

The partial factors γ_r assume values $\gamma_r >> 1$ because they have to take in account the damage of the fibres during placement, the uncertainties in the determination of ultimate strain of fibre and matrix and the statycal fatigue phenomenon. Acceptable values are $\gamma_r = 1.5 \div 2.5$
The value of γ_c, following EuroCodes, is $\gamma_c = 1.5$.

4. Conclusions

In the previous formulations two relations have been found using only equilibrium equations without considering compatibility conditions; these relations correlates the stresses in horizontal and vertical fibres to the shear force. These relations are obtained only by equilibrium equations without using compatibility conditions. It is possible to hypothesise that the compatibility is satisfied only when *plastic adjustments* are allowed at the collapse. In this case the above assumption is no longer valid, because the behaviour of the FRP sheets is elastic-brittle until failure thus not allowing any plastic adjustment.
Another consideration can be made: the theoretically calculated horizontal and vertical tensions are assigned only to the fibres, but, really, the entire laminate (fibres plus matrix) must be considered in order to correctly evaluate them.
It should be more suitable to adopt a resistance criterion which considers the global behaviour of the laminate.

It should be noted that the shear resistant mechanisms such as aggregate interlock and dowel action cannot be taken into account because, in order to be activated, need strains higher than the allowable ones by reinforcing fibres.

Acknowledgements

The Authors wish to aknowledge the Italian Ministry of University Research (MURST) and the Italian CNR for the financial support.

References

[1] *ACI 440-F,* GUIDELINES FOR SELECTION, DESIGN AND INSTALLATION OF FRP SYSTEMS FOR EXTERNALLY STRENGTHENING CONCRETE STRUCTURES - ACI\440-F\440-01B rpt draft of 17/10/1997.

[2] *Chajes M., Januszka T., Mertz D., Thomson T. Jr., Finch W. Jr.,* SHEAR STRENGTHENING OF REINFORCED CONCRETE BEAMS USING EXTERNALLY APPLIED COMPOSITE FABRICS - ACI Structural Journal, V. 92, n.3, 1995.

[3] *La Tegola Ant. La Tegola Alb.,* ACTUAL APPLICATION OF WRAPPING PROCEDURES IN FRP REINFORCEMENT OF CONCRETE STRUCTURES - ACI Field Applications of FRP Reinforcement to Concrete: Parts I and II, Atlanta Georgia, November 9, 1997.

[4] *Arduini M., Nanni A., Di Tommaso A., Focacci F.,* SHEAR RESPONSE OF CONTINUOUS RC BEAMS STRENGTHENED WITH CARBON FRP SHEETS - Third International Symposium on Non-Metallic (FRP) Reinforcement for Concrete Structures, Sapporo, Japan, oct. 1997.

[5] *Collins M., Mitchell D., Adebar P., Vecchio F.,* A GENERAL DESIGN METHOD - ACI Structural Journal, V. 93, n. 1, 1996.

Safety Philosophy for FRP RC Structures

Kyriacos Neocleous[1] and Kypros Pilakoutas[2]

Abstract

This paper examines aspects of safety philosophy that could be incorporated in the design guidelines for concrete structures (RC) reinforced with fibre reinforced polymers (FRP). The majority of current guidelines, though based on modifications of existing codes of practice for steel reinforced concrete structures, do not have an identifiable design and safety philosophy. When dealing with FRP reinforcement, the engineer should be able to prioritise the safety levels of the various modes of failure in order to achieve the desired structural reliability. This prioritisation should be based on an overall safety philosophy. An attempt is made here to establish the overall safety (reliability) level of BS8110. An outline of a new safety philosophy for RC structures is proposed.

Keywords

Safety philosophy; reinforced concrete; partial safety factors; fibre reinforced polymers (FRP), structural reliability.

1. Introduction

New types of corrosion proof reinforcement, such as FRP re-bars, are used by the construction industry when dealing with concrete structures in aggressive environments. Extensive use of any type of reinforcement requires the development of design codes, a process that can take decades to be completed. In the absence of appropriate design codes, professional bodies, or the relevant manufacturing industries, normally provide design guidelines. Currently, there are four main groups of design guidelines: a) from Europe (Clarke J. L. et al., 1996), b) Japan (JMC, 1995), c) America (ACI 440R-96, 1996) and d) Canada (CHBDC, 1996). FIB [TG 3-10] and the European Commission sponsored TMR network "ConFibreCrete" are also working towards the development of such design guidelines.

These design guidelines are mainly provided in the form of modifications to existing codes of practice for concrete construction. The modifications are heavily influenced by the unconventional mechanical properties of FRP reinforcement and incorporate experience from experimental work on concrete elements reinforced with FRP. The brittle linear-elastic behaviour of FRP reinforcement is dominating all design guidelines, and serviceability criteria are applied as rigorously as for steel RC.

In the case of the design guidelines by Clarke et al (1996), the recommendations conform to European codes of practice, such as BS8110 (1997) and ENV 1992-1-1 Eurocode 2 (1992). This approach may seem reasonable, but it may not be entirely appropriate, since the above codes of practice do not have an identifiable design and safety philosophy, as explained in the following. The reliability levels adopted by most codes of practice are: a) not known, and b) not uniform. For the former, codes of practice do not provide information as to how to calculate reliability levels although they are implicitly defined by the application of partial safety factors for loading and materials. In the case of the latter, it is always assumed that for ultimate limit state conditions, a ductile failure will occur due to flexural yielding. The level of over-strength with respect to other modes of failure is neither defined nor can be easily determined. This means that the safety of structural elements will differ according to the level of over-strength and mode of failure.

The real problem to change is the acceptance by the majority of structural engineers that the safety of RC structures is well ensured by the application of the current partial safety factors. However, an initial investigation into the development of partial safety factors has shown that the adopted values of partial safety factors are based on semi-empirical rather than semi-probabilistic methods (Neocleous K, 1997), and do not necessarily lead to economic solutions.

It is evident that the design and safety philosophy of FRP RC structures must be established first in order to assist the development of any future design guidelines and codes of practice. In addition, the new philosophy must be able to accommodate new materials and even new construction techniques. The properties of FRP reinforcement are continuously being developed, and making rigid assumptions will only mean obstacles to future innovation.

In this paper, an initial attempt is being made to establish the overall safety levels adopted in current concrete codes of practice. This is achieved by using the theory of structural reliability for which a brief introduction is given in the following section. Then a new design and safety philosophy for FRP RC structures is discussed.

2. Methodology

In general, structural reliability assessment involves the calculation of the probability of limit state violation, that is the probability of failure P_f (Equation 1). The assessment requires the formulation of an assessment model that represents the structural behaviour for the failure mode for which the assessment is performed.

$$P_f = P(G(R(R_1,R_2...R_i),\ S(S_1,S_2...S_i)) \leq 0) \quad (1)$$

The assessment model is formulated by representing the limit state function, $G(R,S)$, in terms of structural resistance, R, and action-effect, S. Both R and S are modelled by mathematical relationships of random basic variables, R_i and S_i, which represent material properties and actions, respectively. The basic variables are described by their probability density functions to take into account the uncertainties associated in structural design and construction, such as strength variations and dimensional imperfections (Melchers R. E., 1987).

Theoretically, the probability of failure is obtained by solving the resulting integral of the joint probability density function $f_{S,R}(S,R)$ (Equation 2). However, it is not always possible to solve this integral, and thus R and S are assumed independent in order to transform the integral into a convolution integral (Figure 1, Equation 3). The convolution integral can be solved analytically for a few probability distributions, such as for the normal distribution, by applying the concept of the safety index β. In this case, the safety margin, Z, which exists between the R and the S components of the limit state function, is initially evaluated by applying the rule for subtraction of normal random variables (Equation 4). Then, the probability of failure, represented by the shaded area in Figure 2, is estimated by applying the mean and standard deviation of the safety margin into equation 5 (Melchers R. E., 1987).

$$P_f = \iint_{[S>R]} f_{S,R}(S,R)\ dS\ \ dR = \int_0^{\infty} dS \int_0^{S} f_{S,R}(S,R)\ \ dR \tag{2}$$

$$Pf = \int_0^{\infty} dS \int_0^{S} f_R(R) f_S(S)\ \ dR \tag{3}$$

$$Z = R(x_1, x_2 \ldots x_i)\ -\ S(x_1, x_2 \ldots x_i) \tag{4}$$

$$P_f = P(Z \le 0) = \Phi\left(\frac{0-(\mu_R - \mu_S)}{\left(\sigma_R^2 + \sigma_S^2\right)^{0.5}}\right) = \Phi\left(\frac{0-\mu_Z}{\sigma_Z}\right) = \Phi(-\beta) \tag{5}$$

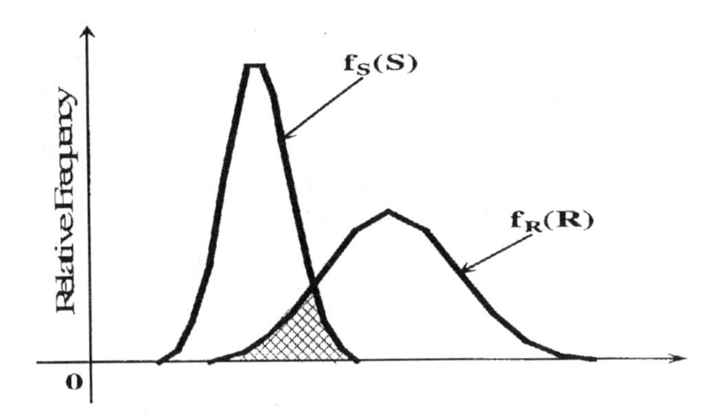

Figure 1. Convolution Integral

In the current study, the structural reliability assessment is carried out by applying a modified "crude" Monte Carlo (MC) simulation method. The basic variables of the limit state function are generated pseudo-randomly according to their probability distributions. It is assumed that all basic variables have truncated normal distributions. In the "crude" MC method, the probability of failure is obtained by dividing the number of failures by the total number of simulations, N, (Equation 6). It is evident that the accuracy of the results is influenced by the number of simulations. In order to improve the accuracy of the results, the "crude" MC method is modified by applying the concept of the safety index β (Equation 5) for the assessment of the probability of failure (Hoogenboom P. C. J., 1996).

$$P_f = \frac{\sum G(R,\ S) \le 0}{N} \tag{6}$$

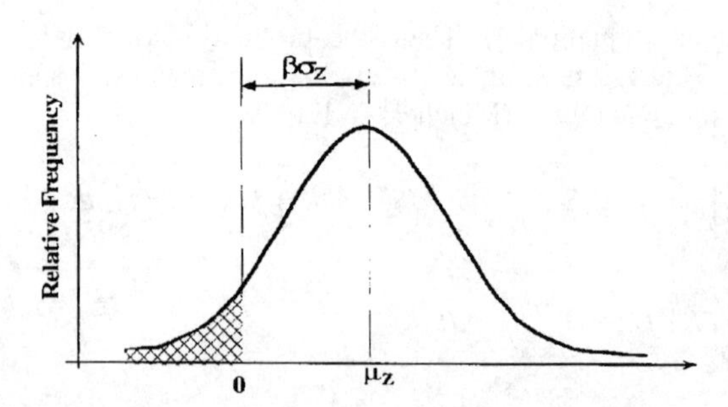

Figure 2. Safety Margin Z of the limit state function $G(R,S)$

3. Description of Experimental and Statistical Data

3.1 Experimental Data

The experimental data, which is used in this study, was taken from work for the Eurocrete project undertaken by the University of Sheffield (Pilakoutas K. et al, 1997). In this paper, two RC beams with steel reinforcement are considered. The dimensions and reinforcement details of these beams are shown in Figure 3. The concrete cover to main reinforcement was 30 mm for both beams. Beam A failed in flexure at a load of 155 kN. The beam exhibited ductile behaviour with concrete crushing after substantial steel yielding. In the case of beam B, a brittle shear failure occurred at a load of 100 kN due to the absence of any shear reinforcement.

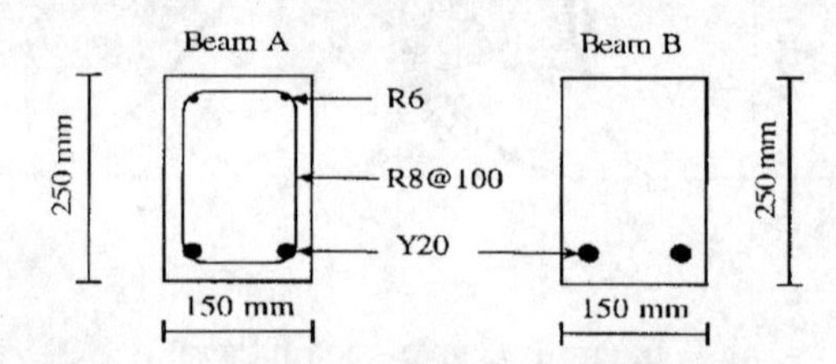

Figure 3. Dimensional and Reinforcement Details of Beams A and B

The test arrangement of both beams is shown in Figure 4. The load was applied to the beam as two point-loads.

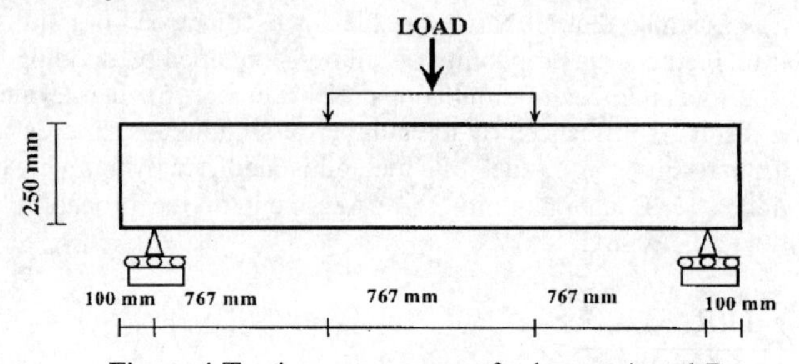

Figure 4.Testing arrangement for beams A and B

3.2 Statistical Data

Statistical data for the dimensional and strength variations, which are required to carry out the structural reliability analysis, were obtained from existing literature. Data from a study by Mirza S. A. et al (1979) was used for the dimensional imperfections of reinforced concrete elements (Table 1). The statistical variation is given in the form of standard deviation. The main problem with this set of data is that some of the values are appropriate for beams of larger dimensions than the ones used in the present study. For example, the standard deviation for the overall depth is given for beams for which the overall depth ranges from 450 to 685 mm. However, it is assumed that these standard deviations can be used in this study as well since the objective is to present the principles of structural reliability assessment. The lower end of the probability distributions for the dimensions were truncated according to the tolerances provided by the CEB-FIB Model Code (1990), (Table 2).

Statistical Variation	Width	Overall Depth	Bottom Reinf. Cover	Effective Depth	Beam Span & Spacing
Nominal Range	280-305	450-685	19-25.4	-	-
Standard Deviation	4.8	6.4	11.13	12.7	17.5

Note: All dimensions in mm

Table 1. Dimensional Variations of In-Situ Beams (After Mirza et al, 1979)

Dimension of Structural Element	Tolerances $\Delta\alpha = (\alpha_{nom} - \alpha_{act})$
$\alpha \leq 200$	$\lvert\Delta\alpha\rvert < 5$
$200 < \alpha \leq 2000$	$\lvert\Delta\alpha\rvert < 3.5 + 0.008\alpha$
$2000 < \alpha$	$\lvert\Delta\alpha\rvert < 17.5 + 0.001\alpha$

Note: All dimensions in mm

Table 2. Tolerances for Concrete Elements (After CEB-FIB Model Code 1990)

The statistical properties of the steel reinforcement, shown in Table 3, are based on data analysed by Mirza S. A. et al (1979a). The majority of this data is provided for steel reinforcement with minimum yield strength of 410 N/mm^2. Despite the fact that steel reinforcement of higher yield strength was used in both tested beams, the same standard deviation is adopted. Similarly, the standard deviation recommended for grade 40 steel was used for the mild steel was used for the shear links.

Statistical Property	Strength of High Yield Steel N/mm^2	Strength of Mild Yield Steel N/mm^2	Bar Diameter ϕ (mm)
Mean	533	309	$0.99*\phi_{nominal}$
Standard Deviation	45	36	$0.024*\phi_{nominal}$
Truncated	-	-	$0.94*\phi_{nominal}$

Table 3. Statistical Properties of Steel Reinforcement (After Mirza et. al. 1979a)

In the case of the concrete compressive strength, a standard deviation of 6 N/mm^2 is adopted. This value is based on a limit recommended by Neville A.M.

(1995). Neville states that for the British mix method, in the absence of any experimental data, the mean value should be 10 MPa greater than the minimum specified value. Thus, this limit is translated into standard deviations by assuming that the minimum strength corresponds to the characteristic strength (Equation 8).

$$\text{Standard Deviation s} = \frac{fcu_m - fcu_k}{1.64} = \frac{10}{1.64} = 6.1 \text{ N/mm}^2 \qquad (7)$$

4. Discussion of the Results for the Reliability Assessment of RC Beams A and B

The structural reliability assessment of beams A and B was performed by using a computer program developed on MATLAB (1992). The basic variables were pseudo-randomly generated by using MATLAB's random-number generator, and the number of simulations was 10^6. Both beams were assessed for flexural and shear failure. The assessment models were formulated according to the design procedure outlined by BS8110 (1997). The values of the partial safety factors were set to unit to obtain the distribution (f_{PR}) of the *predicted* resistance for each beam. The results of the assessment are shown in Figure 5 including the values obtained experimentally. The terms *Predicted Flexure* and *Predicted Shear* refer to the resistance capacity obtained by the structural reliability assessment based on the BS8110. *Design Flexure* and *Design Shear* refer to the design resistance capacity.

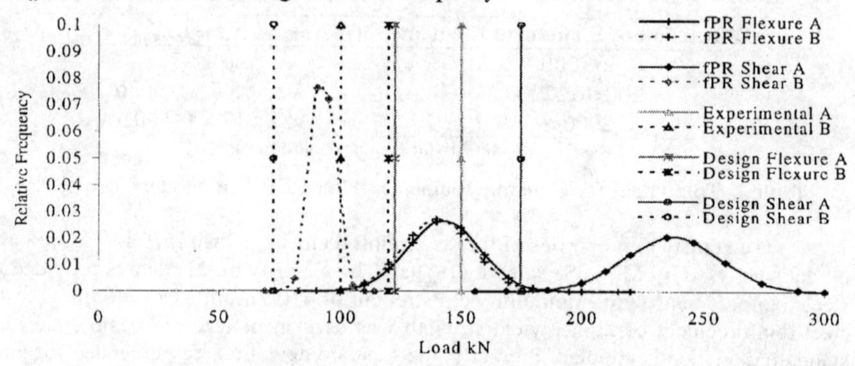

Figure 5. Results of structural reliability assessment for beams A and B

It is evident that the assessment models are reliable, since the values of the experimental loads for both beams are greater than the mean values of the f_{PR}. Ideally, the value of the experimental load should be equal to the mean value of f_{PR}. In addition, it can be seen that the design values lie within the lower end of the f_{PR} distribution. The ratio between the design values of shear to flexural resistance is 1.43, whilst this ratio is even higher if the mean predicted values are used, at 1.58.

The probabilities of design load exceeding f_{PR} for both beams, for each failure mode, are shown in Table 4. The first probability value for beam A adopts the recent modification to γ_{mrein} and the characteristic strength assumed by the code. This probability is an order of magnitude higher than the one calculated using the true characteristic yield strength of British reinforcement. Hence, it can be concluded that the overall safety of flexural design is very much dependant on the steel characteristics and a tight specification of properties should be included in design standards.

Failure Mode	Beam A				Beam B
	fy_k=460 γ_{mrein}=1.05	fy_k=460 γ_{mrein}=1.15	fy_k=490 γ_{mrein}=1.05	fy_k=490 γ_{mrein}=1.15	fy_k=460 γ_{mrein}=1.05
f_{PR} Flexure	8.5E-02	3.2E-02	9.2E-03	5.0-03	(2.68E-02)
f_{PR} Shear	-	-	-	-	5.1E-04

Table 4. Probability of design load exceeding f_{PR} for beams A and B

The probability of the design load exceeding the expected shear capacity for beam B is relatively low. This can be attributed to the brittle nature of this mode, but it can also be seen as an attempt by the code to prioritise the mode of failures. One of the main methods that codes use to achieve the above is through the use of material partial safety factors. However, when shear reinforcement is provided, the overall safety in terms of shear resistance appears to be lowered, as shown for beam A. This can be attributed to the large deviation in the distribution of the yield strength of reinforcement.

The above reduction in overall safety of shear resistance needs further investigation, since it will have a consequence both on the overall element safety as well as on the mode of failure.

It is obvious from the above that current codes are not explicit in their safety philosophy and many aspects of safety need further investigation. In the opinion of the authors, a clear safety philosophy should be declared by the codes. In such a philosophy, a hierarchy of failure modes should be established depending on their consequence of failure. In addition, the overstrength provided to the undesirable modes of failure should be based on a risk assessment exercise.

For most practical cases, a hierarchy of failure modes could be achieved by sets of partial material safety factors that vary according to the situation. Hence, this approach will facilitate the use of new materials and techniques without the need for development of new codes, as in the case now with FRP reinforcement.

5. Conclusions

The safety philosophy of current design guidelines for RC structures has been reviewed. It is concluded that no explicit way of determining safety is provided and new materials can not be used without modifying the codes.

Modification of the exisitng codes, though reasonable as a first step, is not considered the most prudent way forward.

It is proposed that a safety philosophy should be developed based on a prioritisation of the different modes of failure. The desirable mode of failure could be achieved by adopting material partial safety factors suitable for the situation. The overstrength to be designed for the undesirable modes of failure should be appropriate to the accepted level of risk.

Acknowledgements

The authors wish to acknowledge the Overseas Research Student Award of the UK government and the European Commission sponsored TMR Network "ConFibreCrete".

References

ACI 440R-96, "State-of-the-Art Report on Fiber Reinforced Plastic Reinforcement for Concrete Structures", ACI Committee 440, American Concrete Institute, Detroit, Michigan, February, 1996.

BS8110, "Structural use of concrete, Part 1. Code of Practice for Design and Construction", British Standards Institution, 1997.

CEB - FIP Model Code, "Clause 1.4.5 Geometrical Quantities",1990.

CHBDC- Canadian Highway Bridge Design Code, "Section 16: Fibre Reinforced Structures", Final Draft, July, 1996.

Clarke J. L., O' Regan D. P., Thirugnanendran C., "Eurocrete Project, Modification of Design Rules to Incorporate Non-ferrous Reinforcement", Eurocrete Project, Sir William Halcrow & Partners Ltd, 1996.

ENV 1992-1-1 Eurocode 2, "Design of Concrete Structures, Part 1-6:General rules and rules for buildings", European Committee for Standardisation, 1992.

Hoogenboom P. C. J., "Computation of Safety – A Literature Study", Report 03.21.0.22.06, Group of Mechanics & Structures, Faculty of Civil Engineering, TUDelft, July 1996.

JMC, "Guidelines for Structural Design of Reinforced Concrete Buildings Structures", Ed. by Building Research Institute, Japanese Ministry of Construction, July, 1995.

MATLAB, "High-Performance Numeric Computation and Visualization Software: Reference Guide'', The Math Works Inc., August 1992.

Melchers R. E., "Structural Reliability- Analysis and Prediction: Chapter 1- Measures of Structural Reliability'', Pub. Ellis Horwood Limited, 1987, 0-85312-930-4.

Mirza S. A., MacGregor J, "Variations in Dimensions of Reinforced Concrete Members", Journal of the Structural Division, Proceedings of the American Society of Civil Engineers, Vol. 105, No. ST4, April 1979, pp 751-766.

Mirza S. A., MacGregor J, "Variability of Mechanical Properties of Reinforcing Bars", Journal of the Structural Division, Proceedings of the American Society of Civil Engineers, Vol. 105, No. ST5, May 1979a, pp 921-937.

Neocleous K, "Design philosophy of Concrete Structures", Report in progress, Centre for Cement and concrete, Department of Civil and Structural engineering, The University of Sheffield.

Neville A. M., "Properties of Concrete: Chapter 14 – Selection of Concrete Mix Design Proportions (Mix Design)", Pub. Longman Group Limited, 1995, 0-582-23070-5.

Pilakoutas K., Achillides Z., Waldron P., "Non-ferrous reinforcement in concrete structures", Innovation in Composite Materials and Structures, Cambridge, August 1997, pp 47-56.

Numerical assessment of finite element formulations for laminated composite plates

G. Alfano[1], F. Auricchio[2], L. Rosati[1], E. Sacco[3]

[1] - Dipartimento di Scienza delle Costruzioni, Università di Napoli "Federico II" (Italy)

[2] - Dipartimento di Ingegneria Civile, Università di Roma "Tor Vergata" (Italy)

[3] - Dipartimento di Ingegneria Industriale, Università di Cassino (Italy)

Abstract

A family of laminate finite element models have been derived by different formulations based on the Reissner/Mindlin plate model. Extensive numerical simulations have shown that the proposed elements are locking free and exhibit a satisfactory convergence rate. The shear stresses in the thickness have to be evaluated by the integration of the three-dimensional equilibrium equations and an optimal accuracy is obtained if the in-plane strain and the curvature fields are preliminarily regularized by means of a standard nodal projection procedure.

1 Introduction

The fiber reinforced composite materials are characterized by very high longitudinal modulus and low values for the shear and the transversal moduli. Consequently, for flat structures the shear deformation in the thickness appears to be not negligible. On the other hand, laminates present unusual behaviors due to the coupling between extension and bending. Furthermore, the determination of accurate values for the interlaminar normal and shear stresses is very important, since they are responsible for the delamination.

Several laminate theories have been proposed, which consider shear deformation in the thickness and account for the bending-extension coupling [8].

The First-order Shear Deformation Theory (FSDT), presented in [3], is based on the Reissner-Mindlin plate model [1, 2] and found great interest because it can account for the shear deformation in a simple way. Furthermore, the FSDT gives satisfactory results for a wide class of structural problems, even for moderately thick laminates. On the other hand, within this theory, it is particularly difficult the recovery of the interlaminar stresses, which have to be evaluated by integrating the equilibrium equations through the thickness.

In the present work, several laminate finite element based on the FSDT are proposed. They are based on a displacement or on a mixed formulation, are locking free, have not zero energy modes and are able to provide accurate in-plane deformations. Next, the 4- and 9-node full displacement elements are reviewed [5]. Then, the enhanced mixed linked 4-node element is presented [13]. Finally, two 4- and 9-node laminate elements are derived from the MITC family (Mixed Interpolation of Tensorial Components) developed for plate and shells [4, 11].

Numerical results are reported in order to assess the different element performances. Results are compared with the analytical FSDT solution.

2 First-order laminate theory (FSDT)

2.1 The FSDT model

A laminate Ω is considered:

$$\Omega = \left\{(x_1, x_2, z) \in \mathcal{R}^3 \quad / \quad z \in (-h, h) \qquad (x_1, x_2) \in \mathcal{A} \subset \mathcal{R}^2\right\} \tag{1}$$

where $2h$ is the constant thickness and $\mathcal{A}$ is the mid-plane of the undeformed plate identified by the plane $z = 0$. The laminate is made of n layers and the typical kth layer lies between the thickness coordinates z_{k-1} and z_k.

The First-order Shear Deformation Theory of a laminated plate (FSDT) is based on the well-known assumptions, on both the strain and the stress fields [12], that the transverse stress in the thickness of the plate is nil and that straight lines orthogonal to the midplane are inextensible and remain straight after deformation. From now on, Greek indices assume values $1, 2$, while Latin indices assume values $1, 2, 3$. Furthermore, the subscript comma indicates the partial derivative $f_{,j} = \partial f / \partial x_j$ and repeated indices are understood to be summed within their range, unless explicitly stated.

In accordance with the kinematical assumptions the displacement field admits the following classical representation form:

$$\begin{array}{rcl} s_1(x_1, x_2, z) & = & u_1(x_1, x_2) + z\ \varphi_1(x_1, x_2) \\ s_2(x_1, x_2, z) & = & u_2(x_1, x_2) + z\ \varphi_2(x_1, x_2) \\ s_3(x_1, x_2, z) & = & w(x_1, x_2) \end{array} \tag{2}$$

Thus, the strain tensor can be decomposed as: $\varepsilon_{ij} = e_{ij} + z\ \kappa_{ij} + \gamma_{ij}$, where e_{ij} is the in-plane deformation, κ_{ij} is the curvature and γ_{ij} the transverse shear strain.

Taking into account the stress assumption $\sigma_{33} = 0$ and the strain decomposition the stress-strain relations for the FSDT are:

$$\sigma_{\alpha\beta}^{(k)} = \overline{C}_{\alpha\beta\gamma\delta}^{(k)}\ (e_{\gamma\delta} + z\ \kappa_{\gamma\delta}) \qquad \sigma_{\alpha 3}^{(k)} = 2\ C_{\alpha 3\gamma 3}^{(k)}\ \gamma_{\gamma 3} \tag{3}$$

where $\overline{C}$ is the reduced in-plane elastic matrix.

An accurate evaluation for the through-the-thickness shear stress can be recovered using the three-dimensional equilibrium equations [5, 10]. For simplicity, no in-plane loads per unit volume are considered in the following; thus, it results:

$$\widehat{\sigma}_{\alpha 3}^{(k)} = \widehat{\sigma}_{\alpha 3}^{(k)o} - \int_{z_{k-1}}^{z} \left(\sigma_{\alpha 1,1}^{(k)} + \sigma_{\alpha 2,2}^{(k)}\right)\ d\zeta \tag{4}$$

where k is the index relative to the generic laminate layer and:

$$\widehat{\sigma}_{\alpha 3}^{(k)o} = -\sum_{i=1}^{k-1} \int_{z_{i-1}}^{z_i} \left(\sigma_{\alpha 1,1}^{(i)} + \sigma_{\alpha 2,2}^{(i)}\right)\ d\zeta \qquad \text{with } \widehat{\sigma}_{\alpha 3}^{(1)o} = 0. \tag{5}$$

2.2 Variational principles for FSDT

In this section two variational principles for laminated composite plates are introduced. In particular, partial mixed and full displacement functionals are presented. The use of the proposed functionals allows us to deduce the FSDT governing equations and also to develope finite elements.

Initially, the partial mixed functional Π is introduced as:

$$\Pi(u_\alpha, w, \varphi_\alpha, \sigma_{\alpha 3}) = \Pi^b(u_\alpha, \varphi_\alpha) + \Pi^s(w, \varphi_\alpha, \sigma_{\alpha 3}) - \Pi_{ext} \tag{6}$$

where Π_{ext} accounts for the loading and the boundary conditions, Π^b contains the bending and the extensional terms:

$$\begin{aligned}\Pi^b(u_\alpha, \varphi_\alpha) \;=\; & \frac{1}{2}\int_A \mathcal{A}_{\alpha\beta\gamma\delta}\left(u_{\gamma,\delta}+u_{\delta,\gamma}\right)/2\ \left(u_{\alpha,\beta}+u_{\beta,\alpha}\right)/2\ dA + \\ & \int_A \mathcal{B}_{\alpha\beta\gamma\delta}\left(\varphi_{\gamma,\delta}+\varphi_{\delta,\gamma}\right)/2\ \left(u_{\alpha,\beta}+u_{\beta,\alpha}\right)/2\ dA + \\ & \frac{1}{2}\int_A \mathcal{D}_{\alpha\beta\gamma\delta}\left(\varphi_{\gamma,\delta}+\varphi_{\delta,\gamma}\right)/2\ \left(\varphi_{\alpha,\beta}+\varphi_{\beta,\alpha}\right)/2\ dA\end{aligned} \tag{7}$$

while Π^s contains the transverse shear terms:

$$\Pi^s(w, \varphi_\alpha, \sigma_{\alpha 3}) = -\frac{1}{2}\int_\Omega 4\mathcal{S}_{\alpha 3\gamma 3}\sigma_{\gamma 3}\sigma_{\alpha 3}\ dv + \int_A \left[\left(w_{,\alpha}+\varphi_\alpha\right)\int_{-h}^{h}\sigma_{\alpha 3}\ dz\right] dA \tag{8}$$

where $\mathcal{S}_{\alpha 3\gamma 3}$ represents the shear term of the compliance constitutive matrix, i.e. $2\mathcal{S}_{\alpha 3\gamma 3} = 1/(2\mathcal{C}_{\alpha 3\gamma 3})$. The fourth-order in-plane, coupling and bending elasticity matrices are respectively given by:

$$\begin{aligned}\mathcal{A}_{\alpha\beta\gamma\delta} &= \sum_{k=1}^{n}(z_k - z_{k-1})\overline{\mathcal{C}}^{(k)}_{\alpha\beta\gamma\delta} \qquad \mathcal{B}_{\alpha\beta\gamma\delta} = \frac{1}{2}\sum_{k=1}^{n}(z_k^2 - z_{k-1}^2)\overline{\mathcal{C}}^{(k)}_{\alpha\beta\gamma\delta} \\ \mathcal{D}_{\alpha\beta\gamma\delta} &= \frac{1}{3}\sum_{k=1}^{n}(z_k^3 - z_{k-1}^3)\overline{\mathcal{C}}^{(k)}_{\alpha\beta\gamma\delta}\end{aligned} \tag{9}$$

Moreover, in equation (8) the quantity $\int_{-h}^{h}\sigma_{\alpha 3}\ dz = Q_\alpha$ represents the resultant shear stress in the laminate.

The functional (6) is named partial-mixed since both the displacements u_1, u_2, w, φ_1, φ_2 and the shear stresses $\sigma_{\alpha 3}$ are considered as independent fields.

The full displacement functional can be recovered substituting into the shear energy (8) the constitutive relation (3.2), and taking into account that $2\gamma_{\alpha 3} = (w_{,\alpha} + \varphi_\alpha)$. Thus, the following functional is obtained:

$$\tilde{\Pi}(u_\alpha, w, \varphi_\alpha) = \Pi^b(u_\alpha, \varphi_\alpha) + \frac{1}{2}\int_\Omega \mathcal{C}_{\alpha 3\theta 3}\ \left(w_{,\theta}+\varphi_\theta\right)\left(w_{,\alpha}+\varphi_\alpha\right)\ dv - \Pi_{ext}. \tag{10}$$

It is a simple matter to show that the stationary conditions of Π with respect to the parameters $u_\alpha, w, \varphi_\alpha, \sigma_{\alpha 3}$ or of $\tilde{\Pi}$ with respect to the parameters $u_\alpha, w, \varphi_\alpha$ lead to the field and the boundary laminate governing equations. A form of the full displacement functional (10), useful for future developments, is obtained substituting the strain-displacement equation $\gamma_{\alpha 3} = (w_{,\alpha} + \varphi_\alpha)/2$ into the functional (10):

$$\hat{\Pi}(u_\alpha, w, \varphi_\alpha, \gamma_{\alpha 3}) = \Pi^b(u_\alpha, \varphi_\alpha) + \frac{1}{2}\int_\Omega 4\ \mathcal{C}\gamma_{\theta 3}\ \gamma_{\alpha 3}\ dv - \Pi_{ext}. \tag{11}$$

3 Finite element formulations

3.1 Q4r and Q9r elements

The full-displacement 4- and 9-node elements are derived from the functional (10) by interpolating the displacement and the rotation components with the standard bilinear and biquadratic shape functions, see [5, 9] for a full account. In order to prevent locking it is necessary to adopt a reduced integration for the shear energy, that is for the second term of (10).

3.2 EML4 element

The partial-mixed functional (6) represents a suitable basis for the development of a robust laminate finite element. In order to derive satisfactory shear stress profiles via three-dimensional equilibrium equation (4), without any post-processing of the in-plane stress field, an *ad hoc* laminate element is developed. In particular, the in-plane strains are enhanced with a field $\eta_{\alpha\beta}$ [7], so that functional (6) takes the form [13]:

$$\begin{aligned}\Pi(u^{\circ}_{\alpha}, w, \varphi_{\alpha}, Q_{\alpha}, \eta_{\alpha\beta}) \;=\;& \frac{1}{2}\int_A \mathcal{A}_{\alpha\beta\gamma\delta}\,(u_{\gamma,\delta}+u_{\delta,\gamma}+\eta_{\gamma\delta})\,/2\,(u_{\alpha,\beta}+u_{\beta,\alpha}+\eta_{\alpha\beta})\,/2\;dv\;+\\ & \int_A \mathcal{B}_{\alpha\beta\gamma\delta}\,(\varphi_{\gamma,\delta}+\varphi_{\delta,\gamma})\,/2\,(u_{\alpha,\beta}+u_{\beta,\alpha}+\eta_{\alpha\beta})\,/2\;dv\;+\\ & \frac{1}{2}\int_A \mathcal{D}_{\alpha\beta\gamma\delta}\,(\varphi_{\gamma,\delta}+\varphi_{\delta,\gamma})\,/2\,(\varphi_{\alpha,\beta}+\varphi_{\beta,\alpha})\,/2\;dv\;-\\ & \frac{1}{2}\int_A \widetilde{H}_{\alpha\gamma}\,Q_\alpha\,Q_\gamma\;dv+\int_A Q_\alpha\;(w_{,\alpha}+\varphi_\alpha)\;dv-\Pi_{ext}\end{aligned} \qquad (12)$$

where $\widetilde{H}_{\alpha\gamma}$ is the resultant overall shear compliance matrix :

$$\widetilde{H}_{\alpha\gamma}=\sum_{k=1}^{n} 4S^{(k)}_{\alpha3\gamma3}\int_{z_{k-1}}^{z_k} g_\alpha\, g_\gamma \qquad (no\ sum) \qquad (13)$$

with $g_\alpha = \sigma_{\alpha3}(x_1,x_2,z)/Q_\alpha(x_1,x_2)$ denoting the through-the-thickness shear stress shape function.

An isoparametric 4-node composite plate element can be obtained considering the standard isoparametric map [9] and discretizing functional (12) as follows:

- the in-plane displacement is bi-linear in the nodal parameters;
- the transverse displacement interpolation is bi-linear in the nodal parameters and is enriched with quadratic functions linked to the nodal rotations;
- the rotational interpolation is bi-linear in the nodal parameters and is enriched with added internal degrees of freedom associated to bubble functions;
- the shear interpolation is bi-linear and defined locally to each element;
- the enhanced strain is expressed as a function of 13 internal degrees of freedom.

Introducing the above interpolation schemes in functional (12) and performing the stationary conditions for a single element, the stiffness matrix can be obtained. Since the enhanced strain, the bubble rotation and the resultant shear stress are parameters local to each element, they can be eliminated by static condensation. Thus, an element with 5 global d.o.f. per node, named EML4 (Enhanced Mixed Linked 4-node) element, is obtained. A more detailed derivation of the EML4 element can be found in Ref. [13].

It is very important to note that the adopted mixed formulation introduces the resultant stress Q_α as a primary variable. As a consequence, an accurate evaluation of Q_α is expected.

3.3 MITC 4- and 9-node elements

The functional (11) has been used in [4, 11] to derive an efficient locking-free finite element model by expressing the transverse shear strains as interpolation of the values computed by the relation $\gamma_{\alpha 3} = (w_{,\alpha} + \phi_\alpha)/2$ at a set of m "tying" points.

Denoting by ξ_1 and ξ_2 the natural coordinates, the following vectors $\mathbf{g}_1$ and $\mathbf{g}_2$ are adopted as covariant basis:

$$\mathbf{g}_1^t = \left[\frac{\partial x_1}{\partial \xi_1} \quad \frac{\partial x_2}{\partial \xi_1} \right] = [\, J_{11} \quad J_{21} \,] \qquad \mathbf{g}_2^t = \left[\frac{\partial x_1}{\partial \xi_2} \quad \frac{\partial x_2}{\partial \xi_2} \right] = [\, J_{12} \quad J_{22} \,], \tag{14}$$

where $J_{\alpha\beta}$ denotes the $\alpha\beta$ entry in the jacobian of the isoparametric mapping.

The covariant components $\tilde{\gamma}_{\alpha 3}$ at the $k-$th tying point are then given by: $\tilde{\gamma}_{\alpha 3}^k = J_{\alpha\beta}\left(w_{,\beta}^k + \varphi_\beta^k\right)/2, \quad k = 1, 2, \ldots, m$, where the superscript k means that the related quantity has to be evaluated at the $k-$th tying point.

The polynomial shape functions $h^k = h^k(\xi_1, \xi_2)$ are then introduced so as to fulfil the relations $h^k(\xi_1^j, \xi_2^j) = \delta_{kj}$, where (ξ_1^j, ξ_2^j) denotes the coordinates of the $j-$th tying point, and the covariant components are expressed as $\tilde{\gamma}_{\alpha 3}(\xi_1, \xi_2) = h^k(\xi_1, \xi_2)\tilde{\gamma}_{\alpha 3}^k$.

The sets of tying points used to interpolate the component $\tilde{\gamma}_{13}$ are detailed in [4, 11].

3.4 Shear stress profile

It is well known that the constitutive equations $(3)_2$ cannot be used to evaluate the shear stresses in the thickness since they would provide an unacceptable layerwise constant profile. However, it is worth noting that this drawback is not a peculiar feature of the FSDT but it is shared also by more complex and computationally expensive higher order theories.

The optimal way to compute the shear stresses in the thickness is then represented by the integration of the three-dimensional equilibrium equations, that is by using eqs. $(3)_1$, (4) and (5).

Two alternatives are then possible. The former consists in substituting in eq. $(3)_1$ the values of $e_{\gamma\delta}$ and $\kappa_{\gamma\delta}$ directly obtained in terms of the element parameters and the relevant assumed interpolations. The latter requires a post-processing of the in-plane strains and of the curvatures. Actually, it is well known that their finite element solution turns out to be discontinuous at the interelement boundaries and that a better approximation can be achieved by projecting the discontinuous fields onto the subspace of polynomial functions interpolating the nodal values [9], according to a suitably defined inner product.

Numerical investigations revealed that the former procedure can be successfully employed only for the EML4 while for the other elements considered in the paper it is necessary to exploit the latter.

Moreover, in all cases, the stress profile resulting from the direct use of the equilibrium equations (4) can sometimes be not yet satisfactory, due to the fact that in the finite element scheme the equilibrium equations are not locally satisfied. However the accurate evaluation of the resultant shear Q_α leads to the possibility of properly improving the shear stress profiles, solving the following minimum problem:

$$\min \left\{ \left\| \widehat{\sigma}_{\alpha 3}^{(k)} + \left(\int_{z_{k-1}}^{z} \left(\sigma_{\alpha 1,1}^{(k)} + \sigma_{\alpha 2,2}^{(k)} \right) d\zeta - \widehat{\sigma}_{\alpha 3}^{(k)_o} \right) \right\| \right\} \tag{15}$$

subjected to the constraints: $\int_{-h}^{h} \widehat{\sigma}_{\alpha 3} dz - Q_\alpha = 0$; $\widehat{\sigma}_{\alpha 3}|_{\pm h} = 0$, where $\|\bullet\|$ represents a given norm. In particular, the shear stress $\widehat{\sigma}_{\alpha 3}$ is computed as:

$$\widehat{\sigma}_{\alpha 3}^{(k)} = b_\alpha \left[\widehat{\sigma}_{\alpha 3}^{(k)_o} + a_\alpha (z+h) - \int_{z_{k-1}}^{z} \overline{C}_{\alpha\beta\gamma\delta}^{(k)} \left(e_{\gamma\delta,\beta} + \zeta\ \kappa_{\gamma\delta,\beta} \right) d\zeta \right] \tag{16}$$

where no sum on α is performed. Then, the coefficients a_α and b_α are determined enforcing the above constraints as:

$$a_\alpha = -\frac{\overline{\sigma}_{\alpha 3}^{(n)}}{h Q_\alpha} \qquad b_\alpha = Q_\alpha / \left(\overline{Q}_\alpha + \overline{\sigma}_{\alpha 3}^{(n)} h/2 \right) \tag{17}$$

where:

$$\overline{\sigma}_{\alpha 3}^{(n)} = \overline{\sigma}_{\alpha 3}^{(k)}(z)_{z=h,k=n} \qquad \overline{Q}_\alpha = \int_{-h}^{h} \overline{\sigma}_{\alpha 3}^{(k)}(z)\, dz \tag{18}$$

with:

$$\overline{\sigma}_{\alpha 3}^{(k)}(z) = \widehat{\sigma}_{\alpha 3}^{(k)_o} - \int_{z_{k-1}}^{z} \overline{C}_{\alpha\beta\gamma\delta}^{(k)} \left(e_{\gamma\delta,\beta} + \zeta\ \kappa_{\gamma\delta,\beta} \right) d\zeta$$

4 Numerical results

The different laminate finite element models have been implemented in FEAP (Finite Element Analysis Program) [9] and their computational performances have been assessed by several numerical applications.

We here report some results obtained for a square laminated plate with side a. Each layer is characterized by a thickness $a/20$ and the following mechanical properties: $E_L/E_T = 25$, $\nu = 0.25$, $G_{LT}/E_T = 0.5$, $G_{TT}/E_T = 0.2$.

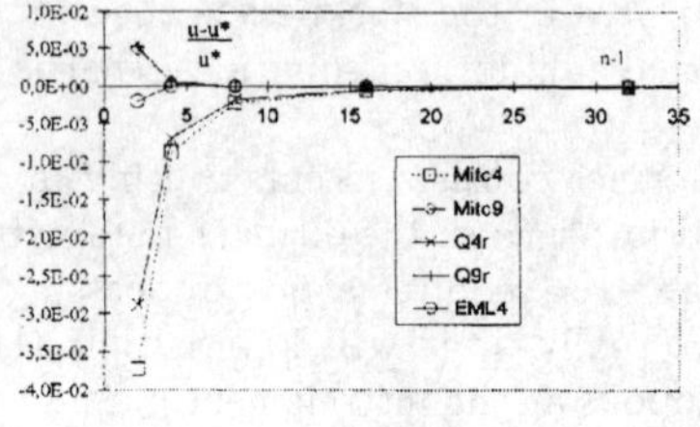

Fig. 1a. 0/90.

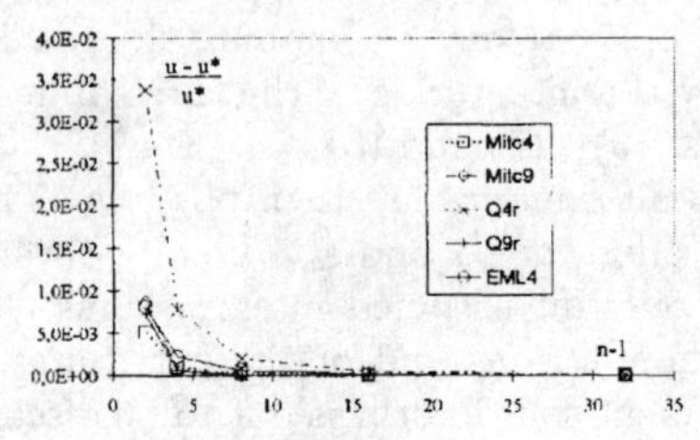

Fig. 1b. 0/90/0.

The plate is simply supported on the boundary and loaded by a sinusoidal transversal load q. Regular meshes have been adopted and only a quarter of the plate have been discretized for symmetry reasons.

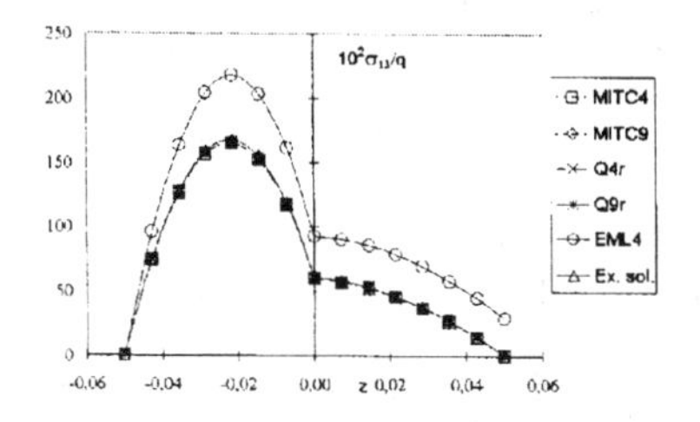

Fig. 2a. $\sigma_{13} = \sigma_{23}$ (0/90) - Non-corrected profile.

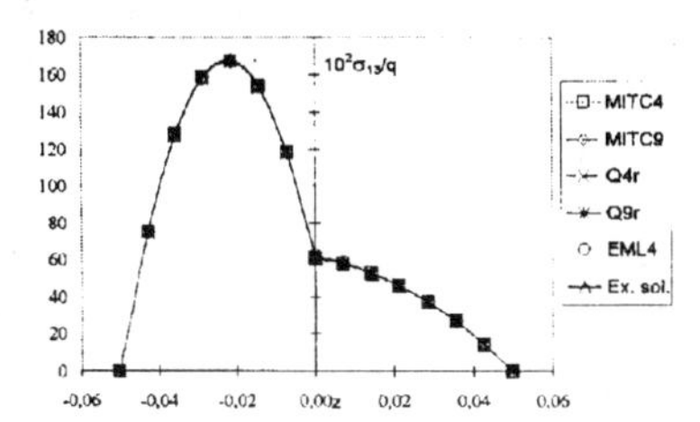

Fig. 2b. $\sigma_{13} = \sigma_{23}$ (0/90) - Corrected profile.

Two layer sequences 0/90 and 0/90/0 have been first considered and in fig. 1 the satisfactory convergence properties of the different elements are shown by plotting the number n of nodes of each edge versus the relative error $(u-u^*)/u^*$ in computing the transversal displacement at the plate center, being u the finite element solution and u^* the exact solution according to the FSDT.

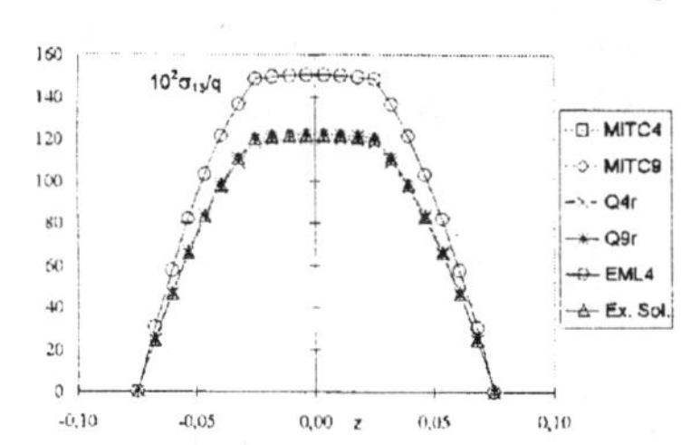

Fig. 3a. σ_{13} (0/90/0) - Non-corrected profile.

Fig. 3b. σ_{13} (0/90/0) - Corrected profile.

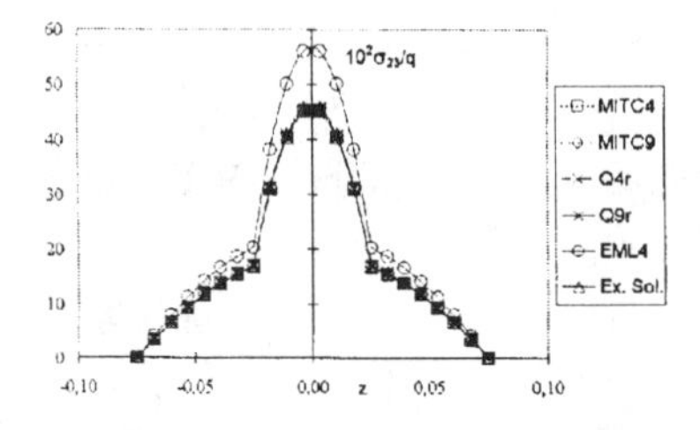

Fig. 4a. σ_{23} (0/90/0) - Non-corrected profile.

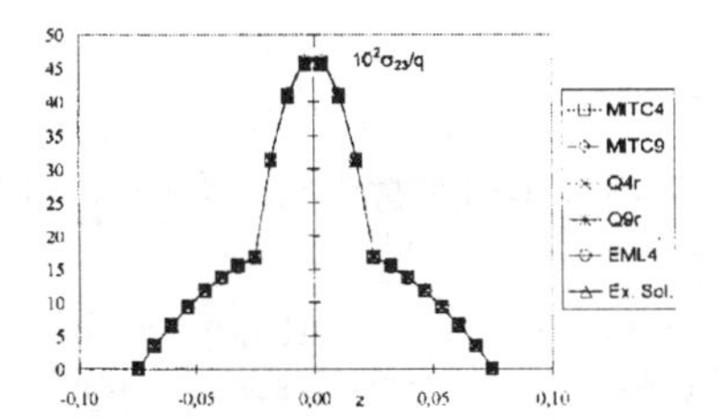

Fig. 4b. σ_{23} (0/90/0) - Corrected profile.

The transverse shear stresses profiles obtained for the different elements are plotted in figs. 2-4 and compared with the analytical FSDT solution. The stresses have been obtained after post-processing the curvatures and the in-plane strains, by the nodal projection procedure, except for the EML4.

Figs. 2a-4a show the results obtained without the correction procedure detailed in section 3.4 while figs. 2b-4b illustrate the results after the correction. Notice that the correction is strictly necessary only for the EML4 element.

Acknowledgments

The financial supports of the Italian National Research Council (CNR) and of the Ministry of University and Research (MURST) are gratefully acknowledged.

References

[1] E. Reissner, 'The effect of transverse shear deformation on the bending of elastic plates', *J. appl. mech.*, **12**, 69-77 (1945).

[2] R.D. Mindlin, 'Influence of rotatory inertia and shear on flexural motions of isotropic, elastic plates', *J. appl. mech.*, **38**, 31-38 (1951).

[3] J.M. Whitney and N.J. Pagano, 'Shear deformation in heterogeneous anisotropic plates', *J. appl. mech.*, **37**, *Trans. ASME* 92/E, 1031-1036 (1970).

[4] K.J. Bathe and E.N. Dvorkin, 'A four-node plate bending element based on Mindlin/Reissner plate theory and a mixed interpolation', *Int. j. numer. methods eng.*, **21**, 367-383 (1985).

[5] J.N. Reddy, *Energy and variational methods in applied mechanics*, John Wiley (1984).

[6] J.N. Reddy, 'A generalization of two-dimensional theories of laminated plates', *Comm. appl. numer. meth.*, **3**, 173-180 (1987).

[7] J.C. Simo and M.S. Rifai, 'A class of mixed assumed strain methods and the method of incompatible modes', *Int. j. numer. methods eng.*, **29**, 1595-1638 (1990).

[8] J.N. Reddy, 'On refined theories of composite laminates', *Meccanica,* **25**, 230-238 (1990).

[9] O.C. Zienkiewicz and R.L. Taylor, 'The finite element method: Voll. I and II', McGraw Hill, 1989.

[10] O.O. Ochoa and J.N. Reddy, *Finite element analysis of composite laminates*, Kluwer Academic Publisher (1992).

[11] M.L. Bucalem and K.J. Bathe, 'Higher-order MITC general shell elements', *Int. j. numer. methods eng.*, **36**, 3729-3754 (1993).

[12] P. Bisegna and E. Sacco 'A rational deduction of plate theories from the three-dimensional linear elasticity', Z.A.M.M. *Zeit. für angew. math. und mech.*, vol. 77, pp. 349-366, 1997.

[13] F. Auricchio and E. Sacco, 'A mixed-enhanced finite-element for the analysis of laminated composite plates', *submitted for the publication* (1998).

Experimental behaviour of concrete beams reinforced with glass FRP bars

M. Pecce*, G. Manfredi*, E. Cosenza*

* Dipartimento di Analisi e Progettazione Strutturale – Università degli Studi di Napoli Federico II – Via Claudio 21, 80125 Naples-Italy

Abstract

The experimental behaviour of beams reinforced with FRP bars is analysed. The bars used are pultruded with E-glass fibers; the surface is treated for obtaining ribs very similar to the steel one for improving the bond to concrete. The beams tested are characterised by different amounts of longitudinal reinforcement and transversal (steel) reinforcement. The experimental results are shown in terms of moment-curvature, crack width, deflections; moreover the numerical evaluation of the maximum stresses in the materials is provided.

1. Introduction

The use of composite rebars in reinforced concrete elements could be a good chance for improving the durability of the structures; however Fiber Reinforced Plastics (FRP) are innovative materials in civil engineering so that many aspects have to be widely analysed before that diffusion and reliability could be effective.

The development of these materials covers various field of engineering. Firstly, the material engineering is involved in studying the possibilities of coupling different types of matrices and fibers varying the percentages of the different components. In this field the research is guided by the required performances of the material referring to its use, however the effective production is already now influenced by the costs of the different types of fibers. Surely the possibility of designing the material and of increasing the production, with the consequent reduction of the costs, could arise the optimisation of the composite materials for the specific application.

On the other side the performances of composite materials have to be analysed in the specific field of application, since the mechanical characteristics are so different from the traditional materials used in the civil structures, that probably all the design approaches, the analytical models and the codes provisions need to be revised.

Considering this aspect it is necessary to carry out many experimental tests in order to point out the meaningful problems and to develop enough data for reviewing the design criteria and models; moreover a complete analysis of the topic is difficult due to the large number of FRP bars on the market and to their continuous evolution. However all the tests can be useful even to individuate the effective applicability or the most suitable application field of the product analysed and the design criteria to be applied.

The flexural behaviour of the concrete beams reinforced with FRP bars is one of the key issues in the development of the applications and experimental tests are actually carried out by many researchers [1-6]
In this paper the results of experimental tests on beams reinforced with FRP bars are reported [7-8]. The FRP bars used for the tests are pultruded with E-glass fibers, in particular the C-Bars™ produced by Marshall Composites. This type of material is interesting since the glass fibers and the pultrusion technology develop to a cheap product, even if the mechanical characteristics can largely improved using other types of fibers. However this type of bars is characterised by a worked surface, with a shape very similar to steel with ribs, for improving bond, that is fundamental for a good behaviour of R.C. structures [9].
The mean characteristics of these FRP bars are the linear behaviour up to failure, and the high ratio of strength to elasticity modulus. The first aspect develops to a brittle behaviour of the structures and makes concrete the ductile component of reinforced concrete; this remark is relevant only if ductility is required as for seismic constructions or design with redistribution.
The second property shifts the design criteria in the serviceability limit states, that check the structure behavioural aspects instead of the strength in order to guarantee functionality and safety during its life. Therefore in the tests the measures and observation were aimed to obtain data about the three well-known verifications in service: check of the materials stress level, deflection limitation, crack width control.

2. Experimental setup

Three simple supported concrete beams reinforced using FRP bars were tested. The span is equal to 340 cm, while the width and the height of the cross section are respectively equal to 50 and 18.5 cm. The load is applied by two equal forces at 120 cm from each support: therefore the bending is constant in the central zone.
Two different amounts of longitudinal reinforcement in tension are designed in order to obtain both the concrete failure (type F1 and F3 beams with 7 ϕ 12.7) and the bars failure (type F2 beam with 4 ϕ 12.7), considering the nominal strength of the materials; in all the beams the reinforcement in compression is due to 2 ϕ 12.7 FRP bars.
Two different arrangement of the transversal reinforcement are realised by steel stirrups only in the zones between the applied forces and the support in order to not influence the cracks development in the constant bending zone. The F1 beam is reinforced only with stirrups in the D-regions, while F2 and F3 beams show distributed stirrups in the variable bending zones. The details of the beams geometry and reinforcement are drawn in Fig.1.
The mean concrete cylindrical strength in compression is about 30 MPa, for all the 3 elements, evaluated by tests on 3 specimens for each beam. The reinforcing bars are C-Bars™ manufactured by Marshall Composites: the C-Bars™ are pultruded bars with fiber glass, and the surface is treated obtaining ribs similar to the steel bars with improved bond; the average mechanical characteristics are tensile strength of 770 MPa and longitudinal elasticity modulus of 42 GPa [10].

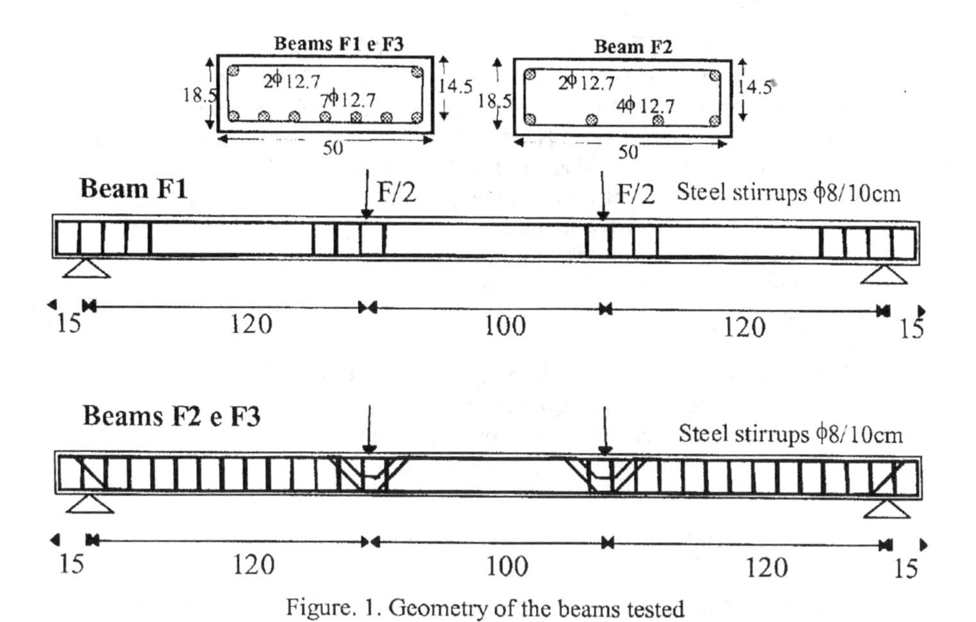

Figure. 1. Geometry of the beams tested

The test was carried out using an electro-hydraulic actuator in displacement control up to the failure of the beam; the load applied was measured by a load cell and the deflection in the midspan by a displacement transducer. Moreover displacement transducers were used for measuring the mean curvature in the centre of the beam on a prefixed base of 14 cm. At last in the middle section of the beam F3, the FRP bars in tension were instrumented with strain gauges positioned before the casting; the crack in the centre was preformed by a thin steel plate for half the height. The strain gauges were 10 and were attached on 3 bars at the 2 sides of the preformed crack, the gauges were protected during the casting so that for a length of about 2 cm for each side of the crack the bond between FRP and concrete was eliminated on the bars with gauges.

3. Experimental results

The experimental results can be divided in two parts: local measures and global measures; the first ones are representative of the behaviour of the central section or zone of the beams and consist of moment-curvature relation, maximum crack openings and maximum stresses in the materials; global aspects are the force-deflection relation and the cracks distribution along the beam.

For analysing the stresses in the materials, a strips model of the section is implemented and constitutive relationships of materials are introduced. For FRP bars the linear behaviour up to failure is considered with the nominal elasticity modulus. On the contrary concrete behaviour is non-linear; the model of Mander et al. [11] is assumed and the secant elasticity modulus of concrete E_c is numerically calculated, according the

formulation suggested by CEB [12]. The cracked model of the section is considered, with no tensile strength of concrete, and assuming the Bernoulli hypothesis.
The first step of the description of the experimental results is represented by the interpretation of the failure mode. The tested beams showed two different modes of failure: in the F1 beam a punching-shear failure in the loaded zone occurred, while beams F2 and F3 achieved a flexural failure with the bars collapse. It should be noted that the beams F1 and F3 were designed for achieving the concrete failure, but the experimental collapse was different.
The punching-shear collapse of F1 beam is due to a low transversal reinforcement, in fact the increase of the transversal reinforcement in the F3 beam leads to a flexural collapse.
The collapse mode of the F3 beam will be discussed in detail in the following; however it is clear that it is necessary a reliable experimental and theoretical assessment of the collapse mode and of the bearing capacity of the structure.

3.1 Section behaviour

The moment-curvature relation of the most stressed section is evaluated considering the strains measured by transducers positioned on the face of the beam at the midspan. The experimental curves are drawn in Fig.2 and represent the mean behaviour of a beam sub-element with a length of 14 cm, therefore these curves take into account the tension stiffening effect of the concrete between cracks. The curvature are evaluated by the measure of transducers 1 and 2, positioned on the face of the beam as shown in Fig.2.
The comparison of the specimens F1 and F2 points out the lower stiffness of the beam with less amount of reinforcement.

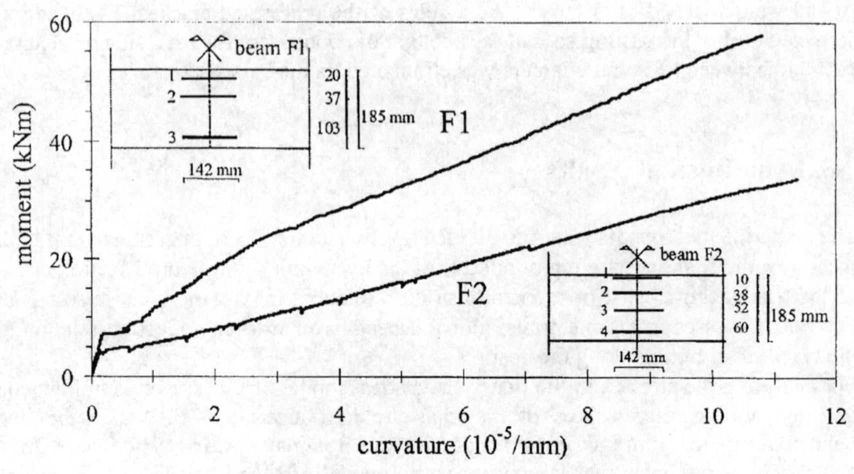

Figure.2. Moment-curvature relation of the central zone.

The transducer positioned at the level of the reinforcement gives the width of the central crack during the test, since in the measurement base only one crack formed; the curves are reported in Fig. 3. The F2 beam with lower tensile reinforcement shows a larger crack opening in comparison with the F1 beam; in particular at failure the crack width of the beam F2 is more than twice the F1 crack width, while the reinforcement is reduced of 43%. Moreover it can be observed that limiting the crack width at 0.3 mm can cause a severe restriction of the serviceability loads and a low fruition of the material strength.
For what concerns F3 beam, the local behaviour in terms of moment-curvature relation and crack width is largely influenced by the local reduction of the bond due to the introduction of the strain gauges on the rebars before the casting: therefore these local measures are not considered.

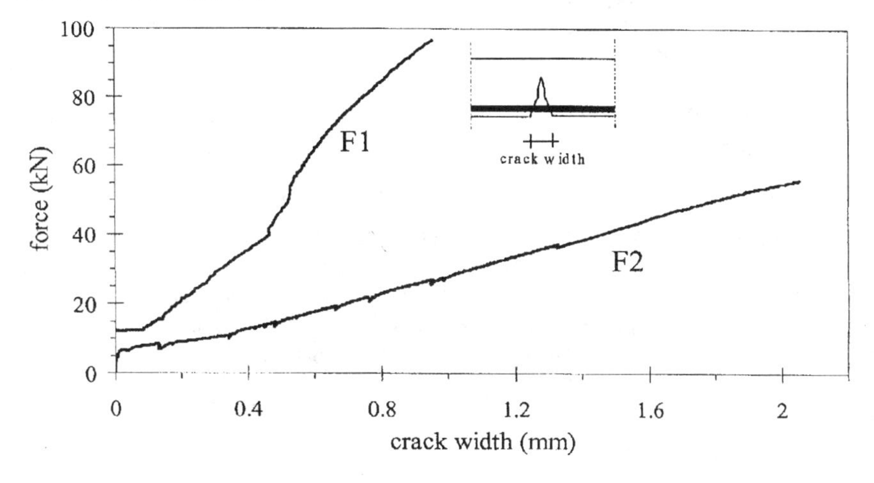

Figure. 3. Crack width of the midspan section.

As above mentioned the material stresses are numerically evaluated; the results for the middle section of the beams are plotted in Fig.4, where the stresses are dimensionless respect to the strength of each material. The curves of concrete are stopped at a strain of 0.005, and the experimental failure is marked with a circle. In the same figure the experimental result obtained by the strain gauges for the beam F3 is drawn, calculating the stress in the bars with the nominal elasticity modulus (42000 MPa) and considering the average of the 10 gauges; the experimental results and the theoretical one are in good agreement confirming the effectiveness of the implemented model in the cracked section.
It is worth to notice that in the beam F2, designed for attaining the bars failure, the ultimate experimental stress of the bars is about the 60% of the nominal strength; therefore it seems clear that the difference between measured and nominal FRP bars strength probably is due not only to the experimental scatter but also to the influence of bending on tensile strength of FRP bars, since the flexural deformation could cause the spalling of the fibers out of the matrix. The force-stresses relations of Fig 4 clear out that the F3 beam, with larger amount of reinforcement, really attained the FRP failure

with the same stress level in the bars of F2 beam while it was designed for reaching the concrete failure.

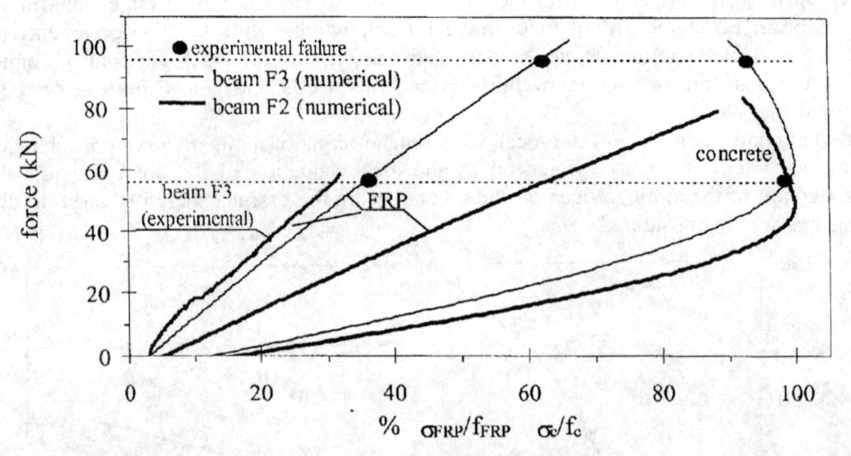

Figure. 4. Numerical and experimental values of the maximum stresses.

This result points out that experimental and analytical study need in order to evaluate the effective strength of FRP bars to guarantee a reliable design, for evaluating the failure mode (concrete or FRP) and the beam bearing capacity.

3.2 Beam behaviour

The beams behaviour is well represented by the force-deflection relation in the middle-span; in fact this relation can point out the beginning of cracking, the tension stiffening effect due to the cracks spacing and bond quality, the possible ductility due to the post-peak behaviour of concrete. In Fig. 5 the force-deflection curves are reported.

The 2 beams with the same longitudinal reinforcement (F1 and F3) show a vary similar trend, except that at the beginning of cracking and at failure. The first aspect is due to the above mentioned presence of the preformed crack in the specimen F3, so that the real uncracked state does not exist in this case. In all the cases the deflection at the maximum load is more than 1/30 of the span, pointing out that the deflection control is surely a significant aspect of the design.

In the last part of the curves it is clear the different failure of the 2 elements; the punching-shear failure in specimen F1 results in a sudden collapse; on the contrary in specimen F3 the flexural crisis occurred, therefore a descending branch of the force-deflection relation was measured, that corresponds to the progressive bars failure. Even though the different failure mechanism, the beams reached the same maximum load and the failure was brittle. In F2 beam the FRP crisis was reached, however no descending branch could be noticed since the bars failure was brittle.

Comparing beams F2 and F3, both failed with the bars collapse, in F3 it is evident the non linear trend in the last branch of the curve due to the non-linear behaviour of concrete especially in the post-peak branch. Therefore in R.C. beams with FRP bars only the ductility due to concrete can be developed, thus the design is favourable if the concrete crisis is provided i.e. the opposite approach of steel reinforced beams.

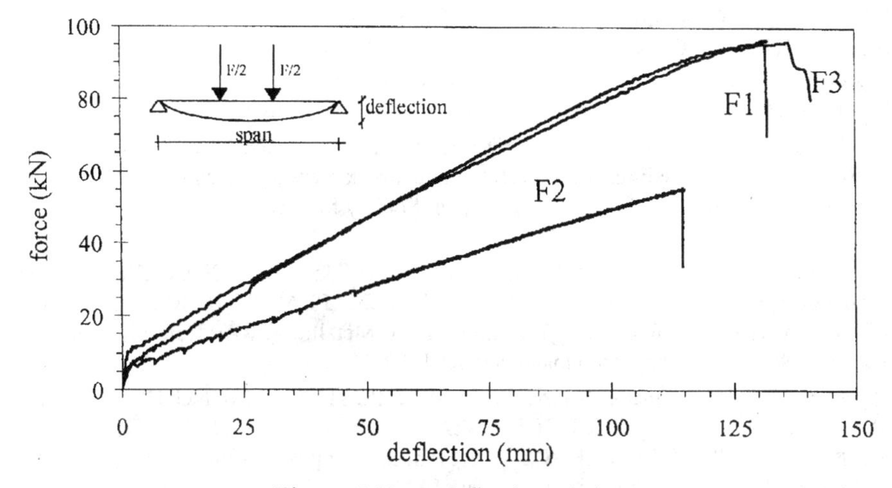

Figure. 5. Force-deflection relation.

At last it can be noticed the deformability of the F2 beam is more than twice of the F1 beam, as already done for the crack openings, providing that the reinforcement amount can be important even for the serviceability verifications; that surely govern design of R.C. beams reinforced with FRP.

4. Final remarks

The experimental results show that the behaviour of the beams reinforced with FRP is characterised by a brittle failure, even in bending, high deformability and crack width, so that the design criteria have to be guided by the serviceability verifications.
Moreover other interesting features can be underlined:

- the strength of the FRP bars in bending results largely lower than the nominal tensile strength; therefore the design have to be referred to the effective value in bending for actually realising the required failure mode and to reach the designed ultimate load;
- the punching and shear strength models have to be revised since the contributes of the various strength mechanisms (interlocking, dowel effect, concrete struts) could be different from that of steel reinforced beams;
- the model of the section assuming the Bernoulli hypothesis is reliable to estimate the stresses in the material, up to the peak stress in concrete is reached.

References

[1] *Abdalla,H., El-Brady, M.M., Rizkalla,S.H.*, DEFLECTION OF CONCRETE SLABS REINFORCED WITH ADVANCED COMPOSITE MATERIALS - Proceedings of 2nd International Conference: Advanced Composite Materials in Bridges and Structures, Canadian Society for Civil Engineering, Montreal, Quebec, Canada, 11-14 August 1996.

[2] *Alsayed, S.H., Almusallam, T.H., AL-Salloum, Y.A., Amjad, M.A.*, FLEXURAL BEHAVIOUR OF CONCRETE ELEMENTS REINFORCED BY GFRP BARS - Proceedings of 2nd International RILEM Symposium: Non Metallic (FRP) Reinforcement for Concrete Structures, August 1995. E&F Spoon.

[3] *Benmokrane, B., Masmudi, R.*, FRP C-BARS AS REINFORCING ROD FOR CONCRETE - Proceedings of 2nd International Conference: Advanced Composite Materials in Bridges and Structures, Canadian Society for Civil Engineering, Montreal, Quebec, Canada, 11-14 August 1996.

[4] *Duranovic N., Pilakoutas K., Waldron P.*, TESTS ON CONCRETE BEAMS REINFORCED WITH GLASS FIBRE REINFORCED PLASTIC BARS - Proceedings of 3nd International RILEM Symposium: Non Metallic (FRP) Reinforcement for Concrete Structures, Sapporo, Japan, October 1997.

[5] *Faza, S.S., Gangarao, H.V.S.*, PRE- AND POST- CRACKING DEFLECTION BEHAVIOUR OF CONCRETE BEAMS REINFORECD WITH FIBER REINFORCED PLASTIC REBARS - I Advanced Composite Materials in Bridge and Structures Conference, Sherbrooke, Canada, 1992.

[6] *Nanni, A.*, FLEXURAL BEHAVIOUR AND DESIGN OF RC MEMBERS USING FRP REINFORCEMENT - Journal of Structural Engineering, ASCE, 1993, Vol.119, No.11.

[7] *Cosenza, E., Greco, C., Manfredi, G. Pecce M.*, EXPERIMENTAL AND THEORETICAL BEHAVIOUR OF R.C. BEAMS WITH FRP REBARS - IABSE Conference New Technologies in Structural Engineering, Lisbon, 3-5 July, 1997.

[8] *Cosenza, E., Greco, C., Manfredi, G. Pecce M.*, FLEXURAL BEHAVIOUR OF CONCRETE BEAMS REINFORCED WITH FIBER REINFORCED PLASTIC (FRP) REBARS - *Proceedings of 3nd International RILEM Symposium: Non Metallic (FRP) Reinforcement for Concrete Structures*, Sapporo Japan, October 1997.

[9] *Cosenza, E., Manfredi, G. Realfonzo R.*, BEHAVIOUR AND MODELLING OF BOND OF FRP REBARS TO CONCRETE - Journal of Composites for Construction, ASCE, 1997, Vol.1, No.2.

[10] *Marshall Industries Composites Inc.*, FRP C-BAR REINFORCING ROD – Technical Data.

[11] *Mander, J.B., Priestley, M.J.N., Park R.*, THEORETICAL STRESS-STRAIN MODEL OF CONFINED CONCRETE - Journal of Structural Engineering, ASCE, 1989, Vol.114, No. 8.

[12] *CEB,* MODEL CODE 90 - Bulletin d' Information n° 213/214, 1993.

The analysis of columns for the design of pultruded frames: Isolated centrally loaded columns

J.T.Mottram*, N.D.Brown*, and A.Lane*,

* Department of Engineering, University of Warwick, Coventry, CV4 7AL, UK

Abstract

Linear elastic bifurcation loads are presented for two pultruded I-profiles in order to investigate the ultimate buckling behaviour of centrally loaded columns. The method of analysis is the Finite Element (FE) method. As the height of a column is reduced from a maximum of 6.3m, specific attention is given to the transition height at which the instability mode of failure changes from global Euler to local flange. It is found using FE and test results that this transition height can be within the practical height range of 3 to 6m, such that ultimate resistance of a column in a frame structure can be governed by local buckling. FE critical buckling load results are compared with data obtained in laboratory testing programmes. It is shown that, despite the shortcomings of the FE modelling, the numerical predictions provide new information in our quest to establish reliable and relevant design approaches for both local and Euler buckling.

1. Introduction

As pultruded profiles of standard section find increasing applications in building and bridge construction, it is natural for design engineers to require design procedures that provide the necessary level of confidence. Fundamental research is, therefore, needed to ensure that published procedures have design expressions that are reliable, safe and lead to economical structures. Standard structural profiles of E-glass fibre reinforced plastic are manufactured economically by the pultrusion process [1]. Their section shapes copy those routinely seen in structural steelwork and so for column members we have thin-walled sections with a wide flange I-profile (i.e. the web's depth and flange's widths are of the same dimension). The range of sections "off-the-shelf" is limited [1]. Previous research on column behaviour has tended to concentrate on section sizes that would not necessarily have applications in buildings [2-4]. The authors' research is, however, concentrating on section sizes that could be used in buildings such that the minimum column height is around 3m [5].

Zureick and Scott [2] have recently published an excellent review on previous work up to 1994. They used their own experimental data to develop a design approach for pultruded columns. The approach, based on Euler (global) buckling, is for *slender* columns of I- and box section with a central load, and accounts for different end boundary conditions by using the effective length concept. Other work (for example see references 3 and 5) has confirmed the *slender* column design approach by way of testing. In all test programmes [2,3,5] the column had simply supported end boundary conditions and bending was restrained about one of the principal axes (usually the major axis for the I-profiles). However, Zureick and Scott gave no details on how to determine whether the column was slender or not.

Barbero and Tomblin [6] used limited laboratory test results on I-profiles to develop a design approach over the full range of column heights. Short (or *stocky*) column tests had been performed [4] by compressing a section in a universal testing machine; the end conditions were now closer to the fully fixed condition. Combined with their test data at higher column heights [3] (where Euler buckling was the failure mode for the *slender* simply supported columns), an interaction equation between local and global buckling was chosen to fit the evidence [6]. The presence of the interaction effect was to lower the calculated buckling load centred around the transition height; in a similar way to how residual stresses influence steel columns. However, Barbero and Tomblin did not conduct sufficient tests on columns falling within the *interaction* height zone to prove its existence. One explanation for why they felt this was the correct physical situation to choose was that as their *short* columns got shorter the load increased before local flange buckling [4] (theoretically the local buckling load can be independent of column height [7-9]). The present authors propose that the increase in load was, in part, due to the restraining effect in the tests and that this observation highlights the importance of correct experimental procedures. The interaction equation relies on the theoretical determination of the column loads for local and global buckling modes. The latter is given by the well-known Euler formula with a correction for shear deformation [2,5,6].

Barbero and Tomblin [6] determined the local buckling load using a numerical solution based on plate theory and the Rayleigh-Ritz method. The correlation between the theoretical predictions and the experimental results was good. Their analysis method has not been made accessible to others and so there is the unresolved issue of how practising engineers can apply the *interactive* design approach. Closed form formulae to determine a column's load for local flange buckling are not available. Several researchers [7-9] have presented *design* charts giving buckling coefficients for uniformly compressed orthotropic plates, having one unloaded edge free and the other restrained (the latter modelling, for example, the junction between the flange and the web of a I-profile). Zureick and Shih [8] also give charts for general orthrotropic I-shaped structural sections. Once the required coefficient is known it can be substituted into a simple closed form expression to calculate the critical buckling load. Such design charts could be used in a design approach for pultruded columns providing the specific geometric and mechanical properties are matched by those given in the charts.

In 1996 a research project on pultruded column behaviour was started at the University of Warwick. In phase one centrally loaded columns were tested using the rig and test procedure described in [5]. The profiles were wide flange sections of E-glass fibre reinforced/poylester having the sizes 203x203x9.53mm, 203x203x12.7mm and 152x152x9.53mm. The sections were manufactured by Creative Pultrusions, PA, USA. New buckling loads for minor-axis buckling have been presented [5] for twelve isolated columns at heights of 3.3, 4.8 and 6.3m. Test results for profiles 203x203x9.53mm and 152x152x9.53mm are reproduced in Figures 4 and 5. Considering previous work (see [4] and [6]), the expected buckling mode of failure for these practical column heights was global. However, for the 203x203x9.53mm profile it was found that local flange buckling was the ultimate state when the height was 3.3m. This section had the highest flange outstand breadth-to-wall thickness ratio of 10 and was, therefore, most prone to the local buckling mode of failure. The ultimate compressive load of 300kN, minimum of two tests when failure was local, was similar to that for Euler buckling (as given by the usual formula), yet the mid-depth deflection at onset of local buckling was < 8mm and stable. Note that Tomblin and Barbero work [4] had indicated that the local buckling load would be significantly higher for the three profiles tested and therefore we had expected the global mode in all twelve tests.

The long preamble on previous work has been necessary to justify the contribution made here. Our new observation that local buckling can occur at a load similar to global buckling and within the practical height range, means that the *slender* design approach of Zureick and Scott [1] and the *interaction* design approach of Barbero and Tomblin [6] are not necessarily reliable or relevant. To enable code-writers to prepare improved design procedures, it is necessary to predict, with confidence, the load at which local buckling occurs. This is not currently possible because of lack of knowledge.

To provide further numerical results the authors have conducted a FE analysis of the problem. Details are given on the model, such that the mesh specification, material properties and boundary conditions correspond to the parameters in the test arrangement at the University of Warwick [5]. Buckling loads for the two profiles with 9.53mm wall thickness are reported for the column height range of 0.5 to 6.3m. Mode shapes for when failure is governed either by local or global buckling are illustrated. Numerical predictions are compared with the test results [2-5], and general observations are made.

2. Finite element modelling

A column was modelled with thin shell parabolic quadrilateral elements using the I-DEAS Master Series 3.0 software. The analysis followed a similar approach to the one given in [10]. Figure 1 shows a typical mesh construction with two elements in the depth of the web and across the width of each flange. The constant sized elements had an aspect ratio of 1.32 and 1.02 for profiles 152x152x9.53 and 203x203x9.53mm respectively, with the longest edges parallel to the axis of the column. To model a column of increased height required the addition of 100mm long sections of elements. The height in a model included 150mm, at each end, which in a test was needed for the steel fixtures [5]. The mid-plane of the flat shell elements was placed at the mid-depth of the wall thickness.

The composite material was modelled as transversely isotropic with three planes of symmetry and for in-plane response by four elastic constants. Taking a local co-ordinate system for a panel in the profile, with '*L*' for Longitudinal and '*T*' for Transverse, the elastic constants were: $E_T = 23$ GPa, $E_L = 8$ GPa, $G_{LT} = 3$ GPa, and $\nu_{LT} = 0.29$. These mechanical properties were the same for web and flange panels. E_L was determined from compressive direct strain reading in the tests [5], but the other three constants were taken to be those of the manufacturer. Elastic constants for pultruded materials are known to vary within, and between, pultruded profiles [2]. It is therefore important that numerical predictions based on the available mechanical property data are not assumed to be exact. Furthermore, this aspect of the research on pultruded profiles needs to be appreciated when analytical results are compared to the *equivalent* experimental buckling loads.

It is important to model the end boundary conditions faithfully. However, to establish whether the FE method was going to provide results that could be used to help to develop a reliable and relevant design approach, it was decided to make the model as simple as possible. Because the column is simply supported, it is necessary for the flanges at each end to be allowed to bend about their major-axis while at the same time the plane section must remain plane (i.e. there is no section warping). To meet the first part of this requirement the boundary conditions were imposed using only the nodes in the web (see figure 1). At one end the nodes on the web were restrained from rotation about the global x-axis and from translation in all three global directions. At the other column end the nodes in the web were restrained from rotation about the global x-axis and from translation in the global y- and z-directions. Loading was applied at the end with the free x-translation. A uniform compression was modelled by a statically admissible *pressure* loading along the length of the web (the flange outstands were not loaded). Invoking the St Venant principle it may be assumed that the inaccuracies in modelling the end conditions died away over a length equal to the column's depth.

3. Results and discussion

The linear buckling option in Master Series 3.0 is based on the Eigenvalue extraction method given in [11]. It predicts a bifurcation buckling load which is the load for which a reference configuration of the structure and an infinitesimally close (buckle) configuration are possible equilibrium configurations. For our column problem the reference configuration is the initial perfect geometry of the model (see Figure 1) and the buckle configuration is similar to either that in Figure 2 (local flange) or in Figure 3 (global Euler).

A comparison between the FE loads (kN) and test loads (kN) are given in Figures 4 and 5 for profiles 152x152x9.5mm and 203x203x9.5mm, respectively. In the figures results are presented for column heights between 0.5 and 6.3m. The mode of failure is labelled. The transition between the local and the Euler buckling mode is clear and without a distinctive *transition* zone. The column heights at transition are 1.75 and 3.18m and the compressive loads 385 and 287 kN, respectively. Note that the average section stress of 92.4 and 51.0 N/mm^2 is significantly less than the material's strength of >210 N/mm^2.

For heights above the transition the correlation is generally close. Zureick and Scott [2] tested profiles from Strongwell [1] and the six columns had an average E_L 0.78 of that of the standard Creative Pultrusions profiles tested in [3] and [5]. It is seen in Figure 4 that the Zureick and Scott test results fall short by 20% or more and this was principally due to the lower E_L. As expected the results from Barbero and Tomblin [3] are much closer because they tested Creative Pultrusions profiles with E_L values near to 23 GPa.

When the column height is below the transition we find that the FE critical load is fairly constant with a slight increase as the column height decreases. However, the test results from [4] show that there is a much higher load at the on-set of local buckling when the column height was similar in length to the buckle's half-wavelength. Theoretical approaches in [6] to [9] treat the restraint conditions at the loaded ends of the buckling plate (or structural section, see [8]) as simple and could explain in part why the local buckling load remains constant for heights between 1.0m and transition. The critical load was determined at the transition height using the analysis of Zureick and Shih [8] and the properties of the two profiles in Section 2. The difference in load between the FE and the other approach was found to < 1.3%. This demonstrates that the FE analysis is accurate in terms of the way practical column heights had been modelled.

The rigid end restraints in the very *short* column tests, at heights of 0.5 to 1.0m, were such that the loaded ends could not behave as simple supports and an additional increase in the buckling load will be due to the extra restraint from the fixed condition [4]. When the column height is increased, the effect of the end restraint conditions becomes less prominent, since it dies away, and this is a reason why the new test result [5] for section 203x203x9.5mm at 3.3m, gives a close correlation to theory.

The FE results in Figures 4 and 5 show that the bifurcation analysis provides buckling loads, and the modes of failure, that correspond well to what is known from the limited number of tests available. To confirm that the FE method could be used reliably by engineers in practice as a design tool, the FE modelling requires further refinement to include:

- bending or twisting deformation of the column.
- variation in profile geometry along the column height.
- variation in material properties along the column height.
- restraints and loading conditions that correspond exactly to those in the tests.

To verify that the FE (or other analytical method) buckling loads are correct, new testing is needed to provide many more results for column heights in the transition zone and with the correct end conditions at shorter heights. There is also the need to use standard test methods to accurately determine the four elastic constants, thereby ensuring the properties in the analysis match those in the tests.

4. Conclusions

The FE method has been used successfully to determine the buckling load and mode of failure for columns of pultruded profile. For slender columns there is generally a good correlation between test results and the numerical predictions. It has been shown that local flange buckling can be the mode of failure for column heights within the practical range. The FE method has been shown to give predictions with sufficient accuracy for its application in the design process. This is an important conclusion, because other analytical methods to determine the local buckling load are not generally accessible. The deficiencies in the FE modelling, in terms of the experimental conditions, are exposed and these are used to discuss current practise when testing columns of very short height. New research is needed before an approved design approach for pultruded columns in frame construction could be prepared by code-writers. However, for column profiles, heights, and end conditions where global buckling governs failure the classical Euler formula can be used with confidence in a design procedure.

5. Acknowledgement

The authors are grateful to Dr B.Shih and Dr A.Zureick for providing their analysis software to calculate the local buckling failure loads of I-shaped structurals [8].

6. References

[1] *Anon.* EXTREN DESIGN MANUAL, Strongwell, Bristol, VA., 1989.
[2] *A.Zureick, and D.Scott,* SHORT-TERM BEHAVIOR AND DESIGN OF FIBER-REINFORCED POLYMERIC SLENDER MEMBERS UNDER AXIAL COMPRESSION, Journal of Composites for Construction, ASCE, 1 4, 1997, 140-149.
[3] *E.Barbero, and J.Tomblin,* EULER BUCKLING OF THIN-WALLED COMPOSITE COLUMNS, Thin-Walled Structures, 17, 1993, 237-258.
[4] *J.Tomblin, and E.Barbero,* LOCAL BUCKLING EXPERIMENTS ON FRP COLUMNS, Thin-Walled Structures, 18, 1994, 97-116.
[5] *N.D.Brown, J.T.Mottram, and D.Anderson,* THE BEHAVIOUR OF COLUMNS FOR THE DESIGN OF PULTRUDED FRAMES: TESTS ON ISOLATED CENTRALLY LOADED COLUMNS, Proceedings of Second International Conference on Composites in Infrastructure, The University of Arizona, 1998, Vol II, 248-260.
[6] *E.Barbero, and J.Tomblin,* A PHENOMENOLOGICAL DESIGN EQUATION FOR FRP COLUMNS AND INTERACTION BETWEEN LOCAL AND GLOBAL BUCKLING, Thin-walled Structures, 1993, 8, 117-131.
[7] *W.M.Banks, and J.Rhodes,* THE INSTABILITY OF COMPOSITE SECTIONS, Proceedings Second International Conference on Composite Structures, Paisley, Elsevier, 1983, 442-452.
[8] *A.Zureick, B.Shih,* LOCAL BUCKLING OF FIBER-REINFORCED POLYMERIC STRUCTURAL MEMBERS UNDER LINEARLY-VARYING EDGE LOADING, Structural Engineering and Mechanics Research Report No. SEM 94-1, Georgia Institute of Technology, Atlanta, August 1994, p.106.
[9] *L.C.Bank, and J.Yin,* BUCKLING OF ORTHOTROPIC PLATES WITH FREE AND ROTATIONALLY RESTRAINED UNLOADED EDGES', Thin-Walled Structures, 24, 1996, 83-96.
[10] *A.R.Vakiener, A.Zureick, and K.M.Will,* PREDICTION OF LOCAL FLANGE BUCKLING IN PULTRUDED SHAPES BY FINITE ELEMENT ANALYSIS, in Proceedings ASCE Speciality Conference, ASCE, NY, 1991, 302-312.
[11] *R.D.Cook, D.S.Malkus, and M.E.Plesha,* CONCEPTS AND APPLICATIONS OF FINITE ELEMENT ANALYSIS, Third Ed., John Wiley & Sons, New York, 1989.

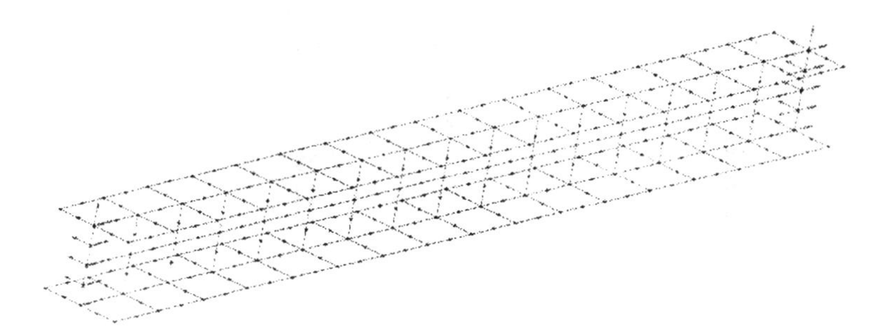

Figure 1. Finite element model of centrally loaded column 152x152x9.53 mm for 1.6m height.

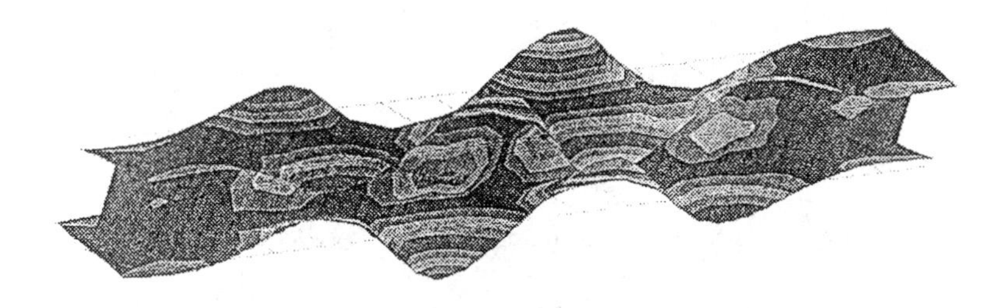

Figure 2. Local flange buckling mode shape for profile 152x152x9.53mm at height 1.6m.

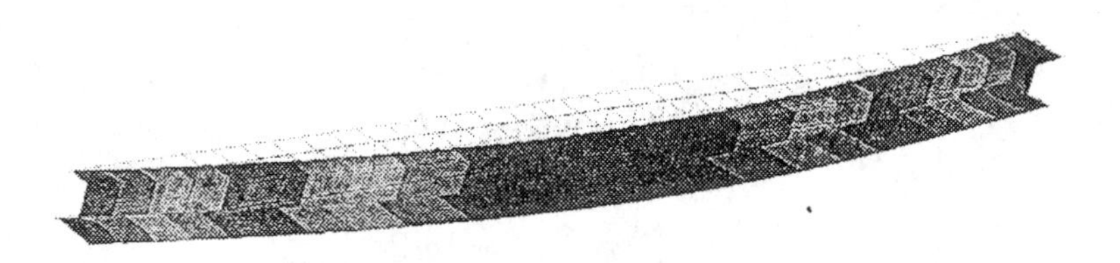

Figure 3. Euler buckling mode shape 152x152x9.53mm at height 3.3m.

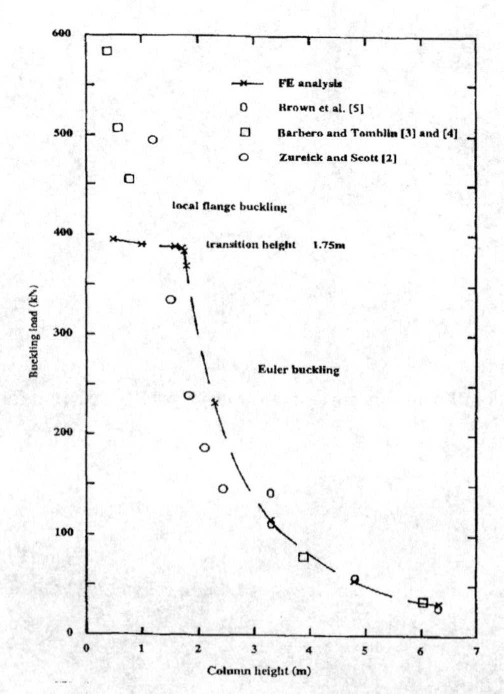

Figure 4. Buckling loads for centrally loaded columns of section 152x152x9.53mm.

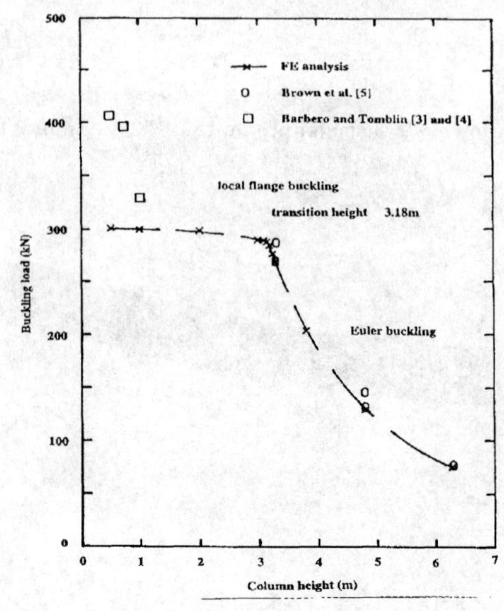

Figure 5. Buckling loads for centrally loaded columns of section 203x203x9.53mm.

Vibration analysis of special orthortopic plate with free edges supported on elastic foundation and with a pair of opposite edges under axial forces

Duk-Hyun Kim*, Jung-Ho Lee**, Won-Suk Lee**

*Korea Composites, 97 Gugidong, Chongrogu, Seoul, 110-011, Korea
Fax (822)379-7118, Tel (822)379-5127
**Department of Civil Engineering, Kangwon Natl. Univ., Chuncheon, Korea

Abstract

A method of calculating the natural frequency corresponding to the first mode of vibration of beams and tower structures, with irregular cross sections and with arbitrary boundary conditions was developed and reported by Kim, D. H. in 1974. In this paper, the result of application of this method to the special orthotropic plates with free edges supported on elastic foundation and with a pair of opposite edges under axial forces is presented. Such plates represent the concrete highway slab and•hybrid composite pavement of bridges. The reinforced concrete slab can be assumed as a special orthotropic plate, as a close approximation. The highway slab is supported on elastic foundation, with free boundaries. Sometimes, the pair of edges perpendicular to the traffic direction may be subject to the axial forces. The plate is subject to the concentrated load/loads, in the form of traffic loads, or the test equipments. Any method may be used to obtain the deflection influence surfaces needed for this vibration analysis. Finite difference method is used for this purpose, in this paper. The influence of the modulus of the foundation, the aspect ratio of the plate, and the magnitudes of the axial forces and the concentrated attached mass on the plate, on the natural frequency is thoroughly studied.

1. Introduction

The problem of deteriorated highway concrete slab is very serious all over the world. Before making any decision on repair work, reliable non-destructive evaluation is necessary. One of the dependable methods is to evaluate the in-situ stiffness of the slab by means of obtaining the natural frequency. By comparing the in-situ stiffness with the one obtained at the design stage, the degree of damage can be estimated rather accurately.
The reinforced concrete slab can be assumed as a special orthotropic plate, as a close approximation. The highway slab is supported on elastic foundation, with free boundaries. Sometimes, the pair of edges perpendicular to the traffic direction may be subject to the axial forces.
Several materials should be tested to find out the best type of pavement material for the future bridge decks, especially advanced composite bridge decks. One solution can be a combination of impregnated woven fibers and toughened polymer with little

abrasion property, forming an integral section with composite deck part. Such pavement will behave as the special orthotropic plate on elastic foundation with free edges. Such pavement may be subject to the axial forces parallel to the traffic direction. Some of the causes of such forces are the longitudinal gradient of the roadway, impact by the moving trucks, temperature changes, and others.
Such plates are subject to the concentrated mass/masses in the form of traffic loads, or the test equipments such as accelerator in addition to their own masses. Analysis of such problem is usually very difficult.
Most of the civil and architectural structures are large in sizes and the number of laminae is large, even though the thickness to length ratio is small enough to allow to neglect the transverse shear deformation effect in stress analysis. For such plates, there are enough number of fiber orientations for which theories for special orthotropic plates can be applied, and simple formulas developed by Kim, D.H. can be used [3].
However, if the plate has boundary condition other than simple supported, obtaining a reliable solution is very difficult. The basic concept of the Rayleigh method, the most popular analytical method for vibration analysis of a single degree of freedom system, is the principle of conservation of energy ; the energy in a free vibrating system must remain constant if no damping forces act to absorb it. In case of a beam, which has an infinite number of degrees of freedom, it is necessary to assume a shape function in order to reduce the beam to a single degree of freedom system. The frequency of vibration can be found by equating the maximum strain energy developed during the motion to the maximum kinetic energy. This method, however yields the solution either equal to or larger than the real one. For a complex beam, assuming a correct shape function is not possible. In such cases, the solution obtained is larger than the real one.
A simple but exact method of calculating the natural frequency corresponding to the first mode of vibration of beam and tower structures with irregular cross-sections and attached mass/masses was developed and reported by Kim, D.H. in 1974. This method consists of determining the deflected mode shape of the member due to the inertia force under resonance condition. Beginning with initially guessed mode shape, exact mode shape is obtained by the process similar to iteration. Recently, this method was extended to two dimensional problems including composite laminates, and has been applied to composite plates with various boundary conditions with/without shear deformation effects and reported at several international conferences including the Eighth Structures Congress[6] and Fourth Materials Congress[12] of American Society of Civil Engineers.
In this paper, the result of application of this method to the subject problem is presented. The effect of the modulus of foundation, the aspect ratio of the plate, and the magnitudes of the axial forces and the concentrated attached mass on the plate, on the natural frequency is thoroughly studied.

2. Method of Analysis

In this paper, the method of analysis given in detail, in the senior author's book[2] is repeated.
The magnitudes of the maximum deflection at a certain number of points are arbitrarily given as

$$w(i,j)(1) = W(i,j)(1) \tag{1}$$

where (i,j) denotes the point under consideration. This is absolutely arbitrary but educated guessing is good for accelerating convergence. The dynamic force corresponding to this (maximum) amplitude is

$$F(i,j)(1) = m(i,j)[\omega(i,j)(1)]^2 \, w(i,j)(1) \tag{2}$$

The "new" deflection caused by this force is a function of F and can be expressed as

$$\begin{aligned} w(i,j)(2) &= f\{m(k,l)[\omega(i,j)(1)]^2\}w(k,l)(1) \\ &= \sum_{k,l} \Delta(i,j,k,l)\{m(k,l)[\omega(i,j)(1)]^2 w(k,l)(1)\} \end{aligned} \tag{3}$$

where $\triangle$ is the deflection influence surface. The relative (maximum) deflections at each point under consideration of a structural member under resonance condition, w(i,j)(1) and w(i,j)(2), have to remain unchanged and the following condition has to be held :

$$w(i,j)(1) \,/\, w(i,j)(2)=1 \tag{4}$$

From this equation, ω(i,j)(1) at each point of (i,j) can be obtained. But they are not equal in most cases. Since the natural frequency of a structural member has to be equal at all points of the member, i.e., ω(i,j) should be equal for all ω(i,j), this step is repeated until sufficient equal magnitude of ω(i,j) is obtained at all (i,j) points.

However, in most cases, the difference between the maximum and the minimum values of ω(i,j) obtained by the first cycle of calculation is sufficiently negligible for engineering purposes. The accuracy can be improved by simply taking the average of the maximum and the minimum, or by taking the value of ω(i,j) where the deflection is the maximum. For the second cycle, w(i,j)(2) in

$$w(i,j)(3) = f\{m(i,j)[\omega(i,j)(2)]^2 \, w(i,j)(2)\} \tag{5}$$

the absolute numerics of w(i,j)(2) can be used for convenience.

3. Numerical Examination

[A/B/B/C/A/A/B]r type laminate is considered. The material properties are ;
E_1 = 38.6 GPa, E_2= 8.27 GPa
ν_{12} = 0.26, ν_{21} = 0.0557, G_{12} = 4.14 Gpa.
The thickness of a ply is 0.000125m. As the r increases, B_{16} , B_{26} , D_{16} and D_{26} decreases and the equations for the special orthotropic plates can be used. For simplicity, it is assumed that A=45°, B=-45°, C=90° and r=22.

3.1 Finite Difference Method (F.D.M.)

The method used in this paper requires the deflection influence surfaces. Since no reliable analytical method is available for the subject problem, the finite difference method (F.D.M.) is applied to the governing equation of the special orthotropic plates.
In finite difference form, the fourth single derivative terms in the biharmonic equation of the special orthotropic plate action require at least five pivotal points for errors of order Δ^2. Hence, it is desirable to transform the fourth-order partial differential equation into three second-order partial differential equations with three dependent variables, namely w, Mx, and My, by considering the equilibrium of a segment of the plate[14].
If F.D.M. is applied to these equations, the resulting matrix equation is very large in sizes, but the tridiagonal matrix calculation scheme used by Kim[14,15,16] is very efficient to solve such equations.

3.2 Accuracy of F.D.M

Since one of the few efficient analytical solutions of the special orthotropic plate is Navier solution, and this is good for the case of the four edges simple supported, F.D.M. is used to solve this problem and the result is compared with the Navier solution.
The aspect ratio used is 1m/1m=1. The mesh size is $\Delta x=0.1$, $\Delta y=0.1$. The accuracy of the F.D.M. solution is justified by comparing two results.

3.3 Influence of the Modulus of Foundation and the Aspect Ratio of the Plate

For this study, the simplest assumption that the intensity of the reaction of the subgrade is proportional to the deflection, w, of the plate, is used. Thus, this intensity is given by the expression kw, where k is the constant called the modulus of the foundation, in Newton per square meter per meter of deflection. The plate geometry and applied load is as given in Figure 1.

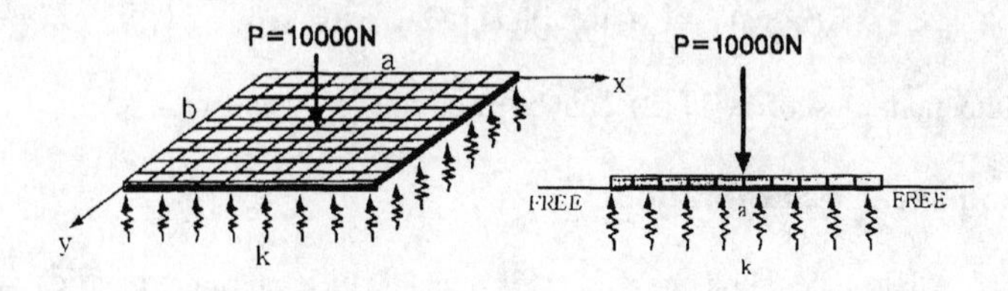

Figure 1. Plate geometry and loading

The laminate orientation is $[45^0,-45^0,-45^0,90^0,45^0,45^0,-45^0]_{22}$, with the same material properties as given previously.
The effect of the modulus of foundation, k, and the plate aspect ratio(a/b) on the deflection at the center of the plate is given in Table 1. Figure. 2 shows the same result as Table 1, in graphics.

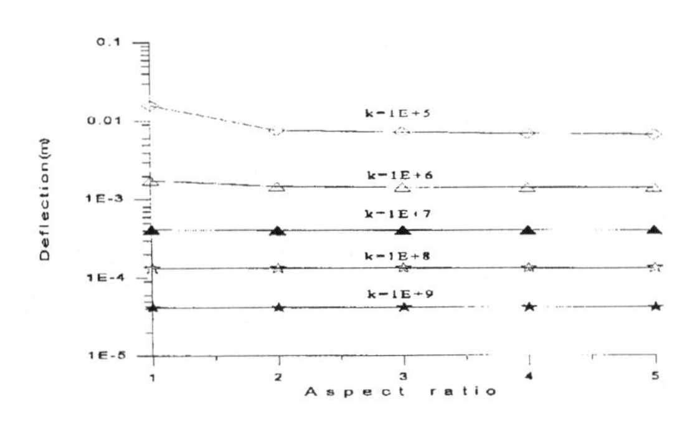

Figure 2. Deflection at the center of the plate with different aspect ratios and k values (Nx=1000 N/m)

Vibration analysis is carried out by the method presented in this paper. The result is given in Table 2. Figure 3 is the graphic presentation of Table 2.

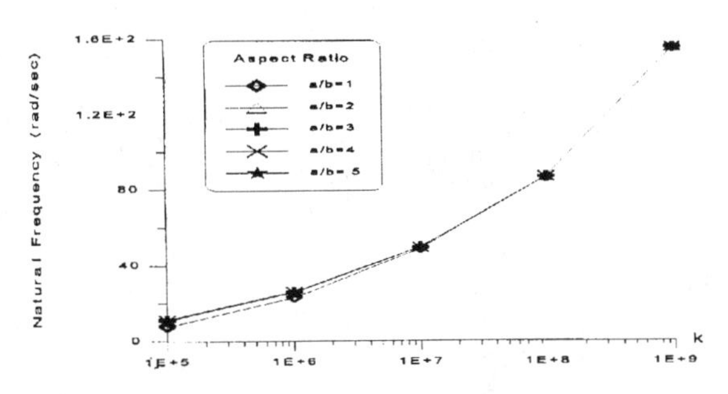

Figure 3. Natural frequency for each k and a/b

3.4 The Effect of the Concentrated Mass

Figure 4. shows the loading conditions of a plate with angle orientation of $[45^0, -45^0, -45^0, 90^0, 45^0, 45^0, -45^0]_{22}$, a=1m, b=1m, k=7MN/$m^3$, Nx=1000 N/m and q=367.8675N/m^2. n is a real number. P(i,j) is at the center of the plate. Tables 3 shows the deflection at the center of the plate with different values of n. Table 4 shows the natural frequency of the plate with different values of n.

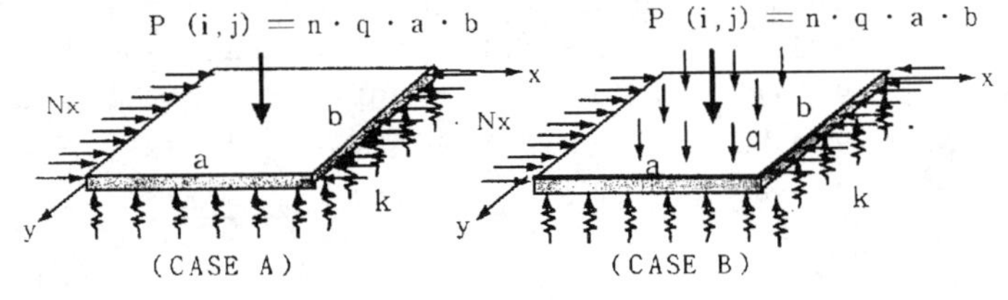

a=b=1m, k=7MN/m^3 , Nx=1000 N/m, q=367.8675N/m^2 , $[45^0,-45^0,-45^0,90^0,45^0,45^0,-45^0]_{22}$

Figure 4. Loading conditions of a plate

Table 1. Deflection at the center of the plate (m) (Nx=1000 N/m)

a/b \ k(N/m³)	k=10⁵	k=10⁶	k=10⁷	k=10⁸	k=10⁹
1	0.157464E-01	0.174296E-02	0.416537E-03	0.132746E-03	0.413241E-04
2	0.763536E-02	0.145253E-02	0.403906E-03	0.132654E-03	0.413241E-04
3	0.716962E-02	0.142352E-02	0.404193E-03	0.132654E-03	0.413241E-04
4	0.693619E-02	0.141226E-02	0.404183E-03	0.132654E-03	0.413241E-04
5	0.673492E-01	0.141228E-02	0.404184E-03	0.132654E-03	0.413241E-04

Table 2. Natural frequency for each case of k and aspect ratio, a/b, (rad/sec) (Nx=1000 N/m)

a/b \ k(N/m³)	k=10⁵	k=10⁶	k=10⁷	k=10⁸	k=10⁹
1	7.71164	22.86935	48.69064	86.63852	155.4520
2	10.86166	25.66662	49.45140	86.66915	155.4520
3	11.28889	26.02688	49.43317	86.66916	155.4521
4	11.21350	26.04478	49.43380	86.66919	155.4522
5	11.42447	26.04412	49.43381	86.66920	155.4522

Table 3. Deflection at the center of the plate (m)

n	CASE A	CASE B	CASE A /CASE B
0		0.5250E-04	
1	0.1865E-03	0.2590E-03	0.7803
3	0.5595E-03	0.6120E-03	0.9145
5	0.9325E-03	0.9850E-03	0.9367
7	0.1306E-02	0.1358E-02	0.9411
10	0.1865E-02	0.1917E-02	0.9581
20	0.3730E-02	0.3782E-02	0.9786

Table 4. Natural frequency (rad/sec)

n	CASE A	CASE B	CASEA /CASE B
0		13.78558	
1	7.322637	6.768390	0.9243
3	4.227726	4.117696	0.9740
5	3.274783	3.223351	0.9843
7	2.767696	2.736574	0.9888
10	2.315621	2.297361	0.9921
20	1.637319	1.630922	0.9961

The effect of the axial force for each case of the aspect ratio on natural frequency is given in Table 5.

Table 5. Natural frequency for each case of the axial forces and the aspect ratios, a/b, (rad/sec) $k=10^7 (N/m^2)$

a/b \ Nx(N/m)	1000	3000	5000	7000	10000
1	48.68064	48.64727	48.61361	48.52819	48.16460
2	49.45140	49.42568	49.37989	49.33506	49.07111
3	49.43317	49.40760	49.39196	49.31756	49.05547
4	49.43380	49.40827	49.38266	49.31836	49.05657
5	49.43381	49.41567	49.38069	49.31832	49.05650

Conclusion

In this paper, the simple and accurate method of vibration analysis developed by, Kim, D. H. is presented. The presented method is simple to use but extremely

accurate. The boundary condition can be arbitrary. Both stiffness and mass of the element can be variable. One can use any method to obtain the deflection influence coefficient needed for this method. The accuracy of the solution is dependent on only that of the influence coefficients needed for this method. One should recall that obtaining the deflection influence coefficients is the first step in design and analysis of a structure. The merit of the presented method is that it uses such influence coefficients values, used already for calculating deflection, slope, moment and shear to obtain the natural frequency of the structure. When the plate has concentrated mass or masses, one can simply add these masses to the plate mass and use the same deflection influence surfaces to obtain the natural frequency.
This method is applied to the special orthotrpic plate with free edges supported on elastic foundation and with a pair of opposite edges under axial forces. Such plate is the case of the most of the concrete highway slab and hybrid composite pavement on bridges. Such pavement may be subject to the axial forces parallel to the traffic direction. Some of the causes of such forces are the longitudinal gradient of the roadway, impact by the moving trucks, temperature changes, and others.
Finite difference method is used to obtain the deflection influence surfaces in this paper.
The effect of the modulus of foundation, the aspect ratio of the plate, and the magnitudes of the axial forces and the concentrated attached mass on the plate, on the natural frequency is thoroughly studied and the result is given in tables to provide a guideline to the design engineers.

References

[1] *Kim, D. H.*, Composite Structures for Civil and Architectural Engineering, Published by E & FN SPON, Chapman & Hall, London, 1995.

[2] *Kim, D. H.*, COMPOSITE MATERIALS FOR REPAIR AND REHABILITATION OF BUILDINGS AND INFRASTRUCTURES, Plenary Lecture at The Third International Symposium on Textile Composites In Building Construction, Seoul, Korea, November 7-9, 1996.

[3] *Kim, D. H.*, A SIMPLE METHOD OF ANALYSIS FOR THE PRELIMINARY DESIGN OF PARTICULAR COMPOSITE LAMINATED PRIMARY STRUCTURES FOR CIVIL CONSTRUCTION, Journal of Materials Processing Technology Vol. 55, Elsevier, London, 1995, pp 242-248.

[4] *Kim, D. H.*, POSSIBILITY OF USING THE CLASSICAL MECHANICS FOR THE PRELIMINARY DESIGN OF LAMINATED COMPOSITE STRUCTURES FOR CIVIL CONSTRUCTION, Proc of KSCE Conference, October, 1991.

[5] *Kim, D. H.*, A SIMPLIFIED METHOD OF VIBRATION ANALYSIS OF IRREGULARLY SHAPED COMPOSITE STRUCTURAL ELEMENTS, First International Society for the Advancement of Material and Process Engineering Symposium (JISSE 1), Tokyo, December, 1989.

[6] *Kim, D. H.*, VIBRATION ANALYSIS OF IRREGULARLY SHAPED COMPOSITE STRUCTURAL MEMBERS – FOR HIGHER MODES, 8th

Structures Congress, ASCE, Baltimore, MD., U.S.A., April 30 – May 3, 1990.

[7] *Kim, D. H.*, VIBRATION ANALYSIS OF IRREGULARLY SHAPED LAMINATED THICK COMPOSITE PLATES, ICCM 8, Honolulu, Hawaii, July, 1991.

[8] *Kim, D. H.*, DESIGN OF COMPOSITE MATERIAL STRUCTURES, China-Japan-USA Trilateral Symposium /Workshop on Earthquake Engineering, Harbin, China, November, 1991.

[9] *Kim, D. H.*, VIBRATION ANALYSIS OF IRREGULARLY SHAPED LAMINATED THICK COMPOSITE PLATES 2, JISSE-2, Tokyo, December, 1991.

[10] *Kim, D. H.*, THE EFFECT OF CONCENTRATED MASS/MASSES ON THE NATURAL FREQUENCY OF VIBRATION OF SPECIAL ORTHORTOPIC LAMINATES WITH VARIOUS BOUNDARY CONDITIONS, Third Asian-Pacific Conference on Computational Mechanics, Sheraton Hotel, Seoul, Korea, September 16-18, 1996.

[11] *Kim, D. H.*, THE EFFECT OF NEGLECTING OWN WEIGHT ON THE NATURAL FREQUENCY OF VIBRATION OF LAMINATED COMPOSITE PLATES WITH ATTACHED MASS/MASSES, The Third International Symposium on TEXTILE Composites In Building Construction, Seoul, Korea, November 7-9, 1996.

[12] *Kim, D. H.*, VIBRATION ANALYSIS OF SPECIAL ORTHOTROPIC PLATE WITH VARIABLE CROSS-SECTION, AND WITH A PAIR OF OPPOSITE EDGES SIMPLE SUPPORTED AND THE OTHER PAIR OF OPPOSITE EDGES FREE, Fourth Materials Congress, American Society of Civil Engineers, Washington, DC, November 10-14, 1996.

[13] *Kim, D. H.*, ANALYSIS OF TRIANGULARLY FOLDED PLATE ROOFS OF UMBRELLA TYPE, 16th Congress of Applied Mechanics, Tokyo, Japan, October, 1966.

[14] *Kim, D. H.*, THE EFFECT OF NEGLECTING THE RADIAL MOMENT TERMS IN ANALYZING A FINITE SECTORIAL PLATE BY MEANS OF FINITE DIFFERENCES, International Symposium on Space Technology and Sciences, Tokyo, Japan, May, 1967.

[15] *Kim, D. H.*, TRIDIAGONAL SCHEME TO SOLVE SUPER LARGE SIZE MATRICES BY THE USE OF COMPUTER, Jour., K.S.C.E., Vol 15-1, 1967.

[16] *Kim, D. H.*, GENERAL THEORY AND APPROXIMATE METHOD OF ANALYSIS OF NON-PRISMATIC FOLDED PLATE STRUCTURES, Jour., K.S.C.E., December, 1971.

[17] *Kim, D. H.*, A METHOD OF VIBRATION ANALYSIS OF IRREGULARLY SHAPED STRUCTURAL MEMBERS, Proceedings, International Symposium on Engineering Problems in Creating Coastal Industrial Sites, Seoul, Korea, October, 1974.

Natural Vibration Of Composite Plates Having Periodical Curvings in The Material Structure

Zafer KUTUG*

* **Civil Engineering Faculty, Department of Mechanics-Yildiz Technical University, 80750 Besiktas / ISTANBUL**

Abstract

In this study, the effect of existence of periodical curvings in the simply supported composite rectangular plate material structure on natural frequency of the plate is examined. Equation of motion of plate is established by using exact equations of theory of elasticity. Galerkin Method is used consistent with boundary conditions of composite plate. Numerical results show that the curvings, no matter how small they are, in the composite plate material structure can not be negligible, in many cases, in determination of natural frequency of the considered composite plate.

1. Introduction

In numerous papers related to the plates made from the layered composite materials, the investigations have been made by using continuum approach in which the pieceswise-homogeneous material of plates has been changed by homogeneous anisotropic material with normalized mechanical properties, and the various refined plate theories have been used. In that case, the layered composite plates have generally been considered as if their layers are ideal, i.e. they are parallel to mid-plane of the plate. But, in practice, layers of plates have some curvings due to producing process or designing. The effect of these curvings on the plate stress was analyzed, for example in [1,2] , and a continuum theory was proposed in [2] for such composites in which the material structure has periodical or local curvings. Additionally, some studies had been performed on the determination of the natural vibration frequency of the plate-strip fabricated from composite material in which the structure has small-scale curvings, for example in [3].

2. Formulation of the Problem and the Equation of Vibration of Plate

Boundaries of simply supported composite rectangular plate are such as $\{0 \le x_1 \le l_1 \,;\; -h/2 \le x_2 \le h/2 \,;\; 0 \le x_3 \le l_3\}$. Therefore, the boundary conditions are

$$u_2 = 0 \quad (x_1 = 0,\, l_1; x_3 = 0,\, l_3), \quad \sigma_{11} = 0 \quad (x_1 = 0,\, l_1), \quad \sigma_{33} = 0 \quad (x_3 = 0,\, l_3) \quad (2.1)$$

and at the upper and lower surface of plate

$$\sigma_{12} = 0 \quad , \quad \sigma_{22} = 0 \quad , \quad \sigma_{23} = 0 \quad (x_2 = \pm h/2) \tag{2.2}$$

conditions are to be satisfied. The equation of motion is established by using the exact equations of the theory of elasticity, i.e.,

$$\frac{\partial \sigma_{ij}}{\partial x_j} = \rho \frac{\partial^2 u_i}{\partial t^2} \qquad i, j = 1,\ 2,\ 3. \tag{2.3}$$

Equations of constitution are such as

$$\begin{Bmatrix} \sigma_{11} \\ \sigma_{22} \\ \sigma_{33} \\ \sigma_{23} \\ \sigma_{13} \\ \sigma_{12} \end{Bmatrix} = \begin{bmatrix} A_{11} & A_{12} & A_{13} & A_{14} & A_{15} & A_{16} \\ & A_{22} & A_{23} & A_{24} & A_{25} & A_{26} \\ & & A_{33} & A_{34} & A_{35} & A_{36} \\ & & & A_{44} & A_{45} & A_{46} \\ & & & & A_{55} & A_{56} \\ & & & & & A_{66} \end{bmatrix} \begin{Bmatrix} \varepsilon_{11} \\ \varepsilon_{22} \\ \varepsilon_{33} \\ \varepsilon_{23} \\ \varepsilon_{13} \\ \varepsilon_{12} \end{Bmatrix} \tag{2.4}$$

where

$$\varepsilon_{ij} = \frac{1}{2}\left(\frac{\partial u_i}{\partial x_j} + \frac{\partial u_j}{\partial x_i} \right) \qquad i, j = 1,\ 2,\ 3. \tag{2.5}$$

and the material parameters are

$$\begin{aligned}
&A_{11}(x_1) = A_{11}^0 \Phi_1^{\ 4} + 2(A_{12}^0 + 2A_{66}^0)(\Phi_1\Phi_2)^2 + A_{22}^0 \Phi_2^{\ 4} \ , \\
&A_{22}(x_1) = A_{11}^0 \Phi_2^{\ 4} + 2(A_{12}^0 + 2A_{66}^0)(\Phi_1\Phi_2)^2 + A_{22}^0 \Phi_1^{\ 4} \ , \\
&A_{33}(x_1) = A_{33}^0 \ , \qquad A_{44}(x_1) = A_{44}^0 \Phi_1^{\ 2} + A_{55}^0 \Phi_2^{\ 2} \ , \qquad A_{55}(x_1) = A_{55}^0 \Phi_1^{\ 2} + A_{44}^0 \Phi_2^{\ 2} \ , \\
&A_{66}(x_1) = (A_{11}^0 - 2A_{12}^0 + A_{22}^0)(\Phi_1\Phi_2)^2 + A_{66}^0 (\Phi_1^{\ 2} - \Phi_2^{\ 2})^2 \ , \\
&A_{12}(x_1) = (A_{11}^0 + A_{22}^0 - 4A_{66}^0)(\Phi_1\Phi_2)^2 + A_{12}^0 (\Phi_1^{\ 4} + \Phi_2^{\ 4}) \ , \\
&A_{13}(x_1) = A_{13}^0 \Phi_1^{\ 2} + A_{23}^0 \Phi_2^{\ 2} \ , \qquad A_{23}(x_1) = A_{13}^0 \Phi_2^{\ 2} + A_{23}^0 \Phi_1^{\ 2} \ , \\
&A_{16}(x_1) = -A_{11}^0 \Phi_1^{\ 3} \Phi_2 + (A_{12}^0 + 2A_{66}^0)(\Phi_1^{\ 3}\Phi_2 - \Phi_1\Phi_2^{\ 3}) + A_{22}^0 \Phi_1 \Phi_2^{\ 3} \ , \\
&A_{26}(x_1) = -A_{11}^0 \Phi_1 \Phi_2^{\ 3} - (A_{12}^0 + 2A_{66}^0)(\Phi_1^{\ 3}\Phi_2 - \Phi_1\Phi_2^{\ 3}) + A_{22}^0 \Phi_1^{\ 3} \Phi_2 \ , \\
&A_{36}(x_1) = (A_{23}^0 - A_{13}^0)\Phi_1\Phi_2 \ , \qquad A_{45}(x_1) = (A_{44}^0 - A_{55}^0)\Phi_1\Phi_2 \ .
\end{aligned} \tag{2.6}$$

The notations used in eqs.(2.1)-(2.6) are conventional. Continuum theory of which Akbarov and Guz' had proposed for such materials is used in the derivation of eqs.(2.6). Assuming the considered plate material consists of alternating two layers each of which is homogeneous isotropic, the terms A_{11}^0, A_{22}^0, A_{33}^0, A_{44}^0, A_{55}^0, A_{66}^0, A_{12}^0, A_{13}^0, A_{23}^0 in eqs.(2.6) are constants of plate material in ideal case which there exist no curvings in the plate structure and determined as follows [4]

$$A_{11}^0 = \mu_1\eta_1 + \mu_2\eta_2 + (\mu_1+\eta_1)\eta_1 + (\mu_2+\eta_2)\eta_2 - \frac{(\lambda_1-\lambda_2)^2}{(\lambda_1+2\mu_1)\eta_2+(\lambda_2+2\mu_2)\eta_1}$$

$$A_{12}^0 = \lambda_1\eta_1 + \lambda_2\eta_2 - (\lambda_1-\lambda_2)\eta_1\eta_2 \frac{(\lambda_1+2\mu_1)(\lambda_2+2\mu_2)}{(\lambda_1+2\mu_1)\eta_2+(\lambda_2+2\mu_2)\eta_1}$$

$$A_{22}^0 = (\lambda_1+2\mu_1)\eta_1 + (\lambda_2+2\mu_2)\eta_2 - \eta_1\eta_2 \frac{\left[(\lambda_1+2\mu_1)-(\lambda_2+2\mu_2)\right]^2}{(\lambda_1+2\mu_1)\eta_2+(\lambda_2+2\mu_2)\eta_1}$$

$$A_{66}^0 = \frac{\mu_1\mu_2}{\mu_1\eta_2+\mu_2\eta_1}, \quad A_{33}^0 = A_{11}^0, \quad A_{44}^0 = \eta_1\mu_1+\eta_2\mu_2, \quad A_{55}^0 = A_{66}^0,$$

$$A_{13}^0 = A_{11}^0 - 2A_{44}^0, \quad A_{23}^0 = A_{12}^0 \tag{2.7}$$

where E_1 and E_2, ν_1 and ν_2, λ_1, μ_1 and λ_2, μ_2; η_1 and η_2 are elasticity modulei, Poisson ratios, Lamé constants and volume ratios, respectively, of two homogen isotropic materials of which composite plate made.

The functions

$$\Phi_1 = \Phi_1(x_1) = \frac{1}{\sqrt{1+\left(a(x_1)\right)^2}}, \qquad \Phi_2 = \Phi_2(x_1) = a(x_1)\Phi_1(x_1), \tag{2.8}$$

where

$$a(x_1) = \varepsilon \frac{\partial f(x_1)}{\partial x_1} \tag{2.9}$$

In eq.(2.9) $\varepsilon \cdot f(x_1)$ indicates the form of the curvings in the plate structure. ε is a parameter defining the curvings order.

Additionally, Kromm's refined plate theory is used [5-6], i.e.

$$u_i(x_1,x_2,x_3,t) = u_i(x_1,x_3,t) - x_2 \frac{\partial w(x_1,x_3,t)}{\partial x_i} + J_i(x_2)\varphi_i(x_1,x_3,t) \quad i=1,3. \; (\Sigma i),$$

$$u_2(x_1,x_2,t) = w(x_1,t) \tag{2.10}$$

where $\varphi_i(x_1,x_3,t)$ is a function defining the average shear strain of the plate section of which the normal is Ox_i (i=1,3.) direction, and the term $J_i(x_2)$ is such that

$$J_1(x_2) = J_3(x_2) = J(x_2) = \frac{1}{4}\left(\frac{h^2}{4}x_2 - \frac{x_2^3}{3}\right) \tag{2.11}$$

If eqs.(2.3) are integrated with respect to x_2 from $-h/2$ to $h/2$, and in addition to this, if eqs.(2.3) are firstly multiplied by x_2 and then integrated with respect to x_2 from $-h/2$ to $h/2$, equations of motion of plate is obtained in terms of forces and moments:

$$\frac{\partial T_{11}}{\partial x_1}+\frac{\partial T_{13}}{\partial x_3}=\rho h\frac{\partial^2 u}{\partial t^2}\ ,\quad \frac{\partial N_{12}}{\partial x_1}+\frac{\partial N_{23}}{\partial x_3}=\rho h\frac{\partial^2 w}{\partial t^2}\ ,\quad \frac{\partial T_{13}}{\partial x_1}+\frac{\partial T_{33}}{\partial x_3}=\rho h\frac{\partial^2 v}{\partial t^2}$$

$$\frac{\partial M_{11}}{\partial x_1}+\frac{\partial M_{13}}{\partial x_3}-N_{12}=-2\rho I_1(h)\frac{\partial^3 w}{\partial x_1\partial t^2}+\rho I_2(h)\frac{\partial^2 \varphi_1}{\partial t^2}$$

$$\frac{\partial M_{13}}{\partial x_1}+\frac{\partial M_{33}}{\partial x_3}-N_{23}=-2\rho I_1(h)\frac{\partial^3 w}{\partial x_3\partial t^2}+\rho I_2(h)\frac{\partial^2 \varphi_3}{\partial t^2} \tag{2.12}$$

where

$$I_1(h)=\int_{-h/2}^{h/2}\frac{dJ(x_2)}{dx_2}dx_2=\frac{h^3}{24}\ ,\quad I_2(h)=\int_{-h/2}^{h/2}x_2J(x_2)dx_2=\frac{h^5}{240}\ ,$$

$$T_{11}=\int_{-h/2}^{h/2}\sigma_{11}\,dx_2\ ,\quad T_{13}=\int_{-h/2}^{h/2}\sigma_{13}\,dx_2\ ,\quad T_{33}=\int_{-h/2}^{h/2}\sigma_{33}\,dx_2\ ,\quad N_{12}=\int_{-h/2}^{h/2}\sigma_{12}\,dx_2\ ,$$

$$N_{23}=\int_{-h/2}^{h/2}\sigma_{23}\,dx_2\ ,\ M_{11}=\int_{-h/2}^{h/2}\sigma_{11}x_2\,dx_2\ ,\ M_{13}=\int_{-h/2}^{h/2}\sigma_{13}x_2\,dx_2\ ,\ M_{33}=\int_{-h/2}^{h/2}\sigma_{33}x_2\,dx_2 \tag{2.13}$$

Introducing eq.(2.5) into eq.(2.4) and using eq.(2.4) in eqs.(2.13),

$$\begin{Bmatrix}T_{11}\\T_{33}\\N_{23}\\T_{13}\\N_{12}\end{Bmatrix}=\begin{bmatrix}A_{11}&A_{13}&0&0&A_{16}\\A_{13}&A_{33}&0&0&A_{36}\\0&0&A_{44}&A_{45}&0\\0&0&A_{45}&A_{55}&0\\A_{16}&A_{36}&0&0&A_{66}\end{bmatrix}\begin{Bmatrix}hu_{,1}\\hv_{,3}\\I_1(h)\varphi_3\\h(u_{,3}+v_{,1})\\I_1(h)\varphi_1\end{Bmatrix}$$

$$\begin{Bmatrix}M_{11}\\M_{33}\\M_{13}\end{Bmatrix}=\begin{bmatrix}A_{11}&A_{13}&0\\A_{13}&A_{33}&0\\0&0&A_{55}\end{bmatrix}\begin{Bmatrix}-2I_1(h)w_{,11}+I_2(h)\varphi_{1,1}\\-2I_1(h)w_{,33}+I_2(h)\varphi_{3,3}\\-4I_1(h)w_{,13}+I_2(h)(\varphi_{1,3}+\varphi_{3,1})\end{Bmatrix}\ . \tag{2.14}$$

obtained, where (,i) indicates the derivation with respect to x_i (i=1,3.). If the mathematical process which is applied to eq.(2.3) is also applied to conditions in eq.(2.1) then the boundary conditions become as follows:

$$T_{11}=0\ ,\quad M_{11}=0\ ,\quad w=0\quad (x_1=0,l_1)$$
$$T_{33}=0\ ,\quad M_{33}=0\ ,\quad w=0\quad (x_3=0,l_3). \tag{2.15}$$

3. Solution of the Problem

For the solution of the considered problem, Galerkin Method is used. Therefore, the displacements of the problem are defined as

$$u(x_1,x_3,t)=X_u(x_1,x_3)e^{i\omega t}=\sum_{m=1}^{N}\sum_{n=1}^{N}u_{mn}\cos\frac{m\pi x_1}{l_1}\sin\frac{n\pi x_3}{l_3}e^{i\omega t}\ ,$$
$$v(x_1,x_3,t)=X_v(x_1,x_3)e^{i\omega t}=\sum_{m=1}^{N}\sum_{n=1}^{N}v_{mn}\sin\frac{m\pi x_1}{l_1}\cos\frac{n\pi x_3}{l_3}e^{i\omega t}\ ,$$
$$w(x_1,x_3,t)=X_w(x_1,x_3)e^{i\omega t}=\sum_{m=1}^{N}\sum_{n=1}^{N}w_{mn}\sin\frac{m\pi x_1}{l_1}\sin\frac{n\pi x_3}{l_3}e^{i\omega t}\ ,$$
$$\varphi_1(x_1,x_3,t)=X_{\varphi_1}(x_1,x_3)e^{i\omega t}=\sum_{m=1}^{N}\sum_{n=1}^{N}\varphi_{1mn}\cos\frac{m\pi x_1}{l_1}\sin\frac{n\pi x_3}{l_3}e^{i\omega t}\ ,$$
$$\varphi_3(x_1,x_3,t)=X_{\varphi_3}(x_1,x_3)e^{i\omega t}=\sum_{m=1}^{N}\sum_{n=1}^{N}\varphi_{3mn}\sin\frac{m\pi x_1}{l_1}\cos\frac{n\pi x_3}{l_3}e^{i\omega t} \tag{3.1}$$

so that the boundary conditions are automatically satisfied. Substituting eq.(2.14) and eq.(3.1) into eq.(2.12), the equation for functions X_u, X_v, X_w, X_{φ_1}, X_{φ_3} is obtained as:

$$\left(hA_{11}X_{u,1}+hA_{13}X_{v,3}+I_1(h)A_{16}X_{\varphi_1}\right)_{,1}+\left(I_1(h)A_{45}X_{\varphi_3}+hA_{55}\left(X_{u,3}+X_{v,1}\right)\right)_{,3}=-\omega^2\rho hX_u$$
$$\left(I_1(h)A_{45}X_{\varphi_3}+hA_{55}\left(X_{u,3}+X_{v,1}\right)\right)_{,1}+\left(hA_{13}X_{u,1}+hA_{33}X_{v,3}+I_1(h)A_{36}X_{\varphi_1}\right)_{,3}=-\omega^2\rho hX_v$$
$$\left(hA_{16}X_{u,1}+hA_{36}X_{v,3}+I_1(h)A_{66}X_{\varphi_1}\right)_{,1}+\left(I_1(h)A_{44}X_{\varphi_3}+hA_{45}\left(X_{u,3}+X_{v,1}\right)\right)_{,3}=-\omega^2\rho hX_w$$

$$\left[A_{11}\left(-2I_1(h)X_{w,11}+I_2(h)X_{\varphi_1,1}\right)+A_{13}\left(-2I_1(h)X_{w,33}+I_2(h)X_{\varphi_3,3}\right)\right]_{,1}+$$
$$\left[A_{55}\left(-4I_1(h)X_{w,13}+I_2(h)\left(X_{\varphi_1,3}+X_{\varphi_3,1}\right)\right)\right]_{,3}$$
$$-\left(hA_{16}X_{u,1}+hA_{36}X_{v,3}+I_1(h)A_{66}X_{\varphi_1}\right)=\omega^2\rho I_1(h)X_{w,1}-\omega^2\rho I_2(h)X_{\varphi_1}$$

$$\left[A_{55}\left(-4I_1(h)X_{w,13}+I_2(h)\left(X_{\varphi_1,3}+X_{\varphi_3,1}\right)\right)\right]_{,1}+$$
$$\left[A_{13}\left(-2I_1(h)X_{w,11}+I_2(h)X_{\varphi_1,1}\right)+A_{33}\left(-2I_1(h)X_{w,33}+I_2(h)X_{\varphi_3,3}\right)\right]_{,3}- \tag{3.2}$$
$$\left(hA_{45}\left(X_{u,3}+X_{v,1}\right)+I_1(h)A_{44}X_{\varphi_3}\right)=\omega^2\rho I_1(h)X_{w,3}-\omega^2\rho I_2(h)X_{\varphi_3}$$

where ω is the natural frequency of the composite plate. Boundary conditions can also be obtained in the same manner:

$$\left[A_{11}\left(-2I_1(h)X_{w,11}+I_2(h)X_{\varphi_1,1}\right)+A_{13}\left(-2I_1(h)X_{w,33}+I_2(h)X_{\varphi_3,3}\right)\right]\Big|_{x_1=0,l_1}=0 \;,$$
$$\left[A_{11}hX_{u,1}+A_{13}hX_{v,3}+A_{16}I_1(h)X_{\varphi_1}\right]\Big|_{x_1=0,l_1}=0 \;,$$
$$\left[A_{13}\left(-2I_1(h)X_{w,11}+I_2(h)X_{\varphi_1,1}\right)+A_{33}\left(-2I_1(h)X_{w,33}+I_2(h)X_{\varphi_3,3}\right)\right]\Big|_{x_3=0,l_3}=0 \;,$$
$$\left[A_{13}hX_{u,1}+A_{33}hX_{v,3}+A_{36}I_1(h)X_{\varphi_1}\right]\Big|_{x_3=0,l_3}=0 \;, \quad X_w\Big|_{\substack{x_1=0,l_1\\ x_3=0,l_3}}=0 \;. \tag{3.3}$$

It can easily be seen that all boundary conditions except the second one of eq.(3.3) are satisfied by the displacement functions in eq.(3.1). If the form of the curvings in the composite plate material is taken as $A_{16}(0)=A_{16}(l_1)=0$ so that second boundary condition of eq.(3.3) is satisfied, the numerical results will be exact otherwise approximate. Introduce the dimensionless coordinates $x=x_1/l_1$, $y=x_3/l_3$ and dimensionless parameters $\lambda=l_1/l_3$, $\overline{\omega}^2=\rho\omega^2 l_1^2/A_{22}^0$.

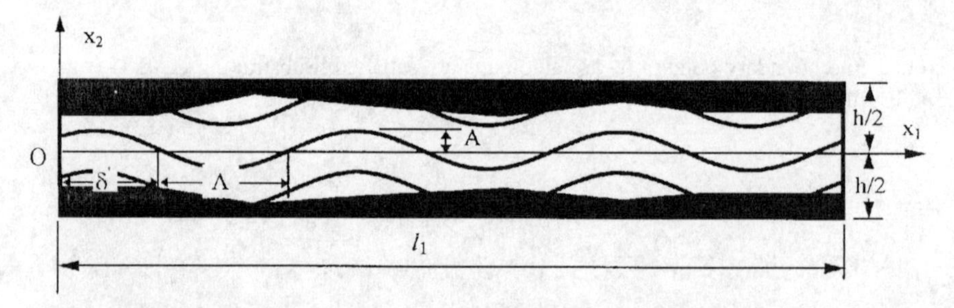

Figure 1. Form of the curvings in the composite plate material

In this case, the form of the curvings in the composite plate material structure (Fig.1) can be given as

$$\varepsilon\cdot f(x)=\varepsilon\cdot\sin(\gamma\pi x+\delta) \tag{3.4}$$

where $\delta=\pi\delta'/l_1$, $\varepsilon=A/\Lambda$, $\gamma=[|l_1/\Lambda|]$. As it is formerly mentioned, if and only if $\delta=\pi/2$ then the eqs.(3.3) are all satisfied and the numerical results will be exact.

Table 1. The results of $\overline{\omega}_I^2$ for small values of ε. ($\delta = \pi/2$)

E_2/E_1	λ l_1/l_3	ε							
		0	**0.01**	**0.02**	**0.03**	**0.04**	**0.05**	**0.06**	**0.07**
10	**0.0**	0.2085	0.2083	0.2079	0.2075	0.2067	0.2058	0.2046	0.2034
20		0.3521	0.3519	0.3513	0.3503	0.3489	0.3474	0.3452	0.3429
50		0.7009	0.7007	0.6999	0.6987	0.6972	0.6950	0.6923	0.6890
100		1.1011	1.1011	1.1011	1.1011	1.1009	1.1005	1.0997	1.0987
10	**0.1**	0.2106	0.2104	0.2101	0.2097	0.2089	0.2079	0.2067	0.2056
20		0.3548	0.3546	0.3540	0.3532	0.3519	0.3501	0.3481	0.3458
50		0.7054	0.7052	0.7044	0.7032	0.7017	0.6995	0.6968	0.6937
100		1.1075	1.1075	1.1075	1.1075	1.1073	1.1071	1.1065	1.1056
10	**0.5**	0.2749	0.2749	0.2745	0.2741	0.2735	0.2728	0.2720	0.2708
20		0.4454	0.4454	0.4448	0.4442	0.4433	0.4419	0.4403	0.4386
50		0.8659	0.8659	0.8655	0.8651	0.8644	0.8634	0.8620	0.8603
100		1.3651	1.3653	1.3661	1.3673	1.3688	1.3708	1.3729	1.3751
10	**1.0**	0.6282	0.6282	0.6282	0.6280	0.6278	0.6276	0.6272	0.6269
20		1.0042	1.0042	1.0040	1.0040	1.0038	1.0034	1.0032	1.0026
50		2.0138	2.0138	2.0144	2.0149	2.0159	2.0171	2.0185	2.0200
100		3.4300	3.4306	3.4323	3.4354	3.4396	3.4448	3.4513	3.4587

Table 2. The results of $\overline{\omega}_I^2$ for high values of ε. ($\delta = \pi/2$)

E_2/E_1	λ l_1/l_3	ε							
		0	**0.1**	**0.2**	**0.3**	**0.4**	**0.5**	**0.6**	**0.7**
10	**0.0**	0.2085	0.1983	0.1753	0.1517	0.1339	0.1218	0.1136	0.1079
20		0.3521	0.3337	0.2876	0.2366	0.1960	0.1681	0.1495	0.1370
50		0.7009	0.6753	0.5915	0.4761	0.3726	0.2964	0.2446	0.2106
100		1.1011	1.0921	1.0087	0.8358	0.6489	0.4991	0.3931	0.3218
10	**0.1**	0.2106	0.2007	0.1776	0.1542	0.1366	0.1245	0.1165	0.1110
20		0.3548	0.3366	0.2909	0.2403	0.2001	0.1724	0.1540	0.1419
50		0.7054	0.6804	0.5976	0.4833	0.3806	0.3052	0.2540	0.2204
100		1.1075	1.0995	1.0185	0.8481	0.6634	0.5151	0.4103	0.3401
10	**0.5**	0.2749	0.2669	0.2479	0.2286	0.2144	0.2050	0.1991	0.1954
20		0.4454	0.4315	0.3948	0.3534	0.3206	0.2989	0.2853	0.2771
50		0.8659	0.8526	0.7956	0.7073	0.6257	0.5669	0.5290	0.5060
100		1.3651	1.3808	1.3552	1.2435	1.1071	0.9962	0.9200	0.8726
10	**1.0**	0.6282	0.6253	0.6169	0.6083	0.6028	0.6005	0.5999	0.6005
20		1.0042	1.0009	0.9874	0.9694	0.9560	0.9489	0.9470	0.9478
50		2.0138	2.0249	2.0310	2.0108	1.935	1.9651	1.9571	1.9571
100		3.4300	3.4864	3.5935	3.6384	3.6288	3.6095	3.5991	3.6001

In numerical investigations, it is assumed that $\gamma = 10$ (in eq.(3.4)), N=20, n=1 (in eq.(3.1)), $\nu_1 = \nu_2 = 0.25$, $\eta_1 = \eta_2 = 0.5$ (in eq.(2.7)) and $h/l_1 = 0.1$. Numerical results are shown in Table 1 and Table 2.

Conclusions

Several numerical investigations have been performed with respect to different parameters such as $\lambda = l_1/l_3$, ε and $E = E_1/E_2$, and $\overline{\omega}$ dimensionless natural frequencies of composite plates have been obtained. According to these results, it can be concluded that

- on the contrary of ideal case, there exist an effect of longitudinal (Ox_1) vibration on the lateral (Ox_2) vibration of the composite plate;
- for same ε and $E = E_1/E_2$ values, dimensionless $\overline{\omega}$ natural frequencies of composite plates increase as λ parameters grow;
- for same $\lambda = l_1/l_3$ and $E = E_1/E_2$ values, dimensionless $\overline{\omega}$ natural frequencies of composite plates , generally, decrease with growth of the values of ε .

Acknowledgements

The author wishes to thank Prof. Dr. Surkay D. AKBAROV due to his valuable suggestions and encouragement.

References

[1] S.D. Akbarov, A.N. Guz' , STATICS OF LAMINATED AND FIBROUS COMPOSITES WITH CURVED STRUCTURES , Appl. Mech. Rev. ASME. V.45,pp.1735, 1992

[2] S.D. Akbarov, A.N. Guz' , ON THE CONTINUUM THEORY IN MECHANICS OF COMPOSITE MATERIAL WITH SMALL-SCALE CURVED STRUCTURES Prikl. Mech. 27, No:2, 3-13, 1991 (in Russian)

[3] Z. Kutug, NATURAL VIBRATION OF THE BEAM-STRIP FABRICATED FROM A COMPOSITE MATERIAL WITH SMALL-SCALE CURVINGS IN THE STRUCTURE. Mechanics of Composite Materials, Vol.32, 502-512. 1996

[4] R.M. Christensen, 1979. MECHANICS OF COMPOSITE MATERIALS, John Wiley and Sons, Inc.

[5] A. Kromm, ÜBER DIE RANDQUERKROFTE BEIGESTUTZTEN PLATTEN , ZAMM, 35,231-242, 1955

[6] A. Kromm, VERALLGENEIERTE THEORIE DER PLATTENSTATIK, Ing. Arch., 21,266-286, 1953.

Decay length of end effects in orthotropic laminates

N. Tullini*, **M. Savoia***

* Distart - Structural Engineering, Faculty of Engineering, University of Bologna, Viale Risorgimento 2, 40136, Bologna - Italy

Abstract

An eigenfunction technique for the estimate of decay length of end effects in multilayered composite laminates is presented. Laminated strips with general layout and interlaminar imperfect bonding are considered. Slower stress decay of end effects is predicted for larger interlaminar slips. For periodic laminates, homogenization technique is used to obtain explicit expressions for effective elastic moduli, decay rate of end effects and microstress distributions at the layer level. Important practical implications of these studies can be found in the fields of analysis, design and mechanical testing of composite materials and structures.

1. Introduction

The study of end (or edge) effects has recently received a great attention in the engineering analysis and design of composite structures. In fact, due to material anisotropy and through-the-thickness layout, many laminate composite structures are very sensitive to local loading, which may often result in delamination phenomena. Hence, the extent of regions where local stresses are important must be known (requiring more detailed analyses based on theory of elasticity), in order to distinguish them from regions where strength of material theory can be safely used.
Theoretical and experimental studies pointed out that decay of local effects in strongly anisotropic structural elements is much slower than that for their isotropic counterparts (see e.g. [1, 2, 3]). For strongly orthotropic strips (as in the case of fiber-reinforced composites), the characteristic decay length has been found to be of the order $H(E_1/G_{12})^{1/2}$, where H, E_1 and G_{12} are the strip height, longitudinal Young modulus and shear modulus. For instance, for unidirectional carbon fiber-reinforced composites, local effects may decay 5 ÷ 7 times slower than for isotropic materials.
The implications of these results in the analysis and design of composite materials and structures are widespread. For instance, in the mechanical testing of anisotropic materials, appropriate specimen sizes and strain gage placements must be adopted in order to disregard the effects of clamping of the extremities [4]. Local stresses are also very important for composite material lap joints and plating systems by means of thin FRP laminae [5].
The most widely used analytical scheme for investigation of Saint-Venant end effects is the eigenfunction technique [1]. General laminations and possible imperfect bonding at the layer interfaces have been considered in [6, 7]. For perfectly bonded isotropic symmetric sandwich strips, it has been shown that very slow decay of end effects occurs when the Young's modulus of the core is small compared with that of face layers [2].

In the present study, an eigenvalue technique to estimate the decay length of end effects for generally laminated orthotropic strips with interlaminar imperfect bonding is presented. Plane elasticity condition is considered. The Airy stress function is taken as the product of an exponentially decaying function in the axial direction and an unknown function (the eigenfunction) over the strip height. It is shown that the presence of interfacial imperfect bonding usually causes slow decay of end effects. The typical case of a laminated strip with periodic layout is addressed making use of homogenization techniques. Explicit expressions for effective elastic moduli and for microstress variations (i.e., at the layer level) are given. The numerical application at the end of the paper shows that the eigenvalues computed for the homogenized material represent the asymptotic values of exact solutions when the number of layers increases, so assessing the validity of homogenization method.

2. Governing equations

A semi-infinite ($x_1 \geq 0$) multilayered rectangular strip in plane strain with total height $H = 2h$ ($-h \leq x_2 \leq h$) is considered. The strip is made of S orthotropic elastic layers of height $h^{(s)}$ ($s = 1, ..., S$), with orthotropy axes coinciding with the reference axes (cross-ply lamination scheme). The long faces of the strip ($x_2 = \pm h$) are traction free and the end section $x_1 = 0$ is subject to a self-equilibrated load distribution. The stress components of the sth layer, satisfying the equilibrium equations with null body forces, are written in terms of Airy stress function $F^{(s)}(x, y)$ as:

$$\sigma_{11}^{(s)} = F_{,22}^{(s)}, \qquad \sigma_{22}^{(s)} = F_{,11}^{(s)}, \qquad \sigma_{12}^{(s)} = -F_{,12}^{(s)} \tag{1}$$

The compatibility equation yields the governing differential equation:

$$F_{,2222}^{(s)} + \overline{E}^{(s)} F_{,1122}^{(s)} + \left(\varepsilon^{(s)}\, \overline{E}^{(s)}\right)^2 F_{,1111}^{(s)} = 0 \tag{2}$$

where:

$$\overline{E}^{(s)} = \frac{R_{66}^{(s)} + 2R_{12}^{(s)}}{R_{11}^{(s)}}, \qquad \left(\varepsilon^{(s)}\, \overline{E}^{(s)}\right)^2 = \frac{R_{22}^{(s)}}{R_{11}^{(s)}} \tag{3}$$

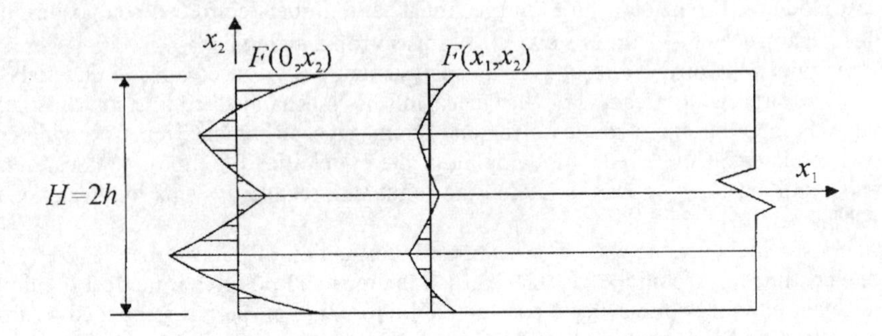

Figure. 1. Decay of end effects in a multilayered strip under plane elasticity.

The form (2) adopted for the compatibility equation is particularly useful to investigate the influence of ddegree of orthotropy of the composite. In fact, for an isotropic material $\overline{E}=2$ and $\varepsilon=0.5$, whereas for strongly orthotropic layers, i.e. when E_1/G_{12} is large (typically for fiber-reinforced composites), $\overline{E}\rightarrow\infty$ and $\varepsilon\rightarrow 0$.
Introducing the dimensionless variables $x = x_1/h$, $y = x_2/h$, the strip domain reduces to $x \geq 0$, $|y| \leq 1$. In the spirit of Saint-Venant's principle, a decaying solution of eqn (2) with boundary conditions representing self-equilibrated load distributions is sought in the form:

$$F^{(s)}(x, y) = h^2\, e^{-\lambda x}\, \psi^{(s)}(y)\,, \tag{4}$$

where λ is the eigenvalue governing the stress decay from the loaded edge, and $\psi^{(s)}(y)$ is the corresponding eigenfunction. The stress components (1) for the sth layer may then be written as

$$\sigma_{11}^{(s)} = e^{-\lambda x}\frac{d^2\,\psi^{(s)}}{d\,y^2}\,, \qquad \sigma_{22}^{(s)} = \lambda^2\, e^{-\lambda x}\,\psi^{(s)}\,, \qquad \sigma_{12}^{(s)} = \lambda\, e^{-\lambda x}\frac{d\,\psi^{(s)}}{d\,y} \tag{5}$$

By integration of compatibility equations, analogous expressions can be obtained for displacement components [6]:

$$u_1^{(s)} = -\lambda\, h\, e^{-\lambda x}\left(\frac{R_{11}^{(s)}}{\lambda^2}\frac{d^2\,\psi^{(s)}}{d\,y^2} + R_{12}^{(s)}\psi^{(s)}\right) \tag{6a}$$

$$u_2^{(s)} = -\lambda\, h\, e^{-\lambda x}\left(\frac{R_{11}^{(s)}}{\lambda^3}\frac{d^3\,\psi^{(s)}}{d\,y^3} + \frac{R_{66}^{(s)} + R_{12}^{(s)}}{\lambda}\frac{d\,\psi^{(s)}}{d\,y}\right) \tag{6b}$$

to within a rigid body displacement. Substituting eqn (4) into eqn (2) yields a fourth-order ordinary differential equation for the eigenfunction $\psi^{(s)}(y)$, which must be integrated by imposing the traction-free boundary conditions on the external faces ($y = \pm 1$) and imperfect bonding at the interfaces y_s ($s = 1, ..., S-1$) [7]:

$$\sigma_{12}^{(1)}(x,-1) = \sigma_{22}^{(1)}(x,-1) = 0\,, \qquad \sigma_{12}^{(S)}(x,1) = \sigma_{22}^{(S)}(x,1) = 0\,, \tag{7a,b}$$

$$\left\{\sigma_{12}^{(s)}, \sigma_{22}^{(s)}\right\} = \left\{\sigma_{12}^{(s+1)}, \sigma_{22}^{(s+1)}\right\} \qquad \text{for } y = y_s \tag{7c}$$

$$\left\{u_1^{(s+1)} - u_1^{(s)}, u_2^{(s+1)} - u_2^{(s)}\right\} = \left\{r_1\sigma_{12}^{(s+1)}, r_2\sigma_{22}^{(s+1)}\right\} \qquad \text{for } y = y_s \tag{7d}$$

where r_1 and r_2 denote interface compliance coefficients. The imperfect bonding is expressed in eqn (7d) by displacement jumps, which are proportional to their associated traction components. For thin flexible adhesive layers, the interface compliances r_1 and r_2 can be written as [8, 9]:

$$r_1 = h^{(a)} R_{66}^{(a)}\,, \qquad r_2 = h^{(a)} R_{22}^{(a)} \tag{8}$$

where $h^{(a)}$, $R_{66}^{(a)}$, $R_{22}^{(a)}$ are the height and elastic constants of the adhesive. When $r_1 = 0$, r_2

$=0$, one recovers the perfect bonding case considered in [6].
Making use of eqns (5, 6), eqns (7) give an homogeneous system of $4S$ equations (eigenvalue problem). Imposing the vanishing of the determinant of the coefficients yields an infinite set of eigenvalue-eigenvector couples. The characteristic decay length d of end effects is conventionally defined as the distance over which stress field decays to 1% of its value at $x = 0$. It is related to smallest eigenvalue λ_1 through the relation $d/h = \ln(100)/\mathrm{Re}(\lambda_1)$. Additional details are reported in [6].
Figure 2 shows the decay rate λ_1 for a bimaterial strip with layers made of Aluminum ($E = 70\,GPa$, $\nu = 0.3$) and of high-strength graphite-epoxy composite (HS1, $E_1 = 137.895\,GPa$, $E_2 = 6.89476\,GPa$, $G_{12} = 4.13685\,GPa$, $\nu_{12} = \nu_{13} = \nu_{32} = 0.25$ and, according to eqns (3), $\bar{E} = 32.811$, $\varepsilon = 0.1322$). The decay rate is computed for different values of thickness of the aluminum layer $f^{(1)} = h^{(1)}/H$ and imperfect bonding between layers is considered. In this case, the first eigenvalue is always real. For an isotropic adhesive, the slip constants χ_1 and χ_2, defined as:

$$\chi_\delta = \frac{r_\delta}{hR_{11}^{(1)}} \qquad (\delta = 1, 2) \tag{9}$$

are related through the relation:

$$\frac{\chi_2}{\chi_1} = \frac{1 - \nu^{(a)}}{2} \tag{10}$$

Hence, $0.25 \le \chi_2/\chi_1 \le 0.5$, corresponding to values of Poisson's ratio of the adhesive in the range $0 \le \nu^{(a)} \le 1/2$. In Figure 2, $\chi_2 = 0.35\,\chi_1$ is considered. As for the order of magnitude of the adhesive compliance coefficients, it is usual to consider $\chi_1 < 1$ for relatively inflexible adhesive layers, $1 \le \chi_1 \le 10$ for intermediate flexibility and $\chi_1 > 10$

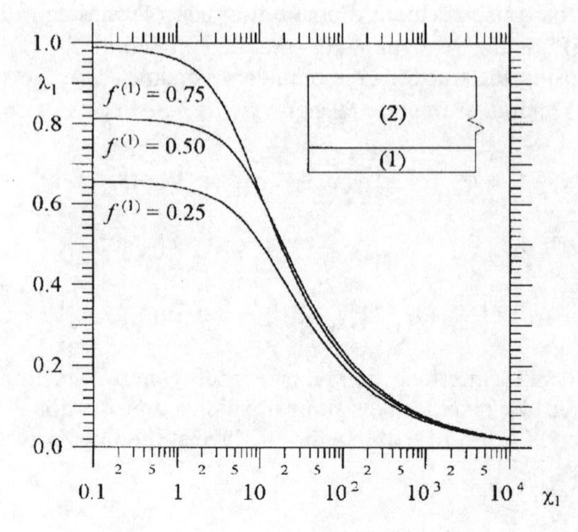

Figure. 2. Bimaterial strip [Al/HS1]: Decay rates λ_1 versus the slip constant χ_1 (with $\chi_2/\chi_1 = 0.35$), for different values of layer thicknesses.

for flexible adhesives.

Small values of λ_1 correspond to large decay lengths of end effects. The figure shows that, when the joint is mainly composed of the composite material HS1 ($f^{(1)} = 0.25$), characterized by a high ratio between axial Young modulus and shear modulus ($E_1/G_{12} = 33.6$), the decay length is longer than in the other cases. Moreover, the decay length tends to infinity ($\lambda_1 \rightarrow 0$) for more and more flexible adhesives ($\chi_1 \rightarrow \infty$). In this case, the characteristics of the adhesive dominate those of the layers, and the decay rate is almost independent of the layer thickness. Conversely, for stiff adhesives ($\chi_1 < 1$) very low debonding occurs and the decay rate is almost constant and close to that of the perfectly bonded strip.

3. The case of multilayered strips with periodic layout

For periodic composite materials, the homogenization method can be used to calculate the effective elastic moduli of an homogeneous material whose overall response is *close* to that of the heterogeneous periodic material when the size of the elementary cell of periodicity tends to zero. It allows also for the a-posteriori computation of local stress variation at the layer level. For more information about homogenization theory the reader is referred to [10, 11, 12].

For a layered strip the effective elastic coefficients can be obtained by homogenization of the fourth-order partial differential operator (2) with periodic interface conditions (7). For the sake of simplicity, the case of perfect interfacial bonding is considered here. For generally anisotropic materials, the governing equations (2, 7a-c) can be rewritten in the compact form ($\alpha, \beta \ldots = 1, 2$):

$$(a_{\alpha\beta\gamma\delta} F_{,\gamma\delta})_{,\alpha\beta} = 0 \qquad x_1 \geq 0, \; -h \leq x_2 \leq h \qquad (11)$$

$$F = F_{,2} = 0 \qquad x_2 = \pm h \qquad (12)$$

where partial differentiation must be considered in the weak sense (virtual force principle), because the coefficients $a_{\alpha\beta\gamma\delta}$ (related to reduced elastic coefficients $R_{\alpha\beta}$) are piecewise constants but traction and displacement continuity (7) must hold at the interfaces (essential and natural boundary conditions, respectively).

The homogenization process is performed by restating problem (11, 12) making use of a variational formulation for $\varepsilon > 0$, (ε being a characteristic small parameter denoting the dimension of the cell of periodicity of the laminate) and solving the limit problem, i.e. for $\varepsilon \rightarrow 0$ (see [6]). The general solution, i.e. for a composite periodic in both directions x_1 and x_2, allows for the definition of homogenized coefficients $\hat{a}_{\alpha\beta\gamma\delta}$:

$$\hat{a}_{\alpha\beta\gamma\delta} = \frac{1}{|Y|} \int_Y (a_{\alpha\beta\gamma\delta} - a_{\mu\nu\gamma\delta} \, w_{\alpha\beta|\mu\nu}) \, \mathrm{d}Y, \qquad (13)$$

where $Y =]0, Y_1[\times]0, Y_2[$ is the representative volume element. Moreover, the functions of homogenization $w_{\alpha\beta}$, appearing in eqn (13), are the (unique) solution of the variational problem (the cell problem):

$$\begin{cases} \text{Find } w_{\alpha\beta} \in H^2_{\text{per}}(Y) \text{ such that } \forall v \in H^2_{\text{per}}(Y) \\ \int_Y (a_{\alpha\beta\gamma\delta} - a_{\mu\nu\gamma\delta}\, w_{\alpha\beta|\mu\nu})\, v_{|\gamma\delta}\, \mathrm{d}Y = 0 \end{cases} \tag{14}$$

where $v_{|\alpha}$ denotes partial derivative with respect to the local coordinate $y_\alpha = x_\alpha / \varepsilon$. Analogously, it can be shown that the Airy stress function admits the following two-scale asymptotic expansion:

$$F^\varepsilon(\mathbf{x}, \mathbf{y}) = u(\mathbf{x}) - \varepsilon^2 F_{,\alpha\beta}(\mathbf{x})\, w_{\alpha\beta}(\mathbf{y}) + O(\varepsilon^4)\,, \tag{15}$$

where $\mathbf{x} = (x_1, x_2)$ and $\mathbf{y} = (y_1, y_2)$ denote the coordinates in the global and the local reference system, respectively.
For a laminated composite, i.e. when the composite is periodic in x_2 – direction only, the homogenization process can be performed analytically. The cell problem (14) reduces to the following differential equation:

$$(a_{\alpha\beta 22} - a_{2222}\, w_{\alpha\beta|22})_{\,22} = 0 \tag{16}$$

to be integrated on the domain of cell of periodicity $]0, Y_2[$. Moreover, the effective elastic moduli and the homogenization functions are given by:

$$\hat{a}_{\alpha\beta\gamma\delta} = \left\langle a_{\alpha\beta\gamma\delta} - a_{\alpha\beta 22}\, a^{-1}_{2222}\, a_{22\gamma\delta} \right\rangle + \left\langle a_{\alpha\beta 22}\, a^{-1}_{2222} \right\rangle \left\langle a^{-1}_{2222} \right\rangle^{-1} \left\langle a_{22\gamma\delta}\, a^{-1}_{2222} \right\rangle \tag{17a}$$

$$w_{\alpha\beta|22} = (a_{\alpha\beta 22} - \left\langle a_{\alpha\beta 22}\, a^{-1}_{2222} \right\rangle \left\langle a^{-1}_{2222} \right\rangle^{-1} / a_{2222} \tag{17b}$$

For an orthotropic laminate eqn (17a) gives the following expressions for the non vanishing effective elastic moduli:

$$\hat{R}_{11} = \hat{a}_{2222} = \left\langle R^{-1}_{11} \right\rangle^{-1}, \qquad \hat{R}_{22} = \hat{a}_{1111} = \left\langle R_{22} - R^2_{12} R^{-1}_{11} \right\rangle + \left\langle R_{12} R^{-1}_{11} \right\rangle^2 \left\langle R^{-1}_{11} \right\rangle^{-1}$$

$$\hat{R}_{12} = \hat{a}_{1122} = \left\langle R_{12} R^{-1}_{11} \right\rangle \left\langle R^{-1}_{11} \right\rangle^{-1}, \qquad \hat{R}_{66} = 4\hat{a}_{1212} = \left\langle R_{66} \right\rangle \tag{18}$$

Hence, the decay rate of end effects for the homogenized material can be evaluated very simply by computing the first nonzero root of the transcendental equation:

$$\frac{\sin \hat{c}_1 \lambda}{\hat{c}_1} \pm \frac{\sin \hat{c}_2 \lambda}{\hat{c}_2} = 0 \tag{19}$$

corresponding to a homogeneous orthotropic single-layer strip [1]. In eqn (19), the constants $\hat{c}_1, \hat{c}_2$ are defined in terms of homogenized material parameters as:

$$\hat{c}_1 = \sqrt{\hat{E}(1 + 2\hat{\varepsilon})}\,, \qquad \hat{c}_2 = \sqrt{\hat{E}(1 - 2\hat{\varepsilon})} \tag{20}$$

with $\hat{E}, \hat{\varepsilon}$ computed from eqn (3) making use of eqns (18). Solution of eqn (19) is reported for instance in [13]. Moreover, making use of (15, 17b), explicit expressions for local stress distributions can be obtained as:

$$\sigma_{11} = F^{\varepsilon}_{,22} = F_{,22}\,\hat{R}_{11} / R_{11}(y_2) - F_{,11}\left[R_{12}(y_2) - \hat{R}_{12}\right] / R_{11}(y_2) + O(\varepsilon) \tag{21a}$$

$$\sigma_{12} = F^{\varepsilon}_{,12} = F_{,12} + O(\varepsilon)\,, \qquad \sigma_{22} = F^{\varepsilon}_{,11} = F_{,11} + O(\varepsilon) \tag{21b,c}$$

Eqns (21b,c) show that transverse shear stress and transverse normal stress are computed uniquely from the stress function F of the homogenized material and, correspondingly, no discontinuities are present at the layer interfaces. On the contrary, the local values of elastic moduli $R_{11}(y_2)$ and $R_{12}(y_2)$ appear in the computation of axial normal stresses, see eqn (21a). Hence, axial normal stress is discontinuous at the interfaces, as is required by the elasticity problem.
In the example of Figure 3, a periodic laminate is considered, whose representative cell of periodicity consists of a sandwich strip $[1-2-1]$ with layer thicknesses $[1/4, 1/2, 1/4]$ of the height of the elementary cell. n_c denotes the number of elementary cells; the total number of layers is $S = 2n_c + 1$, the stacking layout is symmetric and the layers are those of a graphite-epoxy composite, whose elastic constants are (cross-ply lamination):

$$E_1^{(1)} = 127.5\,GPa,\ E_2^{(1)} = E_3^{(1)} = 11\,GPa,\ G_{12}^{(1)} = 5.5\,GPa,\ \nu_{12}^{(1)} = \nu_{13}^{(1)} = 0.35,\ \nu_{32}^{(1)} = 0.25$$

$$E_1^{(2)} = E_2^{(2)} = 11\,GPa,\ E_3^{(2)} = 127.5\,GPa,\ G_{12}^{(2)} = 4.4\,GPa,\ \nu_{12}^{(2)} = 0.25,\ \nu_{31}^{(2)} = \nu_{32}^{(2)} = 0.35$$

Figure 3 depicts the real and imaginary parts of the first five even (λ^e) and odd (λ^o) eigenvalues as a function of the number of cells. The eigenvalues of the homogenized strip are reported by straight lines. It is worth noting that both real and imaginary parts of the exact eigenvalues approach asymptotically those of the homogenized strip as the number of elementary cells n_c increases. Moreover, the convergence turns out to be monotone if the number of cells is not too small, i.e., for $n_c \geq 2j$ ($n_c \geq 2j+1$) for the jth even (odd) eigenvalue.

Acknowledgements

The Authors acknowledge the financial support of the (Italian) Ministry of University and Scientific and Technological Research (MURST - 60%, 1996) and of National Council of Research (CNR-contr. 96.01854.CT11).

References

[1] *I. Choi, C.O. Horgan*, SAINT-VENANT'S PRINCIPLE AND END EFFECTS IN ANISOTROPIC ELASTICITY - J. Applied Mechanics ASME. 1977. Vol. 44, pp. 424-430

[2] *I. Choi, C.O. Horgan*, SAINT-VENANT END EFFECTS FOR PLANE DEFORMATION OF SANDWICH STRIPS. Int. J. Solids Structures. 1978. Vol. 14, pp. 187-195

[3] *C.O. Horgan, J.G. Simmonds*, SAINT-VENANT END EFFECTS IN COMPOS-

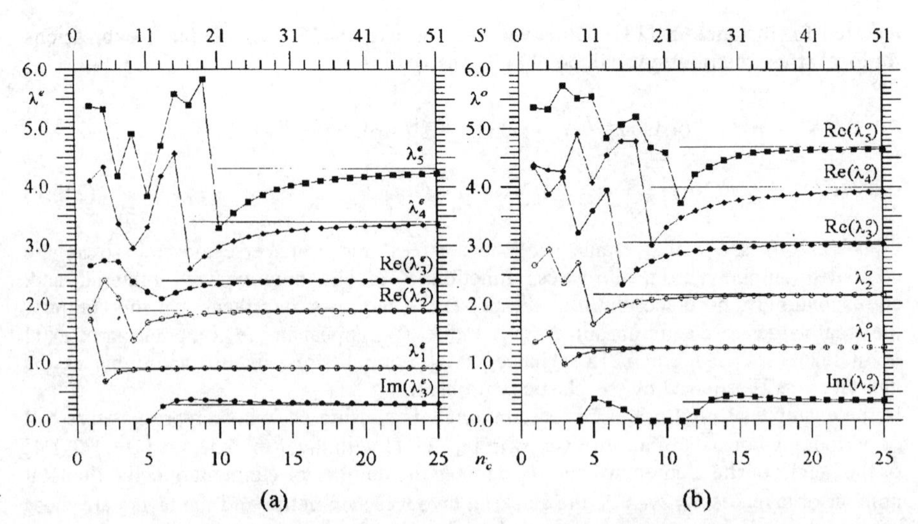

Figure. 3. (a) Even and (b) odd eigenvalues for strips in plane strain vs. number n_c of elementary cells of periodicity. The elementary cells are sandwich strips [1-2-1] with layer thicknesses [1/4, 1/2, 1/4] of the height of the elementary cell.

ITE STRUCTURES - Composites Engineering. 1994. Vol. 4, pp. 279-286

[4] *M.J. Folkes, R.G.C. Arridge*, THE MEASUREMENT OF SHEAR MODULUS IN HIGHLY ANISOTROPIC MATERIALS: THE VALIDITY OF ST. VENANT PRINCIPLE - J. Physics D. 1975. Vol. 8, pp.1053-1064

[5] *N. Plevris, T.C. Triantafillou*, FRP-REINFORCED WOOD AS STRUCTURAL MATERIAL - J. Materials in Civil Engineering. ASCE. 1992. Vol. 4, pp. 300-317

[6] *N. Tullini, M. Savoia*, DECAY RATE OF SAINT-VENANT END EFFECTS FOR MULTILAYERED ORTHOTROPIC STRIPS - Int. J. Solids Structures. 1997. Vol. 34, pp. 4263-4280

[7] *N. Tullini, M. Savoia, C.O. Horgan*, END EFFECTS IN MULTILAYERED ORTHOTROPIC STRIPS WITH IMPERFECT BONDING - Mechanics of Materials. 1997. Vol. 26, pp. 23-34

[8] *A. Klarbring*, DERIVATION OF A MODEL OF ADHESIVELY BONDED JOINTS BY THE ASYMPTOTIC EXPANSION METHOD - Int. J. Engineering Science. 1991. Vol. 29, pp. 493-512

[9] *G. Geymonat, F. Krasucki, S. Lenci*, ANALYSE ASYMPTOTIQUE DU COMPORTEMENT D'UN ASSEMBLAGE COLLÉ - C. R. Acad. Sci. Paris. 1996. Vol. 322 (ser. I), pp. 1107-1112

[10] *E. Sanchez-Palencia*, NON-HOMOGENEOUS MEDIA AND VIBRATION THEORY - LECTURE NOTES IN PHYSICS 127, Springer-Verlag, Berlin - D 1980

[11] *O.A. Oleinik, A.S. Shamaev, G.A. Yosifian*, MATHEMATICAL PROBLEMS IN ELASTICITY AND HOMOGENIZATION, North Holland, Amsterdam - NL 1992

[12] *A.L. Kalamkarov, A.G. Kolpakov*, ANALYSIS, DESIGN AND OPTIMIZATION OF COMPOSITE STRUCTURES, J. Wiley & Sons, New York - USA 1997

[13] *M. Savoia, N. Tullini*, BEAM THEORY FOR STRONGLY ORTHOTROPIC MATERIALS - Int. J. Solids Structures. 1996. Vol. 33, pp. 2459-2484

FREE EDGE EFFECTS OF POLYMERIC LAMINATES IN COMPUTER ANALYSIS

J. LEWIŃSKI, W. M. ORSETTI, A. P. WILCZYŃSKI
Institute of Mechanics and Design, Warsaw University of Technology
02 - 524 Warsaw, 85/206 Narbutta str.

Abstract

The paper presents an extended analysis of the edge effect on free edges of a laminate sample. For this purpose a professional FEM system ADINA is used. As a results graphs of stresses distributions on the laminate sample free edges are presented as functions of thickness of the sample, its width and angle of plies.

1. Introduction

Differences of elastic properties of composite layers, together with boundary conditions on free surfaces, lead to self balanced stress fields, known as free edge effects. An analysis of these effects was presented among many others in [1], [2], [3]. Using the differential equations of equilibrium for a composite element an explanation of existence of additional stresses in the free boundary zone was presented, along with some results of the numerical solution (using the finite difference method). The analysis was given for a four layers specimen under constant elongation in the longitudinal direction. The specimen was symmetric to the mid-plane. The analysis was devoted mainly to estimation of the influence of the layer direction on the values of the additional stresses along the free edges and also the width of the zone of their existence.

Because of practical significance of this problem, leading to delamination of layered composites a more satisfactory analysis was decided, using a professional ADINA stress analysis system. Stress distributions analysis practically in the whole volume of the specimen was performed for the purpose. This included such problems as of thickness and width of the specimen, thickness and number of the layers and stacking angles influence on the free edge effect.

2. Numerical analysis outlines

The analysis was performed on 4 and 8 layered composite specimens, subjected to constant tension σ_{yy} as in fig. 1. Such loading allows easy comparison of stresses while changing the specimen parameters. The layers are stacked symmetrically to the midplane. The orthotropy axes are at $\pm\beta$ angels to the direction of tension as shown an fig. 2. The influence of

- thickness of the specimen h
- width of the specimen b
- number of layers n
- stacking angle β

on the stress and equivalent stress distribution in the mid-section of the specimen was analysed. The used material was epoxy reinforced with Thornel graphite fibre with elastic properties:

$E_{11} = E_{aa} = 129.57$ GPa; $E_{22} = E_{bb} = 9.19$ GPa
$E_{33} = E_{cc} = 9.19$ GPa; $G_{12} = G_{ab} = 5.06$ GPa
$G_{13} = G_{ac} = 5.06$ GPa; $G_{23} = G_{bc} = 3.05$ GPa
$\nu_{21} = \nu_{ba} = 0.308$; $\nu_{12} = \nu_{ab} = 0.020$
$\nu_{31} = \nu_{ca} = 0.308$; $\nu_{13} = \nu_{ac} = 0.020$
$\nu_{32} = \nu_{cb} = 0.476$.

And strength properties in the principal directions:

$R_{1r} = R_{ar} = 1500$ MPa; $R_{1c} = R_{ac} = 1200$ MPa
$R_{2r} = R_{br} = 50$ MPa; $R_{2c} = R_{bc} = 250$ MPa
$R_{12} = R_{13} = R_{ab} = R_{ac} = 70$ MPa; $R_{23} = R_{bc} = 150$ MPa.

These data were given in [4], where a, b, c are as depicted in fig. 2 the principal layer directions. The length of the specimen was $l = 120$ mm. The layer thickness h_w, width b, specimen thickness h and angle β was ranging:

for 4 - layered specimens
$h_w = 0.25, 0.5, 1, 2, 3$ [mm]
$b = 8, 12.5, 25, 50, 75$ [mm]
$h = 1, 2, 4, 8, 12$ [mm]
$\beta = 0, 15, 30, 45, 60, 75, 90$ [°]

for 8 - layered specimens
$h_w = 0.25, 0.5, 1, 2$ [mm]
$b = 25$ [mm]
$h = 2, 4, 8, 16$ [mm]
$\beta = 30°$.

from which suitable combinations were taken. As basic a 4 - layered specimen was adopted, with parameters $l = 120$, $b = 25$, $h = 4h_w = 2$ mm, $\beta = 30°$. All specimens were loaded in tension of $\sigma_{yy} = 100$ MPa. The analysis was performed for points along lines defined in fig. 3, in a cross -section in the mid - length of the specimen. The professional ADINA stress analysis system was employed using 8-node volume elements with finer meshing at the free edges. Each layer additionally was divided into 12 sublayers to increase accuracy.

Some of the obtained results, after a primary selection, which allowed to find maximum stresses in the mid -layers, is presented below.

3. Analysis results

All results are presented as stress distribution graphs. In fig. 4 equivalent stresses accordingly to Tsai - Wu theory are depicted along lines A, AAA, AAB and CCC. It can be seen that maximum values of the equivalent stresses are placed closely to the free edges of the layers adjacent to the mid-plane of the specimen.

Figure 5 gives smooth graphs of effective stresses along CCC line for eight - layered specimens of different thickness. The left column presents values according to Tsai - Wu hypothesis while the right column accordingly to Wilczyński's [7] proposal. These graphs confirm the above conclusion that the maximum values of equivalent stresses are placed in the middle layers of the laminate. However the Wilczyński proposal places their maximum in the symmetry plane while the Tsai - Wu theory gives slightly lower values but in the other side of the layer.

As expected other variables, listed previously, may influence the stress distributions. The analysis has shown, that the free edge effect is stronger for stacking angles of 30÷45°. For further analysis the case $\beta = 30°$ was selected. For this case stress σ_{xx}, σ_{yy}, σ_{zz}, σ_{xy} σ_{yz}, σ_{xz} and equivalent stresses were calculated. Most of these results was presented in [5]. The basic conclusion from these results are as follows:

- the width of the boundary zone is independent of specimen width
- the width of the boundary zone is approximately equal to 1.5÷2 thickness of the composite
- for thick specimen ($h/b \geq 1/4$ for a 4-layered laminate) the boundary zone extends trough the whole width of the specimen.

Additionally all results confirmed a conclusion that the equivalent stress is always greater at the free edges as compared to the specimen interior values. Further analysis will be conducted for the effective stresses which in Authors opinion presents the real effect of stressing of the composite. For this purpose the Tsai - Wu proposal is used. Figures 6 and 7 presents the equivalent stress distribution, depending on the varying thickness, for 4-layered and 8-layered

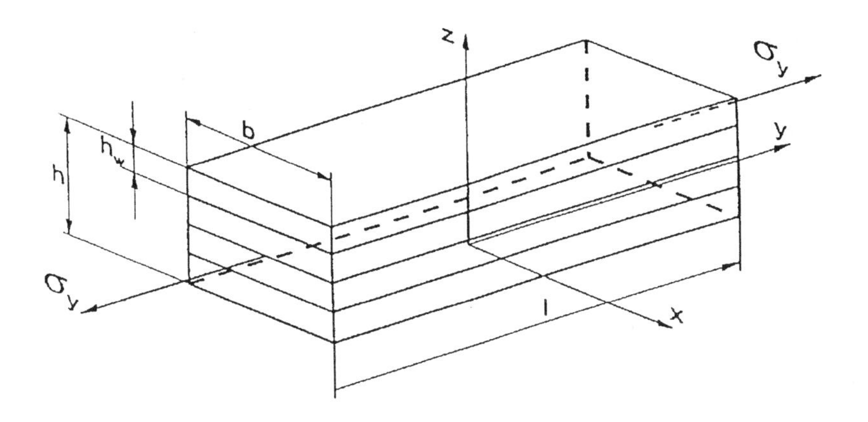

Fig. 1. The laminated specimen

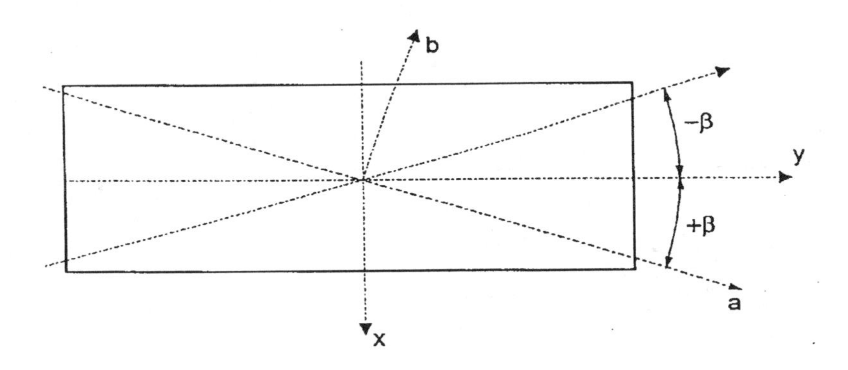

Fig. 2. The stacking angles

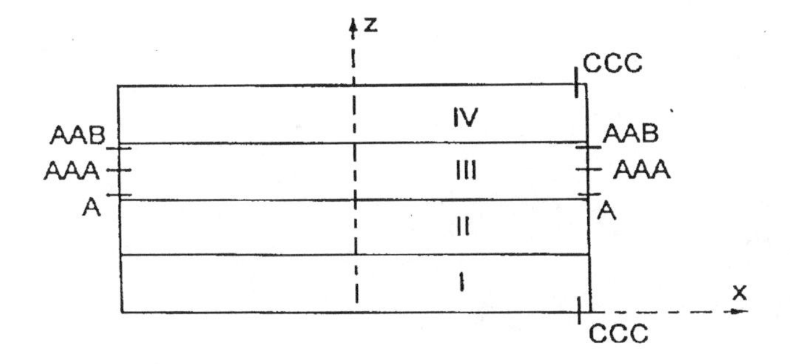

Fig. 3. Positions of analysed stresses in the specimen section

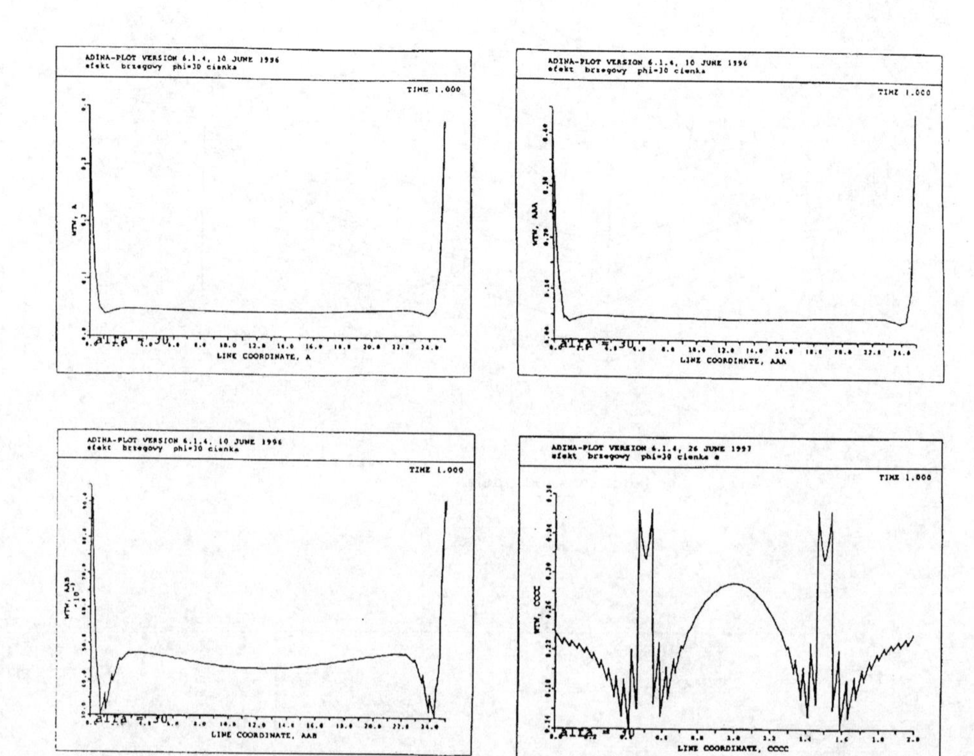

Fig. 4. Equivalent stresses along A, AAA, AAB and CCC lines

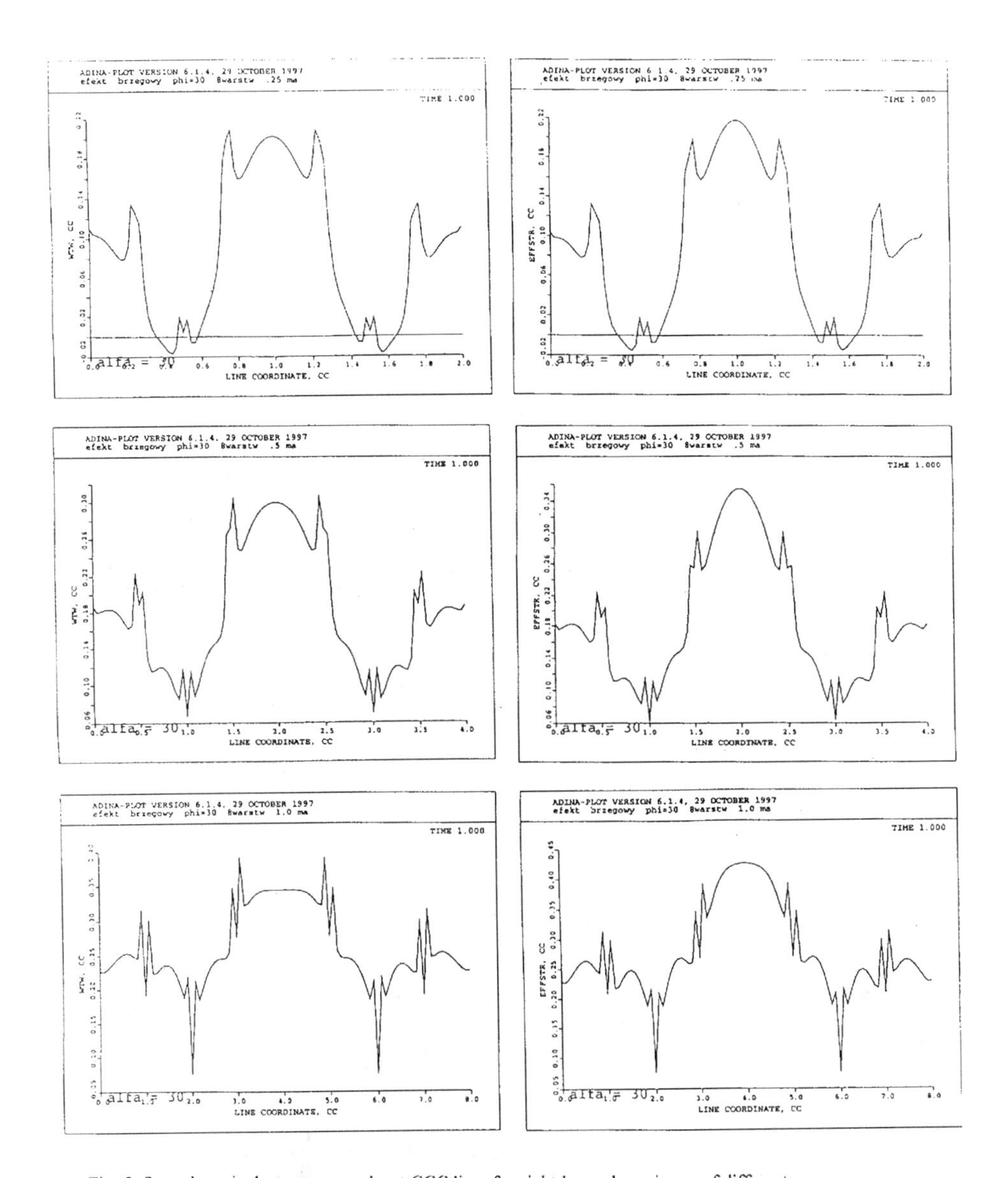

Fig. 5. Smooth equivalent stress graphs at CCC lines for eight-layered specimens of different thickness

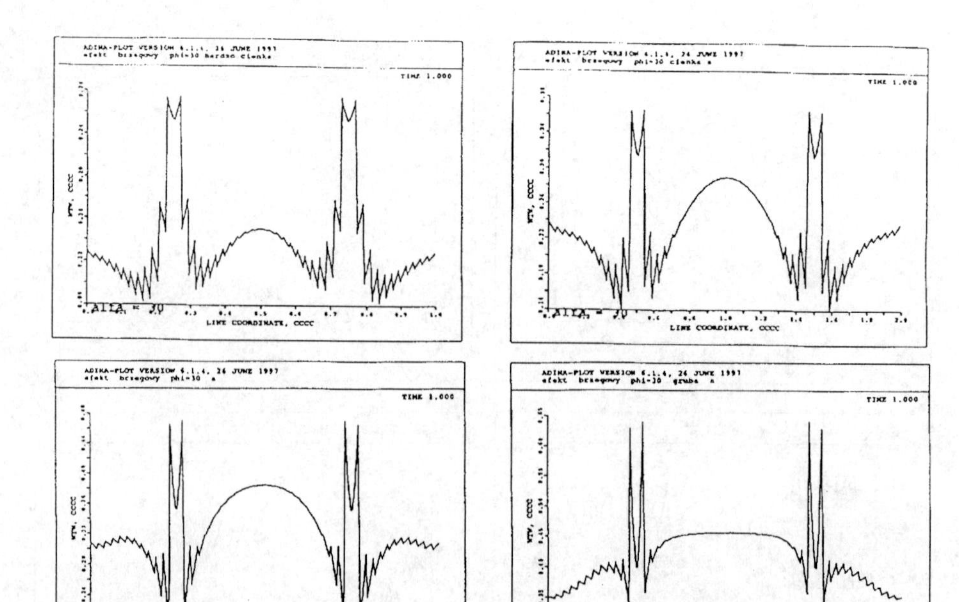

Fig. 6. Equivalent stresses along CCC line in a 4 layers specimen with varying layer thickness

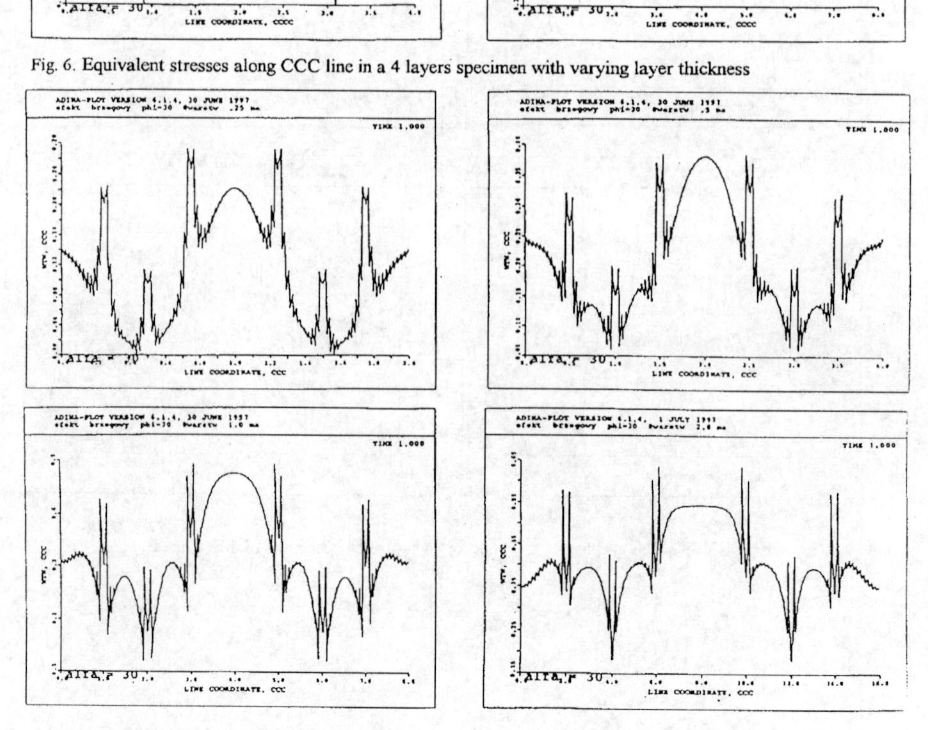

Fig. 7. Equivalent stresses along CCC line in a 4 layers specimen with varying layer thickness

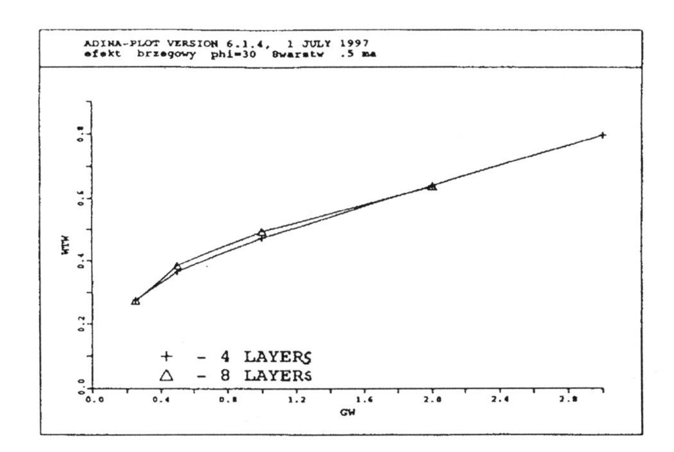

Fig. 8. Comparison of maximum equivalent stresses for specimens 4 and 8 constant thickness layers

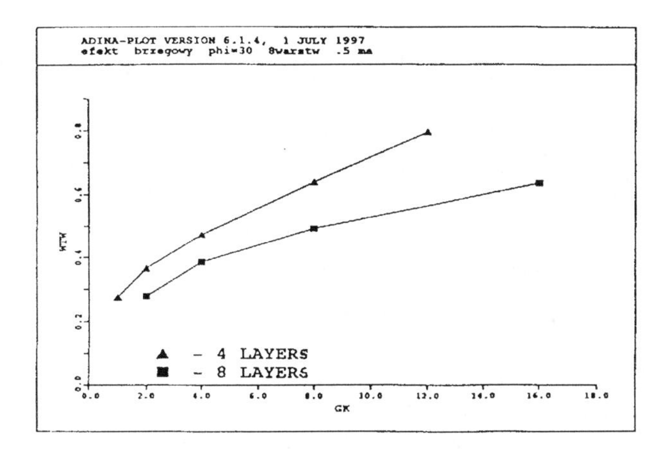

Fig. 9. Comparison of maximum equivalent stresses for specimens with 4 and 8 layers of equal thickness of the laminate

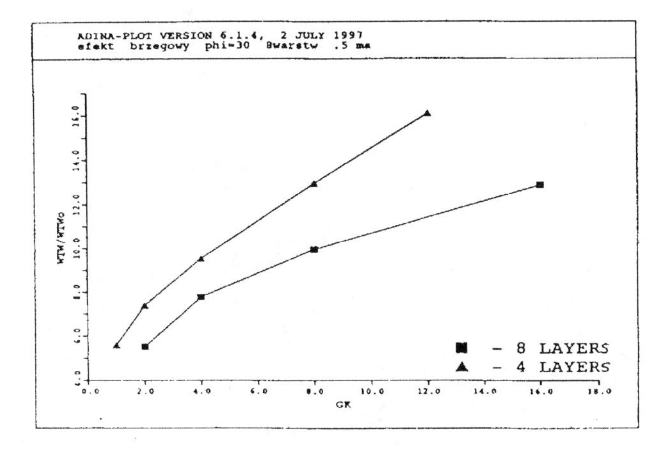

Fig. 10 Maximum equivalent stress to internal equivalent stress for various specimen thickness

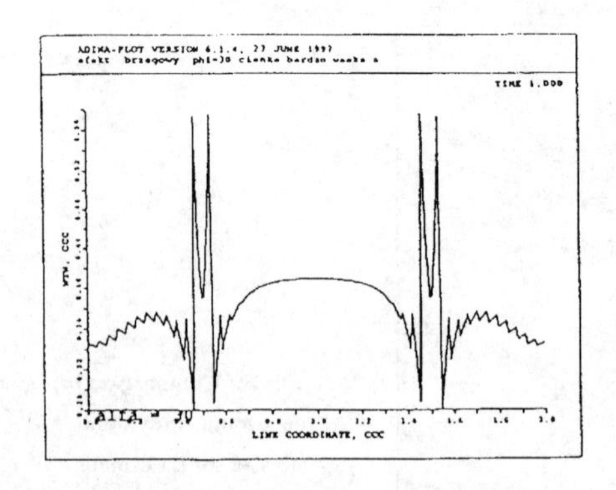

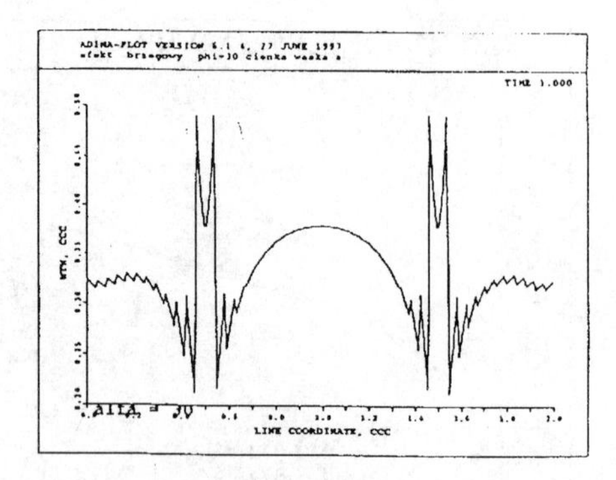

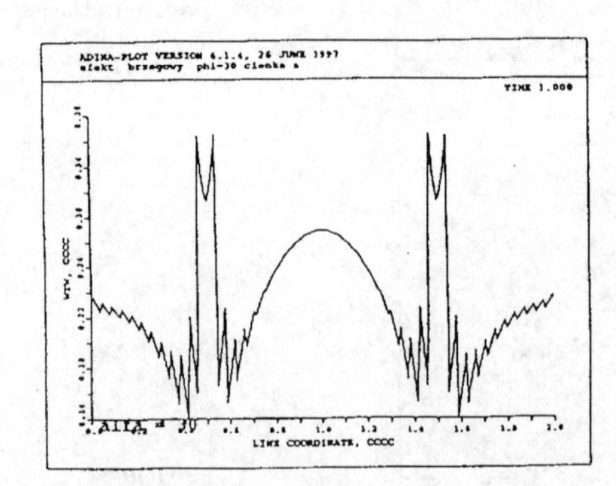

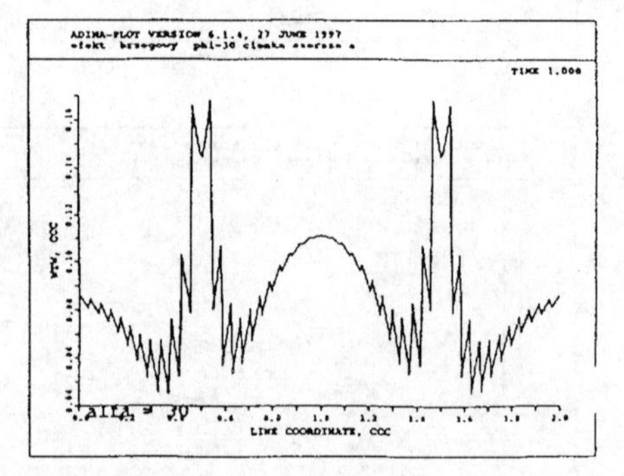

Fig. 11. Equivalent stress distributions along CCC line for a 4-layers specimen for varying width

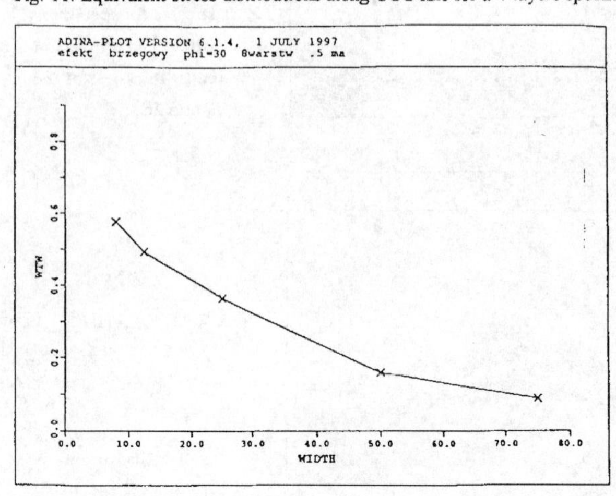

Fig. 12. Maximum equivalent stresses for 4-layers specimen for varying width

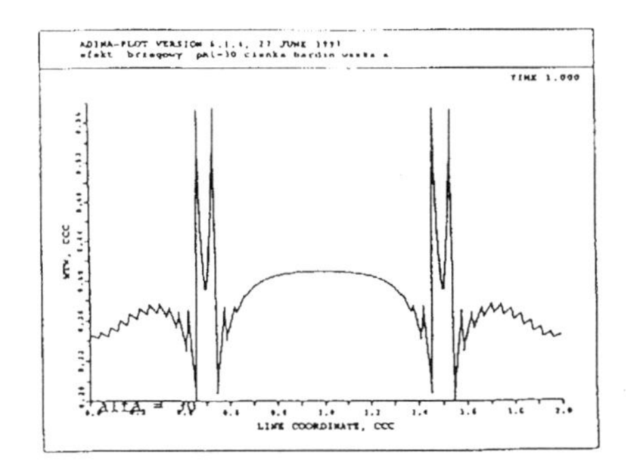

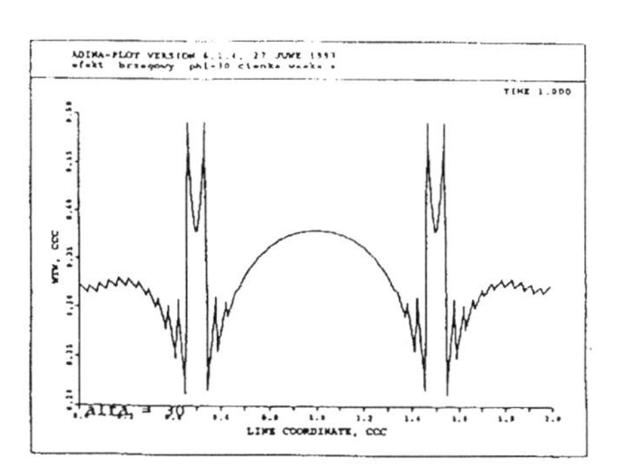

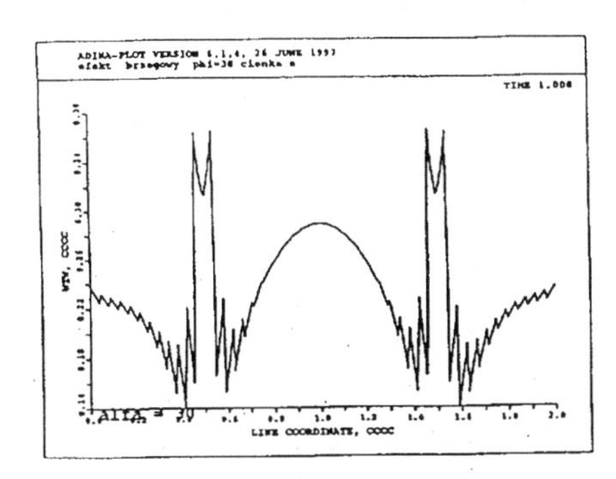

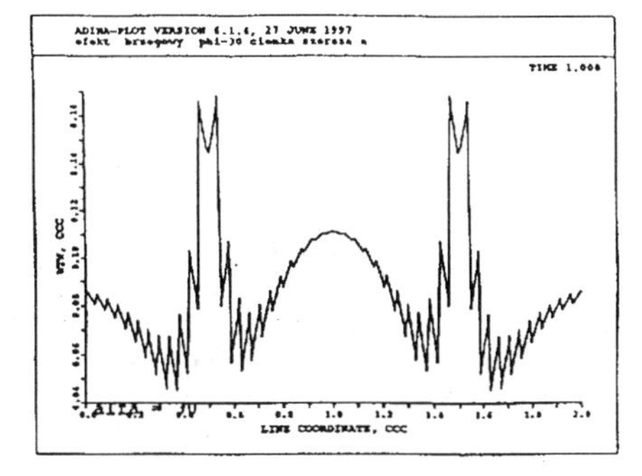

Fig. 13. Equivalent stress distributions along CCC line for a 4 - layers specimen for varying width

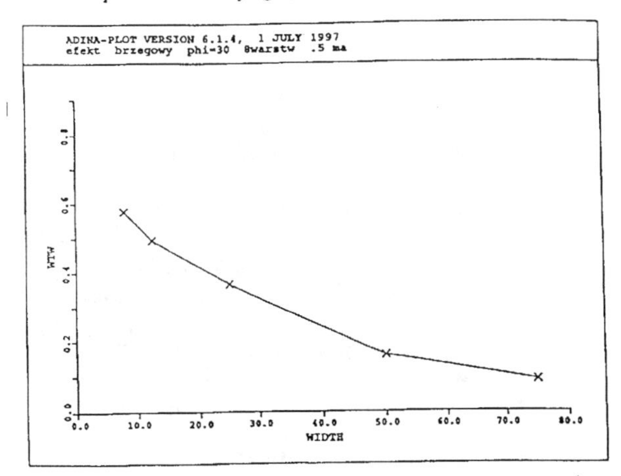

Fig. 14. Maximum equivalent stresses for 4 - layers specimen for varying width (natural scale)

laminates. These two cases enable presentation of calculation of dependence of maximum equivalent stress on thickness of the layers (fig. 8) and of the laminate (fig. 9). It can be seen that an increase of number of layers (from 4 to 8) for equal layer thickness does not affect the maximum equivalent stress values. The equivalent stress increases with the increase of the layer (and laminate) thickness. For a 12-fold increase of layer thickness the equivalent stress increase ratio tends to 3, with a lesser slope for specimens with increased number of layers. Decrease of layer thickness at constant specimen thickness leads to decrease of equivalent stress value. The ratio of effective stress at the free edge to that stress inside of the composite has been found to be 2 to 16, as is shown on fig. 10. A more sophisticated analysis shows this ratio to be dependant on other specimen parameters as well. The average value of this equivalent stress is 0.0495 with ±23% variation, possibly depending also on numerical errors of the calculation.

Analysis of specimen width for 4-layered composites leads to graphs presented on fig. 11 and fig. 12. It can be seen from fig. 12 that a 9.5-fold increase of specimen width leads to a 6.5-fold decrease of equivalent stresses.

4. Final remarks

The numerical analysis of 4 and 8-layered laminate specimen has shown that in the vicinity of the specimen free edges exists a boundary zone, where all components of the stress tensor may occur. This leads to a much higher effective stress in this zone while the maximum of this values are found to be on the layers at the symmetric plane of the composite. This equivalent stress increase with the thickness of the laminate and decreases with the specimen widths and in practice is independent of the number of layers. This effect has a maximum at stacking angles of 30÷45°. The width of the boundary zone is approximately constant at 1.5÷2 of the thickness of laminate. To decrease all these effects if could be recommended to increase the width of the specimen and decrease the thickness of the layers increasing the number of the layers. Nevertheless a significant increase of equivalent stress at the free edges is unavoidable. No quantitative measure of these effects, due to dependence on a multitude of factors could be found. At present, for lack of better means, a estimation presented in [6] could be used for this purpose.

References

1. *Pipes R.B., Pagano N. J.*, Interlaminar stresses in Composite Laminates Under Uniform Axial Extension, Journal of Composite Materials, 1970 vol.4
2. *Puppo A.H., Evenson H. A.*, Interlaminar Shear in Laminated Composites Under Generalized Plane Stress, Journal of Composite Materials, 1970 vol.4
3. *Wang A. S. D., Crosman F. W.*, Some new results on Edge effects in Symmetric Composite Laminates, Journal of Composite Materials, 1977 vol.11
4. *Sharr G.*, Messungen der Elastizitaetszahlen und Waermeausdehnungskoeffizienten von faserverstarkten Kunststoffen, Doctoral thesis, Technische Hochschule, Aachen, 1984
5. *Lewiński J., Orsetti W. M., Wilczyński A. P.*, Analiza komputerowa lokalnych efektów brzegowych w kompozytach warstwowych, II Konferencja Naukowo Techniczna, Polimery i Kompozyty Konstrukcyjne, Politechnika Śląska, Ustroń 96, str. 311-322 (In Polish)
6. *Wilczyński A. P.*, Polimerowe kompozyty włókniste, Warszawa, WNT, 1996 (In Polish).
7. *Wilczyński A. P.*, Use of Tensorial Polynomial Strength Function for Strength Prediction in Laminated Polymeric Composites, Comp. Sc. and Techn., 51, 525, 1994.

Bond of FRP anchorages and splices in concrete beams

Achillides Zenon[1], Pilakoutas Kypros[2] and Waldron Peter[3]

ABSTRACT: This paper examines the bond behaviour of Carbon and Glass FRP single anchorages and spliced bars in concrete beams. Various reinforcement configurations are used in nine beams tested under four-point bending. CFRP bars appear to develop higher bond splitting strength than respective GFRP bars. The higher deformability of GFRP bars in the axial direction appears to play an important role to their lower bond splitting strength. Spliced bars developed higher bond strengths than anchorages. The distribution of bond stresses over single anchorage bars before and after splitting cracking develops, are presented and discussed.

KEYWORDS: Bond, bond splitting strength, FRP bars, splices, splitting cracks, anchorages, embedment length.

1 INTRODUCTION

Bond between concrete and FRP reinforcing bars is one of the fundamental aspects of structural behaviour which needs to be understood before design guidelines for FRP materials can be developed for use in the construction industry. Adequate levels of bond strength and stiffness are required between reinforcement and concrete to transmit forces effectively from one material to the other. Locations of particular concern are the ends of bars, since making bends in FRP bars is not always possible, and splices, where high loads have to be transferred from one bar to an adjacent one. In flexural structural elements, splitting of concrete in the tension zone is the most likely mode of bond failure. This type of failure is substantially different and more dangerous than the pullout mode, since it happens at a much lower bond stress level and the residual bond stress of the reinforcing bar decreases rapidly to zero.

The bond splitting behaviour of FRP bars to concrete is expected to vary from that of conventional steel bars since various key parameters that influence bond performance are different, such as:

- lower FRP modulus of elasticity,
- much lower shear strength and stiffness in the longitudinal and transverse direction,
- high normal strains expected before failure

In order to investigate and understand the bond behaviour of FRP reinforcing bars under flexural conditions, an experimental series of beam tests has been conducted under the EUROCRETE project at the Centre for Cement and Concrete of the University of Sheffield. EUROCRETE was a pan-european research programme, which lasted 4 years having a budget of around 6 millions ECU, aiming to develop FRP reinforcement for concrete structures [1].

This series included four phases of testing over a period of three years and thirty-seven beams were tested during these phases. In the first three phases, 24 beams reinforced with FRP bars were prepared and tested [2]. The beams were not designed to fail in bond since the investigation of the bond behaviour of FRP bars was not the main experimental objective. However, a large number of strain gauges were attached to the main reinforcement of the beams to record the strain values on the bars during the tests. The strain values were used to calculate the developed bond stresses on the bars. Many useful results have emerged from the study of these beams, which contributed to the understanding of bond behaviour of FRP bars [3].

In this paper, the forth phase of the beam testing program is presented. The primary objective of this phase was to examine the mode of bond failure of FRP reinforcing bars under four-point bending loading conditions and the various parameters that influence their bond behaviour. For this purpose, nine medium scale beams (2500 x 250 x 150 mm) were designed to fail in bond, as described in the following section.

2 EXPERIMENTAL PROGRAM

2.1 Material properties

The FRP rebars used in this phase were manufactured by the EUROCRETE project [4]. In this study Glass and Carbon FRP bars of two different diameters, having a rough surface, were used. Their material properties are given in the following table:

	GFRP bar d=13.5 mm	GFRP bar d=8.5 mm	CFRP bar d=8 mm
Young Modulus (MPa)	45000	115000	115000
Direct tensile strength (MPa)	>1000	>1000	>1500
Tensile strength under beam loading conditions (MPa)	700-750	900-950	1300-1380
Elongation at failure under beam loading conditions (%)	1.55-1.65	2.00-2.10	1.13-1.20

Table 1: Properties of reinforcing bars

The links used as shear reinforcement in the beams of this experimental series were specially manufactured filament wound rectangular (200 x 100 mm) GFRP links with 10 x 5 mm cross section. The tensile strength of the used links was measured to be around 425 MPa and their Young Modulus around 47 MPa. They were fixed at the standard spacing of 75 mm centre-to-centre either only in the shear span or, along the whole beam, depending on the arrangement of the main reinforcement. Even though, the amount of shear reinforcement provided falls well below the minimum requirements proposed by [5], this spacing was adopted in the previous phases of testing for similar beams and successfully prevented shear failures.

Compression reinforcement was used in all the beams of this phase, mainly to enhance the concrete compressive strength at the top of the beam. The average concrete values measured in this series of tests are presented in table 2:

Specimens	Day of testing	Cube compressive strength (N/mm^2)	Cylinder indirect tension strength (N/mm^2)
GB29-CB33	28-36	35.0	2.7
GB34-CB37	28-33	45.0	3.6

Table 2: Concrete compressive and tensile strength

2.2 Test Specimens and procedure

Nine medium scale beams (2500 x 250 x 150 mm) were prepared in this phase of testing. Since one of the objectives of this study was to examine the influence of various parameters on the bond splitting behaviour of FRP bars, each beam was prepared to have a slightly different reinforcement arrangement. A summary of the reinforcement arrangement in each beam is shown in figure 1, together with the typical loading arrangement.

Beam	Anchorage length- L (mm, times bar diameter)	Bottom Cover to Diameter Ratio	Main (bottom) - Shear reinforcement	Arrangement of main (bottom) reinforcement 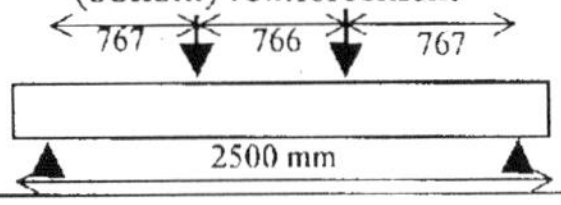 767 766 767 2500 mm
GB29	L=250 (19D)	1.85	3 GFRP bars (13.5mm) - GFRP links 75 mm c/c in the shear span	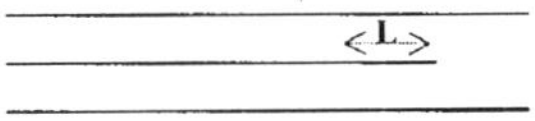 L
GB30	L=300 (22D)	1.85	3 GFRP bars (13.5mm) - GFRP links 75 mm c/c in the shear span	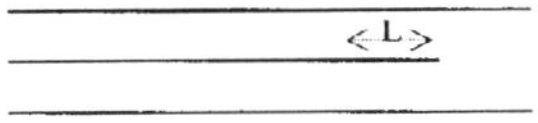 L
GB31	L=300 (22D)	1.85	4 GFRP bars (13.5mm) - GFRP links 75 mm c/c all the way	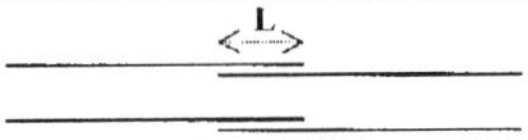 L
CB32	L=300 (38D)	3.13	3 CFRP bars (8 mm) - GFRP links 75 mm c/c in the shear span	 L
CB33	L=300 (38D)	3.13	4 CFRP bars (8 mm) - GFRP links 75 mm c/c all the way	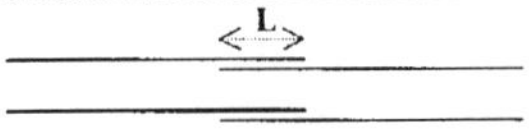 L
GB34	L=370 (44D)	2.94	3 GFRP bars (8.5mm) - GFRP links 75 mm c/c in the shear span	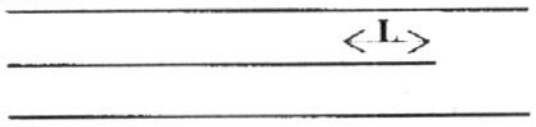 L
GB35	L=300 (35D)	2.94	3 GFRP bars (8.5mm) - GFRP links 75 mm c/c in the shear span	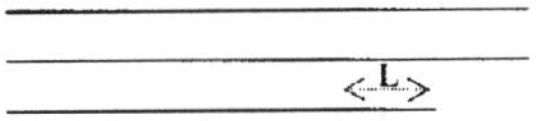 L
GB36	L=300 (35D)	2.94	4 GFRP bars (8.5mm) - GFRP links 75 mm c/c all the way	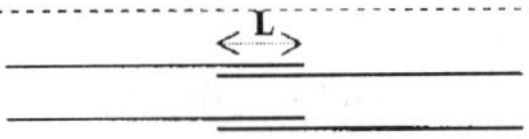 L
CB37	L=580 (72D)	3.13	3 CFRP bars (8 mm) - GFRP links 75 mm c/c in the shear span	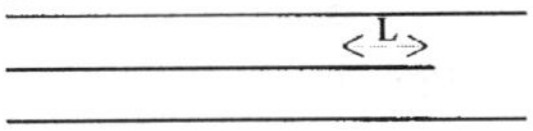 L

Figure 1: Reinforcement arrangement in phase 4 beams

All the beams were instrumented with strain gauges attached on the main reinforcing bars and links. External LVDTs were used to measure vertical and horizontal deflections, as well as surface extensions. This instrumentation was used to monitor and record the response of the beam at all levels of loading.

The load was applied manually by a servo-controlled hydraulic actuator in the displacement control. A load cycle at 15kN was applied before loading to failure. During the test, the developed cracks on the beam were marked and the biggest crack openings were measured with an optical instrument, every 5 to 10 kN. After the completion of each test, the beams were removed from the testing frame and a close examination of the mode of failure was conducted.

2.3 Analysis of measurements

The values from the strain gauges attached on the reinforcing bars were used for the calculation of the developed bond stresses on the bars during testing. The average bond stress between two points, for example (1) and (2), is proportional to the rate of change of strain along that length. From the force equilibrium over a certain length of a round bar in concrete, it can be deducted that:

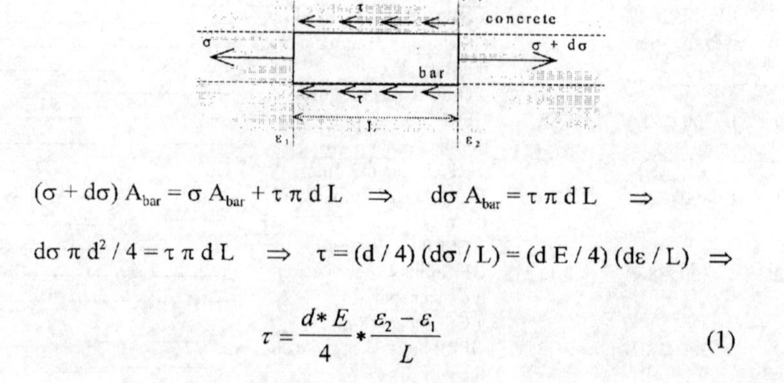

$$(\sigma + d\sigma)\, A_{bar} = \sigma\, A_{bar} + \tau\, \pi\, d\, L \quad \Rightarrow \quad d\sigma\, A_{bar} = \tau\, \pi\, d\, L \quad \Rightarrow$$

$$d\sigma\, \pi\, d^2 / 4 = \tau\, \pi\, d\, L \quad \Rightarrow \quad \tau = (d / 4)\,(d\sigma / L) = (d\, E / 4)\,(d\varepsilon / L) \quad \Rightarrow$$

$$\tau = \frac{d * E}{4} * \frac{\varepsilon_2 - \varepsilon_1}{L} \qquad (1)$$

Hence, by using the experimentally obtained strains, the average bond stress values between successive strain gauges were calculated.

3 EXPERIMENTAL RESULTS

3.1 Bond splitting behaviour

Various types of bond splitting cracks were developed in the beams depending on the arrangement of reinforcement in the cross section. Table 3 shows the types of splitting cracks observed together with the developed bond values in each case. In all cases, the crack initially developed at the very end of the bar and as the load increased, the crack extended along the whole length of the anchorage.

Beam	Type of failure	Embedment length (mm)	Maximum average bond strength (MPa)
GB29	Bond splitting crack under the middle bar	250	τ*= 3.2
GB30	Bond splitting crack under the middle bar	300	τ*= 2.7
GB31	Face and side bond splitting failure	300	τ*= 3.8
CB32	Bond splitting crack under the middle bar	300	τ*= 4.6
CB33	Side bond splitting failure	300	τ*= 5.7
GB34	Bond splitting crack under the middle bar	370	τ*= 3.2
GB35	Bond splitting crack at the side face of beam	300	τ*= 3.0
GB36	Face and side bond splitting failure	300	τ*= 4.1
CB37	Bond splitting crack under the middle bar	580	τ*= 3.7

Table 3: Bond splitting behaviour and average bond strength developed

3.2 Bond Strength values

The maximum average bond strength, τ*, developed over the anchorage or splice, is given in table 3. The maximum strain value of the strain gauge attached at the loaded end of the bar was used in the calculation of τ* over the whole embedment length, by considering equation (1).

By examining the results in the table some useful observations can be made. The first observation is that all the values for the bond splitting strength are lower than the values developed by the same bars in pullout tests which were between 12-15 MPa [6,7]. This is due to the different mode of bond failure in each case. In pull-out tests, the pull-through failure develops by shearing the surface deformations of the bar, whereas in the case of beams, the concrete cover split and the bar slipped through the concrete without any disturbance of the bar surface. This significant difference in bond strength may easily challenge the reliability of pullout tests for assessing the bond behaviour of FRP bars in concrete members, since, for most practical applications, any bond failure of reinforcing bars is likely to be due to splitting.

The second important observation is that the bond splitting behaviour of CFRP and GFRP bars appears to be significantly different. By comparing the bond strengths developed in the single anchorage of beam CB32 with the respective strengths of the anchorages in beams GB34 and GB35, it is obvious that CFRP bars develop higher bond splitting strengths than GFRP bars, under similar conditions. The same observation can be also made in the case of spliced bars in beams CB33 and GB36. It is estimated that CFRP bars develop around 30% higher bond splitting strength than similar diameter GFRP bars.

The above results are in contrast with the results of the pullout tests, previously carried out, which showed that CFRP and GFRP bars developed similar bond behaviour [3]. GFRP bars induce splitting at a lower normal and bond stress, but having a lower modulus of elasticity, at a higher normal strain. Hence, it appears that the deformability of FRP bars in the axial and transverse direction influences significantly their bond splitting behaviour. It can be concluded that pullout tests can not be used reliably to compare the bond behaviour of FRP bars when bond splitting failure is expected in a concrete member and a different test will have to be used.

Another interesting observation from this phase of testing is that the maximum average bond strength developed in the case of spliced bars, for example in beam GB31, was much greater than in the single bar anchorage in beam GB30. In order to determine τ^* in the case of splices, the strain values of the gauges attached at the loaded end of each pair of spliced bars were considered. The value of τ^* calculated over the spliced length was 3.8 MPa at ultimate load level, which was much higher than the value of 2.7 MPa developed in the single anchorage in beam GB30. Similar observations were also made in the case of beams CB32, 33 and GB35, 36 where spliced bars developed significantly higher bond splitting values. A possible explanation for this is that the plane of failure in the case of splices is different from the one in the case of single bars, as shown in table 3. As a result, the influence of stirrups has a different effect on the bond splitting strength of the bars in each case and requires further examination. However, it is considered that reduced bond strengths normally used for the design of splice lengths are not needed in the case of the EUROCRETE FRP bars.

3.3 Strain and bond stress distribution along a single anchorage

The distribution of normal strains along the single anchorage bar of beam GB30 is shown in figure 3, for successive load steps. Similar distribution was obtained along single anchorage bars in the other beams. The slope of each curve is proportional to the bond stress developed on the bar. The bond stress profiles, calculated by using equation (1), are shown in the second graph of figure 3. It is clear that for low load levels the peak bond stress develops at the end of the bar, up to the load level when the bar starts slipping (around 50 kN) and the first splitting crack initiates under the bar.

After the development of the first splitting crack (see load level 60kN of the third graph of figure 3), the peak bond stress migrates from the end of the bar towards the middle of the beam as load increases. The reason for this movement can be the propagation of the splitting cracks along the reinforcing bar. In order to verify this, specially arranged LVDTs were positioned at the bottom face of the beam along the length of the middle bar to monitor the widths of the splitting crack during the test. The crack width propagation along the bar is shown in the third graph of figure 3. It is clear that the splitting crack developed at the bar end at a load just below 60 kN and propagated eventually all the way to the direction of the loading point. The negative values shown for crack widths are due to the poison effect on the concrete which is in tension in the perpendicular direction.

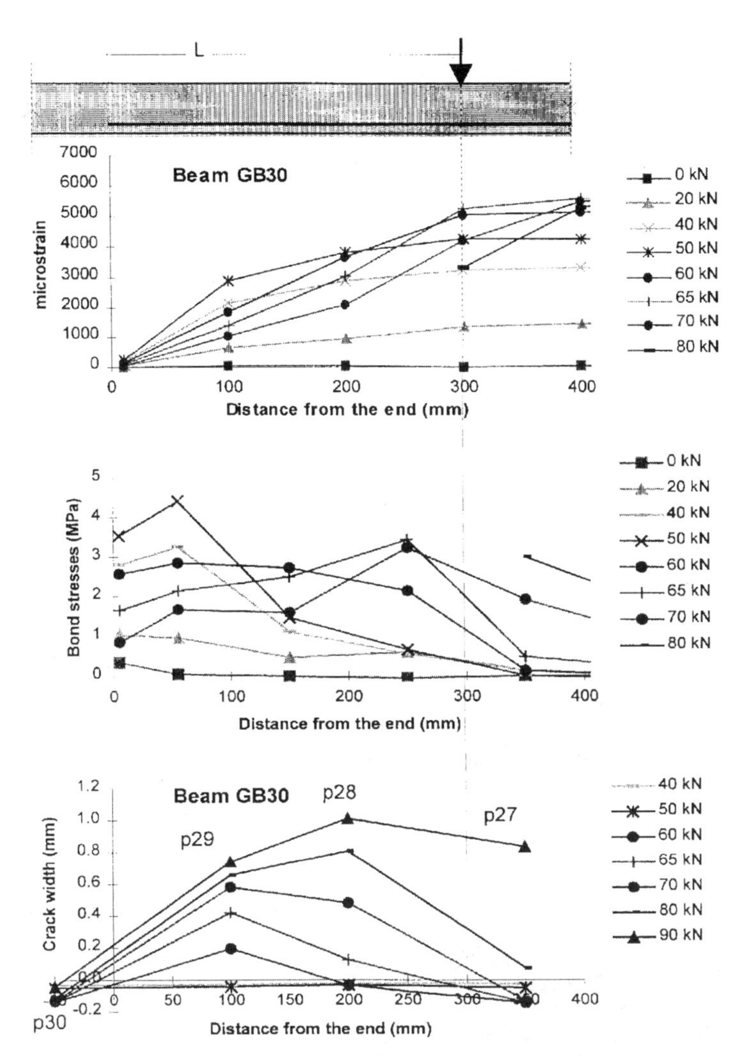

Figures 3: Normal strains, bond stresses and crack widths along the anchorage length

Two additional remarks emerging from observations of figure 3 need to be made. Firstly, the non-zero values of strains and bond stresses at zero load level can be attributed to the fact that these values were recorded at the end of a load cycle which may have locked in some strains. The second remark is associated with the movement of the peak bond stress value along the anchorage length. In beams tested in phases 1-3, the peak bond value was reported to migrate from the loaded end in the middle of the beam, towards the free end of the anchorage at the support, as the load increased [2]. Here, this is not the case since the initial peak bond stress appears at the end of the bar. This difference is due to the different location of the anchorage length with respect to the flexural cracked zone of the beam. In the current case, the whole anchorage length lies within the cracked region of the beam, and hence the bond stresses are less influenced by the propagation of the cracks towards the support, as is the case with bars anchored beyond the support.

4. CONCLUSIONS

The bond splitting behaviour of CFRP and GFRP reinforcing bars is examined in this study. The following conclusions can be made:

- CFRP bars develop around 30% higher bond splitting strength than similar GFRP bars under identical experimental conditions.
- The modulus of elasticity of FRP bars appears to influence their bond splitting strength.
- Traditional pullout tests can not be used to compare the bond strength of different type FRP bars when splitting failure is critical. A new type of test will need to be used when trying to determine the bond splitting characteristics of bars.
- Spliced bars appear to develop significantly higher bond strengths than anchorages and, hence, no reduction factors are required for their design.
- The contribution of shear links to the bond splitting behaviour needs to be examined in more detail.

ACKNOLEDGEMENTS

The authors wish to acknowledge the support of the EUROCRETE partners to this research and the support of the European Commission to the TMR network "Confibrecrete".

REFERENCES

[1] Clarke J.L. and Waldron P., "The reinforcement of concrete structures with advanced composites", The Structural Engineer, Vol. 74, No. 17, Sep. 1996

[2] Duranovic N., Pilakoutas K., Waldron P., "Structural Testing on R.C. Beams, Phase 1-3 Beams", Report Nos. CCC/95/19,20,21,22,33 and CCC/96/34,35 Centre for Cement and Concrete, Dept. of Civil and Structural Engineering, The University of Sheffield

[3] Achillides Z., Pilakoutas K., Waldron P., "Bond Behaviour of FRP Bars to Concrete", 3rd International Symposium on Non-Metallic (FRP) Reinforcement for Concrete Structures, Sapporo - Japan, Oct. 1997

[4] Sheard P., Clarke J., Dill M., Hammersley G., Richardson D., "Eurocrete – Taking account of durability for design of FRP reinforced concrete structures", 3rd International Symposium on Non-Metallic (FRP) Reinforcement for Concrete Structures, Sapporo - Japan, Oct. 1997

[5] Clarke J., "EUROCRETE PROJECT: Modification of design rules to incorporate non-ferrous reinforcement", Confidential EUROCRETE Report, Jan. 1996

[6] Achillides Z., Pilakoutas K. and Waldron P., "Investigation of bond behaviour of FRP bars to concrete", Report No. CCC/96/45A, Centre for Cement and Concrete, Department of Civil and Structural Engineering, The University of Sheffield, Jul. 1996.

[7] Achillides Z., Pilakoutas K., Waldron P., "Structural Testing on R.C. Beams, Phase 4 Beams: Bond Stresses", Report No CCC/97/47A, Centre for Cement and Concrete, Dept. of Civil and Structural Engineering, University of Sheffield

Uplift tests on pultruded GRP column base connections

G.J. Turvey*, C. Cooper*

* Department of Engineering, Lancaster University, Bailrigg, Lancaster, LA1 4YR, UK

Abstract

Details of a test rig for conducting uplift tests on pultruded GRP (Glass Reinforced Plastic) column base connections are presented. The rig was used to carry out tests on four column base connections - one each for two sizes of EXTREN™ 500 Series WF (Wide Flange) section and two arrangements of bolted angle cleats. The columns were first subjected to low uplift loads in order to determine the initial uplift stiffness of the connections. They were then loaded to failure in order to establish the uplift capacities and failure modes of the connections. Details of the uplift load - displacement response are presented for each connection. A simple model for predicting the uplift load capacity of bolted web and/or flange cleat connections is outlined and is used to predict the observed load capacities. It is shown to be in reasonable agreement with the test results, particularly for the larger section column base connections.

1. Introduction

Since 1990, when Bank et. al. [1] published their first series of test results on bolted web/flange cleat connections between pultruded GRP beam and column sections, there has been growing interest in characterising the stiffness and strength of both *conventional* and *novel* forms of beam to column connection between pultruded I and WF-sections. Over the past eight years several series of pultruded GRP beam to column connection tests [2, 3] have been reported which have contributed to the stiffness, strength and failure mode database for such connections. In contrast to the relative wealth of information on beam to column connection response, there appears to be minimal information on column base connection response. Recognition of this situation provided the catalyst for the present investigation of the uplift response of bolted column base connections for pultruded GRP WF-section columns.

Knowledge of the uplift response of column base connections is important in structural frameworks, because uplift may arise as a result of wind pressure/suction on the frame structure. In the column base connection tests reported here, uplift only is considered (bending effects are ignored) and the uplift force is assumed to be directed along the column axis.

2. Column base connection details

Two bolted cleat connection arrangements were tested with two sizes of pultruded GRP WF column section. All of the pultruded sections used in the column base connections were EXTREN™ 500 Series material, manufactured by Strongwell. Two column sections, viz. 102x102x6.4 (all dimensions in mm) and 203x203x9.5, were used. These sections have a 40% (approximately) fibre volume percentage and the polyester resin matrix may include a small percentage of inert filler. 102x102x9.5 angle section cleats and 10mm diameter mild steel bolts were used with the smaller WF column section, whereas 102x102x12.7 cleats and 12mm bolts were used with the larger section. Details of the bolted web and flange cleat column base connections for both WF column sizes are shown in Fig.1. The bolted web cleat column base connection arrangements were as shown in Fig.1, but with the flange cleats removed. All of the bolts in the connections were torqued to 30 Nm. 2mm clearance holes were used in the cleat legs bolted to the steel baseplate; elsewhere close tolerance holes were used. In all cases only the smooth surface of the bolt shank was in contact with the GRP material.

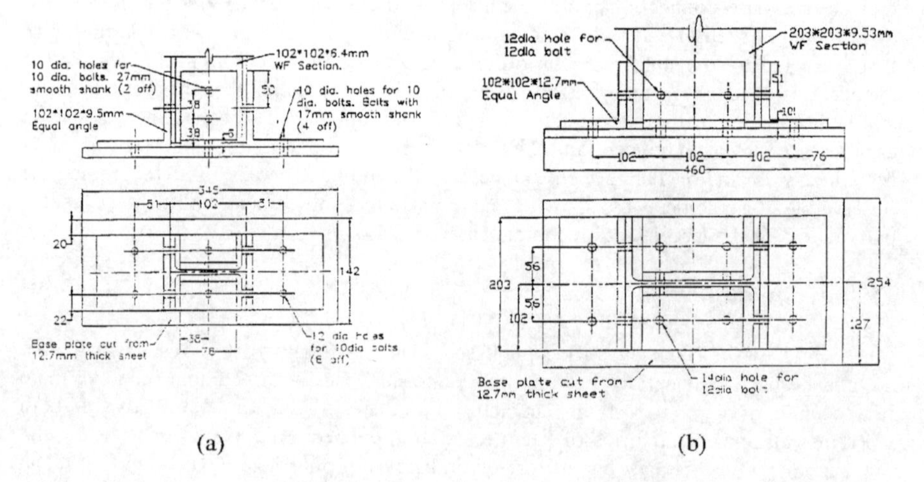

Figure 1. Bolted web and flange cleat column base connections: (a) 102x102x6.4 column and (b) 203x203x9.5 column

3. Uplift test rig

The test rig and instrumentation are shown schematically in Fig.2. The column base connections were tested in an Amsler universal testing machine. The angle cleats were bolted to a steel base plate which, in turn, was bolted to the lower ram of the testing machine. The top of the stub column was reinforced with two 6.4mm thick square GRP plates (with the pultrusion direction parallel to the column axis) bonded on to the column web. A 25mm diameter hole was drilled through the reinforced web so that a loading fork and a hardened steel pin arrangement could be used to attach the top of the column to the upper grips of the testing machine.

The displacement of the stub column relative to the steel base plate was recorded with four 100mm travel linear potentiometers attached to the free edges of the column flanges. In addition, provision was made to monitor the displacement of each cleat relative to the column web and/or flanges with two/four (depending on the number of cleats) 10mm travel linear potentiometers aligned with the vertical centreline of each angle cleat.

4. Low uplift load tests on column base connections

Each column base connection was first subjected to three low uplift load tests in order to establish response repeatability and determine the uplift stiffness. Generally, the loads

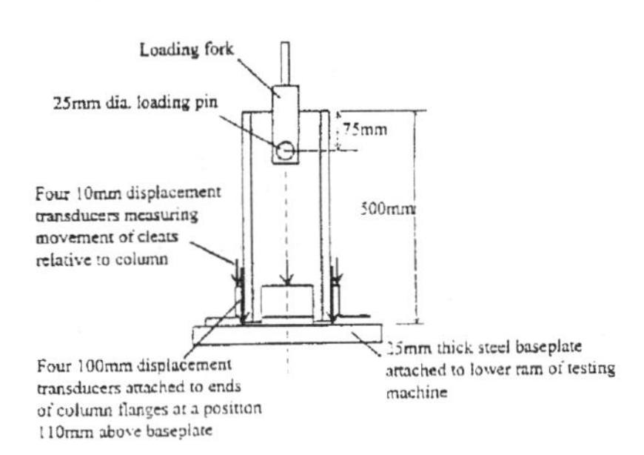

Figure 2. Uplift test setup and instrumentation

were applied in 1kN increments but, in the case of smaller web cleat column base connection, one test was carried out using 0.5kN increments and, in the case of the larger web and flange cleat column base connection, 2kN increments were used. The results of these tests are summarised in Table 1.

After the low uplift load tests on each column base connection had been completed, the connection was then tested to failure. During the course of each failure test the connection was subjected to one unload - reload cycle.

Table 1

Low uplift load test results for web and web and flange cleat column base connections

Connection Type	Max. Uplift Load (kN)	Max. Uplift Displ. (mm)	Initial Stiffness (kN/mm)	Residual Displacement (mm)	Secant Stiffness at 0.1mm (kN/mm)
102x102x6.4 (W)	4	0.100	52.8	0.027	40.0
	6	0.135	62.1	0.014	49.8
	6	0.133	54.8	0.005	47.8
102x102x6.4 (W+F)	8	0.122	92.5	0.022	70.9
	8	0.115	97.9	0.011	73.3
	8	0.120	82.2	0.016	70.9
203x203x9.5 (W)	8	0.191	77.7	0.069	55.2
	8	0.133	85.8	0.016	67.3
	8	0.128	93.3	0.018	69.4
203x203x9.5 (W+F)	20	0.144	-	0.000	171
	20	0.151	-	0.008	163
	20	0.151	-	0.000	164

Note: (W) and (W+F) denote bolted web and web and flange cleats respectively

5. Failure tests on column base connections

In the failure test on the 102x102x6.4 web cleat column base connection the connection was subjected to one unload - reload cycle, which was similar that of the low load uplift tests, i.e. the load was applied in 1kN increments up to 6kN and then the connection was unloaded in two decrements. After unloading the residual displacement was 0.015mm. The connection was then reloaded in 1kN increments up to 8kN following which the increments were reduced to 0.5kN until failure. At loads of 8, 9 and 10kN cracking noises became audible. Fig.4(a), which shows the load - displacement response for this connection, indicates that the connection stiffness started to reduce at a load of 8kN and that the cleats were beginning to be damaged. Delamination cracks in the heels of the web cleats became visible at a load of 10.5kN and became progressively longer and wider as the load increased. When the load reached 12kN the cracking noise became continuous until the load reached 13kN whereupon the connection failed in tension through the heels of the cleats. The top layers of the legs of the web cleats attached to the base plate delaminated and the bolts pulled through them. The uplift displacement at failure was 2.3mm. The failure mode of the web cleats is illustrated schematically in Fig.3(b).

The failure test on the 102x102x6.4 web and flange cleat column base connection also included an unload - reload stage, i.e. the connection was unloaded when the uplift load reached 8kN and then reloaded in 1kN increments to 12kN and thereafter in 0.5kN increments up to failure. The uplift load versus displacement graph, shown in Fig.4(b), indicates that the stiffness began to reduce at a load of 8kN. When the load reached 17kN the connection exhibited a sudden sharp reduction in stiffness and at 18kN a cracking noise was audible. A delamination crack developed in the heels of the flange cleats at 18.5kN. Further load increments caused the delamination cracks in the cleats to become longer and wider until the connection failed when the load reached 21kN. The corresponding uplift displacement was approximately 2mm.

The web cleats failed in flexural tension except for the top lamina which delaminated from the rest of the cleat and failed in shear around the washers as the bolts pulled through (see Fig.3(b)). The web cleats developed a small flexural tension crack in the laminates closest to the baseplate at a position in line with the edge of the washers under the bolt heads (see Fig.3(b)). This crack was caused by high bending stresses at that location.

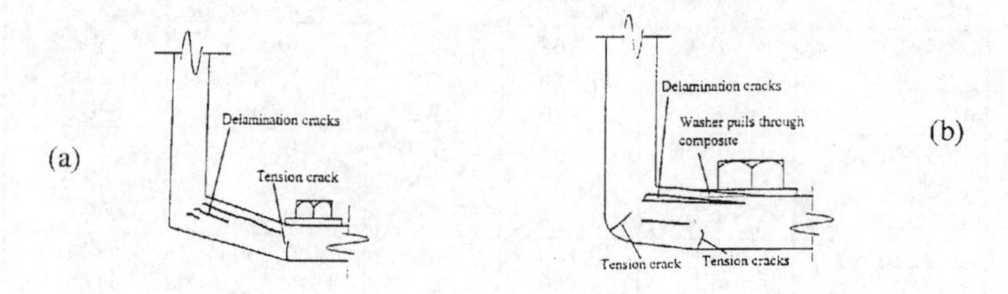

Figure 3. Failure modes of angle cleats: (a) flange cleat and (b) web cleat

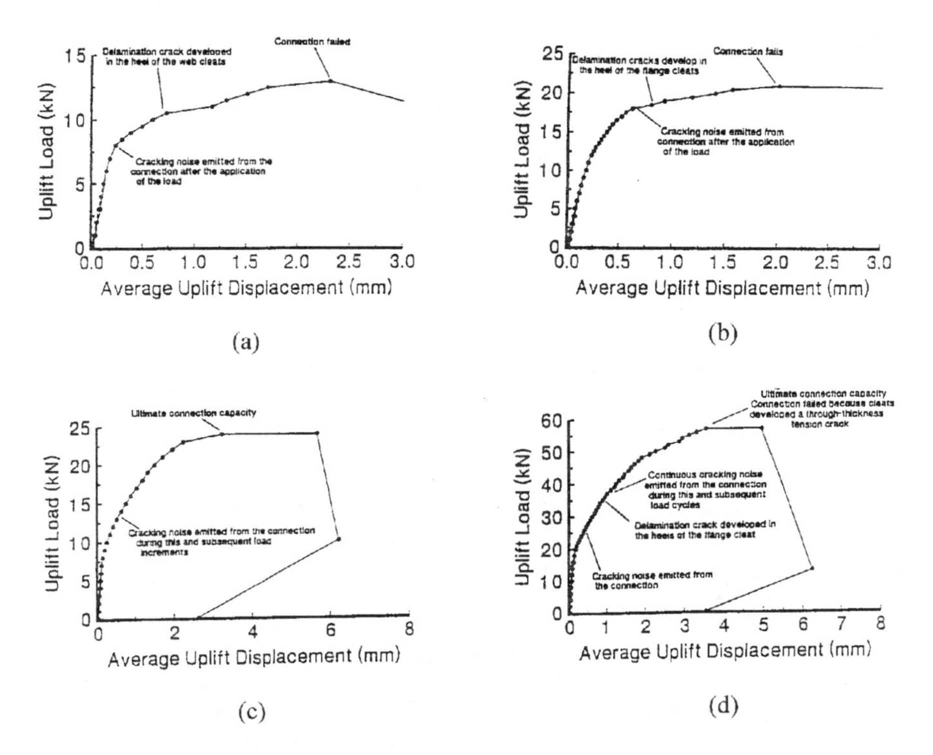

Figure 4. Uplift load versus displacement response for bolted column base connections: (a) 102x102x6.4 column (web cleats), (b) 102x102x6.4 column (web and flange cleats); (c) 203x203x9.5 column (web cleats) and (d) 203x203x9.5 column (web and flange cleats)

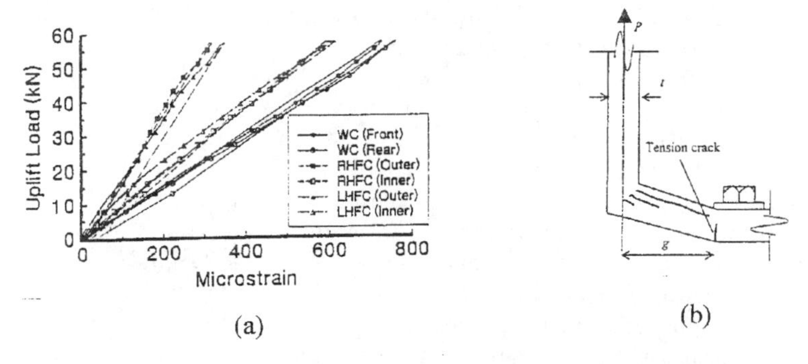

Figure 5. (a) Axial strains recorded on 203x203x9.5 column during uplift test on web and flange cleat column base test and (b) simple cleat flexural tension failure model for determining connection uplift capacity

Visual inspection, prior to testing, of the 203x203x9.5 web cleat column base connection revealed the presence of small delamination cracks in the heels of the cleats. These cracks developed during assembly of the connection as the legs of the cleats were bolted to the steel base plate. Thus the connection suffered minor damage prior to both the low uplift and failure load tests. In the failure test, the connection was loaded in 1kN increments up to a load of 8kN and then unloaded in two decrements. It was then reloaded in 1kN increments and a slight reduction in stiffness was observed as the load reached 8kN. When the load reached 13kN audible cracking noises were noted. Cracking noises were again noted at a load of 15kN and with the application of every subsequent load increment up to failure. These noises, it is believed, were caused by delamination cracks in the web cleats increasing in number and becoming longer and wider. The connection eventually failed when the load reached 24kN and the associated uplift displacement was 3.2mm. The uplift load - displacement response to failure is shown in Fig.4(c). The mode of failure in the web cleats was similar to that illustrated in Fig.3(b). The local curvature change at the edge of the washers in the angle leg attached to the steel base plate caused failure as a through-thickness flexural tension crack. Moreover, delamination in the laminate meant that the laminae were acting independently and this may have accounted for the gradual reduction in connection stiffness.

In the fourth and final test, i.e. the 203x203x9.5 web and flange cleat column base connection, the load was applied in 2kN increments up to 20kN and then the connection was unloaded in two decrements. The connection was then reloaded in 4kN increments up to 20kN and then in 1kN increments to failure. The complete load - displacement response is shown in Fig.4(d). In this test too a reduction in connection stiffness was observed when the load reached 20kN (the load limit of the previous low uplift load tests). When the load reached 23kN the first audible cracking noise was noticed. At a load of 34kN surface cracks were observed along the instep of the angle cleats and when the load reached 37kN (corresponding to an uplift displacement of nearly 1mm) delamination cracks developed in the heels of the cleats. Beyond a load of 38kN the audible cracking noise was continuous. Finally, the connection failed at a load of 57kN with an uplift displacement of 3.5mm.

Throughout the test the heels of the cleats gradually lifted off the base plate as the local change in curvature developed at the edges of the washers under the bolts connecting the angle legs to the base plate. As the cleats lifted delamination cracks in their heels became longer, wider and more numerous. The cleat failure modes are shown in Figs.3(a) and 3(b). The 10mm travel linear potentiometers showed, at failure, that the web cleats moved about 0.56mm relative to the column and that the relative flange cleat movement was between 0.29 and 0.37mm.

This column base connection was also instrumented with a number of uniaxial strain gauges, aligned with the column's vertical axis, in order to determine the strains in the web and flanges of the 203x203x9.5 column section. The strains recorded by each gauge are shown in Fig.5(a). It is evident that the strains on opposite surfaces of the web are similar. However, the strains on opposite surfaces of the flanges are not. The difference in the strain readings appears to be due to flange bending, but the actual cause of this is difficult to ascertain. Overall, the strains in the flanges are smaller than the strains in the web and this seems to indicate a degree of non-uniformity in the strain field at the cross-section where the strains were recorded.

6. Comparison of predicted and actual connection failure loads

The uplift tests have shown that the connection uplift capacity is determined by the strength of the angle cleats, which fail as a result of the development of a flexural tension crack in the leg attached to the steel base plate at a position between the edge of the bolt hole and the edge of the washer (see Fig.3). By assuming that the cleat legs to behave as wide cantilevers, clamped at the edge of the washer, and that the cleat legs attached to the column web are constrained to move vertically, then a simple flexural model, as shown in Fig.5(b), may be used to derive the following expression for the failure load:-

$$P = \frac{bt^3\sigma}{6g} \qquad (1)$$

In Eq.(1) σ denotes the transverse tensile strength of the cleat and b is the width, g is the span and t is the thickness of the cantilever. Tension tests were carried out on coupons cut from the cleat legs and the value of σ was established as 87.4N/mm^2, which is nearly twice the value of 48.3N/mm^2 given in the Strongwell design manual. Using the test and design values of σ in Eq.(1), the uplift load capacities of the individual web and flange cleats in the column base connections have been determined and are listed in Table 2.

Table 2

Uplift load capacities of individual web and flange cleats in column base connections

Cleat Type and Column Size (mm)	b (mm)	g (mm)	t (mm)	P [σ=48.3N/mm^2] (kN)	P [σ=87.4N/mm^2] (kN)
(W) 102x102x6.4	75	10.3	9.5	5.3	9.7
(F) 102x102x6.4	102	35.5	9.5	2.1	3.8
(W) 203x203x9.5	150	32.0	12.7	6.1	10.9
(F) 203x203x9.5	203	32.0	12.7	8.2	14.9

Note: (W) and (F) denote web and flange cleats respectively.

Assuming that all of the cleats fail simultaneously, the uplift capacity of the column base connection may be determined by summing the uplift capacities of each cleat in the connection. The calculated uplift failure loads are compared with test values in Table 3.

7. Concluding remarks

It is evident that the simple cleat failure model predicts the uplift failure loads of the larger section column base connections to within about 10% of the experimental values, whereas for the smaller section connections the predictions significantly over-estimate these values. The reason for this is that the web cleat mode of failure in both of the smaller section column base connections did not really match the model (cf. Figs.3(b) and 5(b)). Failure in the web cleats was not dominated by the flexural tension crack at the edge of the washer, but by shearing at the heel and pull through of the bolt. Hence, an improved model is required for the web cleat in order to achieve improved uplift failure load predictions for the smaller section column base connections.

Table 3

Comparison of experimental and predicted uplift capacities of column base connections

Connection Type	Predicted Uplift Failure Load (P_p) (kN)	Experimental Uplift Failure Load (P_e) (kN)	[P_p/P_e] (%)
102x102x6.4 (W)	19.4	13.0	149
102x102x6.4 (W+F)	27.0	21.0	128
203x203x9.5 (W)	21.8	24.0	91
203x203x9.5 (W+F)	51.6	57.0	91

8. Acknowledgements

The research described in the paper was supported by the UK's Engineering and Physical Sciences Research Council (Grant Ref. No. GR/J41703). The authors are grateful for this support and also wish to acknowledge support from the Department of Engineering, Lancaster University.

References

[1] *L.C Bank*, BEAM-TO-COLUMN CONNECTIONS FOR PULTRUDED FRP SHAPES - Proceedings of ASCE First Materials Engineering Congress - Denver - USA 1990.

[2] *A.J. Bass, J.T. Mottram*, BEHAVIOUR OF CONNECTIONS IN FRAMES OF FIBRE REINFORCED-POLYMER SECTION - Structural Engineer, 1994, Vol.72, No.17

[3] *G.J.Turvey, C. Cooper*, SEMI-RIGID PULTRUDED FRAME CONNECTIONS: TESTS TO DETERMINE STATIC MOMENT-ROTATION CHARACTERISTICS - Proceedings of Seventh European Conference on Composite Materials - London - UK 1996

GFRP- and CFRP-elements for the construction industry

J. F. Noisternig*, D. Jungwirth*

* Department of Central Technics, DYWIDAG-Systems International, P.O.-Box 81 02 68, 81902 Munich, Germany

Abstract

Advanced composite materials, like carbon (CFRP) and glass (GFRP) fiber reinforced plastics offer great potential to the construction industry due to their outstanding properties compared to conventional construction materials. Very high strength and fatigue resistance combined with low density eases the application. But the key problem using advanced composite materials as tensioning elements is how to anchor them. The development of a GFRP-bar as a temporary reinforcing element in rock engineering and tunnelling with a suitable anchoring system lead to successful applications. Another development of an anchoring system for a tendon consisting of CFRP-wires will be dicussed. A CFRP-tendon with an ultimate tensile strength of 5000 kN has been successfully tested under dynamic load according to the PTI recommendation for stay cables.

1. Introduction

Concrete is traditionally reinforced with steel bars and tendons. It is well known that deterioration of concrete structures can in most cases be attributed to corrosion of the reinforcing steel. Advanced composite materials, also known as fiber reinforced plastics (FRP), developed and used primarily in the aerospace and defense industry, have in addition to outstanding strength/weight ratios, also a high degree of chemical inertness to most civil engineering environments which strongly suggests their consideration for such applications. Several developments have changed this picture over the past few years. Advances in the manufacturing of FRP, reduced demands for advanced composites in the high priced defense industry, expansion in the sports industry and prospects for large volume applications in the civil engineering sector led to a rethinking.

There are applications, where FRP-systems have achieved acceptance in civil engineering. Particularly in the rehabilitation of existing structures, CFRP-systems have shown their advantages in many applications. Furthermore, the seismic retrofitting of bridge columns with carbon and glass fiber wraps or pre-formed jackets has been demonstrated to be technically just as effective as and in most cases more economical than conventional steel jacketing [1, 2, 3, 4, 5].

Using tensioning elements consisting of FRP requires new understanding as compared to conventional steel tendons. The key problems are the specific properties and the material behaviour of FRP-elements, which cause great problems in the development of anchoring systems with a high load bearing capacity [6].

2. Requirements for FRP-elements

Using advanced composite materials as reinforcing and tensioning elements in concrete structures requires adequate knowledge of the mechanical properties. Up to now, these elements have in no way been standardized nationally or internationally. Japan, USA and Canada as well as Europe (CEB and FIP) are striving to standardize and characterize materials, applications and design recommendations for FRP. However, some more time will pass until a full set of regulations not only for advanced composite materials, but also for applications and design recommendations will be available.

It has to be clear that in contrast to steel, advanced composites are no homogenous materials and thus each element is an individual material with different properties. Properties of advanced composite elements depend among others decisively on the fiber content. The long-term behaviour and the behaviour under influence of different media are important, besides the mechanical properties under static and dynamic load. Thus the aim of a standardization can only be to give classes for advanced composite elements in dependence of different fiber/matrix combinations. The test methods necessary for a characterization of materials are known from material tests of polymers and have to be standardized for FRP-elements and systems for applications in concrete structures worldwide.

The known requirements to prestressing steel systems (PTI in USA and similar recommendations in Europe) can be taken as a guideline for FRP reinforcing elements and tendons. High efficiency of anchoring systems for FRP-tendons has to be strived for generally to exploit the material as good as possible (among others an economical aspect). Similar to prestressing steel systems, suitability tests for FRP reinforcing elements and tendons are inevitable.

3. GFRP-elements as temporary ground anchors

Anchoring systems for the stabilization of grounds and rocks are a wide field. DYWIDAG-Systems International (DSI) here apply their typical steel threadbars and GEWI-bars. Bars consisting of GFRP are a challenge for the material technology to cover even those fields, where steel has some disadvantages. Developments with regard to the later application of the product have been performed by DSI with GFRP-bars.

GFRP-bars can easily be cut perpendicular to their axis due to their brittleness. At first glance this was considered to be a disadvantage, but using GFRP-bars as a temporary rock reinforcement, it becomes an advantage. In particular, fully grouted rock bolts may be easily destroyed by roller bits or similar tools, which is impossible in the case of steel. Typically, reinforcements placed radially from a pilot tunnel have to be excavated during enlarging the cross-section either by tunnel boring machines or road headers (see Figure 1). Similar applications are common in coal mining, where the mining face in case of difficult stress conditions is reinforced with GFRP-bars which are cut off subsequently by the shearer of a long wall machine. But also in connection with low cost tunnel lining, an inexpensive permanent rock bolt has been asked for for years. Permanent rock bolts and shotcrete are considered sufficient as permanent lining for many tunnels or sections of them. Steel rock bolts require a permanent reliable corrosion protection which increases their costs substantially. For such applications anchoring systems consisting of non-corrosive materials like polymers are an additional advantage. GFRP-bars offer this economic solution.

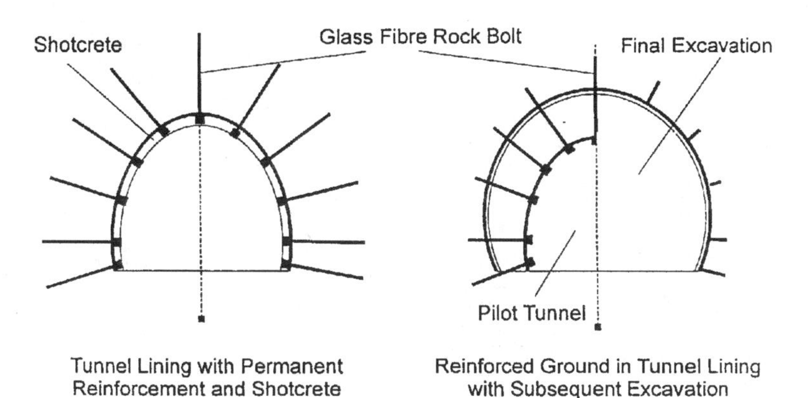

Figure 1. Fields of application for GFRP-bars

It is evident that their use as permanent reinforcement is not restricted to competent rock but may be extended to weak rocks and soils, where soil nailing gains increasing importance. But the applications may grow with the collected positive experiences, covering of the production costs and improvements of the anchoring systems.

Table 1. Mechanical properties of GFRP-bars DYWIDUR

Property	**GFRP-bar ⌀ 22 mm**	**GFRP-bar ⌀ 25 mm**
Bar diameter [mm]	22	25
Density [g/m]	750	930
Breaking load of the bar [kN]	> 300	> 430
Breaking load of anchorage [kN]	110	145
Modulus of elasticity [MPa]	45	40
Bond to cement/resin grout [N/mm^2]	6 - 14	6 - 14

DSI found a solution for such a GFRP-bar with a threaded anchoring system in combination with a polymer nut and plate and a rough surface for a good bonding in the borehole. The anchoring system was the decisive point in this development. A threaded anchoring system with two different thread types has been investigated. The one possibility is to form the thread onto the bar during the pultrusion process, the other possibility is to cut a thread into the bar subsequently. To get an anchoring system completely free of steel, the nut and plate have been developed consisting of thermoplastic Polyamide reinforced with short glass fibers. The mechanical properties of GFRP-bars of diameter 22 mm and 25 mm are shown in Table 1. Both GFRP-bars have the threaded anchoring system.

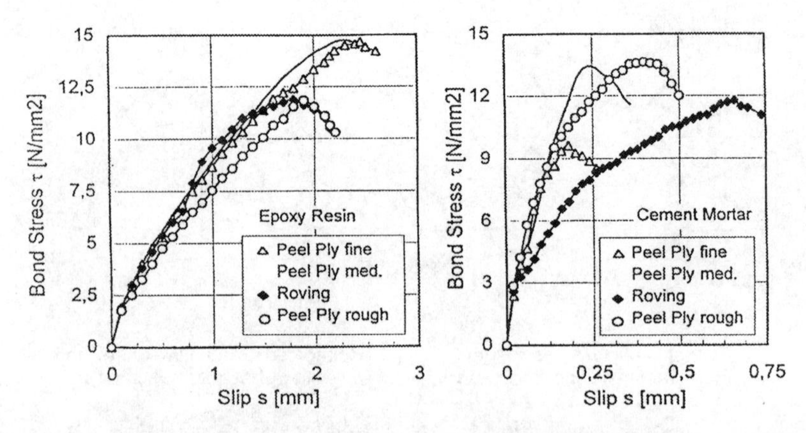

Figure 2. Force-slip behaviour of GFRP-bars DYWIDUR with different surface structures (in resin or cement grout)

The ultimate load bearing capacity of the GFRP-bar can not be achieved with this threaded anchoring system. However, this is not necessary as the GFRP-bar is in most cases glued into the drilled hole over the entire length by a cement or resin grout. Thus, however, good surface roughness of the bar has to provide good bond to transfer the force between bar and surrounding soil via bonding to the grout. The surface roughness is obtained through special peel plies which after removal provide the required roughness or by winding a roving around the bar which also produces a structured surface. The force-slip behaviour of bars with different surfaces in resin and cement grout are shown in Figure 2 [7].

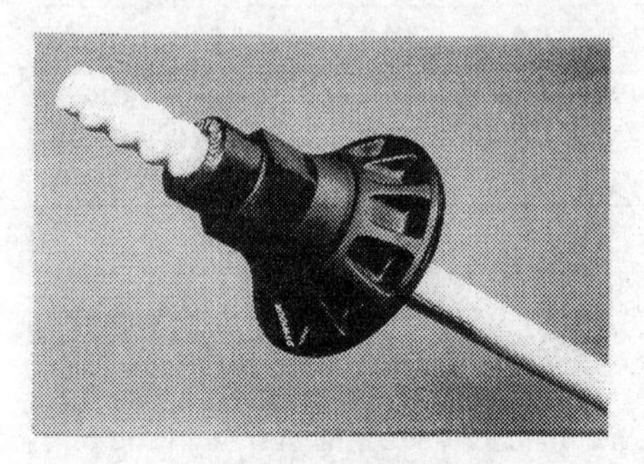

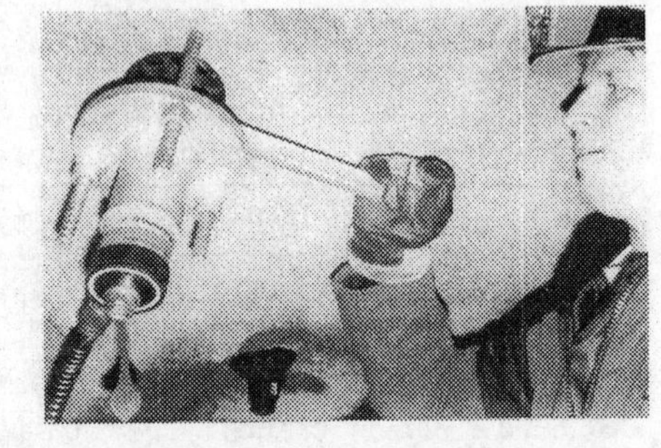

Figure 3. GFRP-bar DYWIDUR (left) and application in a project in Norway (right)

Figure 3 (left side) shows the DYWIDUR-system with the steel free anchoring system. DSI the first time successfully applied GFRP-bars in the Ziegenbergtunnel (Germany). Other applications are excavations in Germany and Austria, a tunnel project in Norway (Figure 3 right side) and in the Netherlands.

4. Tendons consisting of CFRP-wires

Compared to glass and aramide fibers, carbon fibers have some performance advantages, in particular where strength, stiffness and corrosion are important factors. The price/performance ratio of CFRP is becoming more and more attractive in cases where steel tendons failed or needed maintenance due to defects or faults of their corrosion protection. The key problem facing the application of CFRP-tendons and their widespread use in the future is how to anchor them. The outstanding mechanical properties of CFRP-elements are only valid in the fiber direction of the elements. In transverse direction the properties are relatively poor, also the interlaminar shear strength. New designs of the anchoring systems are necessary. DSI is developing a CFRP-tendon consisting of a bundle of CFRP-wires with a potting anchoring system. In Table 2 properties of two developed potting systems and the CFRP-wire Carbon-Stress® from Nedri Spanstaal BV (The Netherlands) which are used in the investigations, are listed.

Table 2. Material properties of potting systems and CFRP-wire Carbon-Stress®

Material type	Property	Value
Potting system S-3	Bending tensile strength	47.0 MPa
	Compressive strength	125.5 MPa
Potting system S-4	Bending tensile strength	46.0 MPa
	Compressive strength	114.5 MPa
CFRP-wire (Carbon-Stress® from Nedri) ⌀ 5 mm	Tensile strength	2800 MPa
	Tensile modulus	160 GPa
	Density	31 g/m
	Fibre fraction	65 - 70 Vol.-%

Analytical and, above all, numerical calculation methods offer, in contrast to experimental investigations and optimizations, timesaving and cost saving proceedings. Numerical calculations of a potting system for a single CFRP-element [6] were used as a basis for further investigations of an anchoring system for 7 CFRP-wires at DSI. These numerical calculations were performed for a better understanding of the complex stress distribution inside the anchoring system. Making use of symmetry of geometry and load, only a three-dimensional quarter of the potting system for 7 CFRP-wires was developed and analyzed with the FE-program MARC. The cross-section of the CFRP-wires were considered as homogeneous orthotropic materials.
To calibrate and verify the numerical model, it was compared with a static test of a tendon with 7 CFRP-wires. The measured strains were used to calculate stresses of the steel hull

and to calibrate and verify the numerical model. Figure 4 shows the comparison of measured and calculated stresses in the steel hull of the CFRP-tendon with 7 wires. As the numerical model describes the load bearing behaviour of the CFRP-tendon quite well, it was used for further calculations. With these calculations, a reduction of the critical stresses in the potting material was possible only by changes in the geometry of the steel hull.

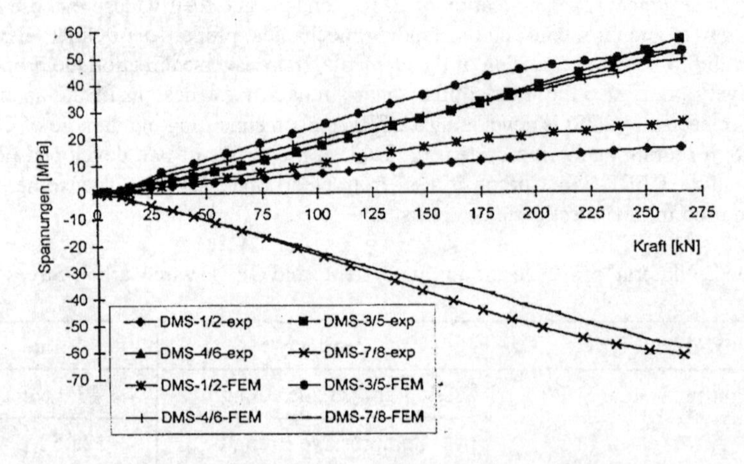

Figure 4. Comparison of calculated and measured stresses in the steel hull of the tendon with 7 CFRP-wires

In addition to the theoretical investigations, experimental tests were performed. In filling studies for the potting material the ideal combination of fillers and epoxy resin as well as their ideal mixing should be determined (see material properties in Table 2). By addition of fillers the viscosity of the potting material is increased, which causes entrapped air inclusions in the material. To prevent these air inclusions the potting material must be pumped into the anchoring system. Pull-out tests on CFRP-wires were conducted to examine the bond behaviour of the CFRP-wires and epoxy resin in dependance of different surface structures of the wires. For this purpose, CFRP prepreg tapes were wound around the CFRP-wires during manufacturing. The winding angle (pitch) and simple or cross-winding of these approx. 5 mm wide CFRP prepreg tapes were varied. The adhesion between the prepreg tape and wire is equivalent to that between wire and potting material, so that nearly no difference in the maximum bond stresses could be observed. From these tests it was concluded that the smooth CFRP-wires should be used for the CFRP-tendon. Using the results of the theoretical investigations, filling studies and pull-out tests, improvements of the potting system are possible. All these led to an anchoring system for CFRP-wires which is shown in Figure 5. It is a conical potting system for a bundle of 7 CFRP-wires. The dimensions of the steel hull (200 mm long with a diameter of 90 mm) result from the calculations on the potting system for CFRP-wires.
With this improved conical potting system and a filled epoxy resin as potting material between 90 and 95% of the theoretical load were reached in static tests. In nearly all tests

the same failure can be observed. Tensile failure of the CFRP-wires occured in most cases at the entrance of the potting system (Figure 6 left side). In tests under dynamic load (maximum stress 0.6 of ultimate tensile strength with an amplitude of 150 MPa) 4 million load cycles were reached without failure. After these tests, in additional static test nearly the same breaking load was reached as before the fatigue test. These tests on tendons with 7 CFRP-wires have shown the high load bearing capacity of the developed potting system.

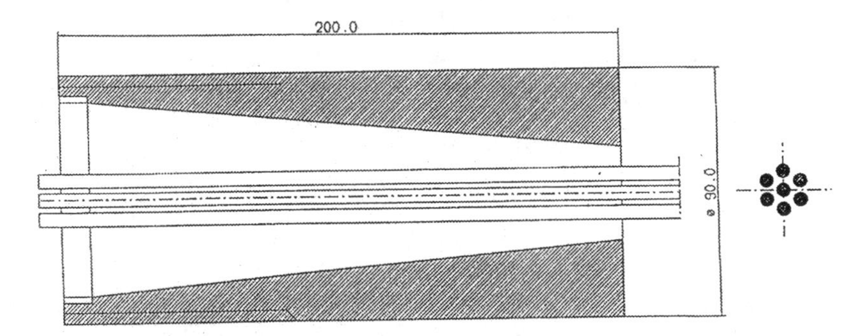

Figure 5. Improved potting system for a tendon with 7 CFRP-wires

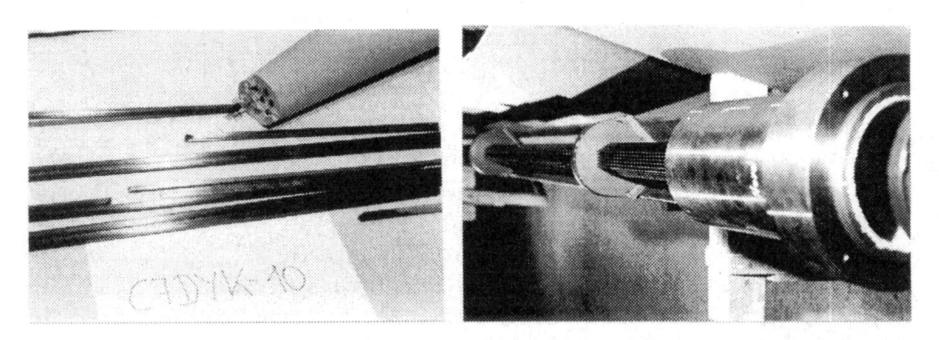

Figure 6. Failure behaviour of a tendon with 7 CFRP-wires (left) and manufacturing of a tendon with 91 CFRP-wires (right)

In the next step this anchoring system has to be transfered to a tendon with a higher load bearing capacity, so that it can be used in the field. The work for a tendon with an ultimate tensile strength of 5000 kN, which means that the tendon has to consist of 91 CFRP-wires, has been started in the end of 1997. Figure 6 (right side) shows the manufacturing of the tendon. This tendon with 91 CFRP-wires has been tested under dynamic load according to the PTI recommendation for stay cables (maximum stress 0.45 of ultimate tensile strength, amplitude of 160 MPa). 2 million load cycles were reached without any failure. In the following test under static load a satisfying load bearing capacity was reached. Next step will be to elaborate all details for a test and application in the field.

5. Concluding remarks

If we compare the price/performance ratio of tendons and consider the corrosion protection systems at steel tendons, GFRP seems to be a cost-effective alternative to temporary steel ground anchors. The use of CFRP-tendons is advantageous, when the conventional prestressing and reinforcing steel has to be protected against corrosion by highly expensive design measures or has to be exchanged early. This especially includes bridges near the sea, prestressing with external tendons and applications in structures in contact with acid and alkaline media. Also long span suspension bridges or cable stayed bridges can only be realized applying CFRP-tendons. It is essential to use pilot projects to learn from mistakes and to convince the owners of structures and the building authorities of the outstanding opportunities of advanced composite materials in construction industry.

Acknowledgment

The authors gratefully acknowledge the financial support from the "Freistaat Bayern" through the project number CA 628 45 0696 "Development of a CFRP-tendon".

References

[1] *H. Saadatmanesh, M. R. Ehsani,* FIBER COMPOSITES IN INFRASTRUCTURE - Proc. of 2nd International Conference on Composites in Infrastructure (ICCI '98) Tucson (Arizona), 5-7 Jan. 1998.
[2] *N. N.,* NON-METALLIC (FRP) REINFORCEMENT FOR CONCRETE STRUCTURES - Proc. of 3rd International Symposium on Non-Metallic Reinforcement for Concrete Structures (FRPCS-3) Sapporo (Japan), 14-16 Oct. 1997.
[3] *U. Meier, R. Betti,* RECENT ADVANCES IN BRIDGE ENGINEERING - US - Canada - Europe Workshop on Bridge Engineering Zurich (Switzerland), 14-15 July 1997.
[4] *N. N.,* TEXTILE REINFORCED CONCRETE - Proc. of 8th International Textile Symposium (Techtextil 1997) Frankfurt (Germany), 12-14 May 1997.
[5] *N. N.,* STRENGTHENING OF STRUCTURES WITH CFRP-LAMINATES (in German) - Proc. of the EMPA/SIA-Conference (D 0128) Zurich (Switzerland), 1995.
[6] *J. F. Noisternig,* INVESTIGATIONS OF THE LOAD BEARING BEHAVIOUR OF ANCHORING SYSTEMS FOR A CFRP-STRAND (in German) - Fortschritt-Berichte VDI Reihe 4 Nr. 133, Düsseldorf (Germany), 1996.
[7] *J. F. Noisternig, D. Jungwirth,* GFRP-BARS AS REINFORCING ELEMENTS IN CIVIL ENGINEERING WITH APPLICATIONS - Proc. of 8th International Textile Symposium (Techtextil 1997) Frankfurt (Germany), 12-14 May 1997.
[8] *J. F. Noisternig, D. Jungwirth,* CFRP-TENDONS FOR STRUCTURAL APPLICATION - REQUIREMENTS AND DEVELOPMENTS - Proc. of 2nd International Conference on Composites in Infrastructure (ICCI '98) Tucson (Arizona), 5-7 Jan. 1998.

Experimental analysis of bond between GFRP deformed rebars and concrete

C. Greco*, G. Manfredi*, M. Pecce*, R. Realfonzo*

* Dipartimento di Analisi e Progettazione Strutturale – Università degli Studi di Napoli Federico II – Via Claudio 21, 80125 Naples-Italy

Abstract

Over the last few years, a number of cases of corrosion of steel reinforcement have resulted in extensive deterioration of reinforced concrete structures.
In order to eliminate corrosion, a new and effective approach consists of the use of corrosion-resistant Fiber Reinforced Plastic (FRP) rebars for concrete structures. If compared to steel rebars, this new type of reinforcement presents some advantages (i.e. lower weight per unit mass, higher durability, etc.) as well as some disadvantages (i.e. lower modulus of elasticity). FRP materials show a brittle behavior, being linear elastic up to the collapse, with tensile strength similar to or greater than the steel one. In this paper results from tests on bond between FRP rebars and concrete are presented and discussed. Tests were performed by using a modified beam-type specimen and bond properties of a new glass fiber reinforced plastic (GFRP) rebars (namely C-Bar™ by Marshall) were investigated.

1. Introduction

Due to the lack of well-established standards, a wide variety of FRP rebars is today commercialized, going from the simple smooth rebars to rebars treated to improve bond characteristics (rebars with surface treatments and/or with deformation of the outer surface). Moreover, within each type, various products made of different materials (fibers and resins) and having different geometry (i.e. spacing and size of ribs), are available. Obviously, all these differences in geometrical and mechanical properties introduce a wide variability, which results in significant differences in bond behavior.
Behavior of both reinforced and prestressed concrete structures is affected to a large degree by bond between the concrete matrix and the reinforcing rebars. For r.c. structures, indeed, the transfer of stresses between the two components, both at the serviceability and at the ultimate state, is strongly dependent on the quality of bond. In fact, resisting mechanisms under bending, shear and torsion are related to the development of an adequate bond; moreover, many serviceability checks, such as control of crack amplitude and of structure deflections, involve the evaluation of the effects of the tension-stiffening which is a phenomenon directly arising from bond behavior. Regarding prestressed concrete structures, a key issue is clearly the transfer of stress from the tendons to the concrete. Therefore, the development of an adequate bond is always a critical aspect of the structural behavior whichever is the type of reinforcement and, then, this is also the case for FRP rebars.

Recently, many experimental studies have been conducted in order to establish both mechanical properties of the commercially available FRP rebars (rather dependent on type and percentage of both fibers and resins) and their bond characteristics. However, the experimental mechanical properties and bond parameters, which were obtained for rebars made of different materials (i.e. glass FRP, aramid FRP or carbon FRP rebars) and having different outer surfaces (smooth, ribbed, sanded), presented large scatters.
Bond mechanisms of FRP rebars are different from those of steel reinforcements because of the different manufacturing of the outer surfaces and of substantial differences in material properties both in longitudinal and in transverse directions. The mechanics of stress transfer by bond between FRP rebars and concrete has been investigated by many Authors [] and two states-of- art were recently published by Cosenza et al. [9] and by Tepfers [10]. In that paper results of many tests were shown by classifying rebars in two main categories: straight FRP rods and deformed FRP rods. "Straight FRP rebars" included smooth rebars and rebars with treatment of outer surface (sanded rebars), while "deformed rebars" included ribbed rebars, braided rebars, indented rebars, twisted rebars, rebars with a spiral glued (with or without surface treatments). As already pointed out, the lack of a standard has resulted in large variations in bond behavior both from qualitative (mechanisms) and from quantitative points of view (bond strength).
From the experimental results [1÷8] it was concluded that bond of FRP reinforcement to concrete is controlled by several factors such as the mechanical and geometrical properties of rebars and the compressive strength of concrete. However, mechanical properties of resin, which the matrix is made of, influence to the largest degree the interaction behavior since they affect strongly strength and deformability of ribs or indentations located on the outer surface.
Bond depends on three different mechanisms: chemical bond, friction and mechanical interlocking. Bond tests consistently evidence that chemical bond (adhesion), initially, is the main resisting mechanism; afterwards, it is replaced by friction and/or mechanical interlocking. Since adhesion between concrete and FRP generally is extremely low, friction and mechanical interlock really become the primary mean of stress transfer. However, interlocking may become less or more important according to different situations [2,4,9]. Namely, the phenomenon is governed by the shear strength and deformability of ribs, which are remarkably lower than those of steel rebars, thus leading to increased slips between rebars and concrete.
In this paper an experimental study of bond between FRP rebars and concrete is presented. After a short discussion on test methods for investigation of bond behavior of FRP rebars, it is introduced the modified beam type test which has been adopted herein to study bond performance of a recently introduced deformed GFRP rebar (namely C-Bar^{TM} by Marshall). Subsequently, trends of bond-slip curves, obtained as the embeddement length varies over a wide range, and failure mechanisms are discussed in order to draw preliminary conclusions about bond performances of such rebars.

2. Testing procedures for the bond assessment

The experimental determination of bond properties in case of FRP rebars is an open question with respect to two issue at least: the first is the possibility of extending the experimental procedures used for steel rebars to the FRP ones, the second is the

assessment of a method to derive from test results a suitable bond-slip law that can be applied in the analysis of structures under actual work conditions.
For what concerns test procedures, the most widely used ones for steel are the following: pull-out tests (with centered or eccentric position of the rebar in the concrete specimen); ring pull-out tests; beam tests (i.e. RILEM type); splice tests. During tests, pull-out or splitting failure can occur depending on cover thickness and confinement.
All the abovementioned types of tests have been conducted by several Authors to evaluate bond strength in case of FRP rebars [1÷8], and substantial differences have emerged from comparison of results obtained adopting various types of test setup; furthermore, some types of test setup do not appear adequate to study the actual behavior.
Comparing bond behavior of FRP and steel rebars it is noted that some properties of FRP rebars can result in remarkable differences compared to deformed steel ones. This is the case of: the lower elastic modulus and the consequent higher tensile elongation, which, along with the higher Poisson ratio, leads to a reduction in the transverse diameter; the lower lateral stiffness that can influence splitting failure in bending conditions; the lower shear strength and stiffness of ribs that can influence the mechanical interlocking.
As a consequence, the following general observations can be made:
the effect of transverse conditions (confinement, cover thickness, etc.) on bond behavior can be very different from that for steel rebars and their influence on the failure mode needs further discussion with respect to the actual work conditions; slip measurements at the free end and at the loaded end can be very different due to the low longitudinal elastic modulus, therefore the embeddement length significantly affects test results; the low transversal stiffness of the FRP bars influence the bond performance according to the test method (i.e. pull-out or beam-tests).
A complete characterization of bond in FRP rebars probably needs a set of different types of bond tests that are representative of the different structural conditions (i.e. tension, bending, etc.) and the assessment of reliable constitutive bond-slip relationships requires an improved interpretation of the test results by using, as an example, identification procedures.
The beam tests with centric placement of the bars, shown in this paper, can represent an upper bound of the bond performances and are representative of the structural behaviour when the splitting collapse is prevented

3. Experimental tests

In the following, results of an experimental investigation of bond properties of a new fiber reinforced plastic composite rebar, namely C-BAR, are presented.
Bond tests were performed in the Department of Structural Analysis and Design of the University of Naples Federico II. They have been conducted on prismatic concrete specimens within which a glass FRP rebar was embedded. The GFRP rebar (the abovementioned C-Bar^{TM}) had a 4/8 inches diameter (≅12.7 mm) and the embeddement length L was ranging from 5 to 20 times the bar diameter d_b, thus obtaining different test arrangements (see Figure 1 for the specimen scheme).
Only the results obtained by performing the first four tests are presented below. Such tests, named test #1, #2, #3 and #4 refer to different embeddement lengths: in particular,

L was equal to $5d_b$ (L=63.5 mm), for tests #1 and #2, to $20d_b$ (L=254 mm) for test#3 and to $10d_b$ (L=127 mm) for test#4.

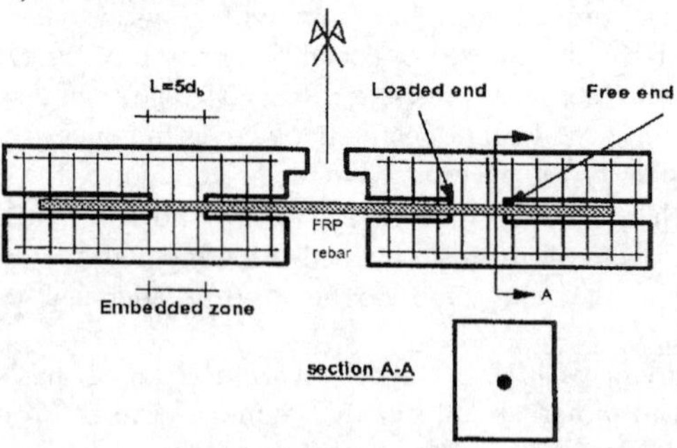

Figure 1 - Specimen scheme

3.1. Experimental setup

Bond tests were performed by using a test machine obtained from a modification of the standard scheme of the beam-test. The machine - shown in Figure 2 - has permitted to apply a tensile force T on the rebar by vertically loading the steel beam under which the concrete specimen was fixed. It can be seen that the test setup permits to conduct contemporaneously two bond tests on the two concrete elements (sub-specimens) connected by the rebar.

The tests were conducted by imposing displacements by means of a mechanical actuator and measuring the corresponding value of the vertical load P with a load cell. Load P was applied on both side of the beam by means of a rigid steel element placed on both sides of the hinge (see Figure 2).

Values of the tensile force T on the rebar has been calculated by the following relationship:

$$T=P \cdot d/(2 \cdot h) \quad (1)$$

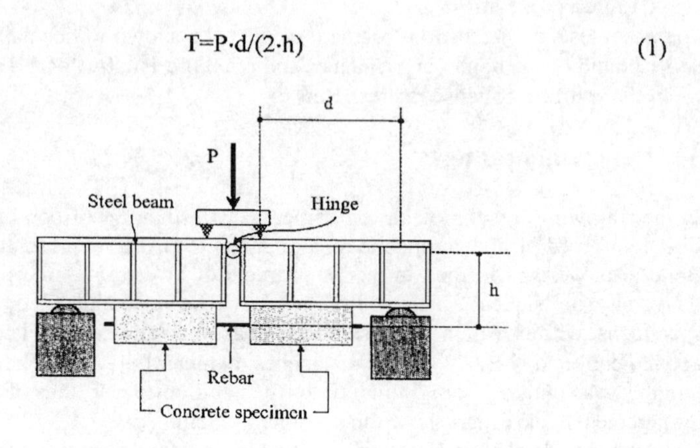

Figure 2 – Test setup for modified beam tests

where d is the distance from the point of application of vertical load to the axis of the beam support; h is the distance between the center of the steel hinge and the axis of the rebar (see Figure 2).
Slips s have been measured by means of four LVTDs placed on both ends (free and loaded) of the two sub-specimens.
Furthermore, a strain gauge was placed on the rebar, at the center of the specimen, in order to measure strains for each value of stress.
As previously said, this test arrangement has permitted to make tests without applying any compression load on the concrete specimen - contrary to what happens in pull-out tests - thus obtaining a good simulation of work conditions of rebars embedded in structural members subjected to bending, when the splitting failure is prevented.

3.2. GFRP rebars and concrete properties

Rebars used in tests are the FRP C-Bar™ produced by Marshall Industries Composites, Inc.. These are obtained by means of a hybrid pultrusion/compression molding production process and are available in four different types of fibers: E-Glass (Type1), Carbon (Type 2), Aramid (Type 3) and a hybrid of Carbon and E-Glass (Type 4). C-Bar reinforcing rods are made in two grades based on the type of deformations and surface characteristics (Grade A and B).
Grade B C-Bar has inclined ribs on the outer surface, intended to improve bond behavior, made of ceramic fibers embedded in uretan modified vinyl ester. The inner core, instead, is composed of unidirectional fibers embedded in recycled PET resin material.
C-Bar reinforcing rods have a surface deformation pattern similar to that of steel rebars and are available in four sizes: #3 (9.7 mm), #4 (12.7 mm), #5 (14.9 mm) and #6 (17.7 mm). The pattern of the outer surface depends on the bar diameter: for #4 rebar, ribs have a minimum average height equal to 1.0 mm and a maximum average spacing equal to 6.1 mm. #4 Grade B/Type 1 C-Bar, used in tests presented in this paper, have a linear stress-strain behavior up to failure with a modulus of elasticity equal to about 42 GPa and an average ultimate tensile strength equal to 770 MPa [13].
The mean compressive strength of concrete R_m has been determined using cubes (150 mm^3). Tests have evidenced different values of R_m: in particular, R_m was equal to 37 MPa in case of test #1, about 40 MPa in case of tests #2 and #3 and 52 MPa for test #4.

3.3. Experimental results

Figures 3 and 4 show the results from tests #1 and #2 respectively. Namely, the experimental bond-slip (τ-s) curves for both sub-specimens are reported. It can be seen that the in both cases only one of the two sub-specimens fully develops bond behavior due to unavoidable asymmetries in the test arrangement. This is evidenced by values of slip s - measured by using the LVTDs placed at loaded and free end – which are much larger on one side than on the other. Values of s at the loaded end are strongly influenced by the elastic deformation of the embedded portion and, therefore, are significantly larger than those found at the free end. Values of bond stresses τ has been calculated by dividing the tensile force T (given by Eq.1) by the area of the lateral surface of rebar in the embedded zone:

$$\tau = T/(2 \cdot \pi \cdot d_b \cdot L) \quad (2)$$

In this way it has been implicitly considered a constant distribution of τ and values obtained with Eq. 2 can be considered as mean values.

It can be seen that satisfactory values of bond strength are obtained in both cases even if values of τ_{max} are rather different. In fact, values of τ_{max} equal to about 11 MPa and about 16 MPa has been obtained in case of test #1 and test #2 respectively. Such a difference can not be explained by the slight increase in the compressive concrete strength for test #2. In any case, values of τ_{max} from both test are similar to the average ones considered in technical literature for deformed GFRP rebars [9], but they are slightly lower than those obtained for C-Bar by other Authors [11,12] as well as than those declared by Marshall Industries Inc. [13]. The good bond performance is confirmed by the low values of the slips s_{max} corresponding to the maximum bond stress, which indicate a stiffer behavior than the average one for deformed FRP rebars [9], being very similar to that of deformed steel rebars ($s_{max} < 1$ mm).

Finally, as observed elsewhere, the contribution of adhesion is almost negligible compared to that presented by deformed steel rebars.

Compared to similar tests on steel rebars, specimens #1 and #2 show a larger difference in the slip measured at the free end and at the loaded end due to the lower Young modulus of the GFRP rebars. Some problems in the definition of a suitable bond-slip relationship for the GFRP bars arise from these findings. In fact, the assumption of constant bond stress can be reliable just for very small embeddement lengths. Therefore, a reduction of the embeddement length (i.e. 3 diameters) can be recommended

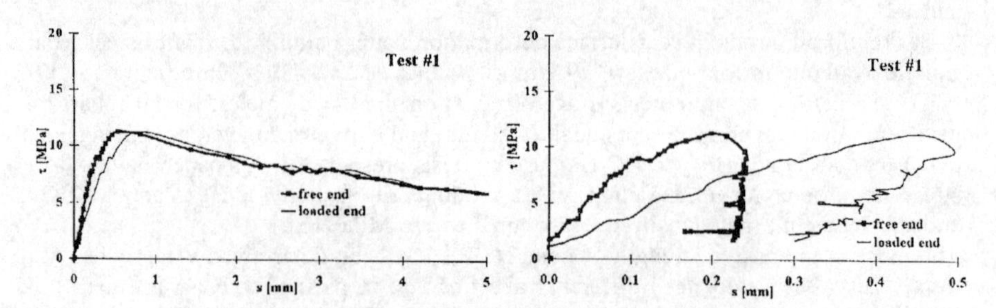

Figure 3 – Test #1: Experimental τ-slip curves

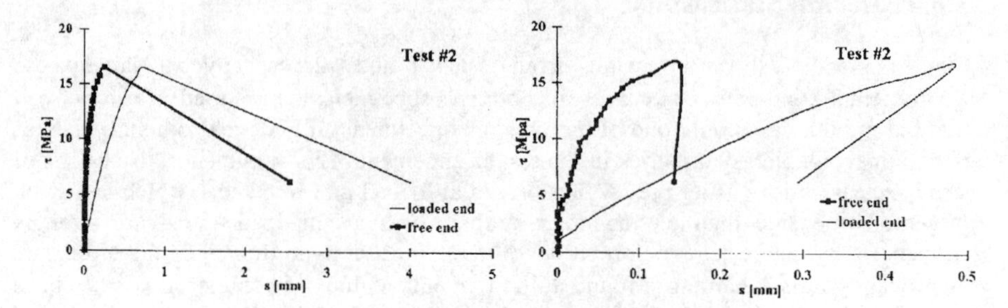

Figure 4 – Test # 2: Experimental τ-slip curves

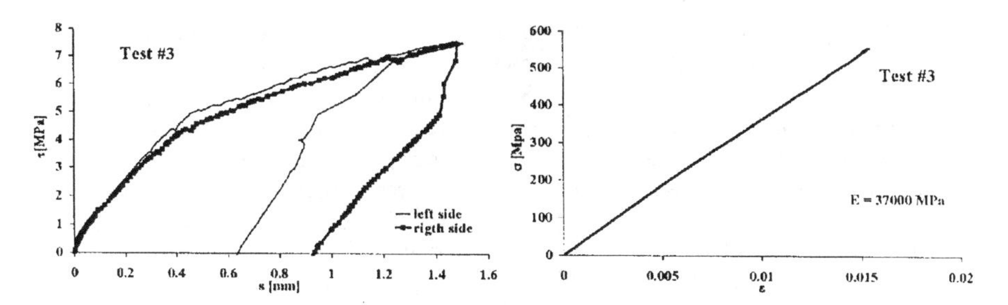

Figure 5 – Test #3: Experimental τ-slip and σ-ε curves

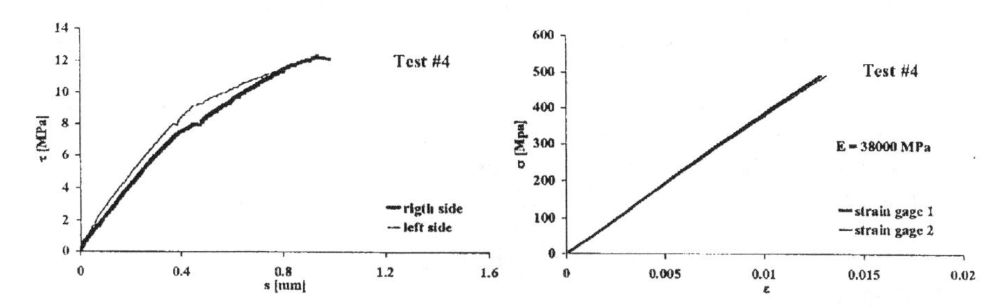

Figure 6 – Test #4: Experimental τ-slip and σ-ε curves

In case of tests #3 and #4, characterized by increased embeddement lengths ($L=20d_b$ and $L=10d_b$ respectively), collapse of the rebar precedes pull out, and slips are practically negligible at the free end. Therefore, Figures 5 and 6 show the bond-slip curves at the loaded ends and the experimental stress-strain (σ-ε) curves of the GFRP rebar. Stress was obtained from the tensile force T (Eq.1), while strain from measurements by the strain gauge (in case of test #4 two strain gauges were positioned).

The measured slips at the loaded end - due to elastic deformation of the embedded portion - are similar for the two sub-specimens. For specimen #3, having longer embeddement length, slips are larger, while bond stresses, evaluated considering a constant distribution along the embeddement length, are clearly lower.

From stress-strain curves the modulus of elasticity of the rebar is found to be about 37÷38 GPa and then very close to that of 42 GPa declared by Marshall Industries Inc..

Failure of the rebar, occurred in both tests #3 and #4, was brittle as expected.

At the collapse, the measured stresses in the rebars were about 550 MPa and about 500 MPa for test #3 and #4 respectively. These values are rather low than the average ultimate tensile strength (770 MPa) given by the producers. This scatter seems not justified by the statistical dispersions on the strength and probably is due to the bending of the bars.

4. Conclusions

In this paper preliminary results from an experimental program aimed at investigating bond properties of deformed GFRP rebars (namely C-Bar™ by Marshall Industries Inc.)

are presented. Tests have been conducted on specimens having embeddement lengths covering a wide range. In case of low embeddement length (5 times the bar diameter) a pull-out failure, without splitting, was evidenced; in case of large embeddement length (10 to 20 d_b) a tensile rupture of the rebar occurred. In the first case, bond strength has been evaluated: excellent bond performances have been evidenced with values of bond strength that have reached values similar to those of steel rebars, though rather scattered. Measured slips were rather low (smaller than slips evidenced by other types of deformed FRP rebars), but a great difference has been noticed by comparing slips obtained at free end and at loaded end. This phenomenon, strictly related to the low elastic modulus, makes difficult the assessment of suitable bond-slip constitutive laws. In fact, being the slip variation along the rebar rather strong, a mean value of bond strength, as evaluated by assuming a constant distribution of bond stresses, can not be considered reliable. Therefore, tests with shorter embeddement length (i.e. 3 d_b) or a statistical back identification of bond properties are needed.
When tensile rupture of the rebar occurred, the elastic modulus and the tensile strength of the rebar was evaluated. Modulus of elasticity was very similar to the declared values (about 40 GPa) while the obtained tensile strengths were much lower of the declared ones, probably due to the slight bending of the rebar during tests.

References

[1] *B. Benmokrane, B. Tighiouart, O. Chaallal.* BOND STRENGTH AND LOAD DISTRIBUTION OF COMPOSITE GFRP REINFORCING BARS IN CONCRETE, ACI Materials Journal, Vol.93, No.3, 1996.

[2] *O. Chaallal, B. Benmokrane.* PULLOUT AND BOND OF GLASS-FIBRE RODS EMBEDDED IN CONCRETE AND CEMENT GROUT, Material and Structures, Vol.26, 1993.

[3] *M.R. Ehsani, H. Saadatmanesh, S. Tao.* BOND OF GFRP REBARS TO ORDINARY-STRENGTH CONCRETE, ACI SP-138, Vancouver, Canada, 1993.

[4] *A. Hattori, S. Inoue, T. Miyagawa, M. Fujii.* A STUDY ON BOND CREEP BEHAVIOR OF FRP REBARS EMBEDDED IN CONCRETE, 2nd RILEM Symposium (FRPRCS-2), Ghent, Belgium, 1995.

[5] *J. Larralde, R. Silva Rodriguez.* BOND AND SLIP OF FRP REBARS IN CONCRETE, Journal of Materials in Civil Engineering, ASCE, Vol.5, No.1, 1993.

[6] *E. Makitani, I. Irisawa, N. Nishiura.* INVESTIGATION OF BOND IN CONCRETE MEMBER WITH FIBER REINFORCED PLASTIC BARS, ACI SP-138, Vancouver, Canada, 1993.

[7] *L.J. Malvar.* TENSILE AND BOND PROPERTIES OF GFRP REINFORCING BARS, ACI Material Journal, Vol. 92, No. 3, 1995.

[8] *A. Nanni, M.M. Al-Zaharani, S.U. Al-Dulaijan, C.E. Bakis, T.E. Boothby.* BOND OF FRP REINFORCEMENT TO CONCRETE - EXPERIMENTAL RESULTS, 2nd International RILEM Symposium, Ghent, Belgium, 1995.

[9] *E. Cosenza, G. Manfredi, R. Realfonzo.* BEHAVIOR AND MODELING OF BOND OF FRP REBARS TO CONCRETE, Journal of Composites for Construction, ASCE, Vol.2, Issue 1, 1997.

[10] *R. Tepfers.* BOND OF FRP REINFORCEMENT IN CONCRETE – A STATE-OF-THE-ART IN PREPARATION, Chalmers University of Technology, Pub.97, Work No.15, Goteborg, Sweden, 1997.

[11] *B. Benmokrane, R. Masmoudi.* FRP C-BAR AS REINFORCING ROD FOR CONCRETE STRUCTURES, 2nd International Conference on ACMBS, Montreal, Quebec, Canada, 1996.

[12] *R. Tepfers, M. Karlsson.* PULL-OUT AND TENSILE REINFORCEMENT SPLICE TESTS USING FRP C-BARS, 3rd International Symposium on Non-Metallic Reinforcement for Concrete Structures, Sapporo, Japan, 1997.

[13] *Marshall Industries Composites Inc..* FRP C-BAR REINFORCING ROD–TECHNICAL DATA, 1995.

Seismic Performance of Welded Frame Joints Retrofitted with Polymer Composites and Adhesively-Bonded Steel Stiffeners

A. S. Mosallam, Ph.D., P.E.
Department of Civil & Environmental Engineering
California State University, Fullerton
Fullerton, California

Abstract

This paper presents the results of a pilot research project on the use of polymer composites and high-strength adhesives for structural repair of damaged steel frame connections. Several steel interior connections were tested under cyclic load to generate information on the energy dissipation capabilities of the different proposed repair techniques. The primary repair system developed in this consists of 3-D braided graphite/epoxy composite connectors attached to the flanges of both beams and columns by adhesives or bolts and both bolted and adhesively-bonded conventional steel stiffeners. Four connection repair details were investigated; *i)* high-strength/high-toughness adhesively bonded composite stiffeners, and *ii)* mechanically fastened composite stiffeners, iii) Adhesively-bonded steel stiffeners, and iv) bolted steel stiffeners. The results of the different repair specimens were compared to those obtained from the *control* welded connection specimens. The control specimens were fully welded and were built according to AISC specifications. In addition, initial results of an on-going linear and non-linear numerical analysis study on the seismic performance of these repair details are reported.

1. Introduction

Proper connection detailing is critical to developing the full strength and ductility of any earthquake-resistant frame structure. Prior to the Northridge (1994), and Kobe (1995) earthquakes, welded steel moment frame (WSMF) structures were considered as one of the most seismic-resistant structural systems. However, over one hundred steel buildings with welded moment connections were found to have experienced beam-column connection fracture during the Northridge Earthquake [1]. In Japan, similar connection damages were also reported. In addition, recent reports indicated that several steel buildings in the Bay Area suffered from similar type of local damages. In this case, connection damages were the results of the 1989 Loma Prieta earthquake, which was only discovered, in the recent years. The two buildings discovered last year, in the Bay Area, were less than 20 years old, and connections suffered cracks in several full-penetration welds at the beam bottom flange attachment to the column. These buildings experienced more than 0.2 ground acceleration during the Loma Prieta quake. Recently, another building suffered from same type of connection damages was reported.

To this date, no one really knows the full extent of these damages and the reason is that no extensive damage survey has been performed yet. It is excepted that as time goes more building damages will be reported. The other end of this problem is that most of the owners are hesitant to check their buildings with their worries on whether or not their insurance claims will be accepted after so many years since the occurrence of the earthquake.

In most cases, the reasons behind these damages are attributed to the failure of the weld metal especially at the flanges. Generally, fractures were initiated at, or near, the *full penetration* (FP) welds between the beam bottom flange and the joined column flange. Fractographic examination of several failed welds also suggests that cracking initiated at this point [2].

Several factors contributed to the damage of the beam-column joints of steel moment frame buildings in the Northridge earthquake [1] & [3]. These included:

- the use of relatively few frame bays to resist lateral demands,
- detailing practices which resulted in large stress concentrations and large inelastic demands at the connection zone,
- poor quality control and assurance in the construction process,
- the inherent inability of material to yield under high tri-axial restraint conditions,
- the large variations in the strengths of rolled shape members relative to specified values.

2. Motivation & Objective

The sever damage experienced by some of the welded steel moment frame structures during the 1994 Northridge Earthquake was the first "*wakeup call*" for the structural engineers, to start reevaluating the design and construction details of such connections. It is almost certain that there are some buildings which have suffered connections damages either as the result of the Northridge earthquake, and its after shocks, or from previous earthquakes such as 1989 Loma Prieta, where these damages are not detected yet, especially those hidden by nonstructural elements. For these reasons, there is a great need for developing different repair and seismic upgrading of connection in order to prevent or minimize damages due to future earthquakes.

In search for optimum and cost effective techniques for rehabilitation of such defective connections, several acceptance requirements were determined to achieve this goal. Prior to the development of the experimental program described in this paper, several criteria were considered including structural, construction and economical requirements.

3. Criteria for an Optimal Connection Repair System

The general acceptance criteria of a successful connection repair system are:

A. Structural Requirements: including the use of a realistic representation of actual pre-earthquake connection details, and the achievement of adequate plastic rotation capacity of the repaired connections (*FEMA calls for a minimum of 0.025 to 0.03 radian.)*

B. Construction Requirements: This include; *i)* ease of erection by eliminating the use of additional welds and by minimizing the need for holes and bolts, ii) maximizing the safety of occupied facilities, and iii) minimizing the erection and repair time.

C. System Cost: In addition, cost effectiveness of the proposed repair/retrofit systems was also considered in the development of these details. According to an ENR report, [4] an estimated inspection costs can be as high as $1,500 per connection, which mostly cover the removal and replacement of fireproofing, while repair costs can reach $20,000 or more per connection, which include removal and finishes replacement. Also, according to EQE report [2], the building value is estimated to be in the order of $125/square foot of floor space for typical seismic connection repair.

4. Development of Innovative Repair Techniques

In this program, the feasibility of using a combination of polymer composites, high-strength adhesives, high-strength bolts and nuts, as well as conventional steel stiffeners as repair systems for welded steel connections was explored. The primary system that was investigated was in the form of attaching three dimensional graphite/epoxy composite stiffener (see Figure 1) to the flanges of both beams and columns. For this system two details were tested; namely, composite connectors with high-strength adhesives only, and composite connectors using steel bolts and nuts. The specimens were tested under both quasi-static and cyclic loading conditions to develop the hysteresis characteristics of each system. The results were compared to the results obtained from the fully welded control specimens.

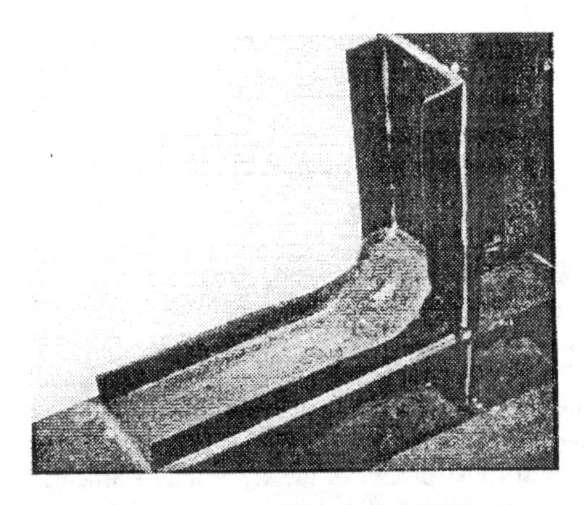

Figure (1): The Composite Stiffener

In the second part of the investigation, a new steel connector were designed and fabricated from A36 steel, and were attached to the flanges using both high-strength/high-toughness adhesives only, as well as steel bolts and nuts. The connection specimens were tested using similar test setup described in Ref. [5]. Both quasi-static and cyclic loading regimes were used to characterize the structural performance of the proposed repair systems. Moment/rotation curves were developed for each connection and the corresponding hysteresis behaviors were recorded. Experimental results for the retrofitted specimens were compared with the results obtained for the control specimens.

5. EXPERIMENTAL PROGRAM

In this pilot project six connection specimens were fabricated using A36 steel and E70XX welds. The specimens were built according to AISC specifications. All specimens, tested in this program, were subjected to a constant amplitude reversal cycle loading conditions (Figure 2) and its low-cycle fatigue behavior was studied, The following repair/retrofit systems were tested; i) adhesively-bonded composite stiffening elements, ii) bolted-only composite stiffening elements, iii) adhesively bonded steel stiffener, and iv) bolted-only steel stiffener. The results were compared with that of the control specimens.

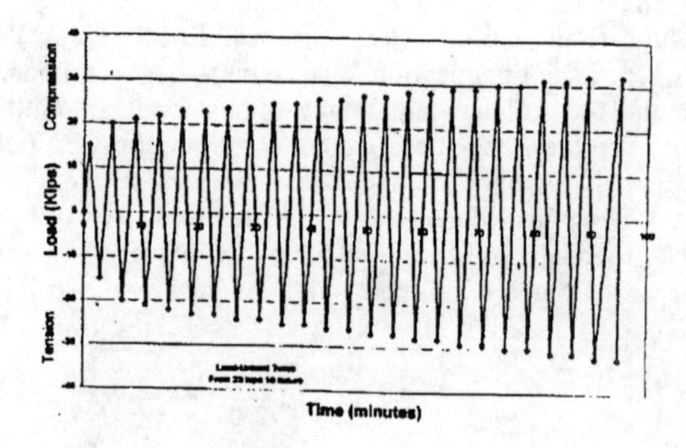

Figure (2): Loading History

A. EXTERIOR BEAM-COLUMN CONNECTIONS

All connections were tested in a closed-loop electro-hydraulic Tinius-Olsen Testing System. All data were collected via a computerized data acquisition system. Each end of the specimen was attached directly to the upper and lower machine heads through a special hinged steel fixture. The relative rotation was measured by two deflection readings at a fixed distance 0.712 m (28") using a rotational rigid metal bar fixture. Strain gages were connected to the connecting elements (3-D composites or steel stiffener) to monitor the behavior and stress history of the connecting elements. The strain data were collected using a data acquisition system connected directly to the testing machine. The reading frequency for all data was one Hz. Details and dimensions of a typical exterior connection specimen is shown in Figure (3).

i) *Fully Welded Exterior Beam-Column Control Connection (Specimen 2A):* Initially, the specimen was subjected to both loading and unloading regimes within the elastic limit of the connections, and up to the point where initial non-

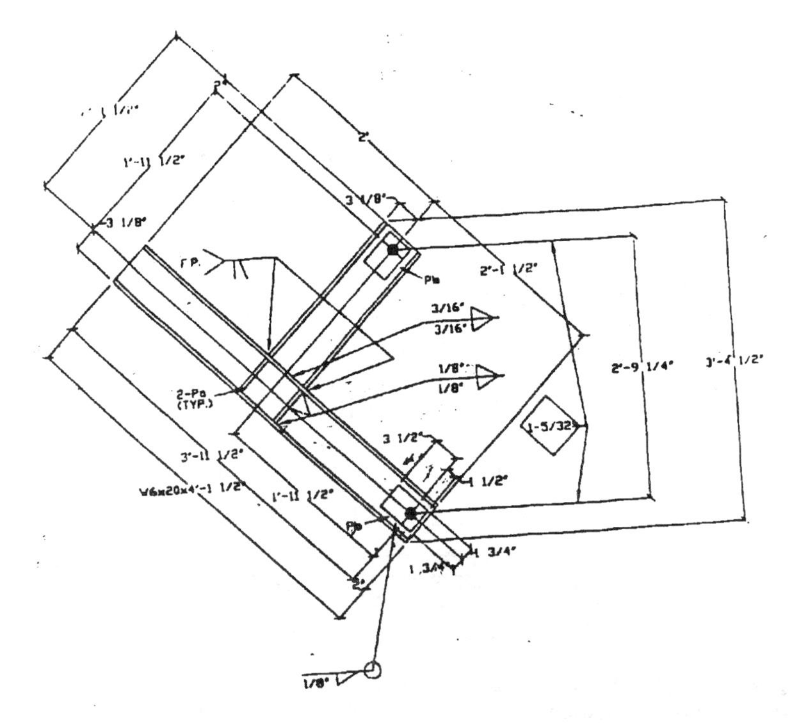

Figure (3): Dimensions and Structural Details of a Typical Exterior Specimen

linearity initiated. The monotonic loading/unloading tests were performed for 5 kips, 10 kips, 15 kips, 20 kips, and 25 kips.

ii) The load/ displacement, (P/δ) and the moment/rotation (M/Θ) behavior were linear for these loading levels. Some non-linearity was observed at the peak of the 25 kips load cycle. The same specimen was also subjected to a full-reversal-cyclic loading with an initial load level of ±22 kips, and up to the ultimate load of ±37 kips. No local buckling was observed. The M/Θ hystirises plot is shown in Figure (4).

ii) Defective Exterior Connection with Adhesively-bonded Polymer Composite Stiffener (Specimen 2B):

- **The Composites Connector:**

For this repair detail, a novel prototype of a birdied AS-4 graphite/epoxy composite stiffeners (Figure (1)) were attached to both flanges of the steel beam and column. The composite stiffeners were especially designed and fabricated for this project. The fiber balance of he stiffener is 0°/ ± 60° and the design fiber volume was 60%. Coupon tests were performed to extract the mechanical properties of the composites stiffener (see Table 1).

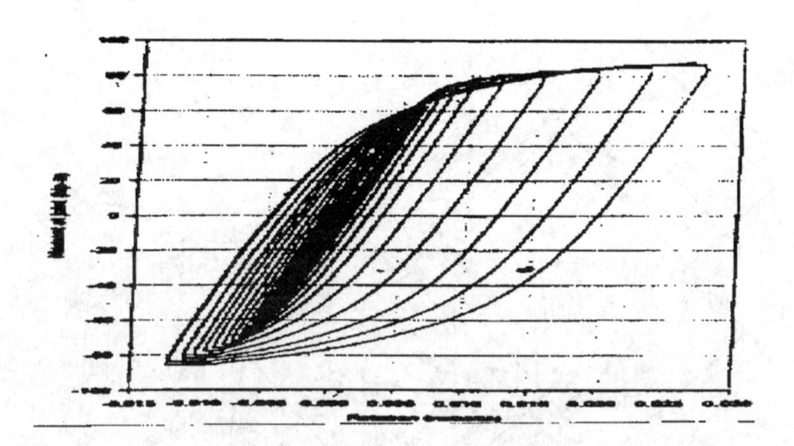

Figure (4): M/Θ Hysteresis of Repair Detail 2A

Table 1. Mechanical Properties of AS-4 Composite Stiffener

Ultimate Stress (ksi)	*Axial Modulus (Msi)*
54.89	6.861

The Adhesive System - In selecting the proper adhesive system for the proposed detail, several performance criteria were considered, including the toughness capacity, as well as shear and axial strengths. In addition, site conditions and ease of application were also important elements in the screening process for the different available adhesive systems. In addition to the structural properties of the adhesive system, constructability requirements such as minimum surface preparation and practical curing time were considered. For this reason, several epoxy and adhesive systems were examined. Several coupon specimens with different commercially produced adhesive systems were tested in pure shear, bending, and axial modes to determine the adhesive properties and to confirm the technical information provided by the manufactures. The results of the selection process indicated that a satisfactory performance could be achieved using a Methacrylate adhesive system. Other adhesive systems suitable for steel structures applications are reported in Ref. [7] & [8]. The stress/strain curve for the adhesive system is shown in Figure (5). In addition to the information provided by the manufacturer, several coupon specimens were prepared and tested to provide preliminary information on the mechanical properties of the adhesive system. The average measured shear and axial tensile strengths were 3.45 ksi and 1.00 ksi, respectively. Table (2) contained some of the mechanical and physical properties of the adhesive system. The adhesive system is composed of two parts, which are dispensed from a mechanical gun simultaneously. According to the supplier's information, no primer, no sanding or grinding of most surfaces are recommended for steel/composites substrate. However, surfaces were wiped cleaned and prepared to ensure good performance. The initial bond starts after 10 to 15 minutes of curing at 22°C (72°F) and develops full strength after 24 hours.

Table 2. Mechanical and Physical Properties of the Adhesive System

Mechanical or Physical Property	Performance Characteristics (cured)
Bond Strength Tensile Shear (ASTM D1002)	24.13 MPa (3.50 ksi)
Peel Strength (ASTM D1876)	35-40 pli
Impact Strength	1.07- 1.12 m-N/mm (20-22 ft-lb/in)
Shore Hardness (D)	78
Elongation	50-75%
Operating Temperature Range	-67°F to + 250 °F
Viscosity:	
Part A	45,000 (cps)
Part B	40,000 (cps)

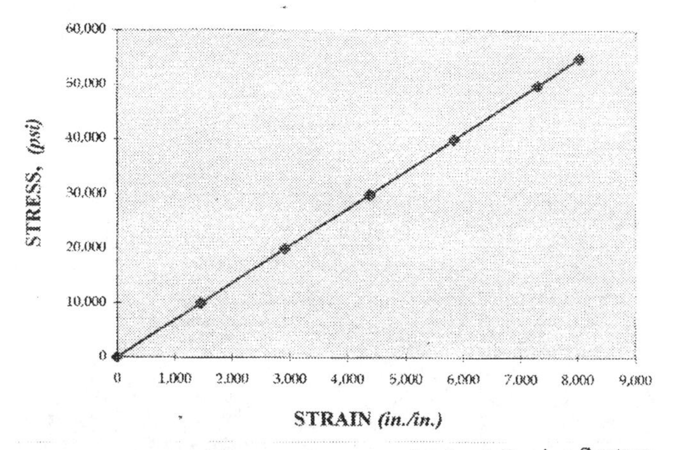

Figure (5): Stress/Strain Curve for the Adhesive System

- **Fabrication** - The following steps were followed in bonding the composite stiffeners to steel surfaces:

- Both contact surfaces were cleaned, roughened, and prepared for the application of the adhesive system.
- The adhesive was then applied to the contact surfaces, and was evenly spread to cover the areas of contact between the composite connectors and the steel surfaces.
- The composite connectors were placed (after preparing the contact surfaces) on the designated positions.
- A clamping force was applied to squeeze out any excessive adhesive, and to provide the compression force required for curing.
- The specimens were cured in the laboratory environment for more than 24 hours as recommended by the manufacturer. However, actual testing of the specimens was done few days later.

- **Testing** – Similar to the control specimen loading, the connection was subjected to both half-and full-cycle repeated loading. At first, the specimen was subjected to half-cycle, monotonic loading conditions. Same loading history was adopted as for the control specimen. Several loading/unloading half-cycles were applied with maximum load levels of 5 kips, 10 kips, 15 kips, and 20 kips, and 23 kips. Unlike the control specimen (2A), permanent deformation was observed at a relatively lower stress level. This permanent deformation increased proportionally as the load increased. The ultimate failure occurred at a load of 23 kips. The vertical displacement (δ) and the relative rotation angle (Θ) at the ultimate load were 0.771" and 0.0337 rad, respectively. As compared to the deformation of specimen 2A, vertical deflection and connection' rotation have increased by 32% and 25%, respectively. The failure was initiated by a combined effect of local compression damage of the composite stiffener's knee and initiation of micro-cracks at the top stiffener adhesives line. As a result, the connection experienced a sudden loss in the rotational stiffness followed by a sudden compression failure of the bottom composite stiffener occurred. The local failure of the adhesives can be attributed to the surface preparation of both composites and steel, and the possibility of uneven thickness of the adhesive film. Failure mode of specimen 2B is shown in Figure (6). The P–δ hysteresis curve of this repair detail is shown in Figure (7).

Figure (6): Failure Mode of Specimen 2B

ii) Defective Exterior Connection with Bolted Composite, (Specimen 2C): In this repair detail, composite stiffeners bolted to the top and bottom flanges of the beam. The average value of the applied torque for all bolts was 50 lb-ft. In the half-cycle loading test, loading history similar to that of specimen 2B was adopted. The behavior of this detail was similar to the behavior of specimen 2B. Permanent deformation was observed at the early stage of the load application. The behavior of the connection up to 10 kips was linear. However, at about 13 kips load level, some non-linearity was observed. During the test, the composite stiffeners were monitored periodically. No cracks were observed at this stage. The ultimate failure of this detail was again due to a compression failure of the composite stiffener at the beam bottom side. It is interesting to note that the ultimate failure load for this specimen was identical to that for the adhesively bonded specimen (P_u=23 kips). This load corresponds to an ultimate moment of 32.00 kip-ft.

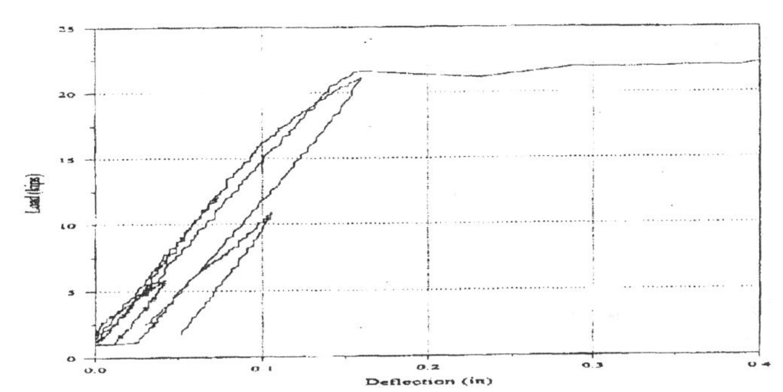

Figure (7): P/δ Hysteresis Curve for Specimen 2B

B. INTERIOR BEAM-COLUMN CONNECTIONS

i) Cyclic Behavior of Fully Welded Interior Beam-Column Control Connection (Specimen 1A) - The load was applied to the specimen through a hydraulic actuator at the mid-span of the connection. Both ends of the specimen were supported using hinged steel fixture. The geometry and physical dimension of the control specimen is shown in Figure (8). Strain measurements were collected at critical locations using both axial strain and Rosette electronic gages. The relative rotation between the beams and the column members were captured using four ±4" LVDT's which were connected directly to the data acquisition system. The load history was similar to those for the exterior connection specimens. As the load approached 45 kips, local buckling of the beam flanges at the two sides of the column (maximum moment zone) occurred. The maximum amplitude of the buckled wave where at about 6" from the two faces of the steel column. In addition, the whole assembly was exposed to an out-of-plane deformation (lateral torsional buckling). This was the ultimate mode of failure of this connection. No damage to weld material was observed. In this case, the plastic rotation (Θ =0.028 rad.) satisfied the recommendation described in Ref. [1].

ii) Cyclic Behavior of Defective Interior Beam-Column Connecting Retrofitted with Adhesively Bonded Steel Stiffener (Specimen 1B) - In this test, test setup identical to specimen 1A was used. In this case, the steel specimen was representing a damaged connection, which was retrofitted using steel stiffener bonded to the 'prepared' steel surface with high strength adhesives. The practical advantages of this detail are the ease of applying the stiffener without the need of welds, hole drilling and bolting. In this

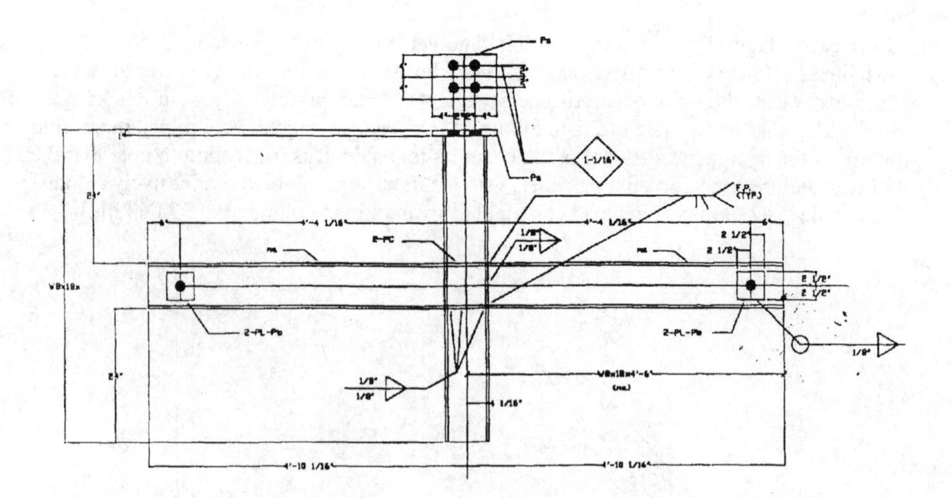

Figure (8): Dimensions and Structural Details of a Typical Interior Specimen

case, same adhesive was used. However, according to the manufacturer recommendations, a special care was taken in the preparation and surface cleaning for steel/steel substrates. This was accomplished using mechanical grinder and chemical solvents. Thin films of adhesives were spread along the clean surfaces of steel columns, beams, and the connecting steel elements. The steel stiffeners were then attached to the marked areas and pressure was applied using spring clamps. Although the specified initial curing time of the adhesive system was 30 minutes, the testing was not performed until full cure of the adhesives was achieved. At the beginning of the test, the connection exhibited a linear behavior. However, in comparison to the control specimen (1A), early signs of nonlinearity in the P/δ hysteresis was observed. This behavior continued to occur until the end of the test. At a load level of about 15 kips, a crackling noise was heard. The first sign of local damages was in the form of hair cracking of the adhesive film at the top stiffener interface (column side). Again, this can be attributed to surface cleaning, adhesive film thickness, air bubbles and other curing related factors. This crack gradually increased until the ultimate failure of the adhesive at a load of 35 kips. Simultaneously, and due to the sudden loss in the connection stiffness, the steel stiffener got detached from the joint and the beam web at this location fractured in shear, Figure (9). The failure was sudden with no warning. No appreciable damages occurred to the other three adhesively bonded steel stiffeners. Results of this test indicated that the retrofitted connection was capable of carrying 78% of the load resisted by the fully welded connection (1A) and about 125% of the calculated ultimate capacity. Figure (10) describes the M/Θ behavior of this connection detail.

iii) *Cyclic Behavior of Defective Interior Beam-Column Connecting Retrofitted with Bolted Steel Stiffeners (Specimen 1C)* - Steel stiffeners were bolted to beam and column flanges. The load was applied according the loading history described

Figure (9): Shear Failure of Beam Web of Specimen 1B

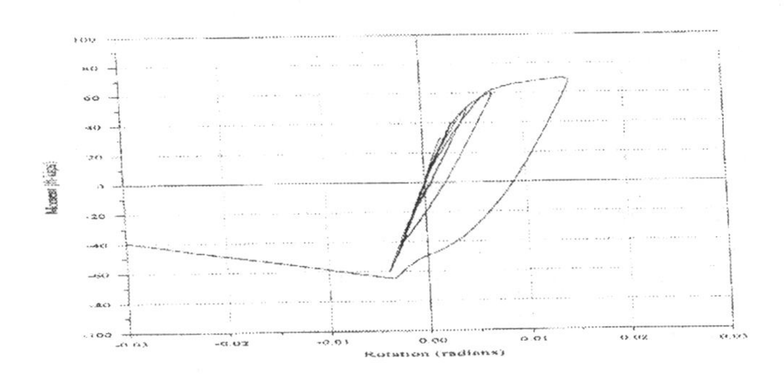

Figure (10): M/Θ Behavior for Specimen 1B Connection Detail

earlier. As for other detail, the M/Θ relation was linear at low stress levels. As the load increased, the connection exhibited some non-linearity. However, in this case no local failure occurred. The ultimate moment capacity was not achieved due to the limited capacity of the actuators. Thus, it can be concluded that this retrofitted detail carried more than 100% of the control specimen (1A). In future experiments the connection size and number of bolts can be reduced to match the control specimen.

6. Numerical Modeling

A comprehensive non-linear numerical analysis, using *ANSYS* software, is in progress. The analysis considered the joint flexibility using the experimental relative rotation stiffness of each connection repair system. Buckling and post-buckling behavior of different connections observed during the experiment was also investigated in this study. In the analysis, the *ANSYS shell-43* plastic shell element was used to model all steel members, while the *solid-45* 3-D solid element was used in modeling the adhesive layers. The non-linear stress-strain behavior of both steel and adhesive were represented by a bi-linear kinematics-hardening rule. The P/δ curve for each connection repair system has been plotted and stress plots at each incremental load described the extent of plasticity and the growth of the plastic region up to the ultimate failure. Figure (11) shows a sample of the initial results of this analysis.

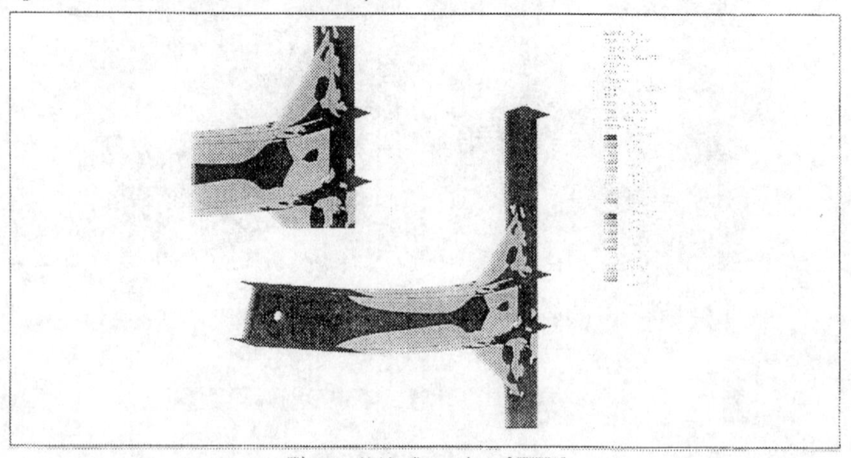

Figure (11): Sample of FEM

7. Idealization of M/Θ Hysteresis:

For design proposes, all M/Θ curves were linearized. Simple linear rotational stiffness coefficients were produced which were used in the analysis. Figure (12) shows a sample of a lineariezd M/Θ hysteresis.

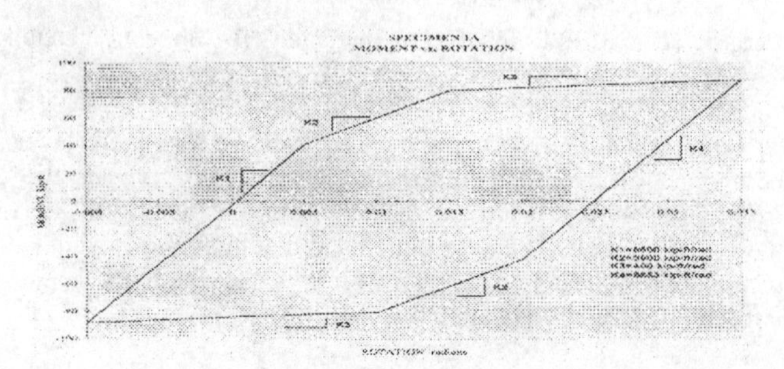

Figure (12): Sample of Lineariezd M/Θ Curves

8. Summary of Results

Figure (13) presents both calculated and experimental ultimate moment capacity of each connection details. From this bar diagram, one can see that in the case of interior connections, all the experimental data have exceeded the calculated values. The increase was from 25% up to more than 61% of the calculated values. For specimen 1C, the connection did not fail, and the estimated ultimate capacity is greater than ±45 kips. In this case, smaller dimension could have been used. Other details are in the process of development including combined connection details, which utilizes both bolts and adhesives. In this case, bolts will be used only at the critical locations. Also, from this plot, one can see that the use of composite stiffeners for exterior connections achieved about 85% of the calculated ultimate capacity. The maximum plastic rotation satisfied the recommended value described in Ref. [1] & [3].

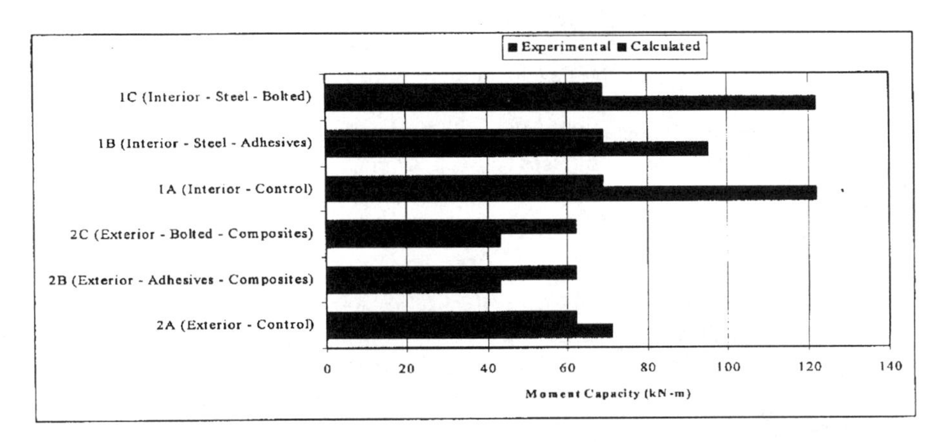

Figure (13) Comparison between Calculated and Experimental Results For Different Systems

9. Conclusion & Recommendations

This study presents the results of a pilot project aimed at developing structural repair/retrofit techniques for steel beam-column connections using both conventional and unconventional systems. Two connection types were investigated; namely exterior and interior beam-column connections. For each test, a control "undamaged" fully welded specimen was tested to compare the results for each proposed repair technique. Test results indicated that all proposed repair/rehab techniques satisfied the ductility requirements recommended by Ref. [1] & [3]. An exception was the lower plastic deformation of the adhesively bonded steel stiffening of connection detail (1B), where maximum plastic deformation was only 0.015 rad. Repair detail of specimen 2B (adhesively bonded composite stiffener) provided the highest ductility, where Θ_p was 0.034 rad. with an increase of more than 25% as compared to the fully welded control specimen (2A), and twice as much as repair detail 1B. This indicates the potential for using polymer composites as a repair and rehab system where ductility is considered essential criteria. In addition, composite stiffeners can offer the structural engineers several attractive features, including lightweight, ease of site application, and minimum construction time and overall cost. However, as it was expected, the composite repair detail failed in a brittle mode, with little warning except for the increasingly crackling sound during the non-liner range. For this reason, several other composites and stiffening details are being investigated to improve the mode of failure. It is expected that this can be accomplish by optimizing fiber architecture and by using combined adhesive/bolted details.

Adhesively bonded steel connection stiffener exhibited sufficient strength and ductility. The adhesive system provides attractive features for construction repair of existing damaged steel structures. These include the ease of applications, minimizing heavy equipment, minimizing or eliminating the need for making holes or using bolts, as well as eliminating the use of field welding and the associated risks of fire to occupied space. Test results indicated that this system can achieve as high as 78%(+) of the control (fully welded) connection capacity. Based on test results and data analysis, further investigation is recommended to develop a more ductile/high strength adhesive system. Also, establishing additional procedure in preparing and treating the steel surface to ensure maximum adhesion is critical. For practical application, a *thin layer of glass mat* should be placed at the contact surfaces between the graphite/epoxy stiffener and the steel to eliminate the activation of the corrosion cycle *(galvanic action)*.

REFERENCES

1. FEMA, (1995). "Interim Guidelines: Evaluation, Repair, Modifications and Design of Welded Steel Moment Frame Structures," Program to Reduce the Earthquake Hazards of Steel Moment Frame Structures FEMA-267, August.
2. Shipp , J., (1994). "*Steel's Performance in the Northridge Earthquake*," EQE Review, Fall, EQE International, 1-6.
3. FEMA, (1997). "Interim Guidelines Advisory No.1 (Supplement to FEMA 267)," Program to Reduce the Earthquake Hazards of Steel Moment Frame Structures, FEMA-267A, March.
4. *Engineering News Records* (ENR), April 15, 1996, pp. 25.
5. Mosallam, A.S. (1994). "*Connections for Pultruded Composites: A Review and Evaluation*", Proceedings, The Third Materials Engineering Conference, ASCE, San Diego, CA, November 13-16, 1001-1017
6. AISC. (1994) *Manual of Steel Construction, Load and Resistance Factor Design*, Second Edition, Chicago, American Institute of Steel Construction,
7. Albrecht, P., and Sahli, A. H. (1986). "*Fatigue Strength of Bolted and Adhesive Bonded Structural Steel Joint*" in Fatigue in Mechanically Fastened Composites, *edt.* J. M. Potter, ASTM STP 927, ASTM, PA, 72-94.
8. Mecklenburg, M. F., Albrecht, P., and Evans, B.M. "*Screening of Structural Adhesives for Applications to Steel Bridges*," Interim Report, Pr. No. DTFH 61-84-R-0027, University of Maryland, College Park, MD, February.
9. Mosallam. A.S. (1998). "Rehabilitation of Steel Structures Using Advanced Composites and Adhesives," Proceeding, 1998 International Composites Expo, Nashville, TN, January 19-12, Session 6-B/1-7.
10. Chakrabarti, P.R., and Mosallam, A.S., (1998). "*Performance of Welded Steel Beam-To-Column Joints Seismic Retrofitted With Polymer Composites And Steel Stiffeners*," in Fiber Composites in Infrastructure, Proceedings of 2nd International Conference on Composites for Infrastructures (ICCI'98), Tucson, AZ, January 5-7, pp. 160-174.

Structural behaviour of concrete elements confined with FRP jackets

Antonio La Tegola*, Orazio Manni*

* Department of Materials Science, University of Lecce, Faculty of Engineering, 73100 Lecce – Italy

Abstract

As a consequence of failures of bridge columns and beams in recent earthquakes, it is adopted the *Wrapping* technique with composite materials in order to reinforce and retrofit structural concrete elements. Theoretical models in order to calibrate the stress-strain curve of continuously wrapped concrete are not sufficiently defined.
Purpose of this paper is to define the constitutive law for concrete transversely confined with FRP jackets, putting in evidence the effectiveness of confinement in the case of rectangular columns.

1. Introduction

The seismic events occurred in this century, determined the collapse of a large number of bridge structures, recently built too; the structural elements was often damaged because of inadequate attention for the structure detail and evidenced a low ductility depending on the insufficient number of steel hoops or spirals near the joints. In fact, it is well known that the discontinuous confinement of reinforced concrete elements prevalently compressed using steel hoops or spirals determines a moderate increase in strength and a significative increase in ductility and toughness. On the other hand, the substantial impossibility to have a diffuse confinement with conventional techniques has promoted the experimental tests and the study of advanced technologies of transverse confinement which adopt non metallic materials having high corrosion resistance, light weight, very high tensile strength and long fatigue life.
It is evident that, when a new technology is introduced, it is necessary to say why it should be used instead of the traditional one. One thing is very clear: the adoption of composite materials in order to reinforce or retrofit reinforced concrete structures permits to have a general saving respect to the use of conventional materials. In fact, the evaluation of the entire cost of the construction process (not only the material cost) is the point to demonstrate the convenience of the system. In the global consideration also the maintenance should be included: in fact, the composite materials have vulnerability to the fire, UV radiations and, in some case, to moisture, but an efficient coating protection using adequate paints is sufficient to assure an acceptable durability and, consequently, to minimise the maintenance cost. For this reason, a large number of applications of "Wrapping" technique are spreading; using this technique, it is possible to transversely wrap prevalently compressed columns with thin carbon or glass flexible straps epoxy-bonded to concrete surface. This methodology of structural retrofitting allows to obtain very good results with reference to the strength and ductility of

reinforced element. The structural behaviour of confined element depends on the column geometry; in fact, the confinement results more efficient for circular columns which are symmetrically hoped by the straps, while, for rectangular columns, it is possible to note a non-symmetric mechanical behaviour which must be considered in the theoretical models.

2. Theoretical model (rectangular columns)

Consider a concrete rectangular column transversely confined with FRP jackets which is axially loaded by an uniaxial compressive stress σ_1.
Adopting the theoretical base related to circular columns indicated in [1], it is possible to generalise the elasticity theory and to applicate step by step the Navier classical relations which define the stress-strain laws in state of triaxial stress. In fact, with reference to a concrete rectangular element confined with FRP jackets loaded by a monotonic history of axial strains ε_1, the total stress state related to a generic strain ε_1, is characterised by a vertical stress σ_1 and two stresses σ_2' and σ_3' acting in the plane perpendicular respect to the direction 1, and confining effectively the concrete element.
Suppose an axial strain ε_1 is imposed. If the concrete element is not transversely confined, in correspondence of the imposed strain ε_1, the transverse strains $\varepsilon_{2l} = \nu_{12}\,\varepsilon_1$, $\varepsilon_{3l} = \upsilon_{13}\varepsilon_1$, will appear, respectively, in the 2 and 3 directions; υ_{12} and υ_{13} are the longitudinal Poisson moduli, respectively, in the 2 and 3 directions. Adopting for concrete the Parabola-Rectangle constitutive law, the following expression may be written:

$$\sigma_1 = \frac{2f^*_{cc}}{\varepsilon_{ck}}\left(\varepsilon_1 - \frac{\varepsilon_1^2}{2\varepsilon_{ck}}\right) \quad for \quad \varepsilon_c \le \varepsilon_{ck} \qquad \sigma_1 = f^*_{cc} \quad for \quad \varepsilon_c > \varepsilon_{ck} \tag{1}$$

with

$$f^*_{cc} = f_{cc} + k\left(\frac{\sigma_2' + \sigma_3'}{2}\right) \tag{2}$$

$\varepsilon_{ck} = 0.2\%$; f_{cc}= average cylindrical strength of unconfined concrete; σ_2' and σ_3' are the effective confinement stresses, respectively, in the 2 and 3 directions; k= experimental parameter, function of concrete and transverse reinforcement characteristics. The relation (2) represents the *Hardening Function* which takes into account the progressive increase of concrete axial compressive strength with increase of the confining lateral stress σ_2.
After allowing the axial strain ε_1, suppose that the element is fictitiously hampered to axially translate (Fig. 1). Transversely, the following compatibility equations are available:

$$\varepsilon_2 + \varepsilon_{rif_2} = \upsilon_{12}\varepsilon_1 \qquad \varepsilon_3 + \varepsilon_{rif_3} = \upsilon_{13}\varepsilon_1 \tag{3}$$

where ε_{rif_2} e ε_{rif_3} are the transverse strains of the concrete element, respectively, in the 2 and 3 directions; for the strain compatibility in correspondence of strap-concrete interface, these strains are equal to the ones of reinforcement in the 2 and 3 directions; ε_2 and ε_3 represent the transverse strains which appear in the concrete specimen fictitiously made impossible to axially translate, because of the confinement pressure explained by the strap, respectively, in the 2 and 3 directions.

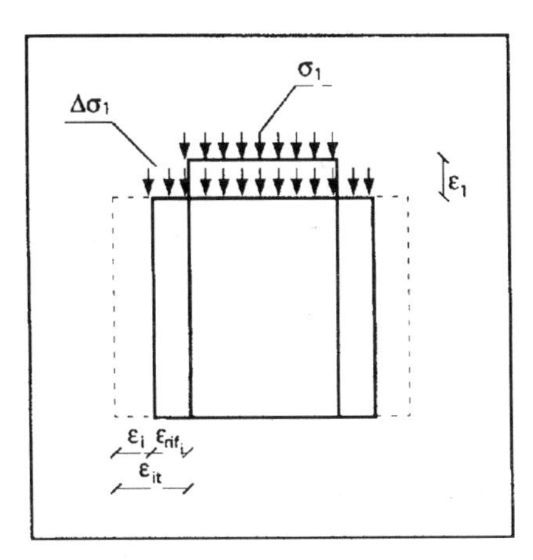

Fig. 1. Scheme of the model

Thus, with reference to the element fictitiously made impossible to axially translate, the transverse application of the confinement stresses σ_2' and σ_3' determines a fictitious axial stress $\Delta\sigma_1$ (see fig. 1). Imposing that the axial strain $\Delta\varepsilon_1$ is equal to zero, the following expression may be written:

$$\Delta\varepsilon_1 = \frac{\Delta\sigma_1}{E_1} - \upsilon_{21}\frac{\sigma_2'}{E_2} - \upsilon_{31}\frac{\sigma_3'}{E_3} = 0 \tag{4}$$

In order to evaluate the lateral confinement stresses σ_2' and σ_3' which appear in the relation (4), it is necessary to take into account the fact that not entire concrete cross-section is efficiently confined (Fig. 2), but only a part of it; a coefficient in order to numerically interpretate this aspect and then to take into account the effectively confined area, can be introduced [2]:

$$k_e = \frac{A_{eff}}{A_{tot}} \tag{5}$$

where A_{tot} = area of the entire concrete cross-section, A_{eff} = effectively confined area.

The unconfined area is defined from second-degree curves having an assigned slope in correspondence of cross-section vertexes; in particular, two different cases may be considered:

a) the initial tangents of curves are inclined of 45°;

b) the initial tangents of curves coincide with the diagonals of the rectangle.

In the case a), considering $b/h \leq 2$, it results:

$$A_{eff} = A_{eff}^{45°} = bh - \frac{1}{3}\left(h^2 + b^2\right) \tag{6}$$

in the case b) it results:

$$A_{eff} = A_{eff}^{diag} = \frac{1}{3}bh \tag{6b}$$

As it can be noted, using the equation (6), the values of k_e tend to decrease increasing the ratio b/h, while k_e assumes constant values using the equation (6b).

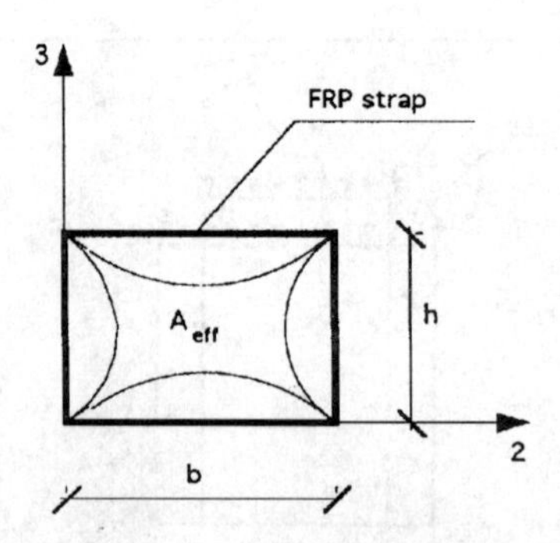

Fig. 2 - Effectively confined area for rectangular columns transversely reinforced with FRP straps

Considering the equilibrium of forces in correspondence of the strap-concrete interface in the 2 and 3 directions, it results:

$$\sigma_2 = \rho_2 \sigma_{rif_2} \qquad \sigma_3 = \rho_3 \sigma_{rif_3} \tag{7}$$

where σ_2 e σ_3 represent the stresses uniformly explained by the reinforcement on the lateral surface of the concrete element, respectively in the 2 and 3 directions; $\rho_2 = 2s/h$ and $\rho_3 = 2s/b$ are the confining reinforcement ratios, respectively, in the 2 and 3 directions; s = thickness of FRP wrap; σ_{rif_2} and σ_{rif_3} represent the stresses of FRP wrap, respectively, in the 2 and 3 directions. Considering the confinement effectiveness coefficient k_e , the following expressions may be written:

$$\sigma_2' = k_e \sigma_2 = k_e \rho_2 \sigma_{rif_2} \qquad \sigma_3' = k_e \sigma_3 = k_e \rho_3 \sigma_{rif_3} \tag{8}$$

Substituting equations (8) in (4), it results:

$$\frac{\Delta\sigma_1}{E_1} = \upsilon_{21} \frac{k_e \rho_2 \sigma_{rif_2}}{E_2} - \upsilon_{31} \frac{k_e \rho_3 \sigma_{rif_3}}{E_3} \tag{9}$$

Moreover, for the concrete element fictitiously hampered to axially translate, it results:

$$\varepsilon_2 = \frac{\sigma_2'}{E_2} - \upsilon_{32} \frac{\sigma_3'}{E_3} - \upsilon_{12} \frac{\Delta\sigma_1}{E_1} \qquad \varepsilon_3 = \frac{\sigma_3'}{E_3} - \upsilon_{23} \frac{\sigma_2'}{E_2} - \upsilon_{13} \frac{\Delta\sigma_1}{E_1} \tag{10}$$

Substituting equation (9) in (10) and taking into account the linear-elastic behaviour of the FRP strap, the first equation of (3) may be written in the following way:

$$\sigma_2' = E_{02}^* \varepsilon_1 \tag{11}$$

where

$$E_{02}^* = \frac{\upsilon_{12} + \dfrac{\upsilon_{13}(\upsilon_{32} + \upsilon_{12}\upsilon_{31})}{1 - \upsilon_{13}\upsilon_{31} + \dfrac{E_3}{k_e \rho_3 E_{rif}}}}{\dfrac{1 - \upsilon_{12}\upsilon_{21}}{E_2} + \dfrac{1}{k_e \rho_2 E_{rif}} - \dfrac{(\upsilon_{32} + \upsilon_{12}\upsilon_{31})(\upsilon_{23} + \upsilon_{13}\upsilon_{21})}{E_2(1 - \upsilon_{13}\upsilon_{31}) \dfrac{E_2 E_3}{k_e \rho_3 E_{rif}}}} \tag{12}$$

with E_{rif}= modulus of elasticity of the transverse reinforcement; the second equation of (3) may be written in the following way:

$$\sigma_3' = E_{03}^* \varepsilon_1 \tag{13}$$

where

$$E_{03}^{*}=\frac{\upsilon_{13}E_3+\frac{E_3E_{02}^{*}}{E_2}(\upsilon_{23}+\upsilon_{13}\upsilon_{21})}{1-\upsilon_{13}\upsilon_{31}+\frac{E_3}{k_e\rho_3E_{rif}}} \tag{14}$$

2.1 Generalised isotropic model

In order to applicate the present model, we can proceed in the following way. After assigning the generic axial strain ε_1 , the corresponding stress σ_1 can be determined adopting the equation (1), if the Hardening Function expressed by the equation (2) is known. Then, we can calculate the effective lateral confining stresses $\sigma_2^{'}$ and $\sigma_3^{'}$ using the equations (11) and (13), the stresses of FRP strap σ_{rif_2} and σ_{rif_3} using the equations (8), the fictitious increase of axial stress $\Delta\sigma_1$ using the equation (9) and, at the end, the total axial stress $\sigma_{1tot}=\sigma_1+\Delta\sigma_1$. Determined the first stress invariant $I_1=\sigma_1+\Delta\sigma_1+\sigma_2^{'}+\sigma_3^{'}$ and considered the quantity $J_1^{*}=\varepsilon_1+\varepsilon_2+\varepsilon_3$, it is possible to define a secant modulus which takes into account the triaxial behaviour of the confined element:

$$E_{1src}=\frac{I_1}{J_1^{*}} \tag{15}$$

In order to obtain the numerical solution of the problem, an internal isotropy of material can be considered:

$$E_1=E_2=E_3=E_{1src} \qquad \upsilon_{12}=\upsilon_{21}=\upsilon_{13}=\upsilon_{31}=\upsilon_{23}=\upsilon_{32}=\upsilon.$$

The modulus of Poisson υ can be considered as a linear function of the first stress invariant until a value equal to r<0.5, then maintaining itself constant until failure:

$$\upsilon=0.1+(r-0.1)\frac{I_1}{f_{cc}} \quad for \quad I_1<f_{cc} \qquad \upsilon=r \quad for \quad I_1\geq f_{cc} \tag{16}$$

If very small ranges for strain history are considered, the calculations of the Hardening Function f_{cc}^{*} , of Poisson modulus υ and of secant modulus E_{1src} may be conducted with a good approximation, considering the parameters related to the previous step.

3.1 Confined concrete ultimate axial strain

In order to evaluate the ultimate compressive axial strain ε_{cu} , it is possible to adopt the rational method proposed by Mander et al. [2], and taken again in [1] with reference to circular columns confined with FRP straps.

Writing the energy balance, it results:

$$U_{cc}=U_{c0}+U_r+U_{rc0} \tag{17}$$

where $U_{cc}=\int_0^{\varepsilon_{cu}}\sigma_1 d\varepsilon_1=$ strain energy capacity of the confined concrete; $U_{c0}=f_{cc}\left(\varepsilon_{c0}-\frac{1}{3}\varepsilon_{ck}\right)=$ strain energy capacity of the unconfined concrete;

If the relation $U_{cc} = \int_0^{\varepsilon_{cu}} \sigma_1 d\varepsilon_1 = \alpha\sigma_{cu}\varepsilon_{cu}$ is considered and the mutual energy is neglected, the confined concrete ultimate axial strain is given by the following expression:

$$\varepsilon_{cu} = \frac{1}{\alpha}\left[\frac{f_{cc}}{\sigma_{cu}}\left(\varepsilon_{c0} - \frac{1}{3}\varepsilon_{ck}\right) + \frac{1}{2}\rho\frac{f_{rif}^{\ 2}}{E_{rif}\sigma_{cu}}\right] \tag{18}$$

where σ_{cu} = strength of the confined concrete; f_{cc}= average cylindrical strength of the unconfined concrete; $\varepsilon_{c0} = 0.0035$= unconfined concrete ultimate compressive strain; $\varepsilon_{ck} = 0.002$; $\rho = (2sb+2sh)/bh$ = transverse reinforcement ratio; f_{rif} = ultimate stress of the FRP strap. The coefficient α which appears in the equation (18) depend on the quality of concrete (f_{cc}), the quantity of FRP strap (ρ) and the type of adopted FRP strap (E_{rif} , f_{rif}).

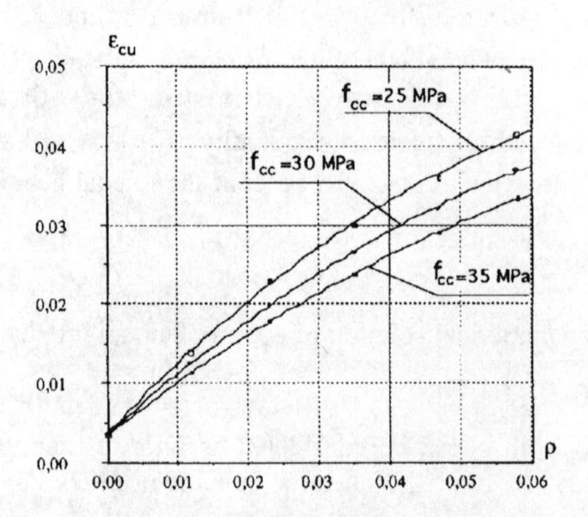

Fig. 3 $\varepsilon_{cu} - \rho$ diagrams related to a rectangular column confined with carbon straps for different confined concrete strengths f_{cc}

With reference to carbon fiber reinforced straps (resin plus fiber) having the following mechanical characteristics: tensile strength $f_{rif} = 2862\ MPa$ and modulus of elasticity $E_{rif} = 172000\ MPa$ [3], the numerical application of the proposed model has allowed to plot the $\varepsilon_{cu} - \rho$ curves indicated in fig. 3. The ultimate compressive axial strain ε_{cu} assumes the following expression:

$$\varepsilon_{cu} = \varepsilon_{c0} + \frac{\alpha}{f_{cc}}\rho + \beta f_{cc}^{\ \gamma}\rho^2 \tag{19}$$

where α, β e γ are three parameters which depend on the type of FRP wrap (E_{rif} , f_{rif}). With reference to the above-mentioned carbon fiber reinforced strap they assume the following values:

$$\alpha = 23.5 \qquad \beta = -\ 607 \qquad \gamma = -1.5.$$

4. Numerical applications of the model

Adopting the proposed model, the $\sigma-\varepsilon$ curves for different confining reinforcement ratios have been plotted. In fig. 4 the $\sigma-\varepsilon$ diagram in non dimensional form related to a rectangular column ($b=400mm$, $h=300mm$) confined with carbon jackets ($f_{rif}=2862\ MPa$ $E_{rif}=172000\ MPa$) [3] is indicated. Increasing the values of the confining reinforcement ratio ρ, we can note a significative increase in the concrete strength σ_{cu} and a very high increase in the ultimate compressive axial strain ε_{cu}.

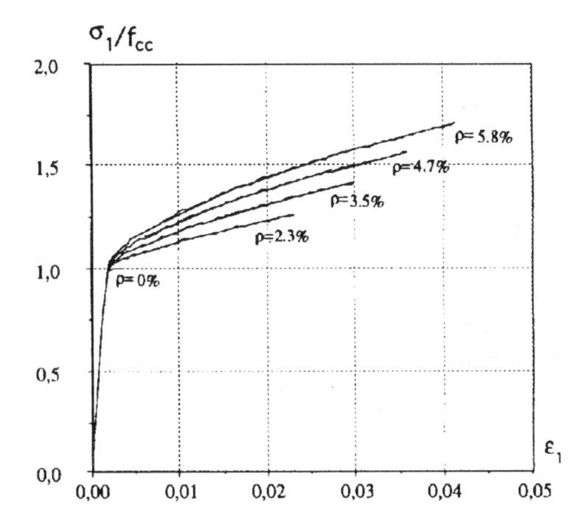

Fig. 4 $\sigma-\varepsilon$ diagram in non dimensional form for rectangular columns confined with carbon fiber straps

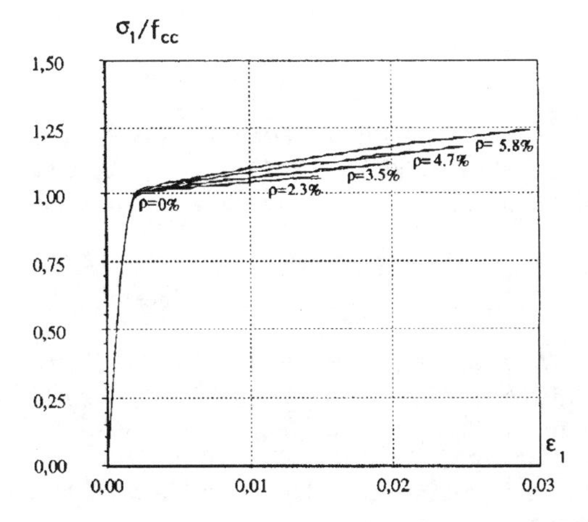

Fig. 5 $\sigma-\varepsilon$ diagram in non dimensional form for rectangular columns confined with glass fiber straps

In fig. 5 the $\sigma - \varepsilon$ diagram in non dimensional form related to a rectangular column ($b = 400mm$, $h = 300mm$) confined with glass fiber straps (resin plus fiber) ($f_{rif} = 1103\ MPa$ $E_{rif} = 48200\ MPa$) [3] is indicated. In this case, the effectiveness of confinement is reduced respect to the case of confinement with carbon wraps, concerning to the strength and ultimate strain.

5. Conclusions

The traditional confinement using steel hoops or spirals improves certainly the structural behaviour of reinforced concrete columns, when these ones are loaded by seismic forces. On the other hand, the production of advanced composite materials has promoted in the last years, the development of the "Wrapping" technique; using this advanced technique, it is possible to transversely wrap reinforced concrete columns with thin carbon or glass flexible straps epoxy-bonded to the concrete surface.
The mechanical behaviour is clearly better than the conventional lateral confinement with steel hoops or spirals and some experimental tests [4] confirm this statement.
In this paper, the authors present a model of constitutive law for concrete rectangular columns confined with FRP jackets which, generalising the elasticity theory, applies step by step the classical Navier relations defining the stress-strain laws in state of triaxial stress. The proposed model adequately describes the qualitative aspects observed in the experimental tests reported in literature; in the future the authors consider the possibility to carry out further laboratory tests in order to calibrate suitably the significant parameters of the model and define a simplified theoretical approach.

Acknowledgements

The authors wish to acknowledge the Italian Ministry of University Research (MURST) and the Italian CNR for the financial support.

References

[1] *La Tegola, A., Manni, O., (1998)* CONSTITUTIVE LAW FOR CONCRETE EXTERNALLY WRAPPED WITH FRP JACKETS, accepted for presentation at Second International Conference on Concrete Under Severe Conditions, CONSEC '98, 21-24 June, Tromsø, Norway.
[2] *Mander, J. B. et al., (1988),* THEORETICAL STRESS-STRAIN MODEL FOR CONFINED CONCRETE - Journal of Structural Engineering, Vol. 114, No. 8, August, pp. 1804-1825.
[3] *Saadatmanesh, H. et al., 1994,* STRENGTH AND DUCTILITY OF CONCRETE COLUMNS EXTERNALLY REINFORCED WITH FIBER COMPOSITE STRAPS, ACI Structural Journal, V. 91, No 4, July-Aug., pp. 434-447.
[4] *Harmon, T.G. et al., 1992,* ADVANCED COMPOSITE CONFINEMENT OF CONCRETE - Proceedings, Advanced Composite Materials in Bridges and Structures Sherbrooke, pp. 299-306.
[5] *Li, M. et al., 1992,* BEHAVIOUR OF EXTERNALLY CONFINED CONCRETE COLUMNS - Proceedings, Materials Engineering's Congress, Atlanta, pp. 677-690.

Local buckling of FRP profiles : experimental results and numerical analyses

M. Pecce*, F. Lazzaro*, E. Cosenza*

* Dipartimento di Analisi e Progettazione Strutturale – Università degli Studi di Napoli Federico II – Via Claudio 21, 80125 Naples-Italy

Abstract

In this paper the experimental results of stub column test on an FRP pultruded profile are reported. The test confirms the failure is governed by local buckling phenomena, that occurs largely before the material strength is reached.
The experimental critical stress is compared to numerical one obtained by a finite element modelling that takes into account the material anisotropy and the shear strain, moreover a simplified model is introduced. The comparison between the experimental results, the refined model and the simplified model concludes the study.

1. Introduction

The pultrusion technique allows to realise profiles in Fiber Reinforced Plastic materials with the shape and the length required at low manufacturing costs.
The benefits of these structural typology are numerous but the most important and typical of composite materials are: the durability and the possibility of design the material choosing the quality of the matrix and the quality and quantity of fibers.
This last characteristic permits to obtain a material with the specific performances required by the destination of the structure. However these advantages are coupled with the necessity of establishing and applying reliable methods of structural design to an innovative material completely different from the traditional ones used for profiles, i.e. steel or aluminium. Moreover actually there are different FRP materials that are continuously changing due to the improving of mechanical performances but also to the variation of the costs of the fibers.
So that it is clear the necessity of studying the structural behaviour of the pultruded profiles for two meaningful aims: to understand the most important behavioural features at serviceability conditions and at failure, but also to define the suitable methods for the experimental and numerical analysis.
One significant characteristic of the FRP pultruded profiles is the orientation of the fibers in the longitudinal direction, therefore the elements are strongly ortotrophic; the strength and the Young modulus in the direction perpendicular to the fibers are largely lower than the ones along the fibers direction.

Moreover, especially when glass fibers, that are cheap, are used, the ratios between strength and elasticity moduli are very large respect to the other traditional materials (concrete and steel). As a consequence the performances of the profiles could be restricted by buckling and deformability problems; in particular local buckling, that could occur in the flanges or in the web of the profile, depends also on the lower elastic constants in the direction perpendicular to the fibers.
At last low values of the shear modulus make shear strain not negligible and make deformability and buckling problems more restrictive.
At now many shapes of Glass-FRP are available on the market and the most important mechanical aspects have been examined by various researchers. The theoretical approach to define the elasticity moduli using the model for laminated materials [1,2] has been widely analysed taking into account the non-homogeneity of the pultruded profiles; the experimental procedures to individuate these characteristics with reference to little specimens of material and to the entire profile have been reviewed in [3].
Global buckling in compressed elements and flexural-torsional buckling in beams [4,5,6] have been studied by analytical and numerical approach, and often the results have been compared with experimental tests. Also local buckling has been analysed by many authors [7,8,9,10] and experimental tests have been made [11,12,13]. However definite design criteria are not yet available, moreover the experimental procedures and the numerical models well-known for the traditional materials cannot readily extended to FRP materials, therefore this field is now still open and needs of contribution.
In the following local buckling is analysed with reference to a wide flange I profile in GFRP; the results of experimental tests are reported, then a refined model is solved by Finite Element Method (FEM) and the test results are compared with the numerical ones.
Furthermore a simplified model is applied. This last approach considers only the behaviour of the flange, as a plate elastically restrained against rotation by the web; the main point of the model is the evaluation of elastic restraint, that has to be defined by experimental tests or, as in this case, by a numerical analysis of the entire profile.

2. The experimental test

Stub column test on a wide flange profile in FRP with E-glass fibers was carried out; the geometrical characteristics of the section are reported in Fig.1a. The profile is produced by MMFG [14] and the nominal mechanics strength in tension is 200 MPa. Three equal specimen 50 cm long were tested up to failure.
In Fig. 1b the relations force-elongation of the 3 specimens are drawn. The curves are very similar; in the first field the profiles have a linear behaviour until the local buckling of the two flanges occur; after this circumstance the behaviour remains linear, but with a different slope because of the reduced stiffness respect to the entire profile.
The failure of the elements is due to the longitudinal crushing of the web in the middle or at the connection to the flanges.
One of the specimen (number 3) was instrumented with 24 strain gauges located along one flange on the two sides. The position of the strain gauges is detailed in Fig.2a. Some results are reported in Fig.2b.
Comparing the force-elongation curves to the force-strain ones, the load at the beginning of local buckling seems different; really it is difficult to estimate a local

phenomenon by global measurement as the elongation of the specimen, but the local measure of the strains is surely more reliable. With reference to experimental trend of the strains, the critical load is about 375 kN, i.e. the critical stress is σ_{cr}=67 MPa. Therefore the critical stress is about the third part of the tensile strength, confirming that local buckling governs the behaviour of the FRP profile not allowing the fruition of the material strength.

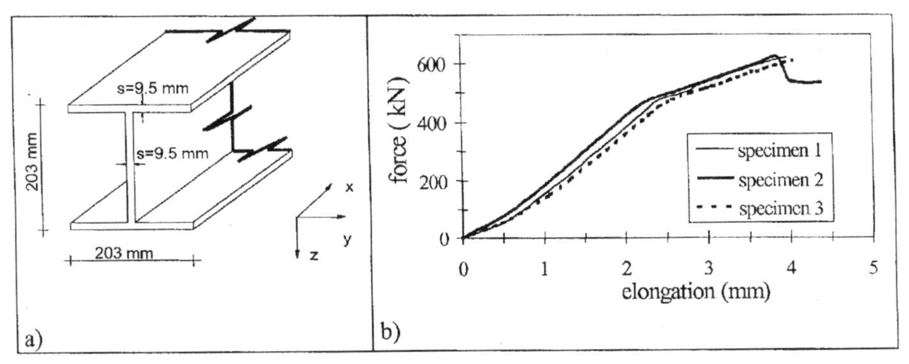

Figure 1. a) Geometry of the profile. b) Force-elongation curves.

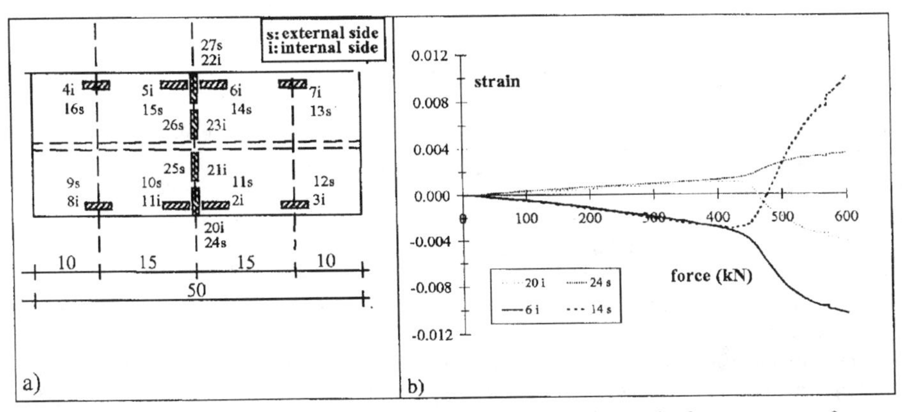

Figure 2. a) Position of the strain gauges on the flange. b) Strain-force curves of some strain gauges.

For what concerns the longitudinal elasticity modulus the tests develop about 22000 MPa, that is similar to the values obtained by tensile tests on specimens cut from the flanges and from the web [3].
It is also interesting that after local buckling occurs the experimental stiffness is about 30% of the initial one, since there is only the contribution of web, that represents about the third part of the entire area, and the material is still in the elastic field.
After failure the specimens were unloaded and the residual strains returned to about zero, confirming local buckling is attained in the elastic field and the material is linear elastic up to failure.

3. The numerical model

In order to model the behaviour of FRP profiles, the finite element method is applied by the code LUSAS [15]. Two different types of finite elements are used in the numerical analysis for considering or not the shear strain: the QTS4 and the QSL8.

The thick QTS4 is a shell rectangular element with 4 nodes situated in the angles, the d.o.f. are 5 for each node: the 3 orthogonal displacement along the 3 orthogonal axes and the 2 orthogonal rotations. This is an hybrid element that prevents the shear locking phenomenon when the thickness of the plate became too small [16].

The semiloof QSL8 is a shell element with 8 nodes situated in the 4 angles and in the middle of the 4 sides; each node is characterised by the 3 displacements along the orthogonal axes, moreover the midside nodes have other 2 d.o.f. that are the 2 orthogonal rotations in the 2 Gauss points along each side [17].

Elements with a ratio between the sides equal to 1 are considered in the following study; the dimensions of the elements were fixed by a numerical analysis of mesh optimisation.
The constitutive relationship of the material is elastic up to failure, and is characterised by 5 elastic constants: the longitudinal and transversal Young moduli (parallel and perpendicular to the fibers), the two relative Poisson ratios, the shear modulus.

4. The numerical-experimental comparison

The comparison of the numerical model with the experimental results is carried out considering the uncertainties of the restraints at the ends of the specimen due to the test machine; in fact the two limit solutions, hinged and fixed at the ends, are analysed. Moreover the solution is obtained with and without the shear strain using the 2 different types of finite elements: QTS4 and QSL8 respectively.
At last the influence of the thickness variation at the web-flange attachment is considered by a simplified modification of the finite element thickness at the attachment.
As regards the elasticity moduli of the material the results obtained by the experimental tests presented in [3] are used, that are the following:

- Young modulus and Poisson ratio parallel to the fibers: E_x=22000 MPa and ν_{xy}=0.3;
- Young modulus and Poisson ratio perpendicular to the fibers: E_y=7500 MPa and ν_{yx}=0.1;
- Shear modulus: G=2400 MPa

In Tab. 1 the theoretical results are summarised for all the cases introduced. It is worth to notice that the experimental result (σ_{cr}=67 MPa) is in good agreement with the numerical ones carried out with the fixed restraints.
The influence of the shear strain is larger for the hinged restraints, moreover it reduces the critical load so that it has to be taken into account in the model, even if it is not significant for the profile considered.

	σ_{cr} (MPa)			
	QTS4		QSL8	
	fixed	hinged	fixed	hinged
	64	53	66	58
	69	58	73	64

Table 1. Results of the numerical analysis.

5. A simplified model

The experimental test showed that local buckling of the flanges occurred, therefore the evaluation of the critical load can be developed by simplified modelling of the flange behaviour. The half of the flange can be considered as a plate substituting the web action with an elastic restraint [9,10], characterised by a stiffness k_φ (Fig.3); in the following the thickness variation at the web-flange attachment is neglected but the shear strain is taken into account.

In order to analyse the problem in a non-dimensional form, the critical stress is divided to the critical stress of a beam with the length of the half-flange b and its stiffness is D_y, that is according to the direction y and it is characterised by the lower elasticity modulus (perpendicular to fibers); the equation (1) is obtained and the experimental result is $N_{cr.ad}=1.20$ ($\sigma_{cr}=67$ MPa, b=203/2 mm, s=9.5 mm, $E_y=7500$ MPa, $\nu_{xy}=0.3$, $\nu_{yx}=0.1$). Also the stiffness k_φ is non-dimensionalised according the expression (3).

$$N_{cr.ad}=\frac{\sigma_{cr}sb^2}{\pi^2 D_y} \quad (1)$$

$$D_y=\frac{E_y s^3}{12(1-\nu_{xy}\nu_{yx})} \quad (2)$$

$$K=\frac{k_\varphi b}{D_y} \quad (3)$$

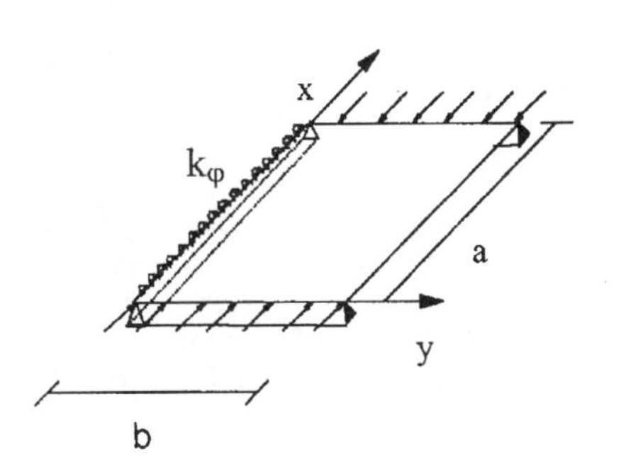

Figure 3. Simplified model of the flange.

In this simplified model the restraint due to the web is very influent on the result. In fact with reference to the dimensions of experimental specimen and assuming the edges fixed, according to the results of the previous paragraph, the two limit cases of $k_\varphi=0$ (the web action is an hinge) and $k_\varphi=\infty$ (the web fix the flange) develop to $N_{cr.ad}=0.4$ and $N_{cr.ad}=1.9$ respectively, with a very large range and uncertainty.
In order to individuate the actual value of the restraint stiffness the numerical model by FEM is used with the QTS4 elements (taking into account the shear strain); analysing the entire profile the value of k_φ is evaluated calculating the ratio between the moment and the rotation in the web-flange joints at the critical condition.
The value of k_φ varies along the profile due to the influence of the edge restraints, but it becomes constant in the central part at about 20 cm from the ends; therefore in the following the value in the central point of the profile is considered.
In particular in Fig. 4 the numerical results of the non-dimensional stiffness K are drawn versus the ratio a/b between the length of the profile and the half dimension of the flange; the 2 limit cases of profile hinged and fixed are reported. It is clear that increasing the profile slenderness the edges effect reduce and for a/b higher than 8 the result is practically independent from the restraints. For what concerns the experimental test, that corresponds to a/b=5, the value of K is included between 1.9 (hinged at the ends) and 2.2 (fixed at the ends).
In order to confirm the reliability of the simplified model the non-dimensional critical load is evaluated by the FEM model considering both the behaviour of the entire profile and the simplified model of the flange; in this last case K is assumed constant along the plate and equal to the value reported in Fig.4.
The comparison of the curves in Fig.5 shows a good agreement between the two models and also with the experimental result, therefore it is confirmed the effectiveness of the simplified model if the web action is correctly evaluated.

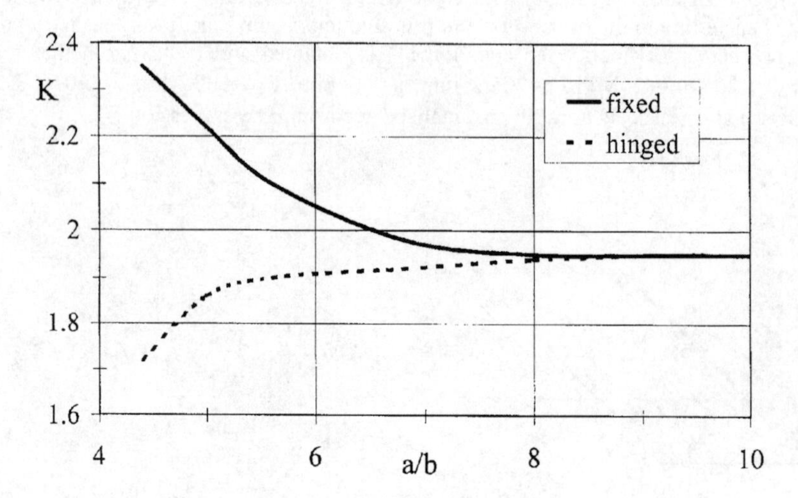

Figure 4. Numerical evaluation of non-dimensional stiffness K.

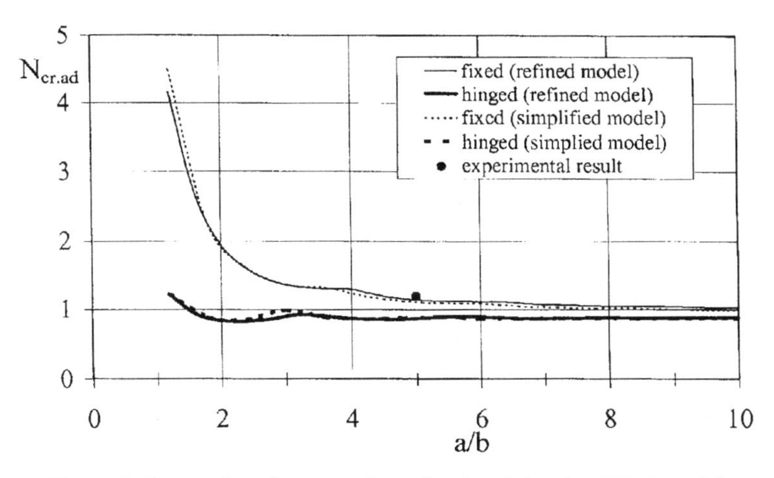

Figure 5. Comparison between the refined and the simplified model.

6. Conclusion

The stub column test described provides that local buckling occurs at a stress level equal to 30% of the strength; therefore it confirms that local buckling has to be considered for defining the failure condition of FRP profiles.
For what concerns the numerical modelling of the profiles behaviour, a finite element model that takes into account the shear strain and the material anisotropy is used. The numerical results are in good agreement with the experimental ones in spite of the uncertainties about the elastic constants of the material and the restraints of the profile in the test machine.
At last the simplified model, which considers the flange as a plate elastically restrained against rotation by the web, allows to evaluate the critical load with about the same reliability of the model of the entire profile if a good estimation of the restraint stiffness is made.

References

[1] *Pizhong, Q., Davalos, F.J.,* MULTICRITERIA OPTIMIZATION MODEL FOR ACM STRUCTURAL SHAPES – Proceeding of the 2st International Conference on Advanced Composite Materials in Bridge and Structures, Canadian Society for Civil Engineering, 1996, Montreal, Canada.

[2] *Nagaraj, V., GangaRao, H.V.S.,* STATIC BEHAVIOUR OF PULTRUDED GFRP BEAMS – *ASCE,* Journal of Composite for Construction, 1997, Vol.1, N. 3.

[3] *Cosenza, E., Lazzaro, F., Pecce, M,* EXPERIMENTAL EVALUATION OF BENDING AND TORSIONAL DEFORMABILITY OF FRP PULTRUDED BEAMS - Proceeding of the 2st International Conference on Advanced Composite Materials in Bridge and Structures, Canadian Society for Civil Engineering, Montreal, Canada, 1996.

[4] *Barbero, E, Fu, S. H., Raftoyiannis, I., LOCAL BUCKLING OF FRP BEAMS AND COLUMNS* – ASCE, Journal of Materials in Civil Engineering, Vol. 5(3),1993.

[5] *Davalos, F.J., Pizhong, Q.*, A ANALYTICAL AND EXPERIMENTAL STUDY OF LATERAL AND DISTORSIONAL BUCKLING OF FRP WIDE-FLANGE BEAMS – *ASCE*, Journal of Composite for Construction, Vol.1, N.4, 1997.

[6] *Mottram, J.Y.*, LATERAL-TORSIONAL BUCKLING OF PULTRUDED I-BEAMS – Composites, Vol.23, N.2, 1992.

[7] *Vakierner, A.R., Zureick, A., Will, K.M.*, PREDICT OF LOCAL FLANGE BUCKLING IN PULTRUTED SHAPES BY FINITE ELEMENT ANALISYS - Proceeding of the ASCE Speciality Conference on Advanced Composite Materials in Civil Engineering Structures, Las Vegas, 1991.

[8] *Bank, L.C., Gentry, T.R., Nadipelli, M.*, LOCAL BUCKLING OF PULTRUDED FRP BEAMS- ANALYSIS AND DESIGN - 49th Annual Conference, Composites Institute, The Society of Plastic Industry, Session 8-D, February 7-9, 1996.

[9] *Bank, L.C., Yin, J.*, BUCKLING OF ORTHOTROPIC PLATES WITH FREE AND ROTATIONALLY RESTRAINED UNLOADED EDGES – Thin-Walled Structures, Vol.24, 83-96, 1996.

[10] *Banks, W.M., Rhodes, J.*, THE INSTABILITY OF COMPOSITE CHANNEL SECTIONS - Proceeding. 3rd Int. Conference Composite Structures, 1983, Ed. I. Marshall, Applied Science Publisher.

[11*] Bank, L.C., Nadipelli, M., Gentry, T.R.*, LOCAL BUCKLING AND FAILURE OF OF PULTRUDED FRP BEAMS - Journal. of Engineering. Material and Technology, April 1994, Vol.116.

[12] *Cosenza, E., Lazzaro, F., Pecce, M.*, LOCAL BUCKLING OF FRP BEAMS – (in italian), 11° Congress CTE, Napoli 7-9 November, 1996.

[13] *Mosallam, A. S., Bank, L.C.*, SHORT-TERM BEHAVIOR OF PULTRUTED FIBER-RINFORCED PLASTIC FRAME - ASCE, Journal of Structural Engineering, July 1992, Vol.118, No. 7.

[14] MMFG Design Manual, Morrison Molded Fiber Glass, Bristol 1989.

[15] LUSAS, Finite Element Analysis System, Finite Element Analysis Ltd., Kingston upon Thames, 1990.

[16] *Criestfield, M.A.*, A QUADRATIC MINDLIN ELEMENT USING SHEAR CONTRAINTS – Computers & Structures, 1984, Vol.18, No. 5.

[17] *Irons, B.M.*, THE SEMILOOF SHELL ELEMENT – Finite elements for thin shells and curved membranes, 1976, cap. 11, John Wiley and Sons.

Vibration analysis of two span continuous special orthotropic plates with elastic intermediate support

Duk-Hyun Kim*, Chi-Moon Won**, Jung-Ho Park***

*Korea Composites, 97 Gugidong, Chongrogu, Seoul, 110-011, Korea
FAX (822) 379-7118, TEL (822) 391-2766
**Department of Construction Engineering, Halla Institute of Technology, Wonju, Korea
***Department of Civil Engineering, Kangwon Natl. Univ., Chuncheon, Korea

Abstract

A method of calculating the natural frequency corresponding to the first mode of vibration of beams and tower structures, with irregular cross sections and with arbitrary boundary conditions was developed and reported by Kim, D. H. in 1974. In this paper, the result of application of this method to the subject problem is presented. The structure considered for this report is two span continuous special orthotropic plates with elastic intermediate support. The use of elastic support as one of the passive control means is common. Any method may be used to obtain the deflection influence surfaces needed for this vibration analysis. Finite difference method is used for this purpose, in this paper. The influence of the modulus of the foundation, the aspect ratio of the plate, and the concentrated attached mass on the plate, on the natural frequency is thoroughly studied.

1. Introduction

A method of calculating the natural frequency corresponding to the first mode of vibration of beam and tower structures with irregular cross-sections was developed and reported by Kim, D. H. in 1974. Since 1989, he has extended this method to vibration analysis of two dimensional problems including composite laminates, and has reported at several conferenes, and it was proved to be very effective for the plates with arbitrary boundary conditions, with irregular sections, with or without axial loads, and with or without attached point masses[5,6,7,8,9].

The problem of deteriorated highway concrete slab is very serious all over the world. Before making any decision on repair work, reliable non-destructive evaluation is necessary. One of the dependable methods is to evaluate the in-situ stiffness of the slab by means of obtaining the natural frequency. By comparing the in-situ stiffness with the one obtained at the design stage, the degree of damage can be estimated rather accurately.

The reinforced concrete slab can be assumed as a special orthotropic plate, as a close approximation. For retrofitting such structures, advanced composite materials are best suited because of many advantages over conventional structural materials[1,2]. The author has identified enough number of fiber orientations which will behave as

special orthotropic plates as the number of plies increases. The structural element of construction is large so that the ratio of the thickness to the length is small enough to ignore the effect of shear deformation. The author has developed simple but accurate method of analysis of such structures[3].
The structure considered for this report is two span continuous special orthotropic plates with elastic intermediate support. The use of elastic support as one of the passive control means is common. The structure is subjected to the concentrated load/loads, in the form of applied loads, or the test equipments. For such plates, obtaining a reliable solution is very difficult.
In this paper, the result of application of this method to the subject problem is presented. The effect of concentrated point mass/masses is also studied.
Since this method has been explained by the senior author already[1], it is not repeated here.
This method uses the deflection influence surfaces. Obtaining the deflection influence surfaces is the first step in any panel design and analysis. Any method can be used to obtain such values.
Because of the special nature of this bridge boundary conditions, the finite difference method(FDM) is used to obtain such influence surfaces.
In finite difference form, the fourth single derivative terms in the biharmonic equation of the special orthotropic plate action require at least five pivotal points for errors of order Δ^2. Hence, it is desirable to transform the fourth-order partial differential equation into three second-order partial differential equations with three dependent variables, namely w, Mx, and My, by considering the equilibrium of a segment of the plate[10].
When the FDM is applied over the all surfaces of the bridge, the resulting matrix size is very huge. However, it is a typical tri-diagonal matrix and can be solved with relative ease[11].
Since one of the few efficient analytical solutions of the special orthotropic plate is Navier solution, and this is good for the case of the four edges simple supported, F.D.M. is used to solve this problem and the result is compared with the Navier solution. The aspect ratio used is 1m/1m=1. The mesh size is $\Delta x=0.1$, $\Delta y=0.1$. The accuracy of the F.D.M. solution is justified by comparing two results.

2. Structure Under Consideration

The bridge considered is as shown in Figure 1.

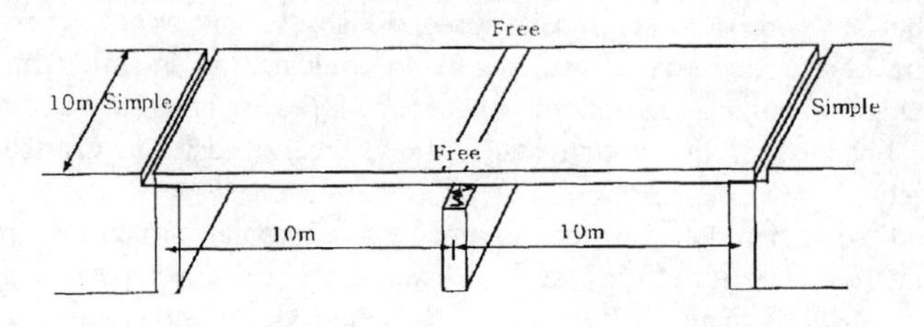

Figure 1. Two span continuous slab bridge

The location of the truck loading is as shown in Figure 2.

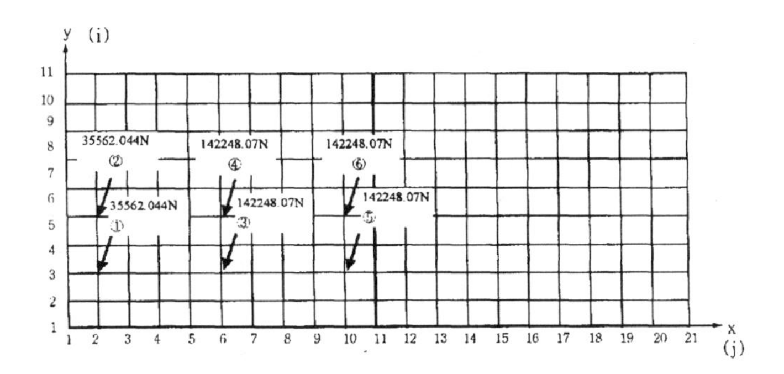

Figure 2. Location of truck loading

2.1 Reinforced Concrete Slab

$\sigma_{ck} = 210\,kg/cm^2 = 20.5942926\,MPa$ and $E_c = 15000\sqrt{\sigma_{ck}} = 21.317118060\,GPa$.
Poissons ratio $\nu_{12} = \nu_{21} = 0.18$ for concrete, Concrete dead weight = $2.5t/m^3 = 24500N/m^3$
Figure. 3 shows the cross section of the slab with unit width.

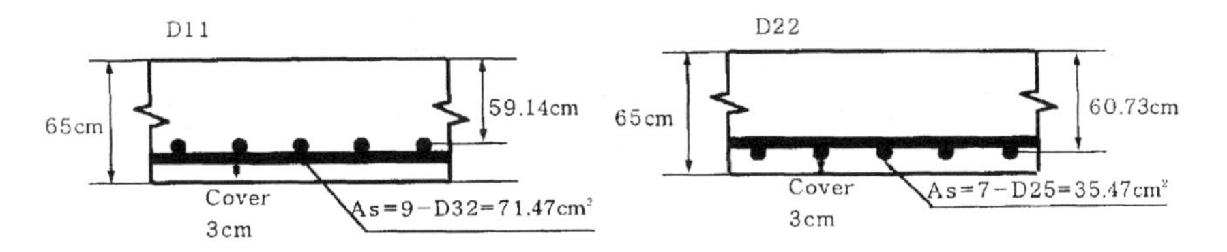

Figure 3. Cross section of the slab with unit width

Three different concepts are adopted for obtaining the stiffnesses, D_{ij}. For all cases the bending extension coupling stiffness, B_{ij}, is assumed as negligible.

Case A. Balanced design using the transformed area for steel in calculating the moment of inertia of the cross-section.

Case B. With $E_c = 15000\sqrt{\sigma_{ck}} = 21.317118060\,GPa$ and $E_s = 199.92\,GPa$, and with concrete $Q_{11} = E_c/(1-\nu_{12}^2)$ and steel $Q_{11}=E_s$, the typical formulas for D_{ij} are used.

Case C. Using the cracked section concept by the maximum moment, the moment of inertia of the cross section is obtained to calculate D_{ij}.

From the reinforced concrete slab section, D_{11}, D_{22}, D_{12}, and D_{66} are calculated as shown in Table 1.

Table 1. The stiffness of each case (N · m)

Stiffness \ Case	Case A	Case B	Case C
D_{11}	273496387.3	275266314.7	277232712.5
D_{22}	163168680.6	171647220.7	176729635.1
D_{12}	90690636.8	90690636.81	90690636.8
D_{66}	206573117.2	206573117.2	206573117.2

To study the influence of D_{22}, D_{12}, and D_{66}, the stiffnesses of six sub-cases are calculated as shown in Table 2.

Table 2. The stiffnesses of six sub-cases, for Case A (N · m)

Stiffness \ Case	Case A-1	Case A-2	Case A-3	Case A-4	Case A-5
D_{11}	273496387.3	273496387.3	273496387.3	273496387.3	273496387.3
D_{22}	163168680.6	209448389.3	243997970.2	273496387.3	273496387.3
D_{12}	90690636.8	90690636.8	90690636.8	90690636.8	0
D_{66}	206573117.2	206573117.2	206573117.2	206573117.2	0

2.2 Sandwich Panel Slab

The core used is the foam core with unit weight of 64kg/m^3. The top and bottom faces are $[ABA]_r$ type laminates with $A = 0°$, and $B = 90°$. The material properties are : $E_1 = 38.6\text{GPa}$, $E_2 = 8.21$ GPa, $\nu_{12} = 0.26$, $\nu_{21} = 0.00557$, $G_{12} = 4.14\text{GPa}$, and h_o=0.00125. Three type are considered:

Type A. $r = 24$, $h = 0.09\text{m}$. The thickness of the core, t_c is 0.368m.
The total thickness of the slab, $t_t = 2\times0.09+0.368 = 0.548\text{m}$.
The weight per unit area is :
core : $64\text{kg/m}^3\times9.8\text{m/sec}^2\times0.368\text{m}\times1\text{m}\times1\text{m} = 230.8096\text{N}$.
face : $1800\text{kg/m}^3\times9.8\text{m/sec}^2\times0.09\text{m}\times1\text{m}\times1\text{m} = 1587.6\text{N}$.
total : $230.8096\text{N}+2\times1587.6\text{N} = 3406.0096\text{N}$.

Type B. r=32, h=0.12
t_c=0.216 m
t_t=2×0.12+0.216=0.456 m

Type C. r=36, h=0.135
t_c=0.168 m
t_t=0.438 m

From the sandwich panel slab section, D_{11}, D_{22}, D_{12}, and D_{66} are calculated as shown in Table 3. The cross section of the Type A sandwich panel slab with unit width is shown in Figure 4.

Table 3. The stiffnesses of sandwich panels (N · m)

Type / Stiffness	Type A	Type B	Type C
D_{11}	276394112	276589504	274260864
D_{22}	178309312	59258948	58760032
D_{12}	20859894	15407327	15277608
D_{66}	39582036	29235688	28989546

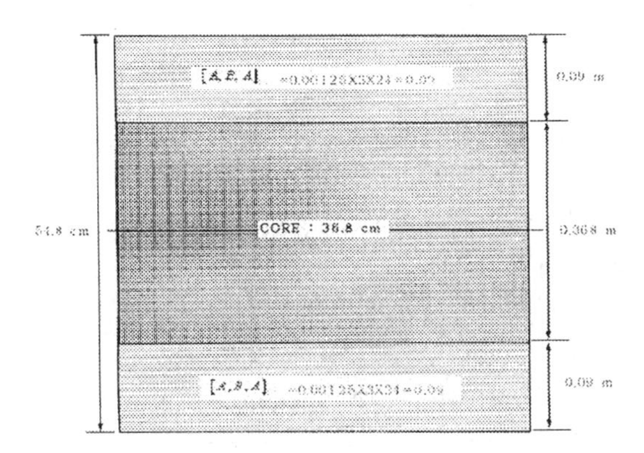

Figure 4. Cross section of Type A sandwich panel slab

3. Result of Vibration Analysis

3.1 Influence of the Modulus of Foundation, k

The influence of the modulus of the foundation on the deflection and the natural frequency of vibration is thoroughly studied by increasing the k values from $14{,}504 \times 10^3$, $14{,}504 \times 10^4$, $14{,}504 \times 10^5$, $14{,}504 \times 10^6$, $14{,}504 \times 10^7$ N/m/m for Type A sandwich panel slab. Table 4 and 5 show the deflection at the concentrated loading points and the natural frequency of vibration, for the Type A sandwich panel slab.

Table 4. Deflection at loading points for the Type A sandwich panel slab (m)

k(N/m²) / Load Point	$14{,}504 \times 10^3$	$14{,}504 \times 10^4$	$14{,}504 \times 10^5$	$14{,}504 \times 10^6$	$14{,}504 \times 10^7$
1	0.2344E-02	0.1373E-02	0.1261E-02	0.1250E-02	0.1248E-02
2	0.2036E-02	0.1195E-02	0.1097E-02	0.1086E-02	0.1085E-02
3	0.8967E-02	0.4470E-02	0.3946E-02	0.3893E-02	0.3888E-02
4	0.7789E-02	0.3894E-02	0.3430E-02	0.3381E-02	0.3376E-02
5	0.8372E-02	0.1844E-02	0.1059E-02	0.9711E-02	0.9619E-02
6	0.7312E-02	0.1638E-02	0.9192E-02	0.8380E-02	0.8294E-02

Table 5. The natural frequency of vibration for Type A sandwich panel slab

k (N/m^2)	Natural frequency (rad/sec)
14,504,000	0.1103856E+02
145,040,000	0.1615736E+02
1,450,400,000	0.1677109E+02
14,504,000,000	0.1682407E+02
145,040,000,000	0.1682975E+02

3.2 Natural Frequency

With the deflection influence surfaces obtained by F.D.M., the presented method is used to obtain the natural frequency for each of the three concrete slab cases and three sandwich panel types.

This calculation is carried out for each of the increasing modulus of the foundation values. The result is given in Table 6.

When k value is $14504 \times 10^6 N/m^2$, the influence of the stiffness D_{22}, D_{12}, and D_{66} on the natural frequency of vibration is studied and the result is given in table 7.

Table 6. Natural frequencies of three concrete slab cases and three sandwich panel types (rad/sec)

k(N/m^2) / Case,Type	$14,504 \times 10^3$	$14,504 \times 10^4$	$14,504 \times 10^5$	$14,504 \times 10^6$	$14,504 \times 10^7$
Case A	0.80059E+01	0.10831E+02	0.10913E+02	0.10912E+02	0.10914E+02
Case B	0.80197E+01	0.10895E+02	0.10978E+02	0.10978E+02	0.10979E+02
Case C	0.80315E+01	0.10951E+02	0.11035E+02	0.11035E+02	0.11036E+02
Type A	0.11038E+02	0.16157E+02	0.16766E+02	0.16824E+02	0.16829E+02
Type B	0.10436E+02	0.15153E+02	0.15713E+02	0.15728E+02	0.15733E+02
Type C	0.10241E+02	0.14904E+02	0.15307E+02	0.15352E+02	0.15357E+02

Table 7. The natural frequency of vibration of Case A concrete slab, with k = 14,504,000,000 N/m^2

Case	Natural frequency
Case A-1	0.1091295E+02
Case A-2	0.1103260E+02
Case A-3	0.1109503E+02
Case A-4	0.1113097E+02
Case A-5	0.1098112E+02

Conclusion

In this paper, the result of application of the simple but accurate method of vibration analysis developed by Kim, D. H. to the subject problem is presented. The method used for vibration analysis is simple to use but extremely accurate. The boundary condition can be arbitrary. One can use any method to obtain the deflection influence coefficient. The accuracy of the solution is dependent on only that of the influence coefficients needed for this method. Finite difference method is used to obtain such influence surfaces.

Three different concepts are used to obtain D_{ij} stiffnesses for the concrete slab. The result is not for off from each other. The differences between the natural frequencies of each cases are only small fractions of a percent.

Three different types of sandwich panels are considered for comparison. Design optimization for the best structural efficiency and the minimum cost requires such comparison. The influences of the magnitudes of different moduli of foundation at the intermediate supports, and different stiffnesses of the slabs on the natural frequency is throughly studied. The influences of D_{22}, D_{12}, and D_{66} are small since the boundaries parallel to the traffic is free.

Reference

[1] *Kim, D. H.,* Composite Structures for Civil and Architectural Engineering, Published by E & FN SPON, Chapman & Hall, London, 1995.

[2] *Kim, D. H.*, COMPOSITE MATERIALS FOR REPAIR AND REHABILITATION OF BUILDINGS AND INFRASTRUCTURES, Plenary Lecture at The Third International Symposium on Textile Composites in Building Construction, Seoul, Korea, November 7-9, 1996.

[3] *Kim, D. H.*, A SIMPLE METHOD OF ANALYSIS FOR THE PRELIMINARY DESIGN OF PARTICULAR COMPOSITE LAMINATED PRIMARY STRUCTURES FOR CIVIL CONSTRUCTION, Journal of Materials Processing Technology, Vol. 55, Elsevier, London, 1995, pp 242-248.

[4] *Kim, D. H.*, A METHOD OF VIBRATION ANALYSIS OF IRREGULARLY SHAPED STRUCTURAL ELEMENTS, Proc. International Symposium on Engineering Problems in Creating Coastal Industrial Sites, Seoul, Korea, 1974.

[5] *Kim, D. H.*, et al., A SIMPLIFIED METHOD OF VIBRATION ANALYSIS OF IRREGULARLY SHAPED COMPOSITE STRUCTURAL ELEMENTS, First Japan International Symposium of the Society for the Advancement of Materials and Process Engineering, Tokyo, December, 1989.

[6] *Kim, D. H.*, VIBRATION ANALYSIS OF IRREGULARLY SHAPED COMPOSITE STRUCTURAL MEMBERS-FOR HIGHER MODES, 8th Structures Congress, ASCE, Baltimore, MD, U.S.A., April 30-May 3, 1990.

[7] *Kim, D. H.*, VIBRATION ANALYSIS OF LAMINATED THICK COMPOSITE PLATES, 3rd EASEC, Shanghai, China, April 23-26, 1991.

[8] *Kim, D. H.*, VIBRATION ANALYSIS OF IRREGULARLY SHAPED

LAMINATED THICK COMPOSITE PLATES II, 2nd Japan International SAMPE Symposium and Exhibition, December 1991.

[9] *Kim, D. H.*, A SIMPLE METHOD OF VIBRATION ANALYSIS OF LAMINATED COMPOSITE PLATES UNDER AXIAL LOADINGS AND WITH ATTACHED POINT MASSES, 4th Japan International SAMPE Symposium and Exhibition, Tokyo, September 25-28, 1995.

[10] *Kim, D. H.*, THE EFFECT OF NEGLECTING THE RADIAL MOMENT TERMS IN ANALYZING A FINITE SECTORIAL PLATE BY MEANS OF FINITE DIFFERENCES, International Symposium on Space Technology and Sciences, Tokyo, Japan, May, 1967.

[11] *Kim, D. H.*, TRIDIAGONAL SCHEME TO SOLVE SUPER LARGE SIZE MATRICES BY THE USE OF COMPUTER, Jour., K.S.C.E., Vol 15-1, 1967.

Composites and construction : design of solar refractive-concentrators prototypes

JOSE IGNACIO PEREZ CALERO
• Dr. Architect, D.. in Physical Sciences. Full Professor in the Mechanics of Continous Media Department, ETS of Architecture at the University of Sevilla, Av. Reina Mercedes, 2, 41012 Sevilla. Spain.

Abstract

In the technology of Refractive Solar Concentrators (R.S.C) are very interesting the design and construction in composites of model tanks for fluids with spherical surfaces. Those models are optimum designs attained in teoric analyses of fundamental parameters. To know the mechanic behavior, in order to perfect functioning, exist the extensometric methodology over reduced models and the computer simulation with Finite Elements Method programs that we use with satisfactory results.

1. Introduction

This paper considers the search for a general methodology-not sufficiently well-known yet- for the design and study of solar concentrators, exclusively based on the refractive phenomenon. The designs desired will be basically simple systems, not very onerous and light, built in many cases in large series and using fluids, especially water as refractive element. We will concentrate especially on spherical designs, either thin or thick, whether working on paraxial zones or not, and with as much opening as it is necessary. We will consider as well any collecting plan that might be interesting, focal or not, and without regards to where the energetic image is.
After the theoretical analysis of models of refractive solar concentrators, as Perez Calero proved (1988), we can conclude that there are optimum designs attained in function of fundamental parameters. These are the objectives of this part of our work.
Moreover, this research line is inextricably linked to the design and later construction of real models of containers for fluids made of plastic materials with spherical surfaces, sometimes combined with plane ones, with the additional problems that this implies, since we have to respond to the need of containing big amounts of liquids within walls of little thickness, (so that we can achieve the correct refractive interphases) and with total transparency, since this is essential to our purposes.
In order to meet the technical exigencies above said, we will resort to composites of policarbonate matrix reinforced with E glass fiber that guarantees absolutely the exigencies requested such as economy and transparency.
It is evident that we need to know the resistent mechanic behaviour of membrane of the models in order to design the zones to be strenghtened with fibers, due to the tight tolerance of strain and strengths compatible with the optimun state of service of the prototypes to be built (generally in large series,

as we say) due to the characteristics of the materials to be used). We will undertake the study on models designed just with matrix; we will analyse their behaviour, and we will deduce the quantity, volumetric relation and interphase with the matrix as well as the distribution of the reinforcing fibers for a posterior re-study, once these be placed, and verification of their perfect work in this second case.

2. Prototypes of models to be investigated

The possible simple designs that we could conceive and construct will belong to the four classic large converging systems, which, naming them according to the incidence surface and exit of the pencil of tridimensional solar beams, are:

Sphere-Plane System	(in electronic treatment ES-PL)			
Plane-Sphere System	("	"	"	PL-ES)
Convergent Meniscus System	("	"	"	ME-CO)
Bi-convex System	("	"	"	BICON)

The two first have one spherical surface and the other plane, corresponding to that of the spheric incidence and that of the plane exit, in the system that heads the list and viveversa in the Plane-Sphere System.
In last two in the list both surfaces are spherical, with their tracing centers on one side or on both, depending on if it belongs to the third or fourth group respectively. Moreover, the Convergent Meniscus System demands a higher thickness in the central axial zones that in the border ones, allowing for two possible alternatives before the solar radiation in concave or convex positions. The two systems mentioned first are also known traditionally together as systems or plane-convex lens or plane-convergent, which makes it neccessary a differentiation in both cases ; this is the reason of the above mentioned nomenclature.

3. Research methodologies

In the research we follow a multiple methodology according to the part of the work that is analysed. For the analysis of the general models we use the mixed one. On the one hand we will develop an analytic-theoretical study in the optic-mathematic field of the following of the pencil of tridimensional beams, incident to the system for any relative position among them, until their intersection with the wished capting planes. On the other hand, the process is followed through computer simulation, according to the several parameters present, both geometric and physic, which allows us to know the general behaviour of any system of our interest. In figure 1 we can see one of the multiple diagrams obtained by this method, that makes it possible to optimize the theorical design of refractive concentrators for a later construction.
As regards to the important related problem of the study of the mechanic behaviour of the simple models of refractive concentrators, they are analised from various

methodologies, generally of great mathematical darkness, and it is due to the complexity of the subject that we think that there are two clear processes of research. On the one hand we have the experimental on scale models of prototypes built of policarbonate, whether with the fixation of extensometric gages in the prefixed points to know the total behaviour of the models (or optionally by photoelastic methods type Moire-Ligtenberg to be applied only on plane faces in the Plane-Sphere System). On the other hand, we have the process of computer simulation by computation with programs based on the Finite Elements Method, which, as it is known, consists on the discretization of the model to be analysed and of the actions that act upon it forming a system of elements linked through equations. It is evident that both methods of research are independent although they can be complementary to set and check the results obtained. In this work we are going to follow both methods with the complementarity pointed out above.

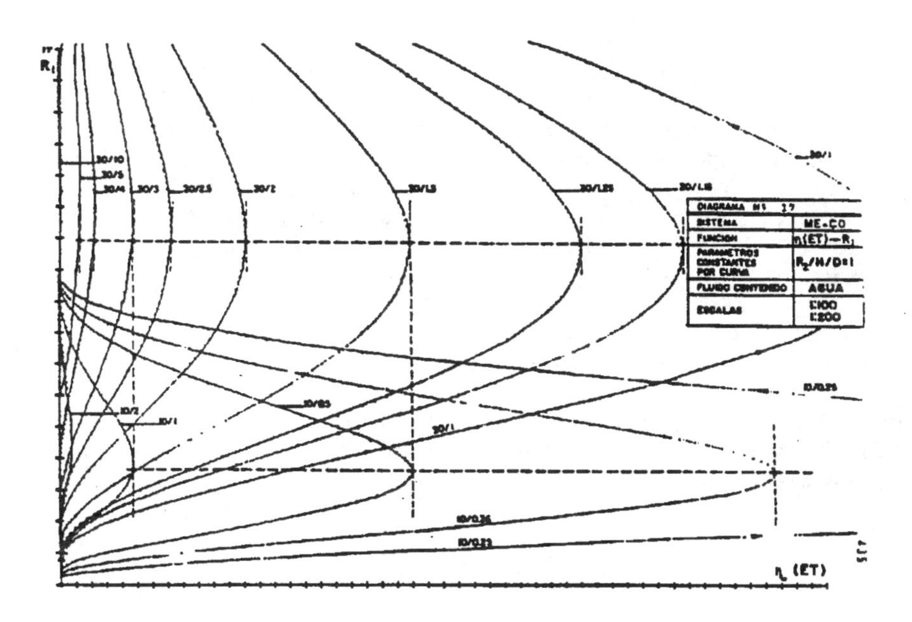

Figure. 1 Diagram of the factor of main concentration

4. Hypothesis of service of the models to be designed

We will use models with optimum design, as we have said before, and among them we will concrete in relation to diameters 59/79, corresponding the first to a hemispherical surface. Evidently, the methodological system is universal although we will focus at the moment on these parameters.
The main angle of service, that is the one that forms the horizontal with the definer plane that contains the maximum circle common to both surfaces, will be of 45 degrees, coinciding with the standard angle of work (in the latitude of Spain) in the refractive solar concentrators (RSC), although the method is valid for any other angle of our interest.
In order to obtain the exact results, the knots will be fixed by a division in meridians every 10 degrees and in paralells every 15 degrees, for the numeric method, and every 30 degrees in meridians and equiangulars in parallels for the experimental research, as can be observed in the enclosed drawings.

The hypothesis of loads supposed has been:

a) Weight considering that of the volume of the fluid contained.

b) Weight plus the wind action from the right to simulate an estimation of greater interest in the practice, due to the real conditions of work of the prototypes studied. The eolic action has been analized taking into account the different models placed in Sevilla (Spain) which gives us an action of 75 kp/m2 and according to the Spanish norm of application and to probabilistic examples. Evidently, the method is valid for any placement.

c) Weight plus wind action coming from the left, for the same reasons and with analogous considerations as those in the example above.

d) Prototype emptied of fluid and exposed to the wind action
from the right. This example responds to real situations of service in certain circumstances.

e) The same as the last case but with the action of the wind
in the opposite direction, that is, from the left.

All the hypotheses have been considered with numeric methodology and only the first one in the study has been by an experimental process.
In the case considered the fluid contained is water (n= 1,33332, d=1.000 kp/m3)
Regarding the anchorage points, we have considered that there is a perimetrical belt that holds the maximum circle common to the two surfaces considered and also fixed bases united to three points perfectly delimitated in all the models.
The containers-models are defined as formed by tridimensional elements of the "shell" type that contain four nodes, with the possibility of acting as membrane or plate, although we are here clearly in the first hypothesis.

5. Design by numeric process

We use the module SAP-80, version 85.02, combination of programs linked by internal files that constitute a data base, every one of them operating upon one or more blocks of data entry, and providing different exit blocks.
The program was implemented in a computer Tandon 486 sx with 8Mb of Ram memory IBM PC, in order to have the most effective discretation.

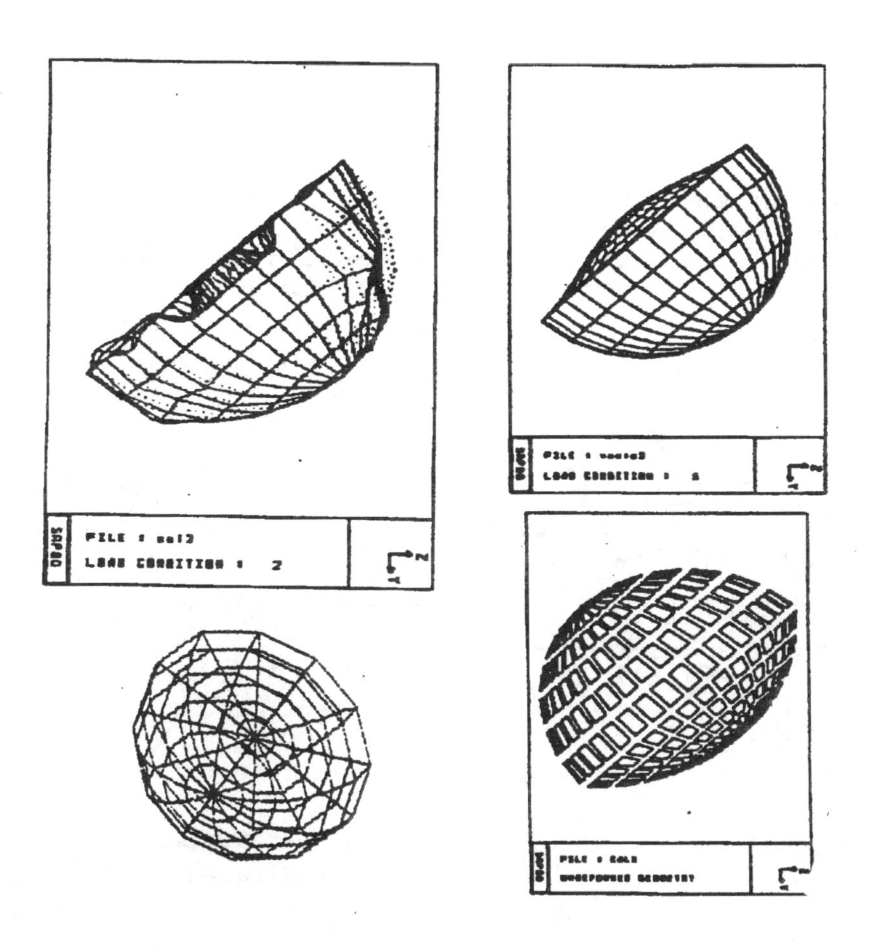

Figure. 2 Informatic graphic treatment

The process of data entry is analised through ASCII files, with 13 blocks of which 6 have been used (Title, System, Joints, Restraints, Quad and Solid) that indicate the nodes and their possibilities od displacement or rotation in tree directions, the tridimensional shells of four nodes and the tridimensional solids of charge of 8 nodes. The exit files analyse and calculate the system for the different hypotheses of supposed load. The program incorporates a modulus called Saplot that makes it possible to visualize the deformations produced in the elements studied, that has been done for a better understanding with a coefficient of rebound of 500 units (Fig. 2).
The analysis of the results from the computer allows us to generate diagrams discretized of efforts, that are the guide, as in detail as we need, for the zones to be reinforced with E glass fiber in a anisotropic behaviour of composite with the policarbonate matrix in the model.

6. Treatment by experimental process

We have used a Strainmeter KYOWA SM-60-D and a Swithching and Balancing Box SS-24-R, with possibilities for 60,120,350 and 500 Ohms and for 1,2 and 4 extensometric gages, which levelled to zero and tared with a piece of the same material allows to proceed to the measurement with everyone of the models.

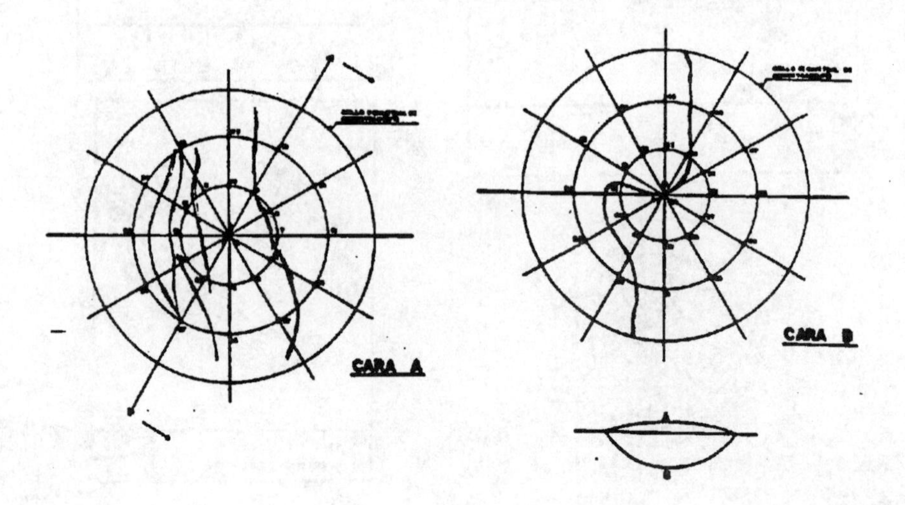

Figure. 3 Breaking lines for side b superior

In the case of the bi-convex model, that with more interesting and useful results regarding the technological appliccations, we include the map of gages on both sides. In order to analise the results of the most characteristic points, we tried first the model in a position of service with side A superior. Experimenting with the opposite position, the sudden breaking of the model is produced, which confirms the unavoidable need of its reinforcement, Fig. 3 showing the breaking lines that were produced, according to the previsions of the non-experimental methodology.

7. Conclusion

The mixed planning undertaken and the potency and versatility of the informatic programs used, as well as the data provided by the experimental process, allow us to reach the end pursued, both numerical and graphically.
In sum, we can assure the design and analysis of the refractive solar concentrates (RSC) of spherical surfaces built into composites (generally made of policarbonate with reinforcement of E glass fibers, allowing us to deduce the quantity, volumetric relation and ideal interphase) counting with strong methodologies of simulation and experimentation to meet optimum designs. In the same way, we can attain the construction of real prototypes with a powerful tool, the Finite Elements Method (FEM) and a powerful way of experimental support, by extensometric gages, that allows us to reach a tridimensional "map" in detail (with a configuration and scale of ranges as tight and adjusted as we need) under the various hypoteses considered, either simple or combined, changeable according to our will. Thus, we can configurate the protopypes of the different systems that we wish to study and use in any case to respond to the technological needs that arise in this area, with enormous possibilities of application in various fields.

References

(1) Oberbach, H. PLÁSTICOS. VALORES CARACTERÍSTICOS PARA EL DISEÑO DE ESTRUCTURAS. Buenos Aires. Argentina. 1978.América L.
(2) Pérez Calero, J.I., A. García Moris. SISTEMA INTEGRAL DE ARQUITECTURA BIOCLIMÁTICA PARA VIVIENDAS SOCIALES. Proceedings XXIII Congreso internacional Comples. Sevilla. España. 1985.
(3) Pérez Calero, J.I. APORTACIÓN AL ESTUDIO DE LOS CONCENTRADORES SOLARES ESTÁTICOS POR REFRACCIÓN: APLICACIÓN A LA CUENCA BAJA DEL GUADALQUIVIR. Tesis Doctoral. Universidad Politécnica. Madrid. España. 1988.
(4) Pérez Calero, J.I. TECNOLOGÍAS DE CONCENTRADORES SOLARES POR REFRACCIÓN EN LOS CAMPOS DE LA ARQUITECTURA Y DEL PLANEAMIENTO URBANO. Proceedings. Congreso internacional Plea.91. Sevilla. España. 1991.
(5) Pérez Calero, J.I. ANÁLISIS MECÁNICO, MEDIANTE EL MÉTODO DE LOS ELEMENTOS FINITOS, DE DEPÓSITOS ACUMULADORES DE SUPERFICIES ESFÉRICAS, CONSTRUIDOS EN COMPOSITES DE POLICARBONATO REFORZADO CON FIBRAS DE VIDRIO. Proceedings Primer Congreso Iberoamericano. de Ingeniería Mecánica Madrid. España. 1993.

(6) Pérez Calero, J.I. DISEÑO Y ANÁLISIS DE REFRACTOCONCENTRADORES SOLARES DE SUPERFICIES ESFÉRICAS CONSTRUIDOS EN COMPOSITES, MEDIANTE EL MÉTODO DE ELEMENTOS FINITOS. Proceeding. VII Congreso Ibérico de Energía Solar. Vigo. España. 1994.
(7) Pérez Calero, J.I. REFRACTOCONCENTRADORES SOLARES: PROPUESTAS PARA LA ARQUITECTURA Y EL URBANISMO. Revista Era Solar. Madrid. España. 1995. nº 65. pp 13-18.
(8) Pérez Calero, J.I. REFRACTOCONCENTRADORES SOLARES PARA LA ARQUITECTURA Y EL URBANISMO: DISEÑO EN MATERIALES COMPUESTOS MEDIANTE MÉTODOS NUMÉRICOS Y EXPERIMENTALES. Proceedings Cobem–Cidim 95. II Congreso Iberoamericano de Ingenieríaa Mecánica. Belo Horizonte. Brasil. 1995.
(9) Pérez Calero, J.I. DESIGN OF REFRACTIVE SOLAR CONCENTRATORS (R.S.C.), BUILT INTO COMPOSITES FOR FINITE ELEMENTS METHODS (F.E.M.) AND EXTENSOMETRIC METTHODOLOGY "In search of the sun" World Solar Congress. Harare. Zimbabwe. 1995.
(10) Pérez Calero, J.I. NUMERICAL AND EXTENSOMETRIC METHODS FOR DESIGN OF REFRACTIVE SOLAR CONCENTRATORS, World Renewable Energy Congress IV. Denver. Colorado. USA. 1996.
(11) Pérez Calero, J.I. LOS COMPOSITES EN EL DISEÑO DE REFRACTOCONCENTRADORES SOLARES PARA APLICACIONES EN ARQUITECTURA Y URBANISMO. I Conferencia internacional sobre materiales compuestos aplicados en arquitectura y construcción. Arquimacon. Sevilla. España. 1996. pp. 281 a 288.
(12) Pérez Calero, J.I. CÁCULO Y DISEÑO INTEGRAL DE PROTOTIPOS DE REFRACTOCONCENTRADORES SOLARES. Proceedings. VIII CONGRESO IBÉRICO DE ENERGÍA SOLAR. Porto. Portugal. 1997.
(13) Quarmby, A. MATERIALES PLÁSTICOS Y ARQUITECTURA EXPERIMENTAL. Barcelona. España. 1976. Gustavo Gili.

A NEW COMPOSITE BUILDING MATERIAL MADE OF WASTE RECYCLED PLASTICS

Olivares Santiago, M.
Doctor arquitecto. Prof. Titular Depto. Construcciones Arquitectónicas I. E.T.S.A. Sevilla. Spain.
Laffarga Osteret, J.
Doctor en Químicas. Prof. Titular Depto. Construcciones Arquitectónicas I. E.T.S.A. Sevilla. Spain.
Galán Marín, C.
Arquitecto. Prof. Asociado Depto. Construcciones Arquitectónicas I. E.T.S.A. Sevilla. Spain.
Martineaux, Ph.
Doctor en Ciencias de los Materiales. Institut des Matériaux Composites. Bordeaux. France.
Alvarez Jurado J.L.
Doctor Ingeniero Industrial. Director Técnico de Almateo S.L. Sevilla. Spain.

Abstract

The aim of this paper is to show some experimental results, within a larger scope investigation, in the field of manufacturing curved specimens and cylindrical pipes made of plastic materials obtained from urban home waste, through a simple recycling process that gives rise to a new material named *polycarbon*.

1. Introduction

Nowadays, our civilisation faces a great problem, as it is the elimination of waste materials. In the urban waste there is a big percentage of plastic materials, which is an urgent problem because of their special organic characteristics that do not allow its degradation and make quite difficult its elimination.

Recycling process of plastic materials does not involve the risk that burning and pirolysis do, so we can get economic materials and with very suitable characteristics for been used in the building industry.

In the Department *Construcciones Arquitectónicas I* of the University of Seville, there exists a deep concern on ecological matters particularly regarding to the relation with Building Science and Technology.

It is not necessary to explain here the full recycling process. But it might be worth to know that due mainly to economic rather than technical reasons it is more convenient to do a full recycling of all waste thermoplastics without previous selection. So investigations about waste plastic recycling were approached under the premise of jumping out the previous cleaning and selection of the different kinds of plastic, so as to reduce manufacturing costs.

Most often consumed thermoplastics are basically of four different types. They four have very similar basic polymers structure:

- Polyethylene (PE)
- Polypropylene
- Polyvinyl Chloride (PVC)
- Polyethylene Terephthalate (PET)

After the first tests on these plastic-recycled materials, we have the first results available. In spite of their limitations because of their plastic nature as an organic material, we can assert that waste plastics can give rise to synthetic construction materials.

Firsts samples were manufactured at high pressure and high temperatures, but the result was that only the outer faces were melted, while the inside plastic mass didn't. Increasing temperatures burned the outermost layer. However, no effect has been detected in the internal core.

Later, at the end of 1993, a new original system based on recycled metallurgical-industry oil heated to recycle thermoplastics waste was developed. This process consisted of a great cubic depot of one-meter side, filled with used hydraulic oil from a sevillian industry. A resistance fixed around the depot heated the oil up to a controlled constant temperature all along the process.

A new material was obtained through this process and we named it *polycarbon* (after poly=many). The appearance was variable, depending on row materials (figure 1). Other experiments using only polyethylene from waste agricultural films were done. These films once cleaned and cut into pieces gave rise to another grey slate-coloured material, so we named it *synthetic slate* (figure 2).

Figure 1.Diferent appearance of *Polycarbon*

Figure 2. Specimens of *synthetic slate*

Further tests on different specimens made of both kinds of recycled products were done in order to establish the differences between them. Apart from the external appearance, their different properties needed separate investigations (figure 3).

This paper reflects part of the works done with the *polycarbon* itself, obtained under the process above mentioned according to the initial ideas of Mr. Alvarez. First experiments were done in the laboratories of the Faculty of Architecture of the University of Seville, and the first results were shown in a paper presented in the *IV Jornadas de Aplicaciones Arquitectonicas de los Materiales Compuestos y Aditivados* [1] and later another paper in *ARQUIMACOM'96* [2].

Figure 3. Polycarbon Specimen tested under presure

Polycarbon is a material whose technical characteristics put in on a level with some good timber. It can be prepared with a very simple and not expensive process.

Later on, more convenient parameters for the recycled plastic mixtures were established. Flat plates of different thicknesses (up to 5 cm.) were obtained. These tiles could be cut, nailed, screwed and drilled with conventional carpenter tools. They can also be glued with several synthetic glues and reinforced with iron or some other materials (figures 4 , 5 and 6).

Figures 4 and 5. Polycarbon cut and nailed

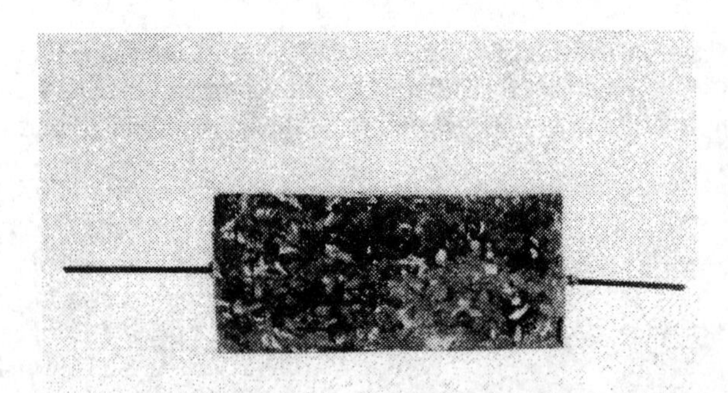

Figure 6. Polycarbon with iron reinforcement

This material has various aspects and colours, so it can substitute another conventional materials. It can substitute wood, at least in several uses, due to its low absorption degree (moulds, urban furniture). According to its high chemical resistance can be useful for building construction, for instance under the shape of security doors, panels, door and window frames. Its low density and acceptable mechanical strength and hardness make it possible to substitute not only wood but also natural and manmade stone, ceramic pavements and roof tiles.

Due to the above characteristics, in our experiments with *polycarbon* we have made plates and sheets that could be used for all different types of building tiles. But the homogeneity of colour and appearance of this *polycarbon* cannot be guaranteed. So it may be used where its multi-coloured granular appearance wouldn't be important or better suitable.

The simplicity of the recycling process makes it possible to focus this investigation on the search of lower manufacturing costs. But soon after the first specimens new difficulties had to be faced, i.e. flat elements, once produced could not be folded.

Several experiments were carried out since 1996, in order to manufacture curved pieces, both opened and closed. This objective will enable *polycarbon* to be more widely used. First results of these experiments are shown in this paper.

2. *Polycarbon* plates manufacturing

First polycarbon plates were produced heating sevillian home waste plastics smashed to bits. This mixture was moulded in iron casts of 60x70 cm. with different thicknesses between 1 to 4 cm. The plastic was agglutinated under heating over 250 °C, atmospheric pressure, for ten to twenty minutes.

The first process, without any extra pressure at all, gave as a result a polycarbon quite oily, as a residuary heating oil remains retained in the inside of the material. This oil can rise to the surface under some conditions. But the product can be suitable for some uses as the quantity of oil lost in each treatment has been measured and it is always under 3-5% of the recycled mass.

It has been experimented that a great part of that oil can be eliminated if the plates are put under pressure while heated. Though we are studying this possibility at this moment, it will undoubtedly arise costs and difficulty of the production means.

As mentioned above, flat plates can be cut, nailed, screwed and glued in order to create different elements for the building industry. However, folding and bending of the pieces is not convenient at all. Thus, a new modification in the moulding system, with curved casts, was experimented.

One of the first uses for curved casts was a recycled plastic pipe. A new iron cast (consisting on two concentric cylindrical tubes, one meter long and two millimetres thick), was manufactured. The mixture of waste plastic bits was introduced between both tubes. The lower edge of the cast was covered with an iron cap. One of the casts was filled up with waste plastic bits, with no extra pressure. And the heating oil was introduced between both cylinders for about fifteen minutes.

First results obtained with these experiments were not satisfactory and showed a great difficulty: the stickiness between *polycarbon* and the inner cylinder once hardened made it very easy dismantling the outer cast but removing the inner one was proved to be impossible without breaking the specimen.

After some tests with different materials, searching for a solution for this problem, we decided to use a lost mould-not able to be reutilized. Small-thick aluminium cylindrical pipes (0,6 mm. thick) were used as a lost mould, under the plastic materials. Once the plastics are hardened, their inner face is a tiny aluminium coating that we also employed as a mould (figure 7).

Figure 7. Polycarbon pipe with a lost Small-thick aluminium mould

Figures 8 and 9. Curved Polycarbon specimens

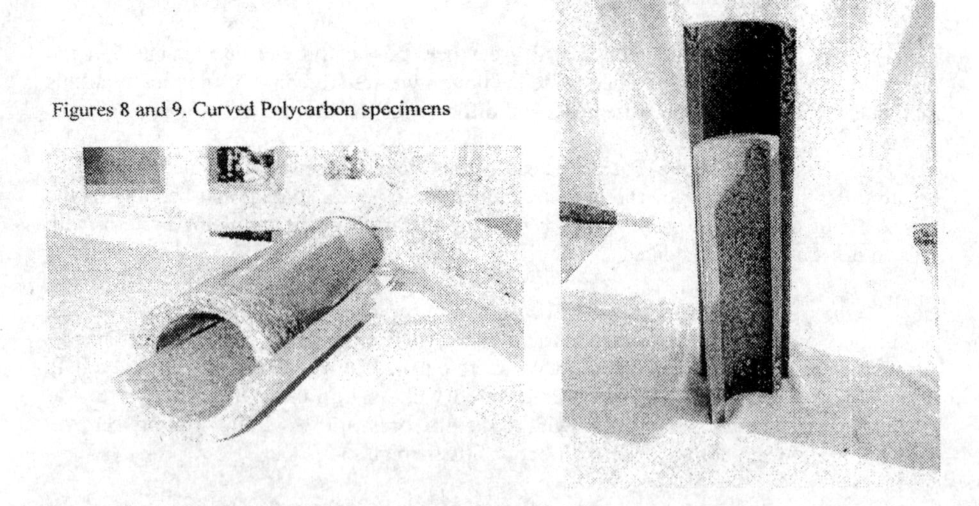

A lot of curved shapes can be obtained using different casts of different diameters and thickness (figures 8 and 9). So through the combination between these and flat elements we will solve various building designs (figure 10).

Figure 10. Moulding posibilities of curved and flat *Polycarbon* specimens

The pipe itself has shown very good qualities when tested to be used under pressure. Besides, the aluminium coat makes the piped water still drinkable. The characteristics of this pipe are similar or even better than other pressure pipes made of conventional materials. So our pipe can take advantage from conventional ones and with a lower price.

The investigation is still on development. The results mentioned before are only the very first steps on it. So ecological, anti-pollution interests and new job and investment opportunities might be combined.

References

[1] *Laffarga Osteret, J. y Olivares Santiago, M.* RECICLADO DE LOS PLÁSTICOS. UNA PRODUCCIÓN ECOLÓGICA DE LOS MATERIALES COMPUESTOS. Libro de Actas de las IV Jornadas sobre Aplicaciones Arquitectónicas de los materiales compuestos y aditivados. Madrid, Abril de 1995.

[2] *Alvarez Jurado, J.L, Laffarga Osteret, J. y Olivares Santiago, M.* MATERIALES COMPUESTOS DE PLÁSTICOS RECICLADOS. Proceedings I International Conference ARQUIMACOM'96. Sevilla, October 1996. pp: 307-310.

NONLINEAR BEHAVIOUR OF FIBRE-REINFORCED CONCRETE STRUCTURES

A. Grimaldi*, R. Luciano**

*Department of Civil Engineering - University of Roma Tor Vergata
**Department of Industrial Engineering - University of Cassino

Abstract

The non linear behaviour of concrete structures reinforced with short diffused steel fibres and concentrated bars is investigated. The influence of fiber reinforcement on the strength and ductility of the concrete under tensile load is studied by modelling the material as a composite with brittle matrix and elastic fibres. Then, by applying the classical principles of limit analysis, upper and lower bounds on the overall tensile strength are obtained. Finally, a two-dimensional model of concrete matrix reinforced with fibres aligned in the tensile direction is presented.

1. Introduction

The possible applications of concrete reinforced by short fibres and the study of its mechanical behavior is an important issue in the field of the Civil Engineering. The most difficult problem is to obtain the constitutive equations of the concrete reinforced with fibres in the linear and nonlinear phase. This material can be considered as a composite made by a matrix of concrete with a brittle behavior and by steel short fibres randomly distributed in the matrix with a volume fraction between 1% and 2%. The aim of such a reinforcement is to improve the mechanical behavior of the material when tensile forces are acting. Infact, a limit tensile strength with ductile behavior is ensured to the material. The tensile uniaxial behavior of the composite is usually schematized by the constitutive relationship represented in figure 1. This behavior has been observed in many experiments (see [5]).
If the material is loaded by an uniaxial strain, it is possible to observe the following characteristic phases of the behavior of the material:
a) Initial linear elastic phase. It is characterized by an elastic modulus and by the peak strength of the concrete σ_{mt}.
b) When the limit strain ε_c of the matrix is attained, the composite changes configuration for values of the strain greater than the limit one. Furthermore, cracks develop in the matrix and the resulting stress is sustained by the composite material made of the cracked concrete and the fibres.
c) The value of the stress σ_r at damaged conditions, represents the minimum value of stress for strains greater than ε_c. As a matter of fact, increasing the applied strain, several damage phenomena can occur: slipping between fibre and matrix or plastic compressive strains in the matrix. The failure condition due to one of the two previous damage phenomena is attained for values of the strain quite greater than ε_c but with ultimate stress close to σ_r. This experimental observation justifies the use of the constitutive behavior shown schematically in figure 1 [5].

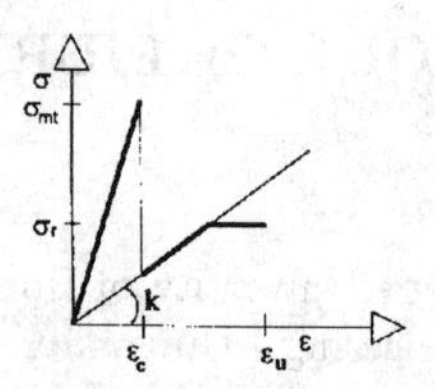

Figure 1: Tensile behavior of the f.r.c.

For this reason, the value of the stress σ_r, represents the maximum stress corresponding to a sufficiently ductile material behavior. Because of such a ductile behavior the performance of structural elements in f.r.c. improves considerably.
As matter of fact, thought the residual stress σ_r is less than the traction strength of the matrix, in the case of structures, for example bending beams, the presence of the fibers can increase the bending strength of the structural element. This behavior is well known from experiments [5] and it is shown in figure 2. In this picture the experimental results obtained for beams subject to bending loads and made with concrete reinforced with fibre volume fractions between 0.5% and 1.5% are reported. It is worth noting that the limit bending moment of the beam grows by increasing the fiber volume fraction and, in several cases, it can became greater than the limit bending moment of beams with unreinforced concrete.
In summary, the use of fibre reinforcement increases the strength and the ductility of concrete structures and, in order to model such a behavior, it is very important to evaluate the residual stress σ_r.
Aim of this work is the evaluation of σ_r by using micromechanics. Initially this task is reached by modelling the concrete as a no tension material [3] and the fiber as linear elastic. The no-tension material is characterized by elastic strain in the compressive stress directions and by cracks in the zero stress directions. Then, by assuming for the matrix a limit value for the compressive stress and a rigid plastic behavior at the interface between fibres and matrix, the global strength of material reinforced by elastic fibres can be obtained by using homogenization methods and the formulations of limit analysis theorems for no tension materials [2], i.e. by using techniques similar to the ones used in [7] and [6].
In the case of periodic arrangement of the fibres, the analysis can be developed by using the finite element method [4] or the cell method [1]. In this work a two-dimensional (2-D) model characterized by rectangular fibres is presented.

2. Strength of the material subject to tensile load

The model adopted in this work is based on the following geometric and constitutive hypotheses: the element of concrete considered in the analysis is reinforced by short fibres parallel to the direction of the imposed strain or stress. The fibres are disposed in a periodic arrangement with staggered pattern (see figure 3). Then the disposition

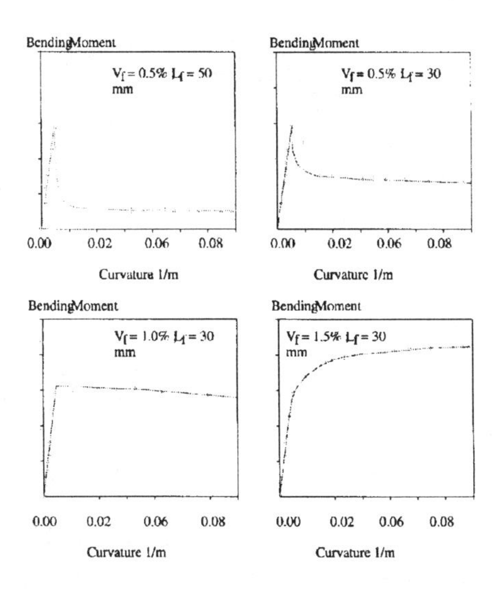

Figure 2: Behavior of beams subject to bending

of the fibres is defined by the fibre length l, the distance between two fibres d and their overlapping s.
In order to determine the overall stiffness of the f.r.c. in the case of a cracked matrix, the constitutive relations adopted for the two components are:
Matrix : non linear elastic relation with no tensile strength defined by the following relations:

$$\sigma \leq 0 \text{ and } |\sigma_i| \leq \sigma^0, \; \varepsilon = B\sigma + \eta, \; \eta \geq 0, \; \sigma \cdot \eta = 0 \tag{1}$$

where σ is the Cauchy stress tensor, ε is the strain, η is the positive strain representing the cracking in the material, B is the compliance tensor of the material in compression, σ_i are the local compression stresses in the matrix and σ^0 is the compression strength of the material.
Fibres: uniaxial linear constitutive relation $\sigma = E_f \varepsilon$
Bonding fibres and matrix : perfect plastic constitutive behavior at the interface with a limit value of tangential stress $|\tau| \leq \tau^0$
For the special geometry and symmetry of the problem, in the following, the unit cell shown in figure 4 is considered, where e_1 is the direction of the fibres. The unit cell is subjected to a mean strain tensor $\mathbf{E}$, with the component $E_{11} > 0$ and the other components equal to zero. Similarly $\mathbf{\Sigma}$ indicates the corresponding mean stress tensor in the unit cell i.e. $\mathbf{\Sigma} = \langle \sigma \rangle = \frac{1}{U} \int_U \sigma dV$, where U is the volume of the unit cell. Further, let $u(x), \varepsilon(x)$ and $\sigma(x)$ denote the displacement, strain and stress

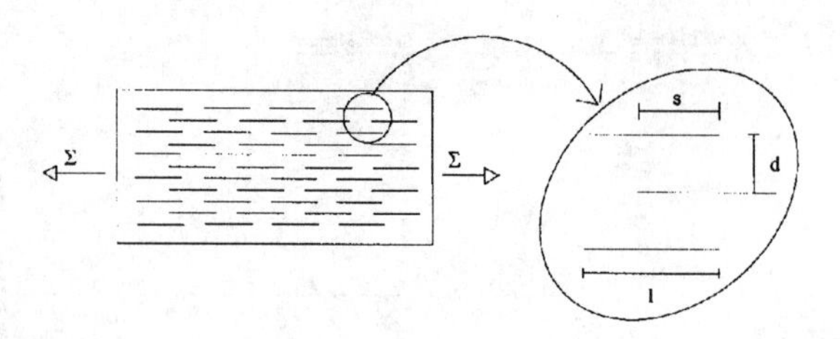

Figure 3: Periodic arrangement of the fibres

field in the unit cell, respectively. Then, for the special geometry of the unit cell and the applied strain, the local displacement field along the boundary of the unit cell is linear i.e. : $u(x) = \mathbf{E}x \; \forall x \in \partial U$. In this case, the mean strain in the unit cell is $\langle\varepsilon\rangle = \frac{1}{U}\int_U \varepsilon dV = \mathbf{E}$.

Increasing the mean strain applied to the material, it is possible to reach a failure configuration. The aim is to evaluate the limit strength Σ_{11}. The failure state is characterized both/either by stresses which are equal to the compressive strength of the matrix and/or by slipping between matrix and fibre. In order to determine the failure mean stress Σ_{11} of the material it is necessary to evaluate the plastic dissipation functional i.e.

$$\mathbf{D}(\varepsilon(u)) = \frac{1}{U}(\sum_i \int_U \sigma^0 \varepsilon_i^- dV) + \int_{\partial V_f} \tau^0 \gamma dA$$

where ε_i^- are the negative principal components of the strain and γ the slipping between fibre and matrix. Therefore, the ultimate failure condition corresponds to the minimum condition of the dissipation functional $\mathbf{D}(\varepsilon(u))$:

$$D(\mathbf{E}) = \min_{u\in C'} \mathbf{D}(\varepsilon(u)) = \min_{u\in C'} \frac{1}{U}(\sum_i \int_U \sigma^0 \varepsilon_i^- dV) + \int_{\partial V_f} \tau^0 \gamma dA \qquad (2)$$

where C' is the space of the admissible displacements, then, the ultimate stress Σ_{11} is equal to:

$$\Sigma_{11} = D(\mathbf{E})/E_{11} \qquad (3)$$

The dual approach is formulated as maximum of the dual functional of $\mathbf{D}(\varepsilon(u))$ defined as:

$$D^c(\mathbf{E}) = \max_{\sigma\in G} \mathbf{D}^c(\sigma, \mathbf{E}) = \max_{\sigma\in G} \left[\langle\sigma\rangle .\mathbf{E} - \chi_1(\sigma) - \chi_2(\tau)\right] \qquad (4)$$

where $\chi_1(\sigma)$ and $\chi_2(\tau)$ are the indicator functions of the sets $\left|\sigma_i^-\right| \leq \sigma^0$ and $\tau \leq \tau^0$ and G is the space of the nonpositive and selfequilibrated stress fields.

Then, the limit mean stress of the material can be evaluated by using the following relation:

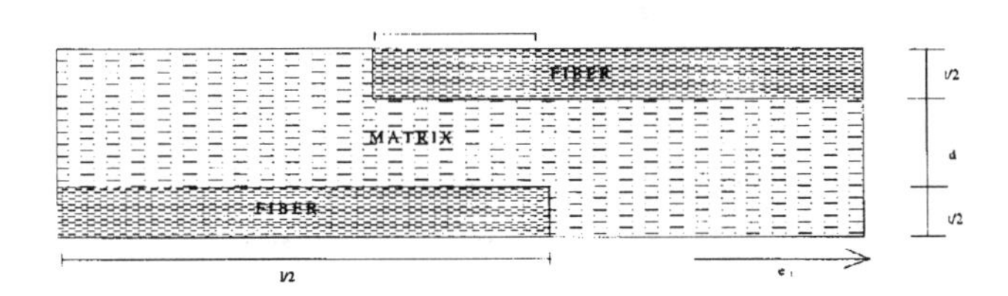

Figure 4: Unit cell

$$\Sigma_{11} = D^c(\mathbf{E})/E_{11} \tag{5}$$

The application of the statical and kinematical limit analysis theorems (2) and (4) provides upper and lower bounds for the tensile limit strength Σ_{11} of the material

3. Applications

A simple application of the results previously presented is the two dimensional model represented schematically in figures 3 and 4. It corresponds to a thin plate of concrete reinforced by rectangular rigid fibres with length l and thickness t. The concrete plate and the fibres have unit width. The unit cell considered in the numerical applications has been analyzed by using a technique similar to the cell method proposed in [1]. In particular, the displacement and stress formulation previously presented have been applied to the unit cell represented in figure 4 by using the displacement distributions depicted in figure 5 and 7 for the dissipation functional formulation and the stress distribution represented in figure 6 for the complementary dissipation functional formulation. These distributions are used to obtain simple expressions for the bounds on the overall limit strength Σ_{11}.
We consider first the displacement mechanism of figure 5, where plastic compressive strain occurs in the central zone subjected to the shear γ^*. The dissipation $\mathbf{D}$ is:

$$\mathbf{D} = \frac{1}{U}\sigma^0\varepsilon^- sd = \frac{\sigma^0}{2U}s(l-s)E_{11} \tag{6}$$

therefore the upper bound Σ_{11}^+ is:

$$\Sigma_{11}^+ = \frac{\mathbf{D}}{E_{11}} = \frac{\sigma^0 s}{2d} \tag{7}$$

In order to obtain another upper bound on the strength it is possible to consider the micromechanism characterized by sliding between fibers and matrix (see figure 7). In this case the dissipation $\mathbf{D}$ is

$$\mathbf{D} = \frac{2}{U}\tau^0(s)u = \frac{\tau^0 s}{d}E_{11} \tag{8}$$

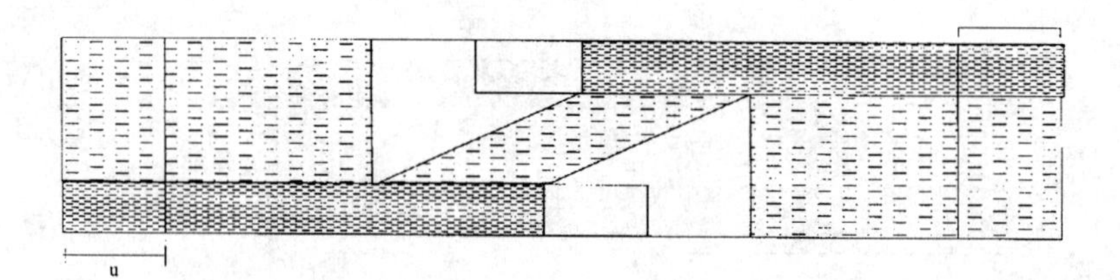

Figure 5: Shear displacement micromechanism

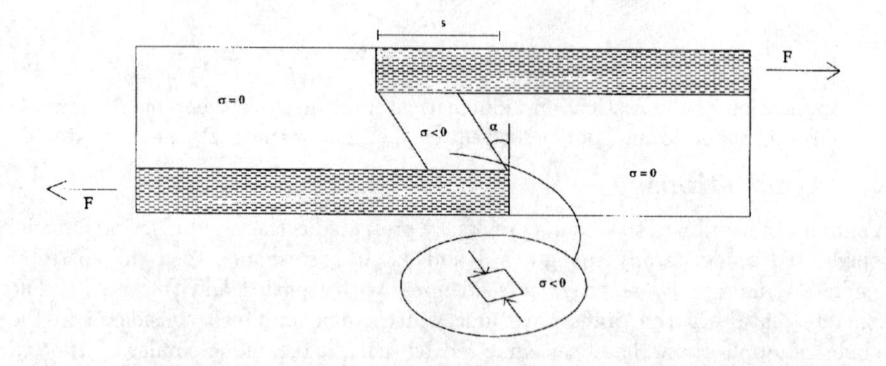

Figure 6: Stress distribution

therefore we get the following upper bound:

$$\Sigma_{11}^{+} = \frac{\mathbf{D}}{E_{11}} = \frac{\tau^0 s}{d} \tag{9}$$

Usually τ^0 is much lower than $\sigma^0/2$ and therefore eq. (9) gives the upper bound Σ_{11}^{+}.

In the following, the evaluation of a lower bound on the strength of the model is obtained by using the static method. In the computations the distribution of stresses shown in figure 6 is considered. The distribution of figure 6 presents stress different from zero only in the central zone inclined of the angle α. According to eq. (4) and assuming $|\sigma^{-}| \leq \sigma^0$ and $\tau \leq \tau^0$ we can write the functional $\mathbf{D}^c$ as:

$$U\mathbf{D}^c = \int_{\partial U} \sigma n.\mathbf{E} x dA = 2Fu = 2\tau(s - d\tan(\alpha))E_{11}\frac{(l-s)}{2}$$

then by performing the following maximization:

$$\max_{\substack{\sigma \in G \\ |\sigma^-| \leq \sigma^0 \\ \tau \leq \tau^0}} U\mathbf{D}^c = \Sigma_{11} E_{11} U$$

it is possible to obtain a lower bound on the strength Σ_{11}. We have to consider two cases: the first case is characterized by failure of the concrete in compression while the second case corresponds to sliding of the fibers. In the first case, $\sigma^- = \sigma^0$ and $\tau = \sigma^0 \frac{\sin 2\alpha}{2} \leq \tau^0$ then, by performing the following maximization with respect to α:

$$\max_\alpha U\mathbf{D}^c = \max_\alpha \sigma^0 s (1 - \frac{\tan(\alpha)}{\tan(\beta)}) \sin(2\alpha) E_{11} \frac{(l-s)}{2} \quad (10)$$

the maximum is attained for $2\alpha = \beta$ where β is such that $tg\beta = s/d$, therefore:

$$\Sigma_{11}^- = \frac{\mathbf{D}^c}{E_{11}} = \frac{\sigma^0 s}{2d}(1 - \frac{\tan(\beta/2)}{\tan(\beta)}) \sin(\beta) \quad (11)$$

which can be rewritten as:

$$\Sigma_{11}^- = \frac{\tau^0 s}{d} \frac{\sin(\beta)}{\sin(\overline{\beta})}(1 - \frac{\tan(\beta/2)}{\tan(\beta)}) \quad (12)$$

by denoting $\sin(\overline{\beta}) = 2\frac{\tau^0}{\sigma^0} < 1$. It is worth noting that the bound (12) is valid for $\tau \leq \tau^0$ and therefore for $\sin(\beta) \leq \sin(\overline{\beta})$ i.e. $\beta \leq \overline{\beta}$.
In the second case, σ^- is less than σ^0 and $\tau = \tau^0$ then, by performing the following maximization with respect to α :

$$\Sigma_{11}^- = \frac{\mathbf{D}^c}{E_{11}} = \max_\alpha \frac{\tau^0 s}{d}(1 - \frac{\tan(\alpha)}{\tan(\beta)}) \quad (13)$$

we obtain a lower bound Σ_{11}^- on Σ_{11}. In this case the maximum is attained for $2\alpha = \overline{\beta}$, then:

$$\Sigma_{11}^- = \frac{\tau^0 s}{d}(1 - \frac{\tan(\overline{\beta}/2)}{\tan(\beta)}) \quad (14)$$

which is valid for $\sigma^- = 2\frac{\tau^0}{\sin(2\alpha)} \leq \sigma^0$ or, in other terms, for $\beta \geq \overline{\beta}$.
In conclusion the results obtained for the two cases: $\beta \leq \overline{\beta}$ (failure for compression in the concrete) and $\beta \geq \overline{\beta}$ (failure for sliding of the fiber) can be summarized as follows:

$$\begin{aligned} \frac{\Sigma_{11}^-}{\Sigma_{11}^+} &= \frac{\sin(\beta)}{\sin(\overline{\beta})}(1 - \frac{\tan(\beta/2)}{\tan(\beta)}) \qquad \beta \leq \overline{\beta} \\ \frac{\Sigma_{11}^-}{\Sigma_{11}^+} &= (1 - \frac{\tan(\overline{\beta}/2)}{\tan(\beta)}) \qquad \beta \geq \overline{\beta} \end{aligned} \quad (15)$$

The equation 15 provides good approximation of the strength Σ_{11} as β grows.

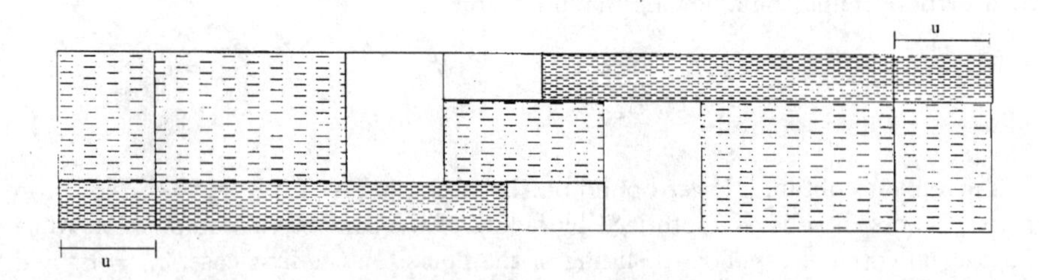

Figure 7: Sliding displacement micromechanism

In conclusion, the methodology proposed in this work is very useful to understand the actual failure micromechanisms of the material and to evaluate the residual stress σ_r as function of the arrangement and the volume fraction of the fibres. It is worth noting that the same procedure developed to obtain the tensile strength of the material for the 2D case can be adopted for the 3D case.

Aknowledgments
The financial support of Ministry of University and Research (MURST) is gratefully acknowledged.

References

[1] Aboudi, J. (1991). Mechanics of Composite Materials, Elsevier.

[2] Como, M. and Grimaldi, A. (1985). A unilateral model for the limit analysis of masonry walls. In Unilateral Problems in Structural Analysis (Edited by. G. Del Piero and F. Maceri), CISM Courses and Lectures, Vol. 288, 25-45 Springer-Verlag.

[3] Giaquinta, M and Giusti, E. (1985). Researches on the equilibrium of masonry structures. Arch. Rat. Mech. Anal 88, 359-392.

[4] Grimaldi, A. and Luciano, R. (1995). Modellazione micromeccanica e valutazione della resistenza a rottura di calcestruzzi rinforzati con fibre. XII Congresso AIMETA, Napoli, 3-6 October.

[5] Lim, T. Y., Paramasivam, P., and Lee, S. L. (1987). Analytical model for tensile behavior of steel-fiber concrete. ACI Materials Journal July/August, 286-298.

[6] Suquet P. (1993). Overall potentials and flow stresses of ideally plastic or power law materials. J. Mech. Phys Solids 41, 981-1002

[7] Willis J. R. (1991). On methods for bounding the overall properties of nonlinear composites. J. Mech. Phys Solids 39, 73-86.

Rehabilitation of corrosion damaged columns with CFRP

M. Blaschko *, K. Zilch**

* Bilfinger + Berger BauAG, Munich, Germany; ** Technische Universität München, Munich, Germany

Abstract

In Germany concrete columns in parking garages are often damaged due to corrosion of the steel reinforcement. The corrosion is induced by deicing salts and moist environment. Concrete columns with damage of this kind have been repaired by wrapping the columns with Carbon Fiber Reinforced Polymers (CFRP) sheets. CFRP sheets are used to replace the corroded horizontal binder reinforcement as well as to protect the concrete against new penetration of deicing salts and moisture and slow down or even stop further corrosion.

1. Introduction

The durability of concrete structures is often limited by corrosion of the steel reinforcement. The corrosion process reduces the load carrying cross section of the steel bar and the increase in volume of rust spalls off the concrete cover. The time until this corrosion process starts is basically related to the thickness of the concrete cover and to environmental conditions. Moist environment and especially deicing salts accelerate this corrosion process.
Concrete columns in parking garages are often damaged due to these effects of corrosion of the steel reinforcement in Germany and in many other countries with snow and ice in winter. Cars bring snow including deicing salts into the garages. Bad drainage keeps chloride polluted water on the foot of the column. According to former specifications the concrete cover is only about 2 cm. In some cases the concrete cover is even smaller because the reinforcement cage moved during the vibrating process. So it takes only about 20 years after constructing till this corrosion process starts.

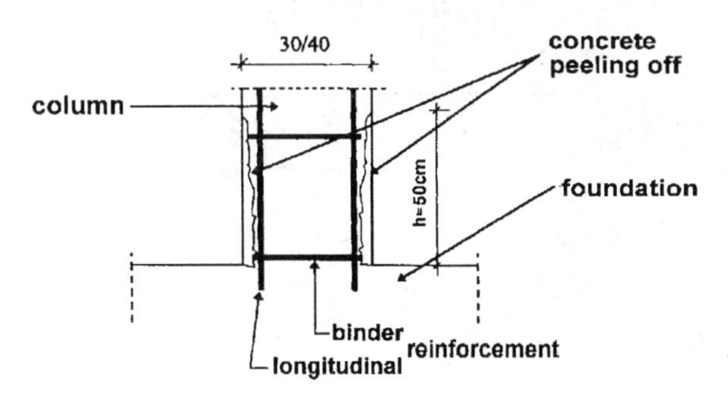

Fig. 1: Typically damaged concrete column

Fig. 1 shows the bottom section of a typically reinforced concrete column about 25 to 30 years after constructing. The damaged areas are from the bottom of the column as far as about 50 cm above the ground.
In most cases only the binder reinforcement, which has got the lowest concrete cover, is really badly damaged. The longitudinal reinforcement is often still in an acceptable condition with fewer corrosion defects.

2. Former rehabilitation methods

Two main targets have to be considered for the rehabilitation process:

- Reconstruct the load carrying capacity of the column
- Stop further damages resp. corrosion

Up to now the shotcreting method has been used in almost all the cases for the rehabilitation of columns [2]. This method includes that damaged and spalled off concrete is removed, e.g. with high pressure water jeting. Corroded reinforcement is replaced by new reinforcement. The column is remodelled with shotcrete. As the bond strength between shotcrete and the original concrete is very high, the total cross section of the column can be seen as one load carrying element. To stop further corrosion not only unbonded but also chlorid containing concrete has to be removed. The column is remodelled with a larger cross section, if there is a lack in concrete cover (Fig. 2). Furthermore the concrete surface is coated with a protecting paint, e.g. on a epoxid resin basis.

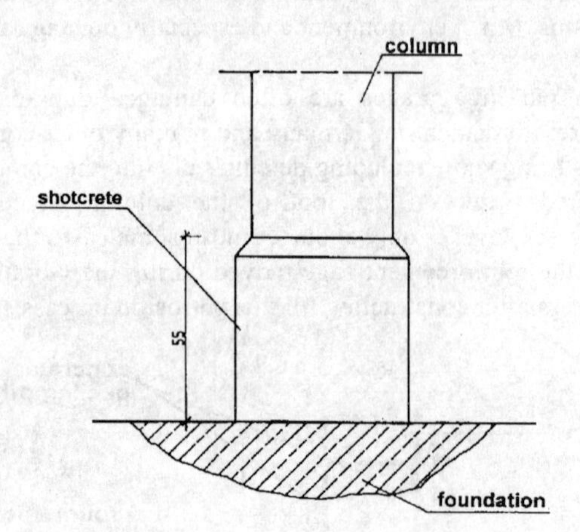

Fig. 2: Concrete column rehabilitated with shotcrete

3. Rehabilitation method with CFRP

A new method for the rehabilitation of corrosion damaged concrete columns is to wrap with Carbon Fiber Reinforced Polymers (CFRP) sheets (Fig. 3). This method has been

used in many cases to strengthen columns especially for seismic loads [1,3,4]. But in the case of rehabilitation the needs and so the application process are different.
During the application process the bondless, spalled off concrete is removed. It is also necessary to remove this concrete because of its high chlorid concentration. The corroded reinforcement is derusted by sand blasting. If the binder reinforcement is badly damaged or the concrete cover ist too small, the binder reinforcement will be taken away. The column is remodelled with PCC mortar to have an alkaline environment for the protection of the steel reinforcement which still exists.

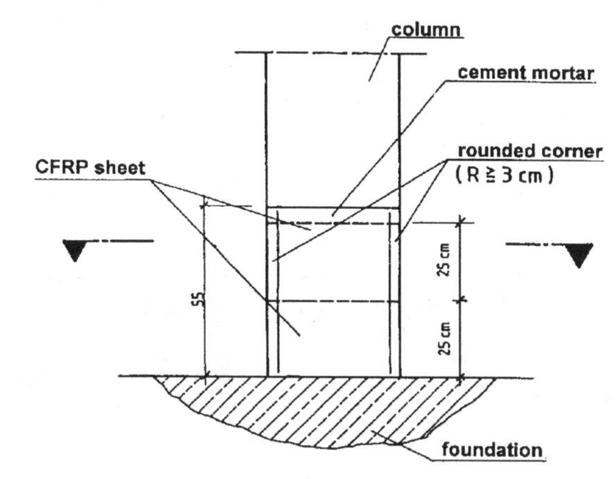

Cross Section:

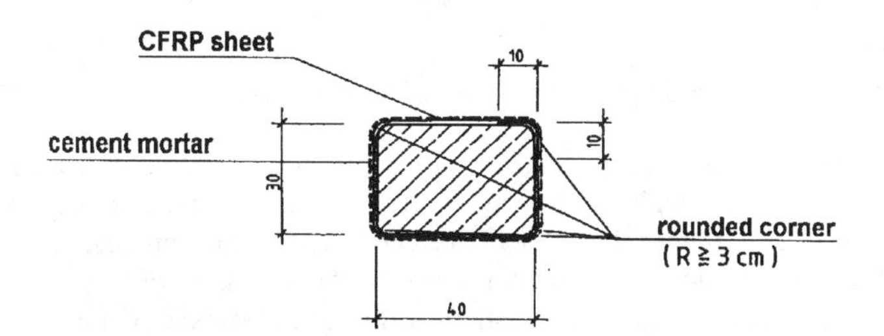

Fig. 3: Concrete column rehabilitated with CFRP

CFRP sheets are wrapped around the column to replace or strengthen the horizontal binder reinforcement. First a primer is applied on the concrete surface. Then the sheets are glued to the concrete body with epoxid resin. This confinement is also necessary to create radial compression forces to make sure that the remodelling mortar carries longitudinal loads without debonding. The edges are rounded by grinding with a radius of about 3 cm to avoid damages of the sheet at sharp edges and to reduce stress concentrations at the corner. Both ends of the sheets are linked together with a total overlapping length of 20 cm at the corner of the column.
The epoxy resin protects the concrete against new penetration of deicing salts and moisture. This is supposed to slow down or even stop further corrosion of the steel

reinforcement. But the CFRP sheet should not be applied over the total height of the column to make an exchange of moisture possible.
Quarzite sand is sprinkled into the still fresh surface of the epoxid resin to form a key for a thin layer of cement mortar which is used to protect the CFRP sheet against small impacts and sabotage. Furthermore the surface is similar to the original surface of the concrete column.

This method was used for the first time in Germany in a parking garage in Munich (Fig. 4). In this case 18 columns with a cross-section of 30 cm by 40 cm were wrapped at a height of 50 to 75 cm.

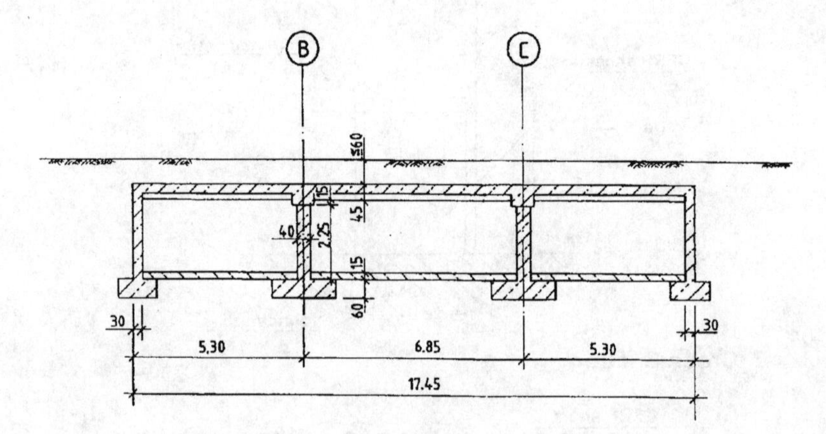

Fig. 4: Cross section of parking garage

This application showed that the CFRP rehabilitation method is very economical. The less concrete has to be replaced the more economical this method is. So it is an important issue, how much chlorid can be accepted in the concrete cover so that corrosion does not start again. This is connected to the moisture content in the column. The moisture content is related to the transferability of the epoxid coating as well as to the boundary conditions of the column. So it seems to be difficult to predict the long-term behaviour of the retrofitted column. Long-term observation of the retrofitted columns in the parking garage is supposed to give this information.

Another important aspect is the fire resistance of the retrofitted column. Because the adhesive becomes weak at temperatures over about 70 °C, the CFRP wrapping does not carry any load soon after the beginning of a fire. A fire protective coating has to be very thick, if it is to give a fire resistance time of 30 minutes or more. In this case the structural safety of the unretrofitted column is greater than one. That means the column is still able to carry load in the case of fire, when the CFRP wrapping is not working. So no fire protecting system had to be applied here. But more research is necessary to evaluate different protecting systems and design rules for fire.

4. Conclusions

The described method of rehabilitation of concrete columns with CFRP is an economical method compared to other rehabilitation methods like shotcreting. This was shown in an application in a parking garage in Munich, Germany.
The application process has to be adapted to the needs of stopping corrosion. So it is e.g. necessary to reduce the moisture content in the column or to protect the CFRP sheet against small impacts from cars. Because of different related corrosion process parameters, it seems to be difficult to predict the long-term behaviour of the retrofitted column. The behaviour of the retrofitted columns in the future has to answer this question.
Another issue of great importance is the fire resistance of a CFRP retrofitted column. More research is necessary to evaluate this.

5. References

[1] *Fardis, M.; Khalili, H.*: CONCRETE ENCASED IN FIBERGLASS-REINFORCED PLASTIC; ACI Journal, Vol. 78, No. 6, November-December 1981, pp. 440-446.
[2] *Held, M.; Blaschko, M.*: INSTANDSETZUNG VON TIEFGARAGEN: PLANEN - AUSSCHREIBEN - AUSFÜHREN; Vortrag bei GEB-Arbeitstagung „Betoninstandsetzung - eine betontechnische Herausforderung“ (in German), München, 07.11.1997.
[3] *Ono, K.; Matsumura, M.*: STRENGTHENING OF REINFORCED CONCRETE BRIDGE PIERS BY CARBON FIBER SHEET; Composite Construction - Conventional and Innovative, Innsbruck, 16.-18.09.97.
[4] *Saadatmanesh, H.; Ehansi, M. R.; Li, M. W.*: STRENGTH AND DUCTILITY OF CONCRETE COLUMNS EXTERNALLY REINFORCED WITH FIBER COMPOSITE STRAPS; ACI Structural Journal, Vol. 91, No. 4, July-August 1994, pp. 434-447.

A NEW POLYMERIC MORTAR MADE OF ALBERO

Olivares Santiago, M.
Doctor arquitecto. Prof. Titular Depto. Construcciones Arquitectónicas I. E.T.S.A. Sevilla. Spain.
Laffarga Osteret, J.
Doctor en Químicas. Prof. Titular Depto. Construcciones Arquitectónicas I. E.T.S.A. Sevilla. Spain.
Galán Marín, C.
Arquitecto. Prof. Asociado Depto. Construcciones Arquitectónicas I. E.T.S.A. Sevilla. Spain.
Martineaux, Ph.
Doctor en Ciencias de los Materiales. Institut des Matériaux Composites. Bordeaux. France.

Abstract

This paper describes a new special polymeric mortar made of *Albero* fines and resin. *Albero* is a special Sevillian name for a non-common lime-sandstone kind of rock. This rock is found just in a very determined area in the province of Seville.

The main objective is to develop a polymeric mortar whose single fine aggregate will be *Albero*. This mortar would be produced in slabs or panels of various sizes and thickness to be used for different applications in the building industry. So the architects will have at his disposal a native Andalusian product with a very adequate aspect. The applications can be more numerous if their physical, chemical and technical characteristics are improved.

1. Introduction

Since the very beginning of the studies about the possibilities and uses of polymeric mortars and concrete (PC) they have been interested in the kind of fines added. Waste fines or scarcely used products are used so as to make the mixture cheaper.

In 1966 some studies about using polymeric mortars with several mineral loads [1] and serpentine sand (asbestos powder less than 1 mm. Size) [2] had already been done. The mixture had less than 14% of UP resin but the tests performed mechanic strength of 100 MPa and 25 MPa for compression strength and bending tests respectively.

After these first results, different experiments with several fine mixtures have been carried out.

2. A rock named *Albero*

In the province of Seville (Spain) is possible to find very different soils and rocks dated on nearly every geological ages [3]. Although some of these rocks might found almost everywhere in Spain, the rock named *Albero* can only be found in Seville. This communication is mainly concerned on this rock. Firstly we will explain the origin of this rock layer.

The geological mass that formed *Los Alcores* (the name of the area in which *Albero* is located) is very peculiar. This geological mass is composed of detritus and fossil limestone, very rich in mollusc shells, Briozoos y equinidus. These fossils are generally bad preserved, withdrawn during the retrogression of the oceans in that era. Limestone lies over a stratum of blue loam, which can be seen in some places.

Another of it characteristics is the abundant presence of quartz grain and the absence of clay. This type of rock is of variable compaction, sometimes sandy and always very rough. Its is ochre or yellow coloured.

Due to its chemical composition with a high proportion of limestone, *Albero* has been used for the very last years as row material for a cement industry and to produce sulphate-resist cements [4].

Figure 1. PC Albero mortar cylindrical specimens

According to the results of the tests done by the authors, in table I are shown the characteristics and properties of *Albero* rocks.

Components in % (weight)		Diverse properties	
Quarcite (such as SiO_2)	5,6 to15,6	Net Weight	1,94
Aluminium composites (such as Al_2O_3)	0,3 to 1,7	Compression resistance (MPa)	61 to 65
Iron composites (such as $Fe_2 O_3$)	0,8 to 1,9	Flexion resistance (MPa)	10 to 12
Limestone (such as CO_3Ca)	80,0 to 88,3	Absorption	0,2 to 0,6
Calcination loses	38,0 to 41,1	Freezing resistance in % of loss of weigh	0,1 to 0,4
		Erosion resistance in mm.	1,2 to 1,6
		Alterations in the surface	Not relevant

Table I. Characteristics of the *Albero* rock used.

In a few words, the *Albero* rock, apart from a great percentage of limestone is composed of a great percentage of quartz with an absence of clay but not a big proportion of iron minerals in different degrees of rustiness. This iron gives the rock its characteristic colour and shade, also named *Albero* yellow in the area.

3. The using of *Albero* rock as a source of building material

Albero, also named *Alcalá* limestone is a building material that has been used for masonry works for centuries. Let us say as an example that the Roman Empire used *Albero* rock in the city of Italica (the capital of the Roman Province of Baetica). Also did the Arabs in the Sevillian Palace Alcazar [5]. But *Albero* rock has also been a very appreciated row aterial for lime and nowadays for common and sulphate-resistant Portland cements. One of the most important Spanish cement factories is placed next to the *Albero* stone quarry in the city of Alcalá de Guadaira (Seville).

Between all this applications one is specially appreciated and used: the *Albero* rock, once crushed with different thickness of grain, it is spread out in roads and path as a pavement material. Its mechanic characteristics, draining properties and dump stability are widely known. *Albero* is also laid as a basement for road construction, and most fine parts are used in pedestrian pathways in parks and gardens.

Thanks to all the qualities above mentioned (stability and draining characteristics mainly), it has always been a very emblematic material in public and private Sevillian spaces.

Albero soil is well known and used not only in Seville and Andalusia but also exported. The pavement of the Plaza de Toros de la Real Maestranza de Caballería (the bullring, one of the most beautiful buildings in Seville) is made of *Albero*, and has been imitated in many other bullrings in Spain. The same material is used in the pavement of the Feria de Abril, pathways and public gardens and the paths of he recent International Exhibition Expo'92.

4. PC Mortars of *Albero* and Resin

PC mortars have been used as an imitation of expensive rocks such as granite and marble for some time ago [6]. There is not too much information about the technical characteristics of these PC just for technical and rivalry reasons.

A study of the possibilities of agglomerate of *Albero* and resin used as a building material, was done (figure 1). This material could be very appreciated by architects and decorators due to the original colour of *Albero*.

As a first step, a chemical and granular analysis of the row material investigated was don. The results are shown in table II, III and IV.

Components	%
Silica (SiO_2)	35.7
Ferric Oxide (FeO_3)	0.7
Alumina (Al_2O_3)	0.6
Calcium Oxide (CaO)	34.9
Magnesium Oxide (MgO)	0.0
Sulphate (SO_3)	0.0
Fire Loss	28.1

Table II. Chemical analysis of the *Albero* rock used (UNE EN 459-2/95).

Liquid limit	Not plastic
Plastic limit	Not plastic
Plastic ratio	Not plastic

Table III. Atterberg limits (NLT 105 AND 106/91).

Test	Result
Absorption	0.37%
Specific weight	1.79
Friction erosion	1.61 mm.
Frost resistance	0.16%
Compression resistance	59,65 MPa
Flexion resistance	11.34 MPa
Impact resistance	96.45 cm.
Loss of weight under temperature changes	0.76 %
Surface changes under temperature changes	No alteration

Table IV. Technical characteristics of the polymeric mortar made of *Albero*.

These works are still on development in the laboratories of the E.T.S. Arquitectura in Seville (figure 2 and 3), but the fist significant results are already available so the author present this paper to this International Conference.

Figure 2. PC Albero mortar cylindrical specimens and iron casts

Different kinds of resins, catalysts and percentages have been tested. Once the most adequate formulation has been stabilised, laboratory tests have been done to determine their technical characteristics.

We can also list some other characteristics of *Albero* polymeric mortar:

- Uniform and permanent colour
- It doesn't suffer from ageing, rustiness, strain or rottenness.
- Can be polished
- Can be drilled, glued, painted , varnished and screwed
- It has no fissures
- It is homogeneous and unalterable under atmospheric agents even sea saltpetre
- It is a very interesting alternative to the use of timber, so from the ecological point of view can be an important material

To summarise, the characteristics of the *Albero* polymeric mortar manufactured and the final appearance of the specimens let us expect that it will be easily used as a coating layer or in pavement tiles. The materiel tested has a good adherence with Portland mortars and concrete surface, been just a few millimetres thick. The main applications can be very widespread and important for the building industry and decoration:

- Vertical surfaces.
- Façade walls.
- Pavements.
- Urban furniture.
- Decoration in general.

Figure 3. PC *Albero* mortar prismatic specimens

References

[1] *Colonna-Ceccaldi, J.* MORTIERS ET BÉTONS DE RÉSINES. Procc. Colloque de la RILEM:Les résines de synthèse dans la construction". Paris, sct. 1967, pp344-354.]

[2] *Hahamovis.J.* & Knezevic. N. PROPIÉTÉS DU BÉTONS DE RÉSINE DE POLYESTER ET DE DÉCHETS D'ASBESTE. Procc. Colloque de la RILEM:"LES RÉSINES DE SYNTHÈSE DANS LA CONSTRUCTION". Paris, sept. 1967, pp 373-379.

[3] *San Miguel de la Cámara, Maximinio.* "GEOLOGÍA. MINERALOGÍA Y NOCIONES DE GEOQUÍMICA". Impresiones C. Bermeo. Madrid, 1957

[4] *Barrios Sevilla, J.* PREPARACIÓN Y ESTUDIO DE UN CEMENTO PAS A PARTIR DE *ALBERO* DE ALCALÁ DE GUADAIRA (SEVILLA). Tesis Doctoral inédita. Universidad de Sevilla, 1.975

[5] *Gonzalez Vila, V.* ARTÍCULO DE LA ENCICLOPEDIA DE ANDALUCÍA. Tomo 2, Cementos

[6] *Sandrolini,F. Y Tassone, P.* NEW ITALIAN (UNI) STANDARDS FOR TESTING POLYMERIC AGGLOMERATED NATURAL STONE FOR WALL AND FLOOR TILES. Procceeding. Symposium RILEM on Propierties and Test Methods for Concrete-Polymer Composites. Ostende, jul. 1995, pp. 47-62.

Mechanical properties of lightweight cement based composites containing polyurethane foam scraps

A. Pompo*, S. Iannace°, L.Nicolais°

* Department of Environmental Physics and Engineering, University of Basilicata, Via della Tecnica 3, 85100 Potenza, Italy

°Istitute for Composite Materials Technology (ITMC-CNR) & Department of Materials and Production Engineering, University of Naples "Federico II", P.le Tecchio 80, 80125 Naples, Italy,

Abstract

The influence of rigid polyurethane foam (PU) particles on mechanical properties of cement based composites is reported in this paper. Three points bending mechanical tests were performed and properties were monitored at 3, 7 and 28 days of water curing. Morphology of fracture surfaces was investigated by SEM analysis.

Flexural strength of the cement based composites decreased with the increase of PU concentration. Modification of crack propagation mechanisms was obtained at high filler concentration. Morphological analysis of fracture surfaces showed a good dispersion of the foamed particles in the cement matrix.

1. Introduction

Lightweight concrete is usually employed in place of ordinary concrete when good insulation properties are requested. By adding lightweight aggregates during mixing, lighter products can be obtained (300-1850 kg/m^3 instead of 2200-2600 kg/m^3 [1]).

Usually, natural (tuff, pumice) or artificial (expanded clay or polystyrene) lightweight aggregates present lower strength than the cement matrix. They also show lower elastic modulus compared to traditional aggregates. Expanded clay is a lightweight filler with good mechanical properties, and it can be used for structural applications [2-6] too. As a matter of fact, bridges and marine structures in Norway are built with this type of lightweight concretes. They contain silica fume and superplasticizer in order to get a high strength lightweight aggregate concrete with compression strengths of 50 MPa [3,7,8].

Lightweight aggregates with poor mechanical properties are usually used to prepare insulated piers and plasters. Expanded polystyrene, which shows a high thermal insulation together with a very low density (12-14 kg/m^3) is an ideal filler to prepare insulated plasters. Another good feature of this material is its low water permeability which yields to products which are less sensitive to humidity if compared to traditional ones. Moreover, they do not influence the mixing process between cement and water as other porous aggregates do. Unfortunately, all these positive features are coupled with problems of segregation [1,3], due to its extremely low density, and matrix adhesion, due to its particular surface morphology, that can be both reduced by using special additives or surface chemical treatments [9].

Rigid polyurethane foam scraps (PU) were used as lightweight aggregates to prepare cement matrix composites. PU represents an interesting class of polymers because in most of their applications, they are used as cellular materials and they possess excellent thermal insulation properties [10].

Due to their thermosetting characteristics, recycling of this polymer is limited [11-16] and the use of these foams as filler in cement based composite can be interesting for several reasons. First, only rough grinding is necessary; foams do not need to become fine powder to be used as filler in these composites. Second, rigid polyurethane foams exhibit good mechanical properties and their characteristics can be considered intermediate between expanded clay and polystyrene beads. Results from mechanical tests and morphological analysis obtained using two different PU grain sizes with Portland cement are here reported.

2. Materials and Methods

PUs were supplied in forms of small pieces (50x50x20 mm³) as industrial waste of structural panels. They were grinded by a mill and two sizes were obtained (named PU-A ad PU-B). Particles were dried mixed with cement powder CEM II/A-L 42.5 R at different percentages, from 0 to 30% by weight of cement. With a constant water/cement ratio of 0.5 by weight and a standard planetary mixer used for 5 minutes, 20x20x100 mm³ composite samples were prepared. They were kept 24 hours in a 100% RH and then in 25°C water for 3, 7 and 28 days.

Particulate size distribution was analyzed by using a computer image analyzer. Particles were digitized obtaining the bi-dimensional projection of their three-dimensional shape. They were then interpreted as elliptical particles having a maximum and a minimum axis. The equivalent diameter can also be obtained from a hypothetical circle having the same area than the ellipse.

An Instron 4204 was used to test mechanical properties on samples having different aging times and PU percentages. Three points bending tests were performed in the following conditions: crashed speed of mm/in and span of 80 mm. Three samples were considered for any experimental point.

After mechanical tests, fracture surfaces of broken samples were analyzed by means of a Hitachi scanning electron microscope.

3. Results and Discussion

Distributions for both bead sizes PU-A and PU-B are reported in figure 1. Cumulative distribution of the minimum and the maximum axis of the ellipse are compared to that of the equivalent diameter. For each system the three curves do not overlap because the particulate shape is not spherical. The median diameters, calculated at F=0.5, are different: d=2.30 mm for PU-A and d=1.03 mm for PU-B.

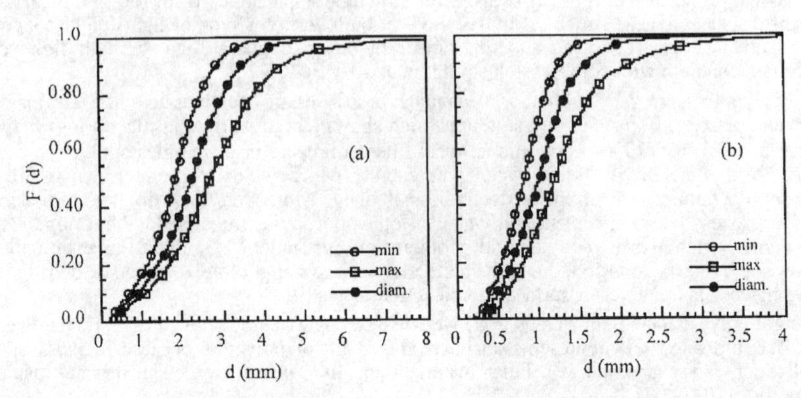

Figure 1: Cumulative distribution function of the two particulate systems: a) PU-A and b) PU-B.

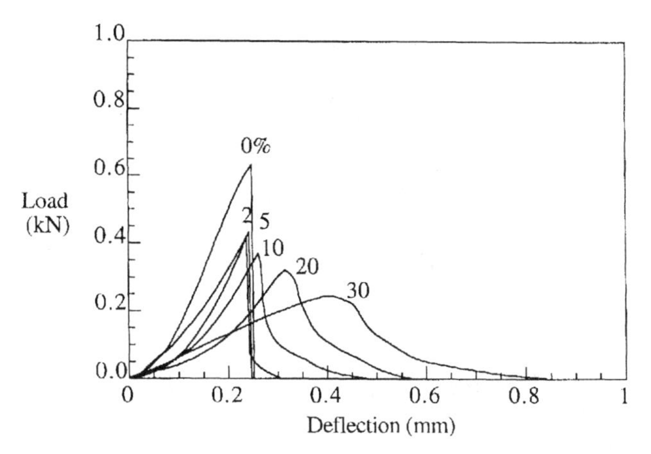

Figure 2: Effect of composition on mechanical properties of PU-A composites (curing time=3 days)

Selected experimental load-deflection curves, evaluated on samples containing PU-A are reported in figures 2, 3. These bending tests were performed on samples cured in water for 3 and 28 days. Similar behaviour was observed on samples cured for 7 days, except for the absolute values which are higher for longer curing times. Figure 4 shows the mechanical behaviour of the other system (PU-B) after 28 days.

By increasing the amount of PU, the strength of the composite systems lowered in both systems A and B. Moreover, in composites with larger particulate size (PU-A), an increase of PU content yields to an increase of the maximum deflection. This behaviour is absent in PU-B based composites.

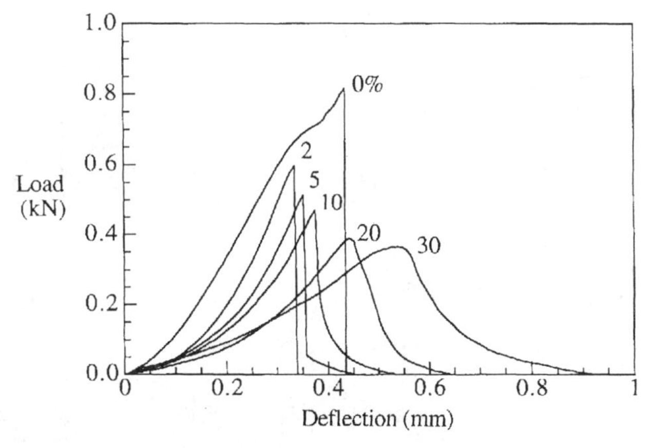

Figure 3: Effect of composition of mechanical properties for PU-A composites (curing time=28 days).

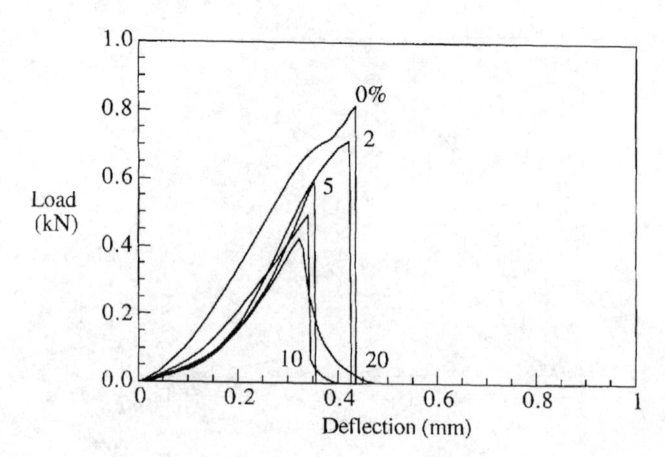

Figure 4: Effect of composition of mechanical properties for PU-B composites (curing time=28 days) .

At certain compositions, particles modify fracture mechanism of the cement matrix composite. This variation can be observed from load-deflection curves, by comparing the curves tract after the load peak. With low particles percentage, crack propagation is not stable and brittle fracture with catastrophic failure occurs. Under these circumstances load goes instantaneously to zero after the peak. By increasing the PU amount in the composite, crack propagation becomes slower suggesting the occurrence of crack stabilization effects. Transition from brittle to ductile fracture mechanism is taking place and this behaviour is followed by a modification of the load-deflection tract after the peak [17]. As a matter of fact, a gradual load reduction as function of deflection is replacing the catastrophic behaviour observed at low particle concentration.

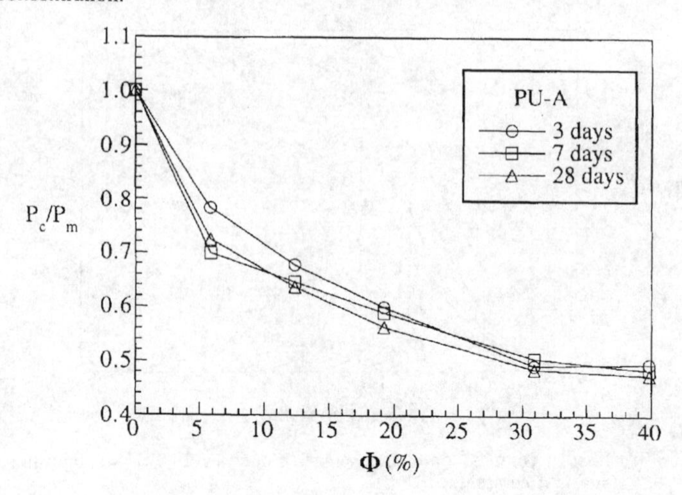

Figure 5: Effect of PU-A concentration on maximum strength

Transition from brittle to ductile fracture mechanism is also affected by filler size. 10% of bigger size particles (PU-A) was enough to modify the load-deflection tract after the peak, while a bigger amount of the smaller PU fillers (PU-B) was necessary to obtain this effect. The different particle size distribution determine different values of the maximum packing fraction ϕ_{max}. PU-A particles have lower ϕ_{max} than PU-B particles, due to their larger size. The filler is responsible of the crack mechanism modification, so that, filler with higher values of ϕ_{max} needs higher volumetric fraction of particles to alter the crack propagation phenomena. The different packing factor would explain why the fracture behaviour of system B is still brittle-like at 10% of composition while it is already ductile-like for system A.

Peak values of load-deflection curves of the composite materials (Pc), divided by the peak value of the cement matrix (Pm) is reported in figures 5 and 6 as function of volumetric fraction of PU-A and PU-B particles, respectively. Normalized composite strength values (Pc/Pm) are reported in these figures for each curing time (3, 7 and 28 days). Values decrease with increasing filler content in both systems A and B while curing time has a negligible effect. Strength reduction is a common behaviour in particulate composite materials and it is usually due to a weak adhesion between the continuos and the dispersed phase. With the hypothesis that no adhesion occurs between PU particles and cement matrix, the filler would act as simple voids and the composite strength reduction becomes only a function of the volumetric fraction and the packing factor ϕ_{max} . The lower strength displayed by system A is therefore due to the lower packing factor of the PU-A particles.

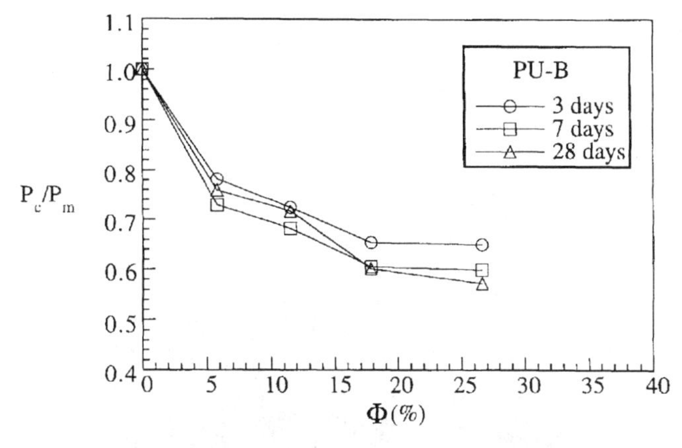

Figure 6: Effect of PU-B concentration on maximum strength

Lightweight fillers results in lighter materials, and it is interesting to compare the specific strength (S) of these composite systems for different compositions. S is calculated as the ratio between the material strength and its density ($S=P/\rho$). The ratio between the specific strength of the composite and that of the matrix (Sc/Sm) is reported in figure 7. An increase of filler composition produces a reduction of the ratio Sc/Sm, which is more pronounced for system A since, as previously discussed, bigger particles (PU-A) lead to a stronger decrease of mechanical strength. The normalized values indicate that the specific strength of the 40% (v/v) PU-A based composite is only 30% less than that of the matrix itself while absolute values were 55% lower, as shown in figure 5.

Fracture surface of a PU foams particulate composite is shown in figure 8. As previously discussed, particle shape is not spherical. Grinding of rigid PU foams produced particles with an irregular external surface. This particular geometry is the consequence of the breakage of closed cells and their exposition on the external surface. The high irregular shape of these particles determines a very high specific surface.

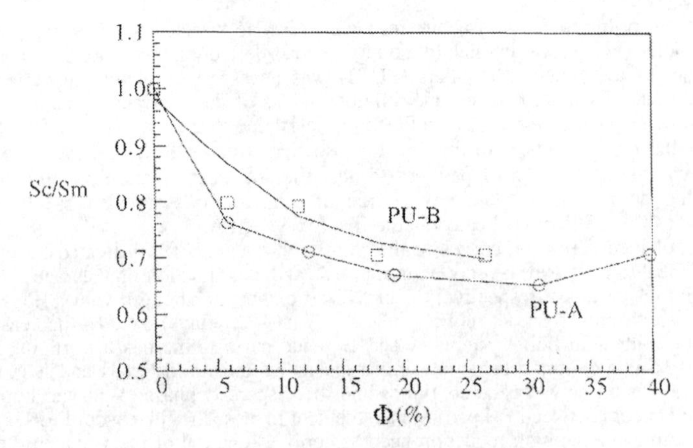

Figure 7: Effect of PU concentration on the specific strength

If all the exposed cells are filled by the cement matrix, than a high surface contact between the two phases is obtained, yielding to composite materials with improved properties. Moreover, the irregular shape contributes to avoid migration of the low density foamed materials, as occurrs in other foamed systems such as polystyrene beads. In the latter case, the very low density and the hydrophobic nature of the polymer phase can lead to segregation phenomena with poor dispersion of the foam particles.

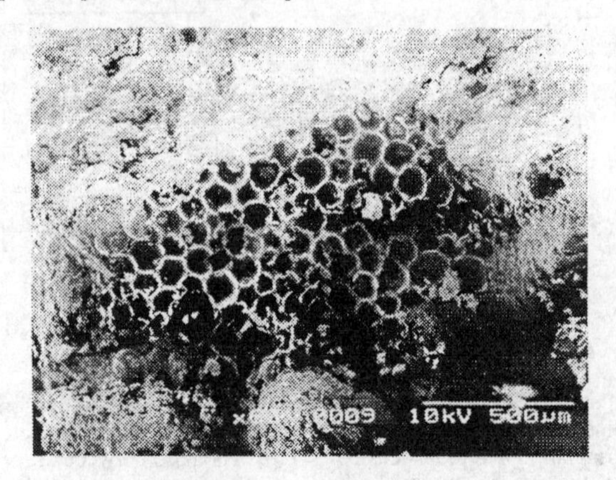

Figure 8: Fracture surface of the PU/cement composite

4. Conclusions

Grinded rigid polyurethane (PU), were used as lightweight aggregates to prepare lightweight cement matrix composites. Three point bending mechanical tests were performed on samples with different concentration of fillers. Bending strength of these composites decreases with the increase of PU. Moreover, transition from brittle-like to ductile-like fracture mechanism was observed when filler concentration increased.

Grinding of PU foams yields to particles with a high specific surface, because of the exposure of the cellular structure. Good interpenetration between phases and good dispersion of the filler in the continuous matrix can be therefore obtained.

5. Acknowledgments

Authors wish to thank Dr Milena Marroccoli (Cementi Lucania spa) and Dr PierCarlo Porta (Tecnopol srl) for materials supply. Special thanks to Clara Mercogliano and Annarita Rega for providing assistance with specimens preparation and testing

References

[1] *A. Castagnetti*, I CALCESTRUZZI LEGGERI, Ed. I.T.E.C. Milano, 1974
[2] *T.W. Bremner, T.A. Holm and V.F. Stepanova*, IL CALCESTRUZZO LEGGERO. UN MATERIALE DI COMPROVATA QUALITÀ PER DUE MILLENNI, L'Industria Italiana del Cemento, 10(1995)569-576
[3] *S. Chandra and L. Berntsson*, DETERMINATION OF MIXTURE PROPORTIONS OF STRUCTURAL LIGHTWEIGHT AGGREGATE CONCRETE, Real World Concrete, Proceedings of R.N. Swamy Symposium, Milwaukee,USA, p.209 (1995)
[4] *Sw.M. Tshalakowa, Hr.J. Popivanov and H.Z. Zlatanov*, LIGHTWEIGHT COMPOSITION FOR THERMAL ISOLATING COMPOSITES AND PRODUCTS, 5° CANMET/ACI. International Conference on Fly Ash, Silica Fume, Slag and Natural Pozzolans in Concrete, Suppl. Papers, Milwaukee, USA, p.741 (1995)
[5] *T.A. Burge*, HIGH STRENGTH LIGHTWEIGHT CONCRETE WITH CONDENSED SILICA FUME, Fly Ash, Silica Fume, Slag and Other Mineral by-Products in Concrete", Vol.II, SP 79-39, p731 (1983)
[6] *S. Chandra and L. Berntsson*, TECHNICAL NOTES: INFLUENCE OF POLYMER MICROPARTICLES ON ACID RESISTANCE OF STRUCTURAL LIGHTWEIGHT AGGREGATE CONCRETE, International Journal Cement and Composite Lightweight Concrete, **5**, 127 (1983)
[7] *T.A. Hammer and M. Sandvik*, USE OF HIGH PERFORMANCE LIGHTWEIGHT AGGREGATE CONCRETE IN NORVEGIAN MARINE STRUCTURES, 2° CANMET/ACI Int. Symposium on Advances in Concrete Technology", Suppl. Papers, Las Vegas, USA, p.301 (1995)
[8] *P. Soroushian, M. Nagi and J.W. Hsu*, OPTIMIZATION OF THE USE OF LIGHTWEIGHT AGGREGATES IN CARBON FIBER REINFORCED CEMENT, ACI Materials Journal, **89**, 267 (1992)
[9] *R. Sri Ravindrarajah and A.J. Tuck*, PROPERTIES OF HARDENED CONCRETE CONTAINING TREATED EXPANDED POLYSTIRENE BEADS, Cement and Concrete Composites, **16**, 273 (1994)
[10] PLASTIC RECYCLING, PRODUCTS AND PROCESSES, R.J. Ehrig ED., Hanser Publisher, Munich
[11] *F. Simioni, M. Modesti e S.A. Rienzi*, RRIM RICICLATI NELLA PRODUZIONE DI ISOLANTI TERMICI RIGIDI, Macplas, **151**, 121 (1993)
[12] *G. Cossi*, ESPANSI PER ISOLAMENTO A UTECH '94. NOZZE D'ARGENTO, Macplas, **161**, 128 (1994)
[13] *G. Modini*, LA GERMANIA INSEGNA, Materie Plastiche ed Elastomeri, 54
[14] DAL RECUPERO AL RIUTILIZZO, Macplas, **184**, 113 (1996)
[15] RESIDUI ESPANSI, Macplas, **184**, 42 (1996)
[16] PANNELLI ISOLANTI IN ESPANSO POLIURETANICO RIGIDO RICICLATO PER UN ULTERIORE CONTRIBUTO AL RISPARMIO ENERGETICO, Seleplast, **9**, 41 (1996)
[17] *B.L. Karihaloo, A. Carpinteri and M. Elices*, FRACTURE MECHANICS OF CEMENT MORTAR AND PLAIN CONCRETE, Advanced Cement Based Materials, **1**, 92(1993)

PREDICTION OF THE DURATION OF THE RELAXATION BEHAVIOUR OF A COMPOSITE CABLE (KEVLAR 49) IN RELATION TO TEMPERATURE

S. KACI*, Z. TOUTOU*

ABSTRACT

This study deals with the determination of the duration of the relaxation behaviour, in relation to temperature, of composite cables intended for use as prestressing tendons.

The material studied is a braided cord of Kevlar 49 impregnated with a thermo-setting resin of polyester.

Based on rheological model, (Maxwell type), predicting correctly the experimental behaviour of the relaxation modulous in uniaxial tension obtained over a duration of 120 hours and a stress level of 50 % of the ultimate stress, a relaxation law taking into account the effect of temperature is proposed. This law is used to predict the duration of the total relaxation of the material for different temperatures (25 °C, 35 °C, 55 °C, 65 °C). As a result, it is noticed that a temperature increase of 40 °C reduces the duration of the total relaxation by about 70 % of its initial value.

Keywords : Composite, Kevlar 49, rheological model, relaxation, temperature, duration of the total relaxation.

INTRODUCTION

The use of composites cables as substitutes for steel tendons in prestressed works could be a good alternative in reason of their mechanical properties [1], of their substantial rheological behaviour, also well in static behaviour, notably in relaxation, that in dynamic behaviour, of their lightness and easiness of stake and especially of the problem of corrosion which doesn't exist.

An experimental program [1], consisting in to test the material in relaxation, for different stress level and for different temperatures, was carried on composite cables of Kevlar (type 0.95 TWA2). The obtained results were already been published [2], [5].

In addition, this experimental behaviour was simulated using a Maxwell model type, taking into account the effect of temperature. This study is also published in the acts of an international conference [7].

Many other studies were carried on these composite cables of Kevlar rather experimental than numerical [3], [4], [6], [8], [9], [10].

Thus, as a continuity of these several studies, and in order to get a good knowledge of their long time behaviour, we propose in this paper, a method for predicting the duration of the total relaxation behaviour of this composite material.

DURATION OF THE TOTAL RELAXATION BEHAVIOUR

The period of time at which the relaxation phenomenon doesn't increase is called the duration of the total relaxation.

The experimental viscoelastic behaviour of the studied composite cables which are tested in uniaxial tension, over a duration of 120 hours, a stress level of 50% of the ultimate stress and for different temperatures, was correctly described by a Maxwell three sells, related in parallel with a linear spring, model. Thus, the relaxation modulus of the material, for a given temperature T, is described by the following equation:

$$E(t)=E_{\infty}+E_1.e^{-t/\tau_1}+E_2.e^{-t/\tau_2}+E_3.e^{-t/\tau_3} \quad (1)$$

The numerical results obtained from the simulation of the experimental curves of the relaxation modulous for different temperatures (25°C, 35°C, 55°C and 65 °C), are summarised in table I, and the corresponding plots are shown in figure 1.

Table I : Numerical results obtained from the simulation of the experimental measurements.

Temperature [°C]	E_{∞} [GPa]	E_j [GPa]	Error	τ_j [mn]
25	91.90	3.21 3.12 2.53	0.067	5.16 118.34 2501.02
35	90.56	3.30 2.34 2.77	0.104	6.98 119.97 2013.89
55	88.98	3.02 2.24 2.07	0.071	6.56 80.32 1887.30
65	88.69	2.73 2.09 2.27	0.021	2.23 31.76 669.91

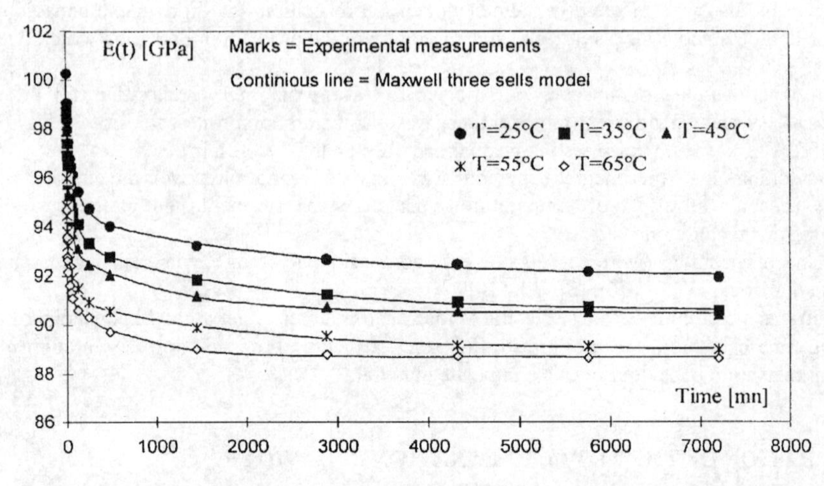

Figure 1 : Comparison between experimental plots and numerical results of the relaxation modulus of composite cables (type 0.95 TWA2), at different temperatures.

PREDICTION OF THE DURATION OF THE TOTAL RELAXATION BEHAVIOUR

Let $\Delta E(t)$ be the relaxation of the material at instant t such as:

$$\Delta E(t) = E(0) - E(t) \tag{2}$$

With E(0) represents the initial value of the relaxation modulus given by:

$$E(0) = E_\infty + E_1 + E_2 + E_3$$

And E(t) is the relaxation modulus at instant t.

Let ΔE be the total relaxation, in term of modulus, of the material, given by the following relationship:

$$\Delta E = E(0) - E(\infty) \tag{3}$$

With : $E(\infty) = E\infty$

So ; $$\Delta E = E_1 + E_2 + E_3 \tag{4}$$

If we suppose that the relaxation phenomenon of the material is achieved when we get 95 % of the total relaxation, and then the relaxation phenomenon becomes stationary, the corresponding time will be deduced like following :

$$\Delta E(t) = 95\%\ \Delta E \tag{5}$$

$$\Leftrightarrow \quad \Delta E = E(0) - E(t) = 0.95\ (E_1 + E_2 + E_3) \tag{6}$$

$$\Leftrightarrow \quad \Delta E = E_1 + E_2 + E_3 - E_1.e^{-t/\tau_1} - E_2.e^{-t/\tau_2} - E_3.e^{-t/\tau_3} = 0.95 * (E_1 + E_2 + E_3) \tag{7}$$

$$\Leftrightarrow \quad 0{,}05\ (E_1 + E_2 + E_3) = E_1.e^{-t/\tau_1} - E_2.e^{-t/\tau_2} - E_3.e^{-t/\tau_3} \tag{8}$$

Supposing that the effect of terms with τ_1 and τ_2 is neglected near the term with τ_3, it follows then :

$$e^{-t/\tau_3} = \frac{0.05 * (E_1 + E_2 + E_3)}{E_3} \tag{9}$$

We put: $$x = \frac{0.05 * (E_1 + E_2 + E_3)}{E_3} \tag{10}$$

It comes $$t = -\tau_3 \ln x \tag{11}$$

The equation (11) represents the time at which we have 95% of the total relaxation of the material and called also the duration of the total relaxation behaviour. Thus, using this relation, we have calculated the different duration of the total relaxation corresponding to different temperatures (25°C, 35°C, 55°C and 65 °C). The corresponding results are summarised in table II.

Table II : Effect of temperature on the evolution of the duration of the total relaxation.

Temperature. [°C]	relaxation modulus E(t) [GPa]	Time corresponding to 95% of ΔE [hour]
25	$91.90+3.21e^{-t/5.16}+3.12e^{-t/118.34}+2.53e^{-t/2501.02}$	72.64
35	$90.56+3.30e^{-t/6.98}+2.34e^{-t/119.97}+2.77e^{-t/2013.89}$	62.61
55	$88.98+3.02e^{-t/6.56}+2.24^{e-t/80.32}+2.07e^{-t/1887.30}$	54.46
65	$88.69+2.73e^{-t/2.23}+2.73^{e-t/31.76}+2.27e^{-t/669.91}$	20.72

INTERPRETATIONS

From the obtained results, we deduce that a temperature increase of 40 °C, reduces the duration of the total relaxation phenomenon about 70% of its initial value which is obtained at temperature of 25°C. This substantial reduction of the duration of the total relaxation, confirms that temperature is an accelerating factor of the relaxation phenomenon of Kevlar 49 composite cables. So, a fracture of these composite cables may happen without increasing loads.

CONCLUSION

In this paper a method of predicting the duration of the total relaxation behaviour of composite cables of Kevlar 49 (type 0.95 TWA2), is described. This method is based on the numerical results obtained from the simulation of their experimental behaviour.

In fact, using a maxwell three sells rheological model, the experimental plots were correctly simulated and this result allowed to identify three relaxation times temperature dependant. These relaxation times are affected by temperature change ; they decrease rapidly with increasing temperature. This property is caused by the fact that at high temperature all configurationnal modes of motion can freely occur.

Using an approximating method, we have observed that a temperature increase of 40 °C reduces the duration of the total relaxation behaviour of Kevlar composite cable about 70 % of its initial value estimated at 25 °C. So, it is important to take into account the effect of temperature on Kevlar composite cables relaxation if they were used to work as prestressed tendons in prestressed works. In fact, a fracture at long time could occur without increasing load by the effect of temperature.

SYMBOLS

E_∞ = the relaxation modulus when time increases indefinitely (GPa),
E_1 = the relaxation modulus corresponding to the first mechanism (GPa),
τ_1 = the relaxation time of the first mechanism (mn),
E_2 = the relaxation modulus corresponding to the second mechanism (GPa),
τ_2 = the relaxation time of the second mechanism (mn),
E_3 = the relaxation modulus corresponding to the third mechanism (GPa),
τ_3 = the relaxation time of the third mechanism (mn),
$E(t)$ = the relaxation modulus at instant t (GPa).
t = time (mn)

REFERENCES

1. Kaci, S., Câbles composites pour la précontrainte: Etude de la relaxation, Thèse Doctorale de l'Université de Bordeaux1, (France), Juillet 1989.
2. Kaci, S., Câbles composites pour la précontrainte: Etude de la relaxation , Proceeding of Advanced Composite Materials in Bridges and Structure, 2nd International Conference, (Canadian Society for Civil Engineering), pp 89-100, (Sherbrooke, Canada), October, 1992.
3. Kaci, S., Study of dynamic behaviour of tension of prestressed composite cables (temperature-related) , Proceeding of Ninth International Conference of Composite Materials, (I.C.C.M/9), Vol. III, pp 293-300, University of Zaragoza, (Madrid, Spain), 12-16 July 1993.
4. Kaci, S. & Khennane, A., Study of the creep behaviour of composite cables for prestress, Proceeding of Third International Conference on Deformation and Fracture of Composite, University of Surrey, pp 644-652, (Guilford, U.K), 27-29 March 1995.
5. Kaci, S., Experimental study of mechanical behaviour of composite cables for prestress, A.S.C.E, journal of engeneering mechanics, vol. 121, NO. 6, pp 709-716, June 1995.
6. Kaci, S. & Khennane, A., Temperature effect on the relaxation modulous of composite cables for prestress, Proceeding of Seventh European Conference on Composite Materials, Vol. 1, pp 159-166, (London, U.K), 14-16 May 1996.
7. Kaci, S. & Toutou, Z., Modélisation du comportement viscoélastique des câbles composites pour la précontrainte, Proceeding of Advanced Composite Materials in Bridges and Structure, 2nd International Conference, (Canadian Society for Civil Engineering), pp 335-340, (Montréal, Quebec, Canada), Août, 1996.
8. Kaci, S. & Khennane, A., Temperature effect on the creep and dynamic behaviors of Kevlar prestressing cables, Canadian, Journal of Civil Engineering, N° 24, Juin 1997
9. Toutou, Z., Contribution à l'étude du comportement viscoélastique des câbles composites pour la précontrainte, Thèse de Magister, Institut de Génie Civil, Université de Tizi-Ozou, Algérie, Juillet 1997.
10. Kaci, S. & Toutou, Z., Study of the viscoelastic functions of composite cables for prestress, Proceeding of Fifth International Conference on Automated Composites, pp .435-442, (Glasgow, UK), September 1997.

The Effect of UV Light on Advanced Composite Material from the Tech21 Bridge

Dr. Manoocheher Zoghi
Laura Casper & Teresa Mueller
University of Dayton
Department of Civil Engineering and Engineering Mechanics
300 College Park
Dayton, OH; USA 45469-0243

The alarming state of disrepair of the United States' infrastructure, along with the high cost of traditional construction have challenged civil engineers to investigate new methods to repair and replace the deteriorated structures. The Tech21 team has responded to this challenge by initiating an infrastructure replacement project that involves the exclusive use of advanced composites materials. The Tech21 team is comprised of a variety of advanced composites experts. Members of the team are: Butler County Engineer's Office (BCEO); Martin Marietta Materials (MMM); LJB Engineers and Architects, Inc. (LJB); the US Air Force Wright Laboratory's Material Directorate (WL/ML), the University of Dayton (UD), and the University of Dayton Research Institute (UDRI).

The Tech21 bridge is located in Butler County, Ohio; USA. Installed in May of 1997, it is a two-lane vehicular bridge, 33 feet in length, and 24 feet in width. It was designed for a maximum load of 24,000 pounds. This design was controlled by a maximum deflection of L/800. As a result, the bridge has a very high factor of safety in regard to strength. The bridge is made entirely of continuous E-glass reinforcement in a polyester matrix.

The most unique characteristic of the Tech 21 bridge is the extensive health monitoring program that has been implemented. The program includes both short-term and long-term monitoring. The short-term monitoring consists of a live-load test program that will continue through July of 1998. The results of these tests will be used to verify the design calculations and to set formal load ratings for the bridge. The long-term testing program utilizes an extensive set of permanent sensors that were installed in the bridge during fabrication. These sensors monitor various health issues such as creep; the integrity of adhesive bonds; and the possible loss of structural performance due to changes in material properties associated with freeze/thaw cycling, load cycling, and environmental exposure. One factor that the sensors cannot monitor is the possible loss of structural integrity due to exposure to UV light. It is well-known in the composites world that advanced composites can suffer degradation when exposed to UV light. In order to prevent such degradation on the Tech21 bridge, a UV protection coat was applied to the surface of the bridge at the manufactuer's location. However, the actual performance of the UV protective coat is unknown.

In order to determine the anticipated performance of the UV protective coat, samples of the bridge material with the protective coat have been obtained from MMM. These samples will be subjected to an accelerated UV exposure of 25 years. The specimens are alternately exposed to ultraviolet light alone and to condensation alone in a repetitive cycle. This cycling simulates the deterioration of the material caused by water as rain or dew and the ultraviolet energy in sunlight. The exact determination of the cycle was extrapolated from the cycles UDRI typically uses to accurately simulate the climatic conditions in the Southwestern United States and in Florida,. Performing this extrapolation for the climatic conditions of Southwestern Ohio yields an actual exposure time to testing exposure time of about 2.5 years to one week. The experiment will run for ten weeks, therefore a 25 year time span will be considered.

The results of this experiment will prove or disprove the durability of the applied UV protective coat and therefore the durability of the entire structure. This will provide civil engineers with the confidence to use the coat in future designs or with the incentive to investigate better means of protecting the advanced composite material from UV light and moisture. The Tech21 bridge has the potential to provide the global civil engineering community with critical data to validate the long-term durability of fiber reinforced plastics as a bridge material.

BEHAVIOUR OF FRP RC FLAT SLABS WITH CFRP SHEAR REINFORCEMENT

Abdel Wahab EL-GHANDOUR[*1], Kypros PILAKOUTAS[*2] and Peter WALDRON[*3]

ABSTRACT: This paper reports on the work of an experimental programme aiming to study the structural behaviour and design of FRP reinforced concrete flat slabs with CFRP shear reinforcement. The results show that failure of the slabs takes place as a result of bond slip of the flexural FRP reinforcement which was not totally prevented by the CFRP shear reinforcement. Bond slip may be facilitated by the formation of cracks directly above the main bars and could be prevented by reshaping the shear reinforcement to provide additional anchorage to the flexural bars or reducing the spacing and area of the flexural bars.

KEYWORDS: FRP reinforcement, shear reinforcement, concrete slabs, punching shear, bond slip

1 INTRODUCTION

In recent years, there is an increasing research effort in developing an understanding for the use of fibre reinforced polymer (FRP) rods as high performance non-corroding reinforcement for concrete structures. The reason for this is that FRP materials could eliminate durability problems which lead to structural degradation and consequent costly repairs and loss of use. The use of FRP reinforcement in flat elements is particularly interesting since the concrete cover can be reduced and, hence, the overall thickness of the element. However, thin 3-D elements may suffer from punching shear due to the effect of concentrated loads, especially at supports. Work at the University of Sheffield, initiated under the pan-European research project EUROCRETE and now continuing as part of the TMR European Research Network "Confibrecrete", aims at developing a fundamental understanding of the behaviour of FRP materials in concrete. This paper will cover some of this work on flat slabs.

The authors have presented in a previous paper [1] aspects of the structural behaviour and design of two slabs SG1 and SC1 reinforced with Glass and Carbon flexural bars respectively, without shear reinforcement. It was suggested that the flexural design of the slabs could not be based on traditional approaches used for steel reinforced slabs, due to lack of potential ductility along failure surfaces. Furthermore, it was found that the punching shear design as modified by Clarke [4] appeared to give good predictions of the capacity of slabs unreinforced in shear. However, in the two slabs tested, the failure attained was not due to punching, but due to bond slip as a result of concrete splitting. Concrete splitting may be avoided if suitable reinforcement is used. Such reinforcement could be part of the shear reinforcement. This paper will describe the testing of the second two slabs, SGS1 and SCS1, similarly reinforced in flexure as SG1 and SC1, but with carbon FRP shear reinforcement. The most important results from the four tests will be presented in this paper.

2 DESIGN OF THE SLABS

The slabs were tested under symmetric point loads applied at eight locations on a circle of diameter 1.7 metres, as shown in figure 1. The slabs were designed in such a way that flexural failure could not exceed a load of 320 kN and punching shear resistance was checked at this load level [2].

2.1 FLEXURAL DESIGN

Taking into consideration that FRP is a brittle material having a sudden tensile failure, it is reasonable to assume that the flexural failure of these slabs, either governed by crushing of concrete in the compression zone or by rupture of the FRP reinforcement, will be brittle. Hence, the flexural failure of the slabs cannot be predicted by using the yield line theory as in the case of expected ductile behaviour. Therefore, the assumed plane of failure ($X_{22.5}$ - $X_{22.5}$) in the weaker flexural direction (figure 1), appears to be reasonable. According to this assumed plane of failure, the ultimate design moment of the CFRP reinforced slabs, M_{design}, was found to be 88 kNm [2].

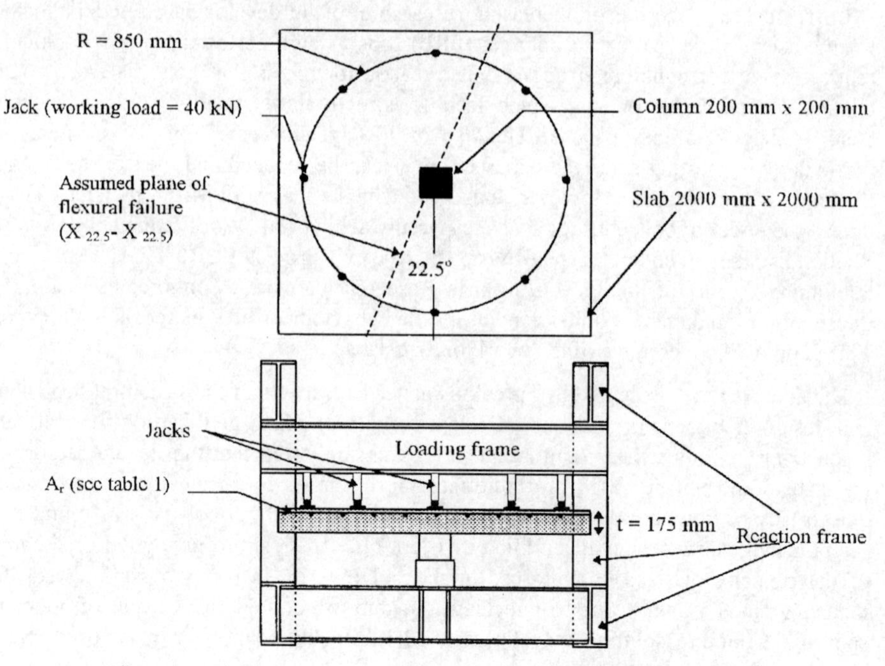

Figure 1 : Slab geometry and schematic representation of test set-up

The flexural design of the slabs was then carried out through simple section analysis techniques assuming equilibrium of forces and compatibility of strains, but without material partial safety factors. The results are shown in table 1.

Table 1 Slabs and reinforcement details

Slab	Concrete strength (MPa) (Cube)	Flexural Reinforcement (Spacing = 200 mm) GFRP Longit.	GFRP Transv.	CFRP Longit.	CFRP Transv.	Shear CFRP 1 x 25 mm
SG1	41.6	11 ϕ 8.5	11 ϕ 8.5			
SC1	43.4			9 ϕ 8.5	9 ϕ 8.5	
SGS1	46.1	11 ϕ 8.5	11 ϕ 8.5			256 legs
SCS1	36.1			9 ϕ 8.5	9 ϕ 8.5	224 legs

2.2 PUNCHING SHEAR DESIGN MODIFICATION

All the checks for punching shear as well as the design of punching shear reinforcement for SGS1 and SCS1 were made according to the BS 8110 [3] code requirements, taking into consideration two modifications to incorporate non-ferrous reinforcement proposed by Clarke [4] for the EUROCRETE project. The first modification is used when determining the shear capacity of concrete, v_c, since the area of the longitudinal FRP tension reinforcement, A_r, is transformed into an effective area, A_e, for use in the existing expressions, by multiplying by the modular ratio. The second modification limits the ultimate strain in the shear reinforcement to 0.0025 (which is only slightly greater than the maximum strain implied by the British Standards which limits the stress for steel to 460 N/mm^2). Therefore, when designing the punching shear reinforcement, the characteristic strength of the shear reinforcement is taken to be equal to $0.0025E_r$, where E_r is the modulus of elasticity of the shear reinforcement.

3 EXPERIMENTAL DETAILS

Four slabs were designed, manufactured and tested at the Heavy Structures Laboratory of the University of Sheffield [2]. The slabs were given codes SG1, SC1, SGS1 and SCS1. The first two slabs (SG1 and SC1) were reinforced in flexure by glass and carbon FRP bars, respectively, but had no shear reinforcement. The second pair of slabs (SGS1 and SCS1) were similarly reinforced in flexure as SG1 and SC1, but had carbon FRP shear reinforcement. This reinforcement was based on the "Shearband" system shown in figure 2 and developed at Sheffield [5], by using CFRP strip. The shear reinforcement was specially manufactured of high strength CFRP flat strip of thickness 1 mm and width 25 mm, for EUROCRETE. The dimensions of the slabs were chosen to be similar to those in tests undertaken previously at Sheffield [6]. The four slabs used were 175 mm thick and 2000 mm square with a 200 x 200 x 200 mm column stub located centrally below the plate. The slabs were reinforced with symmetrical top flexural reinforcement mats made of EUROCRETE round bars having a rough surface which results in good bond characteristics [7]. The characteristics of the slabs are shown in table 1.

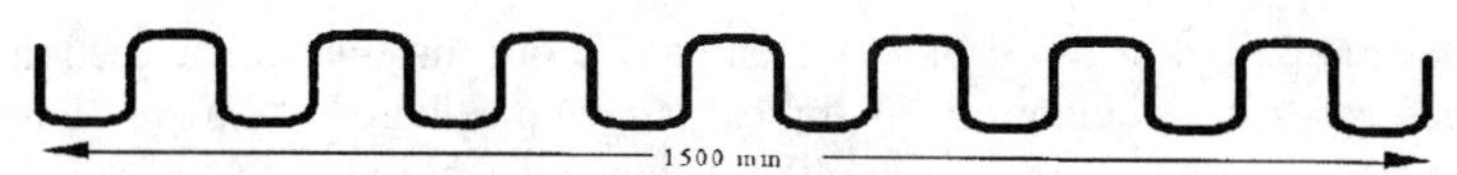

Figure 2 : CFRP Shearband

The spacing of shear reinforcement in both slabs was the same, but slab SGS1 contained two more strips than slab SCS1. The shear reinforcement in both slabs SGS1 and SCS1 was placed above the top flexural reinforcement, a feature unique to the "Shearband" system. A normal weight concrete was used, of nominal strength C40. The concrete cover used was 25 mm. For control purposes, specimens were obtained and tested consisting of cubes, cylinders and prisms. The concrete strengths obtained on the day of testing are shown in table 1.

Extensive measurements were made of strains on key locations on the flexural and shear reinforcement. LVD transducers were also used to measure the vertical and horizontal deflections as well as concrete strains. The width of cracks was not measured during the test, but the development of the cracks was marked on the top surface [2].

4 EXPERIMENTAL RESULTS AND DISCUSSION

4.1 DESCRIPTION OF EXPERIMENTS

A loading procedure comprising of two loading cycles was adopted. The slabs were firstly loaded up to 150 kN, then subjected to one unloading - reloading cycle and then loaded to failure. The applied load was manually controlled.

The crack development in all four slabs followed a similar pattern. The first cracks opened up on the top surface in the form of flexural cracks above the column, in the weaker direction, at around 100 kN for both SG1 and SCS1, 130 kN for SGS1 and 150 kN for SC1. A noticeable drop in load was caused as a result of first cracking. After the initial load stage, cracks propagated from the middle outwards and reached the slab edges. Subsequently, with increasing loads, more cracks developed and advanced radially from the column faces towards the slab edges, along the four axes of symmetry of the slab (central X and Y axes and two diagonals). Cracks parallel to the X and Y axes opened up at loads of 130 kN and 190 kN for SG1 and SC1 respectively, and at a load of 180 kN for both SGS1 and SCS1, whilst cracks parallel to the diagonal axes opened up at higher loads. By the time the applied load reached 150 kN and 220 kN for SG1 and SC1 respectively and 195 kN for both SGS1 and SCS1, only a few new diagonal cracks developed and most of the cracks were already formed. After that, with increasing load, the width of the cracks closer to the column increased substantially.

Slabs SG1 and SC1 failed after undergoing substantial deformations due to a suspected bond slippage of their flexural bars, the fact that accelerated their failure at loads less than their flexural capacity. This suspected bond slip failure resulted in a sudden increase in the widths of the flexural cracks on the top of the slabs as well as compressive concrete crushing close to the column. The maximum recorded loads for these two slabs were 170 kN and 229 kN, respectively. Penetration of the column into the slab was noticed after the end of the tests. No clear evidence of punching shear failure or punching shear cracks was found on the top surface of the slabs.

Slab SGS1 also failed after undergoing substantial deflections due to suspected bond slippage of its flexural bars just before reaching the maximum expected flexural load. The maximum recorded load for this slab was 198 kN. Penetration of the column into the slab was noticed after the end of the test. As in the first pair of slabs, no clear evidence of punching shear failure or punching shear cracks was found on the top surface of the slab. The widths of the flexural cracks close to the column on the top surface were considerable and concrete compressive crushing was observed at the bottom side of the slab.

Slab SCS1 failed in a combined flexural and suspected bond slip failure after reaching a reduced flexural load, possibly due to bond slip, as well as, low concrete strength. The maximum recorded load for this slab was 200 kN, lower than that of SC1 without shear reinforcement. From the crack pattern of this slab, it was noticed that the flexural crack that caused failure followed closely the plane of flexural failure ($X_{22.5}$ - $X_{22.5}$) assumed in the design stage of the slabs (section 2.1). Concrete compressive crushing was observed on the bottom side of the slab, following the diagonal top flexural crack. Again, no clear evidence of punching shear failure occurred in this slab even though penetration of the column into the slab was noticed after the end of the test.

4.2 RESULTS AND DISCUSSION

Due to the large amount of data collected for each slab, it is not possible to include them all in this paper. Hence, only a selection of important data is presented for the four tested slabs. The load versus average deflection history at a radius of 850 mm is shown in figure 3. A typical history of the load versus strain for the flexural reinforcement near the column is shown in figure 4. A typical history for the load versus strain for the shear reinforcement at 150 mm from the column face is shown in figure 5. Finally, the load history versus concrete strain on the top of the slab is shown in figure 6.

Considering figure 3, it is obvious that the post-cracking portions of the load-deflection curves of all slabs show no big increase in the load up to failure, and the stiffness after cracking is substantially lower than the one that would normally be expected from RC slabs. Nevertheless, it is obvious that the post-cracking behaviour of the SC slabs is better than that of the SG slabs, since carbon bars have a higher elastic modulus. Considering that deflections at failure serve as a measure of deformability, then figure 3 shows larger deformability for the SG slabs due to the lower elastic modulus of Glass.

In terms of capacity, figure 3 shows a significant increase in capacity of slabs reinforced with CFRP flexural bars (34.7 % load enhancement is achieved), with the exception of slab SCS1 which had a lower concrete grade. Furthermore, figure 3 shows a slight increase in the capacity of slab SGS1 reinforced with shear reinforcement (16.5 % load enhancement is achieved), which can be attributed to the effect of shear reinforcement in delaying bond slip.

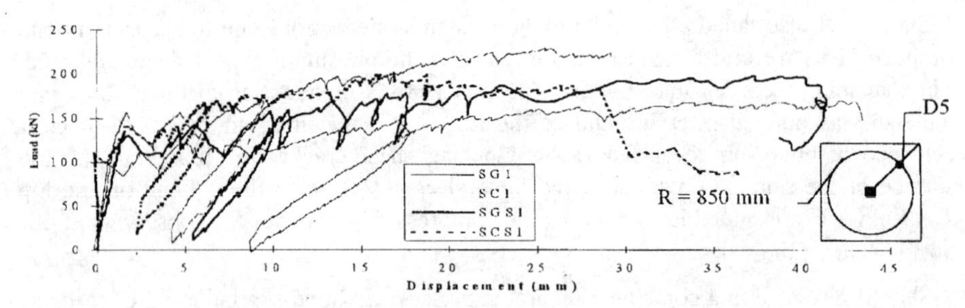

Figure 3 : Full History of Load-Deflection Relationship D5

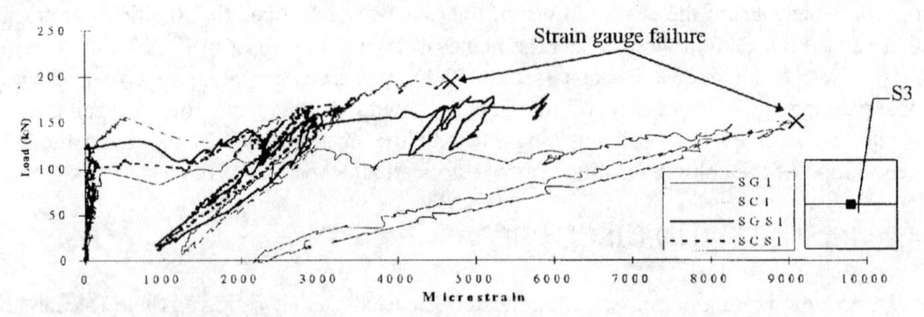

Figure 4 : Full History of Load-Strain Relationship S3

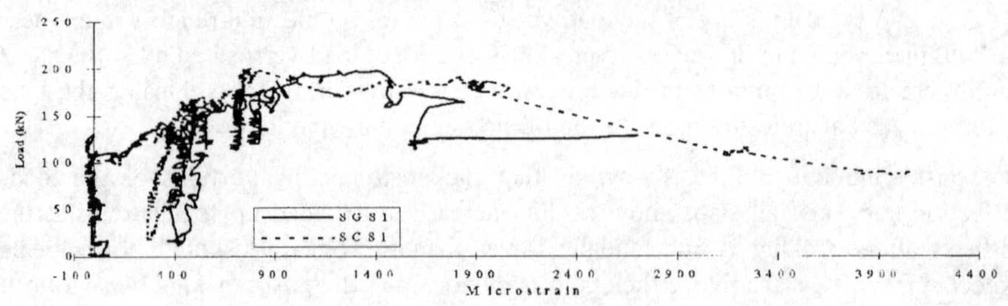

Figure 5 : Full History of Load-Strain Relationship SS5

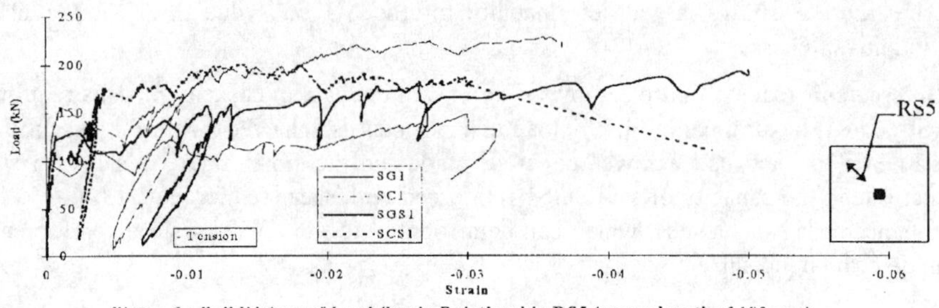

Figure 6 : Full History of Load-Strain Relationship RS5 (over a length of 180 mm)

Considering figure 4, it can be seen that the strains in the reinforcement at failure are around 4800 and 9000 microstrain in CFRP and GFRP bars, respectively, corresponding to stresses of 528 N/mm^2 in CFRP bars and 360 N/mm^2 in GFRP bars. This indicates that only relatively low stresses were developed in the reinforcement at failure compared to their strengths, revealing the possibility of bond slip occurring at failure.

From figure 5, it can be seen that the maximum strains in the shear reinforcement just before failure are around 1400 and 1900 microstrain in slabs SGS1 and SCS1, respectively, corresponding to stresses of 154 N/mm^2 and 209 N/mm^2 in the CFRP shear reinforcement. This indicates that only relatively low stresses were developed in the shear reinforcement at failure compared to the strength of its material. This could be attributed to the low failure loads of slabs SGS1 and SCS1, as mentioned previously, as well as the high amount of shear reinforcement provided in both slabs.

The occurrence of bond slip at failure is also supported by the curves of figure 6, where it is obvious that the average strains on the top concrete surface at failure were of the order of 0.03 to 0.05, which are an order of magnitude greater than the maximum strains in the bars at failure. This provides firm evidence that bond slip failure has occurred in the flexural bars of all four slabs.

In all cases, it was observed that the main cracks developed directly above the longitudinal reinforcement, partly because there exists a plane of weakness. This may be the underlying reason of splitting and bond slip, since the flexural bars have been demonstrated to have good bond characteristics [7].

As a result, one of the main advantages of using FRP reinforcement in slabs, that is the reduction of the cover thickness, may lead to other problems and requires re-examination. The problem of bond slip, which was expected to be less in slabs SGS1 and SCS1 with shear reinforcement, did not go away. The ineffectiveness of the shear reinforcement in totally preventing bond slip in both slabs SGS1 and SCS1 could be attributed to its shape. Reshaping the "Shearband", in order to minimise the distance between the vertical legs to which the top flexural bars will be anchored, is thought to provide a solution to the dominating problem of bond slip. Another solution may be to decrease the spacing and reduce the area of the flexural reinforcement. Both solutions will be tried by the authors as part of the ongoing research at the Centre for Cement & Concrete of the University of Sheffield.

5 CONCLUSIONS

The first three slabs SG1, SC1 and SGS1 failed due to bond slip of the flexural bars at loads less than their expected flexural and punching shear capacities. The fourth slab SCS1 also failed far below its expected capacity, due to combined flexural and bond slip failure, as a result of the low concrete strength of its batch.

The flexural design of FRP reinforced flat slabs cannot be based on traditional approaches used for steel reinforced slabs, due to lack of potential ductility.

The provision of the shear reinforcement system in slab SGS1 was effective in increasing its load capacity. This increase in capacity was small which means that the provided shear reinforcement did not totally eliminate bond slip of the flexural bars, but only retarded its occurrence.

Reshaping the "Shearband", and decreasing the spacing and cross-sectional area of the FRP bars are thought to provide solutions to the dominating problem of bond slip.

ACKNOWLEDGEMENT

The authors wish to acknowledge the support of the EUROCRETE Partners, the DTI / EPSRC, LINK Structural Composites Programme and the European Commission for funding provided to the TMR Network "Confibrecrete".

REFERENCES

[1] El-Ghandour, A. W., Pilakoutas, K. and Waldron, P., "Behaviour of FRP Reinforced Concrete Flat Slabs," Proceedings of the Third International Symposium on Non-Metallic (FRP) Reinforcement for Concrete Structures FRPRCS-3, Sapporo, October 1997, Volume 2, pp.567-574.

[2] El-Ghandour, A., Pilakoutas, K. & Waldron, P., "Behaviour of FRP Reinforced Concrete Flat Slabs, Series I: SG1 to SCS1," Report No. CCC/97/0060A, Centre for Cement and Concrete, Department of Civil and Structural Engineering, the University of Sheffield, November 1997, 66 pp.

[3] BS 8110 : 1985, "Structural Use of Concrete," Parts 1:3, British Standards Institution, 1985.

[4] Clarke, J. L., "Modification of Design Rules to Incorporate Non-ferrous Reinforcement," Report No. 1, Sir William Halcrow & Partners Ltd, January 1996, pp.13-14.

[5] Li, X. and Pilakoutas, K., "Alternative Shear Reinforcement for RC Flat Slabs, Series II: PSS-E to PSS-H," Report No. CCC/94/0039A, Centre for Cement and Concrete, Department of Civil and Structural Engineering, the University of Sheffield, March 1996, 35 pp.

[6] Li, X. and Pilakoutas, K., "Alternative Shear Reinforcement in Flat Slabs," Report No. CCC/94/0030A, Centre for Cement and Concrete, Department of Civil and Structural Engineering, the University of Sheffield, March 1995, 42 pp.

[7] Achillides, Z., Pilakoutas, K. and Waldron, P., "Bond Behaviour of FRP Bars to Concrete," Proceedings of the Third International Symposium on Non-Metallic (FRP) Reinforcement for Concrete Structures FRPRCS-3, Sapporo, October 1997, Volume 2, pp.341-348.

Effect of fiber coatings on low cycle fatigue properties of B/AL laminated composites

Shao-Lun Liu* and Xi-Peng Han*

*Lab of mechanical properties,Institute of Aeronautical Materials,Beijing 100095,China

Abstract: An experimental study of the effect of fiber coatings on low cycle fatigue life was conducted on B/Al LD_2 $[0]_6$ laminated composite under strain control conditions. The results showed that low cycle fatigue life of B/Al LD_2 $[0]_6$ laminated composite with fiber coatings was 7 times as large as those without fiber coatings,a reasonable explanation was giran for coating effects.

KEYWORDS: Coating, Metal matrix composite, Low cycle fatigue, Boron fiber, Aluminum alloy

1.Introduction

Fiber reinforced metal matrix composite (MMC_f) is considered to be attractive as structural materials in the fields of aeronautical and aerospace industries because of their higher strength-to-density ratio, higher stiffness to weight ratio, improvements in fatigue crack growth resistance and, principally, their superiority to polymer matrix composites(PMCs)in high temperature applications. Therefore, as metal matrix composites technology matures, one of the most urgent problems is characterization of durability and damage tolerance behaviors for the wide use of MMC in aeronautical, aerospace and automotive structures. An understanding of mechanical behavior and especially fatigue properties of MMC_f is of significant importance for their increasing application.

This paper presents characterization of LCF behaviors of B/Al LD_2 $[0]_6$ laminated composites, by means of Coffin-Manson formula, and verified the fatigue damage assessment proposed by Johnson[1]. Finally, this study analyses the effects of fiber coatings on LCF life of B/Al LD_2 $[0]_6$ laminates composites.

2.Materials and testing

The materials used in this study was boron fiber reinforced aluminum alloy LD_2 matrix $[0]_6$ laminated composites. The laminates had a nominal fiber volume fraction of 45% with the fibers uniaxially oriented in the loading direction. The fiber diameter was 0.140mm.The B/Al laminated composites used two kinds of fibers, one kind was fiber with Coatings（see fig.1）, another was fibers without Coatings.

LCF tests were conducted on flat specimens with dimensions of thickness t=1mm and width w=10mm and length L=75mm. The specimens were cut from the unidirectional panel using an electric-discharge machining (EDM) technique. The specimens were polished manually by light filing after EDM cutting to remove any damage associated with the machining.

Tabs were bonded onto the specimen grip section to provide a cushion between the sharp grip surface and the specimen surface to prevent damage to the underlying filaments.

Experimental Procedures

Low cycle fatigue tests were carried out on a MTS-809 fatigue machine with hydraulic grips. The strain gage section is 12.7mm length in this case, the active region of a strain gage must be placed accurately within the gage length. In all cases, specimens were cycled under constant strain amplitude controls with a triangle waveform. The strain ratio R=0.1, and frequency f=0.33HZ were used. The environment was lab air. The cyclic stress-strain curves of B/Al LD_2 $[0]_6$ laminated composite specimen were plotted by x-y recorder.The details of LCF Test method for MMC_f see literature[2].

3.Results and discussion

Characterization for LCF of MMCf

According to mixed law, the elastic modulus of composite materials is

$$E_c = E_f V_f + E_m(1-V_f) \quad (1)$$

where E_f: elastic modulus of fiber

E_m: elastic modulus of matrix

V_f: fiber volume fraction.

Based on Coffin-Manson formula which is applicable to metallic materials, low cycle fatigue behavior of MMC_f can be characterized as

$$\Delta\varepsilon_t / 2 = \frac{\sigma_f'}{E_f V_f + E_m(1-V_f)}(2N)^b + \varepsilon_f'(2N)^c \quad (2)$$

The LCF Test results of B/Al LD_2 $[0]_6$ laminated composite with fiber coatings.

The LCF test results of B/Al LD_2 $[0]_6$ laminated composite specimen P-6 with fiber coatings is showed in Fig.2. Log-log plot of $\Delta\varepsilon/2$ versus $2N_f$ obtained in specimens with fiber coating is given in Fig.2. Through fitting test data by the formula(2),LCF equation for B/Al LD_2 $[0]_6$ laminated composite with fiber coatings P-6 can be obtained as following

$$\Delta\varepsilon_t / 2 = 0.0025(2N_f)^{-0.023} + 0.0024(2N_f)^{-0.082} \quad (3)$$

It can be seen from Fig.2 that the shape for LCF curves of the laminates P-6 is greatly different from that of the matrix materials, aluminum alloy LD_2, and the elastic life curves $\Delta\varepsilon_e / 2 - 2N_f$ and the plastic life curves $\Delta\varepsilon_p / 2 - 2N_f$ absolutely don't intersect. Also there is no transitional life, because plastic deformation of B/Al LD_2 $[0]_6$ laminated composite was too small to consider.

<u>The LCF test results of B/Al LD_2 $[0]_6$ laminated composite without fiber coatings.</u>

Fig.3 presents the LCF test results of B/Al LD_2 $[0]_6$ laminated composite PR-2 without fiber coating. Log-log plot of $\Delta\varepsilon / 2$ versus $2N_f$ is shown in Fig.3. Through fitting the test data by formula (2),the LCF equation of B/Al $[0]_6$ laminated composite without fiber coatings can be obtained as following.

$$\Delta\varepsilon_t / 2 = 0.0022(2N_f)^{-0.026} + 0.0008(2N_f)^{-0.010} \qquad (4)$$

It can be seen from Fig.3 that the elastic-life curve and the plastic-life curve don't intersect at all. The elastic-life line and the total life line overlap. This obviously indicated that plastic deformation only occupied very small portion within total deformation. Therefore, it suggests that considering LCF behavior of B/Al LD_2 $[0]_6$ laminated composite as elastic is reasonable.

<u>The effect of fiber coatings on LCF life of B/Al laminated composites.</u>

In order to compare the effects of fiber coating on LCF life of B/Al LD_2 $[0]_6$ laminated composites, Fig.4 gives comparisons of LCF life of B/Al LD_2 $[0]_6$ laminated composites with and without fiber coatings. It can be seen obviously from Fig.4 that the LCF life of laminated composites with fiber coatings is 7 times as large as that without coating.

Due to the thermal expansion coefficient mismatch(CTE) between the fiber and the matrix, high residual stresses exist in metal matrix composite systems upon cool down from processing temperature to room temperature. The residual stress in interface results in fatigue crack initiation and growth in matrix under cyclic loading. An interface material can be placed between the fiber and the matrix to reduce the high tensile residual stresses in the matrix. The experimental results have proved that interplace materials, i.e. coating, can reduce and even diminish the residual stresses as long as the modulus, thickness and the thermal expansion coefficient of coating materials are rightly selected[3,4,5]

The cycle stress-strain curve with hysteresis loops for B/Al metal matrix composites without fiber coatings are showed in Fig.5 and Fig.6, respectively. The reduction in the width of the hysteresis loops indicates cyclic hardening, whereas the increase the in the width of the hysteresies loops indicates cyclic sorting. It can be seen from Fig.5 and Fig.6 that B/Al LD_2 $[0]_6$ laminated composites without fiber coatings shows cyclic sorfting behavior, and that with fiber coatings shows cyclic hardening behavior.

The test results also showed that coatings not only gave long LCF life of B/Al

LD_2 $[0]_6$ laminated composites, but also greatly improved its tension and compression properties[6].

4.Conclusions

1. Low cycle fatigue life of B/Al LD_2 $[0]_6$ laminated composite with fiber coatings was 7 times as large as those without fiber coatings.
2. Fiber coatings greatly improved LCF properties for B/Al LD_2 $[0]_6$ laminated composites.
3. B/Al LD_2 $[0]_6$ laminated composites without fiber coatings showed cyclic softing behavior,but that with fiber coatings shows cyclic hardening behavior.

Acknowledgments

The authors thank Mrs. Xiu-Fen Guo for her doing fatigue test, and Guang–Yi Sun for his providing B/Al $[0]_6$ materials in this study.

References

[1] Johnson, W.J., Fatigue testing and damage development in continuous fiber reinforced metal matrix Composites, in Metal Matrix Composites: Testing, Analysis and failure modes, ASTM STP1032, 1989.

[2] S.L.Liu, X.P.Han and X.F.Guo, Low Cycle Fatigue Behavior of B/Al LD_2 $[0]_6$ Laminate,Fatigue'96 ,Vol.III,pp1493–1496,1996.

[3] Carman, G.P., R.C. Averill, K.L. Reifsnider and J.N. Reddy, Optimization of fiber coatings to minimize stress concentrations in composite materials. *J. composite materials*, 1993,27: 589-612.

[4] Popejoy, D.B. and L.R. Dharani , Effect of fiber coating and interfacial debonding on crack growth in fiber-reinforced composites. *Theoretical and Applied Fracture Mechanics*, 1992,18: 73-79.

[5] S.L.Liu and D.Chen,Optimal Design of MMC_f Granular Coat,Journal of Inner Mongolia Polytechnic University,China, Vol.16,No.3,1997（in chinese）

[6] Liu, S.L. and J.Q. Zhang, Test study on tension and compression behaviors of B/Al LD_2 $[0]_6$ laminated composites（in chinese）.95' Annual meetings of Beijing Aeronautics and Aerospace Association (BAAA). Beijing, China. 1995, Oct.(in Chinese),unpublished.

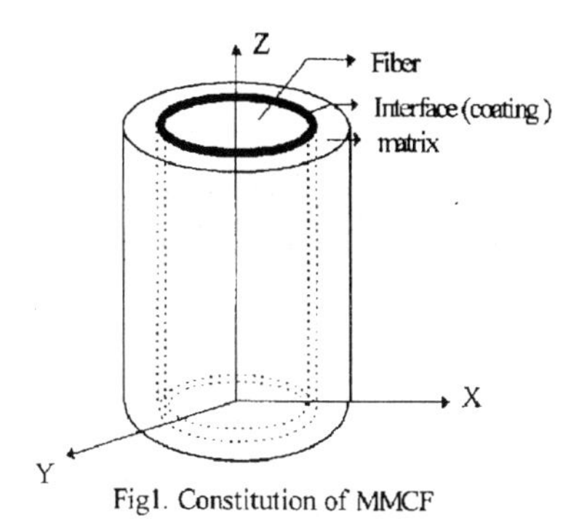

Fig1. Constitution of MMCF

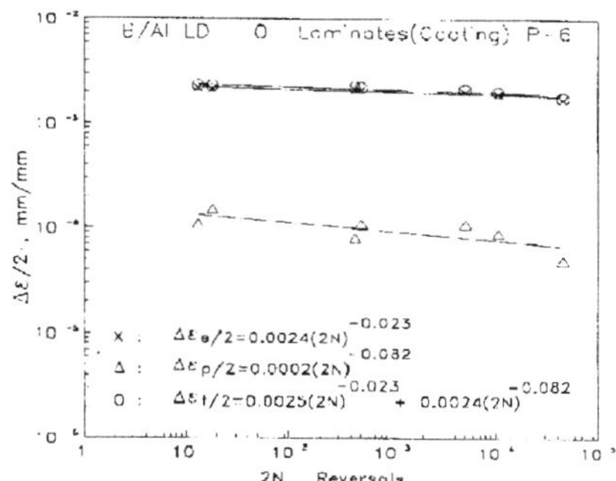

Fig.2 strain-life relation of B/Al LD_2 $[0]_6$ laminated composite P-6 without fiber coatings

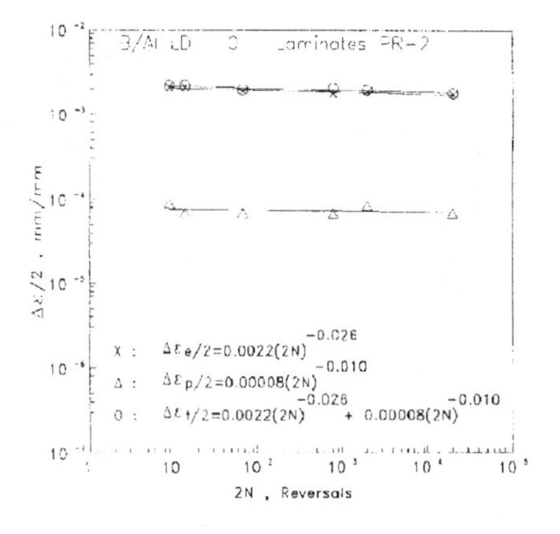

Fig.3 Strain-life relation of B/Al LD_2 $[0]_6$ laminated composite PR-2 without fiber coatings

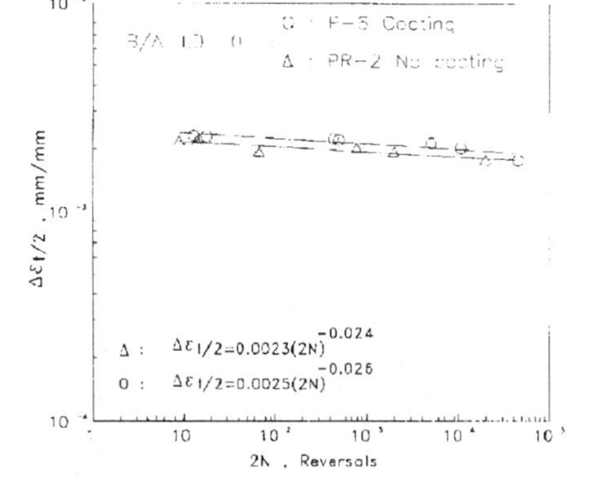

Fig.4 comparison of LCF life of B/Al LD_2 $[0]_6$ laminated composite with and without fiber coatings

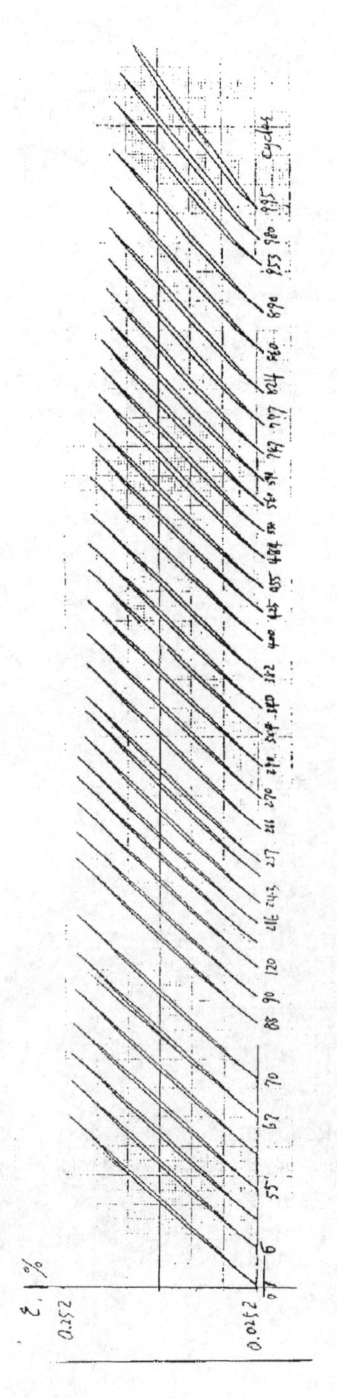

Fig 5. Stress–strain behavior with hysteresis loops at various cycle number for B/Al LD_2 $[0]_6$ laminated composite with fiber coatings

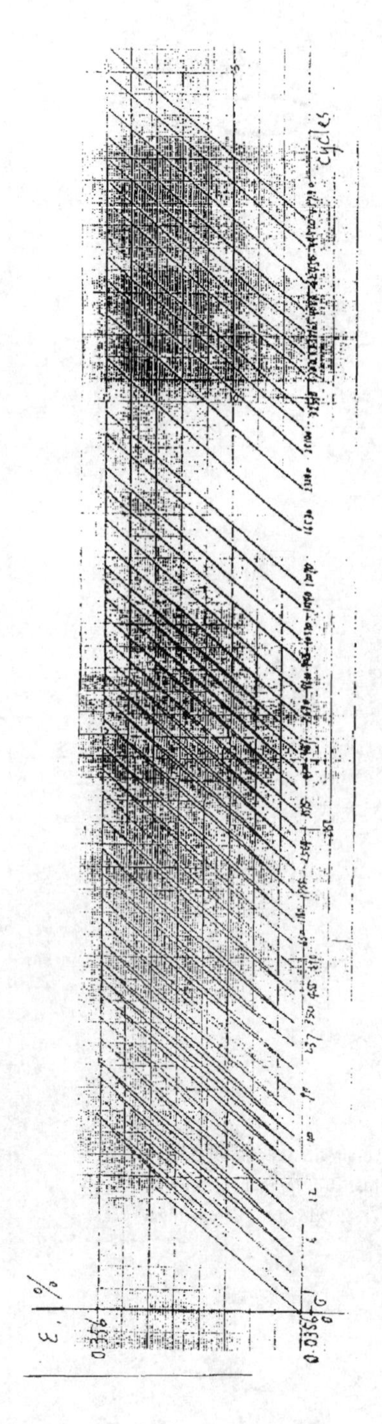

Fig 6. Stress–strain behavior with hysteresis loops at various cycle number for B/Al LD_2 $[0]_6$ laminated composite without fiber coatings

SYMPOSIUM 4

High Performance PET Fibers Via Liquid Isothermal Bath High-Speed Spinning: Fiber Properties and Structure Resulting from Threadline Modification and Post Treatment

Jiunn-Yow Chen, Paul A. Tucker, John A. Cuculo

Fiber and Polymer Science Program, College of Textiles,

North Carolina State University,

Raleigh, North Carolina 27695-8302, U.S.A.

Abstract

Poly(ethylene terephthalate) fibers with improved mechanical properties and dimensional stability were spun in the range of 2000-5000 m/min via controlled threadline dynamics by a liquid isothermal bath (LIB) spinning process, followed by post drawing and annealing. Fiber properties of the as-spun fibers and post-treated fibers of each process were compared. Two commercial tire cords were also included. Unlike unperturbed spinning, the LIB as-spun fibers show unique structural properties of high amorphous orientation, low crystallinity, high strength, and high initial modulus. Moreover, noncrystalline chains are further extended during post treatment. The post-treated LIB fibers exhibit mechanical properties with tenacity higher than *ca.* 9 g/d, initial modulus higher than 120 g/d, and ultimate elongation less than *ca.* 10%. They also demonstrate superior stability with thermal shrinkage less than 6% and LASE-5 higher than 5 g/d. The overall properties are not obtainable by either the traditional spin-draw process or any modified process which produces low shrinkage tire cord. Evidence is presented that strongly indicates the existence of a "third phase", referred to as the taut-tie noncrystalline phase *(TTNC)*, in addition to the traditional two-phase model, *i.e.* crystalline and random amorphous phases.

1. Introduction

Considerable research has been done to improve mechanical properties and dimensional stability of PET fibers. In a traditional spin-draw process, as-spun fibers are taken up at a relatively low-speed producing fibers in a random amorphous state with very low orientation and low crystallinity. Post drawing and annealing gradually develop an oriented structure with appropriate high crystallinity. These traditionally spun fibers possess high strength, but relatively poor high temperature dimensional stability. The latter characteristic is undesirable in industrial end-use applications. In improving the thermal shrinkage of PET industrial yarn, a high modulus low shrinkage (HMLS) tire cord yarn was developed and commercialized. Dimensional stability was improved, but strength and initial modulus were unexpectedly not similarly improved.

When high-speed spinning, a one-step process (OSP), was introduced with promise of high productivity and cost reduction, the expected benefits unfortunately did not materialize. The high-speed spun fibers showed limited mechanical properties, with lower strength, lower initial modulus and greater elongation than those of traditional spin-draw fibers [1].

One of our main goals in high-speed spinning is learning how to control threadline dynamics via threadline modifications of both stress and temperature profiles by introduction of a tension enhancing liquid isothermal bath (LIB) in the spinline [2-6]. In this study, the liquid isothermal bath (LIB) is placed near the structure development zone in the threadline with the expectation of increasing the amount of extended chains in as-spun fibers. To develop the full potential of these fibers, the oriented as-spun fibers are subjected to post drawing and annealing. The taut-tie noncrystalline chains are further rapidly developed and crystallized along the fiber axis.

2. Experimental

Melt Spinning Process and Post Treatments

(PET) chips with intrinsic viscosity (IV) of 0.97 dL/g and viscosity molecular weight M_v of *ca.* 29,440 were used in this study. Unless otherwise specified, spinning denier was set at 4.5. Figure 1 illustrates the two process setups. One is the liquid isothermal bath (LIB) process and the other, the unperturbed process. In the LIB process, the liquid bath was positioned such that the bottom of the bath was 100cm from the spinneret. The liquid medium, 1, 2-propandiol, was heated at 175°C. The depth of liquid in the bath was kept at 45 cm for 2000-4000 m/min and 30 cm for 4000-5000 m/min take-up speed. Downstream, the spinline was cooled by ambient air at 23°C and taken up by high-speed godet rolls.

Samples of as-spun fibers were then selected and in a continuous post treatment process passed over two heaters between three rolls and then to a wind-up. The temperature of the first step, drawing, was set at 180°C and the second step, annealing, at 220°C. The feed roll was controlled at a constant speed of 10 m/min, limited simply by the capability of the drawing equipment. The as-spun fibers were drawn mainly to a near maximum draw ratio in the first step and only to a minimal draw ratio in the second step to retain threadline tension.

Characteristics of As-Spun and Post-Treated Fibers

Characterization of the fibers followed generally traditional methods including birefringence, tensile testing, (LASE-5)[7], DSC, boil-off shrinkage (BOS) and thermal shrinkage (177°C), density and crystallinity, WAXS [8,9] and fraction of taut-tie noncrystalline phase [10,11,12].

3. Results and Discussion

Properties of As-spun and Post-treated Fibers

Effect of take-up speed on as-spun and post-treated fibers

Birefringence versus take-up speed of as-spun and post-treated fibers is shown in Figure 2. Unperturbed as-spun fibers have, as expected, low birefringence *ca.* 0.07-0.10. Drawing to their near maximum draw ratios of 1.7 for the 3000 m/min take-up, and 1.5 for the 5000 m/min take-up spun fibers raised the fiber birefringence significantly to *ca.* 0.19-0.20. LIB as-spun fibers have high birefringence of 0.17-0.21. After drawing these fibers, at the near maximum draw ratios of only 1.2 for 3500 m/min, and 1.1 for 5000 m/min take up these fibers exhibit ultra-high birefringence of *ca.* 0.22-0.23. It is believed that the LIB as-spun fibers exhibit high orientation because of the effect of the threadline modifications on both stress and temperature profiles via the liquid bath. These modifications alter first the threadline dynamics and in turn the macromolecular chain extension along the fiber axis. Thus, the morphology and mechanical properties are significantly improved.

Some of the tensile properties of as-spun fibers of both LIB and unperturbed spinning processes and their respective post-treated fibers are shown in Table 1. In general, both tenacity and initial

modulus are respective functions of take up velocity. The highest tenacity, 9.6 gpd, of the LIB as-spun fibers was obtained at take-up speed of 5000 m/min. After post-treatment at draw ratio of 1.1, tenacity increased to 10.3 gpd. The highest initial modulus of 147.5 gpd was obtained from post-treated LIB fibers, which was taken up at an intermediate speed of 3500 m/min with a draw ratio of 1.2. The high values of LASE-5 for LIB fibers is also evident.
Elongation at break of the LIB fibers, see Table 1, is 14.8% at 3500 m/min and falls to 10.5% at 5000 m/min. The general finding is, the higher the take-up velocity, the lower the elongation at break. The post-treated LIB fibers have low elongation at break, in the range of *ca.* 8-11%, while the post-treated unperturbed is 16% or higher. The unperturbed as-spun fibers have crystallinity (X_v) from 7.8% to 39.3% as take-up velocity increased from 2000 to 5000 m/min. It has been found that the crystallinity of the LIB as-spun fibers is gradually suppressed going from 2000 m/min to 5000 m/min speed. At 4000 m/min and above, the crystallinity is less than that of unperturbed fibers. In traditional high-speed spinning, a fiber with low crystallinity, all other things equal, generally has a higher extensibility than that of fiber with a high crystallinity. In the case of unperturbed fibers, the near maximum draw ratios are 1.7 and 1.5 for the fiber spun at 3000 m/min and 5000 m/min, respectively. However, in the case of LIB fibers, the near maximum draw ratio is only 1.1-1.3 (see Table 2). Although the LIB as spun fibers have a low crystallinity of 29.1% at the 5000 m/min speed, the draw ratio is still lower than that of the unperturbed fiber made at the same speed and having a high crystallinity of 39.3%. Thus, it appears that the extensibility is not correlated simply to crystallinity (X_v). In the literature, a third phase, or so-called intermediate phase, or oriented mesophase, or oriented amorphous phase or taut-tie molecular phase has been discussed by many researchers [10, 13-17]. The above mentioned unique combination of properties, *i.e.* low crystallinity, low elongation and high birefringence, might be looked upon as first evidence that the non-crystalline chains of the LIB as-spun fibers are highly ordered or oriented. In addition, with low extensibility (*i.e.* low draw ratio) exhibited during post treatment and at the same time showing low crystallinity and high birefringence, a second bit of evidence presents itself strengthening the notion that LIB fibers have taut-tie non-crystalline chains in the as-spun stage and in their respective post-treated fibers. These two pieces of evidence, if not proof, at least suggest the presence or existence of a taut-tie noncrystalline phase, referred to as a third phase, in addition to the two phases, *i.e.* random amorphous and crystalline phases, commonly accepted as existing in the traditional two-phase model.
Further, boil-off shrinkage (BOS) of unperturbed as-spun fibers starts high, above 60%, at 2000 m/min and drops off rapidly to *ca.* 2% at 5000 m/min while crystallinity increases rapidly and significantly, from *ca.* 8% to *ca.* 39% over the same wind-up speed range. This follows expected behavior reported by Vassilatos *et al.* [18] and Heuvel and Huisman [19]. On the other hand BOS of LIB as-spun fibers drops from *ca.* 13% to *ca.* 10% over the same wind-up speed range of 2000 m/min to 5000 m/min as crystallinity also drops, from 35.3% to 29.1%. The reason for this drop in crystallinity with increasing wind-up speed for LIB fibers has been discussed in a previous publication [4]. The BOS/crystallinity behavior represents a surprising and unique departure from conventional behavior. This suggests that BOS of LIB fibers is not simply dependent upon crystallinity, as is the case in traditionally high-speed spun fibers.
It is obvious that the post-treated LIB fibers have not only higher initial modulus and higher strength but also higher LASE-5 than either of the two commercial tire cords. (See Table 1) The post-treated LIB fibers have LASE-5 of 5.03-5.78 gpd, while the two commercial tire cords have 2.94 gpd and 3.31 gpd, respectively. In addition to superior LASE-5, the post-treated LIB fibers

show superior lower thermal shrinkage, *ca.* 5%, than any of the conventional or low shrinkage commercial tire cords. Notice that both LASE-5 and thermal shrinkage are considered the two main parameters of dimensional stability. From the results, the post-treated LIB fibers have indeed excellent dimensional stability, superior to that of the traditional spin-draw fibers and the two commercial tire cord fibers.

As stated in a previous publication [4], the greater the liquid level, the higher the spinline stress. This situation produces higher birefringence in the as-spun LIB fibers than is found in the unperturbed fibers. The difference in tenacity between as-spun and post-treated fibers is 2.4 gpd, from 7.1 gpd to 9.5 gpd, at LIB depth of 20 com. At a higher LIB depth of 30 cm, the difference is only 0.7 gpd, from 9.6 gpd to 10.3 gpd. The same tendency is also shown in initial modulus. The difference is 21.7 gpd, from 117.2 gpd to 138.9 gpd, at LIB depth of 20 cm while it is only 1.5 gpd, from 139.4 gpd to 140.9 gpd, at LIB depth of 30 cm. LASE-5 also shows the same trend. The difference in LASE-5 between as-spun and post-treated fibers is 1.11 gpd, from 4.26 gpd to 5.37 gpd, at LIB depth of 20 cm, while it is only 0.41 gpd, from 5.07 gpd to 5.48 gpd, at LIB depth of 30 cm. According to the results, as LIB depth was increased from 20 cm to 30 cm, the properties of the as-spun fibers move closer to those of their respective post-treated fibers. This is an important and suggestive finding. It lends credence to the statement that the LIB spinning process has the potential to produce high performance fiber via high-speed spinning in one step instead of the traditional two steps.

Structure of As-spun and Post-Treated Fibers

In the LIB spinning process, both stress and temperature profiles were modified via controlled threadline dynamics [2] leading to the unique properties of low crystallinity and high birefringence and implying that the occurrence of folded chains has been reduced and that of extended chains has increased. Moreover, with the properties of high birefringence, low crystallinity and the propensity of low draw ratio, the existence of taut-tie molecular chains in the non-crystalline region seems reasonable and acceptable. Orientational nucleation then proceeds rapidly to induce orientational crystallinity during post drawing and annealing. Meanwhile, taut-tie molecular chains are further developed and crystallized.

Crystallization during spinning of the unperturbed as-spun fibers proceeds following the concepts of stress-induced crystallization [20]. At take-up speeds of 2000-3500 m/min the cyrstallinity (X_v) of as-spun LIB fibers is higher than that of unperturbed fibers. However, when the take-up speed is increased to 4000 m/min and above, spinline stress is increased and crystallinity of the LIB spun fibers is significantly suppressed when the unperturbed fibers are compared at each take-up speed. At 5000 m/min the crystallization of the unperturbed fiber is 39.3% while that of the LIB fiber is only 29.1%. X-ray analysis confirms the reduced crystal size for the LIB fibers. [4]

After post treatment, the apparent crystal sizes of L_{010}, L_{100}, and L_{105} of the LIB post-treated fibers are still smaller than those of post-treated unperturbed fiber. The post-treated LIB and unperturbed fibers have essentially the same crystallinity (X_v) of *ca.* 55%. The same crystallinity level and the small crystal dimensions imply that the number of crystals in the post-treated LIB fiber is greater that that of post-treated unperturbed fiber. This means that the distance between crystals is shortened, *i.e.* shorter noncrystalline chains are formed. In addition, the unique high amorphous orientation factor (f_a) of *ca.* 0.85-0.90 (see Table 2) is higher than that of unperturbed fiber f_a of below 0.70. It implies that the noncrystalline chains are highly extended. These data and the high amorphous orientation factor (f_a) strongly support the assumption that the noncrystalline region makes a dominating contribution toward the high orientation detected. This provides more evidence for the likely existence of a taut-tie noncrystalline phase in the LIB as-spun and post-treated fibers.

The Relationships Among Structural Properties

Both crystalline and noncrystalline parts contribute to orientation. Table 2 illustrates that crystalline orientation factor (f_c) of all the post-treated fibers have reached *ca.* 0.92 to 0.96. The amorphous orientation factor (f_a) of the post-treated LIB fibers is *ca.* 0.85 or higher while it is relatively low at *ca.* 0.72 or less in the low shrinkage tire cord and the post-treated unperturbed fibers. It seems that the overall orientation is dominated by the noncrystalline contribution.

The fraction of taut-tie noncrystalline phase (*TTNC%*) is calculated on the basis of a parallel-series three-phase model proposed by Takayanagi *et al.* [10]. In the as-spun fibers, the fraction of taut-tie noncrystalline phase of the LIB fibers is significantly higher than that of unperturbed fibers as shown in Table 3. When the LIB depth was increased from 20cm to 30cm at take-up speed of 5000 m/min, the fraction of the taut-tie noncrystalline phase is proportionally increased. As shown in Figure 3, a linear relationship is found between the fraction of the taut-tie noncrystalline phase and initial modulus of PET fibers.

4. Conclusions

LIB fibers, both as-spun and post treated, demonstrate superior mechanical properties and excellent dimensional stability. They also exhibit unique characteristics possessing high amorphous orientation, high fraction of taut-tie noncrystalline phase with low crystallinity, and exhibit high strength and high initial modulus in the as-spun state. The high fraction of taut-tie noncrystalline phase is produced by the stress-enhanced LIB spinning system. Further, with post drawing and annealing the extended chains of LIB as-spun fibers are further rapidly developed and crystallized along the fiber axis. In addition, the enhanced fraction of taut-tie noncrystalline chains with shorter length play a dominant role in the structure responsible for the unique combination of superior properties. It is found that overall orientation is dominated by the noncrystalline contribution. The calculated fraction of the taut-tie non-crystalline phase agrees with and strongly supports evidence for the existence of a third phase in addition to those generally accepted in the traditional two-phase model. It is also speculated that initial modulus is primarily determined by the fraction of the taut-tie noncrystalline phase present in the PET fibers.

References

[1] T. Kawaguchi, Chap. 1 of *"High-Speed Fiber Spinning,"* Ed. By A. Ziabicki, and H. Kawai, Wiley-Interscience (1985).

[2] J. A. Cuculo, P.A. Tucker, and G. Y Chen, *J. Appl. Polym. Sci., Appl. Polym. Symp.*, 47, 223 (1991).

[3] C.Y. Lin, P. A. Tucker and J. A. Cuculo, *J. Appl. Polym. Sci.*, 46, 531 (1992).

[4] G. Wu, Q. Zhou, J. Y. Chen, J. F. Hotter, P.A. Tucker and J. A. Cuculo, J. Appl. Polym. Sci., 55, 1275 (1995).

[5] Q. Zhou, G. Wu, P.A. Tucker and J. A. Cuculo, *J. Polym. Sci., Polym. Phys.*, 33, 909 (1995).

[6] G. Wu, P. A. Tucker and J. A. Cuculo, *Polymer*, 38, 5, 1091 (1997).

[7] P. B. Rim, and C. J. Nelson, *J. Appl. Polym. Sci.*, 42, 1807 (1991).
[8] L. E. Alexander, *"X-Ray Diffraction Methods in Polymer Science,"* 191, p. 335, reprint ed., Krieger (1985).
[9] J. H. Dumbleton, *J. Polym. Sci., Part A-2*, 6, 795 (1968).
[10] M. Kamezawa, K. Yamada, and M. Takayanagi, *J. Appl. Polym. Sci.*, 24, 1227 (1979).
[11] C. L. Choy, M. Ito, and R. S. Porter, *J. Polym. Sci.* , *Polym. Phys.*, 21, 1427 (1983).
[12] T. Thistlethwaite, R. Jakeways, and I. M. Ward, *Polymer*, 29, 61 (1988).
[13] J. Shimizu, N. Okui, and K. Kikutani, Chap. 15 of *"High- Speed Fiber Spinning,"* Ed. By A. Ziabicki, and H. Kawai, Wiley-Interscience (1985).
[14] A. Peterlin, Chap. 10 of *"Ultra-High Modulus Polymers,"* Ed. By A. Ciferri and I. M. Ward, Applied Science (London) (1979).
[15] T. Sun, A. Zhang, F. M. Li and R. S. Porter, *Polymer*, 29, 2115 (1988).
[16] H. A. Hristor and J. M. Schultz, *J. Polym. Sci., Part B: Polym. Phys.*, 28, 1647 (1990).
[17] G. Wu, J. –D Jiang, P. A. Tucker and J. A. Cuculo, *J. Polym. Sci., Part B: Polym. Phys.*, 34, 2035 (1996).
[18] G. Vassilatos, G. H. Knox and H.R.E. Frankfort, Chap. 14 of *"High-Speed Fiber Spinning,"* Ed. By A. Ziabicki, and H. Kawai, Wiley-Interscience (1985).
[19] H. M. Heuvel and R. Huisman, Chap. 11 of *"High-Speed Fiber Spinning,"* Ed. By A. Ziabicki, and H. Kawai, Wiley-Interscience (1985).
[20] A. Ziabicki, Chap. 2 of *"High-Speed Fiber Spinning,"* Ed. By A. Ziabicki, and H. Kawai, Wiley-Interscience (1985).

Table 1. A comparison of the mechanical properties of liquid isothermal bath (LIB) as-spun and their respective post-treated fibers with fibers made by the unperturbed processes under different conditions, and with two commercial tire cords.

Take-up speed (m/m) and process	Draw ratio (DR)	Temperature of 1st heater (°C)	Temperature of 2nd heater (°C)	T/E/M (gpd/ %/gpd)	LASE-5 (gpd)
3500 LIB[a]	-	N/A	N/A	8.3/14.8/128.8	3.64
3500 LIB/DA	DR=1.1	180	N/A	9.1/ 9.7/134.1	5.03
〃	DR=1.1	180	220	9.1/10.7/138.9	5.49
〃	DR=1.2	180	220	10.0/ 9.8/147.5	5.78
5000 LIB[b]	-	N/A	N/A	9.6/10.5/139.4	5.07
5000 LIB/DA	DR=1.1	180	220	10.3/ 8.7/140.9	5.48
5000 unperturbed[c]	-	N/A	N/A	4.1/67.5/ 62.5	1.23
5000 unperturbed/DA	DR=1.5	180	220	5.7/16.1/116.8	3.13
Conventional tire cord[d]	unknown	-	-	9.5/16.6/ 96.1	2.94
Low shrinkage tire cord[d]	unknown	-	-	7.4/16.5/ 87.9	3.31

[a]: LIB depth=45cm, spinning denier=6dpf; [b]: LIB depth=30cm, spinning denier=4.5dpf
[c]: spinning denier=4.5dpf; [d]: from Ref.7; N/A: not applicable

Table 2. Draw ratio (DR), crystallinity (X_v), apparent crystal size, crystalline orientation factor (f_c) and amorphous orientation factor (f_a) of post-treated LIB and unperturbed fibers, and commercial tire cord.

Take-up speed(m/min) & spin-draw process	Draw ratio (DR)	X_v (%)	Apparent crystal size (Å) L_{010}	L_{100}	$L_{\bar{1}05}$	f_c	f_a
2000 LIB/DA	1.3	53.4	57.5	43.3	62.1	0.945	0.848
3500 LIB/DA	1.2	55.3	58.6	41.4	60.0	0.956	0.896
5000 LIB/DA	1.1	55.3	52.5	41.6	58.0	0.946	0.901
Low shrinkage tire cord	unknown	45.3	49.8	37.4	54.9	0.936	0.723
3000 unperturbed/DA	1.7	58.8	63.2	43.1	59.0	0.940	0.679
5000 unperturbed/DA	1.5	55.4	76.5	54.2	58.9	0.926	0.641

Table 3. The effect of LIB depth on individual fraction of taut-tie noncrystalline phase (*TTNC%*), initial modulus, and crystallinity (X_v) of the LIB as-spun fibers spun at take-up speed of 5000 m/min.. Unperturbed (w/o LIB) as-spun fiber are also included for comparison.

LIB depth (*cm*)	Fraction of taut-tie noncrystalline phase (%)	Initial modulus (*gpd*)	X_v (%)
20	10.69	117.2	32.3
25	12.21	129.7	27.7
30	13.31	139.4	29.1
w/o LIB	4.06	62.5	39.5

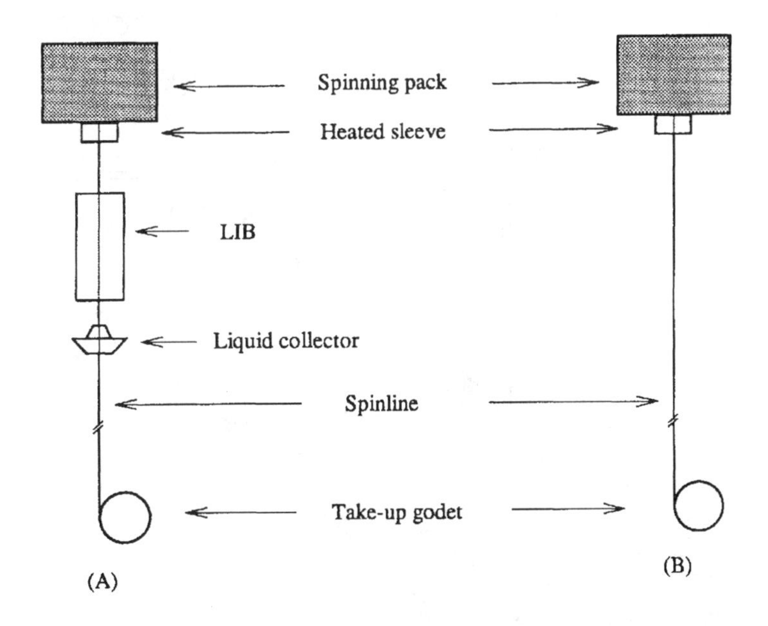

Figure 1. Schematic diagram of melt spinning via two different processes: (A) liquid isothermal bath (LIB) spinning process; and (B) unperturbed spinning proces

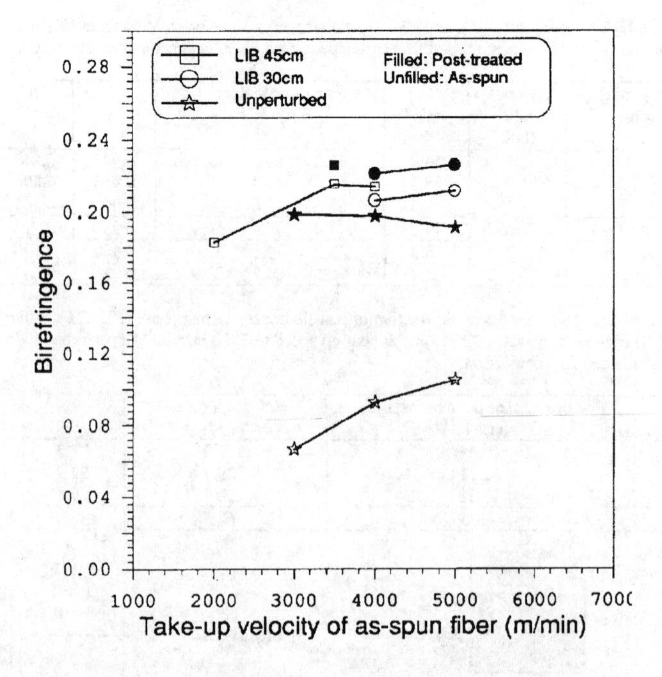

Figure 2.

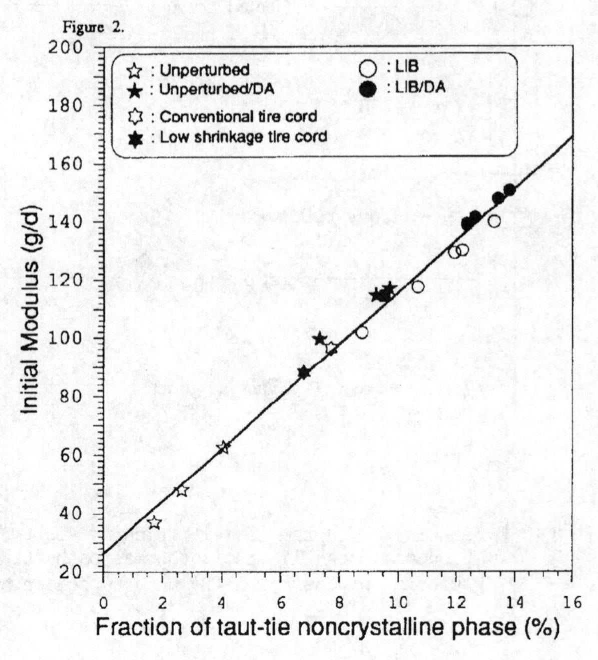

Figure 3

Aerotiss® 3.5D - A breakthrough in braiding

CAHUZAC Georges
Aerospatiale - Branche Espace et Défense
Industry Business Unit
BP 11 33165 St Médard en Jalles - France
Tel: 33 5 56 57 36 77 Fax: 33 5 56 57 35 48

Abstract

AEROSPATIALE has succeeded in developing a new type of textile structure named Aerotiss 3.5D . It's a multilayer fabric with 3 directions of fibers . This interlock fabric has good mechanical properties. This product is made on a mandrel , circular or with any kind of shape , by example a blade for a wind power generator or a boat mast . It also can be cut to obtain a flat fabric.
The machinery is a new type of circular braiding machine which minimizes the ratio between the size of the part and the size of the machine. It's a cost effective machine because it works like a conventional braiding machine by exchanging bobbins between rotating wheels.

Introduction

AEROSPATIALE has developed during the last twenty years several processes for making thick textile structures labeled **Aerotiss®** . Its last development has been in the field of braiding machinery . Braiding is well known as an industrial process for making little complex items by laying successively many layers . But its use for making thick fabric in one step had not been successful because of the poor quality of the pattern which can be created by the mean of a multilayer braiding machine and because of the magnitude effect between the size of a part and the size of the machine. While 3D multilayer Interlock braiders have been used by many companies in USA from few years and some new ones have just been made in Europe, they are not able to braid our textile architecture (fig. 1) which needs a braiding machine designed for that purpose.

Aerotiss 3.5D architecture

This textile structure is a multilayer fabric which has 3 main directions 0° and ± 60° with light slopes and few voids into the unit cell. And this fabric has a increased damage tolerance and energy absorption due to the interlocking of all the layers .
It will be manufactured on a close mandrel but it can also be cut and flattened . That fabric is very suitable for making the wall of a part like a flange or a spar.
Multilayer interlock braids will be used as an economic way to have close items made directly at their final shape on a mandrel and ready for RTM process . Any kind of shape is possible from a tube to a propeller blade .

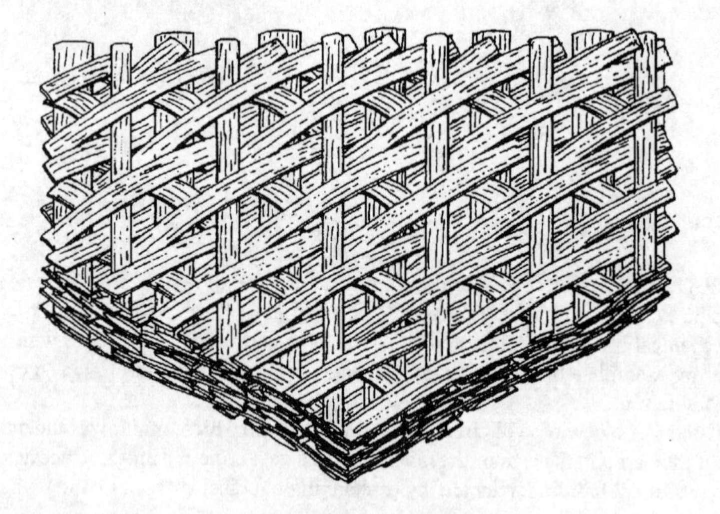

Fig. 1 - Aerotiss 3.5D fabric

The aerotiss 3.5D fabric shown has 7 layers of 0° yarns , 3 layers of 60° yarns and 3 layers of -60° yarns. It's easy to see the light waves of the 60° yarns while the 0° yarns are straight. Calculated mechanical characteristics are very good because they are directly linked to the angle of the yarns.

This type of structure can be done on a patented type of circular braiding machine . The main features of that machine are the position of the bobbin carriers and the 0° yarns which are situated between 4 bobbin carriers.

Bobbin carrier design

The main difference between existing braiding machine and this new type is the pattern of the bobbin carriers which reproduces the pattern of the 3.5D fabric.

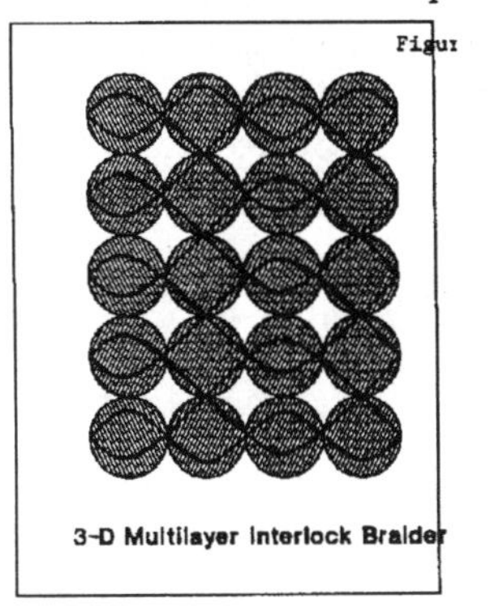

Fig.2 Traditional track paths and Horn Gears

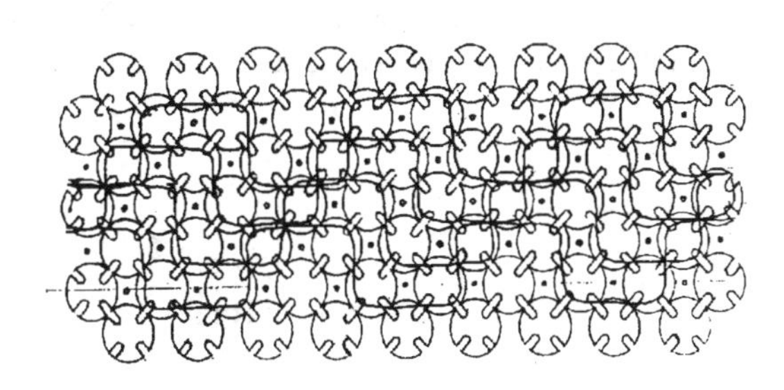

Fig.3 Disposition of the bobbin carriers and path for the 3.5D fabric

That disposition allows to have the bobbins moving along circular or radial paths depending of the positions of the needles. The even levels of bobbins are moving in one direction and the odd levels in the opposite direction. We obtain the same degree of freedom as on a weaving loom to create different types of fabric with three directions of yarns. That explains why it's possible to create good fiber architectures from such a design.

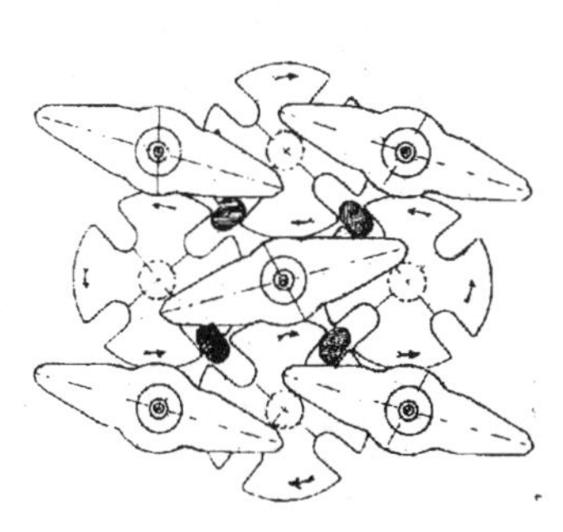

Fig.5 detail of the bobbin carriers and needles . O° yarns are situated into a hole in the needle center.

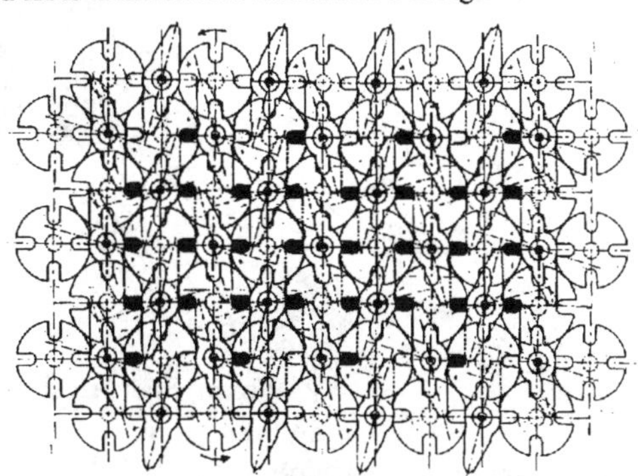

Fig.4 Position of the needles allowing a transverse crossing

Standard size 3.5D braiding machine

It's an example of braiding machine which can be made by using the 3.5D architecture. the aerotiss 3.5D fabric will have 10 layers of yarns, 4 layers at 0°, 3 layers at +60° and 3 layers at -60° . Depending of the size of the used yarns , diameters between 50 mm and 200mm and thickness between 1.5 and 7 mm can be done of that machinery.

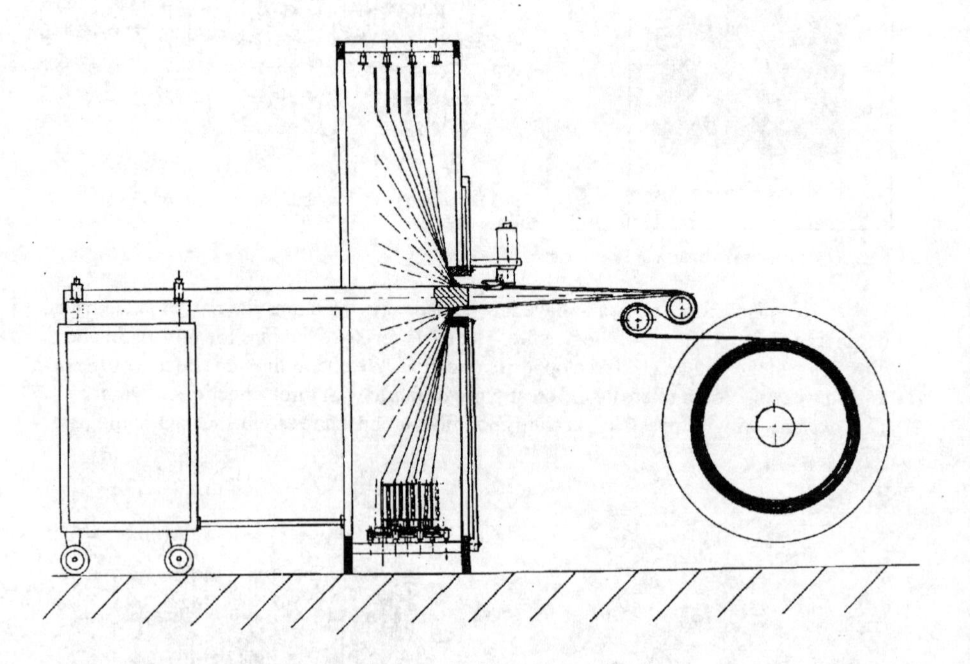

Fig.6 - Standart size 3.5D braiding machine

Diameter of that machine : 4 m

120 rows of 5 and 4 bobbin carriers

420 bobbins of 0° yarns

720 moving bobbins, 360 bobbins of +60° and 360 bobbins of -60° yarns.

That braiding machine will be the greatest one in EUROPE.

We are looking for partners and companies interested in developing such a machine.

We are also interested by any potential application for that type of machine and product.

Huge size 3.5D braiding machine

Making a big item on a braiding machine looks like an oxymoron. But it's possible by using a huge machinery. This braiding machine is not more complicated than one with a standard size but needs parts made in big numbers. Its axes is vertical to have an easy access to put all the bobbins into the machine before starting a production run . The fabric is rolled up on a mandrel situated on the shop floor. Making a particular item on a specific mandrel , like an exit cone of a nozzle, is possible by moving vertically the central mandrel.

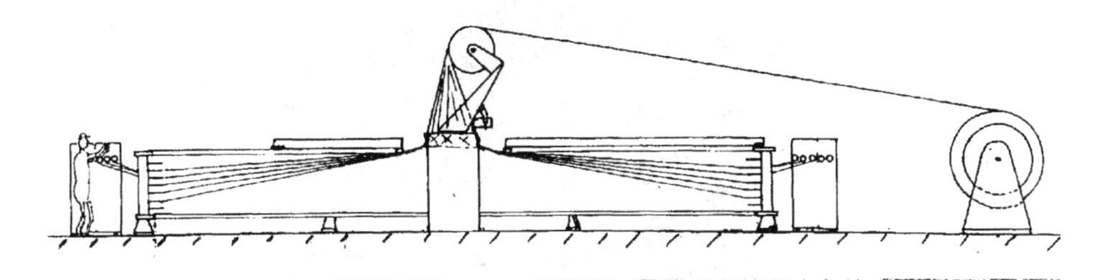

Fig.7

Main characteristics of this machine are :

a mean diameter around 13m,
3510 bobbin carriers ,
2970 straight yarn bobbins
and 5400 moving bobbins.

The fabric made by this braiding machine will be a 3m width fabric with a 3 mm thickness and made of 21 layers , 11 at 0°, 5 at +60° and 5 at -60° . Production rate will be around 40 m an hour.

Conclusion

This new type of braiding machine is a real breakthrough in the field of machinery for making composite materials . It can produce complex items in a one step process . Aerospatiale is looking for partners interested in developing that technology or finding a interest in using items made by this process for their own products .

Fig.8 Aerotiss 3.5D carbon fabric

REFERENCES

1. D. BROOKSTEIN -Albany International Research Co. - ECCM6 - 3D braided composites- Design and Applications

Electrically conductive glass fibre reinforced epoxy resin

M. Kupke*, K. Schulte*, H.-P. Wentzel**

* Technical University Hamburg-Harburg, Polymers & Composites
21071 Hamburg, Germany

** Daimler-Benz Aerospace Airbus, EVT, 28183 Bremen, Germany

Abstract

The research on an industrially manufactured, electrically conductive glass fibre reinforced epoxy prepreg for aviation applications is reported. In a co-operative effort between Technical University Hamburg-Harburg (TUHH) and Daimler-Benz Aerospace Airbus (DASA) a new glass-epoxy composite with both electrical and good mechanical properties was successfully developed.
The electrical conductivity was achieved adding carbon black as a conductive filler into the epoxy matrix and this at a very low level of content. The range of possible applications for this new material is not only limited to aviation. It can also be used in other transport systems or in electric and electronic devices.

1. Motivation

Glass fibre reinforced plastics (GFRP) are an electrically insulating material and are most widely used in numerous applications, including the aircraft industry (e. g. radome, fairings). To avoid electrical charges or electrical and magnetic fields from disturbing communications or electronic devices, metallic fabrics, foils, or antistatic or conductive coatings are used at present together with GFRP components. Conductive coatings must be stored, painted and discarded to manufacture and to repair components, additionally there are ageing problems and there is an increase in weight.
The development of an electrically conductive, non ageing matrix system for GFRP with an electrically resistivity less than 10^8 Ωcm and mechanical properties equal to present matrix systems should make it easier and cheaper to produce, maintain and repair components. Figure 1 shows the present state and the future design of sandwich components with electrically conductive or antistatic GFRP.

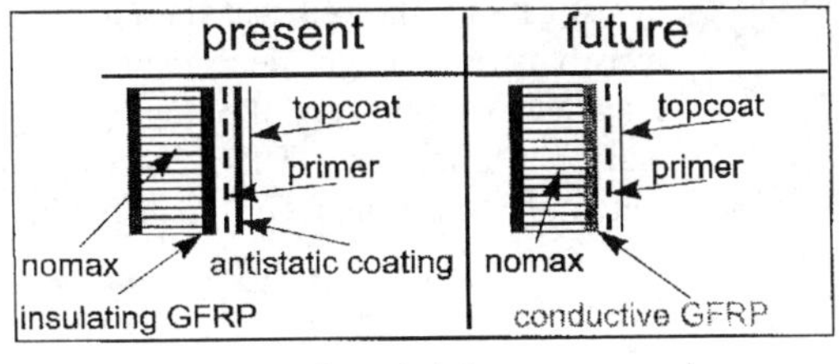

Figure 1: Design of sandwich components in present and future.

With an electrically conductive GFRP a conductive coating is not anymore necessary and present problems with the coatings, like painting and drying, that need much time and manpower, can be avoided.

2. Previous investigations

There are several possibilities to develop an antistatic / electrically conductive polymer matrix:

- intrinsically conductive polymers (ICP) like polypyrrole or polyacethylene,
- increase of ionic conductivity of standard polymers,
- an electrically conductive filler like carbon black.

Investigations have shown, that intrinsically conductive polymers lack in ageing, would lead to reduction in long term properties [1]. This is not acceptable for aircraft components. The ionic conductivity of standard polymers cannot be increased to the levels demanded, (only above the glass transition temperature T_g), that otherwise leads to insufficient mechanical properties in the components [2]. Therefore the best choice at present seem to be conductive fillers of carbon black. It has been shown in earlier works, that carbon black can form an electrically conductive network at a very low level of content [3, 4]. The dependence of specific electrical conductivity to the content of carbon black is schematically shown in figure 2.

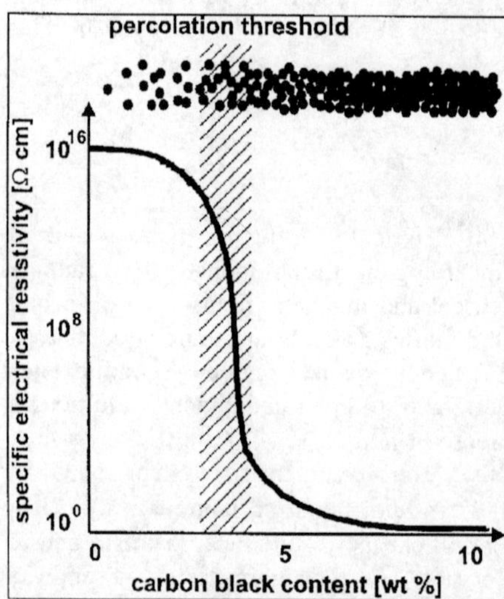

Figure 2: Qualitative dependence of electrical resistivity to carbon black content. Hatched part is called *percolation threshold*.

The hatched part with the greatest slope is called *percolation threshold*. It could be shown, that it is possible to shift the percolation threshold to lower contents of carbon black with increasing ionic conductivity [4, 5]. In an optimised system of epoxy resin (Araldit LY556 / HY932, Ciba Geigy) with carbon black (Printex XE2, Degussa) a specific resistivity of less than 10^8 Ωcm was achieved with a content of only 0.1 wt% (= 0.06 vol%) carbon black and additional 0.003 wt % copper-chloride ($CuCl_2$). Copper-chloride was added to increase the ionic conductivity. The mechanical properties of GFRP are not influenced if the content of carbon black does not exceed 1.5 wt%. However, the investigations showed that the electrical resistivity is strongly dependent on the manufacturing process, the impregnation and consolidation. This makes the transfer of know how from the laboratory stage into industrial manufacturing difficult.

3. Industrially manufacturing of electrically conductive GFRP

The goal was the industrial production of prepregs and laminates with a defined electrical resistivity but retaining the mechanical properties.
Materials used were

- matrix: epoxy resin, Rütapox VE 3828, Bakelite AG
- reinforcement: glass fibre fabric, US-Style 7781,
- conductive filler: carbon black, Printex XE2, Degussa AG.

The matrix is a high performance, aviation qualified epoxy system used for glass fibre prepregs. The carbon filler is a well known and already characterised high conductivity carbon black in previous investigations. As a result of these, carbon black contents of 1.3 wt% and 1.5 wt% was chosen to achieve the lowest level in resistivity without influencing the mechanical properties.
Within the manufacturing process for electrically conductive GFRP three steps had to be developed and optimised:

1. the dispersion of carbon black in the epoxy resin,
2. the impregnation of glass fibre fabrics / prepreg manufacturing,
3. the lamination and consolidation of composite sheets.

The first manufacturing step was performed together with *Altropol Kunststoff GmbH, 23617 Stockelsdorf, Germany*. Using the adjusting processing variables, it was possible to produce a reproducible and excellent epoxy polymer with a fine and homogeneous dispersion of carbon black (named *Neukadur EN 403 / 404* for *1.3 wt% / 1.5 wt%*, respectively). The quality of dispersion is very important for the following manufacturing steps. The prepreg manufacturing was performed at *Stesalit AG, 4234 Zullwil, Switzerland*. No additional problems arose from manufacturing prepregs with an epoxy that contained carbon black fillers. It was as easy as the production of the traditional prepregs. The manufacturing of laminates at *DASA, 28183 Bremen, Germany* was as well without any problems. The lamination and consolidation was equal for both the filled and unfilled prepregs. The laminate sheets were produced in an autoclave process. In order to control the build up of the conductive network the variation of the electrical resistivity of the carbon black material was in situ monitored during the autoclave process.

4. Properties of industrially manufactured electrically conductive GFRP

Three different kinds of electrical conductivity were measured in order to verify if the specific values have been achieved:

- specific resistivity through bulk material in two directions (figure 3)
- specific surface resistivity in two directions (DASA specification, figure 4)
- specific surface resistivity in two directions (figure 6)

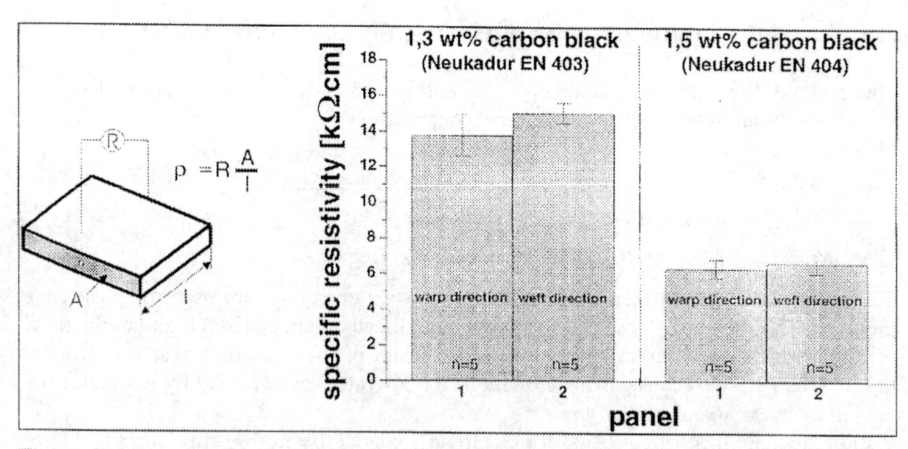

Figure 3: Measurement of specific resistivity in warp and weft direction for 1.3 wt% and 1.5 wt% of carbon black.

It is illustrated in figure 3, that the specific resistivity is in the range of kΩ. The specific resistivity of GFRP with 1.3 wt% carbon black is about two times higher than that of the only slightly more filled material (1.5 wt%). This shows, that the carbon black content is near the percolation threshold.

Figure 4 shows the surface resistivity according to a DASA specification.

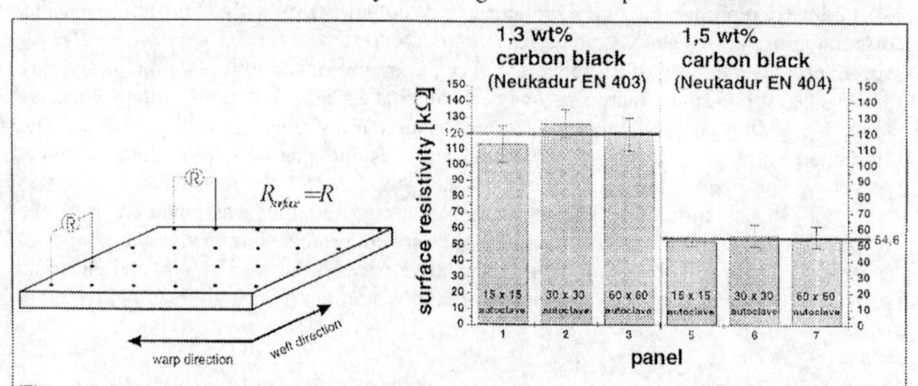

Figure 4: Measurement of surface resistivity according to DASA specification in warp and weft direction for 1.3 wt% and 1.5 wt% carbon black. The average values of all measurements, warp and weft direction, are shown. The distance between measurement points are 5 cm in reality. Panel's size is given in [cm x cm].

The resistivity of the laminate with 1.3 wt% carbon black in its matrix is again two times higher than in the laminate with 1.5 wt% carbon black. No significant influence of the panel's size could be observed. The achieved results of about 120 and 55 kΩ respectively are very satisfying and much better than expected. With this method it is possible to plot the surface resistivity distribution of a panel as shown in figure 5.

In figure 5 are additionally shown three different methods of manufacturing

a) Impregnated and laminated by hand in laboratory,

b) industrially impregnated to prepregs, then consolidated in a laboratory press at 1.5 bar.

c) industrially impregnated, then consolidated in an autoclave process at 2.0 bar.

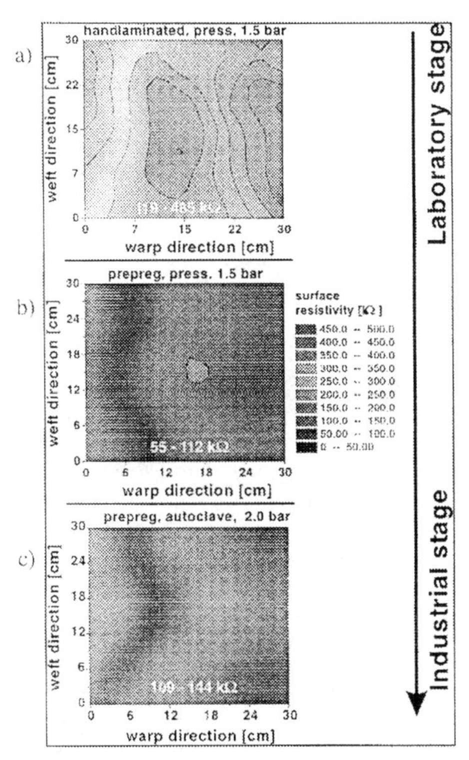

Figure 5: Resistivity distribution for differently manufactured panels.
Shown carbon black content is 1.3 *wt%*.

The size of the panels is always 30 x 30 cm. The carbon black content is 1.3 wt% in each case. From top to bottom the improved in the development from the laboratory to industrial stage is demonstrated. Figure 5a shows an inhomogeneous resistivity distribution. Due to handlaminating marks of the shear motion can be seen. The level of average resistivity is relatively high. In figure 5b the resistivity distribution is much more homogeneous due to the improved impregnation process with less shear motions during prepreg manufacturing. The level of the average resistivity is very low. The most homogeneous distribution due to a homogeneous pressure is achieved in an autoclave (figure 5c). The average resistivity is higher then for the prepreg-laminate consolidated in a press, due to the higher pressure in the autoclave. The dependence of resistivity to consolidation-pressure was already found in previous investigations [6].

With an alternative method a more realistic specific surface resistivity was measured (figure 6).

The values of specific surface resistivity are about two times lower than those according to DASA specification. The resistivity of the GFRP with the 1.3 wt% carbon black filled matrix is again two times higher than the one with 1.5 wt% carbon black. However, the electrical properties of the newly developed material sufficient for numerous applications and can be adapted to the individual value needed.

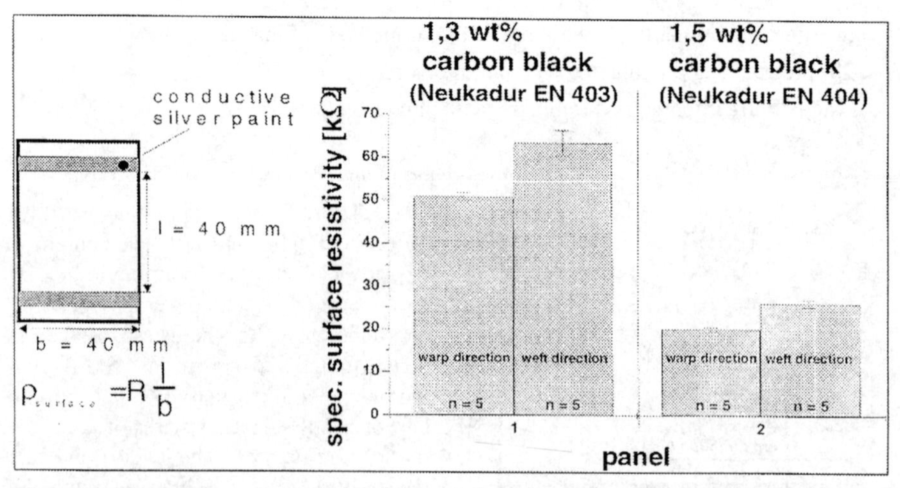

Figure 6: Measurement of specific surface resistivity in warp and weft direction. n: number of specimens

The mechanical properties were determined in tensile tests (DIN 29971) and in three point bend tests (DIN EN 2377). The mechanical test were performed in order to show if there are any differences in properties between the carbon black filled and the traditional matrix systems. In order to avoid any influence of the manufacturing process only those specimen were compared that were consolidated in a joined autoclave process. It was the aim of the investigation to retain the mechanical properties of the unfilled prepreg systems also in the carbon black filled composite. It can be seen from figure 7 that this goal could easily be achieved, when comparing the mechanical properties even an increase in the mechanical properties can be observed.

It can be assumed that the fine dispersed carbon black particles (diameter $\overline{d} \approx 20nm$) act in a nanoscale as a reinforcement phase. However, it is astonishing that an increase in the fracture strain and the interlaminar shear stress can be observed. In further investigations it has to be made clear, if the distinctive structured surface of the primary carbon black particles which results in the low percolation level and high conductivity respectively is the reason for this.

5. Applications

The possible applications of this new material are not only within the aircraft industry, where GFRP's are used mainly for fairings, radome or the fin of the vertical and horizontal stabilizer. It is also possible that there are applications within the automotive industry, where antistatic and electrical properties together with an improved optic (black colour and no patch-like structure due to matrix-rich zones), are of importance. Also in the electro and electronic industry components made of GFRP can be produced, mainly those in which the insulating properties of conventional GFRP is of disadvantage.

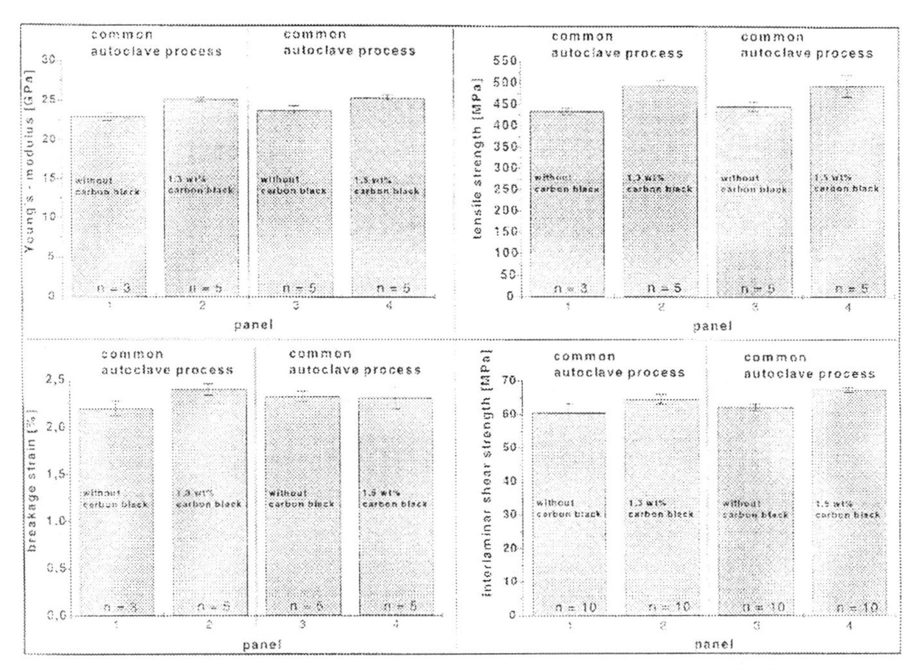

Figure 7: Measured mechanical properties of both filled and unfilled GFRP. The number of specimens is given with *n*. To avoid influences by manufacturing respectively curing process, specimens with common autoclave processes were compared. Young's modulus, tensile strength and breakage strain were measured in unidirectional tensile test according to DIN 29971. The interlaminar shear strength was measured in a three point bending test according to DIN EN 2377.

Electrically conductive epoxy systems are not only reduced to that presented in this publication. In previous investigations it was shown that the production process is transferable to other epoxy systems. One has to state, that with normal epoxy systems the production of a conductive polymer is much more easy as in the one introduced in this publication.

In addition to glass fibre or aramide fibre reinforced components the application can be increased also to parts manufactured of unreinforced resins, as they are used at present in the electronic industry.

With a variation of the amount of carbon black, higher or lower electrical properties can be achieved. In case of higher electrical resistivity the mechanical properties will not be influenced, however, in case of very low electrical resistivity one has to expect a reduction in mechanical properties due to the higher amount of carbon black.

References

[1] H. Münstedt, Kunststoffe 79, 1989, page 6

[2] R. Schüler. *Entwicklung polymerer Verbundwerkstoffe mit elektrischer Leitfähigkeit.* PhD thesis, Technical University Hamburg-Harburg, December 1994

[3] B. Jachym in: *Carbon Black-Polymer Composites.* E. K. Sichel (Hrsg.), New York: Marcel Dekker, 1982

[4] G. Oblieglo and U. Herrmann, Kunststoffe 74, 1984, page 97

[5] R. Schüler, J. Petermann, K. Schulte, H.-P. Wentzel. *Agglomeration and electrical percolation behaviour of carbon black dispersed in epoxy resin.* Journal of Appl. Polymer Science, 1997, pp. 1741 - 1746

[6] R. Schüler, A. Mulkers. *Elektrisch leitfähige Epoxidharze.* Co-operative effort DASA-TUHH, interim report (not published), Technical University Hamburg-Harburg, 1995

COMPACTION AND RELAXATION OF DRY AND SATURATED TEXTILE REINFORCEMENTS: THEORY AND EXPERIMENTAL RESULTS

F. Robitaille, R. Gauvin

CRASP - Center for Applied Research on Polymers, Mechanical Engineering Department, École Polytechnique de Montréal, C.P. 6079, Succ. "Centre ville", Montréal, Québec, Canada H3C 3A7

Abstract

Textiles represent, by far, the most common reinforcements in thermoset composites production. When subjected to normal compaction forces, textile reinforcements are usually regarded as non-linear elastic materials. However experimental evidence shows that their behaviour is influenced by factors such as the number of layers, the presence of a resin, etc; stress relaxation was also observed. In a context where pre-production process simulations become the norm, knowledge of the compaction behaviour is required, given the effect of the reinforcement's architecture on properties such as the permeability. This paper presents the main results of a program undertaken to define this behaviour. Firstly, the conclusions of a survey of published results are presented. Secondly, the results of a comprehensive experimental program on the effects of selected processing parameters are reviewed. Finally, the phenomenon of fibre reorganisation is discussed.

1. Introduction

Many studies of the mechanical behaviour of textiles were published. Some papers present results of compaction experiments performed on textile reinforcements. These works show curves of the compaction pressure P as a function of the fibre volume fraction v_f or porosity ϕ , as these are major production parameters. Most authors comment on the effect of relevant processing parameters, and some propose semi-empirical compaction models for idealised fibre assemblies [1].

On the other hand, analytical studies where the geometry of the assemblies is thoroughly defined were published [2,3,4]; here most authors assimilate the textiles to continuous media and propose expressions for the terms of a stiffness matrix. This approach has not been used to describe the phenomena observed in composites production. The geometrical information required by these models may explain this; furthermore, the objectives motivating the two approaches differ.

Finally, other works show the effect of tow compaction on the in-plane properties of the textiles (extension, shearing and drape) [5]. The authors of these works define the geometry of the textiles at the level of the tows, instead of considering individual fibres. While such definitions contain much less information than the ones established for the analytical works mentioned above [3,4], they may allow the description of phenomena observed on the reinforcements, at the macroscopic level; the presence of tows, stitch threads and interlacing patterns is not considered in models such as [1]. In this work it is shown that the dimensions of the tow's sections evolve during the compaction, with an effect on properties such as v_f. After discussing some elements of the analytical works, the evolution of the tow geometry is considered.

2. Overview of analytical and semi-empirical works

Van Wyk [2] proposed a compaction model for fibre assemblies where the fibre segments extending between two permanent contacts with other fibres are assimilated to beams deforming in flexion only. The model is based on the beam bending equation and the average distance between contact points is obtained from consideration of the geometry of the assembly of randomly-oriented fibres. The expression of the model is:

$$P = CT \cdot E \cdot \left(\frac{1}{V^3} - \frac{1}{V_o^3} \right) \tag{1}$$

where E is the fibre's modulus, V and V_o are the initial and current total volumes of the fibre assemblies and CT is a fitting parameter. The distance between two successive fibre contacts b is proportional to V, resulting in a rigidifying behaviour. Komori and Makishima defined distribution functions for the segment orientations [6]; Lee and Lee [3] based their expressions for the initial values of the moduli of fibre assemblies on these functions and van Wyk's work. The authors use the following equation for the average segment length b:

$$b = \frac{V}{2\, L_T D} I \tag{2}$$

where L_T is the total fibre length, D is the fibre diameter and I is an average of projected lengths; I is a function of the segment orientations and does not vary otherwise. The other basic ideas behind Lee and Lee's expressions are the same as in van Wyk's work. Pan and Carnaby [7] included the slippage at fibre contacts in their compaction model for arbitrarily oriented assemblies. The contact behaviour is given by:

$$C_p \geq C_n\, \mu + WF_o\, b \tag{3}$$

where C_p and C_n are the parallel and tangent force at the contact of two fibres, μ is a coefficient of friction and WF_o is a constant force per unit length. Equation (3) allowed the representation of the compaction hysteresis in homogeneous fibre assemblies. The cited works and others allowed fibre bending and slippage to be identified as key phenomena behind the compaction of fibre assemblies. In all these works the proportionality between the segment length b and the volume of the assembly V is respected, apart from deviations resulting from fibre reorientation during compaction; P is therefore proportional to v_f^3.

Gutowski et al [1] published a well-known model for the compaction of aligned fibres. This work is also based on fibre bending; however the authors introduced a higher bound on v_f, termed v_a and related to the stacking of parallel cylinders. Hence the model is adequate for parallel fibres such as found in prepreg tapes; however when used as a fitting equation for textile reinforcements, the resulting lower v_a values are more difficult to justify.

Karbhari and Simacek [8] proposed a model for the mechanical behaviour of assemblies of unlubricated fibres. The authors consider Coulomb's law of friction for the behaviour at the fibre contacts, and mention the probable effect of the geometry of the fibre network and resulting interference on the modelled behaviour.

3. Review of experimental works

Published experimental results for the compaction and relaxation of random mats and textile reinforcements were fitted to two power laws by Robitaille and Gauvin [9]:

$$\boldsymbol{v_f} = \boldsymbol{A} \cdot \boldsymbol{P}^{\boldsymbol{B}} \tag{4}$$

$$(\boldsymbol{P}/\boldsymbol{P}_o) = 1 - \boldsymbol{C} \cdot \boldsymbol{t}^{(1/\boldsymbol{D})} \tag{5}$$

where $\boldsymbol{P_o}$ is the pressure initially applied and t is the time. $\boldsymbol{A}$ is the fibre volume fraction at unit pressure, $\boldsymbol{B}$ is the compaction stiffness index ($\boldsymbol{B}<1$; rigidity raises faster as $\boldsymbol{B}$ is lowered), $\boldsymbol{C}$ is the pressure decay at 1 s, ($1/\boldsymbol{D}$) indicates the concavity of the relaxation curves. Two additional parameters are reported in [1]: $\boldsymbol{M}$ is a representative rigidity equal to the slope of a straight line drawn between $\boldsymbol{P} = 0.2\ \boldsymbol{P_{max}}$ and $\boldsymbol{P} = \boldsymbol{P_{max}}$, $\boldsymbol{P_{max}}$ being the maximum pressure applied, equal to $\boldsymbol{P_o}$ (fig. 1) . $\boldsymbol{P_{300}}$ is a pressure ratio equal to ($\boldsymbol{P}/\boldsymbol{P_o}$) at 300 s . These parameters were chosen for their capacity to clearly illustrate key phenomena and general trends seen with the reinforcements. Also, while remaining simple these equations are more flexible than some of the models discussed above, where the relations between some variables are dictated by geometrical hypothesis.

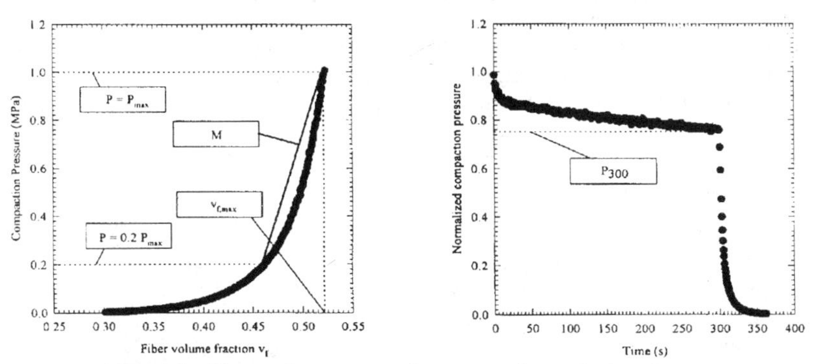

Figure 1. Observed parameters for compaction and relaxation curves

The following results were found for compaction curves. As the number of layers rises, parameters $\boldsymbol{M}$ and $\boldsymbol{A}$ rise, while the maximum fibre volume fraction $\boldsymbol{v_{f,max}}$ and parameter $\boldsymbol{B}$ decrease; compaction curves ($\boldsymbol{v_f},\boldsymbol{P}$) progressively shift to lower $\boldsymbol{v_f}$ values, and higher rigidities are seen close to $\boldsymbol{v_{f,max}}$. As the number of consecutive compaction cycles rises, $\boldsymbol{M}$, $\boldsymbol{A}$ and $\boldsymbol{v_{f,max}}$ rise and $\boldsymbol{B}$ decreases; compaction curves are shifted to higher $\boldsymbol{v_f}$ values. For relaxation curves, $\boldsymbol{P_{300}}$ decreases with higher number of layers and raises with higher applied pressure. The trends were identified from more than 300 tests performed on 16 reinforcements. Typical results appear in figure 2, along with the variability observed for parameters M and $\boldsymbol{B}$; the values were obtained from sets of 2, 3 or 4 curves produced in the same experimental conditions.

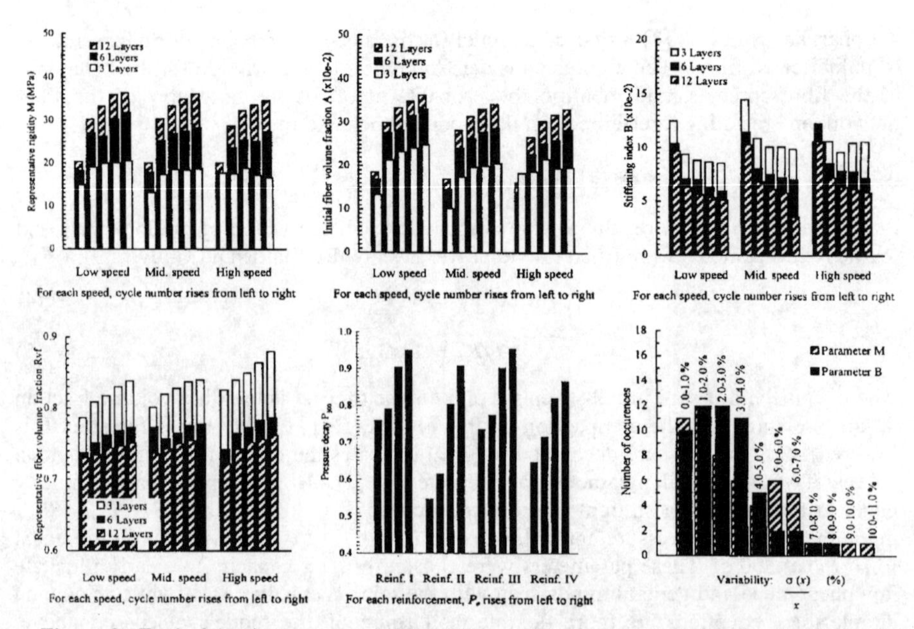

Figure 2. Typical results: parameters $\boldsymbol{M}$, $\boldsymbol{A}$, $\boldsymbol{B}$ and $\boldsymbol{Rv_f}$ ($v_{f,max}$) for varying numbers of layers and cycles; parameter $\boldsymbol{P_{300}}$ for varying initial pressure Po; variability of $\boldsymbol{M}$ and $\boldsymbol{B}$.

Most fitted $\boldsymbol{B}$ values are in the (0.201-0.341) range for random mats, (0.097-0.199) for textiles, and (0.041-0.057) for aligned fibres. The equivalent $\boldsymbol{B}$ value for van Wyk's model is 0.333 . Fitted $\boldsymbol{B}$ values for mats correspond with the model because their initial v_f values are low; fitted $\boldsymbol{A}$ ranges are (0.001-0.03) for mats and (0.04-0.35) for textiles.

4. Methodology and experimental results

In some processes such as bag moulding, the porosity of the reinforcement is allowed to vary after resin injection. An experimental program was undertaken in order to quantify the effect of a fluid on the compaction behaviour, and to produce supplementary data for relaxation [10]. Statistical experiment design techniques were used. Complete loading cycles (load at constant rate, hold at constant v_f, retrieval at constant rate) were applied to textiles #1, #2 and #3, respectively a twill weave (Bay Mills 154), a plain weave (Bay Mills 196) and a non-interlacing bidirectional (JB Martin 82001A). The processing parameters investigated are the maximum applied pressure $\boldsymbol{P_{max}}$ (1.0, 1.5,

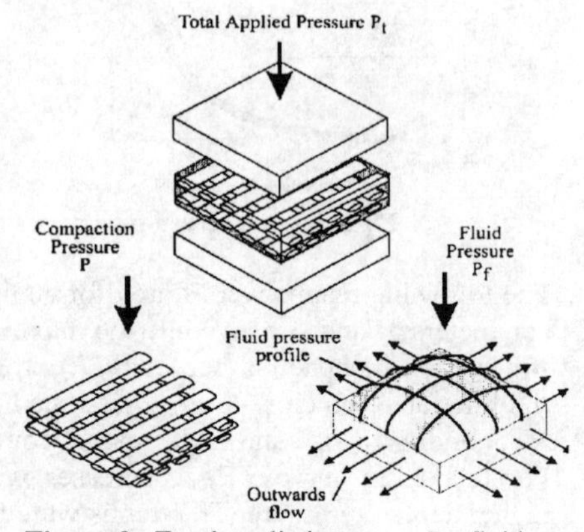

Figure 3. Total applied pressure Pt, fluid pressure Pf and compaction pressure P

2.0 MPa), compaction rate ***CR*** (2, 60 mm/min), holding time t_{hold} (1, 300 s), number of layers ***NOL*** (textile #1: 6 & 12; textiles #2 & #3: 12 & 24), cycle number ***CN*** (1, 2, 3, 4, 5), and immersion in distilled H_20 ***SAT***. In these experiments, the measured variable is the compaction pressure ***P***; in the case of saturated reinforcements, ***P*** was obtained by the subtraction of the fluid pressure P_f from the total applied pressure P_t (fig. 3), using platens instrumented with pressure transducers. In addition to the parameters mentioned above, the compaction energy E_I and energy loss E_L (%) were calculated.

All the trends previously mentioned were observed. The effects of the number of cycles were weaker; experimental data shows this to be due to the thicker stacks used in this study. However the use of SED techniques allowed to make quantitative comparisons, and to obtain a detailed portrait of the behaviour. The results were expressed as:

$$A = 0.186 - \frac{2.25e^{-3}\,NOL}{2} - \frac{6.50e^{-3}CR}{2} + \frac{2.44e^{-3}P_{max}}{2} + \frac{2.63e^{-3}\,t_{hold}}{2} + \frac{5.35e^{-2}\,CN}{2} - \frac{6.56e^{-3}\,SAT}{2} + \textit{interaction terms} \quad (6)$$

where the variables ***NOL***, ***CR***, etc. are equal to -1 or +1 for the lower and higher values mentioned above, and the coefficient are determined from the experimental results and for each observed parameter, such as ***A*** in the polynomial illustrated in equation 6.

The coefficients of the polynomials for parameters ***B***, ***C*** and E_L appear in figure 4. Experimental evidence show that the most important parameter affecting the behaviour in compaction is the cycle number ***CN***. It was mentioned in the previous section that ***B*** is related to the structure of the material; figure 4 shows that it is mostly on successive cycles that the structure changes, or is rearranged. Figure 5 shows that the two parameters controlling the relaxation are the compaction rate ***CR*** and presence of a fluid ***SAT***. More relaxation is seen after 1 s for dry textiles compacted at high speed. Figure 6 shows a more subtle result: E_L is seen to decrease as the number of layers ***NOL*** rises. Also, fig. 2 shows that $v_{f,max}$ ($= Rv_f$) is reduced as ***NOL*** rises, indicating that the energy loss is higher for textiles compacted to higher volume fractions v_f. This is confirmed by the raise in E_L resulting from a raise in the maximum applied pressure P_{max}, shown in fig. 6 . However this figure also shows that E_L becomes lower on successive compaction cycles ***CN***, although $v_{f,max}$ raises on successive cycles. These apparently

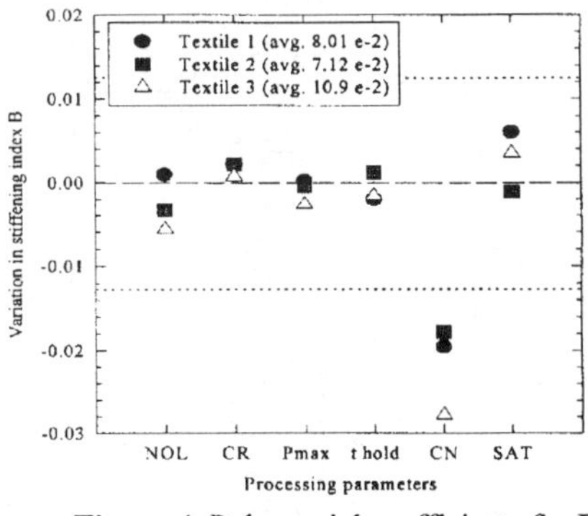

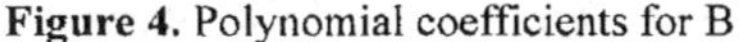

Figure 4. Polynomial coefficients for B

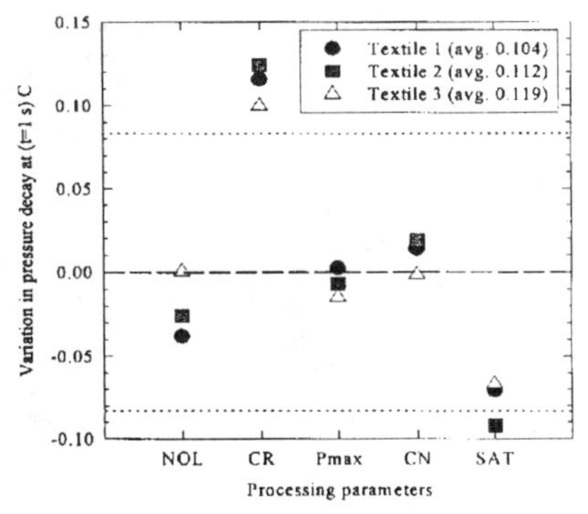

Figure 5. Polynomial coefficients for C

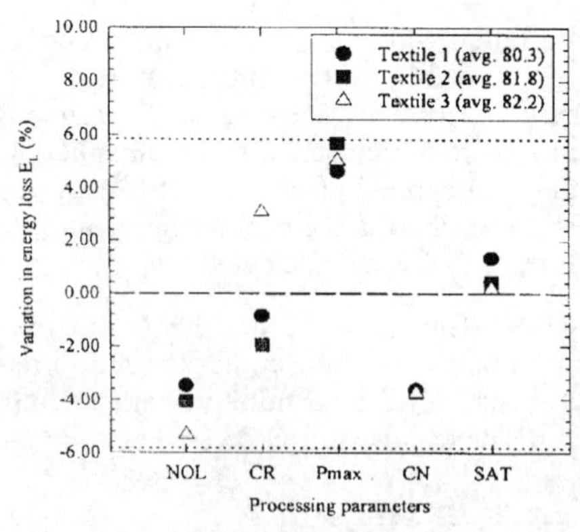

Figure 6. Polynomial coefficients for E_L

contradictory conclusions show the impact of fibre reorganisation. On the first cycle, the recoverable energy progressively reduces and reorganisation occurs through fibre friction. This results in a raise of v_f at which the pressure begins to build-up, A, in a raise, albeit smaller, of the maximum fibre reached during compaction $v_{f,max}$, and in a diminution of the energy loss E_L.

Finally, the results show that the fibre reorganisation that occurs during relaxation at constant v_f has no impact on the geometry of the fibre assembly and on successive compaction cycles.

5. Fibre reorganisation

An experimental program was undertaken [11] to gain a better understanding of fibre reorganisation. Series of 51 successive compaction cycles (t_{hold} = 0 s) were applied to 4 textiles; dry, distilled H_2O- and silicon oil- saturated stacks were tested. Figure 7 shows compaction cycles 1, 11, 21, 31, 41 and 51 for the JB Martin 80153A non-interlaced

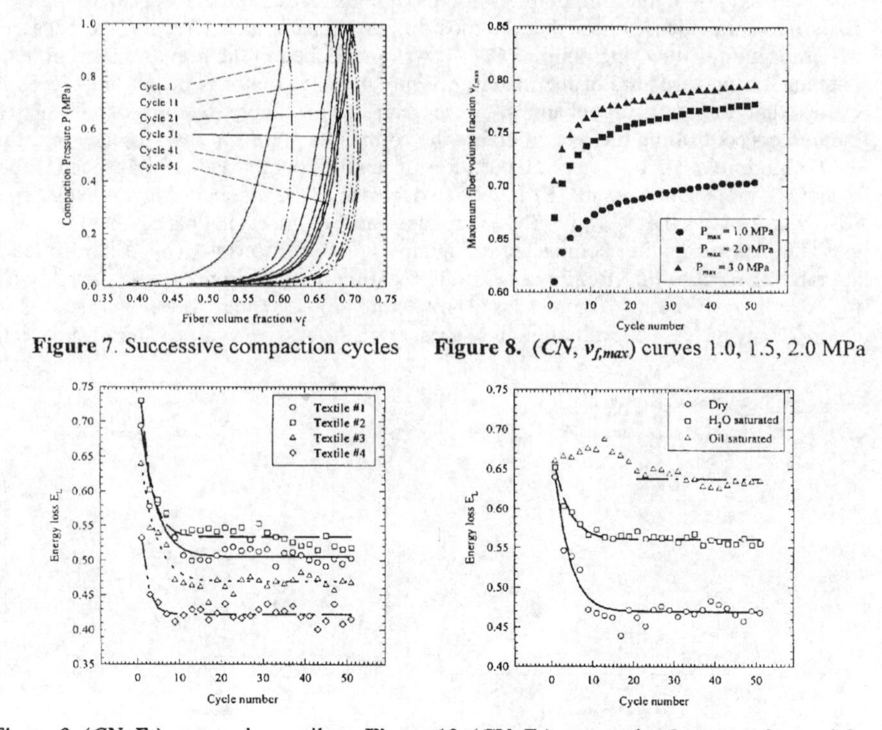

Figure 7. Successive compaction cycles

Figure 8. (CN, $v_{f,max}$) curves 1.0, 1.5, 2.0 MPa

Figure 9. (CN, E_L) curves, dry textiles

Figure 10. (CN, E_L) curves, dry & saturated material

fabric in the dry state; $v_{f,max}$ is seen to raise. Figure 8 shows that this increase is fast for the 10 first cycles and then becomes slower; the rates do not depend of the applied pressure P_{max}. Figure 9 shows how E_L evolves with CN for dry materials; the energy loss decreases for the first 10 cycles and then takes a constant value, indicating that most reorganisation occurs during these cycles. Figure 10 shows that the changes in behaviour associated to immersion in different fluids appear only after the first 10 cycles. Therefore, macroscopic phenomena such as tow widening define the initial behaviour. Afterwards, microscopic phenomena such as friction at fibre contact become preponderant. Figure 11 show (22x) the tow width before (a) and after (b) 51 cycles for the same material; these photographs show important levels of tow widening.

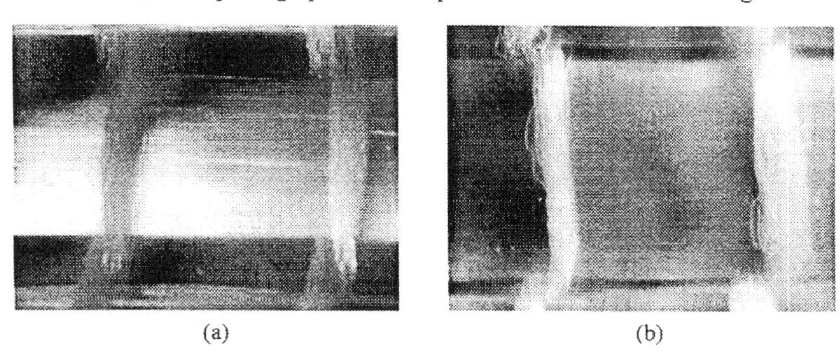

(a) (b)

Figure 11. Tow width (22x) before (a) and after (b) undergoing 51 consecutive cycles

6. Conclusion

The compaction and relaxation behaviour of textile reinforcements is dependent on the processing parameters; however the parameters that have the most important effect on each phenomena are not the same. Tow widening also has an influence on the mechanical properties of fibre assemblies. Quantitative models must include this phenomenon, which often has a more important effect than the interactions at fibre contacts.

References

[1] *Z. Cai, T. Gutowski*, THE 3D DEFORMATION BEHAVIOR OF A LUBRICATED FIBER BUNDLE, Journal of Composite Materials, 26, 1207-1237, 1992.

[2] *C.M. van Wyk*, NOTE ON THE COMPRESSIBILITY OF WOOL, Journal of the Textile Institute, 37, T285-T292, 1946.

[3] *D.H. Lee, J.K. Lee*, INITIAL COMPRESSIONAL BEHAVIOR OF FIBRE ASSEMBLY, in Objective measurement: application to product design and process control, Edited by the Textile Machinery Society of Japan, 613-622, 1988.

[4] *J.I. Curiskis, G.A. Carnaby*, CONTINUUM MECHANICS OF THE FIBER BINDLE, Textile Research Journal, 55, 334-344.

[5] *G.A.V. Leaf, K.H. Kandil*, THE INITIAL LOAD-EXTENSION BEHAVIOUR OF PLAIN-WOVEN FABRICS, Journal of the Textile Institute, 71, 1-7, 1980.

[6] *T. Komori, K. Makishima*, NUMBER OF FIBER-TO-FIBER CONTACTS IN GENERAL FIBER ASSEMBLIES, Textile Research Journal, 47, 13-17, 1977.

[7] *G.A. Carnaby, N. Pan*, THEORY OF THE COMPRESSION HYSTERESIS OF FIBROUS ASSEMBLIES, Textile Research Journal, 59, 275-284, 1989.
[8] *P. Simacek, V.M. Karbhari*, NOTES ON THE MODELLING OF PREFORM COMPACTION: I- MICROMECHANICS AT THE FIBER BUNDLE LEVEL. Journal of Reinforced Plastics and Composites, 15, 86-122, 1996.
[9] *F. Robitaille, R. Gauvin*, submitted to Polymer Composites, 1997.
[10] *F. Robitaille, R. Gauvin*, submitted to Polymer Composites, 1997.
[11] *F. Robitaille, R. Gauvin*, submitted to Polymer Composites, 1997.

Diels-Alder Based Co-reactants for Bismaleimides

B. Dao and T.C. Morton

CSIRO, Molecular Science, Private bag 10, Clayton South MDC Victoria, 3169, Australia

Introduction

Bismaleimides have been one type of matrix resin that have shown promise for high temperature applications; with additional benefits because of their low-cost, ease of synthesis and relatively mild cure conditions. New co-reactants are of interest because many of the existing systems suffer from excessive brittleness and poor long term aging properties [1].

The maleimide ring is a well known dienophile for Diels-Alder addition [2] and there have been several attempts to use suitable bis dienes as curing agents for bismaleimides [3,4,5]. A potential diene not so far investigated is 2-naphthol and its derivatives, which readily undergo Diels-Alder addition with N-phenylmaleimide and bismaleimides. This study was carried out to determine if this reaction could be utilized in curing bismaleimides and to determine what effects the bicycloctanone structure, which is produced on curing, would have on thermal stability.

To answer to the first question, the bismaleimide curing behavior of 7-allyloxy-2-naphthol (MAN, Figure 1, Structure I.) has been investigated. A second target compound bis[7-(2-hydroxynaphthyloxy]bisphenol A (Figure 1, Structure II) has proven difficult to synthesize in sufficient quantities (in a pure state) but some results will be presented. The second question was addressed by placing an appropriate bis-functionality on a naphthol adduct which would allow its co-reaction with bismaleimides to form a thermosetting polyimide matrix (Figure 1, structures III and IV).

Results and Discussion

Satisfactory cures were obtained with 7-allyloxy-2-naphthol and appropriate bismaleimides. However the laminate coupons made using this system had lower thermal stability than those of a comparable system using the commercial co-reactant Matrimid 5292B. Typically a neat resin bar on curing at 250°C gave a Tg at 1Hz of 247°C (tan δ) and an modulus of 23.4 GPa and a temperature of break of modulus of 218°C. Percentage weight losses on aging were 0.73, 2.32 and 3.40% after 1, 7 and 14 days at 250°C in air respectively.

The diallylyl ether adduct, Compound III, when cured with a suitable bismaleimide, yielded a neat resin with Tg of 299°C, tensile strength of 69MPa and a tensile modulus of 3.8GPa whereas a cure with using the common 3,3'-diallylbisphenol A (Matrimid 5292B) exhibited a Tg of 265°C, a tensile strength of 56MPa and a modulus of 3.0 GPa. Composite coupons made with the new system gave a Tg of 296°C and 0.2, 1.2 and 2.0% weight loss when aged in a hot air oven at 250°C. This compares with weight losses of 0.41, 1.55 and 2.24% of laminates made using Matrimid 5292B as the co-reactant. The fact that these laminate coupons had better thermal performance than the system using commercial co-reactant showed that the presence of the Diels-Alder adduct structure in the resin backbone was not detrimental to the normal performance of ene cured bismaleimides. On the other hand the thermal stability limitations of the bicyclooctenone structure did show up in the more thermally stable phenylethynyl end capped resin materials. Compound IV was highly processible and made very satisfactory laminate coupons but thermal stability was markedly compromised. Carbon fibre composite coupons (5 Ply) made using Compound IV were well compacted and void free, had a Tg of 361°C (DMTA, figure 12) and exhibited thermal weight changes of +0.21, -0.39 and -1.59% when aged in a hot air oven at 250°C.

References

[1] Morgan R J, Shin E E and Lincoln J E 1996 *28th International SAMPE Tech Conf.* **28** 213
[2] Carruthers W 1990 *Tetrahedron Organic Chemistry Series* **8** Chapter 1
[3] Stenzenberger H D and Konig P 1991*High Perform. Polym.* **3** 41
[4] Diakoumakos C D and Mikroyannidis J A 1992 *J Polymer Sci: Part A: Polymer Chem.* **30** 2559
[5] Hawthorne D G , Hodgkin J H, Jackson M B and Morton T C 1994 *High Perform. Polym.* **6** 249.

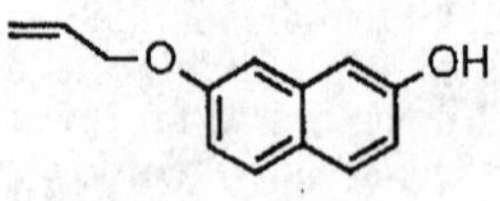

O
OH
I MAN

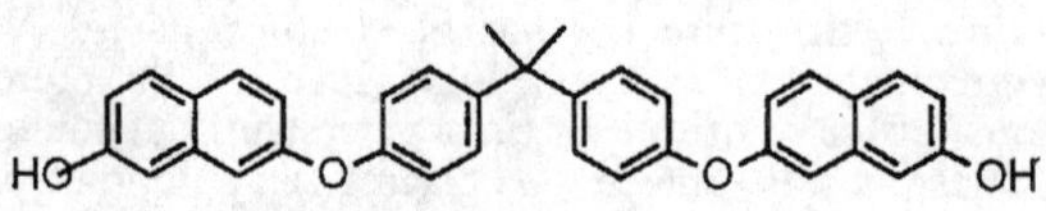

HO
O
O
OH
II

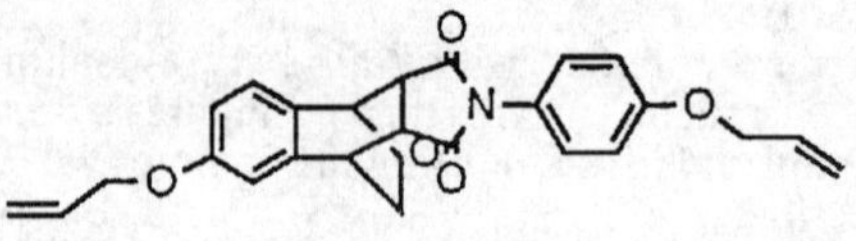

O
N
O
O
O
O
III

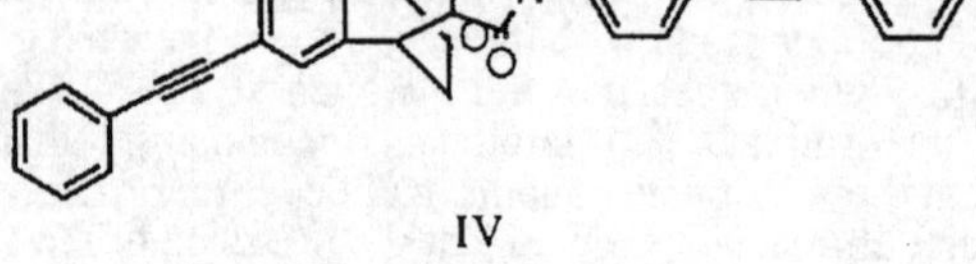

O
N
O
IV

Figure 1

RECYCLED CARBON FIBER COMPOSITE AS REINFORCING ELEMENT IN NEW THERMOPLASTIC COMPOSITE.

Toni Pisanikovski *, Jan Wahlberg (Lund university),
J-E Ståhl (Lund university)

*) Lund University, Production & materials engineering, Box 118, S-221 00 Lund, Sweden

Abstract

As the us of carbon fiber reinforced plastic (CFRP) and other thermoset based composites increase the demand for more environmentally friendly processes and life cycles. The last few years these kinds of materials have started to appear in more everyday applications and the volume increases every year.

The objective of the study was to investigate the possibility of re-using CFRP as reinforcing elements in a polypropylene (PP) based composite. The reason PP was chosen was that the material already is used in large amounts as a matrix in different glass fibre composites such as GMT (Glass Mat Reinforced plastic). The idea was to develop a material that was by a flow moulding process and had equivalent or superior to the GMT material.

Introduction

The principle of reusing reinforcing elements of a cured CFRP composite and not only use the elements as filler but as reinforcement in a new thermoplastic composite system has been investigated. The project is funded by Swedish institutes and companies i.e. NUTEK, FMV, SAAB, Karlskronavarvet AB and RAPID Granulator.

The main objective was to develop a material system with randomly oriented short fibers reinforced recycled elements. The reinforcing elements are in this case not strictly fibers but a system of matrix and fibers. This means that in the recycling process not only the fibers but also the matrix is reused in the same process. The demand of such recycling process increases together with the elements internal and surface morphology. The work presented in this paper can be divided into:

- Cutting process and production of reinforcing elements.
- Surface morphology
- Pull-out test
- Applications

The goal is to produce a material with a thermoplastic matrix only reinforced with recycled elements from carbon/epoxy composites. In the trials the work was concentrated on a polypropylene matrix. The goal was to produce a material with properties equal to conventional GMT. The production of the new material was performed by compounding the PP with the reinforcing elements. The produced material contained 35%V_f randomly oriented elements. Due to the high V_f the extruder, a Brabender DSK 42/7, was modified so that the pressure in the compression section and the length of the metering section was decreased. The produced material was then used as conventional GMT material in a flow-molding operation.

Cutting process

The most economical cutting process is using a granulator with a rotating cutting knife, figure 1.1. The equipment used was delivered by Maskin AB RAPID in Bredaryd Sweden [5,7]. The granulator used in the trials was equipped with force sensors on the cutting edge and a computer for output registration. By controlling the cutting sequence the production of reinforcing elements with high properties was possible. Because of the complex system many properties have to be optimised and the most important factors to take in count are:

- Cutting force and power output.
- Morphology.
- Tool and machine wear.
- Working environment

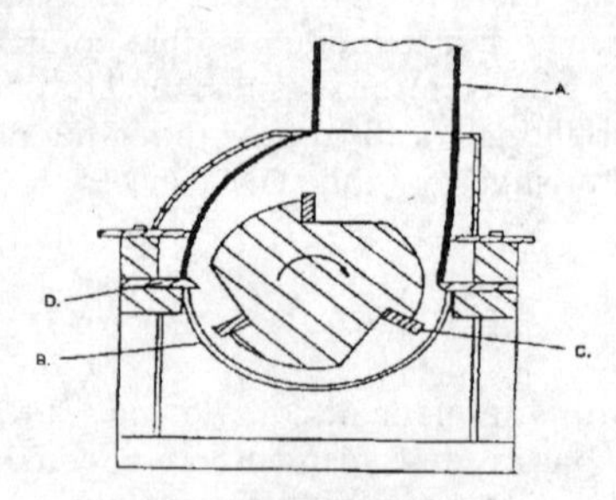

Figure 1.1 Principal drawing of granulator, three rotating knifes and two firm. A. Inlet, B. Outlet, C. Rotating knife, D. Fix knife.

The granulated polymeric material can be used in many different processes and applications. This means that the material have to be classified and weight against the process. In the classification the cutting force both the radial and tangential, power, sound, dust, size distribution and tool wear. The process classification is connected with the produced element type and the specifications for that system. The developed element types were divided into three groups in which the demand increase for each group.

- Volume reduced elements, produced material is usually energy recycled.
- Process elements, usually thermoplastic materials
- Structure elements, usually thermoset based composites that are material recycled

The main demand on the volume reducing element group is that the cutting operation is performed economical. This means that the cutting sequence is optimised so that the highest economical output is obtained. In the cutting economy the most important factors are; Wear, down-time and tool change. The system is applied on elements cut for i.e. energy recycling. The process element class, include the element morphology which means that the elements dimensions and geometry is controlled as well as the surface morphology. The surface is characterised against i.e. surface smoothness and surface cleanness. The structure element class the demands are increased to involve the whole element volume. In this case the element is characterised against damages and changes in the element. In this article/study the elements produced will be classified as structure elements. This means that the carbonfibre/epoxy composite will be recycled as new reinforcement in a thermoplastic based composite. In the trials a T300 carbon fibre pre-preg based composite from the aviation industry will be used.

The materials were cut in a standard RAPID 2645 KU to which a cyclone was connected. The cyclone was used to control the size distribution and to decrease the number of air-born dust particles on the element surface. The outlet in the granulator is changed so that the size distribution is matched against the theoretical critical length. For the standard cutting process most of the reinforcing elements could be used in a com-pounding operation. In the cutting process 1/5 of the produced material couldn't be used as reinforcement in following processes. The material was instead cut to a volume-reduced element and used as filler in other composite systems.

Because the recycled material is to be used as reinforcement the produced elements had to have a low internal damage level. Besides of the damages, the surface morphology had to be suitable for the matrix system used, in this case polypropylene. This means that the element surface had to have a low particle level on the surface. The elements were therefore automatically both cleaned and size distributed in the cyclone. In figure 1.2 an example of a material processed with a feed-backed cyclone.

a. 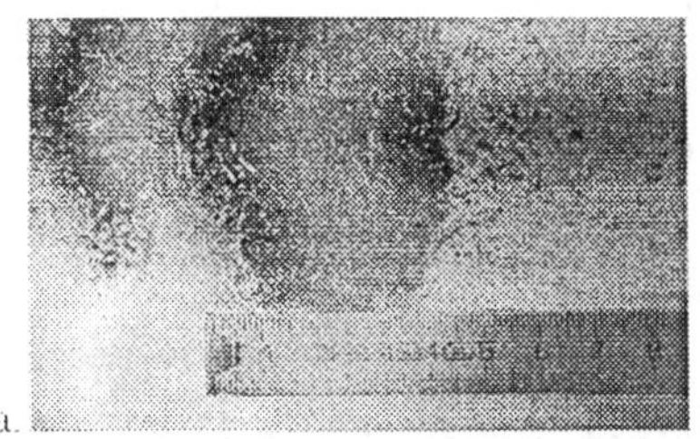

b. 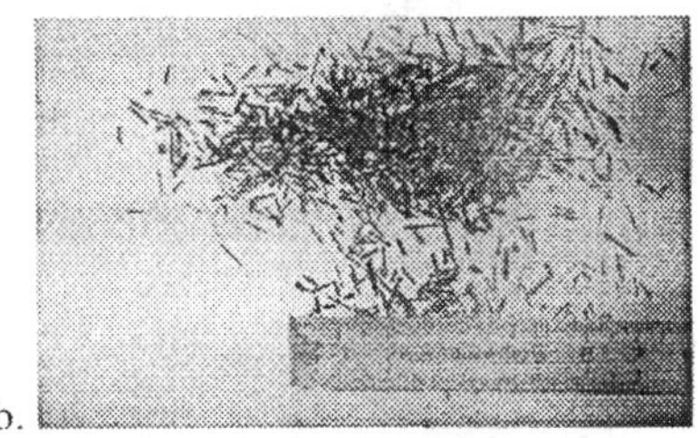

Figure 1.2 a. Material processed without a feed-backed cyclone. b. Material processed and sized distributed with a feed-backed cyclone.

Besides of the contamination level of the particle surface the degradation damages have to be low after the cutting process. The most common damages after the cutting sequence are:

- Fibre break
- Multiple break
- Debonding

In figure 1.3 both a reinforcing element with high level of damage and one with low is shown.

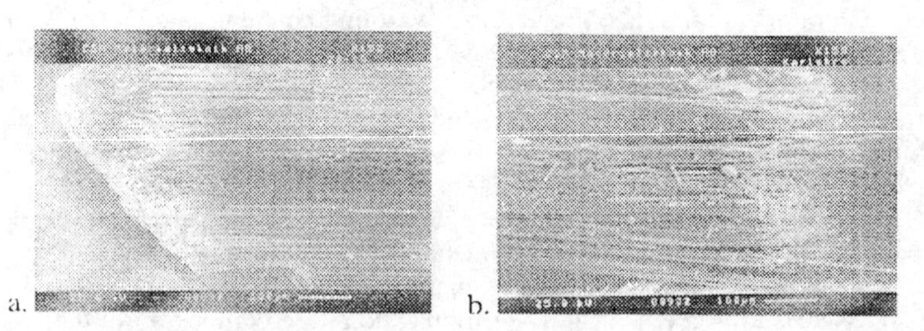

Figure 1.3 a. Reinforcing element with low degradation. b. Element with high level of degradation

Surface morphology and matrix modification

The surface structure after the cutting operation can be described as a mixture of carbonfibre and epoxy [7]. From figure 1.4 it can be observed that the new thermoplastic matrix has to bond both the fibre and the old epoxy matrix. The pre-trials showed that some modification of the system was needed because of the low bonding. In order to assume good adhesion some part of the system had to be changed beside of treating the element with an cyclone. The priority used is: 1. Modification of matrix 2. modification of reinforcing elements 3. both modification of matrix and reinforcing element.

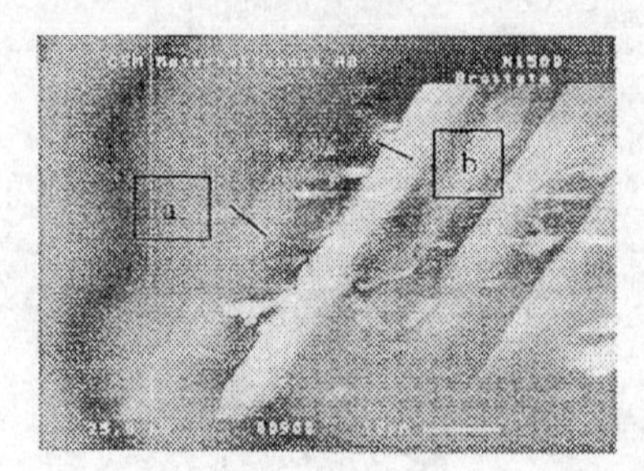

Figure 1.4 Reinforcing element surface. a. Carbonfibre. b. Epoxy

The thermoplastic matrix was modified with a maleic anhydride that was pre compounded in an extruder together with the polypropylene. In the trials commercial pre-modified polymers were also tested i.e. polybond 3002, 3150. As base matrix compound a polypropylene Duplen NS 10 N was used. In addition to matrix modifications some changes were made on the reinforcing element.

- Adding modifiers to the surface.
- Change the surface morphology.

To increase the compatibility elements were coated both with polyvinyl alcohol, PVA (Mol.m 72000) and maleic anhydride bonded to the element surface. Figure 1.5 show a coated reinforcing element on which the coating has bonded. The goal was to make the element surface more susceptible to the polypropylene.

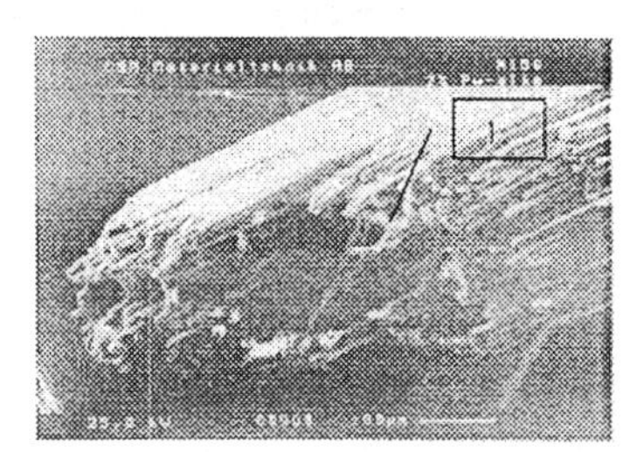

Figure 1.5 Coated reinforcing element, 1. PVA coating.

By etching parts of the element with an acid the surface morphology changed. The main goal was to increase the matrix possibilities to form a bond with the reinforcing element. At the same there is an increased mechanical anchoring between the polypropylene and the reinforcing element. The etching substance was chosen so that only parts of the matrix in the reinforcing element were influenced. By creating a sponge like structure the mechanical properties of the reinforcing element weren't changed. The structure is created after introducing acrylic acid to the surface.

a. 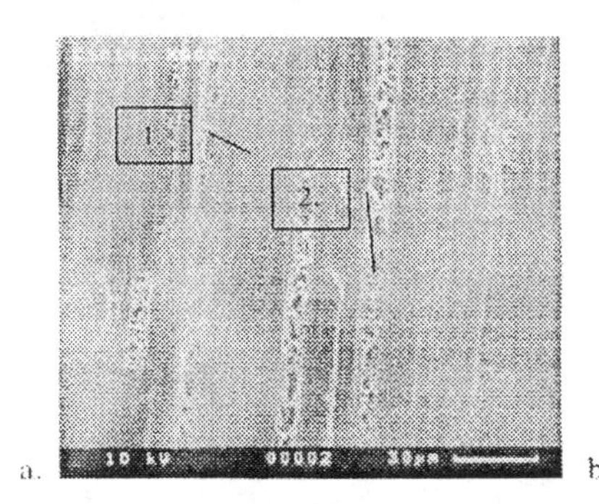

b. 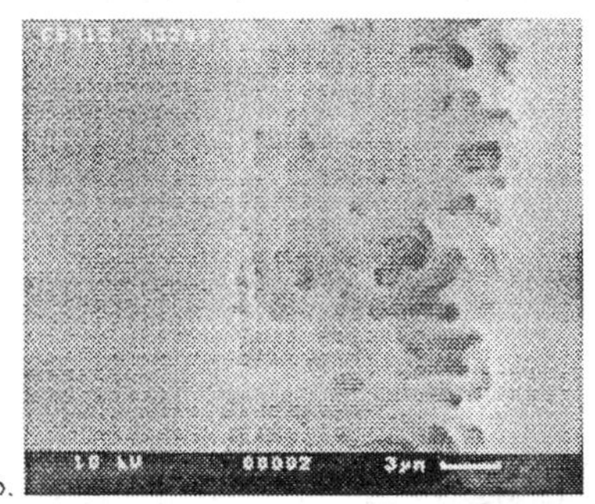

Figure 1.6 Reinforcing element etched with acrylic acid. a) 600x area 1. unaffected carbonfibre. 2. etched epoxy matrix. b) Magnified etched epoxy

Element Pull-out test

The pull-out test was performed both to evaluate the influence of different production parameters but also for the evaluation of modifiers [7]. The used pull-out test controlled production parameters as: holding-time, pressure, temperature and cooling rate. The main demand on the process was that the sample during sample preparation was pressurised both during melting and cooling with a constant pressure and that the embedded length could be varied. The equipment used is shown in figure 1.7. The equipment is heated with electrical rods and cooled with water (1. and 2.). The matrix is placed in the pre-heated cavity (5.). The matrix sample is produced in an injection moulding process in which the modifiers are added. The elements were cut and measured to fit the slot (4.). The lid and the piston (1. and 2.) were tightened against the body (3.) and before to sample is pressurised the element was protruded into the molten matrix through the slot in the body to the right embedded length.

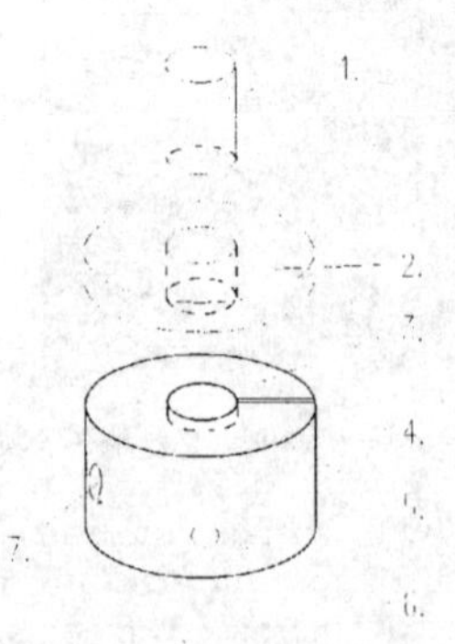

Figure 1.7 Sample preparation equipment for element pull-out test.

The results from the pull-out test is summarised in table 1.8 from which the modified polypropylene contain 10% polybond 3002. The result is presented as a relation between the maximum pull-out force and the embedded area (F/A).

Material	Element	F/A	Pressure [Mpa]
PP, Duplen NS 10 N	No modification	9,61	10
PP, Duplen NS 10 N	No modification	9,12	20
PP, Duplen NS 10 N, 10% polybond 3002	No modification	14,5	10
PP, Duplen NS 10 N, 10% polybond 3002	No modification	13,5	20
PP, Duplen NS 10 N, 10% polybond 3002	PVA coating	0	10
PP, Duplen NS 10 N, 10% polybond 3002	PVA coating	0	20
PP, Duplen NS 10 N, 10% polybond 3002	Etched element surface	17,3	10
PP, Duplen NS 10 N, 10% polybond 3002	Etched element surface	17,2	20
PP, Duplen NS 10 N, 10% polybond 3002	Etched element surface, coated with PVA	18,2	10
PP, Duplen NS 10 N, 10% polybond 3002	Etched element surface, coated with PVA	18,0	20
PP, Duplen NS 10 N, 10% polybond 3002	Maleic acid coating	20,5	10
PP, Duplen NS 10 N, 10% polybond 3002	Maleic acid coating	19,8	20

Table 1.8 Summarised results from element pull-out test

The maximum transferred force for the non-modified sample show a level. And from the force-displacement curve, figure 1.9, it can be observed that the force does not decrease dramatically after the bond in the interface fails. The etched specimen gave an 80% increase in transferred force in comparison to the non-modified. The force-displacement curve is presented in figure 1.9. The largest difference between the two curves is that the force-build up in the etched sample is non-linear. Which can be explained with the shearing of the thermoplastic matrix that is bonded and anchored mechanically to the element, figure 1.10. The force level is after total failure identical between the non-modified and the etched sample.

From the samples with PVA coating there was a problem with the bond between the PVA and the element. This led to that the PVA was sprayed to an element with an etch surface instead. The force displacement curve for the sample show a linear force build-up and an 89% increase in transferred force figure 1.9. By coating the element with maleic anhydride a 113% increase of the transferred force was observed. And as for the PVA modified element the maleic anhydride coated show a shared linear force build up.

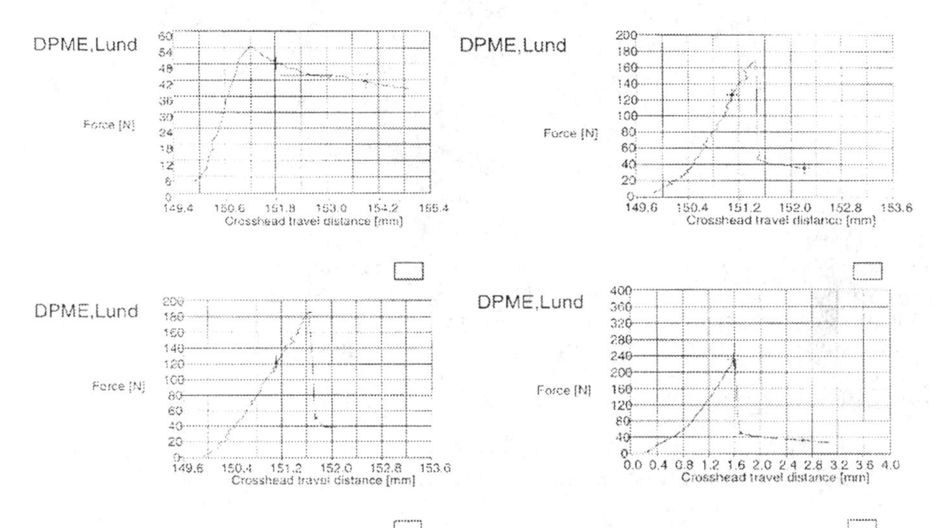

Figure 1.9 Force-displacement curve for a) non-modified sample. b) Etched element with Polybond modified matrix. c) Etched element with PVA coating and Polybond modified matrix. d) Maleic anhydride coated element with polybond modified matrix.

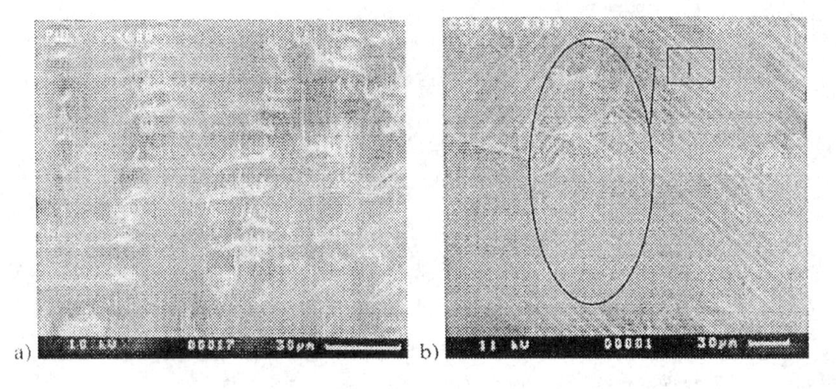

Figure 1.10 a) Element with bonded and mechanically anchored polypropylene. b) In the non-modified sample is there only low amounts of sheared of matrix, 1. area with polypropylene.

Application

From the extruded material a semi-processed plate was manufactured. Which could be processed as a GMT material. And from in a three point bending test the materials showed a young modulus of 6GPa which can be compared with 4GPa for a convential GMT material. In figure 1.11 a disk manufactured with 30V_f recycled CFRP in the same figure there is a sample of a processed specimen with 5V_f reinforcing elements just to show the element distribution after processing.

a) 

b) 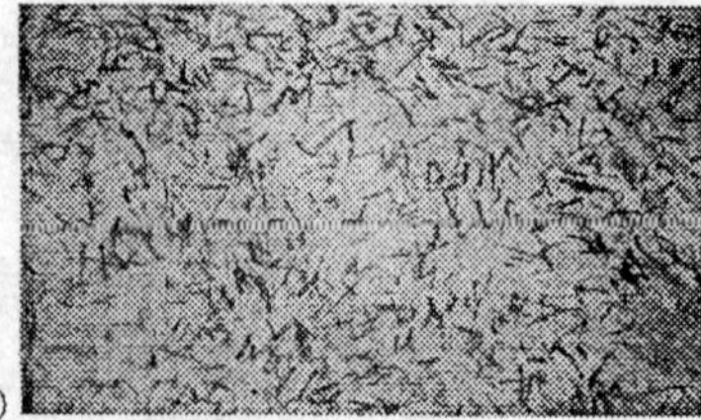

Figure 1.11 a) Press moulded disk with 30 V_f recycled reinforcing elements. b) Disk with 5 V_f reinforcing elements

A larger study over the mechanical properties is necessary in the future work. This study have to be combined with an optimised material system. The study so far have shown that it's possible to produce a new composite material with recycled reinforcing elements, both with good mechanical- and processing properties.

Acknowledgements

The work has been funded by the Swedish board for technical development (NUTEK), SAAB, Karlskronavarvet AB and FMV. The authors gratefully acknowledge the financial, material and technical contribution given by the partners in the project *Miljö- och kretsloppsanpassning av produktionsmetoder för högpresternade kolfiberkopositer[8].*

References

[1] Gottberg, J.P
Grundlagen der scheidzerkleinerung hochpolymerer weichstoffe, Kunstofftechnik, no. 9, sid. 419-423, 1970

[2] Gottberg, J.P
Scheidvorgang und arbeitsbedarf bei der zerklieinerung hochpolymerer weichstoffe. kunstofftechnik no. 10, sid. 242-251, 1971

[3] Gottberg, J.P
Hochpolymere weichstoffe beim scheidvorgang, kunstofftechnik, no. 10. Sid. 7-65. 1970.

[4] Jönsson, H.
Återvinning av GMT. Inst. för mekanisk teknologi och verktygsmaskiner LTH, LU, 1995

[5] Pisanikovski, Toni
Granuleringsmeknik för sammansatta material. Inst. för mekanisk teknologi och verkstadsmaskiner LTH, LU, 1997

[6] Ståhl, Jan-Eric.
Från fiber till komposit. Inst. för mekanisk teknologi och verktygsmaskiner LTH, LU, 1995

[7] Ståhl, Jan-Eric.
Verstadstekniska tillverkningsmetoder, del 1. Inst. för mekanisk teknologi och verktygsmaskiner LTH, LU, 1995

[8] Ståhl, Jan-Eric.
Miljö- och kretsloppsanpassning av produktionsmetoder för högpresternade kolfiberkopositer. Inst. för mekanisk teknologi och verktygsmaskiner LTH, LU, 1997

[9] West, G.A.
Radical reassesment of rotary knife milling design. ANTEC, sid. 842-847, 1985

Microstructural studies of the compression of woven fabrics in the processing of polymer composites

R.A.Saunders, C.Lekakou and M.G.Bader

School of Mechanical and Materials Engineering
Department of Materials Science and Engineering
University of Surrey
Guildford, Surrey GU2 5XH

Abstract

This paper focuses on the compression of fibre reinforcements during the processing of polymer composites, covering a range of fabrics, namely a plain weave, a twill, a satin and a noncrimped, stitch-bonded fabric. Laminates of these fabrics impregnated with polyester were produced under different degrees of maximum compression, sectioned through their thickness and subjected to microstructural analysis in order to (a) investigate the mechanisms of deformation in the compression of fabric assemblies and (b) compare the compression of the different types of fabrics in terms of resulting average area porosity (resin rich areas), area pore structure, average area voidage and voids for the different types of fabrics. The investigated mechanisms of deformation included elimination of the resin rich layer between fabrics, nesting of fabrics, deformation of the yarn waveform in woven fabrics and deformation of individual yarns.

1. Introduction

Compression of the reinforcement/resin system is always present in the processing of composites in order to determine the thickness of product and to enable consolidation of plies. Compression of fibre reinforcement has significant effects on the permeability of reinforcement during processing and on the properties of the composite product. The scope of the present study is to carry out microstructural studies on laminates of woven fabrics with the aim of (a) analysing the compression in terms of deformation mechanisms inside the fibre reinforcement and (b) examining the change of volume and area porosity as well as pore structure during the compression of different types of weaves.

In a previous microstructural study of this type, Yurgartis et al [1] manufactured laminates reinforced with plain woven cloths and measured inclination angles and crimp angles to describe yarn shape, and angle match between inclination angles of adjacent yarns to assess yarn nesting. A three-dimensional reconstruction of a yarn section from available data illustrated local distortions and irregularities of the waveform. However, it was possible to draw sinusoidal fits for the waveforms through the data points for the various yarns. Jortner [2] observed in a similar study that across ply compaction during lamination of an assembly of resin impregnated plain woven cloths does not change significantly the weave's wavelengths. Simacek and Karbhari [3] reported mean measurements of geometric parameters of a unit cell in an anisotropic plain weave where compression was performed from an uncompacted state to a maximum fibre volume fraction of 0.58. Matsudaira and Qin [4] carried out an empirical analysis of their experimental data regarding the compressional behaviour of fabrics.

2. Materials and experimental techniques

Three different types of E-glass woven fabrics, purchased from Fothergill Engineered Fabrics, were used: a plain weave Y0212, a twill weave Y0185 and a 5 harness satin weave Y0227. A noncrimped stitch-bonded, E-glass fabric was also used, consisting of four unidirectional layers of aligned rovings with a 0, 45, -45 and 90° orientation with a corresponding tex of 1243, 290, 290 and 482 g and a corresponding number of ends for each of the layers of 2.5, 7.0, 7.0 and 5.2 per 10 mm, respectively. A polyester resin Crystic 471 PALV unsaturated polyester from Scott Bader Company Ltd was employed in this study, with methyl ethyl ketone peroxide (MEKP) as initiator and cobalt naphthanate as accelerator.

The compression experiments were carried out on an Instron 1195 testing machine at constant compression speed. The reinforcement was impregnated with the curing resin system. The wet reinforcement was then compressed at a constant compression speed of 1 mm/min until a maximum target pressure had been reached. At that point further compression was stopped and the crosshead remained in place until the laminate had completely cured in-situ. A sample was then sectioned from the centre for subsequent microstructural analysis.

In the presented studies, the weft yarns are the longitudinal yarns in the examined sectioned samples whereas the warp yarns are represented by elliptic cross-sections. The mosaic of pictures was converted into a digital image which was further analyzed by using UTHSCSA *Image Tool*. First the areas of the warp yarns cross-sections were measured. By then making the assumption of elliptic yarn cross-section, measurements were made of the height, h, and the width, w, of the warp yarns together with the coordinates of the centre of the measured ellipses.

Yarns were modelled as sinusoidal waveforms. It would not have been possible to examine the yarn across its thickness since its thickness and other features of its perimeter would depend at which location across its width the yarn was sectioned. So, the centrelines of the longitudinal weft yarns were drawn and each centreline was assumed to form a sinusoidal wave which would represent the yarn waveform. Three best fit lines were then drawn through various types of points of the centreline for each yarn: (a) maxima, (b) minima and (c) mid-points (see Fig.1). The distance, 2a, between the lines of maxima and minima represented twice the amplitude of the yarn waveform. The distance between the crossings of the centreline with the mid-point line represented half the wavelength, λ, of the yarn waveform. The distance, d, between the mid-point lines of consecutive weft yarns represented the distance between plies.

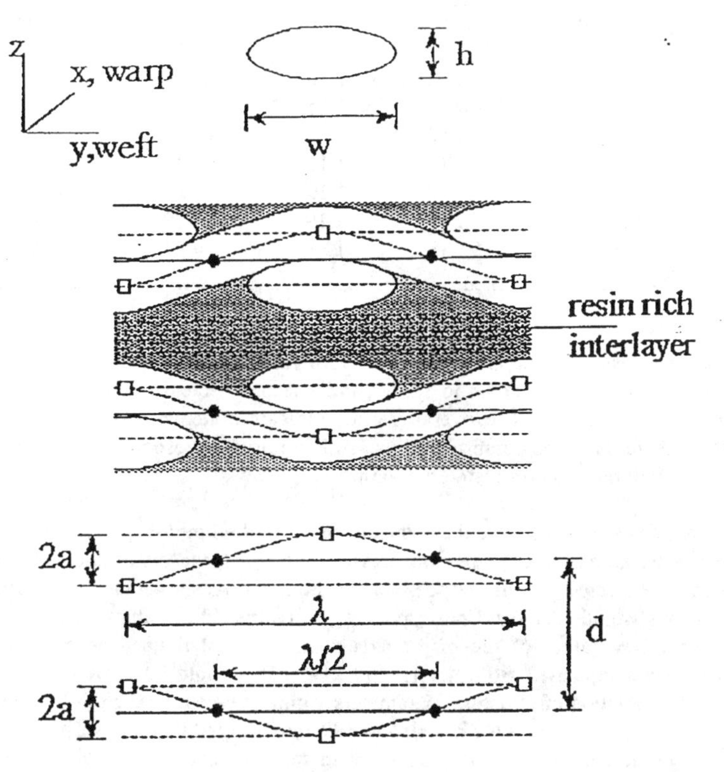

Fig.1. Diagram presenting the microstructural measurement technique and measured parameters for plain woven cloths.

3. Microstructural analysis: modes of deformation

This section includes the microstructural studies of the through thickness sections of laminates compressed under different pressures and cured in-situ. The studied laminates contained 20 layers of plain woven cloth assembled warp to warp and weft to weft. The aim of the microstructural studies was to investigate at microlevel the compression of resin impregnated cloths as a combination of four modes of deformation: (a) elimination of the resin rich layers between plies; (b) nesting of cloths; (c) deformation of the yarn waveform; (d) compression and deformation of the yarn cross-section.

Table I presents the change of the various mean geometrical parameters as measured from the mosaic of the assembled micrographs.

Table I: Average data from the microstructural studies of 20 ply laminates.

P_{max} (kPa)	d (mm)	A(mm^2)	w (mm)	h (mm)	2a (mm)	λ(mm)
4.4	0.52	0.270	1.16	0.31	0.31	3.06
88	0.44	0.252	1.15	0.28	0.27	2.94
265	0.40	0.257	1.13	0.29	0.28	3.03
442	0.39	0.237	1.19	0.28	0.29	3.07
619	0.38	0.237	1.19	0.28	0.28	3.01
884	0.37	0.237	1.19	0.29	0.26	2.89
1768	0.34	0.233	1.31	0.27	0.25	2.98

The mean cross-sectional area, A, of the warp yarns experiences a first decrease as the pressure is increased from 4.4 to 88 kPa and a second decrease as the pressure is raised from 265 to 440 kPa. The overall results indicate a stepwise compaction of fibrils within the yarns as the compression pressure is raised corresponding to an increase of the fibre volume fraction inside the yarns.

The yarn cross-sections were then considered to be of elliptic shape. If flattening of the yarns occurred as the compression pressure increased, the distribution of major axis would be expected to shift respectively to higher values whereas the distribution of minor axis would display a corresponding shift to smaller values. However a small shift to lower values of the major axis is in fact first noticed as the compression pressure was increased from 4.4 to 400 kPa. This could be attributed to the wide statistical variation of structural parameters within the examined cloth specimens or to effects of cloth nesting. One has also to take into account that the warp yarns may not be exactly at a right angle or even all of them at the same angle with respect to the plane of cross-section. On the other hand, a certain decrease in the values of minor axis is observed as the compression pressure is raised from 4.4 to 88 kPa indicating yarn compression. This is consistent with the observations about changes in the area of yarn cross-section. A definite increase of the major axis and a small decrease of the minor axis are observed for an increase of the compression pressure from 880 to 1760 kPa indicating an elongation of warp yarn cross-section at high pressures.

The yarn waveform was considered sinusoidal, described by the relation

$$y = a\sin\left(\frac{2\pi x}{\lambda} + \phi\right) \qquad (1)$$

The mean amplitude of the yarn waveform decreased from 2a=0.31 mm under 4.4 kPa to 2a=0.25 mm under 1768 kPa, as is displayed in Table I. The changes in the waveform amplitude were observed to be randomly distributed between plies. No trend could be concluded in the change of the wavelength of the yarn waveform for increasing maximum compression pressures in which case any observed changes were considered within the standard deviation of the statistical variation of wavelengths in the employed cloth. The phase angle ϕ followed a random distribution under all compressions indicating that the cloths were laid up at random phase differences between each other.

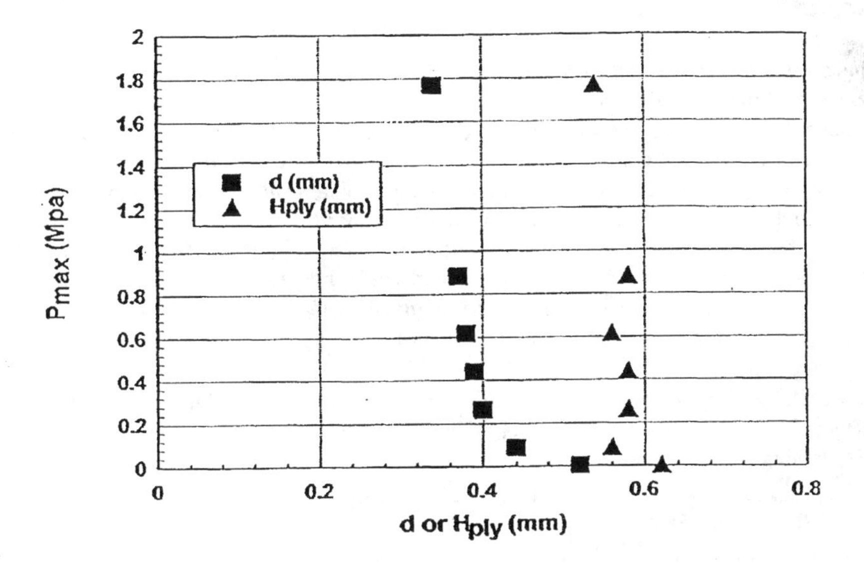

Fig.2. Compression of assemblies of 20 resin impregnated cloths at 1 mm/min up to P_{max} and microstructural studies of the cured laminates: a comparison between changes in the average distance between plies and the average ply thickness as a function of P_{max}.

A comparison between the average ply thickness and average distance between plies gives further indications about certain modes of compression. From Fig.1, the average ply thickness, H_{ply}, is taken as

$$H_{ply}=\max(2h,4a) \tag{2}$$

The average distance, d, between consecutive plies is defined as the distance between consecutive midpoint lines each of which is formed by drawing a line of best fit through the midpoints of each weft yarn centreline. In theory, if a number of dry cloths were perfectly assembled and fitted together, warp to warp and weft to weft at ϕ=0, d should be equal to H_{ply}. Fig.2 presents the average ply thickness and the average distance between plies as functions of the target compression pressure for the different cured laminates. If H_{ply} is smaller than d a resin rich layer should exist between cloths (see Fig.1) whereas if H_{ply} is larger than d the cloths have nested with each other without a distinct resin free layer. In this case it seems that nesting generally occurs as is shown in Fig.2. A certain amount of nesting occurs on average between consecutive cloth layers even at low compression and as the compression increases the extent of nesting increases since the change in d is greater than the change in H_{ply}.

4. Microstructural analysis: Comparison of compression of different fabrics

This section focuses on the comparison of the compressed microstructures of different types of fabrics, where microstructural analysis was carried out in laminates sectioned through their thickness and the following variables were evaluated: fibre area fraction, pore area fraction and void area fraction. The pore (or porosity) area fraction includes the sum of the resin rich areas and the voids and indicates the pores initially available for resin flow. At the resolution at which the measurements were carried out, the bundles were considered as solid fibres and that is also assumed in the values of fibre area fraction. The area fractions of pores and voids indicate macro-pores and macro-voids between bundles. The measurements given in this section do not include any micro-pores and micro-voids inside the bundles.

Table II. Area fractions of the fibre, porosity and void content of a number of different fabric structures (assemblies of twenty plies) in sectioned polyester (Crystic 471) laminates.

Fabric type	P_{max} (kPa)	α_f	α_p	α_v
PW	88	0.77	0.23	0.040
PW	884	0.92	0.08	0.017
TW	88	0.76	0.24	0.006
TW	884	0.85	0.15	0.006
5HSW	88	0.73	0.27	0.001
5HSW	1768	0.90	0.10	0.001
NCSB	88	0.64	0.36	0.004
NCSB	884	0.78	0.22	0.005
NCSB	1768	0.88	0.12	0.001

Table II presents a summary of the calculated average values of the measured variables for selected maximum target pressures and for the four types of examined fabrics, i.e. plain weave (PW), twill (TW), 5 harness satin (5HSW) and noncrimped,stitch-bonded (NCSB) fabric.

The plain and twill weave exhibit similar pore area fraction at a compression pressure of 88 kPa. However, the yarns of the plain weave are larger than those of the twill weave and this, coupled with nearly twice the number of ends and picks for the twill weave, results in the plain weave having fewer but larger pores and, hence, wider flow paths. Table II demonstrates that as the compression pressure is increased to 884 kPa, the pore area fraction of the plain weave decreases to 0.08 whereas the pore fraction of the twill weave is reduced to 0.15. This can be attributed to the capability of adjacent plain woven fabrics for better nesting.

The 5 harness satin weave also exhibits a similar pore area fraction to that of the plain and twill weave (see Table II). The pores are much smaller than those of the plain and the twill weave. Although the 5 harness satin weave has the smallest pores, it also has the least tortuous flow paths. The decrease of the pore area fraction of the 5 harness satin with increasing pressure is as expected due to the low crimp of its yarns. However the pore fraction is not as low as in the plain weave due to the reduced ability of the 5 harness satin for nesting between adjacent layers because of the low level of crimp and no symmetry.

Laminates of noncrimped stitch-bonded fabric allow higher fibre volume fractions [6] than woven fabrics, have no crimp and are more stable than simple weaves regarding distortion. The relatively high value of pore area fraction at a compression pressure of 88 kPa (see Table II) is due to the large dimensions of rovings, a lower number of ends and picks, non-unidirectional fibre orientation between layers and low crimp and therefore no nesting. As the pressure is increased to 1768 kPa, the layers of reinforcement adjust together well due to the low crimp, ensuring that there are no rich resin areas and reducing the pore area fraction dramatically leaving almost no pores for resin flow. The reason for the pore area fraction not decreasing further is the remaining pores between rovings in a single ply. These would be difficult to reduce further as the plies are stitched together inhibiting the movement of rovings and also nesting is greatly reduced because of the low crimp.

Control of voids is very important in the production of polymer composites. Void formation is related to resin impregnation and more specifically to viscous flow and wetting properties of the resin related to the reinforcement. The formation of voids can be attributed to many factors, the main one in the current studies being the entrapped air from the mixing of the resin system and the layup stages. The size of voids is affected by the available pore size and their position along the flow path. The plain weave laminate at the relatively low compression pressure of 88 kPa has the highest void area fraction and the largest sized voids. As the compression pressure is raised to 884 kPa, the voids are forced out of the system along the in-plane flow paths. The tortuosity of the flow path is an important factor in the process of voids movement and elimination. The plain weave fabric has the most tortuous flow path and, hence, it is more difficult for the entrapped air to flow out of the system. The twill weave has the next most tortuous flow paths and there is a corresponding reduction in the void area fraction. The 5 harness satin weave is next, where there is a corresponding reduction of the void area fraction toward zero. The noncrimped stitch-bonded fabric has large pores at low pressures, leading to medium voidage. However, at high compression, air escapes though the low-crimp, non-tortuous path yielding low voidage.

5. Conclusions

From the microstructural studies it can be envisaged that the compression of woven cloths can be modelled as a three mode compression. In compression mode 1 the fibre cloths nest closer by slipping while under compression. When the cloths are impregnated with resin an equivalent amount of resin is expected to be expelled. Nesting is related to both type and geometrical parameters of reinforcement and resin rheology and, in the presented experimental study, nesting was present even at low pressures of 4.4 kPa. In compression mode 2 the fibre yarns are deformed by decreasing the amplitude of yarn waveform in which case the thickness of individual plies is reduced. In compression mode 3 the fibre yarns are individually compressed and deformed. Another mode in the compression of laminate is the resin flow through the fibres The microstructural data of the present study indicated that mode 1 is dominant over a wide range of pressures in the low and intermediate pressure regime whereas modes 2 and 3 become significant at high pressures.

Research on the comparison of the compression behaviour of different types of fabrics focused on the compression of assemblies of glass fabrics of the following types: plain weave, twill, satin and noncrimped, stitch-bonded fabric. Polyester laminates were made under different degrees of compression and subjected to microstructural studies. The type of fabric was a deciding factor in its influence on compression. Ease and perfection of nesting between layers was combined with the ease of resin removal to maximise compression. Microstructural studies of laminates produced under different degrees of maximum compression revealed porosity (resin rich areas), area pore structure and sizes and average voidage and void sizes for the different types of fabrics.

Acknowledgements

The authors gratefully acknowledge the support of EPSRC initially in the form of a PhD studentship to the first author and also in the form of grant GR/J76972.

References

[1]. Yurgatis S.W., Morey K. and Jortner J. 'Measurement of yarn shape and nesting in plain-weave composites', Comp.Sci.Techn., 46(1), 1993, pp.39-50.
[2]. Jortner J. 'Microstructure of cloth-reinforced carbon-carbon laminates', Carbon, 30, 1992, pp.153-164.
[3]. Simacek P. and Karbhari V.M. 'Notes on the modelling of preform compaction: I-Micromechanics at the fibre bundle level', J. of Reinf. Plastics and Comps., 15(1), 1996, pp.86-122.
[4]. Matsudaira M. and Qin H. 'Features and mechanical parameters of a fabrics mechanical behaviour', J. of the Textile Institute, 86(3), 1995, pp.476-486.
[5]. Bader M.G. and Lekakou C. 'Processing for laminated structures', Chapter 8 in 'Composites Engineering Handbook', Ed. P.M.Mallick, Marcel Dekker, New York, 1997, pp.371-479.

Modelling approach of microbuckling mechanism during cure in a single fibre carbon epoxy composite

Christian Jochum, Jean-Claude Grandidier & Michel Potier-Ferry

Laboratoire de Physique et de Mécanique des Matériaux (LPMM), URA CNRS 1215,
Institut Supérieur de Génie Mécanique et Productique,
Université de Metz, Île du Saulcy, F-57045 METZ CEDEX 01.

Abstract

In this work, the mechanism of fibre positioning imperfections observed in carbon epoxy composites is studied. A first simple approach has allowed to show that these imperfections do not appear during the final cooling stage of the cure process, but that they are the consequence of a fibre microbuckling phenomenon created by the resin shrinkage during the crosslinking reaction. In order to have a good description of this fundamental mechanism, both experimental and theoretical studies on single fibre composite specimens are done. The structure is described through a model of beam on foundation whose response is determined experimentally by DMTA analysis.

Keywords

Undulations Defects - Cure Process - Microbuckling during cure - Single fibre Composite Specimen - Crosslinking - Volume variation

1. Outline and motivation

Nowadays, most of works have allowed the comprehension and modelling of the compressive failure mechanisms in long-fibre reinforced laminate composites. Authors have shown that the appearance of a fibre microbuckling phenomenon triggers the compressive failure. This mechanism is controlled by the physical non-linearity of the matrix and by the presence of initial fibre positioning imperfections. Additional recent works have established that this mechanism is also affected by some structural parameters of the problem (stacking sequence, ply thickness and loading). For example Drapier et al [1] have derived an intermediate scale model which takes into account all these parameters. However, the efficiency of this approach is directly conditioned by the knowledge of the initial fibre positioning imperfection characteristics (wavelength, amplitude). Of course, it is possible to develop measurement techniques for this imperfection but, from a scientific point of view we found that it is more strategic to study the origin of this fibre initial waviness.

2. Mechanism identification

An very realistic observation of fibre positions within a long-fibre reinforced ply composite has been established by Paluch [2]. In fact, he has rebuilt the three dimensional progress of carbon fibres within the ply by a succession of cross cuttings (figure 1).

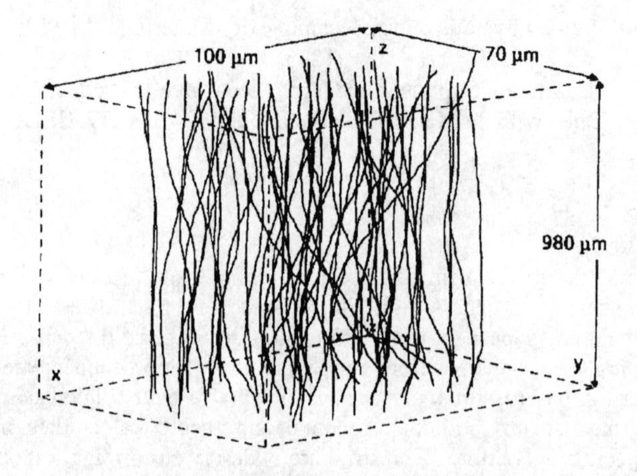

Figure 1 : Perspective view of carbon T300 fibres.

From this observation it results that the carbon fibre defects are very closed to a sinusoidal form. This has also been observed on single fibre composite specimen (Ahlstrom [3] and Grandsire [4]) and it suggests us that the fibre waviness results from a microbuckling instability mechanism which appears during the hot phase of the composite cure cycle (figure 2).

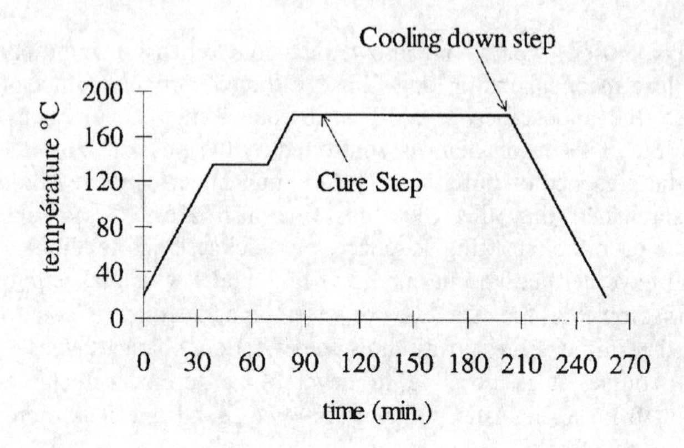

Figure 2 : Classical epoxy resin cure cycle.

Some authors like Grandsire [4] have supposed, that the instability appears during the final cooling phase (figure 2). As a theoretical point of view, the mechanism which generates the instability is essentially due to the difference of axial thermal dilatation between the epoxy matrix and the fibre (high strength carbon fibre), whose thermal dilatation coefficients are around 11E-5°C^{-1} and around -0.02E-5°C^{-1} respectively. Therefore, during the cooling phase, the fibre dilates axially while the matrix contracts. According to the coherence of the fibre-matrix interface, a compressive loading is generated on the fibre. As this phenomenon occurs at a hot temperature, we can consider that the support provided by the matrix to the fibre is still limited and can not avoid the fibre microbuckling. We assume that the length of the cure step allows the relaxation of internal stresses by a viscous flow of the matrix. During the final cooling phase, the matrix is not subjected to chemical reactions and has its definitive characteristics because of a quasi total crosslinking.

The long-fibre reinforced laminate composite thermal stress is estimated by a simple model presented more in details in Datoo [5]. The microstructure is represented by the alternative superposition of two beams, one for the matrix and the other for the fibre. During the cooling, axial deformations in fibres and in the matrix are supposed to be identical, this allows to estimate the compressive stress generated in the fibre.

To evaluate the microbuckling risk, the thermal compressive fibre stress during cooling is compared with the necessary stress to generate an instability (figure 3). In this approach, stiffness and thermal coefficient changes are taken into account as well as the matrix behaviour change at the glass temperature transition (Tg) crossing. Above Tg temperatures, the resin is into a rubber state and the critical microbuckling stress (noted σ_{fc}) can be calculated with the Rosen [6] model which gives the following relationship :

$$\sigma_{fc} = \frac{G_m}{f(1-f)}$$

where G_m is the elastic shear modulus of the matrix and f the volume fraction of the fibre. In this approach the modulus change with the temperature.

Below Tg temperatures, the plastic matrix behaviour appears and the critical stress is estimated thanks to the model of kink band proposed by Budianski and Fleck [7] :

$$\sigma_{fc} = \frac{1}{f} \cdot \frac{G}{1+n\left(\frac{3}{7}\right)^{\frac{1}{n}}\left(\frac{\bar{\phi}/\gamma_{Yc}}{n-1}\right)^{\frac{n-1}{n}}}$$

where G is the composite shear modulus and $\bar{\phi}$ the initial incline angle of the fibre. The composite behaviour is approached by a Ramberg-Osgood law whose hardening coefficient is noted by n and the elastic shear strain limit by γ_Y. In comparison with these works, we have taken into account the temperature influence on parameters n, G, γ_Y evolutions during the whole cooling. These evolution laws have been characterised by video-controlled shear tests (G'Sell C., Jacques D., Favre J.P[8]).

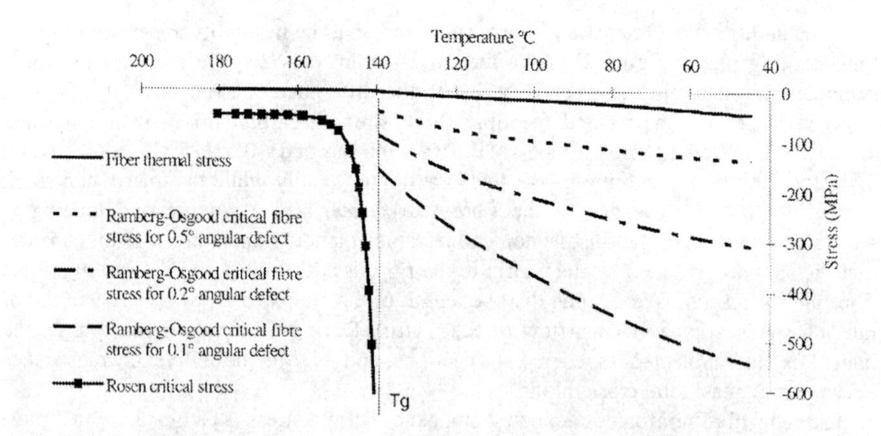

Figure 3 : Comparative stress layout during cooling.

Figure 3 shows thermal stress and necessary instability stress evolutions versus temperature. It clearly appears that it is unlikely to observe any instability during the final cooling phase.

To prove that the final cooling phase is not at the origin of the microbuckling, we have compared two single fibre composite specimens, each one is made with only one carbon fibre T300 embedded in an epoxy resin LY 556. They were carried out with different cure cycles but having the same final cooling stage. Results are displayed on following figures and pictures 1 and 2 :

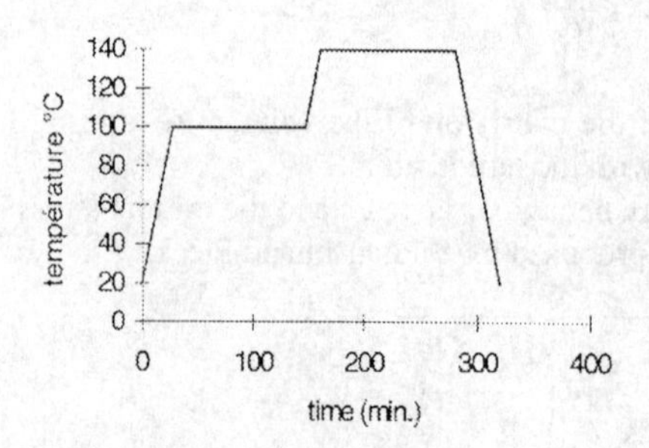

Figure 1 : cure cycle n°1 : *2h at 100°C + 2h at 140°C*

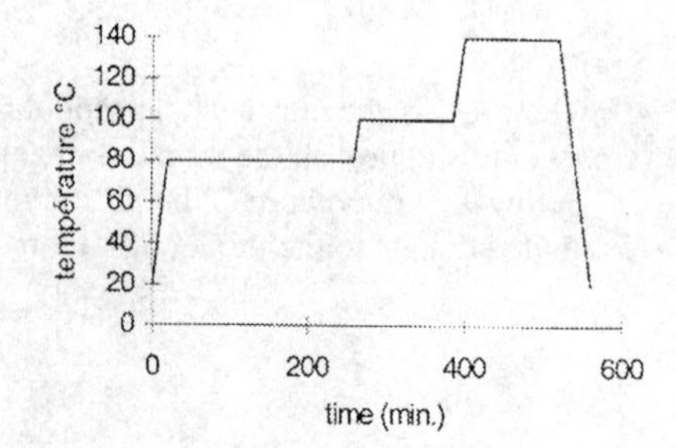

Figure 2 : cure cycle n°2 : *4h at 80°C+ 2h at 100°C + 2h at 140°C*

Picture 1 : fibre microbuckling

Picture 2 : no fibre microbuckling

It appears clearly that the instability is essentially due to the resin crosslinking reaction since the two cycles chosen here have the same final cooling phase and that the microbukling only occurs in case of a fast cure cycle.

3. Mechanical characterisation of the crosslinking reaction

In order to specify the physical and mechanical behaviour of the resin during the crosslinking chemical reaction, measurements of the shear stiffness variation have been carried out with a dynamic and mechanical thermal analysis (D.M.T.A.) system under a configuration of torsion.

An automatic control for the torsion plate to have a vanishing axial force has also allowed us to obtain the volume variation of the resin during its crosslinking with different cure rates. Using different frequency analysis, we can finely characterise the resin viscoelastic behaviour during the chemical reaction.

Typical curves for 2Hz and 3Hz are shown on figure 4. We observe that the elastic modulus change quickly (figure 4-a) and grows from zero to around 10 MPa within 10 minutes. During this interval of time (from 35 to 55 minutes), the resin presents a viscous behaviour as shown on figure 4-b with a rubber behaviour after 50 minutes. During the whole crosslinking phase, the resin is subjected to a shrinkage as shown on figure 4-c. At the end of this stage, the matrix dilates in important proportions. This dilatation is however cancelled during the final cooling phase.

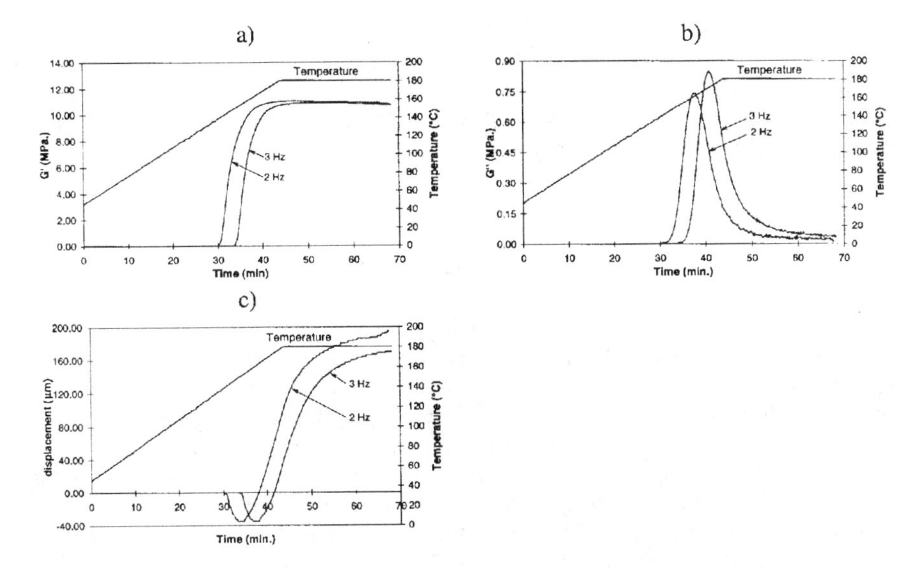

Figures 4-a , 4-b, 4-c : Modulus G', G'' and displacement changes during the crosslinking reaction of the resin for a 180°C cure with a cure rate of 3°C/min at two frequencies.

By changing the cure rate, the mechanical behaviour remains similar (figure 5-a and 5-b), only the time to obtain 10MPa. is changed. Figure 5-c confirms the resin shrinkage dependency with the chemical reaction rate. The more stronger the reaction is the more important the shrinkage is.

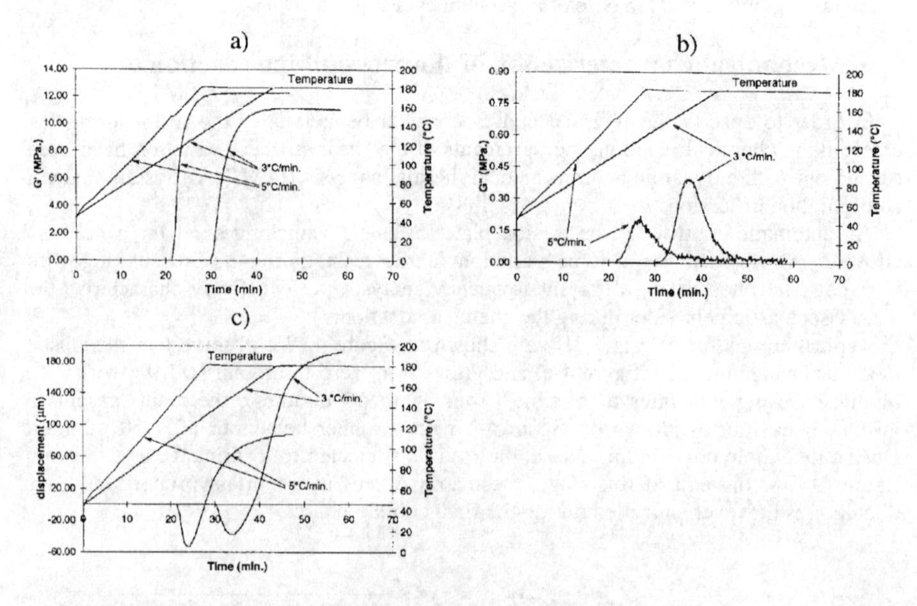

Figures 5-a , 5-b, 5-c : Modulus G', G'' and displacement changes during the crosslinking reaction of the resin for a 180°C cure at two different cure rate of 3°C/min and 5°C/min with the same frequency.

4. Mechanism modelling

According to previous results, we can explain the fibre defect generation by a warm phase microbuckling phenomenon. During the cure process, the volume variation (shrinkage) generated by the chemical reaction of crosslinking induces a compression loading in the fibres. Since the matrix is in a viscous state, it provides only a limited support to the fibre. This mechanism is not obvious to study because, on one hand, the mechanical characteristics of the resin change strongly during the chemical reaction, and on the other hand, in a laminate composite the thickness thermal gradient generated by the exothermic aspect of the reaction induces different crosslinking rates at each point of the material. Moreover, datas about volume shrinkage generated by the crosslinking reaction are rare in the literature and must still be confirmed in order to set a predictive modelling.

Thus, in order to grasp correctly the mechanism we decided to study the cure of a single fibre composite specimen. The interaction between fibre and resin can simply be described by a model of beam on foundation (figure 6).

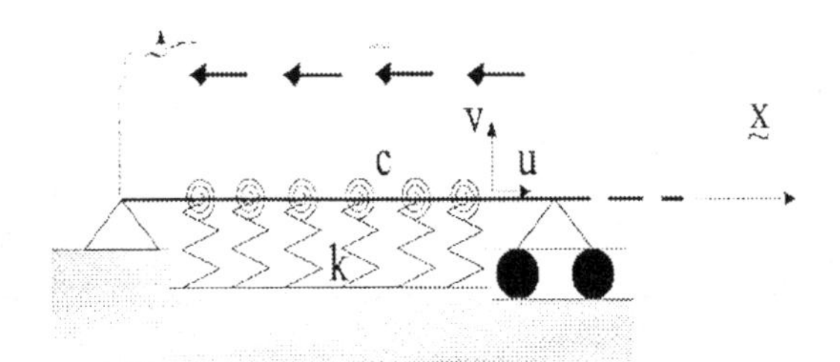

Figure 6 : model of beam on foundation.

The reaction of the resin is represented by two springs with viscoelastic stiffness noted k(t) and c(t). They are respectively representative of the y-axis return force and of the reaction moment created by the resin when the fibre moves transversely. Assuming that the behaviour of the material is incompressible, stiffness is an explicit function of the elastic modulus G' and loss modulus G''. The instability can be seen in this first approach as being the solution at each moment of the minimum of the potential energy (noted P) and whose expression is :

$$P = \frac{1}{2}\int_0^l E_f I_f \left(\frac{d^2v}{dx^2}\right)^2 dx + \frac{1}{2}\int_0^l k(t)\, v^2\, dx + \frac{1}{2}\int_0^l c(t)\left(\frac{dv}{dx}\right)^2 dx + \varepsilon_r(t)\,\frac{1}{2}\int_0^l (x-l)\left(\frac{dv}{dx}\right)^2 dx$$

The function v(x) denotes the fibre transverse displacement. E_f, I_f and S_f are respectively the Young modulus, the quadratic moment and the section of the fibre. The resin shrinkage is represented by the function of crosslinking deformation $\varepsilon_r(t)$. In this approach we assume that the prebuckling is linear and that the load-history does not affect the physical values. To solve this equation functions k(t), c(t) and $\varepsilon_r(t)$ have still to be determined.

5. Conclusions and perspectives

On the basis of these experimental results, the instability can be characterised as function of the cure cycle. In order to improve our approach, the influence of the history on the behaviour of the resin and the effect of the temperature will be taken into account in the model of beam on foundation. This approach provides a reliable basis to the comprehension and the prediction of the appearance of waviness imperfections occurring in laminates.

References

[1] *Drapier S., Gardin C., Grandidier J-C. Potier-Ferry M.,* 1997. "THEORETICAL STUDY OF STRUCTURAL EFFECTS ON THE COMPRESSIVE FAILURE OF LAMINATE COMPOSITES", Comptes Rendus de l'Académie des Sciences, tome 324 IIb, pp 219-227

[2] *Paluch B.*, 1994. "ANALYSE DES IMPERFECTIONS GEOMETRIQUES AFFECTANT LES FIBRES DANS UN MATERIAU COMPOSITE A RENFORT UNIDIRECTIONNEL", La Recherche Aéronautique. 6, 431 - 448.

[3] *Ahlstrom C.*, 1991. "INTERFACE FIBRE DE VERRE / MATRICE POLYEPOXY-INTRODUCTION D'UNE INTERPHASE A PROPRIETES CONTROLEES", thèse de l'Institut National des Sciences Appliquées de Lyon, Juillet 1991

[4] *Grandsire I.*, 1993. "COMPRESSION DES COMPOSITES UNIDIRECTIONNELS : METHODES D'ESSAIS ET APPROCHE MICROMECANIQUE", thèse de Doctorat de l'ENS Cachan, Avril 1993.

[5] *Datoo M.H.*, 1991. "MECHANICS OF FIBROUS COMPOSITES", ed. Elsevier Science Publishers Ltd.

[6] *Rosen*, 1964. " MECHANICS OF COMPOSITE STRENGTHENING", Fiber Composite Materials, American Socety Metals Seminar, Metal Parks, Ohio, 37-75.

[7] *Budianski B., Fleck N.A.*,1993. "COMPRESSIVE FAILURE OF FIBER COMPOSITES", Journal of Mechanic and Physic of Solids, 41(1), 183-211.

[8] *G'Sell C., Jacques D., Favre J.P.* "PLASTIC BEHAVIOUR UNDER SIMPLE SHEAR OF THERMOSETTING RESINS FOR FIBRE COMPOSITE MATRICES", Journal of Materials Science 25 (1990) 2004-2010.

A New Tool for Fibre Bundle Impregnation: Experiments and Process Analysis

A. Lutz[1], K. Velten[2], M. Evstatiev[1]

[1]Institut für Verbundwerkstoffe GmbH, D-67663 Kaiserslautern, Germany

[2]Institut für Techno- und Wirtschaftsmathematik, D-67663 Kaiserslautern, Germany

Abstract

Highly viscous matrix materials are common in thermoplastic composites manufacturing. A recently developed impregnation technique based on melt impregnation using a porous impregnation wheel overcomes problems associated with the high viscosity of the thermoplastic matrix and with the very thin fibre filaments. It allows for long effective impregnation times, features uniform matrix flow even in the presence of a highly nonuniform fibre distribution and makes use of non-newtonian effects to reduce matrix viscosity. Fibers and molten polymer matrix will get in contact at the outer surface of the impregnation wheel, which is the centre-piece of the impregnation tool. An analysis of the method based on experiments and a quantitative description of the process is presented.

1. Introduction

The most important difficulty in manufacturing continuous fibre reinforced thermoplastic composites is to gain a high degree of impregnation. This is especially true for the connection of the highly viscous and glutinous thermoplastic melt and the very slender filaments of the reinforcing material. The viscosity of the thermoplastic melt is, depending on the selected matrix, often more than two or three orders of magnitude higher in comparison to thermosets.

To overcome these problems and to gain an easy and comfortable impregnation process a new kind of impregnation tool was developed which is called: impregnation wheel (Fig. 1). This kind of impregnation tool is described in more detail in [1] and [2]. The impregnation wheel consists of a porous ring, which allows the molten thermoplastic matrix to penetrate through it and through the fibre bundle, which is in physical contact with the outer surface of the ring along half of its perimeter (up to 220°). The fibre bundle is pulled between a supporting rail system placed at the edges of the wheel. By increasing the diameter of the impregnation wheel, the duration of the impregnation step can be largely extended. In addition the effect of uniform matrix flow which is described in section 2 supports the impregnation process.

The latter allows to produce high quality impregnated fibre bundles. It is also possible to vary the fibre volume fraction within a certain range by controlling the amount of molten polymer. The station has to be placed next to a commercially available extruder that provides the molten polymer thus feeding the latter into the impregnation tool.

The aim of this work was to analyze the non-newtonian permeabilities of different porous materials and based on this practical knowledge to optimize the impregnation situation.

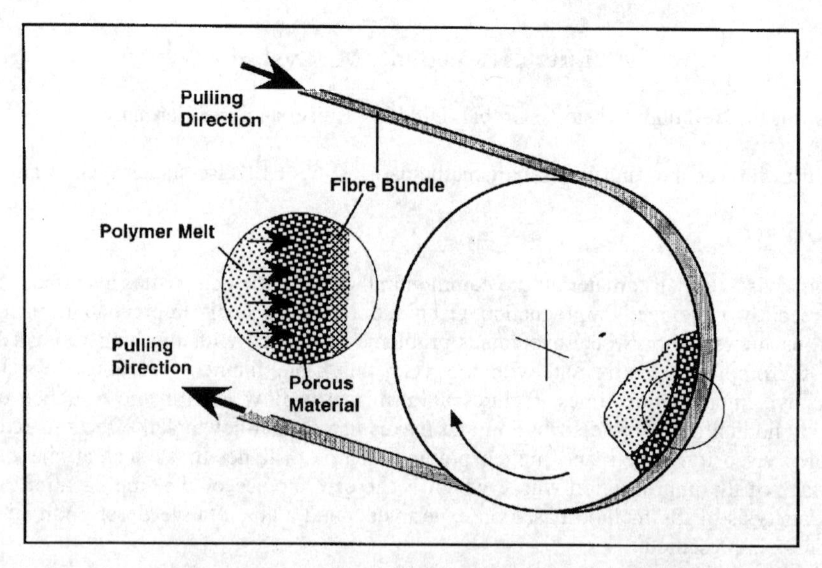

Fig. 1: The new impregnation tool (impregnation wheel)

2. The principle of uniform matrix flow

It is fact that a fibre bundle always shows a deviation in its thickness as a function of its width which means that also the flow resistance is a function of ist width. Low flow resistances of fibre bundle sections lead to nonuniform impregnation since the molten matrix tends to penetrate the fibre bundle only in positions where the bundle thickness is low. Experiments with the combination of porous body and fibre bundle showed that depending on the flow resistance of the porous body different impregnation conditions can be achieved. E.g. in the case of sufficiently high flow resistances of the porous material the matrix flows uniformly through the fibre bundle and fully impregnated prepregs or tapes can be produced.

As already mentioned the flow resistance of the porous material should be higher than that of the fibre bundle in order to achieve uniform flow of the molten polymer through fibre bundle. A uniform flow front can be realized by adjusting the permeability of the porous material so that the flow resistance (pressure drop) is higher than that of the fibre bundle (Figure 2). From Figures 1 and 2 it also can be deduced, that the geometrical aspects of the fibre bundle (tex number (number of layers of individual filaments)) and the width of the spread bundle affect the impregnation process. If a fibre bundle or roving with a high tex number is chosen, the thickness of the fibre bundle increases (if the same width in the supporting rail system is used). This means that for fibre rovings with a higher tex number also the permeability of the porous body has to be decreased (a higher h_{Body}) or that the porosity of the permeable body has to be reduced (steeper rise of the pressure drop as a function of the material thickness or pore size distribution). Another alternative is that the width of the rail system is increased (decreasing the fibre bundle thickness). In order to ensure a uniform flow through the fibre bundle and to achieve a high impregnation degree all these mentioned actions can also be combined, making the impregnation tool very flexible in its design. Additionally the non-newtonian effect of the matrix is used to reduce the viscosity to a minimum value because of the high shear rate while the molten polymer penetrates through the porous material.

However, it was not a goal at this study to determine the permeability of a real sized fibre bundle yet. As a first step, we wanted to investigate the flow behaviour of non-newtonian thermoplastic melts penetrating through different porous media. For our study, we used materials with different pore geometries (see Fig. 3).

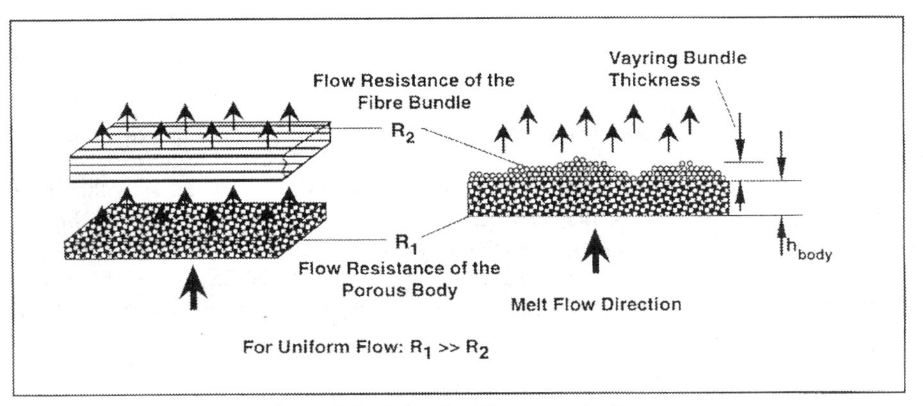

Fig. 2: Principle of uniform matrix flow

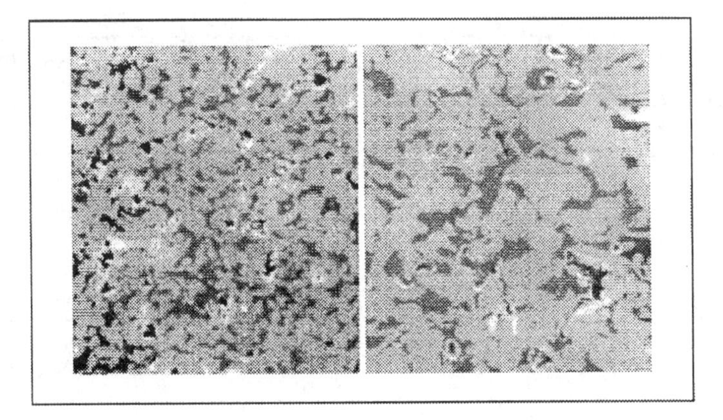

Fig.3: Example of two porous materials with different pore-size distribution (cross section, same scale).

3. Experiments and quantitative process analysis

To optimize this manufacturing process, we used several different porous materials in our investigations, which were characterized by different pore sizes (see Fig.3) and surface properties. The goal was to compare the permeabilities of the porous materials. We used two simple flow experiments: In the first experiment, the power law parameters n and m of the polymer were determined. The necessary data were obtained based on the results of capillary measurements where the pressure difference Δp [Pa] and the volume flow rate Q $\left[\frac{m^3}{s}\right]$ was given. A relation between p and Q involving the power law parameters n and m can be obtained from the following equation [3]:

$$\Delta p = \frac{2ml}{r}\left(\frac{3n+1}{n\pi^3}\right)^n Q^n$$

The power-law parameters n and m can now be determined by a fit of this equation.

To determine the non-newtonian permeability of the porous materials, another flow experiment was used: A thermoplastic melt flows under a pressure difference through the porous material placed in the mounting (Fig. 4A). In [3] it is discussed how the Darcy permeability can be obtained from this experiment. Based on the Darcy permeability, the fibre bundle impregnation time can be determined as follows:

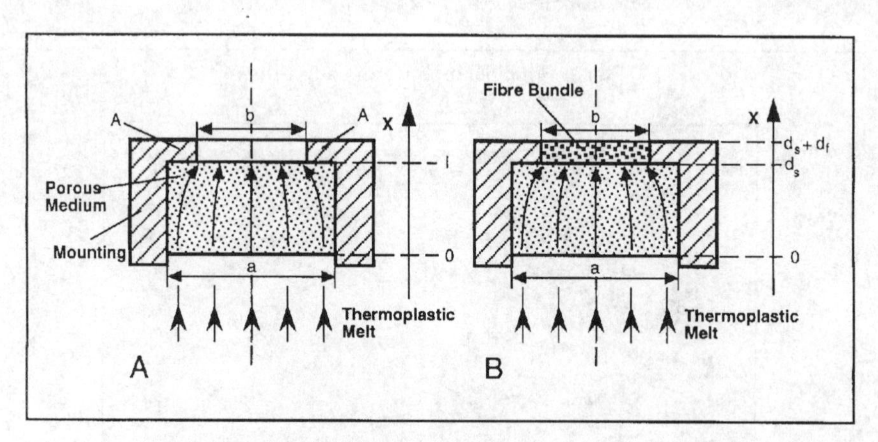

Fig. 4: Experimental set-up for pressure-drop experiments

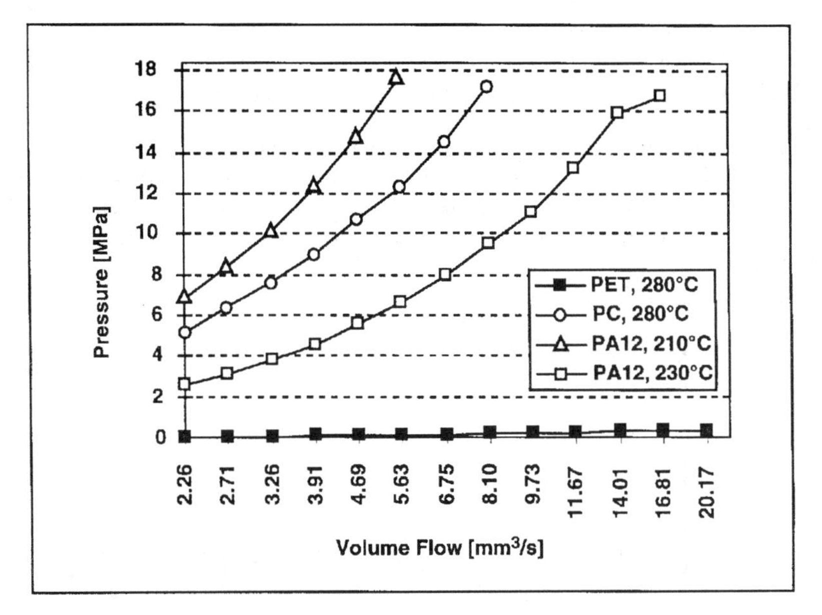

Fig.5: Example for gained measurements of pressure-drop experiments while penetrating a porous body with different thermoplastic melts

Let d_s be the thickness of the porous medium, d_f the thickness of the fibre bundle and $d_s \leq \xi(t) \leq d_s + d_f$ the coordinate position of the flow front in the fibre bundle (Fig.4B). At time t=0 (when the fibre bundle touches the impregnation wheel) the condition

$$\xi(0)=d_s$$

is fulfilled. At any time t>0, the movement of the flow front is determined by the flow rate:

$$\frac{d\xi(t)}{dt} = q(\xi(t),t)$$

The driving force of the flow is a constant applied pressure difference,

$$\Delta p = p(0) - p(\xi(t))$$

Using the methods presented in [4] and [5], the impregnation time t_{imp} (which is given by $\xi(t_{imp})=d_s+d_f$) can be calculated from the previous equations:

$$t_{imp} = \frac{n}{n+1}\left(\frac{b}{a}\right)^{n+1}\left(\frac{m}{\Delta p}\right)^{\frac{1}{n}} K_f^n \left(\left(\left(\frac{a}{bK_f}\right)^n d_f + \frac{\Gamma}{K_s^n} d_s\right)^{\frac{n+1}{n}} - \left(\frac{\Gamma}{K_s^n} d_s\right)^{\frac{n+1}{n}}\right)$$

Here, K_s and K_f are the permeabilities of the porous material and the fibre bundle and Γ is a (constant) factor reflecting the geometry of the mounting in Figure 4. In [3] it is discussed how K_s , K_f and Γ can be obtained from flow experiments.

E.g., the last formula has been used to estimate the influence of the fibre bundle permeability K_f on the impregnation time. A situation has been considered where all parameters were known from measurements except for the fibre bundle permeability, K_f. Figure 6 shows the dependence of the impregnation time on the fibre bundle permeability.

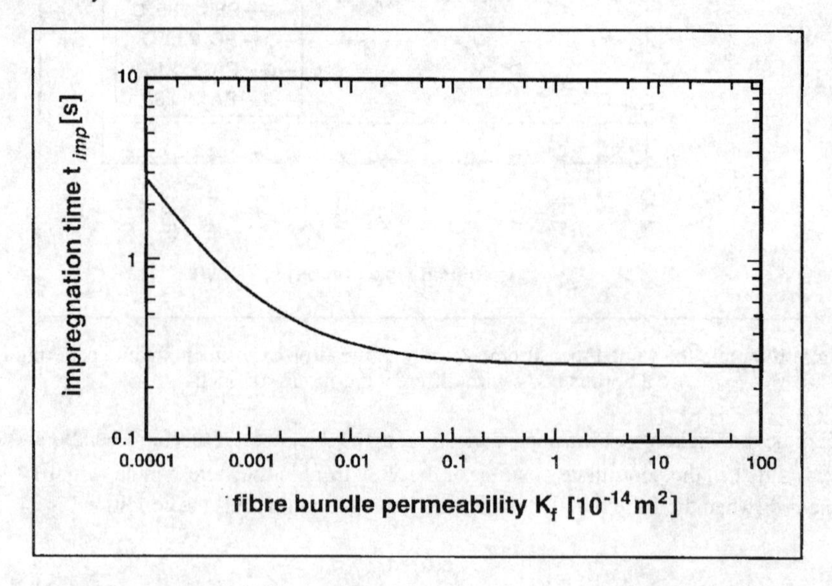

Fig. 6: Impregnation time t_{imp} as a function of the fibre bundle permeability K_f

As can be seen, the impregnation time is almost independent of K_f if $K_f > 10^{-15} m^2$. This is due to the fact that K_s was approximately $4 * 10^{-14} m^2$ in this case: as K_f approaches the same order of magnitude, it becomes less important for the determination of the impregnation time since the diameter of the fibre bundle d_f is much smaller compared to the diameter of the porous material, d_s.

However, the Figure shows that values of K_f below 10^{-15} m^2 (which can occur in dense zones of the fibre bundle) have a significant effect on the impregnation time and should thus be avoided.

4. Conclusions

A new process for fibre bundle impregnation based on a porous impregnation wheel has been presented. Long impregnation times, uniform matrix flow, low impregnation pressure and moderate fibre tension with less risk of filament damages characterize the new process.

A particular advantage of the process is its flexibility: various constructive changes can be used to adjust the process requirements of different manufacturing environments. E.g., extended impregnation time can be achieved if necessary by the use of a larger wheel diameter. Also the porous material can be optimized with respect to the process conditions.

A quantitative model of the process has been presented which allows for the determination of the impregnation time as function of the various process parameters. E.g. the impact of the fibre bundle permeability on the impregnation time has been estimated. The quantitative description of the process is of particular value in the practical application of the method since it allows to estimate the effects of changes in the process parameters.

5. Acknowledgements

The work was financially supported by the Deutsche Forschungsgemeinschaft (DFG FR 675/23-1). Further thanks are due to the FONDS DER CHEMISCHEN INDUSTRIE, Frankfurt, for support of Professor Friedrich's personal research activities in 1998.

6. References

[1] Lutz, A.; Harmia, T.; Friedrich, K.; Halonen, E.-R.: Development of a new impregnation-tool for manufacturing void-free, melt-impregnated continuous fibre bundles; 17th International Conference & Exhibition SAMPE EUROPE, May 28 - 30, Basel, (1996)

[2] Lutz, A.; Funck, R.; Harmia, T.; Friedrich, K.: A New Impregnation Tool for On-Line Manufacturing of Thermoplastic Composites; 11th International Conference on Composite Materials; Gold Coast, Queensland, Australia; July 14th - 18th; (1997); p. IV 113 - IV 120

[3] Velten, K., Lutz, A., Friedrich, K.: Quantitative characterization of porous materials in polymer processing; Paper submitted for publication in Composites Science and Technology, December (1997).

[4] Velten, K.: Quantitative analysis of the resin transfer molding process; Journal of Composite Materials; (1998); In press.

[5] Z. Cai: Simplified mould filling simulation in resin transfer moulding; Journal of Composite Materials; 26 (17); (1992); p. 2606 - 2630

The Role of Cost Estimation in Design for RTM

G. Veldsman, A.H. Basson[1]
Department of Mechanical Engineering,
University of Stellenbosch,
Stellenbosch, South Africa

Abstract

Manufacturing cost estimation models can be key elements in expanding the use of composites. Cost estimation is an essential part of Design for Manufacturing. The use of a cost estimation model can overcome some of the limitations of relying only on design experience in selecting materials and production methods for a particular part. Cost estimation models for low production volumes must reflect the increased importance of cost elements such as design cost, tooling cost and set-up cost. RTM is often used for low to medium production volumes. Cost estimation models for RTM should therefore accurately take the tooling cost, and production labour and consumables costs into account, in addition to the part's material cost. The cost estimation model must help the designer to identify the most economical tooling designs for a particular part. This is particularly important in RTM because there usually are various feasible mould designs for a given part.

1. Introduction

Due to the relative newness of composite materials, and their manufacturing processes relatively few designers know how to design an economically viable Fibre-Reinforced Thermoset (FRT) product. Boothroyd et al. [1994] showed in a survey conducted among designers that only 21 % of designers surveyed had a great or fair amount of knowledge of thermoset polymers. In discussions held with designers, it was found that most still see composites as only suitable for a few niche markets, mainly due to material costs. Recent studies tend to contradict this notion, showing that composite material products can be highly competitive on a cost and performance basis, especially when produced with newer methods, such as resin transfer moulding. Examples are the body parts of the Lotus Elan and Nissan SANI, as well as parts for the kit-plane industry [Bonner and Teeter, 1993]. Veldsman [1995] and Hutcheon [1989] have given further examples. The deciding factor in these instances was that a detailed cost analysis was done and it was found that these products could be economically produced making use of FRT and Resin Transfer Moulding (RTM).

The discussion above therefore indicates that composite products can be expected to become more common, but that designers' unfamiliarity with the materials and processes are limiting their use. One of the main factors that influences the choice of material and manufacturing process, is cost. Cost estimation models can help to show designers with little experience in composite when this type of material should be considered.

[1] Correspondence should be addressed to Prof. A.H. Basson

The main focus of this paper is to consider the role of cost estimation during the early phases of design for Resin Transfer Moulding processes. The paper also outlines the requirements for a cost estimation procedure.

The term "cost estimation model" is used here to indicate a set of empirical equations that relate the direct production costs to design parameters of the product.

2. The Role of Manufacturing Cost Estimation Models during Design

It is generally accepted that up to 50% of the avoidable costs [Corbet and Crookall, 1986], and over 70 % of the final product costs [Ferreirinha et al., 1993], are determined during design (Figure 1). This demonstrated that the best time to minimise product cost is during design. Sheldon et al. [1993], Mileham et al. [1993], and Boothroyd et al. [1994] all emphasised the importance of cost estimating during design and the resulting benefits.

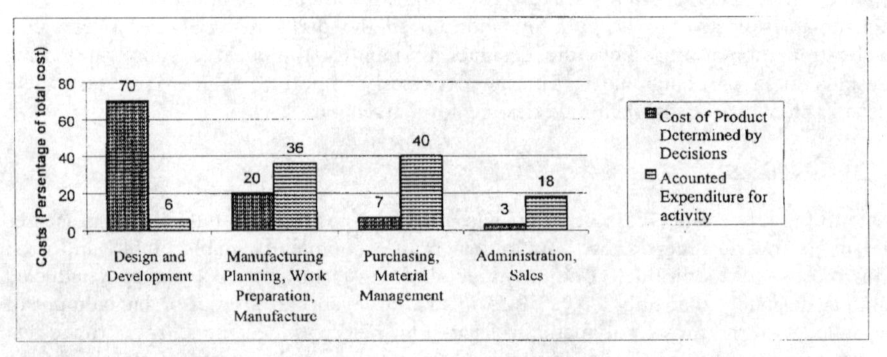

Figure 1: Influence of Company Sections on Manufacturing Cost of Product vs. Accounted Expenditures [Ferreirinha et al., 1993]

The importance of Design for Manufacture is generally recognised today. Central to Design for Manufacture is the selection of the best material and manufacturing processes for a particular product, and the optimisation of the product's design for the chosen material and processes. Factors such as production volumes and rates have to be taken into account. The designer usually has to choose between a number of viable materials and processes. In the final instance, cost will play a major role in the decision, in addition to timescales, technical risks, availability, strategic considerations, etc. It is therefor essential that the designer can assess the cost implications of his decisions [Heitger and Matulich, 1986].

Figure 2 shows a breakdown of the design process into different phases. The selection of material and manufacturing process has to be made, at the latest, during the Embodiment Design stage. However, Design for Manufacture should play a major role during final concept selection in the conceptual design phase. As stated above, these selections will be greatly influenced by cost. One of the challenges in formulating a manufacturing cost estimation model, is that a designer must be able to use it early in the design process. In the absence of formal cost models, designers rely on their own or other designers' experience (which usually include cost models that have not been explicitly formulated). Therefore we can state that cost models, whether they are explicitly formulated or developed through "experience" play a very important role in the selection of the material and manufacturing process of a product and that these cost models should significantly influence the selection of concepts [French and Widden, 1993].

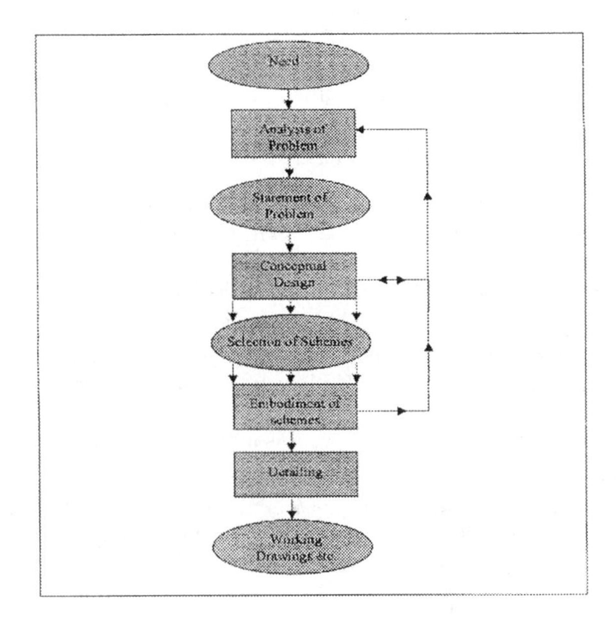

Figure 2: Block Diagram of Design Process [French and Widden, 1993]

How does an explicitly formulated cost model compare to one implicitly developed as part of experience? Engineers have remarkable capabilities in "generalising" their experiences, e.g. to select appropriate aspects of previous experience and apply it to a new problem. An explicit cost model will therefore be of little use to a designer when he is working within a field where he has developed great experience. It does, however, overcome some of the inherent limitations in personal experience, for example:

- Cost models can be shared by many designers simultaneously.
- Cost models enable designers to make good decisions far sooner in their careers, thus speeding up the learning curve.
- If suitable cost models are available, a designer can consider more materials and processes.
- Cost models can be used by different people to compare different designs and processes.

- Cost models help overcome the tendency of human beings to make choices with which they feel comfortable because of previous experience, even though they cannot justify the choice.

Cost estimation models are not only used to select materials and processes, but also to optimise designs for selected manufacturing processes. The cost models typically build up the cost estimate by considering different elements involved in the manufacturing operations. The designer can therefore easily identify which aspects in the design incur the largest part of the cost. Once these aspects have been identified, the designer can investigate alternatives where the potential benefits are the largest. Trade-offs between, for example many simple components vs. a few more complex components, can accurately be made with the help of cost estimation models.

3. Cost Estimation for Low Production Volumes

The authors' research serves some South African industries that are typical of a manufacturing sector characterised by small production volumes. The small production volumes, usually associated with niche markets, influences the optima because the balance between various cost elements is different to that encountered in medium and larger production volumes.

One example of a cost element that has a substantial influence on the final product cost, is the cost of the design process itself. Because of this, design tools with a more favourable cost:benefit ratio are sought. Cost estimation models therefore have to be kept as simple as possible to be able to serve this industrial sector.

Another cost element that is more prominent in small production volume environments, is tooling costs. In some cases, the production volumes envisaged at design time are even lower than the production life of the tooling. Cost estimation procedures therefore have to make particular provision for the inclusion of tooling costs.

Small production volumes are also characterised by a high set-up time component in overall costs. Increasing emphasis on agile manufacturing approaches is leading to similar cost shifts even in higher production volume environments. It is therefore increasingly important that cost estimation models accurately reflect set-up costs so that the products being designed for these situations can be optimised appropriately.

4. Cost Estimation in the Composites Industry

Little has been published with regards to cost estimation models for composite manufacturing. This may be due to the following:

- Companies consider cost information to be confidential.
- The processes are fairly new compared to processes such as metal forming or injection moulding.
- Research institutions tend to focus on cost estimation models for processes that are widely used and for which cost data are readily available. These tend to be the better established processes, particularly those used for high production rates and volumes.

The most significant publications discussing cost estimation in composites are by Gutowski [1994], (this article describes the work done by Foley [1993], Kim [1991], and Tse [1992]), Scott and Heath [1992], Bonner and Teeter [1993], and Veldsman et al. [1996]. ACCEM [1976] gives empirical correlations of labour times for detailed recurring labour operations in the lay-up of composite pre-pregs. This process is mostly used for the manufacture of advanced structural products (e.g. in the aerospace industry) and not for medium to high production volume consumer goods. Only Veldsman et al. [1996] and Foley [1993] considered low to medium volume manufacturing processes. Foley [1993] looked at RTM, filament winding and at the automated tape layer process with the focus on designing a new process FRTM (Flexible RTM). Veldsman et al. [1996] looked at cost modelling for RTM with the focus on making use of parameters that are readily available to the designer at the beginning of the design process.

5. Cost Estimation and Design for RTM

From the previous discussion it is clear that there is a need for cost estimation models in the medium production volume range (500 to 40 000 parts annually). These volumes are well suited to (RTM) [Rudd and Kendall, 1992]. The process may even be used for production volumes lower than 500 parts annually due to the many advantages of the process over hand-lamination and pre-preg mouldings [Foley 1992].

Under RTM we include all processes where a rigid mould system is used and the following general procedure is followed: dry reinforcement is placed in the mould, the mould is closed, the reinforcement is wetted out by either injecting or sucking in a thermoset resin, the part is left to cure, and finally the part is ejected (see Potter [1997] and Veldsman [1995] for different variants).

As stated earlier, cost estimation should start during Conceptual Design and have a significant influence during Selection of Schemes (Figure 2). At this stage in the design process the focus should not fall on exactly determining the final product cost, but rather on identifying the main cost elements that the designer can influence. These costs are mostly the direct costs [Potter, 1997]. The designer has a limited influence on the overhead costs and these are usually determined by the type of company and by the company structure.

During cost estimation, the designer has to estimate the tooling cost, production labour and consumables costs, and the product material cost. The combination of these costs forms the direct product cost. The cost of the tooling plays a major roll in determining whether the product can be produced economically. Figure 3 shows a typical cost breakdown of a RTM product. The influence of the mould goes beyond the contribution of the mould cost to the product's cost. The mould design also strongly affects the production economics: its weight and ease of handling, the way the reinforcement has to be loaded, the clamping, the resin porting control, the injection pressures it can withstand, the ease of cleaning, the release agent application, provision for part ejection and part transport, etc. are all major factors influencing the costs involved during the production stage. The mould design's effect can be further amplified by its effect on the scrap rate.

Considering the various aspects in Figure 3 that are affected by the tooling design, it is clear that the tooling has a strong influence over two thirds of the final product cost. This shows why it is important that tooling design and component design proceed in parallel.

An immediate problem in estimating the tooling and production costs, is that there are usually more than one feasible tooling design available to the designer. For example, Potter [1997] lists 9 different materials suitable for RTM tooling (GRP, aluminium, steel, nickel plated, etc.), each of these having 19 different properties (specific gravity, tolerance limitations, coefficient of thermal expansion, size limitations etc.) that may affect the suitability of the material for a particular application.

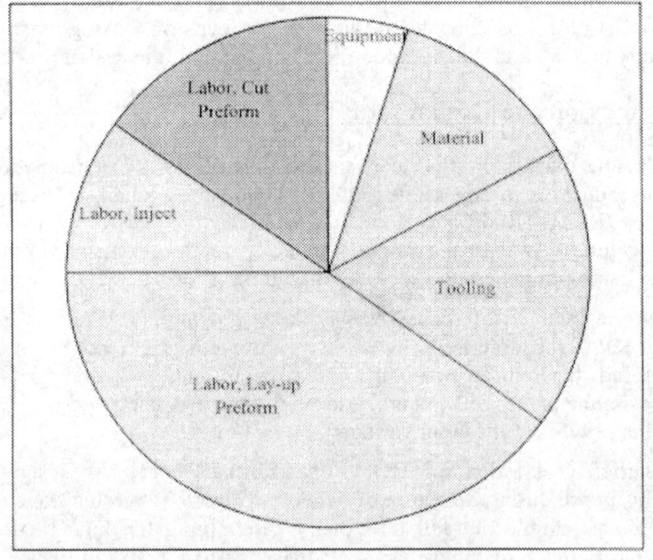

Figure 3: Typical Cost Breakdown for the RTM Process - 1000 Parts/Year [Foley, 1992]

The selection of a particular tooling design option is dependent on various customer and designer specifications. Even when the feasible design options have been identified, it would be too time consuming for the designer to manually calculate the manufacturing cost for each of the mould designs.

A cost estimation model would, however, have to provide for all the feasible tooling design options. The cost model should ask the designer for the necessary design parameters, then, if possible, identify tooling designs that are feasible for the particular application, and estimate the cost for each design. The designer can then review the results produced by the cost estimation models, and confirm the suitability of the lowest cost option identified by the cost estimation models. The designer will further be able to quantify the cost penalties incurred by using tooling designs that do not have the lowest cost, but that may have other advantages, e.g. strategic.

The cost estimations need to be accurate enough to be used for making design decisions. Because of the limited amount of information available to the designers when he uses the cost estimation models, the models cannot be accurate enough for final manufacturing cost predictions.

For the designer to be able to make best use of RTM cost estimation models during the conceptual and embodiment design phases, the models must also have the following characteristics:

- The number of inputs required from the designer must be kept to a minimum.
- The inputs must be parameters readily available to the designer. If inputs related to the production process are required, the designer must be provided with the necessary information to make informed choices.
- The cost estimation models must calculate the cost involved taking into account all the cost factors that can be directly influenced by the design, such as the influence of the tooling design on the production cost.
- The output must be in such a format that the designer will be able to realise which design parameters played the biggest part in determining the costs.
- All the necessary documentation must be recorded for design reviews.
- The system must be user friendly and easy to learn to use.
- The system must be able to be configured for the designer or company's specific needs.

6. Conclusions

The more widespread use of composites may be hindered by lack of experience and, therefore, uncertainty in the mind of the designer of the costs involved. Cost estimation models can be a key element in expanding the use of composites, because they can be used by designers with little experience in composites to determine whether composites are an economically feasible option for a particular design.

The use of cost estimation models can overcome some of the limitations of relying on design experience in selecting materials and production methods for a particular part. Their qualitative nature and the ability to be distributed are two particular advantages.

Cost estimation models for RTM have to take the mould cost, and production labour and consumables costs into account, in addition to the part's material cost. The cost estimation models must help the designer to identify the most economical tooling designs for a particular part and to quantify the additional costs involved in using other tooling designs that may be chosen for non-economic reasons.

A number of requirements that a cost model should comply with, have been listed.

Such manufacturing cost estimation models will make it possible for the designer to design for minimum cost without compromising product quality. The system will also give the designer a better understanding of the RTM process, especially to the role the mould plays in determining final product cost as well as the influence it has on production. The system will also increase the competitiveness of the process as well as that of the company.

References

ACCEM, 1976, Advanced Composite Cost Estimating Manual, Northrop Corporation Technical Report AFFDL-TR-76-87, Vol. 1.

Bonner, H.M., Teeter, D.B., 1993, "Resin Transfer Molding Applications in the Kitplane Industry", Proceedings 38th International SAMPE Symposium, pp. 488 - 495.

Boothroyd, G., Dewhurst, P., Knight, W., 1994, Product Design for Manufacture and Assembly, Marcel Dekker Inc., New York.

Corbett, J., Crookall, J.R., 1986, "Design for economic manufacture", Annals of CIRP, Vol. 35, pp. 93-98.

Ferreirinha, P., Hubka, V., Eder, E.W., 1993, "Early Cost Calculation: Reliable Calculation, Not just Estimation", Design for Manufacturability, DE-Vol. 52, ASME, pp. 97-104.

French, M.J., Widden, M.B., 1993, "Function-Costing: A Promising Aid to Early Cost Estimation", Design for Manufacturability, DE-Volume 52, ASME, pp. 85 -90.

Foley, M.F., 1992, Techno-Economic Synthesis of a Cost Effective Composite Manufacturing Process, Ph.D. Thesis, Dept. of Mech. Eng., MIT.

Heitger, L.E., Matulich, S., 1986, Cost Accounting, McGraw-Hill Book Company.

Hutcheon, K.F., 1989, The application of RTM to the motor industry, M.Phil. Thesis, University of Nottingham.

Kim, C.E., 1991, Composites Cost Modeling: Complexity, Master's Thesis, Dept. of Mech. Eng., MIT.

Potter, Kevin, 1997, "Resin Transfer Moulding", Chapman & Hall.

Rudd, C.D., Kendall, K.N.,1992, "Towards a Manufacturing Technology for High-Volume Production of Composite Components", Proceedings of Institute of Mechanical Engineers, Vol. 206, pp. 77 - 91.

Scot , F.N., Heath, R.J., 1992, "RTM for Civil Aircraft Manufacture", SAMPE European Chapter, pp. 235 - 247.

Sheldon, D.F., Huang, G.Q., Perks, R., 1993, "Specification and Development of Cost-Estimating Databases for Engineering Design", ASME Design for Manufacturability, DE-Vol. 52, pp. 91 - 96.

Tse, M.,1992, Design Cost Model for Advanced Composite Structures, Master's Thesis, Dept. of Mech. Eng., MIT.

Veldsman, G., Basson, A.H., Van der Westhuizen, J., 1996, "Composite Mould Design and Cost Estimation for Resin Transfer Moulding", Proceedings ICCST/1, Durban, South Africa, pp. 571 - 576.

Veldsman, G., 1995, Aspects of Design for Manufacturability in RTM, M.Eng. Thesis, Dept. of Mech. Eng., University of Stellenbosch, South Africa.

Vernon, I.R., 1986, Realistic Cost Estimating for Manufacturing, Society of Manufacturing Engineers, Dearborn, Michigan, 1986.

Impregnation of fibres and fabrics due to capillary pressure in resin transfer moulding

S. C. Amico and C. Lekakou
School of Mechanical and Materials Engineering, University of Surrey,
Guildford, Surrey, GU2 5XH - UK.

Abstract

The capillary pressure of glass-fibre yarns and fabrics was measured on the basis of the Wilhelmy principle. Height and weight independent measurements of the impregnation of the reinforcement by silicone oil and epoxy resin were each analysed by two different fitting procedures, which systematically underestimated equilibrium positions due to the required long times to produce reasonable estimations. Long-run experiments produced results of capillary pressure in agreement with the theoretical predictions according to the Young-Laplace equation, with a form factor for fabrics in the range of 2.3-2.7.

1. Introduction

Resin Transfer Moulding (RTM) is a low pressure process in which a liquid resin is injected into a closed mould containing pre-placed fibre reinforcement. After the mould is filled, time is allowed for curing and, when the part is strong enough, it is demoulded. The advantages of the RTM process are well established, being referred to as a versatile process, able to efficiently produce fibre-reinforced polymeric composites of different shapes and high structural performance characteristics for intermediate volume production runs [1].

The chemical and rheological properties of the resin, the orientation of the fibrous reinforcement and the surface characteristics between fibre and matrix play a major role in resin impregnation [2]. The permeability of the reinforcement and the capillary pressure (P_c), which is the main driving force to act inside fibre tows, have to be fully understood in order to optimise the use of RTM, reducing, for instance, the presence of voids which deteriorate the mechanical properties of the final part.

In this paper, an effort was made to measure P_c within yarns and one layer of glass-fibre fabric through experiments based on the Wilhelmy principle, in which a sample is partially immersed into a test fluid and the liquid uptake is related to P_c.

2. Capillary effects in RTM

While many aspects of RTM are common to other composite manufacturing methods, the flow of resin through long distances of a dry preform of complex structure is regarded as the most unique characteristic affecting RTM [3].

Flow models usually consider the flow to occur in a porous medium, based on Darcy's law, expressed by equation (1) for unidirectional flow.

$$\frac{Q}{A} = \varepsilon \cdot \upsilon = -\frac{\kappa}{\mu} \cdot \frac{\Delta P}{\Delta x} \qquad (1)$$

where: Q is the volumetric flow rate; A is the cross-sectional area of the medium; ε is the porosity of the preform; υ is the interstitial velocity; κ is the permeability of the preform; μ is the fluid viscosity and $\Delta P/\Delta x$ is the fluid pressure drop per unit length of the specimen. ΔP may be expressed as the sum of P_m (mechanical pressure), P_g (gravitational pressure), P_v (vacuum pressure) and P_c (capillary pressure).

Various factors are known to influence RTM, such as different types, styles, orientations or surface treatments of the fibres, non-uniformities in bundles and fibres, mould design, part geometry, processing conditions and even the test fluid [4].

Although the applicability of this law has been questioned [5, 6], Darcy's law is still the most used model to describe the flow of resin through reinforcements and, as mentioned by Parnas *et al* [1], many of the discrepancies about this model may be due to neglecting the effects of the microscopic flow (flow of resin within the fibre tow) on the interpretation of the macroscopic flow (in the pore spaces between fibre tows). Under certain processing conditions, the microflow and the fibre wet-out manifest as a significant driving force for resin impregnation and need to be considered for manufacturing good quality composites. Besides, once pressures involved in the injection procedure affect distinctly micro and macroflow, differences in the velocity of the flow front may be responsible for entrapping air either within or between the tows.

3. Capillary Pressure

The dynamic liquid-fibre interaction process involves wetting, liquid uptake in the porous structure (due to capillary forces) and, for some fibres, liquid absorption, making the wetting measurements of complete fibrous assemblies more difficult to analyse. P_c has been theoretically estimated using the Young-Laplace equation, first applied to idealised capillary tubes. In a reinforcing fibrous preform, usually anisotropic, there is a distribution of pore sizes and shapes and the direction of pore path may also vary. Therefore, a dimensionless shape or form factor (F), dependent on the flow direction, was introduced to merge all different anisotropic/geometric configurations, resulting in the following relationship for one-dimensional resin flow:

$$P_c = \frac{F}{D_f} \cdot \frac{(1-\varepsilon)}{\varepsilon} \cdot \sigma \cos \theta \tag{2}$$

where: D_f is the diameter of a single fibre filament; σ is the surface tension of the wetting fluid and θ is the contact angle between the liquid and the solid. If a dry unidirectional fibrous preform is being impregnated, F is considered 4 for axial impregnation (flow along the fibre axis) and 2 for transverse impregnation (flow normal to the axes of the fibres). For complex fibre alignment such as the woven fabric preform, F may be only determined indirectly by experimental evidence [2, 7].

The capillary action has commonly been analysed by experiments where a fabric sample is partially immersed in a test fluid and the liquid uptake (height rise) due to capillary forces is related to surface tension, contact angle and permeability, according to the Wilhelmy principle [8]. A similar approach described in the literature for analysing data for a rising liquid experiment has been followed [9]. Combining equation (2) with equation (1), and substituting the interstitial velocity by the velocity of the rising front of the liquid (dh/dt) and the length of the flow by the height, one would have:

$$\frac{dh}{dt} = \frac{\kappa}{\mu\varepsilon} \cdot \left(\frac{P_c - \rho g h}{h} \right) = \frac{\kappa . P_c}{\mu\varepsilon} \cdot \frac{1}{h} - \frac{\kappa \rho g}{\mu\varepsilon} = a \cdot \frac{1}{h} - b \tag{3}$$

In an analogous way, equation (3) can be written in a weight based form ($w = h.\rho.A$), as shown below by equation (4).

$$\frac{dw}{dt} = \frac{\kappa \rho^2 A^2 P_c}{\mu\varepsilon} \cdot \frac{1}{w} - \frac{\kappa \rho^2 A g}{\mu\varepsilon} = a_w \cdot \frac{1}{w} - b_w \tag{4}$$

where A is the cross-sectional area and ρ is the density of the infiltrating fluid.

Upon integration of equation (3), the height of the liquid can be found for a particular time (see equation (5)) and also, the equilibrium time can be estimated for $h = 0.99h_e$.

$$t = \left(\frac{-h_e}{b} \right) \cdot \left[\ln\left(1 - \frac{h}{h_e} \right) + \frac{h}{h_e} \right] \tag{5}$$

After data for the height rise with time has been collected for sufficient time to approach equilibrium (h_e), two different approaches can be followed: 1) a non-linear parameter fitting procedure can be used to fit the data to equation (5), or 2) dh/dt can be plotted against 1/h and the fitting line, with slope $a = \kappa.P_c/\mu.\varepsilon$ and intercept $b = -\kappa.\rho.g/\mu.\varepsilon$, which can be used to estimate Pc, according to equation (6). Following the nomenclature of the reviewed literature, the first method mentioned will be called the integral method and the second, the differential method.

$$P_c = -\rho.g. \left(\frac{\text{slope}}{\text{intercept}} \right) \tag{6}$$

4. Experimental Procedure

- **Single Glass-fibre Yarns**

Capillary experiments, as shown below (figure 1), were used to follow the axial capillary impregnation within a fibre yarn in two ways: 1) height rise in the yarn; 2) weight increase of the assembly (fibre + fibre support).

Basically, the fibres were carefully removed from the original cloth, cut in the desired length and attached with adhesive tape to the fibre support and to the weight in such a way that the weight would be below the level of liquid inside the beaker. Then, the liquid level was slowly increased, until a meniscus around the yarn was formed and, at that time, the current values for height and weight were considered as initial values and the stopwatch was started. As the liquid uptake occurred, weight and height readings of the column of liquid were continuously taken as a function of time. The experiments were conducted isothermally at room temperature and the used electrobalance and ruler could follow weight and height changes of 0.1 mg and 1 mm, respectively.

Whereas some investigators placed the fibres in glass capillaries tubes [7, 8, 9], in this work the capillaries were avoided in order to prevent possible edge effects inside the capillaries. On the other hand, extra care had to be paid to avoid trimming of the yarn, which would change its flow characteristics, and its cross sectional area became an extra variable to be measured.

The fibre yarn used was obtained from a strand of plain weave cloth (see table 1) used as supplied. The fluids were silicone oil (with a black die) and epoxy resin (without curing agent). Density, viscosity (measured by a Brookfield viscometer) and surface tension (measured by a DuNuoy ring apparatus) of the fluids were measured (table 2). Contact angle (θ) values were determined by the single fibre pullout test [8] using a dynamic contact angle analyser (DCA - 322 CAHN), where a balance measures the wetting force exerted on a fibre (F_w) slowly lowered in a beaker containing the fluid. F_w is correlated to θ according to: $\cos\theta = F_w/\sigma.P$, where P is the perimeter of the fibre.

Table 1: Properties of the glass fabric used

Areal density (A_w)	Density of E-glass (ρ_{glass})	Nominal fabric thickness (t)	Number of filaments per yarn (n)	Filament radius (R_f)	Approximate yarn radius (R_b)
0.546 kg/m^2	2560 kg/m^3	4.8 x 10^{-4} m	2094	5 x 10^{-6} m	3 x 10^{-4} m

Table 2: Properties of the wetting fluids

Liquid	Density (kg/m^3)	Viscosity (mPa.s)	Surface tension (N/m)	Contact angle with E-glass (°)
Silicone oil	855	120	21 x 10^{-3}	38
Epoxy resin	1132	2160	44 x 10^{-3}	56

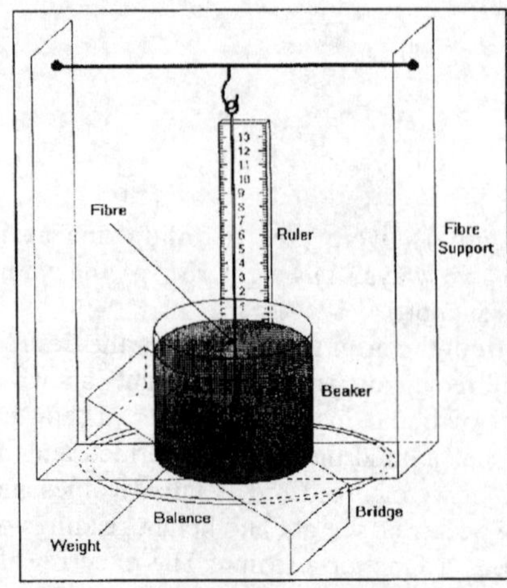

Figure 1: Experimental apparatus to measure weight and height changes in fibres

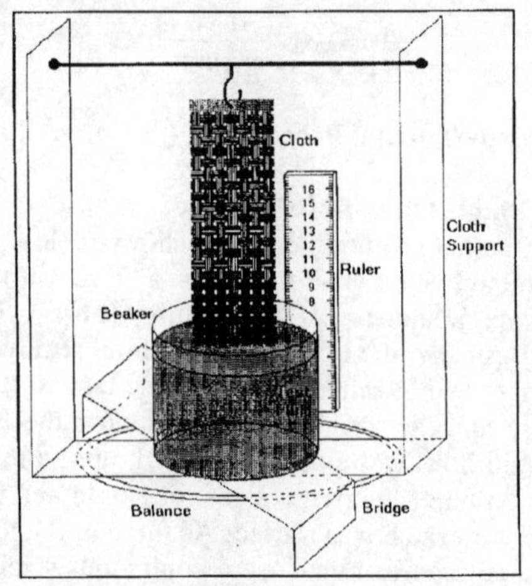

Figure 2: Experimental apparatus to measure weight and height changes in fabrics

- **Plain Woven Glass-fibre Fabric**

The same experimental apparatus was used (figure 2). In this case, however, no weight attachment was necessary since the fabric was already reasonably aligned transversely to the liquid surface. Manual stitching of the sides and the bottom extreme of the cloth had to be carried out to prevent the fabric from fraying, destroying the fabric architecture and loosing mass due to detachment of the lower yarns. Tape was not used in the sides of the fabric, because it proved to drive the flow in the edges.

The same equations as for single yarns were used. The porosity of the fabric (ε_f) and of the yarn (ε_y) were theoretically estimated and are, respectively, 0.42 and 0.56. For the fabrics, only the height data was found to be useful due to the inherent difficulty to cut and submit a standard width of the fabric for testing.

Results and Discussion

- **Single Yarns - Silicone Oil**

The experiments produced curves with a shape similar to the one shown in figure 3 for the readings of height and weight increase with time. The equilibrium state was not reached since the values were still varying at the end of the experiment.

The integral fitting procedure produced a good representation of the data, as can be seen in this figure. The data dh/dt against 1/h (derivative method) also varied linearly (see equation (3)), although scatter was observed in some experiments.

The height readings for the experiments conducted for the silicone oil are shown in figure 4. This figure shows two main features that are applicable to all capillary experiments. Firstly, the fitting procedure occasionally correlates the final readings more accurately, which is expected to influence the output values.

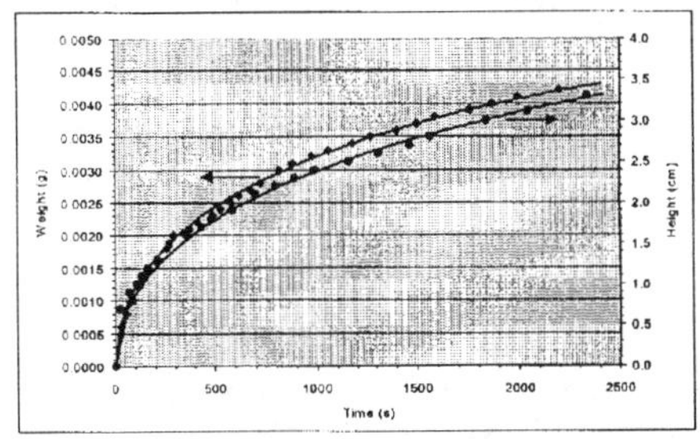

Figure 3: Height (•) and weight (♦) readings and respective fitting curves for a typical experiment

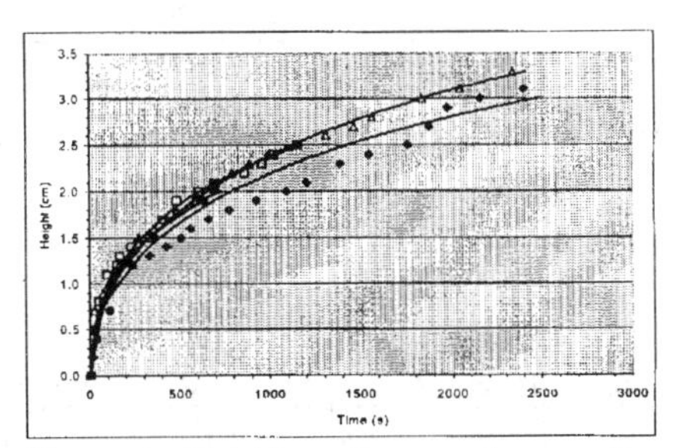

Figure 4: Height readings and respective fitting curves for 4 different experiments (•), (♦), (□) and (Δ), using silicone oil

Secondly, as mentioned in the literature [9], the duration of the experiment influences the estimation of h_e. Examination of figure 6 (specially curves •,□ and Δ) shows that although the fitting curves are practically coincident, the final values for the equilibrium height were 1.55, 3.5 and 5.2 cm, respectively. In other words, the further the height is from the equilibrium position, the more underestimated will the h_e value be. The estimation of t_{99} was also an indication that the duration of the experiment should be longer and, therefore, the period of the experiments was increased.

Average results for the two different methods of data analysis with silicone oil as the impregnating fluid are shown in table 3, with a reasonable agreement between them. One important result from this table is that the pore area of the fibre yarn, or total area of voids - A_{Tv}, is higher than initially predicted ($\approx$ 0.0012 cm^2). This might have happened due to handling of the yarn when hanging it in the support. As a result, the actual value for the porosity increases from 0.42 to 0.55 and the theoretical values for h_e, w_e and P_c (according to equation 2) will also change - see table 4. The theoretical value for h_e was found by the relationship $h_e = P_c/\rho.g$ and for w_e, $w_c = \rho.A_{Tv}.h_e$.

Although the values for h_e are at the same order of magnitude as the ones found by Batch *et al* [9] for DOP oil impregnating continuous roving glass fibres reinforcements (with similar ε), they are 14-16 times lower than the theoretical ones. Thus, it was proposed by these authors that in case equilibrium was not reached due to the long times involved, the theoretical value for the equilibrium height should be used instead, and only b would be estimated by both methods. This half-empirical methodology produced more consistent results and significantly larger h_e values than the previous ones but it was not followed in this work due to its strong influence on the estimations.

For the weight measurements, an equivalent equilibrium height (h_{eq}) can be predicted according to $h_{eq}=w_e/(A_{Tv}\rho)$. The estimated values for h_{eq} according to the integral method agreed quite well, showing a better coherence than the derivative one. In other words, equivalent results were obtained for both height and weight (with independent measurements).

Table 3: Results for the two methods of data analysis for the yarn impregnated by silicone oil

	Integral method			Derivative method			
Height	h_c (cm)	P_c (Pa)	t_{99} (h)	h_e (cm)	A_{Tv} (cm^2)	P_c (Pa)	t_{99} (h)
Average value	4.0	360.7	4.5	4.5	0.0020	374.6	4.5

Weight	W_e (g)	h_{eq} (cm)	P_c (Pa)	W_e (g)	h_{eq} (cm)	P_c (Pa)
Average value	0.0057	4.0	360.7	0.0049	2.8	252.5

Table 4: Theoretical values according to the Young-Laplace Equation

	$A_{Tv} \cong 1.2 \times 10^{-7}$ m^2 $\varepsilon_y = 0.42$ and F = 4	$A_{Tv} \cong 2.0 \times 10^{-7}$ m^2 $\varepsilon_y = 0.55$ and F = 4
P_c (Pa)	9141	5416
h_e (cm)	109.0	64.6
w_e (g)	0.1102	0.1104

- **Single Yarns - Epoxy Resin**

Table 5 shows the final results of the experiments conducted with epoxy resin, with a good agreement for both fitting methods. Although the experiments were carried out for up to 2.5 hours, the estimated equilibrium times were much longer ($\approx$ 19 h). The high viscosity epoxy is also responsible for high A_{Tv} values (0.0083 cm^2). It seems that due to the low load imposed to the yarn, as the fluid flows through the pores, it opens them, increasing the total flowing area and, consequently, the porosity (0.83). Table 6, including the theoretical values, has the same features discussed in the previous section, with values for h_e and w_e 9 and 20 times bigger than the estimated ones, respectively. Although the final values for the weight data for both methods showed a reasonable agreement (table 5), the h_{eq} values are much lower than h_e. This might be a consequence of further difficulties in setting the balance reading to zero as soon as the yarn was put in contact with the high viscosity resin.

Table 5: Average results for the two different methods of data analysis for the epoxy resin

	Integral method			Derivative method			
Height	h_e (cm)	P_c (Pa)	t_{99} (h)	h_e (cm)	A_{Tv} (cm^2)	P_c (Pa)	t_{99} (h)
Average value	2.0	222.2	17.4	2.2	0.0083	250.7	19.8
Weight	w_e (g)	h_{eq} (cm)	P_c (Pa)	w_c (g)	h_{eq} (cm)	P_c (Pa)	
Average value	0.0055	0.6	66.6	0.0057	0.6	66.6	

Table 6: Theoretical values according to the Young-Laplace Equation

	$A_{Tv} \cong 1.2 \times 10^{-7}$ m^2 $\varepsilon_y = 0.42$ and F = 4	$A_{Tv} \cong 8.3 \times 10^{-7}$ m^2 $\varepsilon_y = 0.83$ and F = 4
P_c	13591 Pa	2016 Pa
h_e	122.4 cm	18.2 cm
w_e	0.1663 g	0.1706 g

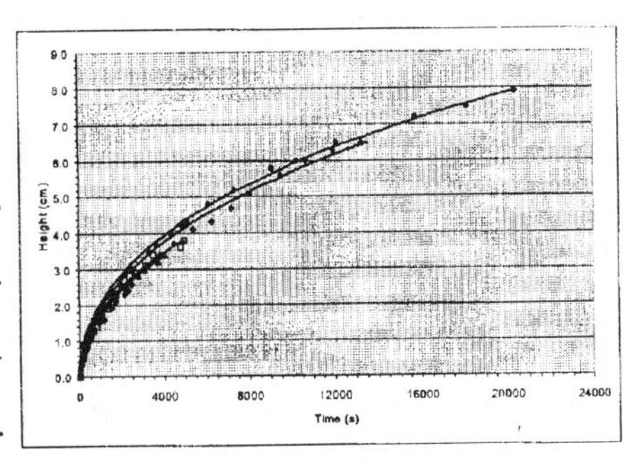

Figure 5: Height readings and respective fitting curves for 4 experiments (Δ), (□), (♦) and (•)

- **Plain Woven Fabric - Silicone Oil**

Figure 5 presents the height increase curve. The final results (integral method: h_e = 18.2 cm, b = 1.1 x 10^{-4} cm/s, P_c = 1549.7 Pa and t_{99} = 176.5 h; derivative method: h_e = 20.7 cm, P_c = 1734.3 Pa and t_{99} = 219.1 h) are approximately 2.5 times lower than the theoretical predictions (P_c = 3901 Pa and h_c = 46.5 cm - for ε = 0.56 and F = 3 - arbitrarily chosen).

- **Plain Woven Fabric - Epoxy Resin**

Considering the difficulties in finding h_e, it was decided to leave the experiment to run for a period much longer - more than 2 months . Still equilibrium was not reached (figure 6), with a resin rise of 0.2 cm per day. However, the results were in the expected range for h_e and, consequently, P_c. The theoretical value (table 7) indicates an F value in the range of 2.3-2.7.

Table 7: Results for the long-run experiment

Integral method		Derivative method	
h_e (cm)	P_c (Pa)	h_e (cm)	P_c (Pa)
39.5	4385.6	46.3	5142.8

Theoretical value for h_e (ε =0.56)	for F = 2: 34.8 cm	for F = 4: 69.6 cm

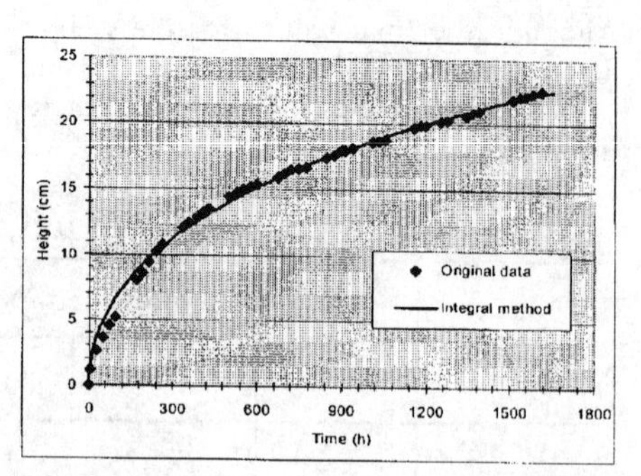

Figure 6: Long-run experiment with epoxy resin

6. Conclusions

The experimental results of the capillary experiments showed that independent height and weight measurements provided results with a reasonable agreement.

The two different fitting methodologies tried (derivative and integral) produced consistent results although the final values were underestimating Pc due to the long required equilibrium times. When the experiment was left to run for sufficiently long time, the results were consistent with the theoretical predictions according to the Young-Laplace equation, and the form factor was in the range of 2.3 - 2.7 for a sample of plain weave, an intermediate value between the values for flow axial and transverse to the fibres.

Acknowledgements

This work is part of a PhD degree sponsored by CAPES - Brazil.

7. References

1. PARNAS, R. S.; SALEM, A. J.; SADIQ, T. A. K.; WANG, H-P and ADVANI, S. G. *Composite Structures*, 27, 1994, pp. 93-107.
2. AHN, K. J.; SEFERIS, J. C. and BERG, J. C. *Polymer Composites*, 12 (3), 1991, pp. 146-152.
3. PARNAS, R. S. and SALEM, A. J. *Polymer Composites*, 14 (5), October 1993, pp. 383-394.
4. STEENKAMER, D. A.; WILKINS, D. J. and KARBHARI, V. M. *Journal of Materials Science Letters*, 12, 1993, pp. 971-973.
5. GAUVIN, R.; CHIBANI, M. and LAFONTAINE, P. Journal of Reinforced Plastics and Composites, 6, October 1987, pp. 367-377.
6. PARNAS, R. S.; HOWARD, J. G.; LUCE, T. L. and ADVANI, S. G. *Polymer composites*, 16 (6), 1995, pp. 429-445.
7. BAYRAMLI, E. and POWELL, R. L. *Colloids and Surfaces*, 56, 1991, pp. 83-100.
8. HSIEH, Y-L. *Textile Research Journal*, 65 (5), 1995, pp. 299-307.
9. BATCH, G. L.; CHEN, Y-T and MACOSKO, C. W. *Journal of Reinforced Plastics and Composites*, 15, 1996, pp. 1027-1051.

MODELLING AND NUMERICAL SIMULATION OF NON-ISOTHERMAL MOULD FILLING IN RTM PROCESS

G. Di Vita
Centro Italiano Ricerche Aerospaziali,
Capua (CE), Italy

M. Marchetti, V. Bonafede
University of Rome "La Sapienza",
Aerospace Department, Rome, Italy

Abstract

The aim of this work is to present some recent developments in modelling and simulation of the resin transfer moulding process.
A satisfactory numerical description of the process was reached through mathematical models that allow us to simulate the most important phenomena occuring during RTM: the flow of resin under pressure in the mould and the preform impregnation; the prediction of the preform permeability, considered as a stratification of unidirectional or mat type layers; the heat transfer between the mould walls, the fiber preform and the resin; the chemical transformations occuring in the resin, that cause a change of viscosity affecting the resin flow.
A set of differential equations can be written. The strongly non-linear character of the resulting system may be reduced by decoupling some of the phenomena and dependent variables, and solving separately the relative models.
A simulation code, on the basis of the mentioned models, was developed. The numerical algorithms employed in the code are outlined in the paper, and some relevant results are presented: complex shape mould filling simulations, with injection port placed in any part of the mould and multiple resin fronts; temperature, degree of cure and viscosity distribution along the mould thickness as a function of the injection temperature and the mould walls temperature.

1. Introduction

The resin transfer moulding process, in which a pre-catalyzed resin is injected under pressure into a reinforcement preform in a closed die mould where the resin cures, allows the production of high quality composites by use of a simple and economical type of technology. This process is regarded as an important method for the net-shape production of a wide range of composite products.
At present much of the tooling, preform design, and process development for a new RTM part is done by trial and error. This methodology leads to expensive modifications of the mould during the set-up phase, and can also lead to alterate the preform design to eliminate impregnation problems. The use of computer simulation codes enables us to considerably reduce the costs of the preliminary parameters imposition phase.
The fiber impregnation in a closed mould depends on a number of parameters. The

part geometry, the injection pressure, the location of the injection gate and of the air vent, which should ideally be located at the points of the mould that fill last in order to allow the air to escape. Another parameter that will influence the mould filling is the structure of the fiber preform. Fiber preform with different geometries or fiber arrangements will offer different resistances to the flow. Lastly, resin viscosity will vary throughout the mould due to its dependance on temperature and degree of cure. The varaiation of fluid viscosity can strongly influence the mould filling pattern [1]. To successfully predict the mould filling behaviour, all these parameters, and the nature of their coupling, must be correctly represented.

2. Modelling Resin Transfer Moulding

Prediction of resin motion through the fibers during mould filling as well as temperature, degree of cure, and viscosity is the objective for the resin transfer moulding process modelling.

2.1. Resin Flow Model

The moulds for RTM usually consists of two top and bottom parts which are tightly pressed together and sealed around the edges during the injection process. One or more air vents are located around the mould edges (Fig. 1). One of the important tasks in mould design is to position the injection ports and the air vents so that the mould will be filled completely without using excessive pressure. Other tasks include estimating mould distortion and minimizing required clamping forces. Successful completion of these tasks requires knowledge of the resin flow, especially the pressure distribution, the resin front profiles and positions at different stages of the injection process.

The resin flow in RTM is usually modelled as a flow through porous media. This particular flow is governed by Darcy's low, a well-known experimental expression, later theoretically derived, which states that the resin velocity is proportional to the pressure gradient, to the inverse of the resin viscosity and to the considered porous medium permeability [2].

RTM components are tipically thin shell structures, the thickness being much less than the other overall dimensions. A scaling analysis [1] shows that in this type of geometry the transverse flow is small compared to the in-plane flow. Therefore, the flow may locally be modelled as two-dimensional, considering only the in-plane flow. The permeability can be characterized as a function of the porous medium structure and porosity. In general, a porous medium can present anisotropic permeability properties. This is particularly true for fibrous porous materials such as wood, fabric and fibrous preforms. In this case the scalar permeability must be replaced by the permeability tensor $\mathbf{K}$, and the Darcy's law can be written as:

$$\vec{u} = -\frac{\mathbf{K} \cdot \vec{\nabla} p}{\mu} \tag{1}$$

where $\vec{u}$ is the velocity vector, $\vec{\nabla} p$ the pressure gradient and μ the viscosity.

Nevertheless, in a homogeneus porous medium the permeability is orthotropic. Therefore it is always possible to find an orthogonal coordinate system which diagonalizes the permeability tensor. The directions of the axes of this coordinate system are

referred to as the principal directions.
Applying the mass conservation for the resin flow:

$$\nabla \cdot (\rho \vec{u}) = 0 \tag{2}$$

where ρ is the density, and assuming that the resin density and the viscosity are invariable during mould filling, the Eqn. (1) and (2) can be combined to yield:

$$\nabla(\mathbf{K} \cdot \vec{\nabla} p) = 0 \tag{3}$$

The boundary of the integration domain Γ is subdivided in 2 parts (Fig. 1): Γ_1 where the the pressure is known (the mould inlet and the resin front) and Γ_2 where the pressure gradient is known (mould solid walls).

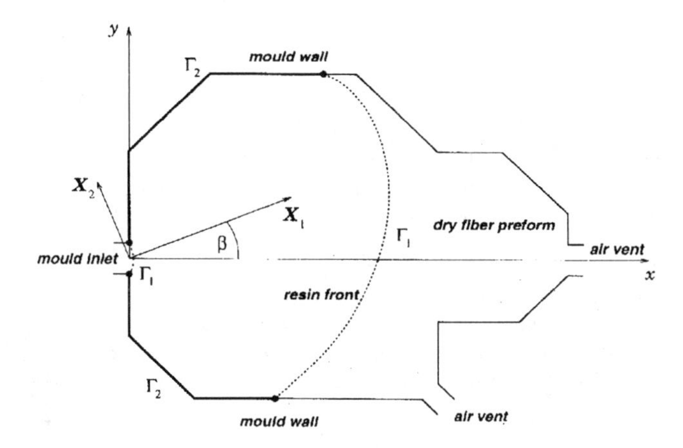

Figure 1 - Sketch of the mould integration domain and its boundaries.

Therefore, the boundary conditions for the problem can be expressed as follows:

$$\begin{array}{lll} \Gamma_1) & \text{resin front} & p = 0 \\ \Gamma_1) & \text{mould inlet} & p = p_0 \\ \Gamma_2) & \text{solid wall} & \frac{\partial p}{\partial n} = 0 \end{array} \tag{4}$$

In order to solve Eqn. (3) along with boundary conditions (4), the boundary element method can be employed [3,4,5]. This method is based upon the concept of reducing the dimensions of the problem by generating approximating equations that have unknown parameters associated with only the boundary of the computational domain. Said X_1 and X_2 the principal direction of permeability of the preform and considering them as reference system, the permeability tensor can be regarded as diagonal, and the Eqn. (3) can be written:

$$\nabla\left(K_1 \frac{\partial p}{\partial X_1} + K_2 \frac{\partial p}{\partial X_2}\right) = 0 \tag{5}$$

that, for constant permeability components, reduces to the Laplace's equation:

$$\nabla_0^2 = K_1 \frac{\partial^2 p}{\partial X_1^2} + K_2 \frac{\partial^2 p}{\partial X_2^2} = 0 \tag{6}$$

This equation can be manipulated and discretized on the boundary of the integration domain [3,4,5,6] to get a linear system of equations where the unknowns are the pressures on the Γ_2 boundary and the pressures gradient on the Γ_1 boundary.
Once the system has been solved the displacement of the resin front at each time step can be calculated.

2.2. Thermal Model

In the RTM process the mould filling does not take place under isothermal conditions. In general, the mould is heated and cool resin is injected into the mould. As the viscosity of the resin is dependent on the temperature, the knowledge of the temperature distribution inside the mould would be useful. In order to calculate the temperature field, the energy balance equation have to be solved.
The nature of the problem and the considered part geometry allowed to make several simplification hypotesis for the physical model:
1) A time-variant bidimensional domain, obtained making a section of the mould with a xz plane at the $y = 0$ symmetry plane was considered (Fig. 2).
2) The temperature on a "front", defined as the set of resin-points that are injected into the mould at a time (having, in other words, the same age), is considered constant along the whole front.
3) The convection terms along the z axis and with the lateral walls, as well as the conductive heat fluxes with the lateral walls, are considered negligible if compared to the conductive heat fluxes with the upper and lower walls and to the convective heat flux in the direction of advancement of the resin front (x).

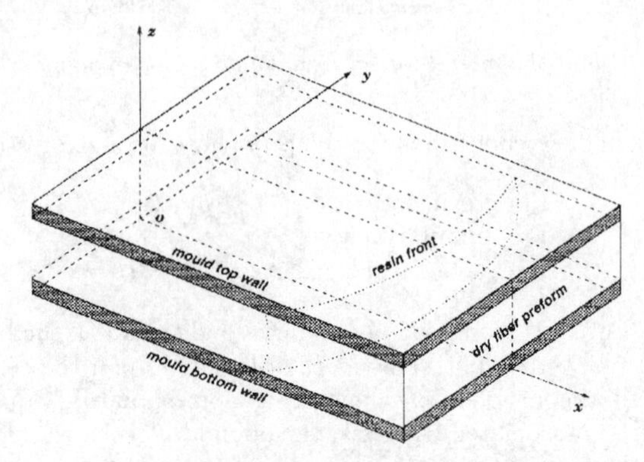

Figure 2 - Mould scheme and computational domain for the Thermal Model.

Under these hypotesis the energy balance equation for a unit volume can be written as follows:

$$\rho_c C_c u \frac{\partial T}{\partial x} + \rho_r C_r \frac{\partial T}{\partial t} = \frac{\partial}{\partial x}\left(K_x \frac{\partial T}{\partial x}\right) + \frac{\partial}{\partial z}\left(K_z \frac{\partial T}{\partial z}\right) + \rho H_r \frac{d\alpha}{dt} \tag{7}$$

ρ_c, C_c, ρ_r and C_r are, respectively, instantaneous density and specific heat for the composite and the resin, K_x and K_z the thermal conductivity of the composite in the x and z direction; u is the resin front velocity component on the x axis for the point on the plane $y = 0$, that is given from the resin flow model, T the temperature, t the time, H_r the total heat of reaction.
The last term of the Eqn. (7) represents the heat developed during the polymerization reaction until the time t for a unit volume.
α is the degree of cure, defined as the fraction of the total heat of reaction (H_r) developed at the time t:

$$\alpha(t) = H(t)/H_r \tag{8}$$

The balance equation contains two unknown quantities: the temperature T and the degree of cure α (the resin front velocity is known from the resin flow model). In order to complete the model an expression for the reaction rate is needed:

$$\frac{d\alpha}{dt} = f_1(T, \alpha) \tag{9}$$

Solving the set of the two Eqs. (7) and (9), the temperature T and the degree of cure α at each time and point of the domain can be achieved. Finally, knowing the rheological behevoiur of the considered resin in an analytical form:

$$\mu = f_2(T, \alpha) \tag{10}$$

the istantaneous viscosity at each point of the domain can be calculated, and the viscosity in the resin flow model can be upgraded.

With regard to the initial conditions necessary to integrate the two Eqs. (7) and (9), it has to be noted that at each integration step the resin front, and all the points that are located on the previous resin front positions, move forward replacing the old value of temperature and degree of cure, that are necessary to compute the actual values. The initial conditions for each "new front", defined as the forward boundary of the resin injected during an integration step, are $T = T_{in}$ and $\alpha = 0$, where T_{in} is the resin injection temperature.

The boundary conditions concern the injection port, the metallic external (top and bottom) walls and the part of the mould that is not already reached from the resin. If T_m is the mould temperature and the border of the integration domain $\Omega(x, z)$ is:

$$\partial\Omega = \partial\Omega_{inlet} \cup \partial\Omega_{wall} \cup \partial\Omega_{front} \tag{11}$$

the boundary conditions can be written as follows:

$$\begin{aligned} T &= T_{in} \qquad && \forall(x, z) \in \partial\Omega_{inlet} \\ T &= T_m && \forall(x, z) \in \partial\Omega_{wall} \cup \partial\Omega_{front} \end{aligned} \tag{12}$$

The development of appropriate models for the reaction rate will be discussed in the kinetic model section. Moreover, an appropriate model for the composite properties (ρ_c, C_c, K_x and K_z), as a function of fiber volume fraction, temperature and degree of polimerization must also be provided [1,6].

2.3. Kinetic Model

While the composite geometry and the instantaneous position of resin and reinforcement are given by the Resin Flow Model, and the temperature distribution is given by the Thermal Model, this model provides polymerization rate, degree of cure and heat of reaction of the resin as a function of position and time.

Several kinetic models have been developed to describe the polymerization of thermoset matrices [7,8,9,10]. Considering that the exact chemical composition of the resins is generally unknown, for commercial systems only empirical models can be used.

For instance, an extensively used empirical model for polyester and epoxy based resin systems is the following:

$$\frac{d\alpha}{dt}(T,\alpha) = K(\alpha_{max} - \alpha)^n \tag{13}$$

where K is the reaction kinetic constant, that is temperature dependent, n the order of reaction and $\alpha_{max}(\leq 1)$ the maximum conversion value. In this case the maximum rate of degree of cure is obtained when $\alpha = 0$. If the maximum rate of degree of cure is obtained at $\alpha > 0$, there are 2 kinetic constants and 2 pseudo-reaction orders (m and n) that have to be included in the model.

2.4. Rheological Model

This model provides the viscosity of the resin as a function of position and time. With respect to the rheological behaviour of commonly used reactive systems, two main contributions must be considered. The first one is associated with the molecular mobility changes as a consequence of temperature variations (physical effect); the second one is related to the molecular structural changes induced by the cure reaction (chemical effect). Several theoretical and empirical models have been developed [7,8,9,10]. Again, only empirical models were considered in this work as useful tools to describe the rheological behaviour of commercial systems with unknown composition.

3. Numerical Results

Two examples of simulation of the resin front advancement during filling of a rectangular mould is given in Fig. 3.
For the case on the left side, the ratio between the permeabilities in the $x \equiv X_1$ and $y \equiv X_2$ directions is 1.8. The time step between successive resin fronts is 0.5 sec, and the filling time is 9.5 sec. As can be noted, the permeability of the reinforcement, the shape of the mould, the resin viscosity, and the inlet pressure values adopted during the simulation are such that the lower corners of the mould will not be properly and completely filled by the resin.
A very interesting difference can be noted with the simulation on the right side, where the ratio K_1/K_2 is 0.7. The time step between successive resin fronts is always 0.5 sec, but the filling time is 19.5 sec.
Other relevant examples of simulations are presented in Fig. 4.
On the left side the principal axis of anisotropy X_1 is not coincident with the horizontal x axis.

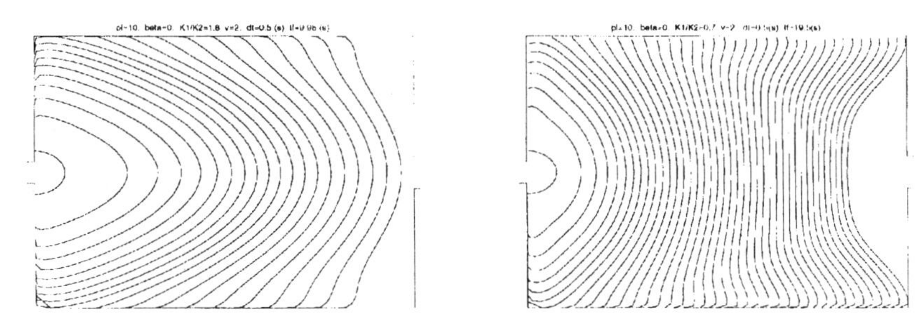

Figure 3 - Mould filling simulations: a) $K_1/K_2 = 1.8$; *b)* $K_1/K_2 = 0.7$.

In the case presented on the right, where the resin is injected in a complex shape mould, the reiforcement impregnation is not satisfactory: multiple resin fronts are generated and the mould is not completely filled. After such an analysis the designer could change some of the process conditions, and, for example, place at least two air vents on the lateral walls of the mould.

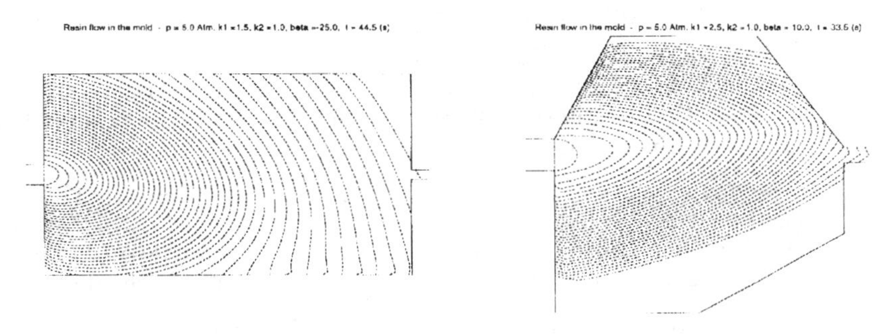

Figure 4 - Mould filling simulations: a) $K_1/K_2 - 1.5$, $\beta = -25°$; *b)* $K_1/K_2 = 2.5$, $\beta = 10°$.

Conclusions

Mathematical models for the description of non-isothermal resin impregnation of fibrous preforms into irregular geometry moulds was developed and presented. The resin flow model is based on the application of the Darcy's law for porous media into a bi-dimensional domain. The temperature field inside the mould is computed solving the energy balance equation. Once the temperature distribution is known, the degree of cure and viscosity of the resin can be calculated making use of opportunely developed kinetic and rheological models. A numerical code based on the aforesaid models was developed. At each time step the pressure, velocity, temperature, degree of cure and viscosity fields are upgraded, regarded the other variables as constant, and the resin front posisions are recalculated. The developed code is also suitable to handle complex mould shapes, multiple resin fronts and any position for the resin injection port. The use of the code will allow the technological process designer to anticipate problems such as dry spots, premature resin gel stage achievement or tardy resin cure, formation of voids which may occur at merging flow fronts. The designer can then optimize parameters such as injection pressures, placement of gates and air vents, preform architecture, resin injection temperature and mould imposed temperature.

Further improvements to the global simulation system can be made removing some of the simplification hypotesis. Infact, the two dimensional simulation method is adequate for most RTM part geometries only if the parts are thin. Moreover, the hypotesis of invariance of the temperature on a "front", defined as the set of resin-points that are injected into the mould at a time, can be removed, allowing to take count of the heat conduction in the y direction.

References

1. S. G. ADVANI, "Flow and rheology in polymer composites manufacturing" Composites Materials Series, Vol. 10, Elsevier, 1994.
2. R.E. COLLINS, "Flow of fluids through porous materials", Reinhold Publishing Company, New York, 1961.
3. D.E. BESKOS, "Boundary Element Methods in Mechanics", Vol. 3, North-Holland, 1987.
4. M.K. UM, W.I. LEE, "Numerical Simulation of Resin Transfer Molding process using Boundary Element Method", Proceedings of the 35th International Sampe Symposium, 1990.
5. E. BRUGNOLI, G. DI VITA, M. MARCHETTI, A. MOSCATELLI, "Two dimensional simulation of resin front advancement in resin transfer moulding by use of the boundary element method", in ESDA-92, Design Analysis, Machinability and Characterization of Composite Materials, Instanbul, Turkey, 1992.
6. V. BONAFEDE, "Computer aided structural-technological of advanced composite materials manufactured by the resin transfer moulding" (in italian), MSc thesis, Università degli studi di Roma "La Sapienza", 1994.
7. W.I. LEE, A.C. LOOS, G.S. SPRINGER, "Heat of reaction, degree of cure, and viscosity of Hercules 3501-6 resin", pp. 510-520, in Journal of Composite Materials, Vol. 16, 1982.
8. A.C. LOOS, G.S. SPRINGER, "Curing of Epoxy Matrix Composites", pp. 135-169, in Journal of Composite Materials, Vol. 17, 1983.
9. M.R. DUSI, W.I. LEE, P.R. CIRISCIOLI, G.S. SPRINGER, "Cure kinetics and viscosity of Fiberite 976 resin", pp. 243-261, in Journal of Composite Materials, Vol. 21, 1987.
10. J. M. KENNY, A. APICELLA, L. NICOLAIS, "A model for the thermal and chemorheological behaviour of thermosets. I: Processing of epoxy-based composites", on "Polymer Engineering and Science", Vol.29, n.15, 1989, pp.973-983.
11. M. OPALICKI, J. KENNY, "Row materials characterization and selection for system testing", First annual report of the research project Brite/Euram n. BREU-CT91-0451, 1992.

MACHINING CHARACTERISTICS OF CARBON/PHENOLIC ABLATIVE COMPOSITES

P. S. Sreejith[1], R. Krishnamurthy[2], K. Narayanasamy[3], S.K. Malhothra[4].

1. Research Scholar, 2. Professor, 3. Associate Professor, Department of Mechanical Engineering
4. Chief Design Engineer, Composite Technology Centre.

Indian Institute of Technology, Madras – 600 036, India
E-mail: mctools@iitm.ernet.in

Abstract

Machining of composite materials is an area still full of open questions, pertaining to assessment of machining and chip formation compared to metal cutting. Due to the peculiar nature of composite materials, precise analysis lacks progress yet. In this paper, data pertaining to face turning of carbon/phenolic ablative composites parallel to the fibre orientation by coated carbide tool inserts are illustrated. The discussion include mechanism of material removal, chip formation and temperature of machining. The aim of the study was to generate data and to create a basis for understanding the process of machining these composites.

1. Introduction

Mostly, polymeric composites are either fabricated/moulded; hence it is very difficult to achieve the desired design tolerances/near nett shape. In order to achieve near nett shape and also to attain the required dimensional tolerances or finish, it is necessary to go for machining of these composites. Only limited data on machining of these composites are available; it is therefor necessary to carry out machining studies for proper understanding of the cutting mechanisms and also to develop useful data on machining of such composites.

In view of high hardness and abrasiveness of carbon fibres, in carbon/phenolic composites, machining studies were attempted with coated carbide tools. Retention of cutting edge is a problem for both HSS and carbide tools [1]. High wear resistance and the ability to use at higher cutting speeds are claimed as the significant advantages of coated carbide tools [2].

Polycrystalline diamond tools are most frequently specified for lighter finishing cuts at high speeds, since their usable cutting edge length is very less [3]. Data on machining performance of titanium nitride (TiN) coated, K20 type cemented carbide tools are presented in this paper.

2. Experimental Equipments and Procedures

The work-piece material used in the investigation was filament wound carbon/phenolic ablative composite in the form of thick cylindrical shell of thickness 25mm. and having 0° fibre orientation. TiN coated tungsten carbide (WC) square inserts (SPUN 120308) were used for machining.
The machining trials were carried out on a Precision high-speed VDF lathe, provided with positive infinitely variable, stepless speed regulation. A three component piezoelectric crystal type of dynamometer (Kistler type 9441) measured the force components along with charge amplifiers and analogue indicators. Temperature at the tool work-piece interface was measured by a non-contact optical pyrometer. The chips produced during machining were examined using a scanning electron microscope (SEM) model Jeol JSM – 5300.

3. Results and Discussion

The performance of the cutting tool was evaluated in terms of cutting force and temperature.

Cutting Forces

Typical observations of the influence of the cutting conditions on the cutting force components are illustrated in Fig.1. The values of the cutting force components presented within ±1% scatter limits. The force components after an initial rise decline in magnitude. At lower speed, the cutting wedge tends to plough on the work surface resulting in higher order force. Then the cutting becomes steadier as the cutting speed increases with a consequent reduction in the force components. Further observed rise in the force components with higher order cutting velocity can be attributed to possible deformation of the cutting wedge which is a thermal associated form of tool wear. The same trend of variation was retained at higher depth of cut.

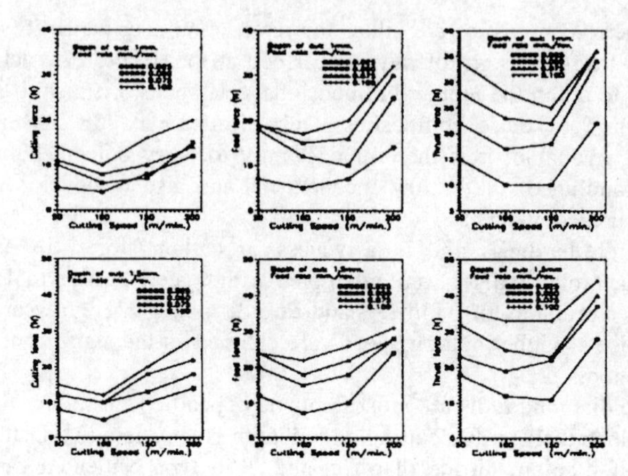

Figure 1.Typical Variation of Machining Forces with Speed

From the illustrations on force components, it can be observed that there exists a critical velocity of 100m/min. with TiN coated tools at which minimum cutting forces exist. Out of the three components, the thrust force is seen to be the highest. This may be due to the lack of form stability of the cutting wedge especially at higher order cutting temperatures, coupled with greater hardness of the material causing increased resistance to indentation of the tool wedge.

Chip Formation

Machining of the material produced small chips due to perfect shearing/rupturing of the carbon fibres with little deformation. The machining resulted in a series of cracks with instantaneous rupture of the fibres ahead of the cutting wedge leading to the formation of a chip as illustrated in Fig.2. The possible reduction in hardness of the cutting tool with increasing temperature result in the deformation of the cutting nose and flank, the instability of the wedge geometry can also be attributed to the abrasive environment prevailing during machining of carbon/phenolic composites.

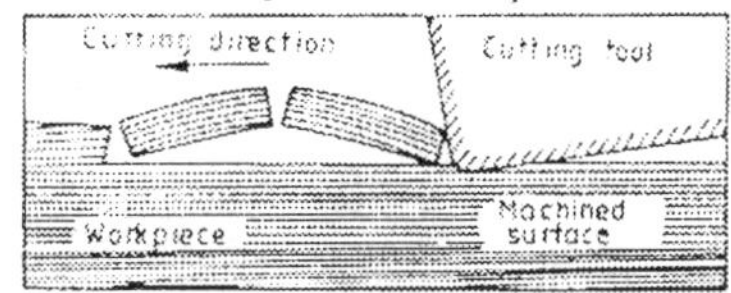

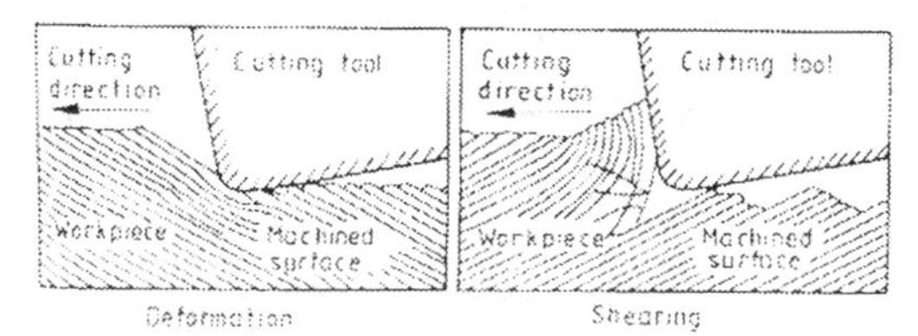

Figure 2 Mechanism of Chip Formation

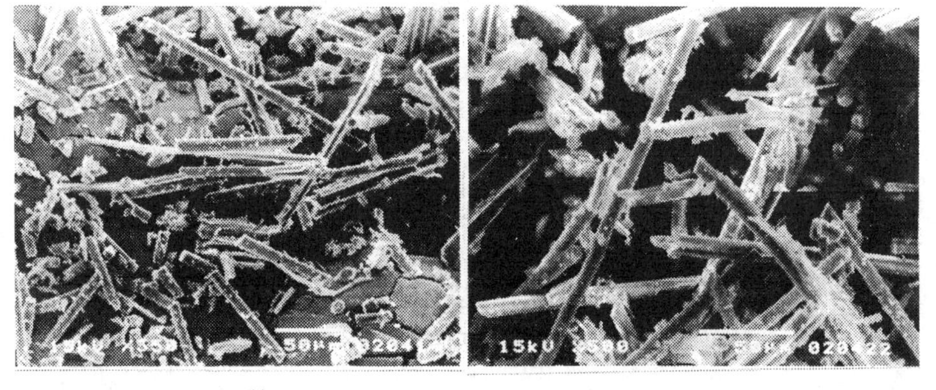

Figure 3 SEM Photographs of Chips

The chips were small fragments of the composite material formed by a series of fractures as well as rupture of carbon fibres as seen from SEM photographs [Fig.3]. The fibres have undergone sharp fracture with very little deformation of the matrix material. Because of the weak inter-laminar bond strength between the fibres and the matrix, the fibres have easily loosened without much of the matrix material sticking to them. The layer lattice nature of graphite, enables the fibres to be easily pulled out of the matrix. This in turn will produce better surface texture of machined component.

Cutting Temperature

Fig.4 shows the influence of cutting parameters on temperature over the cutting zone. Here it can be seen that the temperature registers an increasing trend when this polymeric composite was machined with a velocity greater than the respective critical velocity. The temperature registers a smaller value of magnitude at the critical velocity of 100m/min., since the force components are minimum at this speed.

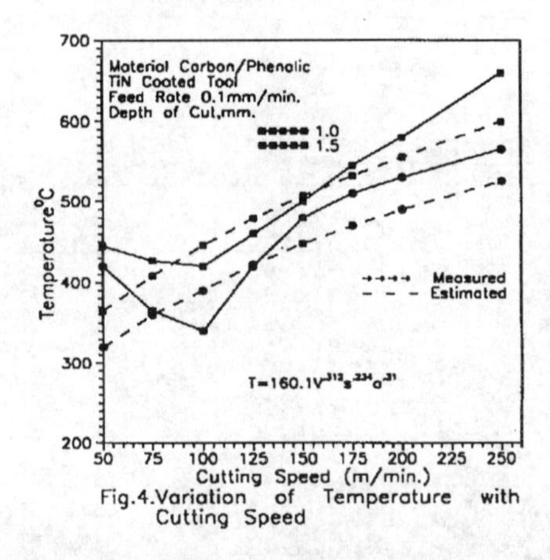

Fig.4.Variation of Temperature with Cutting Speed

In the machining process, the cutting tool experiences higher temperature due to the higher stress rate associated with the process. Hence as the cutting speed increases, the strain rate and consequently the cutting temperature increases; also the increased frictional heating over the rake and flank portion, also contribute to the rise in temperature.

Using the observed data, statistical modelling of temperature, based on log linear multiple regression analysis was developed. The model explains, the significance of the cutting conditions for temperature developed during machining.

The functional form of the estimated model is

$$\theta_c = V^{0.312} s^{0.234} a^{0.31} \tag{1}$$

where 'θ_c' is the temperature, 'V' is the cutting speed, 's' is the feed rate and 'a' is the depth of cut.

The machining temperature θ_c is usually influenced by the strain rate and thermal conductivities of both work-piece and cutting tool materials and as such it can be expressed as

$$\theta_c \cong K(V)^b \quad (2)$$

where 'b' is the material dependent exponent. The higher value of the exponent of cutting speed confirms the above result. Further higher order dependence on the feed rate is attributed to the deformation of the cutting wedge enhancing the chip strain, as seen proportionately larger variation in force components with changes in feed rate.
It is also interesting to note that the depth of cut is also a significant parameter, unlike the case of machining of homogeneous engineering materials. The fluctuations in the cutting force experienced by the cutting tool during machining of polymeric composites induces higher order tool-tip oscillations; this in turn will produce frictional heating which will increase the cutting temperature. Hence when machining polymeric composites, it is necessary to maintain low order magnitude of tool-tip overhung and also it calls for a machine tool of relative higher order stiffness.
Fig.4 also illustrates the variation of temperature for the cutting conditions for the actual experimental values and predicted values by the equation (1). From the graphs, it can be observed that the values show good correlation.

4. Conclusions

Results of face turning of carbon/phenolic ablative composites using TiN coated tools were presented. From the investigations, the following conclusions can be made.

- A critical velocity of 100m/min. exists in the machining of carbon/phenolic composites at which the cutting forces are minimum.
- The mechanism of chip formation is by a series of cracks ahead of the cutting tool producing a sharp fracture of the fibres.
- Temperature increases as cutting speed or feed rate increases except at the critical velocity.
- The equation relating temperature and cutting parameters shows good correlation.

References

[1]. *G. Caprino, I. De Iorio, L. Nele, L. Santo*, EFFECT OF TOOL WEAR ON CUTTING FORCES IN THE ORTHOGONAL CUTTING OF UNIDIRECTIONAL GLASS FIBRE REINFORCED PLASTICS – Composites, Part A, 27A, pp. 409-415, 1996.

[2]. *W. Konig, R. Fritsch, D. Kammermeier*, NEW APPROACHES IN CHARACTERISING THE PERFORMANCE OF COATED CUTTING TOOLS – Annals of CIRP, Vol.41/1, pp. 49-54, 1992.

[3]. *Robin P. Bergstrom*, MACHINING TOUGH ABRASIVE PLASTICS – Manufacturing Engineering, pp. 61-63, Nov. 1983.

The Use of PCD Tools for Machining Fibre Reinforced Materials

F. Klocke, C. Würtz

Fraunhofer Institute for Production Technology IPT
Steinbachstr. 17, 52074 Aachen, Germany

Abstract

The machining of fibre reinforced plastic components is normally limited to finishing processes as the manufacturing process itself already produces a near-net shape. Milling and drilling are the most commonly used finishing techniques after laser beam and water jet processing. Tool wear caused by the reinforcing fibres is an important factor when costing the machining processes. Polycrystalline diamond and carbide tools are the most common. The development and improvement of tools used for FRP machining is a current focus of research at the Fraunhofer Institute for Production Technology. This article provides an overview of the potential uses of PCD in FRP machining processes and also compares PCD tools to carbide tools in terms of both economics and quality.

1. Introduction

Fibre reinforced plastics (FRPs) are mainly used in aircraft and automobile constructions and, to some extent, in the chemical industry. The applications concerned are not generally mass-produced but are manufactured in small to medium-sized series. Despite the superior performance of these materials, their use is limited in industrial applications for various reasons. Many of these restrictions are production-related. Any attempt to improve the situation must stand up to economic scrutiny as well as satisfy the demands made by the manufacturing and machining processes on the material itself in order to fulfil the desired product characteristics. Post-processing is often very cost-intensive because of the complex construction geometries and large degree of tool wear, particularly in the case of FRP components. Product planning and development stages should therefore take the machining process into consideration. The dimensions of a component, the necessary machining accuracy and the accessibility are all factors which are reflected in the machining equipment, process and tools. Such cost factors are often underestimated during the product planning and development stages. It is therefore very important that one have the necessary technical understanding and incorporate it into the planning process.

Further more the manufacturing processes used for machining FRP components must be robust in order to ensure a constant standard of quality: Changing the material parameters often means that less robust processes and the necessary tools be modified to suit the particular machining task. Different fibre weave structures or fibres from different manufacturers, for instance, can often require a new processing strategy.

The machine tool plays an important role in the machining of FRP components in terms of the factors mentioned here [1]. Carbide tools and, particularly relevant for carbon fibre reinforced plastics, those made from polycrystalline diamond have been proved successful. In the latter case, the diamond cutter is soldered onto a carbide support which allows for small tool diameters even with multiple-toothed tools. This is particularly important when milling slots or grooves with small inside radii. Larger tool diameters are achieved by attaching the cutters to a tool holding system as illustrated in Figure 1.

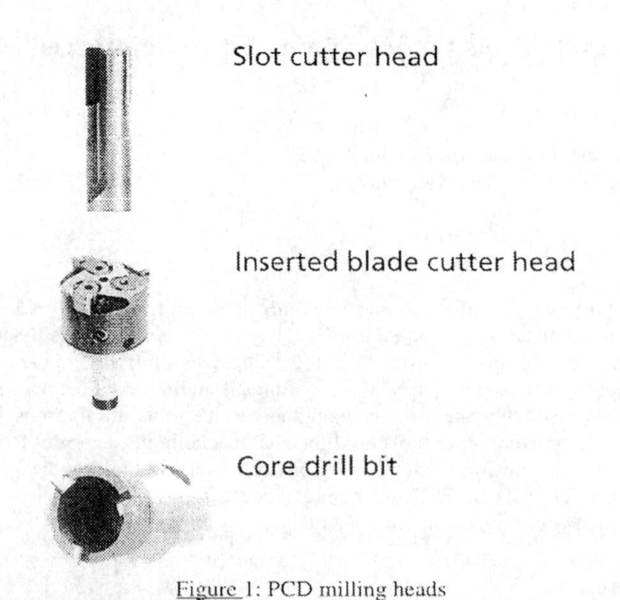

Figure 1: PCD milling heads

2. Wear Analysis of PCD Tools

The design and modification of machine tools follows an analysis of the wear mechanisms. The tool wear which occurs during the machining of FRP components is characterised by the abrasive effects of the reinforcing fibres. Carbon fibres in particular can lead to ridging along the cutting edge, depending on the fibre orientation. In comparison to the carbides, the diamond grains are affected more by the cobalt bonding phase dissolving than by the friction between the workpiece and the cutter [2]. This causes less wear to the tool flank than observed with carbide tools. What is more obvious with PCD tools is that the cutting edge begins to round after only a few metres milled path: this can be partly attributed to manufacturing-related defects. After further milling, the wear behaviour of the PCD tool stabilises and results in a cutting edge life which is significantly longer than for carbide tools (see Fig. 2).

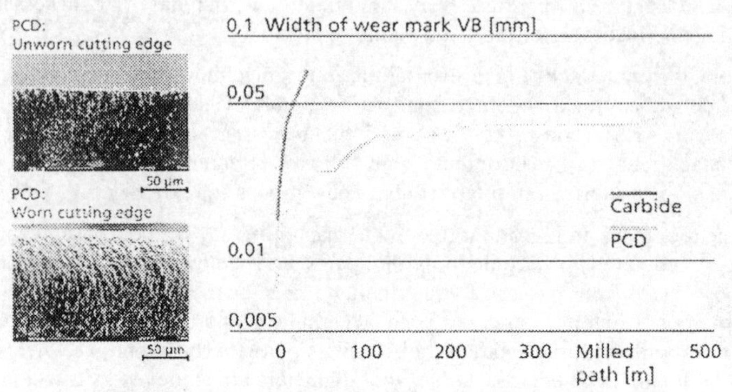

Figure 2: Wear curves for PCD and carbide tools

Polycrystalline diamond materials with a grain size of 1 - 2 μm can be used for machining of FRP components [3]. The cutting edge radius of a PCD cutter is between 6 and 7 μm depending on the grind. Tools with coarser grains, and thus with larger cutting edge radii as a result, produce slightly poorer quality and have less wear resistance (see Fig. 3).

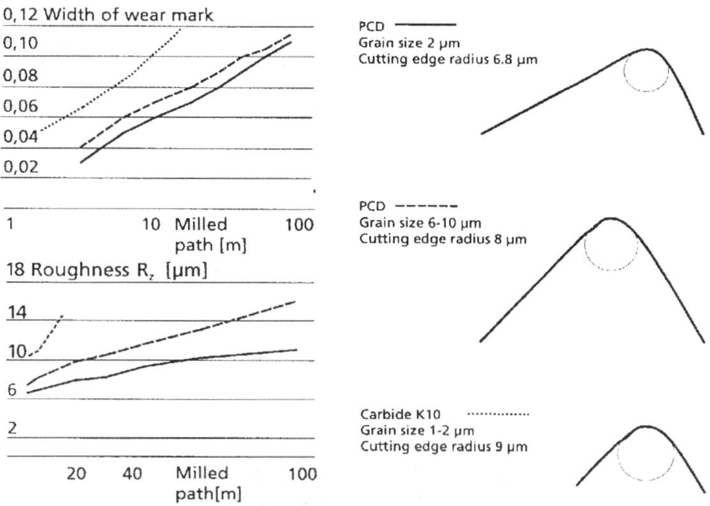

Figure 3: Comparison of different grain sizes and cutting materials

This can be mainly attributed to the superior quality of fine-grained polycrystalline diamond cutters which is also reflected in superior cutter surface properties and lower notching. Despite the coarser material being harder, these factors have a crucial effect on the machining quality. Figure 3 shows that the surface quality achieved using a finer grain is 25% better than that for a coarser grain over the same milled path. Compared to carbides, the smaller cutting edge radius and notching as well as the hardness have a positive effect on tool wear.

The investigations have shown that, in terms of cutting edge life, the usual criteria which concentrate on the width of the wear mark are not appropriate for FRP machining. The tools are better evaluated from a qualitative point of view. The decreasing cutter sharpness or, rather, the increasing rounding of the cutting edge, means that fibres at the surface layers of the laminate are no longer cut properly. In contrast, the superior quality of PCD cutters and the significantly superior wear resistance mean that protruding tufts of fibres are only observed after several hundred metres milled path.

3. Reducing Tool Stress

Important factors affecting the use of PCD tools and their wear behaviour include the cutter micro-structure and the design of the cutter and tool geometry. The machining processes used for FRPs involve large process dynamics which are a result of the high processing speeds and the component geometries. Tight processing contours and thin laminates mean that, at high spindle revolutions, the component vibrations must be absorbed by appropriate clamping. If this is not the case, the quality as well as the wear resistance suffer. Furthermore, restricted access to the components often requires the use of 'finger' milling tools with small diameters [4]. These are very sensitive to the process dynamics: the clamping, the tool and the process itself must be adjusted in relation to one another in order to carry out the machining task. The elastic properties of FRP materials can now and then cause tool flank regrinding as the cutter leaves the material (see Fig. 4).

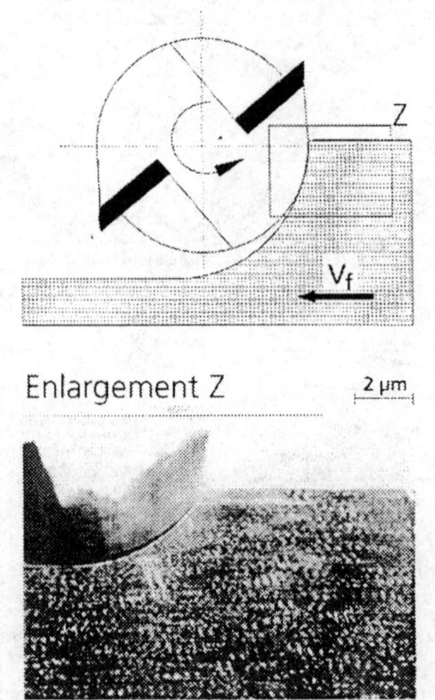

Figure 4: Regrinding of PCD tools during FRP milling

This abrasive movement not only increases the process dynamics but also increases the thermal stress on the tool. The low thermal conductivity of fibre reinforced materials makes this problem particularly evident in tools with small diameters. Almost all of the process heat which is generated must be absorbed by the tool. If the capacity to absorb the heat is too low, an increase in friction inevitably leads to even higher thermal stress in the tool which in turn leads to more tool wear and poorer operational safety. Cooling, either from the outside or from the inside of the tool can reduce this problem. External cooling relies on compressed air or another gaseous or liquid coolant absorbing the heat. The problem in the case of PCD tools with small diameters are that the high spindle revolutions often only allow for coolant feed at high pressure. In the case of internal cooling, the coolant is fed through the tool: this is more effective but requires that the spindle be modified to accommodate the coolant feed. Experiments have shown that changes in the construction of the tool itself can also reduce the frictional stress: the addition of a second clearance angle significantly reduces the friction and thus the thermal stress within the tool. One must, however, be aware that larger clearance angles reduce the cutter stability. The tool must therefore be designed around component-related specifications in order to guarantee the highest cost effectiveness and quality.

4. Cost Analysis

PCD tools are expensive to buy but this can be compensated for through the machining process and the cutting tool life. If one considers a double-bladed, straight-grooved PCD finger milling tool with a diameter of 8 mm, the PCD version would be 12 times more expensive than the carbide tool (K10). On the other hand, the wear resistance of PCD tools and their ability to withstand higher processing speeds makes them the a more attractive option. The following comparison between the processing costs for PCD and carbide tools involved in machining two different FRP materials is based on cost per milled metre. The following costs will be considered:

- Interest and repayments
- Depreciation
- Wages
- Rent
- Energy costs
- Maintenance costs
- Tool costs

We have assumed a cost of 450,000 DM for the milling machine which includes the casing around the machine as well as an extraction unit which sucks out the dust emitted during machining. The operating rate in a one-shift factory is 80%. The plant is to be written off over 6 years and the interest rate is 10% per annum. The wages are set at 60 DM/hour which corresponds to the hourly rate for a machine operator [6]. The overheads are set at 220%.

The use of PCD tools for machining glass fibre reinforced thermoplastic resulted in a cutting edge life of approx. 1900 m at a cutting velocity of 800 m/min. The cutting edge life of the carbide tool, on the other hand, was only around 100 m at a cutting velocity of 400 m/min. Despite the PCD tool being much more expensive (650 DM compared to 60 DM for a carbide tool), the more favourable processing conditions reduce the machining costs: they are 90% higher for carbide tools. This comparison is less positive in the case of carbon fibre reinforced duroplastics for which the tool life is shorter: at slightly lower cutting velocities, PCD tools only provide an 18% saving compared to the carbide tools.

The tool life does not, however, only depend on the tool and the process parameters. The type of fibre as well as the laminate thickness have a significant effect on tool life, particularly in the case of carbon fibre reinforced materials. Component vibrations intensify this so that the clamping becomes an even more vital factor, particularly in the case of thin-walled components. The angle at which the fibres are cut also plays an important role. Tufts of fibres at the cut edge are more likely to form if the fibres are cut at 90° than at 0°. The effect of various factors on tool life and thus on the costing of the process should be taken into consideration when machining FRP components.

Research carried out by the Fraunhofer IPT has shown that significant savings can be made. Around 40% of the specific costs (DM/m) arising during the milling of FRPs can be attributed to tooling costs. These costs can be reduced in a variety of ways, as already mentioned. The longer tool life and the superior properties of PCD tools at higher cutting velocities should be taken advantage of. The example in Figure 5 shows that an increase in tool life leads directly to an almost 9% reduction in costs. An increase in cutting parameters will, however, lead to a reduction in machining costs of up to 18%.

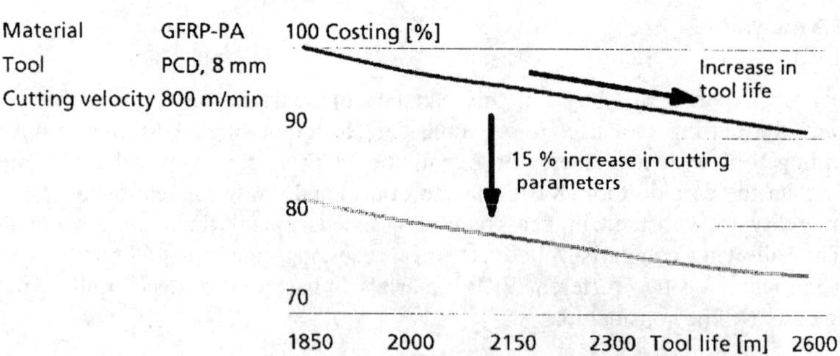

Figure 5: Cost reductions achieved by adjusting cutting parameters

That these cutting velocities can be achieved is not only a question of process design but is also directly linked to the ways in which the chosen tool can be used. The high process dynamics often cause the PCD tools with small diameters to fail - the small PCD plates break off - and the maximum tool life is not reached. The aim must therefore be to design and construct tools in such a way that they are able to cope with the dynamic stress profile as well as the abrasive nature of the reinforcing fibres. Research at the Fraunhofer IPT focuses on the design and construction of tools (with an 8 mm diameter or less), specifically on where the PCD plates are soldered on to the tool so that they can be used economically at high cutting speeds. Attempts are being made to redesign machining processes from an economic point of view, particularly in the field of fibre reinforced materials but also wood.

5. Outlook

Cutting processes performed on fibre reinforced plastics are strongly dependent on the tools implemented, both in terms of quality and economics. This is not only related to the wear which the tools are subjected to when cutting reinforcing fibres but also, to a large extent, dependent on parameters specific to the component itself. Tools should therefore be modified to suit the individual machining tasks. The correct choice of cutting material is also important. Polycrystalline diamond is an economical alternative to carbide despite the higher purchase costs: the tool life is longer and PCD tools can withstand higher processing speeds. Operational safety is, however, a higher priority: tools with smaller diameters are subject to high thermal stress which must be compensated for through equipment-oriented tooling concepts.

References

[1] Klocke, F.; Pfeifer, T.; Eversheim, W.; Weck, W. and others
Produktionstechnik für Bauteile aus nichtmetallischen Faserverbundwerkstoffen, Teil 2: Produktion,
Ingenieur Werkstoffe 6(1997) Nr. 3/4,
Springer VDI Verlag, Düsseldorf, 1997

[2] Rummenhöller, S.
Werkstofforientierte Prozeßauslegung für das Fräsen kohlenstofffaserverstärkter Kunststoffe,
Doctoral Thesis at the RWTH Aachen, Aachen, 1996

[3] Hohensee, V.
Umrißbearbeitung von faserverstärkten Kunststoffen
Dissertation an der Universität Hannover, Hannover, 1992

[4] Klocke, F.; Würtz, C.
Tool design for CFRP milling,
European Machining, Turref-RAI plc, GB, 1997

[5] Klocke, F.; Würtz, C.;
Technologische und ökologische Aspekte beim Fräsen von faserverstärkten Kunststoffen,
3. Nationales Symposium der SAMPE Deutschland e.V.
Fertigungstechnologie der Faserverbundkunststoffe in Industrie und Forschung, Aachen, 1997

[6] Spur, G.; Liebelt, St.
Bearbeitung von glasmattenverstärkten Thermoplasten (GMT),
ZWF 92 (1997) 4
Carl Hanser Verlag, München, 1997

Machining of Particle-Reinforced Aluminium Metal Matrix Composites

J.T. Lin and D. Bhattacharyya
Composites Research Group, Department of Mechanical Engineering, University of Auckland, New Zealand

Abstract

Continuous turning of composite materials, Comral™-85 (6061/Al_2O_3/20p) and *Duralcan*™ (A359/SiC/20p), using polycrystalline diamond (PCD) inserts has been studied in this paper. The test conditions included cutting speeds varying from 75 to 700m/min and feed rates from 0.1 to 0.4 mm/rev while the depth of cut was kept constant at 0.5 mm.

The tool life and machining force data are analysed using regression techniques and generalised Taylor's models have been developed to describe the tool performance on these composites. The derived tool wear-machining force equation can be used to indirectly monitor the development of tool wear during a machining operation, thus minimising the wear measurement time and increasing the total useful tool life. The results show that the equation derived from the feed force data is better suited to monitor tool wear than the one derived from the cutting force. The alternative way of indirectly monitoring the tool wear by studying its relationship with power consumption is also discussed.

Introduction

With the increasing usage of metal matrix composites (MMCs) in various applications such as the aerospace, the automotive and the sports related industries, the machining of such materials has become an important subject of study. As reported in earlier publications [1 - 5], owing to the addition of reinforcing materials which are normally harder and stiffer than the matrix, the machining becomes significantly more difficult than that of conventional materials. Among many types of MMCs, the most popular are aluminium alloys reinforced with ceramic particles since they cost less but provide high strength, stiffness and fatigue resistance with only a minium increase in density over the base alloy [6-7]. In this study, the two materials used are Comral-85, a standard 6061 aluminium matrix reinforced with a nominal 20 vol% of alumina-containing MICRAL-20® microsphere reinforcement, and *Duralcan*, that consists of A359 aluminium base reinforced with a nominal 20 vol% of silicon carbide (SiC) particles. The inherent weakness of the alumina reinforcing particles within the Comral-85 causes its inferior

---

Duralcan™ is a registered trademark of Alcan Aluminium Ltd.
Comral™ is a registered trademark of Comalco Aluminium Limited.

fracture property [6] and hence in this study, much of the emphasis would be put on the *Duralcan* material.

Regression analysis techniques [8] has been employed to find the generalised Taylor equations or to establish the tool wear/machining forces and the tool wear/power consumption relationships.

Table I Summary of experimental conditions

Workpiece material	Direct-Chill Cast A359/SiC/20p (175 mm x 400 mm long diameter billets) & Comral-85 T6 Heat Treated bars (50 mm x 500 mm long diameter bars)
Tool insert	TPG322 COMPAX 1500 (PCD) & TPG321L1 PAX 20 (PCD)
Tool holder	TARP-16-3hR175.2-2525-16 (Sandvik Coromat) Back rake angle : 6°
Cutting parameters	Cutting speed : 300, 500 & 700 m/min & 74, 200 & 250 m/min Feed rate : 0.1, 0.2 & 0.4 mm/rev & 0.16 mm/rev Depth of cut : 0.5 mm Side cutting edge angle : 0°
Tool dynamometer	Kistler 3-D force dynamometer (Type 9441)

Experimental Procedure

The MMC bars were machined using polycrystalline diamond (PCD) inserts (25 μm) at different speeds (75 ~ 700 m/min) with varying feed rates (0.1 ~ 0.4 mm/rev) and a constant depth of cut of 0.5 mm. Table I summaries the details of the experimental conditions. The use of PCD inserts in this study is due to their excellent performance compared with ceramic tools when machining aluminium MMCs [1,2,9].

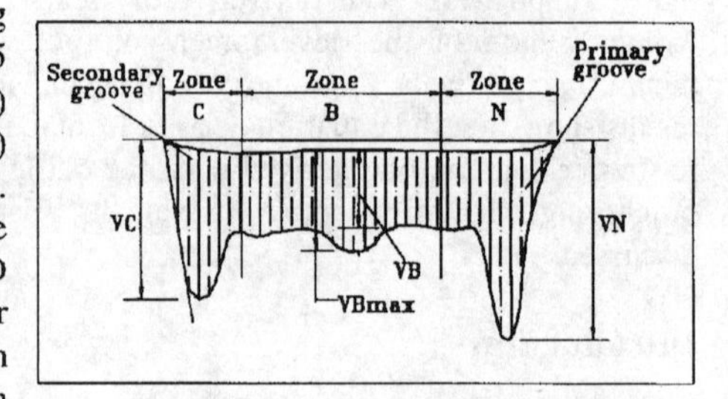

Figure 1 View on major flank

The tool wear (flank wear, VBmax, as shown in Fig. 1) was measured at predetermined intervals by using an Olympus optical microscope with measuring grids on the eye-piece. The tool performance was judged primarily by the flank wear growth and the workpiece surface finish. To facilitate tool regrinding, the maximum flank wear of 0.25 mm was chosen as the tool life limit for both Comral-85 and *Duralcan* respectively. The cutting and feed forces were monitored at every pass by using the piezoelectric 3-D force dynamometer Kistler 9441 and record-ed with an oscilloscope.

Results and Discussion

Tool Wear and Tool Life

The typical wear growth curve during machining Comral-85 is shown in Fig. 2. It is clear that the tool wear grows faster (ie. the tool life is shorter) with increasing speed, as expected from the traditional Taylor's model. By using the technique of linear regression analysis, the Taylor's equation of machining Comral-85 can be expressed as

$$VT^{1.079} = 8462 \qquad (1)$$

with a confidence level of 99.9%, where T (in minutes) is tool life, V (m/min) is cutting speed and 8462 is a constant representing the cutting speed that gives a one-minute tool life [10]. Note that the feed rate and depth of cut remain constant at 0.16 mm/rev and 0.5 mm respectively.

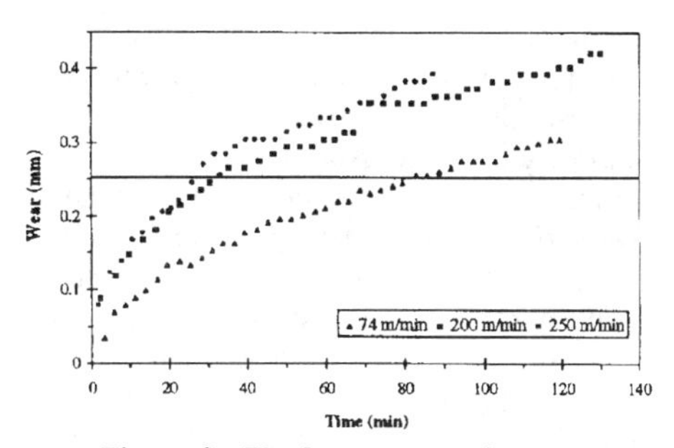

Figure 2 Tool wear growth curves when machining Comral-85 at three different speeds.

For machining *Duralcan* material, some typical wear growth curves are shown in Fig. 3. The general Taylor's equation may be derived as,

$$T = 4.82 \times 10^8 V^{-2.733} f^{-0.637} \qquad (2)$$

with a confidence level of 99.8%, where T is tool life (mins), V and f are cutting speed (m/min) and feed rate (mm/rev) respectively. It shows that the tool life decreases if the cutting speed or feed rate is increased. On the other hand, if the tool life is considered in terms of material volume removal, which makes more sense from the industrial point of view, a different picture emerges, Fig. 4. This interesting point can be explained by rewriting the general Taylor's equation as

$$VR = 2.41 \times 10^8 f^{0.36} / V^{1.73} \qquad (3)$$

This equation clearly shows that the material volume removal (VR) increases with feed rate despite a decrease in tool life (time) while the effect of speed remains the same.

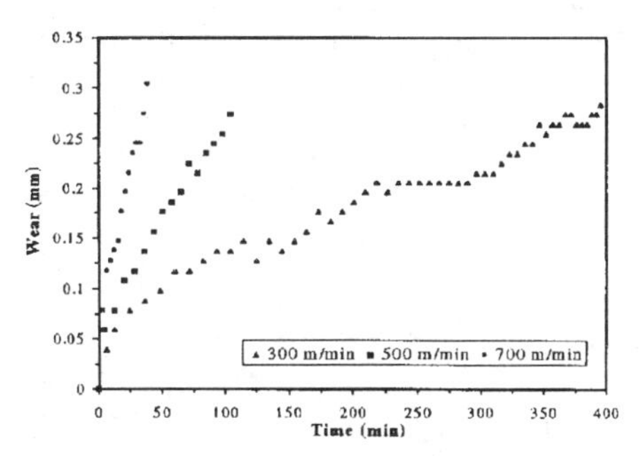

Figure 3 Tool wear growth curves when machining *Duralcan* at three different speeds with feed rate of 0.1 mm/rev.

Tool Wear and Machining Forces

Similar to tool wear, the machining forces also steadily increase with time, Fig. 5. The similarity of variation in the tool wear and machining forces suggests that the measurement of machining forces during the machining process may be used as an indirect way of monitoring tool wear if a proper relationship can be found between these two parameters. That can save time by avoiding the removal of tool for wear assessment process and keep the working condition of the tool more consistent.

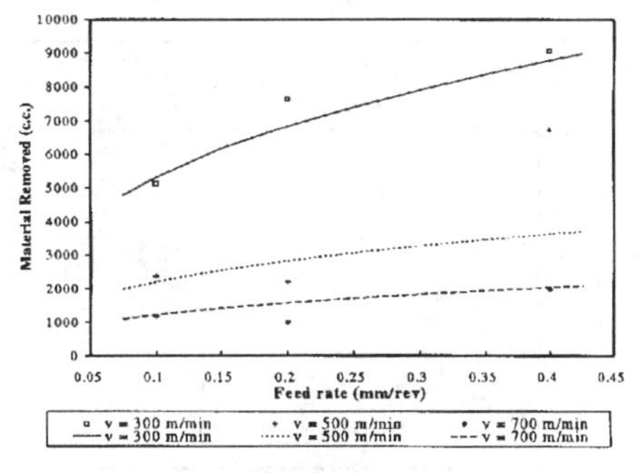

Figure 4 The comparison of experimental tool life with the value from the Taylor equation for increasing feed rate, in terms of material removal.

In order to find the relationship between the tool wear and the two machining forces, linear regression analyses were carried out on the experimental data under different cutting conditions. In the analyses, two forms of data fit, a simple linear form and power law, have been implemented. The results are shown in Figs. 6 and 7 for feed force and cutting force respectively. From the regression results, it appears that the feed force-tool wear equation can describe the experimental behaviour more accurately than the equation between cutting force and tool wear. On the other hand, the power law relationship has been more accurate than the linear equation in matching the experimental data, which can be judged by the value of R^2. In general, though, the accuracy of both these relationships is good, and under any particular cutting condition, either of them can be used to predict the tool wear with sufficient accuracy.

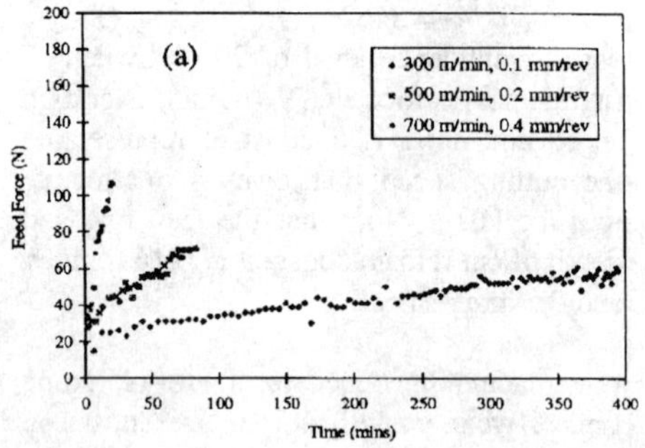

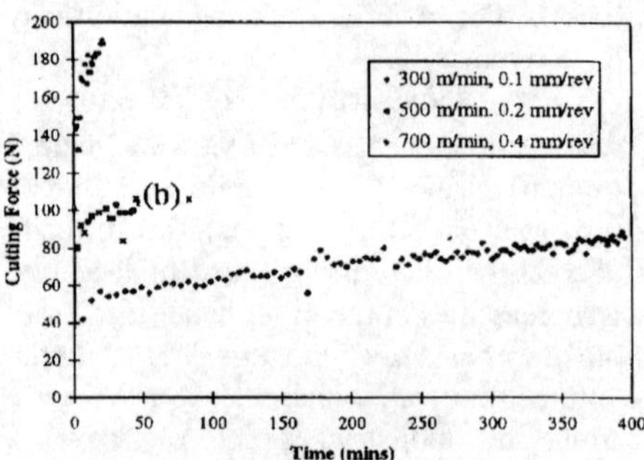

Figure 5 Machining forces versus time in three different cutting conditions; (a) feed force; (b) cutting force.

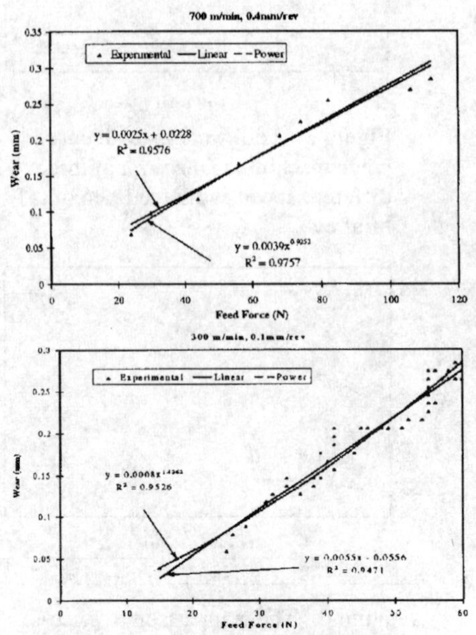

Figure 6 Regression results of the tool wear against feed force in different cutting conditions.

The technique of linear regression has been used to analyse the experimental data from all the different machining conditions. The general equations for the machining forces and tool wear are obtained as following:

$$W = 10^{-2.439} \times V^{-0.149} \times f^{-0.197} \times F_x^{1.15}, \quad R^2 = 0.929 \qquad (4)$$

and

$$W = 10^{-6.162} \times V^{0.188} \times f^{-1.226} \times F_z^{2.017}, \quad R^2 = 0.877 \qquad (5)$$

where W is tool wear in mm, V is cutting speed in m/min, f is feed rate in mm/rev,

Fx and Fz are feed force and cutting force respectively, in Newton (N). When compared with the experimental data for some of the cutting conditions, the value generated by the general equations can predict the trend with a reasonable accuracy even though it is not as accurate as those predicted by the individual equations. It is interesting to note that when the ± 10% error zone is put in, the band covers most of the experimental data, as shown in Figs. 8 and 9.

It is also evident that even in the general equation, the feed force-tool wear relationship gives a better indication of tool wear than what is achieved from cutting force-tool wear relationship. The reason for this may be attributed to the greater fluctuation of the cutting force compared to the feed force. This is caused by the higher probability of high speed impacts of the reinforcing particles on the tool surface.

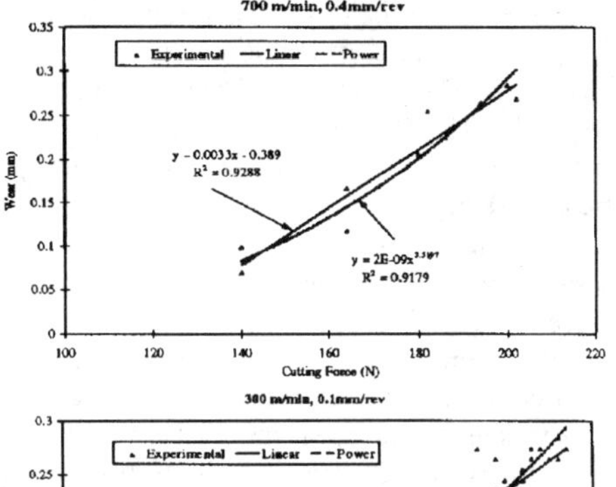

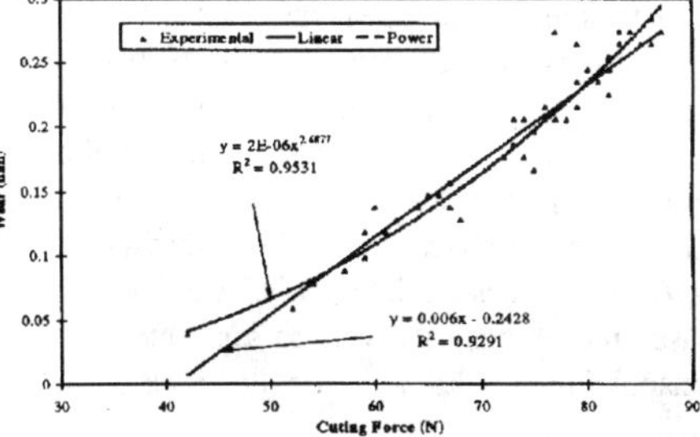

Figure 7 Regression results of the tool wear against cutting force in different cutting conditions.

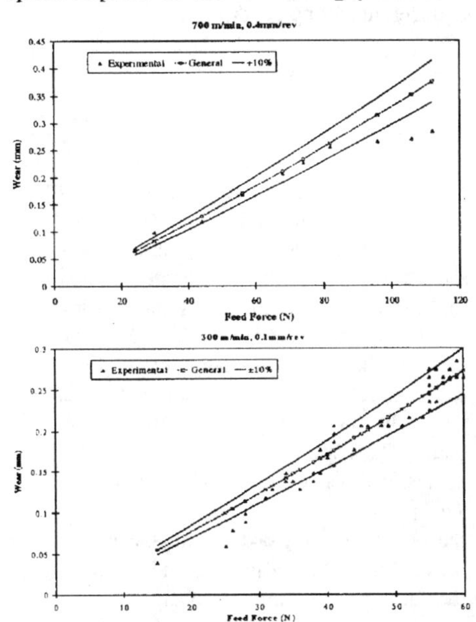

Figure 8 General regression results of the tool wear against feed force in different cutting conditions.

Tool Wear and Power Consumption

Apart from the machining forces, the power needed during machining could also be another suitable index for monitoring too life. Fig. 10 clearly shows that the power consumption for machining the *Duralcan* MMC increases with increasing tool wear. When the cutting conditions become harsher, the change is much more obvious. This leads to another possibility of using the power reading to monitor the progress of tool wear during machining. In order to achieve this goal, a relationship between tool wear and power consumption has to be found. Once again, the technique of linear regression has been employed to derive such relationship; both linear and quadratic forms can be used to describe the relationship between tool wear and power consumption with similar accuracy. The results have been shown in Fig. 11. The R^2 value in this analysis is not as good as that in the force-wear relationship; it increases as the cutting speed and feed rate are increased, especially when machining at very high speed and feed rate. The lower accuracy of the regression results is due to the scatter of data points when machining at lower speed and feed rate. However, even though the R^2 value of this analysis is not very high, the equations may still be used to monitor the tool wear quite effectively as long as the maximum allowable tool wear is the only threshold. Of course these relationships need to be used more conservatively to guarantee the machining quality and the desirable tool life.

For general purposes, considering all the different cutting conditions, the power consumption-wear relationship is obtained as,

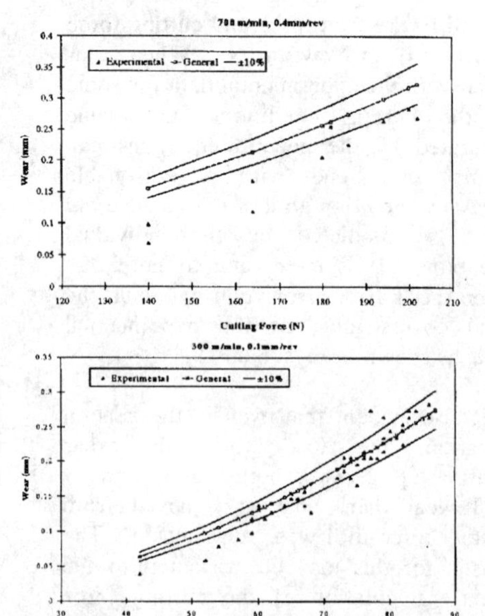

Figure 9 General regression results of the tool wear against cutting force in different cutting conditions.

$$W = 10^{3.594} \times V^{-1.888} \times f^{-1.121} \times P^{1.825}, \ R^2 = 0.914 \qquad (6)$$

where W is tool wear in mm, V is cutting speed in m/min, f is feed rate in mm/rev, P is the power consumption in kilowatts. The comparison between the general regression results with ± 10% error bar and the experimental data for some cutting conditions are shown in Fig. 12. The pictures show that the magnitude of wear calculated from the regression results is generally higher than the experimental data. The confidence level (0.95) shown in this analysis is not as high as expected. The possible reasons for this are firstly, the power reading taken from the power meter was not very precise and secondly the amount of power consumption is directly proportional to the cutting force which does not have a very satisfying regression due to its fluctuation during

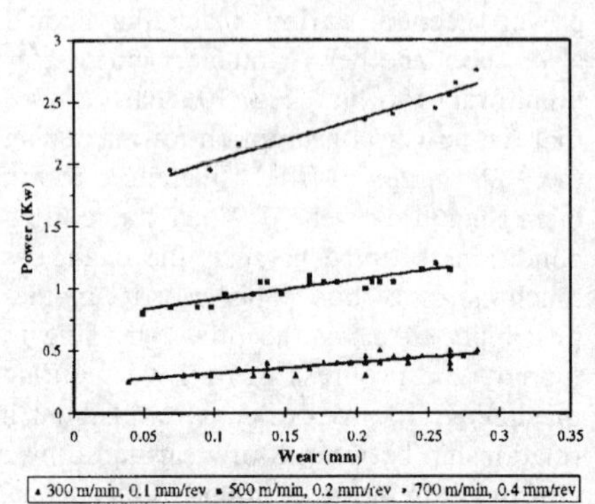

Figure 10 Power consumption against tool wear for machining *Duralcan* MMC at three different conditions.

machining. Therefore, the lower confidence level for the power-wear relationship could be expected from the result of cutting force-wear relationship.

However, from economical point of view, Eq. (6) might not be the ideal way to describe the tool wear progress because the higher value of tool wear predicted means the tool will be changed well before it reaches the pre-determined life. On the other hand, if the tool is replaced for resharpening to lengthen its useful life, it is better to change it before it is too severely damaged. This implies that a more conservative approach is taken for the tool wear prediction if the tool wear/power consumption relationship is used for monitoring tool wear progress.

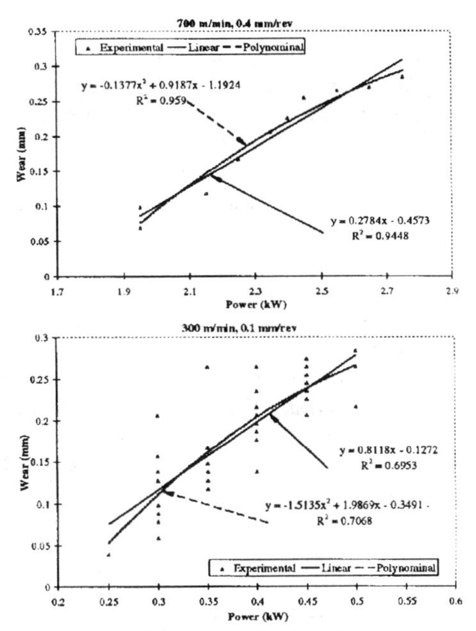

Figure 11 Regression results of the tool wear against power consumption.

Concluding Remarks

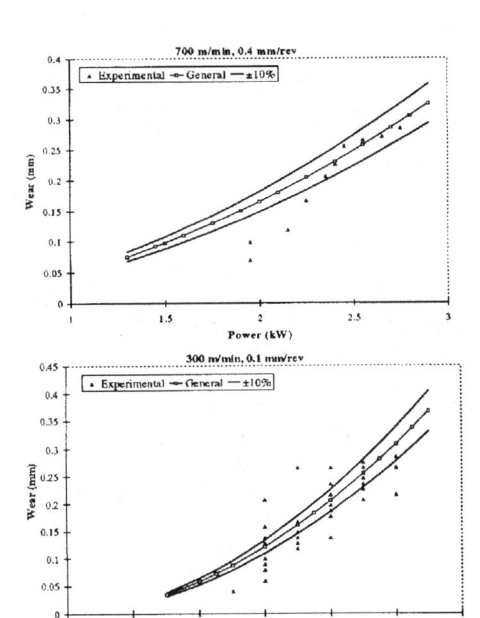

Figure 12 General regression results of the tool wear against power consumption.

The following conclusions can be drawn from this study:

1. For machining Comral-85 composite material, the tool life is shortened by the increase of cutting speed as expected from the traditional Taylor's model. With a linear regression analysis, the Taylor's equation can be obtained, and this relationship can be used to predict tool life with sufficient accuracy when machining Comral-85 metal matrix composite with different cutting speeds keeping feed rate and depth of cut unchanged.
2. As for machining *Duralcan* material, PCD inserts wear more rapidly with the increase of cutting speed and/or feed rate. Any increase in cutting speed shortens the tool life no matter whether the tool life is calculated in terms of time or the material volume removal. But when the feed rate is

increased, the tool life increases in terms of volume removal although the time life decreases as expected from the traditional tool life equation. This interesting phenomenon, important to the practical users, can be explained by rewriting the Taylor equation to incorporate material volume removal.

3. The primary wear mode of machining the particulate reinforced aluminium metal matrix composite is confirmed to be the flank wear on the tool tip, and the main wear mechanism is abrasive wear caused by the reinforcing particles.
4. By using cutting and feed forces, tool wear can be monitored without interrupting the machining operation. For any single cutting condition, either a power law equation or a linear equation can be used to describe the progress of tool wear with sufficient accuracy and the feed force appears to give a better response in the monitoring exercise.
5. The power consumption may also be used as an indirect way of monitoring the tool wear progress. The power-wear relationship may prove to be more useful because it provides a reasonable safety factor to ensure that the tool change takes place well before the tool wear limit is reached and thus avoids the possibility of catastrophic damage to the PCD tool.

References

[1] L. Cronjäger, D. Meister, MACHINING OF FIBRE AND PARTICLE-REINFORCED ALUMINIUM - Annals of C.I.R.P. Vol. 41/1 (1992) 63-66.

[2] N. Tomac, K. Tønnessen, MACHINABILITY OF PARTICULATE ALUMINIUM MATRIX COMPOSITES - Annals of C.I.R.P. Vol. 41/1 (1992) 55-58.

[3] K. Weinert, A CONSIDERATION OF TOOL WEAR MECHANISM WHEN MACHINING METAL MATRIX COMPOSITES (MMC) - Annals of C.I.R.P. Vol. 42/1 (1993) 95-98.

[4] Z. J. Yuan, L. Geng and S. Dong, "ULTRAPRECISION MACHINING OF SICW/AL COMPOSITES", Annals of C.I.R.P. Vol. 42/1 (1993) 107-109.

[5] J. T. Lin, D. Bhattacharyya and C. Lane, MACHINABILITY OF A SILICON CARBIDE REINFORCED ALUMINIUM METAL MATRIX - Wear, Vol. 181-183 (1995) 883-888.

[6] D. Bhattacharyya, M.E. Bowis and J.T. Gregory, THE INFLUENCE OF ALUMINA MICROSPHERE REINFORCEMENT ON THE MECHANICAL BEHAVIOUR AND WELDABILITY OF A 6061 ALUMINIUM METAL MATRIX COMPOSITE - Machining of Composite Materials, T.S. Srivatsan and D.M. Bowden ed., ASM International, Materials Park, OH., U.S.A. (November, 1992) 49-56

[7] M. J. Couper and K. Xia, DEVELOPMENT OF MICROSPHERE REINFORCED METAL MATRIX COMPOSITES - Metal Matrix Composites - Processing, Microstructure and Properties, N. Hansen, D. Juul Jensen, T. Leffers, H. Lilholt, T. Lorentzen, A.S. Pederson, O.B. Pederson and B. Ralph ed., RISO National Laboratory, Roskilde, Denmark (1991) 291-298.

[8] R.T. Leslie and G. Lorenz, TOOL-LIFE EXPONENTS IN THE LIGHT OF REGRESSION ANALYSIS - National Standards Laboratory Technical Paper No. 20, CSIRO, Melbourne, Australia (1964).

[9] C. Lane, THE EFFECT OF DIFFERENT REINFORCEMENTS ON PCD TOOL LIFE FOR ALUMINIUM COMPOSITES - Machining of Composite Materials, T.S. Srivatsan and D.M. Bowden ed., ASM International, Materials Park, OH., U.S.A.(Nov. 1992) 17-27.

[10] M.C. Shaw, METAL CUTTING PRINCIPLES - Clarendon Press Oxford (1984) 6.

Drilled hole damage and residual fatigue behaviour of GFRP

A. Bongiorno, E. Capello, G. Copani, V. Tagliaferri

Dipartimento di Meccanica - Politecnico di Milano - Via Bonardi 9, Milano 20133 Italy

Abstract

Drilling operation of composite laminates leads to a decrease of mechanical characteristics [1,2]. This problem becomes particularly critical when holes are designed to house junction elements which stress the hole section with concentrated and cyclic loads. Fatigue failure can occur and fatigue life is connected to the damage generated in the drilling phase. While many authors have studied the fatigue behaviour of composites [3-7], the experimental work carried out to study the fatigue life as a direct result of the damage generated by drilling is really limited [8].

In the present work several holes have been drilled in different process conditions, thus generating different kinds and levels of damage. Fatigue bearing tests have been carried out. The fatigue behaviour has been quantified in terms of hole deformation and *peel up* and *push down* delamination. Moreover, microscopic observations of the hole damage have been carried out to study and understand the mechanism of damage propagation during the tests. Based on these results, a theoretical model about the physical mechanism of damage propagation has been proposed.

Major results show that the presence of defects in the central point of the hole section deeply affects the fatigue behaviour, while delamination plays a minor role.

Introduction

Fibre Reinforced Plastics (FRP) laminates are well suited to substitute traditional materials in several applications where high stiffness and strength, low weight and good corrosive resistance properties are requested. On the other hand, the difficulty in machining operations due to their mechanical anisotropy often limits their application and diffusion.

Drilling is the most common composite machining operation in order to house mechanical fasteners. Drilling operations generate several kinds of damage around the hole, among which delamination, intralaminar cracks, fibre and matrix debonding and thermal alterations are the most relevant [9]. Poor hole quality leads to a decrease of fatigue resistance properties, especially when the hole is subjected to concentrated loads [8].

The present work is divided into two steps. The first one aims to the comprehension of how drilling conditions affect the hole quality. Detailed results are reported in [10] and [11], while the present work is focused on the second step, that is the relation between hole damage and fatigue behaviour.

In the drilling phase of the work the investigated parameters were the ratio between feedrate (V_t) and revolution speed (V_r), the presence of a support under the specimen (to avoid material inflection) and the pre-heating of the twist drill. As described in literature [1, 2, 9-11], these parameters deeply affect the hole quality.

In the fatigue test phase a tension-tension bearing load has been adopted in order to produce load conditions similar to the real ones. Damage growth has been observed and measured at several stops before the test ended at 10^5 cycles.

Starting from measurements, microscopic observations and from the results of statistical analyses, a theoretical model about the physical mechanism of damage propagation has been proposed. This model can be used to carefully and consciously select the process parameters that optimise the fatigue behaviour of drilled holes.

Experimental set up

GFRP Material

Tested material was a plastic thermosetting matrix composite (polyester resin) reinforced with glass woven fibres (300 g/m²). The nominal thickness was 2.2 mm obtained stacking together seven laminae. Two 1200×1200 mm² plane panels have been obtained by hand lay-up and have been cut into rectangular specimens (200×50 mm – see fig. 1) by Abrasive Water Jet machining. Cutting parameters of AWJ have been chosen in order to avoid any damage near the generated kerfs.

Drilling process

Specimens have been drilled using a 5 mm diameter HSS twist drill. The experimental set up of the drilling process is reported in figure 2. The following process parameters have been investigated:

- Vt/Vr ratio
- presence of a pre-drilled support under the specimen to avoid inflection
- pre-heating of the twist drill.

A 2^3 Design Of Experiments technique has been adopted and four replicas for each experimental condition have been performed. The experimental plan and the levels of investigated drilling parameters are reported in table 1. The quantitative description of hole quality has been performed in terms of delamination and obtained diameter. Some specimens have been sectioned along the hole diameter in order to observe with a microscope the generated surface and internal damage, aiming to a qualitative description of the hole quality.

Fatigue test

Bearing cyclic tests have been performed using an Instron 8501 hydraulic machine. The experimental set-up is reported in figure 3 while figure 4 represents the tension-tension applied load. Cycling has been interrupted in correspondence of fixed stops to analyse and measure the damage propagation with the same method described for the drilling process.

Results analyses

Quantitative analysis

The drilling phase: After the drilling phase, the measured delamination and hole diameter data have been used to perform an analysis of variance (ANOVA). This statistical technique can be used to identify the process parameters that significantly influence the measured results (table 2).

Push down delamination is significantly affected by the Vt/Vr ratio, the presence of the support and by the twist drill pre-heating.

Peel up delamination is influenced only by Vt/Vr ratio.

Diameter of holes drilled with pre-heated twist drill is bigger than the nominal one because of the retraction of matrix due to its thermal vitrification.

Adequate discussion of these results is reported in [10] and [11].

Fatigue tests: Considering the fatigue tests, diagrams of damage growth show a rapid propagation of all damages at the initial stage of the test, followed by a second phase in which trend significantly decreases (see figure 5). After the fatigue tests have been stopped at 10^5 cycles, an ANOVA has been performed on the collected delamination and hole deformation measures.

Hole deformation has been considered as the index of fatigue resistance, while statistical analysis on the other variables has been used to understand the damage propagation mechanism.

Both Vt/Vr ratio and drill pre-heating affect the hole diameter measured at 10^5 cycles, while incremental data are influenced only by Vt/Vr ratio. This because higher is Vt/Vr ratio adopted in drilling, wider is the extension of delamination and internal cracks. High drill temperature causes a higher deformation at the end of the test because it generates holes with a bigger diameter, but it does not influence final damage propagation. Surprisingly, the presence of a support under the specimen does not affect the fatigue behaviour. All these results are explained by the proposed model for fatigue damage growth mechanism.

Qualitative analysis

Microscopic analysis of hole section after the drilling phase showed that for the low level of Vt/Vr ratio (0.11 mm/rev) the internal surface is regular (figure 6a), the only visible damage is a small delamination localised in correspondence of extreme layers. On the contrary, in hole sections obtained with the high value of Vt/Vr ratio (0.50 mm/rev), delamination is wider and involves several layers even in the central part of the section (figure 6b). Moreover, intralaminar cracks and the debonding effect between fibres and matrix is marked. All damages are deeply extended into the material.

The pre-heating of the twist drill causes matrix retraction and a deep fibre-matrix debonding (figure 6c). This kind of damage involves all laminae of composite material.

Independently from the kind of damage present after the drilling phase, microscopic observations of hole sections after the fatigue tests have showed that the hole deformation is generated by the bending of two groups of laminae in opposite directions (figure 7c). The entity of hole deformation is affected by the presence of internal defects rather than by delamination.

Fatigue damage growth mechanism

Combining the previous observations, a two steps mechanism of fatigue failure can be proposed. The first step is the crack generation, while the second is the crack propagation. Drilling operation can generate cracks that compromise the fatigue behaviour, since the first step of the mechanism is eliminated.

Consider a specimen which presents both delamination and some damage localised along the hole section (figure 7a). The laminae which initially resist to the load pin are placed between the delaminated layers. In fact, the delaminated laminae offer poor resistance since they are free to bend under the load (fig. 7a). According to the fatigue mechanism of damage propagation, as the cyclic test proceeds the load pin causes the propagation of the cracks localised in the central part of the section (figure 7b). This internal delamination separates the material thickness into two layers which are free to bend in opposite directions leading to a macroscopic hole deformation (fig. 7c). Therefore, the fatigue behaviour of the hole is strictly connected to the presence of internal damage rather than delamination. This has been confirmed by the identified relationships between process parameters, induced damage and fatigue life.
Vt/Vr ratio: hole sections after fatigue tests of specimens drilled using two Vt/Vr ratios are represented in figure 8 and 9. It can be seen that the section obtained with Vt/Vr = 0.50 mm/rev appears very irregular and cracks are extended into the material. Moreover, intralaminar cracks extend deep into the material (figure 10). Considering that the *push down* delamination is extended in the material, intralaminar cracks can be found if the energy required to cross the laminae between the crack and the delamination is lower than the energy required for the interlaminar propagation.
On the other hand, a Vt/Vr = 0.11 mm/rev generates a substantially undamaged section where internal cracks are not present. Considering the two steps fatigue failure model, low values of Vt/Vr imply that the crack generation step has to occur before the crack can grow into the material, generating the hole deformation.
The twist drill pre-heating: The twist drill pre-heating is the cause of evident damages into the entire hole section. These thermal alterations, which do not affect a wide zone of material, act like small cracks which can immediately propagate leading to a rapid hole deformation since the first cycles.
The support under the specimen: Considering the proposed mechanism, the absence of statistical influence of the support on hole deformation can be explained. In fact, the ANOVA has shown that the specimen support only affects the *push down* delamination, while it does not influence internal damages which are responsible of the elimination of the first step of the fatigue model.

Conclusions

The aim of the present work is to indicate the drilling conditions which lead to an optimum fatigue behaviour. Based on the obtained results, following conclusions can be drawn.

- It has been observed that the presence of defects in the central part of the hole section negatively affects the fatigue behaviour, while the delamination plays a minor role. In fact, small cracks localised in the middle of the section can immediately propagate into the material. Then the composite laminae are free to bend under cyclic load and a wide deformation of the hole occurs.
- Therefore, the optimal drilling parameters should be selected in order to avoid the generation of damage in the middle of the hole section.
- The Vt/Vr ratio deeply affects the internal damage. Cracks and debonding have been observed. This kind of damage (and the consequent fatigue behaviour) seems to be mainly determined by the Vt/Vr ratio, while the other parameters play a minor role. Therefore, an effective improvement of fatigue behaviour can only be obtained

if low Vt/Vr ratios are selected. In this condition, a limited temperature of the twist drill can further improve the fatigue behaviour.

- The presence of a support only reduces *push down* delamination and does not affect the internal hole conditions. Therefore, it has no influence on the fatigue behaviour.
- Even if the pre-heating of the twist drill has a positive effect on the *push down* delamination, it generates a thermal damage on the whole section. Therefore, for a proper fatigue behaviour, an increase of twist drill temperature should be avoided.
- The proposed damage propagation mechanism is valid only for the applied load conditions. Different load conditions may lead to different mechanisms.

Finally, it has been observed that the presence of voids inside the manufactured composite negatively affects the fatigue behaviour and could vanish the positive effects of a careful selection of machining parameters. Thus, special care should be taken in the manufacturing of composite laminates to avoid these defects [12].

REFERENCES

[1]V. Tagliaferri, G. Caprino, A. Diterlizzi, "Effect of drilling parameters on the finish and mechanical properties of GFRP composites", *Int. J. Mach. Tools Manufact.* Vol 30 No 1 1990.

[2]A. Diterlizzi, G. Caprino, V. Tagliaferri, "Damage and residual compressive strength of GFRP drilled with conventional tools", *Proceedings of the seventh international conference on composite materials.*

[3]J. Xiao and C. Bathias, "Fatigue damage and fracture mechanism of notched woven laminates", *Journal of composite materials* Vol 28 No 12 1994.

[4]A. Ozturk, "The influence of cyclic fatigue damage on the fracture toughness of carbon-cerbon composites", *Composites* Part A 27 A 1996.

[5]T.Fujii, T. Shina and K. Okubo, "Fatigue notch sensitivity of glass woven fabric composites having a circular hole under tension-torsion biaxial loading", *Journal of composite materials* Vol 28 No 3 1994.

[6]C. Dahlen and G.S. Springer, "Delaminaton growth in composites under cyclic loads", *Journal of composite materials* Vol 28 No 8 1994.

[7]N.H. Tai, C.C.M. Ma and S.H. Wu, "Fatigue behaviour of carbon fibre/PEEK laminate composites", *Composites* Vol 26 No 8 1995.

[8]Erik Persson, Ingvar Eriksson and Leif Zackrisson, "Effects of hole machining defects on strength and fatigue life of composite laminates", *Composites* Part A 28 A 1997.

[9]G. Caprino, V. Tagliaferri, "Damage development in drilling glass fibre reinforced plastics", *Int. J. Mach. Tools Manufact.* Vol 35 No 6 1995.

[10]A. Bongiorno, E. Capello, G. Copani, V. Tagliaferri, "La generazione del danno nella foratura di GFRP", *Internal report of the Mechanical Department of the Politecnico di Milano.*

[11]E. Capello, V. Tagliaferri, "Induced damage and structural behaviour of drilled holes in GFRP laminates", *Proceedings of the 1998 ESDA Conference, July 1998.*

[12]E. Capello, V. Tagliaferri, "Technological analysis of hand lay-up technique in order to produce voids free thermosetting composites", *AITEM September 1997.*

levels / factors	Low	High
V_t / V_r (mm/rev)	0.11	0.50
Specimen support	absent	present
Pre-heating	Ta	180 °C

Table 1 – The experimental plan

Measured results	Significative factors		
	V_t / V_r	Support	Pre-heating
Push down delamination	+	–	–
Peel up delamination	+	no	no
Hole diameter	no	no	+

Table 2 – Results of the ANOVA for the drilling machining (α=5%)
(+ or – indicates the main effect of the considered factor)

Measured results	Significative factors		
	V_t / V_r	Support	Pre-heating
Absolute deformation	+	no	+
Incremental deformation	+	no	no
Absolute push down delamination	+	-	-
Incremental push down delamination	no	+	+
Absolute peel up delamination	+	no	no
Incremental peel up delamination	no	no	no

Table 3 – Results of the ANOVA for the fatigue test (α=5%)
(+ or – indicates the main effect of the considered factor)

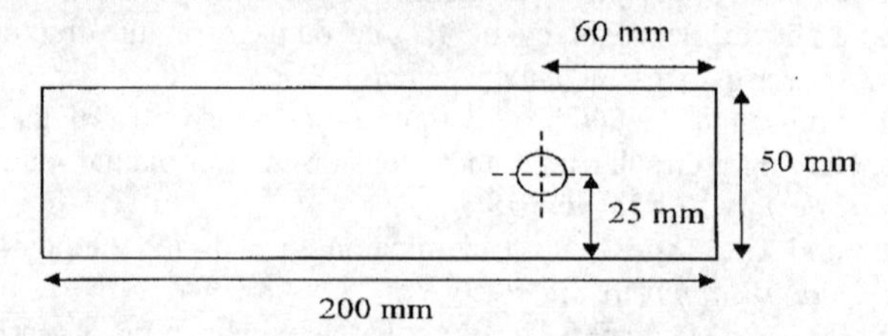

Figure 1 – The specimens

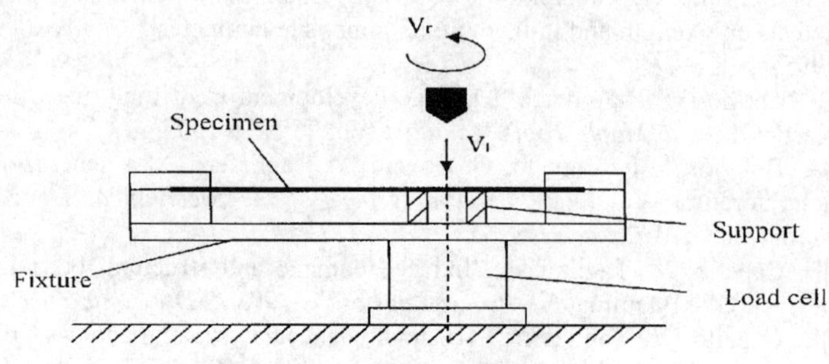

Figure 2 – The drilling set up

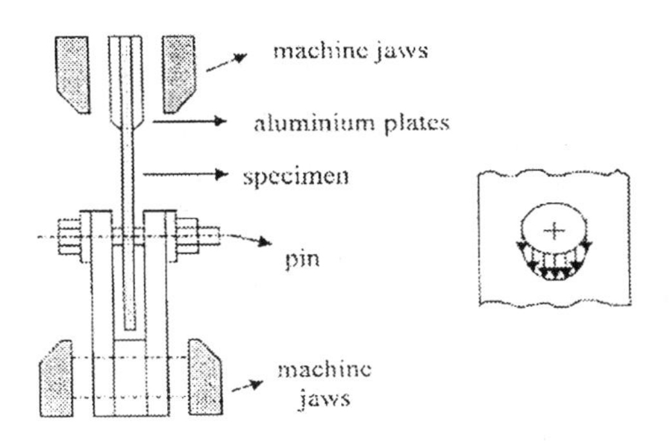

Figure 3 – The fatigue test set up

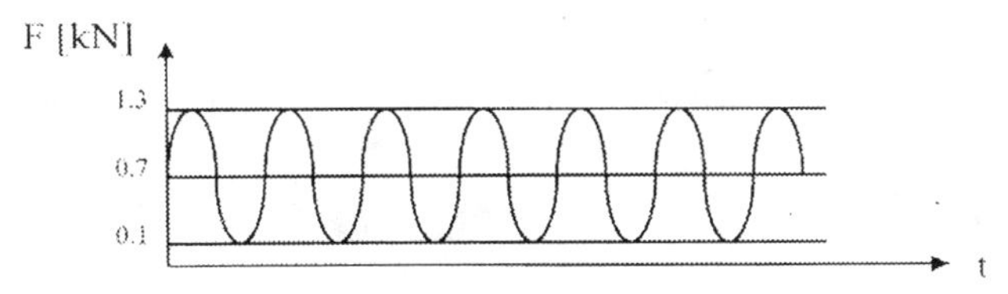

Figure 4 – The applied load in the fatigue test

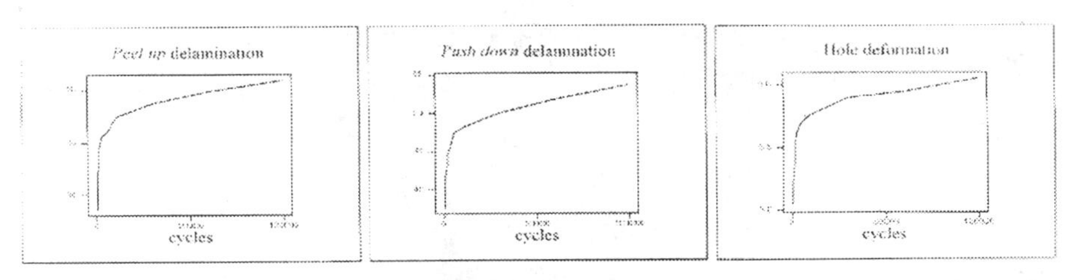

Figure 5 – Damage propagation in cyclic load

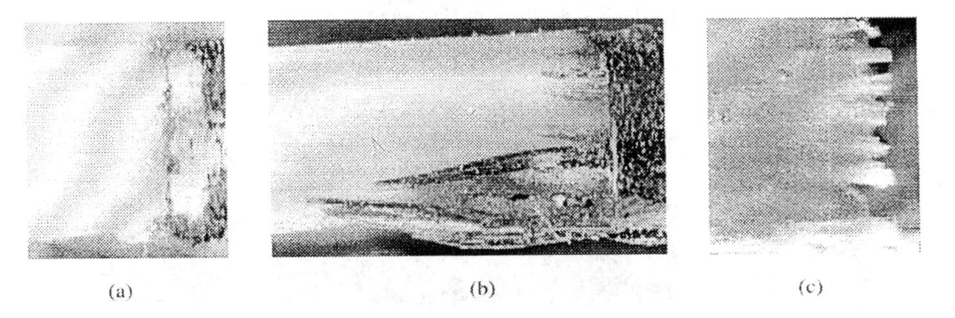

Figure 6 – Microscopic observations after drilling

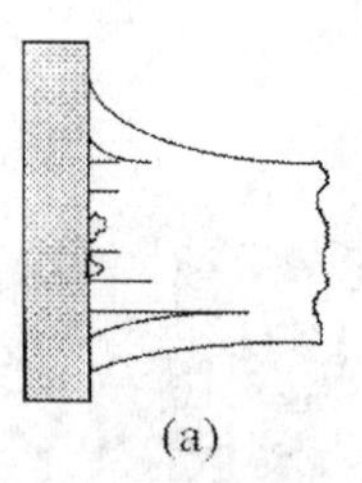

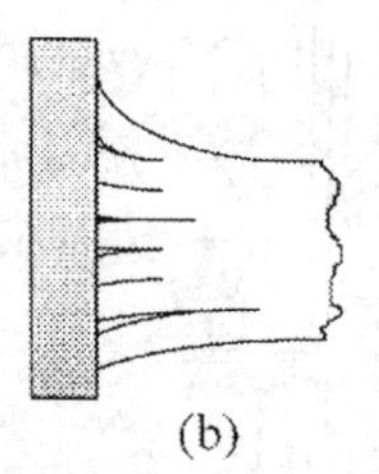

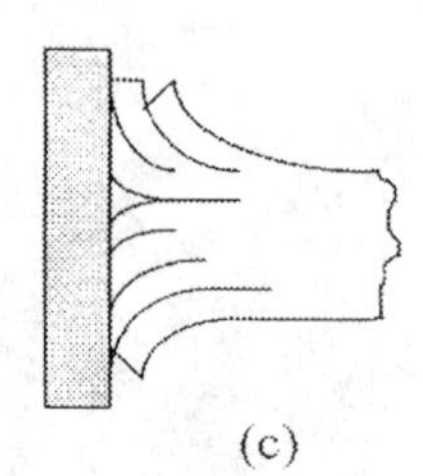

Figure 7 – Damage propagation during cyclic load

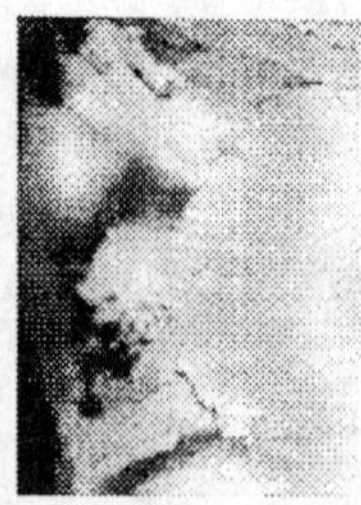

Figure 8 – Hole section after fatigue test (V_t / V_r =0.5 mm/rev)

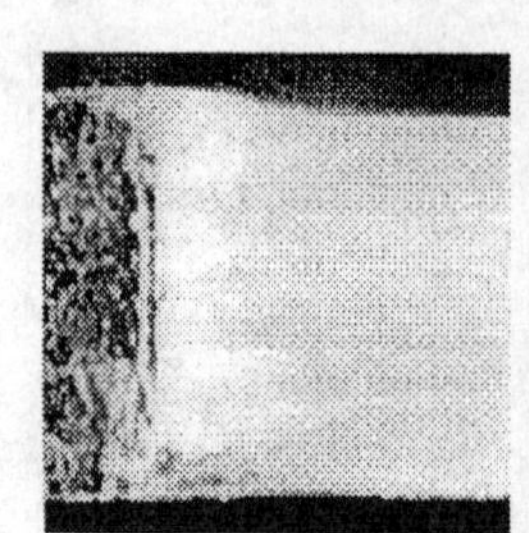

Figure 9 - Hole section after fatigue test (V_t / V_r =0.11 mm/rev)

Figure 10 – Intralaminar cracks (V_t / V_r =0.50 mm/rev)

Dynamic analysis of drilling forces of GFRP characterised by different structure

I. De Iorio, C. Leone*, V. Tagliaferri**

Dipartimento Scienze ed Ingegneria dello Spazio, Università degli Studi di Napoli, "Federico II", Napoli, Italy.
* Dipartimento di Ingegneria dei Materiali e della Produzione, Università degli Studi di Napoli, "Federico II", Napoli, Italy.
**Dipartimento di Meccanica, Politecnico di Milano, Milano, Italy.

Abstract
Aim of the work is to analyse the forces due to the interaction between the fibres/matrix and the tool in drilling of glass fibre reinforced polymers (GFRP). Experimental analysis has been developed machining monodirectional and fabric GFRP. Spectrum analysis of the cutting force are done under different cutting conditions and fibre layout. The results shown that the instantaneous forces can be described by means of an analytical model that correlate the force with the cutting parameter and the layout structures. This model can be employed to predict the force to induce the delamination damage.

1. Introduction

Some works have proposed studies on the drill mechanisms of fibre reinforced polymers by means conventional tools to evaluate the influence of the cutting parameters on the structural material damage [1-6]. The principal cause of damage has been correlated with the bit drill action modelled as a punch that to apply a force on the ultimate lamina to produce a delamination in the down side of the material [5-6]. In other experimental works the influence of the cutting parameters on the average value of the thrust force and torque are reported [7-13]. These approach consider the composite materials as an homogeneous structures and do not take in to account the local interaction phenomena between the tools and the fibres. The local phenomena can be relevant when the material is strongly anisotropic and/or non-homogeneous. In [14] the effect of the non-homogeneity in the thickness of the material on the cutting force are described. Experimental results shown that the non-homogeneity of the material is the cause of instantaneous increase of the cutting force. This forces can produce a delamination under conditions that are not predicted by the principal model proposed in literature. An experimental study on the influence of the layout structure on the drilling force is reported in [15] where the interaction matix/fibre-tool are studied for different combination of fibre layout and cutting parameter. The results show that the cutting force are affected by high oscillation induced by the presence of the reinforce. The amplitude and the frequency of the oscillation are function of the layout structure and the cutting parameter. In the range of the cutting parameters adopted an delamination control system based on the monitoring of the average thrust force are not sufficient to prevent the damage.
Aim of this work is to show the influence of the local phenomena due to the material lay out on the cutting forces. For this reason the interaction phenomena between tool and material is analysed in drilling operation of monodirectional and fabric GFRP. The analysis were done in the time dominion and in the frequency dominion. Observations of the instantaneous variation of the cutting force are described under different cutting conditions. The obtained results shown that the instantaneous variation of the cutting force, due by the local interaction tool-material, can be predicted with an simple analytic model that to correlate the force with the cutting parameter and the layout structures. This model can be adopted in a control system of the delamination damage.

2. Experimental

Materials: Two type of GFRP was used for the drilling tests, an epoxy-matrix reinforced by monodirectional with a fibre volume ratio, Vf, of 40% and an epoxy-matrix reinforced by fabric with a

fibre volume ratio, Vf, of 55%. The samples dimension were 40x60 mm2 with a thickness of about 7 mm. A panels of only resin system was made as reference material.
Experimental apparatus: The tests was conduced using the experimental apparatus described in [7]. It consist in a drilling machine fed by a piezoelectric transducer able to reed the thrust force, F, the torque, T, and the axial position of the tool. It is also equipped with a system to control the cutting speed, Vr, and the feed speed, Va, of the spindle. A digital I/O card inserted in a personal computer allows the data processing.
Procedures: For every materials were performed drilling test for different cutting condition, setting the cutting speed at 1500, 2000, 3000, 4000 and 6000 RPM and changing the feed speed at the value of 5 and 10 mm/s. The cutting tool employed were normally HSS bit drill (Din 388) of 5 mm in diameter. To limits the wear effects five holes were done for every tool. All the tests were scanned at a frequency of 4 kHz with an low pass filter (180 kHz). The cutting force were evaluated, for all the drilling time and in two reverse of the tool with cutting edge completely engaged. The spectrum analysis were done for all the time when the with cutting edge completely engaged for all the cutting condition.

3. Results

In figs 1a e 1b are reported the thrust force and the torque as a function of the time during the drilling of monodirectional composites for Vr = 3000 RPM and Va =10 mm/s. We can notice that when the cutting edge are completely engaged the amplitude of the oscillation of the cutting force are higher than the mean value. In this case is very difficult to evaluate a reference value to characterise the thrust force. The same oscillation are present during the drilling of a fabric composites but aren't present for the only resin system, as observed in [15].

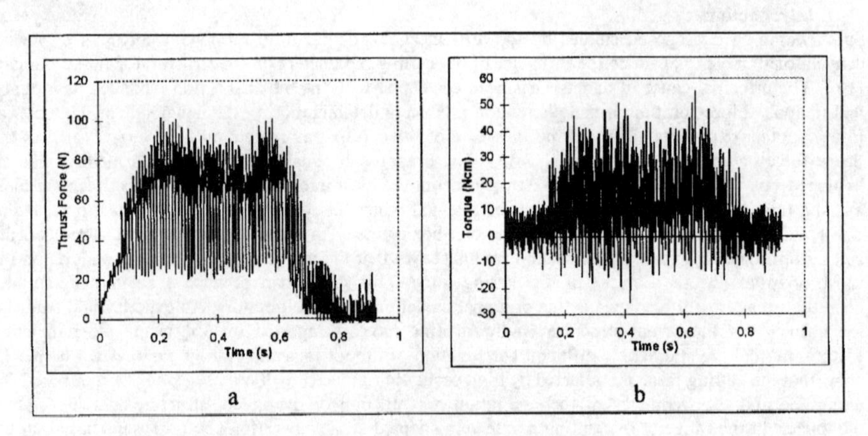

Fig: 1, Thrust force and torque vs time. Monodirectional composites.

To avoid the influence of the non-homogeneity along the thickness on the thrust force and the torque analysis of the signal test analysis are done for two reverse in the regime conditions. In the figures 2 and 3 are reported the thrust force and the torque vs time for different layout structure.
We can observe that in two reverse are present four periodic oscillations and eight periodic oscillations respectively for the monodirectional composites (figures 2a and 3a) and for the fabric composites (figures 2b and 3b). The resin panels (2c and 3c) show only high frequency oscillation, with amplitude less than that observed for both the GFRP. Spectrum analysis were done in order to show the fundamental harmonic component for the different cutting condition.
The analysis were done on the signal, with the mean value less, during the regime, in this matter there are not present the harmonic related to the drilling time.

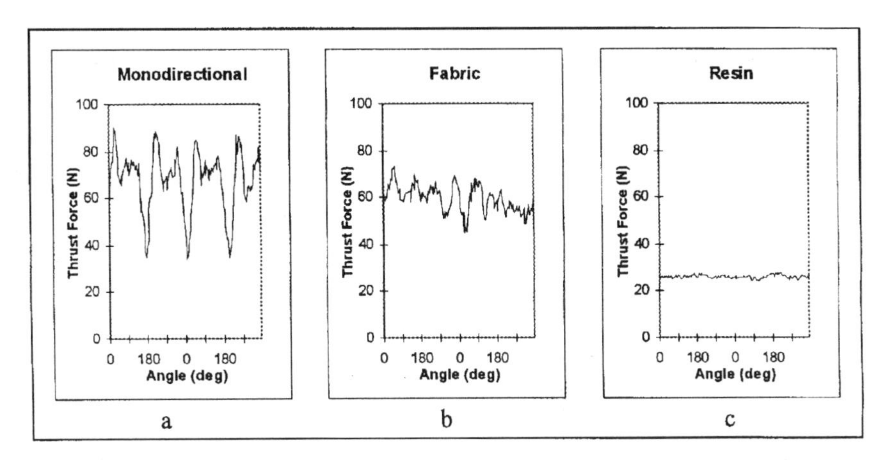

Fig: 2, Thrust force in two reverse, for different composites, Vr = 3000 RPM and Va =10 mm/s.

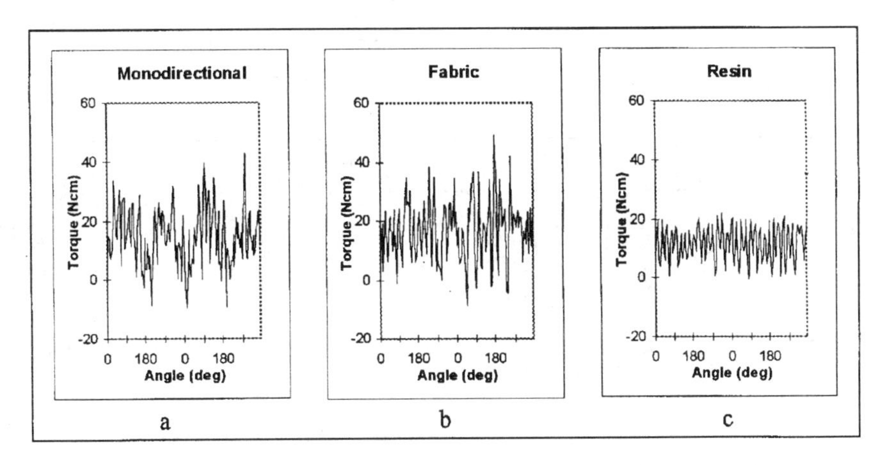

Fig: 3, Torque in two reverse, for different composites, Vr = 3000 RPM and Va =10 mm/s.

In the figures 4-5 the spectrum of the monodirectional composites force signals are reported for different cutting condition
It is possible to observer that are present different fundamental harmonic component at multiple value of the frequency correlated at the cutting velocity. For the thrust force are present a fundamental harmonic at the double of the frequency rotation Fr, that is equal to the rotation speed in reverse for second. The amplitude of this component increase increasing the cutting velocity. Also for the torque are present a component at double of Fr, but there are a large band of noise at high frequency (700÷800 Hz).

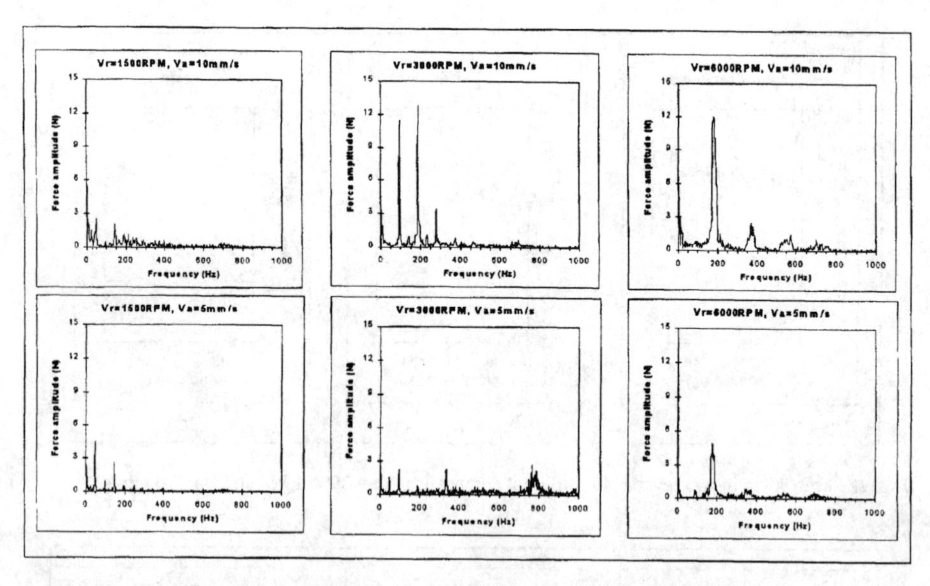

Fig: 4, Thrust force spectrum in different cutting condition. Monodirectional composites.

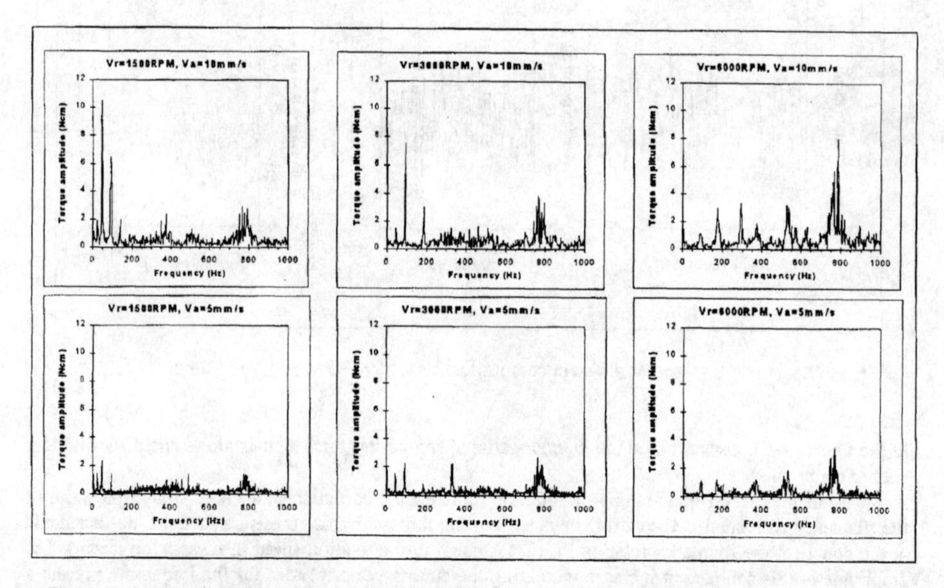

Fig: 5, Torque spectrum in different cutting condition. Monodirectional composites.

In the figs. 6-7 are reported the spectrum of the force of the fabric composites under the same experimental conditions. In this case the principal harmonic component are present four time the frequency rotation Fr and the amplitude are less than the monodirectional. The torque present different component and the noise at 700÷800 Hz and, for high cutting speed, are visible the Fr frequency that are masquerade in the monodirectional composites.

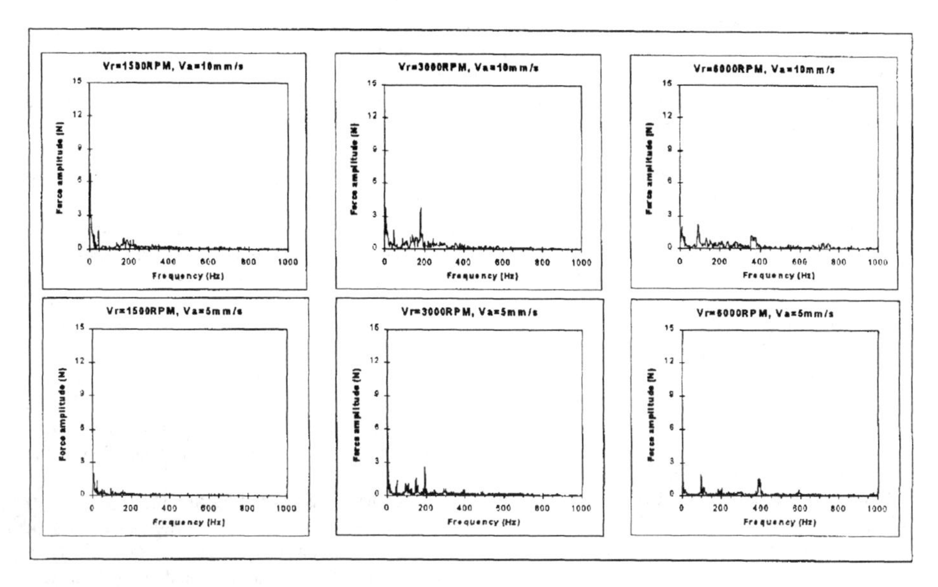

Fig: 6, Thrust force spectrum in different cutting condition. Fabric composites.

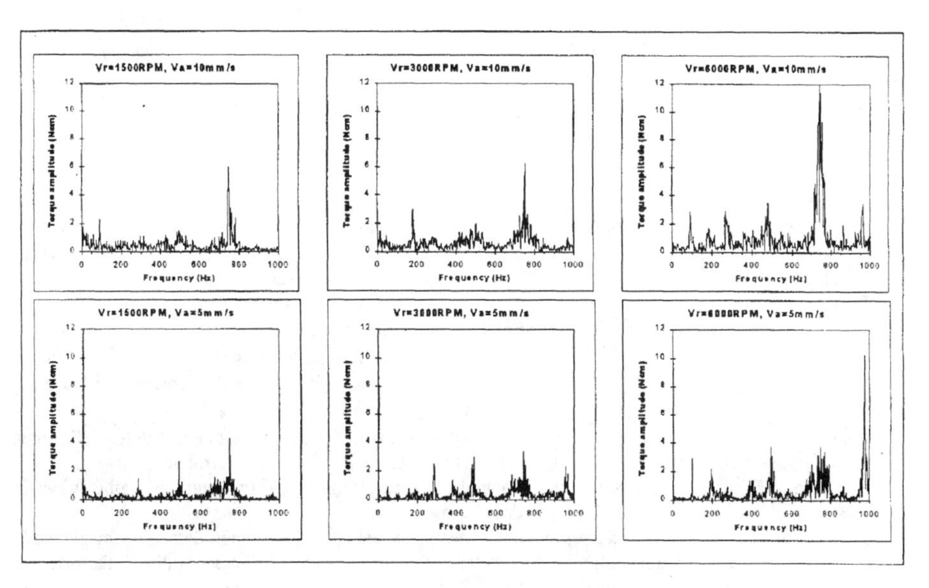

Fig: 7, Torque spectrum in different cutting condition. Fabric composites.

In the resin system case the fundamental component have an amplitude that don't are comparable with booth the GFRP materials.
During GFRP drilling the cutting force show a frequency correlated to the layout structures and the cutting speed. The amplitude of the thrust force oscillations are lower for the fabric structures than the monodirectional and increasing as the cutting speed increase. Moreover we can notice that the fundamental frequency have a π period for the monodirectional GFRP and a $\pi/2$ period for the fabric.
The presence of a π oscillation in the drilling of monodirectional are due to the rupture of the fibre loaded along the axis during the contact tool-fibre. Similar condition has been observed in [15]. Analogous observation is possible to do about the fabric GFRP, in this case the buckling load condition between tool and fibre to occur four turn in one reverse. For this reasons the frequency results double than the monodirectional. The amplitude of the oscillation observed in fabric case is lower than the monodirectional since the length of the fibre submitted to the buckling load is shorter. The presence of the noise at 700-800 Hz is due to the instrumental mechanical noise and to the chip phenomena formation.

4. Analytical Model

The maximum thrust force is strongly influenced by the local distribution of the fibre and for this reason it is not an robust parameter to prevent the delamination damage especially for the monodirectional composites in which is maximum mechanical anisotropy. The average value of the thrust force is correlated with the feed rate in the monodirectional layout but the noise introduced by the local interaction tool-fibres do not permit any correlation between the maximum thrust force, that induce delamination, and the cutting parameters.
On the basis of experimental observations a model has been developed assuming that the forces can be described by means of a Fourier series as the sum of a constant component (mean value) and by some harmonic components. By means of the eq. 1) and eq. 2) the thrust force (F) and the torque (T) can be predicted as a function of coefficient k and ε, correlated to fibre layout, and the relative position tool-fibres (θ).

$$F = Fa\,(k_{0f}+k_{1f}(sen2\theta+\varepsilon_f)+k_{2f}(sen4(\theta+\phi_{1f}+\varepsilon_f)) \qquad \text{eq. 1)}$$

$$T = Ta\,(k_{0t}+k_{1t}(sen2\theta+\varepsilon_t)+k_{2t}(sen4(\theta+\phi_{1t}+\varepsilon_t))+k_{3t}(sen8(\theta+\phi_{2t}+\varepsilon_t))) \qquad \text{eq. 2)}$$

Were:

Fa Average thrust
k_{0f} Mean coefficient.
k_{1f} I harmonic's coefficient.
k_{2f} II harmonic's coefficient
ε_f Initial phase angle
ϕ_{1f} Phase angle between the I and the II harmonic

Ta Average torque
k_{0t} Mean coefficient.
k_{1t} I harmonic's coefficient.
k_{2t} II harmonic's coefficient
k_{3t} III harmonic's coefficient
ε_t Initial phase angle
ϕ_{1t} Phase angle between the I and the II harmonic
ϕ_{2t} Phase angle between the I and the III harmonic

The coefficients have been evaluated on the basis of experimental tests. In particular the K_{if}, (K_{it}) have been evaluated as the A_{if}/Fa (A_{it}/Ta) ratio were A_i is the amplitude of i -order harmonica
As example in figs 8 are reported the experimental thrust force signal and the predicted value evaluated by means the eq. 1 in the case of monodirectional layout.
We can observe an agree between the predicted thrust force and the experimental data. The torque model (fig. 9) describe the real signals only for low value of the cutting speed. This fact is due to the presence at high cutting speed of high frequency harmonic do not described by the model.
Instantaneous thrust force can be predicted by the model and its critical value that to introduce delamination can be evaluated.

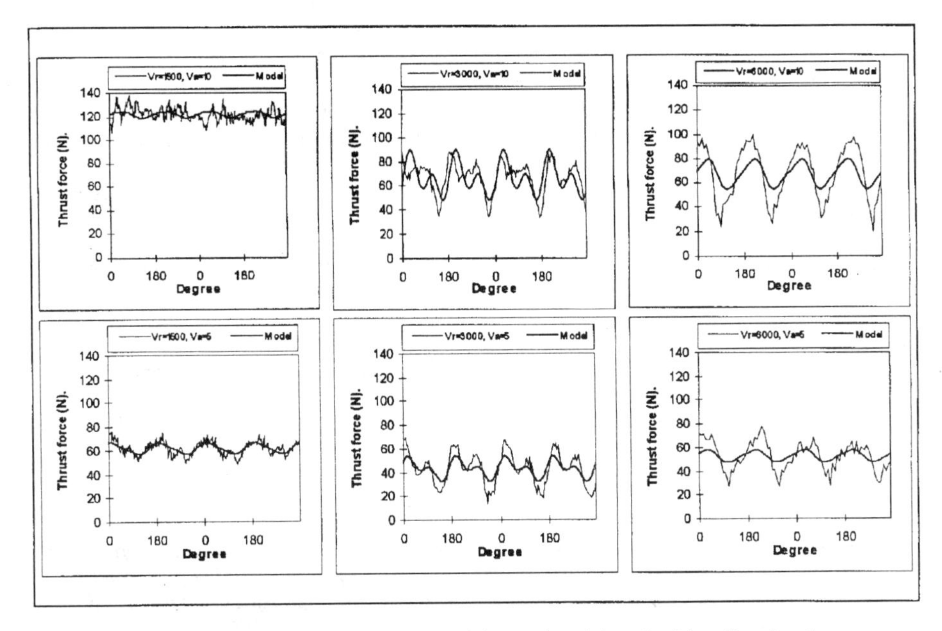

Fig. 8, Confronts between thrust force experimental data and model results. Monodirectional composites.

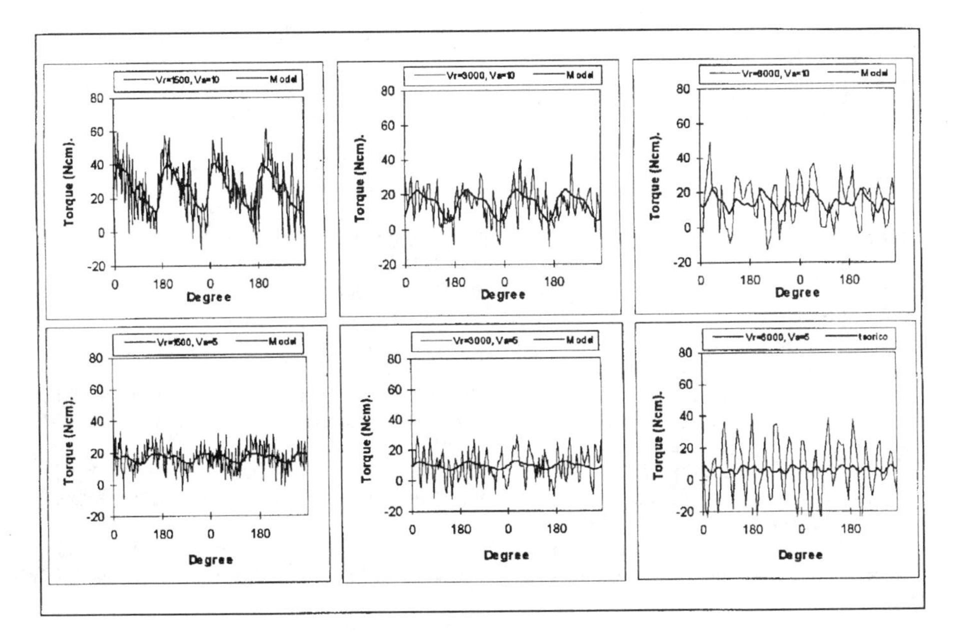

Fig. 9, Confronts between torque experimental data and model results. Monodirectional composites.

5. Conclusions

- The mechanical anisotropy of material is one of the a principal factor that affect the oscillation of the thrust force.
- The GFRP drilling force show a frequency correlated to the layout structures and the cutting speed.
- The instantaneous thrust force can be predicted by a simple mathematical model optimised on the basis of experiment performed to characterise the material.
- The proposed model do not permit to evaluate the torque because it is affected by the presence of high frequency harmonic.

In the future we will apply the model and the results described in [6,8] to evaluate the critical thrust force under dynamic regime.

References

[1] Koenig W., Wulff C., Grass P. and Willerscheid H., "Machining of Fiber Reinforced Plastics", Annals of CIRP, vol. 34, no. 2, 1985, pp. 538-548.

[2] Koenig W. and Grass P., "Quality Definition and Assessment in Drilling of Fiber Reinforced Thermosets", Annals of CIRP, vol. 38, no. 1, 1989, pp. 119-124.

[3] Tagliaferri V., Caprino G. and Diterlizzi A., "Effect of Drilling Parameters on the Finish and Mechanical Properties of GFRP Composites", Int. J. Mach. Tools Manufact., Vol. 30, no 1, 1990, pp. 77-84.

[4] Caprino G., Diterlizzi A. and Tagliaferri V., " Damage in Drilling Glass Fiber Reinforced Plastics", Advancing with Composites, Acts of Int. Conf. on Comp. Mat., May 10-12, Milan, 1988, pp.493-503.

[5] Bhatnagar N., Naik N.K. and Ramakrishnan N., "Experimental Investigations of Drilling on CFRP Composites", Processing and Manufacturing of Composite Materials, Acts of The Winter Ann. Meet. of ASME, Atlanta, Dec. 1-6, 1991, pp. 77-92.

[6] Ho-Cheng H. and Dharan C.K.H., "Delamination During Drilling in Composite Laminates", J. of Engineering for Industry, August 1990.

[7] Di Ilio A., Tagliaferri V. and Veniali F., " Cutting Mechanisms in Drilling of Aramid Composites", Int. J. Mach. Tools Manufact., Vol. 31, 1991, pp. 155-165.

[8] Jain S. and Yang D.C.H., "Effect of Feedrate and Chisel Edge on Delamination in Composite Drilling", Processing and Manufacturing of Composite Materials, Acts of The Winter Ann. Meet. of ASME, Atlanta, Dec. 1-6, 1991, pp. 37-51.

[9] Doerr R., Greene E., Lyon B. and Taha S., "Development of Effective Machining and Tooling Technique for Kevlar Composite Laminates", Fabrication of Composite Materials: Source Books, ASM, Metal Park, 1985.

[10] Di Ilio A., Paoletti A., Tagliaferri V., Veniali F., "Progress in Drilling of Composite Materials" Proc. of ASM Materials Week '92 Conference, Chicago 2-5 November 1992.

[11] Koplev A., Listrup A. and Vorm T., "The Cutting Process, Chips, and Cutting Forces in Machining CFRP", Composites, Vol. 14, no 4, October 1983.

[12] R. Komanduri, B Zhang and C. M. Vissa., 1991, "Machining of Fiber Reinforced Composites", Processing and Manufacturing of Composite Materials, PED-Vol. 49/MD-Vol. 27, pp. 1-36.

[13] Caprino G., Santo L., De Iorio I., “Chip formation mechanism in machining unidirectional carbon fibre reinforced plastics”, Proceeding of the Third Conference A.I.TE.M., Salerno, Italy, 17-19 September 1997, pp. 165-172.

[14] De Iorio I., Leone C. and Tagliaferri V., 1994, "Influence of composite structure on drilling mechanism", Proceeding of the Conference on Engineering Systems Design and Analysis PD Vol 64-2, pp 43-49. London.

[15] De Iorio I., Leone C. and Tagliaferri V., "Analysis of GFRP lay out structures effect on drilling forces" Third Int. Conference "Integrated Design & Process Technology" SPDS, Berlin, Germany, July 6-9, 1998.

THE INFLUENCE OF SHAPE AND CONTENT OF REINFORCEMENT ON GRINDABILITY OF METAL MATRIX COMPOSITES

Antoniomaria Di Ilio, Alfonso Paoletti

Department of Energetics, University of L'Aquila, 67040, Monteluco di Roio, L'Aquila, ITALY

Abstract

One of the main problems arising when machining metal matrix composites is the low machinability due to abrasion of cutting tools by the reinforcement and poor quality of the surface finishing. Among traditional machining, grinding operation is playing an important role both for finish and heavy-duty operations.
In this work the influence of shape, orientation and content of the reinforcement on the grindability of metal matrix composites is analysed by means of experimental tests carried out with grinding wheels made with conventional abrasives. Investigations deal with grinding forces and degradation of the grinding wheel surface, acquired during the machining process and ground surface roughness. The effects of workpiece material composition on grinding wheel wear and ground surface quality is described considering some grindability indices. Ground surface and chip morphology have been also analysed by SEM.

1. Introduction

Composite materials and composite structures are widely used in today's industrial applications [1]. Particularly, as the speed of aerospace vehicles pushes the supersonic envelops, aerodynamic heating and structural strength and weight are becoming even greater design factors. Proper vehicle performance, therefore, requires the use of light weight materials capable of maintaining excellent strength characteristics at elevated temperatures. With increasing interest in new high performance aircraft and turbine engines over the last several years, the materials scientists have focused their attention on Metal Matrix Composites (MMCs) as one of a number of potential candidates for high temperature structural applications. MMCs are new materials consisting of a metal matrix reinforced with a ceramic constituent which greatly improves their mechanical properties, such as yield stress, Young's Modulus and tensile strength. Even if MMCs are often fabricated with near net-shape techniques, such as precise casting and forging [2] and spray techniques [3], which do not require large material removal by machining, however also in these cases, final machining operations are always necessary and, consequently, high wear resistance tool materials are often recommended [4-8]. The implementation cost of MMCs is still high compared to that of non reinforced materials and particularly machining is among the contributing factors to the final cost of MMC

components [9]. Grinding process is considered as a final machining in the production of components requiring fine tolerances and smooth surfaces, but it can be also used in heavy-duty machining. Generally, finish grinding is more costly than other machining processes when the unit volume of material removal is taken into account. However, the development of methods for more precise casting and forging, which are closer to the final configuration of component, has led to consider grinding more economical as a single process for machining directly to the final dimension, as it eliminates the need for prior traditional machining processes, such as turning or milling [10]. Aim of this work is to enhance the knowledge about the grindability of MMCs, since grinding of metal composites, with few exceptions, has received little attention still now [11, 12]. In [13] a comparison of the grinding performance of wheels based upon conventional abrasives and superabrasive wheels, such as cubic boron nitride and diamond, has been made in machining of metal matrix composites.
In this work we want to establish the influence of geometry, orientation and volume fraction of reinforcement on the grindability of the material. To this purpose, grinding tests have been carried out using alumina abrasive wheels. Grinding forces, wheel worn area and surface finish of ground workpieces have been measured and their trends have been analysed as a function of specific material removal, for different shape and content of reinforcement.

2. Experimental

Experimental tests have been carried out on a horizontal surface grinder. The specimens were clamped on a two component piezoelectric dynamometer in order to acquire the normal and tangential component of the grinding force. The morphology of the active surface of the grinding wheel has been monitored by using an on line system for images acquisition [14]. A stylus profilometer has been used to evaluate the surface finish of the workpiece. The grinding wheel used is based upon conventional abrasive, while constant cutting parameters have been adopted for all tests, as reported in Table I. Different kinds of MMCs, made with an Al alloy matrix and reinforced with SiC powders with different volume fraction and whiskers have been investigated. Specimens 23 mm length and 13 mm width have been cut from extruded rods in direction parallel and normal to extrusion direction (Table II). For each test, one hundred plunge cut grinding passes with a depth of cut of 0.01 mm have been performed. Force components, flat area percentage and surface roughness have been measured during a single up-grinding pass. The wheels have been dressed using a single-point diamond dresser, tilted to the wheel radius with 15°, adopting one coarse pass with 3 μm radial depth and one finish pass of 1 μm, followed by five spark out passes.

Table I. Characteristics of grinding wheel and grinding parameters

Grinding wheel identification	: 32A 46-IV	Depth of cut, a (mm)	: 0.01
Diameter / Width (mm)	: 220 / 25	Workpiece speed, V_W (mm/s)	: 300
Abrasive	: Al_2O_3	Wheel peripheral speed, V_S (m/s)	: 20
Grit size	: 46		
Bond type	: Vitrified	Lubrication	:Yes/No

Table II. Workpiece materials employed for tests

Material	Denomination	Hardness (HRB)
Al-2009 / SiC-**15P** Parallel to extrusion direction	**15P-p**	83.4 ± 1.0
Al-2009 / SiC-**20P** Parallel to extrusion direction	**20P-p**	62.4 ± 1.7
Al-2009 / SiC-**20P** Normal to extrusion direction	**20P-n**	67.6 ± 1.5
Al-2009 / SiC-**25P** Parallel to extrusion direction	**25P-p**	72.6 ± 1.0
Al-2009 / SiC-**15W** Parallel to extrusion direction	**15W-p**	62.1 ± 1.4
Al-2009 / SiC-**15W** Normal to extrusion direction	**15W-n**	70.5 ± 1.1
Al-2009 / SiC-**20W** Normal to extrusion direction	**20W-n**	95.9 ± 0.7
Al-6061 / SiC-**25P** Normal to extrusion direction	**6061_25P**	52.4 ± 0.6
Al-7075 Normal to extrusion direction	**7075**	47.4 ± 0.6

3. Results and discussion

Material grindability is not dependent on a specific technological property, but is a function of a set of different characteristics which cannot be correlated each other.
In order to compare the behaviour of different cases under test, the weighted average values of normal and tangential components of grinding forces, flat area, and roughness, have been calculated defining the following *grindability indices* [15]:

$$I(x) = \frac{\sum_i (x_i \cdot V'_{wi})}{\sum_i V'_{wi}} \qquad (1)$$

where x is F'_n, F'_t, A and R_a respectively, V'_w is the specific material removal and index "i" refers to pass number. The results have been reported as a function of the ground material type.

Grinding forces
The measured grinding forces have been normalised with respect to the workpiece width. For each up-grinding pass, the representative force value has been obtained evaluating the average value of the correspondent trace. In Figure 1 the trends of normal and tangential components of grinding forces as a function of specific material removal, evaluated as above mentioned, are reported.

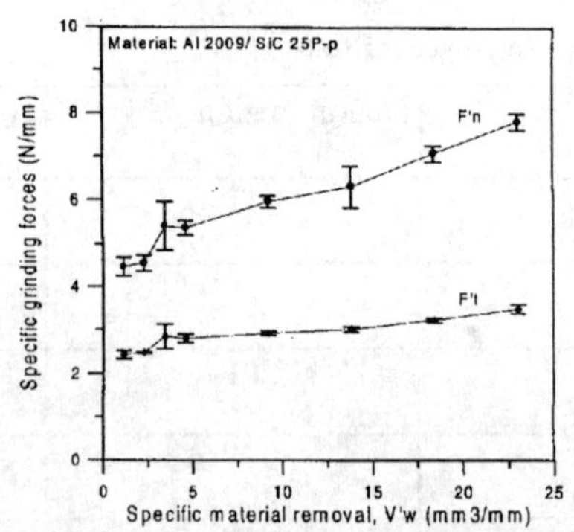

Figure 1. Specific grinding forces as a function of specific material removal

As can be seen, the tangential component of grinding force shows only a slight increase with specific material removal. On the contrary, the normal component exhibits a significant growth. Therefore, it may be assessed that the progressive wheel clogging does not influence significantly the tangential component of the grinding force. As far as the normal component is concerned, the grinding wheel seems to be very sensitive to clogging. Figure 2a, b show the weighted average values of normal and tangential components of specific grinding force for different workpiece materials.

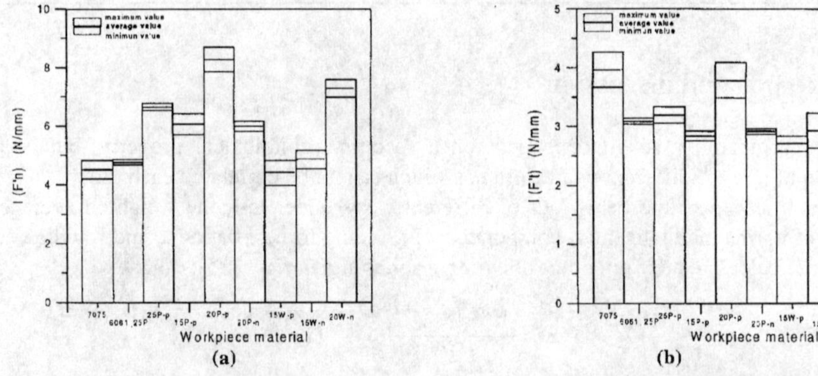

Figure 2. Weighted average values of normal (a) and tangential (b) components of specific cutting force for different workpiece materials

The normal and tangential components of the grinding forces are affected by orientation, shape and volume fraction of the reinforcement and by the type of matrix. For particulate composites having the same matrix, namely Al-2009/SiC-15P, Al-2009/SiC-20P, Al-2009/SiC-25P, both the components of the grinding forces are inversely related to the material hardness. On the contrary, materials reinforced with whiskers, namely Al-2009/SiC-15W and Al-2009/SiC-20W, show opposite tendency. This different behaviour is also found if it is considered the reinforcement orientation. In fact, while for particulate composite Al-2009/SiC-20P, the lowest force values have been obtained perpendicularly to the extrusion direction, for whiskers composite Al-2009/SiC-15W, the lowest force values have been reached parallel to the extrusion direction. However, the reinforcement, both powder or whiskers, contributes to reduce the tangential component of force. As can be seen in Figure 2b, the highest value of the tangential force has been obtained for non reinforced aluminium.

Flat area

The decrease of grinding wheel cutting ability is due to the formation of flat areas produced by adhesion of chips on the active surface of the tool.

The trend of flat area percentage exhibits significant increase after few initial passes, thereafter it rapidly tends to a constant value (Figure 3). However, the trends are characterised by large oscillations that can be attributed to the uneven distribution of the clogged area on the active surface of the wheel. Figure 4 shows the weighted average values of flat area percentage for different workpiece materials.

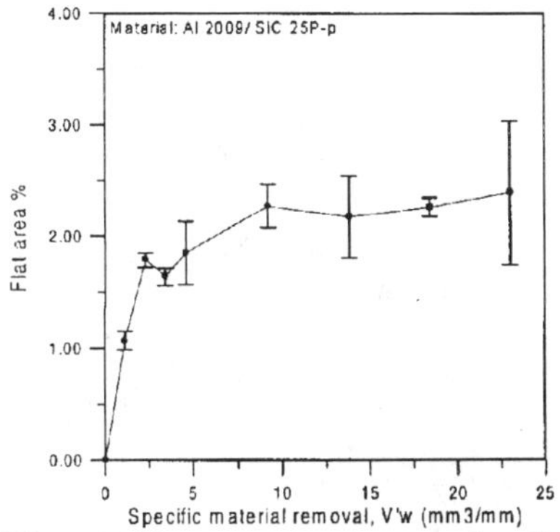

Figure 3. Flat area percentage as a function of specific material removal

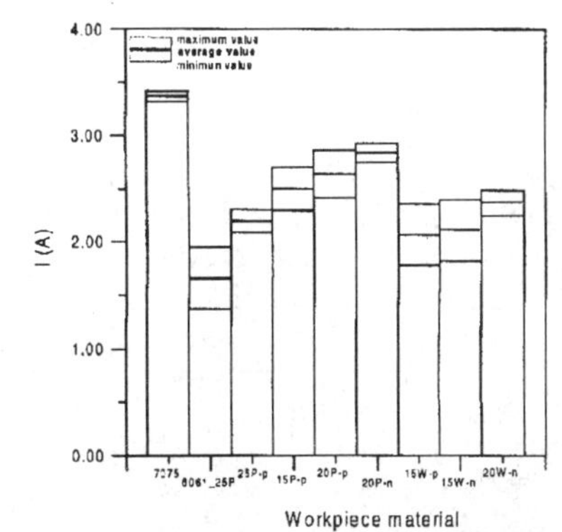

Figure 4. Weighted average values of flat area percentage for different materials

Grindability indices concerning flat area percentage of grinding wheel for different workpiece materials are fairly close to those concerning specific tangential force. The highest flat area percentage has been revealed during grinding of non reinforced aluminium, while the composite Al-6061/SiC-25P seems to show the lowest tendency to clog the wheel.

Surface texture

The morphology of the ground surfaces is characterised by the presence of side flow ploughing marks and scratches and by areas which evidence high plastic deformation and lack of ridges, as can be seen in Figure 5.

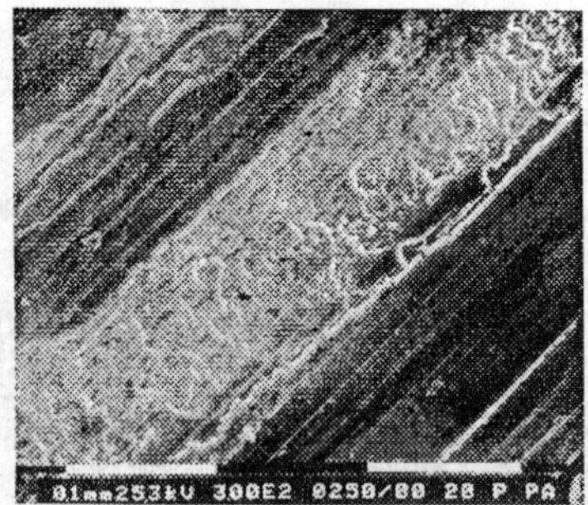

Figure 5. Surface morphology of composite Al-2009/SiC-20P

It has been noted that the zones with high plastic deformation decrease for materials which exhibit a higher hardness value. The surface roughness shows an increasing trend as specific material removal increases (Figure 6). Increasing specific material removal, Skewness parameter R_{sk} , that is a measure of the symmetry of the profile curve with respect to the mean line, also shows a growing trend for all types of materials under test, as depicted in Figure 7. This fact means that, just after wheel dressing, the machined surface is characterised by a full profile, while as wheel wear proceeds there is tendency to exhibits an empty profile. In particular, for the composite Al-2009/SiC-15W, ground along extrusion direction, this parameter is characterised by the largest slope.

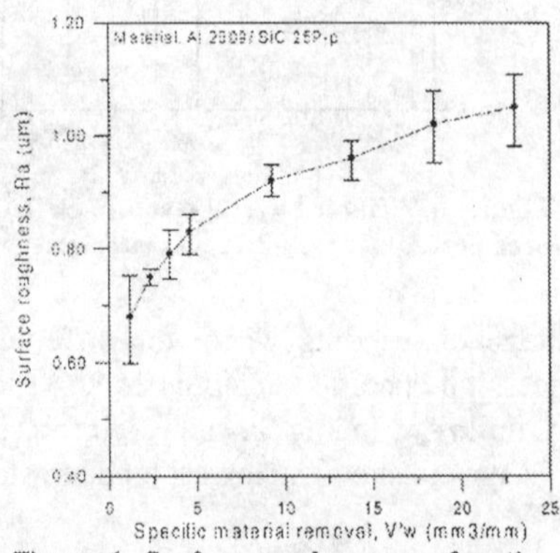

Figure 6. Surface roughness as a function of specific material removal

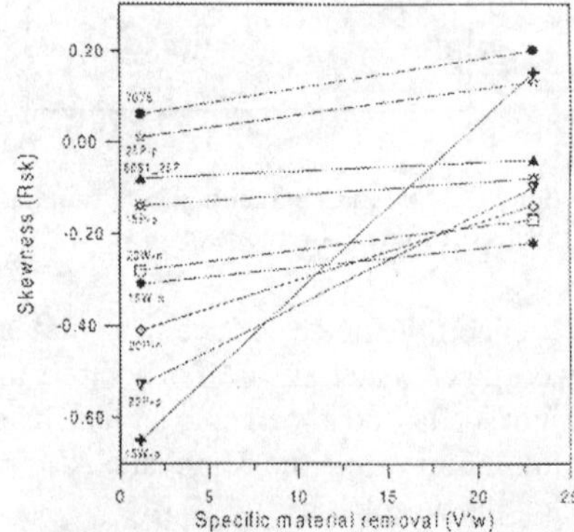

Figure 7. Skewness as a function of specific material removal

Figure 8 shows the grindability index for surface roughness for different workpiece materials.

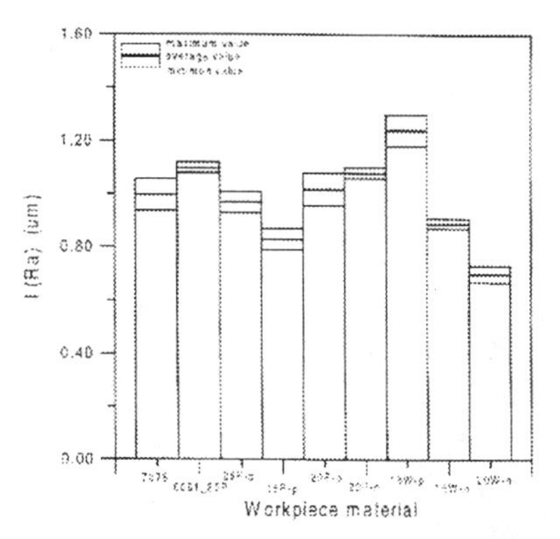

Figure 8. Weighted average values of surface roughness for different workpiece materials

Both for particulate and whiskers composites having the same matrix, the surface roughness follows a decreasing trend as a function of material hardness. However, the above mentioned correlation falls into failure when it takes into account the constituents of composite material, matrix and reinforcement, as well as reinforcement orientation. In fact, the highest roughness value has been obtained for composite Al-2009/SiC-15W, ground along the extrusion direction, notwithstanding its higher hardness with respect to both composite Al-6061/SiC-25P and non reinforced aluminium Al-7075.

<u>Chip morphology</u>

SEM observations of the debris have shown the presence of curled chips with lamella structure both for non reinforced aluminium and for composite materials (Figures 9a, b). However, while the former exhibits dense lamellae, reinforced aluminium debris show irregular and broken lamellae, probably due to the presence of hard reinforcement. It has been evaluated that the lamella spacing is about 2 µm for non reinforced aluminium and about 4 - 6 µm for composite Al-2009/SiC-20P. For whiskers composites the debris are smaller than those produced by particulate composites.

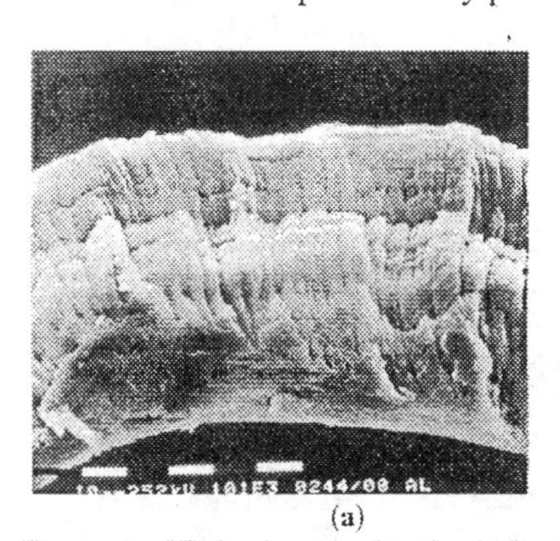

(a)

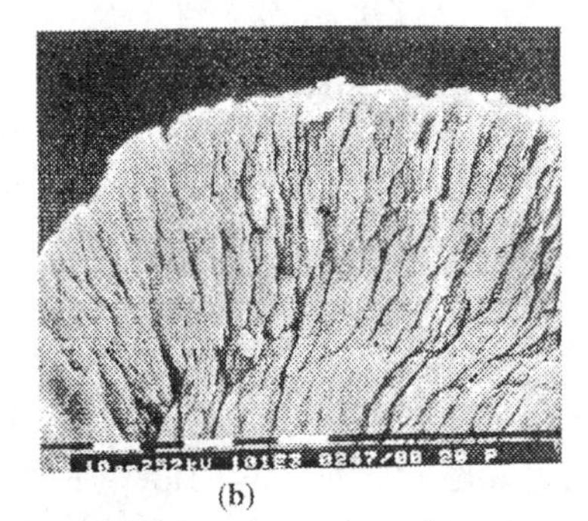

(b)

Figure 9. SEM micrographs of grinding chips of non reinforced aluminium (a) and composite Al-2009/SiC-20P (b)

Influence of grinding fluid

Grinding fluid provides cooling of the workpiece, thus lowering the temperature in the grinding zone.

The use of grinding fluid allows to obtain a lower plastic deformation of the ground surface, so the surface texture is entirely characterised by the presence of side flow ploughing marks and scratches, as shown in Figures 10a, b.

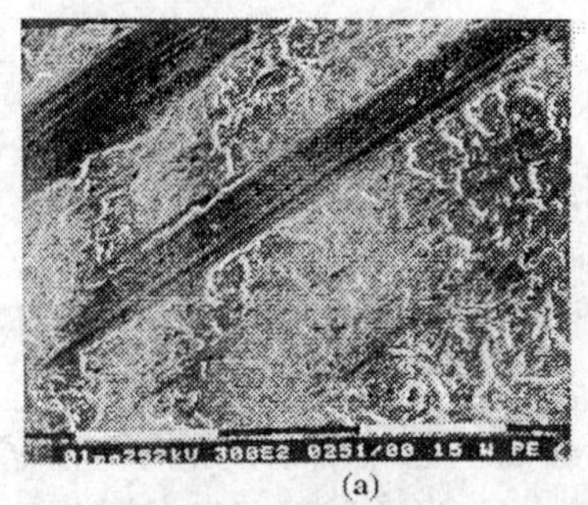

(a)

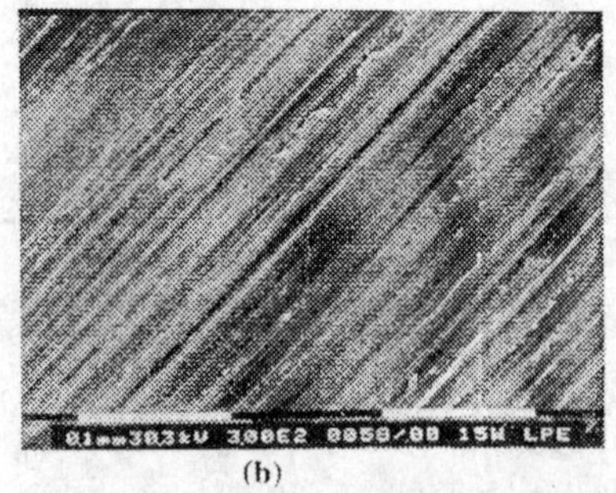

(b)

Figure 10. Surface morphology of composite Al-2009/SiC-15W obtained with dry conditions (a) and with grinding fluid (b)

Furthermore, the use of grinding fluid yields lower grinding forces, lower clogging of the wheel and better surface finish. Table III shows a comparison among the *grindability indices* obtained during grinding of composite Al-2009/SiC-15W adopting dry and wet conditions.

Table III. *Grindability indices* of composite Al-2009/SiC-15W

Grindability indices	Dry condition	Grinding fluid
$I(F'_n)$ [N/mm]	4.8 ± 0.3	4.4 ± 0.3
$I(F'_t)$ [N/mm]	2.9 ± 0.3	2.2 ± 0.3
I(A)	2.1 ± 0.4	0.9 ± 0.4
$I(R_a)$ [µm]	0.89 ± 0.02	0.81 ± 0.03

4. Conclusions

When grinding MMCs, the main cause of wheel degradation is the obstruction of the wheel due to the tough matrix material. These results indicate that the ductile matrix is the constituent material that mostly influences the grindability of the composites, rather than the hard reinforcement. Grindability indices have been settled taking into account grinding forces, wheel wear and workpiece surface roughness in order to compare the behaviour of different materials. The results indicate that composites generally exhibit a certain free-cutting tendency which make them easier to be ground with respect to non-reinforced lights alloys. However, the grindability of these materials does not depend on a specific technological property, but is a function of a set of different characteristics which are not simply related each other. In fact, geometry, orientation and volume

A STUDY OF PHYSICAL AND STRUCTURAL FEATURES OF HIGH POROUS RESIN BONDEND GRINDING STONE

1. Yoji Tomita*, H. Eda*, J . Shimizu*

*Faculty of Engineering– Ibaraki University of Japan, Nakanarusawa machi, 4-12-1, Hitachi-shi, Ibaraki-ken, Japan

Abstract

In this paper, we are reporting our study referring to the development of resin bonded grinding stones which can be applied to the ultra fine precision processing of the various kinds of work piece selected from a ductile material to a brittle material. As the method for processing, a lapping method which uses a lapping plate composed by resin bonded grinding stones instead of a cast iron plate is used. The important factors which compose a synthetic grinding stone and gives major influence on machining features are summarized in (1) abrasives, (2) bonding materials and (3) a structural feature of grinding stone. We have investigated the best combination of resins, structural properties and physical properties of grinding stones which are suited for the processing of various kinds of work piece.

1. Introduction

Recently, the resin bonded grinding stones having soft and fragile feature are becoming more popular in the region of high precision processing using planetary type lapping machine [1, 2]. This kind of processing method is called as the grinding stone lapping, and the grinding stone which can be used to this purpose is basically belonging to a kind of conventional grinding stone such as vitorified type or resinoid type, however the physical properties differ from that of mentioned types. Further, this kind of grinding method is recognized as a method which promises good surface roughness however the obtained removal rate is low. The feature and the physical properties of the grinding stone are mainly controlled by following three factors, that is, (1) abrasives ; concentration and grain size of abrasives, (2) a kind of bonding materials ; combination of various kinds of resin, (3) a structural feature ; quantitative balance of abrasives, resin and pore (porosity). In this paper, we are reporting the development of the best grinding stones suited to the grinding stone lapping for aluminium alloy, chrome plated metal, silicon wafer, soda-lime glass and phenol resin by changing the above mentioned three fundamental factors of grinding stone and carried out actual grinding tests. The obtained test results are reported in this paper.

2. The feature of the grinding stone lapping

Before the preparation of a grinding stones and the actual tests by prepared grinding stone, the features of grinding stone lapping are summarized as follows [1]. That is :

1) The grinding stone to be used is comparatively soft and fragile and easily be worn out so as to be the abrasive grain easily loosed, and has a good feature for self sharpening of a grinding stone.
2) The processing speed is about 100 ~ 1000m/min and is comparatively slow compared with that of conventional grinding by the harder type stone (1000 ~ 12000m/min).
3) Large quantity of machining fluid is used. The average quantity of machining fluid is about 100 ~ 1000 L/min. m^2.
4) The temperature of the processing point of an abrasive is not so high, it becomes 5 ~ 10°C higher than the room temperature.
5) The grinding process is carried out by a pressure control method and ductile mode.

3. Preparation of a grinding stone

From the results of our previous study [1, 2, 3], it becomes clear that the resin bonded grinding stone which uses polyvinyl alcohol resin and a thermosetting resin such as phenol resin or melamine resin gives good result for the ultra precision processing by the grinding stone lapping, because these resins give a proper softness and fragility to a grinding stone. And also, it has become clear that a high porous structure gives good structural features to a grinding stone. While polyvinyl alcohol (PVA) resin is a soft and elastic type resin, phenol resin is recognized as a hard and stiffened type resin and melamine resin is recognized as a hard and fragile type resin. The experiments to clarify the features of grinding stone are carried out, and we confirmed that the PVA resin has a proper elasticity and acts like as a cushion material which weakens the partial high pressure of a working point of abrasive loaded to the work piece, further it works as a dispersion agent to disperse abrasives homogeneously in a matrix of grinding stone at the manufacturing process. Phenol resin is a hard type resin and has a good dimensional stability, which relief the exceeded elasticity of PVA resin. Melamine resin is a hard and fragile type resin and can reduce the bonding strength without affecting the hardness of grinding stone when it is used together with other kinds of resin. Mixing use of the melamine resin brings a grinding stone a feature of easy worn out, which relates to a good self sharpening feature of the grinding stone.
As the previous tests to determine the effect of resin to the physical properties of grinding stone, above mentioned three resins are mixed by typical representative ratio, then the grinding stones are prepared. The physical properties of the obtained grinding stone are measured. Hardness and bonding strength are the representative physical properties of the grinding stone. Fine particles of silicon carbide (SiC) is used as the abrasive, and the abrasive content is fixed to 60% by weight.
This kind of grinding stone is manufactured by a wet process [3]. The aqueous solution of monomer or pre-polymer of resin is added into the aqueous solution of PVA resin. Fine particles of abrasive, formaldehyde (hardner), sulfulic acid (catalyser) and pore forming chemicals are added to the solution, then the mixture is agitated well so as to form a homogeneous and high viscose slurry. The slurry is poured into a mold, then the mold is placed into about 353K controlled water bath for around one day to complete the chemical reaction. Removed from the mold, the solidified slurry is washed and dried, then heat treated at the hardening temperature of the thermosetting resin.
The physical properties and structural properties of obtained grinding stones are summarized in Table 1.

Table 1 preparation test of grinding stone

Test number	1	2	3	4	5	6	7
Abrasive conc wt%	59.2	60.3	60.1	58.6	58.7	61.0	59.9
PVA resin wt%	13.3	18.8	20.0	0.0	19.7	9.8	9.7
Phenol resin wt%	13.8	20.9	0.0	20.2	10.6	20.3	10.4
Melamine resin wt%	13.7	0.0	19.9	21.2	11.0	8.9	20.0
Porosity vol%	69.8	72.0	71.4	69.1	74.1	72.7	73.0
Pore size micron	25	25	25	28	28	23	25
Young modulus Mpa	2430	2580	1880	3100	2350	2800	2880
Bonding strength	0.37	0.15	0.42	0.21	0.33	0.28	0.35
Hardness H_{RR}	-82	-55	-210	20	-115	-26	-10

As clearly understand from Table 1, the fundamental physical properties of grinding stone such as Young' s modulus, hardness and bonding strength is largely effected by bonding materials (resin).
In a case of the grinding stone lapping process, grinding ability is closely related to the self sharpening of the grinding stone, and it must be altered concerning the physical properties of a work piece to be processed. When the feature of grinding stone is conformed to the physical property, high and continuous grinding force can be expected.

4. Development of grinding stone for various applications

1. Aluminium alloy

Aluminium alloy is a typical ductile material, and recognized as a soft metal. The surface is mainly processed by ductile mode. Ground surface causes plastic deformation by sharpened edges of fixed abrasives in the matrix of a grinding stone. The test grinding stones whose abrasive concentration and resin (polyvinyl alcohol resin, phenol resin and melamine resin) combination ratio are altered are prepared and the physical properties of them are measured. For the present application, a soft and fragile type grinding stone whose abrasive concentration is about 60wt% (10vol%), Young' s modulus is about 2500Mpa and bonding strength is about 0.4mm is suitable. For the preparation of the bonding material which satisfy the above mentioned requirement, it is clarified by our investigation that the best combination ratio is approximately, PVA resin : 20wt%, melamine resin : 15wt%, phenol resin : 5wt%.

Table 2 Developed grinding stone for aluminum alloy

Abrasive size micron	41 #400	20 #800	16.3 #1000	8 #2000	5.3 #3000	3 #4000
Abrasive conc wt%	63.4	61.1	60.8	58.4	59.3	54.4
PVA resin wt%	16.6	18.6	18.6	21.1	20.2	24.3
Melamine resin wt%	15.3	15.0	15.3	15.3	15.2	15.3
Phenol resin wt%	4.7	5.3	5.3	5.2	5.3	6.1
Porosity vol%	63	70	70	74	74	74
Pore size micron	30	25	25	25	25	22
Young' s modulus Mpa	2800	2550	2500	2300	1850	1550
Bonding strength mm	0.27	0.35	0.40	0.35	0.43	0.47

The grinding stones of various abrasive size are developed and features of them are summarized in Table 2, and the results obtained by actual grinding test are shown in Figure 1. As afore mentioned, since the work piece is processed by plastic deformation caused by sharpened edges of fixed abrasives, the appearance of ground surface becomes luster and by the microscopic observation, uniform trace marks of one direction marked by abrasives are visible. Figure 2 is a SEM observation of the ground surface which shows the typical pattern processed by ductile mode.

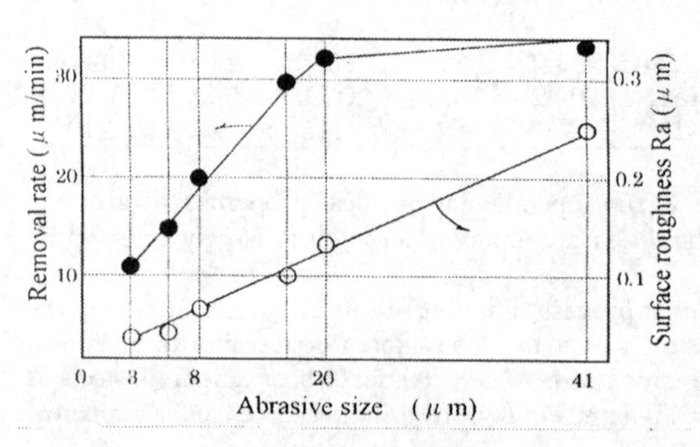

Figure 1 Test results by developed grinding stones

Figure 2 SEM picture of ground surface of aluminum alloy

2. Chrome plated metal

The surface of chrome plated metal is too hard to be processed by a soft grinding stone developed for the purpose of aluminum alloy. As the abrasives, white aluminum oxide (Al_2O_3 = WA) is preferably used because the affinity of it is better than that of silicon carbide (SiC). And as the bonding materials, hard and stiffened type resin is required. That is, in this case, since the surface is processed by plastic deformation caused by sharpened edges of fixed abrasives, the harder abrasives and the stiffened bonding materials are necessary. Accordingly the processing pressure is higher, therefore the good dimensional stability is required to the grinding stone.

The best mixing ratio of resin of the developed grinding stone is approximately ; PVA resin : 0wt%, melamine resin : 15wt%, phenol resin : 15wt%, while WA abrasive content is about 70wt% and porosity is 50vol%. The physical properties of developed grinding stone are ; Young' s modulus is 3500 Mpa, bonding strength is 0.50mm and Rockwell' s hardness is $50H_{RR}$.

The representative date of this grinding stone is summarized in Table 3.

3. Silicon wafer

Silicon wafer is recognized as a typical brittle material, and currently is processed by a conventional lapping method using WA or FO loose abrasives and cast iron plate. In this case, the surface of silicon wafer is processed by fine fracture caused by loose abrasives and the appearance of the processed surface is frosty and dull. While, since the grinding stone lapping method is basically based on plastic deformation caused by fixed abrasives, the processed surface is quite different from that of obtained by loose abrasives. Since the work piece is a hard and brittle material, an abrasive to be used is desirably a hard type such as diamond or aluminum oxide (WA) . However, when these hard abrasives are used, deep scratches remain caused by sharpened edges on the surface of work piece further the deep work damaged layer is generated. To avoid these problems, in this paper, SiC abrasive which is softer than diamond and harder than silicon wafer is used. Further, to improve the grinding force, the concentration of abrasives is increased to the maximum level (approximately 80wt%) and bonding strength is weakened. The physical properties of developed grinding stone for this application are ; Young' s modulus is 3000 Mpa, bonding strength is 0.20mm and Rockwell' s hardness is $40H_{RR}$. The best mixing ratio of resin to accomplish the above mentioned physical properties of grinding stone is approximately ; PVA resin is 5wt%, melamine resin is 10wt% and phenol resin is 5wt%, while WA abrasive content is about 80wt% and porosity is 45vol%. The representative date of grinding stones developed for this application are summarized in Table 3.

Table 3 developed grinding stone for silicon wafer

Abrasive size micron	41	20	8
Abrasive content wt%	80.3	80.5	82.5
PVA resin wt%	4.8	4.7	4.5
Melamine resin wt%	10.8	9.2	8.2
Phenol resin wt%	6.1	5.6	4.8
Porosity vol%	44.9	45.7	45.4
Pore size micron	22	25	24
Young' s modulus Mpa	3350	3020	2430
Bonding strength mm	0.11	0.17	0.26
Hardness H_{RR}	47.8	42.2	22.2

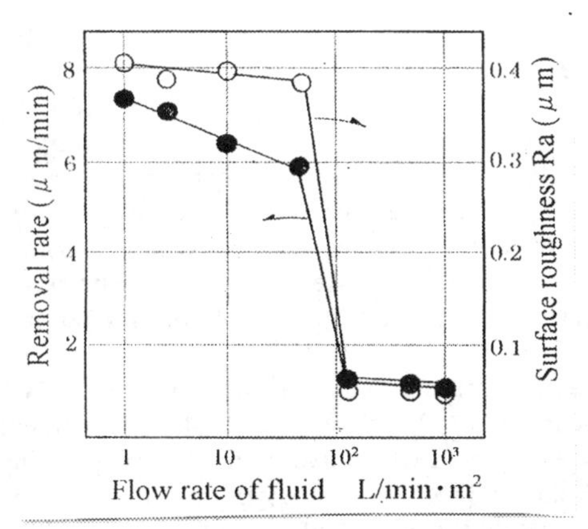

Figure 3 Influence of flow rate to grinding results

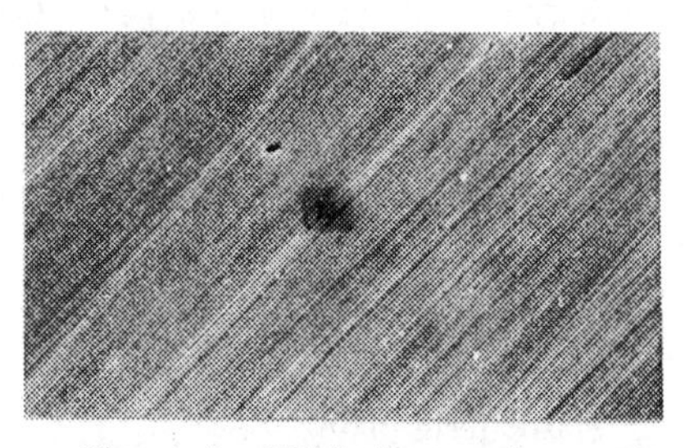

Figure 4 SEM picture of ground surface by high flow rate (x 1000)

Figure 5 SEM picture of ground surface by low flow rate (x 1000)

Using the developed grinding stones, actual grinding tests are carried out on an as-cut silicon wafer. The appearance of ground surface is luster and uniform trace marks of one direction marked by fixed abrasives can be observed by microscopic observation, while the removal rate is very slow compared with a conventional lapping method. To obtain a frosty and dull surface and to improve a removal rate, the actual tests to change the grinding condition are carried out, and we have clarified that the amount of grinding fluid effects remarkably to the grinding results [4]. The amount of flow rate of grinding fluid is altered from 1 to 1000 L/min.m^2, and the obtained results are shown in Figure 3. As clearly understand from this figure, the surface roughness and removal rate changes remarkably at 40 ~ 100 L/min.m^2 region. At smaller flow rate than this region, a frosty and dull surface and high removal rate can be obtained. When the amount of flow rate is bigger than this region, grinding surface is washed by sufficient fluid and the surface of work piece is processed by ductile mode by fixed abrasive. And when the amount of flow rate is smaller than this level, removed abrasives from the grinding stone remain on the surface and act as loose abrasives and the process is carried out by brittle mode which causes fine brittle fractures on the surface of work piece. The SEM observation of ground surface is shown in Figure 4 and Figure 5.

4. Glass

Soda-lime glass is used as a work piece. Since this material is also a kind of brittle material, the grinding stone which is similar to that of silicon wafer is suited. However, a soda-lime glass is softer than a silicon wafer, therefore, the developed grinding stone is slightly different.

The best mixing ratio of resin of the developed grinding stone is approximately ; PVA resin : 5wt%, melamine resin : 15wt%, phenol resin : 5wt%, while WA abrasive content is about 75wt% and porosity is 50vol%. The physical features of developed grinding stone are ; Young' s modulus is 2800 Mpa, and bonding strength is 0.25mm. The representative date of this grinding stone is summarized in Table 3.

The influence of grinding fluid flow rate is tested likely to the case of silicon wafer, and very similar result have obtained. The remarkable changing point from ductile mode to brittle mode is observed at flow rate of 10 to 50 L/min.m^2. At smaller flow rate than this region, a frosty and dull surface and high removal rate can be obtained [5].

5. Plastic resin plate

As a work piece, a plate of phenol resin (Bakelite) is used. In comparison with other materials, since Bakelite is a very soft and fragile material, when it is processed by fixed abrasives, surface becomes very defective by deep scratches and appearance becomes very bad. Therefore, in this case the surface must be processed by the loose abrasives. We developed soft (not elastic) and fragile type grinding stone whose abrasive content is very high.

The best mixing ratio of resin of the developed grinding stone is approximately ; PVA resin : 2wt%, melamine resin : 0wt%, phenol resin : 3wt%, while SiC abrasive content is about 95wt% and porosity is 50vol%. The physical features of developed grinding stone are ; Young' s modulus is 500 Mpa, and bonding strength is 0.05mm. The representative date of this grinding stone is summarized in table 7. This type is a flaky type grinding stone and easily generates loose abrasives by a light friction caused by a light contact with a work piece. Said loose abrasives work on the grinding surface.

5. Discussion

1) In general, the mechanism of grinding stone lapping is understood to be ductile mode caused by fixed abrasive, and the appearance of ground surface is luster and fine and uniform trace marks of one direction marked by abrasives can be observed by microscopic observation.
2) Since the effect of working point of abrasive is finite, an abrasive whose sharpened edge is dulled must be removed smoothly and a new abrasive must be regenerated. The continuance of the grinding ability of an abrasive is closely depended to the physical properties of the work piece. The hardness and bonding strength of grinding stone are controlled by changing the combination ratio of resins, in accordance with the continuance of the abrasive.
3) By changing the grinding condition, it is possible to change the grinding mechanism remarkably. Under the condition of low flow rate of the grinding fluid, many removed abrasives from the grinding stone remain on the grinding surface and act as loose abrasives, which cause fine brittle fractures of work piece. The ground surface of work piece is dull and frosty. That is, the processing mechanism changes from ductile mode to brittle mode at this point. This effect can be effectively obtained by the grinding stone which has high abrasive concentration and low bonding strength.
4) The phenomenon mentioned above brings good processing effect on the materials which lacks of ductility such as silicon wafer or glass. However, on the ductile materials such as aluminum alloy or other metals, it does not bring good effects because the loose abrasives are stuck or buried into the surface of work piece and deteriorate the surface quality of work piece.
5) The representative date of developed grinding stone using 16 micron size abrasives for each applications are summarized in Table 3 with the approximate processing results.

Table 3 Developed grinding stone for each applications

Application (kind of work piece)	Aluminum Alloy	Chrome Plated Metal	Silicon Wafer	Soda-Lime Glass	Phenol Resin
Abrasive	SiC	Al_2O_3	SiC	Al_2O_3	SiC
Abrasive wt%	60(10)	70(20)	80(30)	75(25)	95(35)
Resin wt%	40(30)	30(30)	20(25)	25(25)	5(15)
Porosity vol%	60	50	45	50	50
Pore size micron	20	20	20	20	20
Hardness H_{RR}	-	50	40	-	-
Y' modulus Mpa	2500	3500	3000	2800	500
Bonding St mm	0.40	0.50	0.20	0.25	0.05
Roughness Ra Micron	0.10	0.05	0.05 [0.3]	0.05 [0.5]	0.15
Removal rate Micron/min	30	20	1 [6]	1 [8]	30

Remarks ( ) : vol%
[ ] : processed by low flow rate of grinding fluid

The SEM observation showing structural feature of developed high porous resin bonded grinding stone is shown in Figure 6.

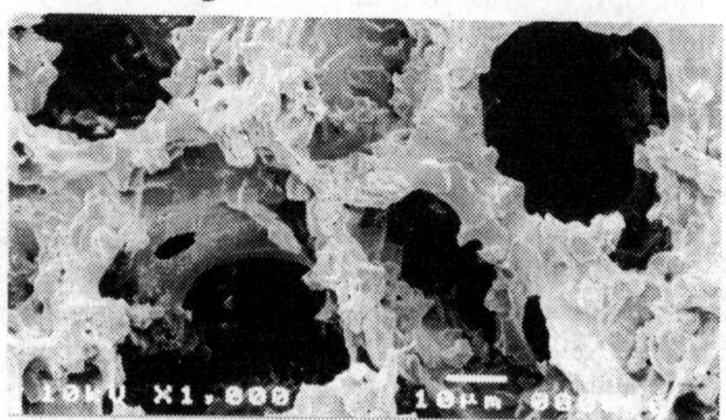

Figure 6 SEM picture of grinding stone

6. Conclusion

1) The processing method by grinding stone lapping can be applied to the ultra precision processing for various kinds of materials from ductile materials and brittle materials.
2) The resin bond grinding stone is used for this application, and the physical and structural properties of the grinding stone must be changed with accordance to the feature of materials to be processed.
3) The most important physical properties of grinding stone are hardness (stiffness) and bonding strength, which are effected by (1) the materials and content of the abrasives, (2) combination ratio of bonding resin and (3) the structural features of the grinding stone.
4) Basically, the processing mechanism of grinding stone lapping is ductile mode and by changing the grinding condition, the grinding mechanism changes to brittle mode.

References

[1] *Y. Tomita, H. Eda,* A STUDY OF THE ULTRA PRECISION GRINDING PROCESS ON A MAGNETIC DISK SUBSTRATE – Wear, 1 July 1996, Vol 195

[2] *Y. Tomita, H. Eda, Y. Yamamoto, M. Nakamura,* STUDY OF ULTRA PRECISION PROCESSING OF SILICON WAFER BY GRINDING STONE – Proceedings of 2nd International Conference of Precision Engineering (ICMT/2) Singapore, 22-24 November 1995.

[3] *T. Sato, Y. Tomita,* COMPOSITE WHETE STONE FOR POLISHING SOFT METAL – United States Patent 4,750, 915, June 1988.

[4] *Y. Tomita, H. Eda, M. Nakamura,* A STUDY FOR ULTRA PRECISION PROCESSING OF SILICON WAFER BY LAPPING STONE – Journal of the Society of grinding Engineers, 1 March 1996, Vol 40, No.11

[5] *N. Kurata, H. Eda, Y. Tomita,* STUDY ON PRECISION MACHINING OF GLASS SUBSTRATE - Proceeding of International Conference on Manufacturing Milestones toward the 21st Century (MM21) Tokyo, 22-25 July 1997.

Production technology for molding a composite fixing head for a G.R.P. anchoring rock bolt

Luigi Giamundo*, G. Maddaluno*, V. Landolfi, N. Giamundo***

* ATP
** PROTECO

Abstract

The use of composites for the building industry and civil engineering has been limited by the lack of truly industrial mass production technology.

As a matter of fact, the many advantages in performance of the composites are vanished by the high production cost related to the most used production technologies. Pultrusion is an exception in so much as it offers a good compromise between cost and performance. It is however true that pultruded composites, most of the times, cannot be used "as produced", but they need to be post- worked or to be assembled with other pultruded shapes or with accessories. This is the real reason why often production technologies are not adequate to produce competitive applications.

Since long, GRP pultruded rods and tubes have found massive application in tunnelling and mining, mainly for consolidation and pre-consolidation during tunnel excavation. Such intensive use is justified by the excellent resistance and life of the product. However, one big problem, which is still limiting some applications, comes from the necessity to supply these bars with special heads that can transfer large axial loads; it is not possible to apply traditional solutions based on threaded nuts and plates.

This presentation refers to the development work for a composite head and its specific production process.

1. Production definition

The main reason why for pultruded bars it is not possible to utilise threaded nuts is the fact that the stress applied on a composite thread is supported mainly by the resinous matrix and by the fiber/matrix interface; let us not forget that with pultruded bars the reinforcement fibers are oriented almost totally in the direction of the axis. For this reason the loads are not very well transferred to the fibers.

Otherwise, it can be affirmed that a cone/counter cone system, derived from those used for metal ropes, should be a valid system to transfer loads to the continuos reinforcement fibers. In such case, the rod is inserted into a bushing which has conical hole in the centre in which are seated a serious of small cones to lock the sliding of the bar in one direction.

The behaviour mechanism can be synthesised as follows:

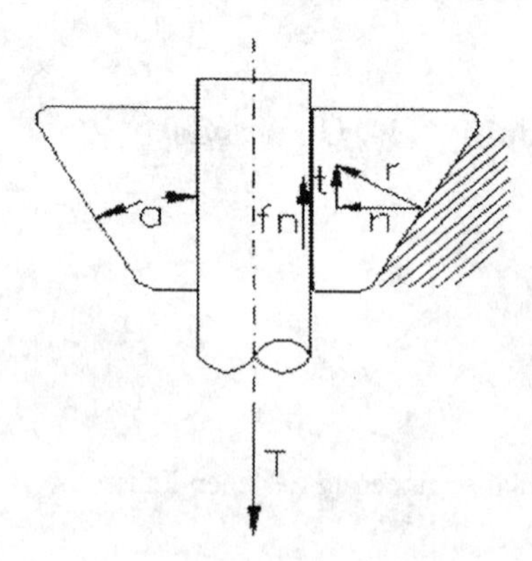

Figure 1 - Cone system

The reaction "r" on the contrast cone surface, which derives from the axial force "T" , has a component "n" normal to the axis of the bar which causes a friction = fxn, inversely proportional to the amplitude "a" of the cone angle, and directly proportional to the friction coefficient between the pultruded bar and the cone.

As "T" increases, so will "n" and the friction reaction. This reaction acts in the direction of the fibers and it can be supported very well by the material.

A correct definition of the product implies the following:

- Dimensioning of the resistant section;
- Definition of the geometry of resistant section;
- Material choice;
- Design.

The pictures 1, 2, 3 and 4, show the result of the designing respectively with reference to the bar, to the locking cones, to the plate which acts as counter-cone, to the accessories, and to the assembly.

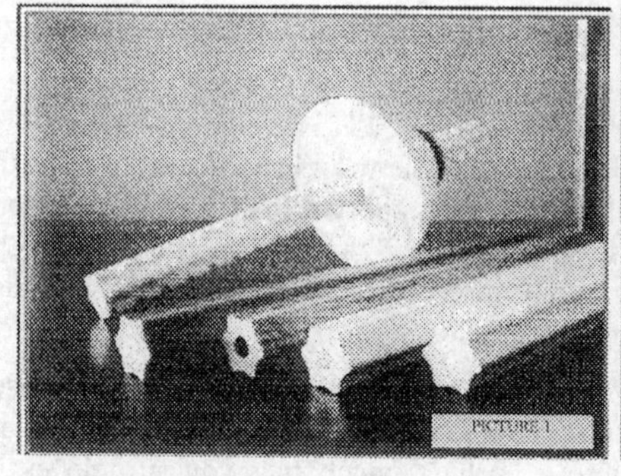

PICTURE 1

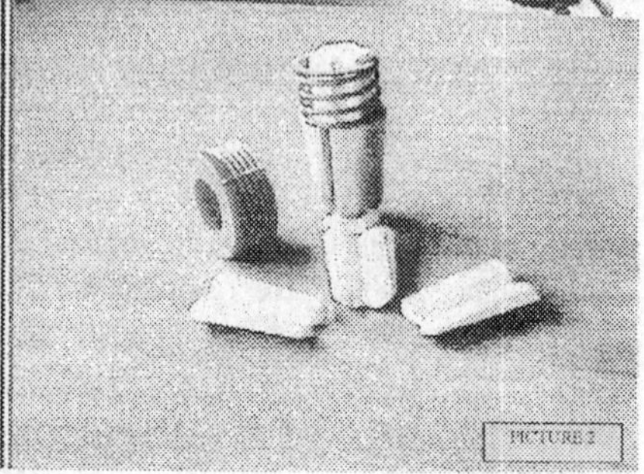

PICTURE 2

At this point we would like to illustrate the problems that we have coped with for the development of the head.

Marketing data indicated the necessity to develop a product with a contrast base(head) at least 100 mm. diameter; with maximum height 55 mm. (dictated by necessity to cover it with subsequent material application) and with load capacity up to 10.000 DN (at break) induced by a bar with a section equivalent to a 22 mm. diameter circular rod.

We thought that it would be very important to develop an easy to use system (head and cones) such as to favour a very easy application by simple sliding of the head over the rod up to the contrast wall. As it can be seen from picture (4), this was obtained thanks to a metallic spring held in place by a cap.

From a structural point of view, the design made by F.E.M. analysis resulted into the manufact shown in picture (3).

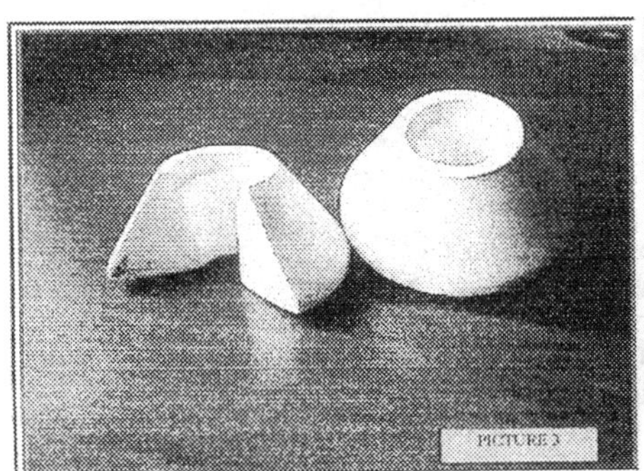

PICTURE 3

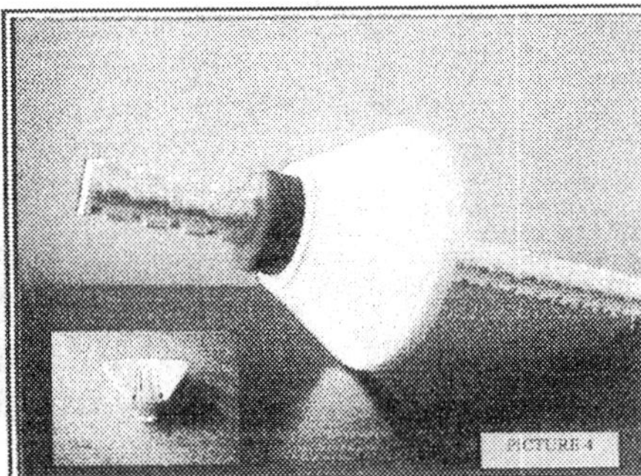

PICTURE 4

Calculations indicated that the problem could be solved by use of a composite with a polyester matrix and continuos glass roving "wound" around the "hole" with a winding angle "0" (zero), and with a reinforcement content at least 50%.

The reason why the orientation of the fibers should be applied by winding was that the stress inside the hole is similar to a very high internal pressure.

2. Development of production technology

Considering the above data it was taken into consideration the "Filament Winding" technology, but two types of problems became evident: first, difficulties in obtaining a good geometrical manufact, second, and worse, economical; as a matter fact the price of a part made with traditional filament winding would be much above the limits indicated by marketing, for the following reasons:

- High production equipment costs;
- Slow production cycles;
- Low flexibility in the choice of raw materials, with consequent limits in cost reduction.

ATP srl has developed a "mixed" production technology which we will try to explain with reference to figure 2.

If on a conical mandrel we would place, by winding, some continuous glass fibers (roving) taking care to tension the fibers in a way to obtain a cylindrical shape, and then we compress this "cylinder" in the direction of the axis by means of 2 parallel planes, the fibers will partially unwind and assume a shape indicated in the figure 2.

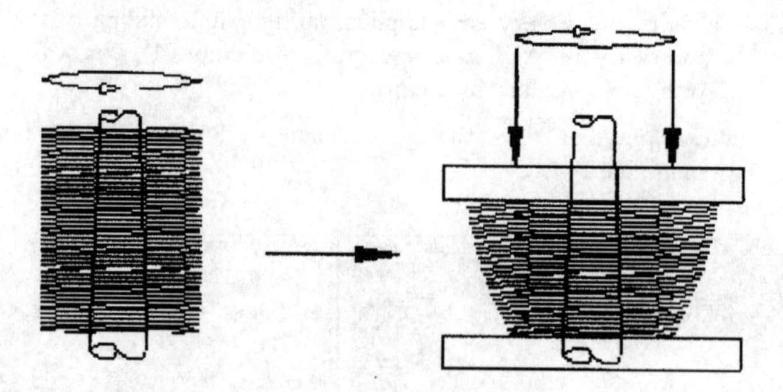

Figure 2 - Mixed production technology

Hence the possibility to produce the final manufact by compression moulding on a pre-form similar to the one shown in the above figure 2.

The complete production cycle developed is shown below:

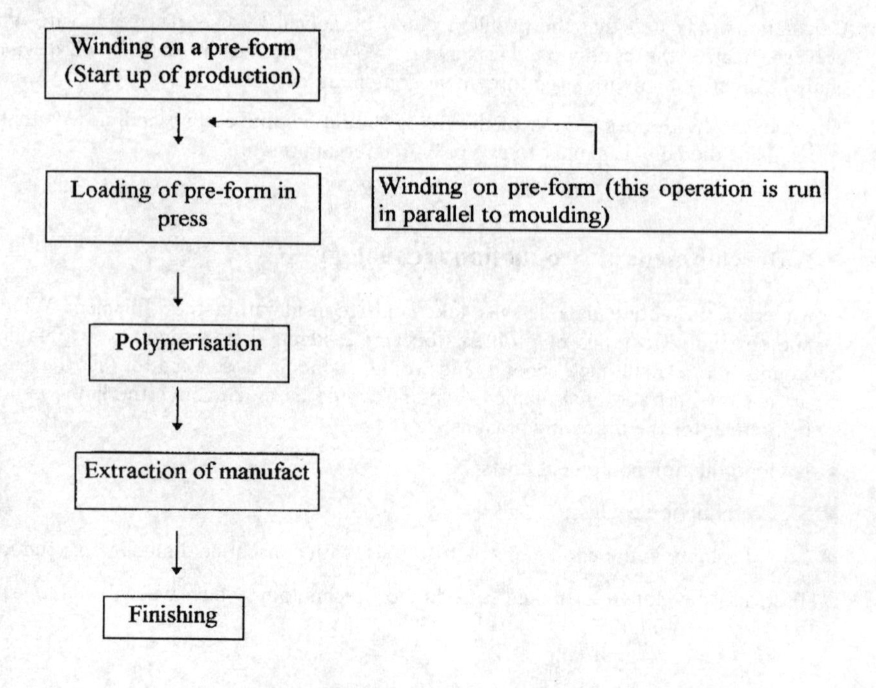

The following technical problems were dealt with:

- A method to obtain a pre-form without filament tensioning.
- Development of a resinous matrix which would have enough lubrication and would let an easy transport of the "WET" reinforcement and at the same time sufficient stability so that the pre-form would not loose its programmed geometrical form for several minutes.
- Design and detailed definition of the mould positioning cycles.
- The optimisation of the polymerisation cycle (the target was to obtain curing cycles below 5 minutes even in presence of large thickness variations).

Picture 5

The pre-forms are obtained by the equipment shown in picture 5.

Feeding of continuous glass fiber roving impregnated with resin is made with a speed which is synchronised, by a mechanical equipment, to the rotation speed of the mandrel. Cycle time for one pre-form is about 1 minute.

Picture 6

Two pre-forms are needed to feed the mold holding plattens as shown in picture 6.

Details of production cycle are shown in picture 7 - loading of pre-forms - and in picture 8 - demolding.

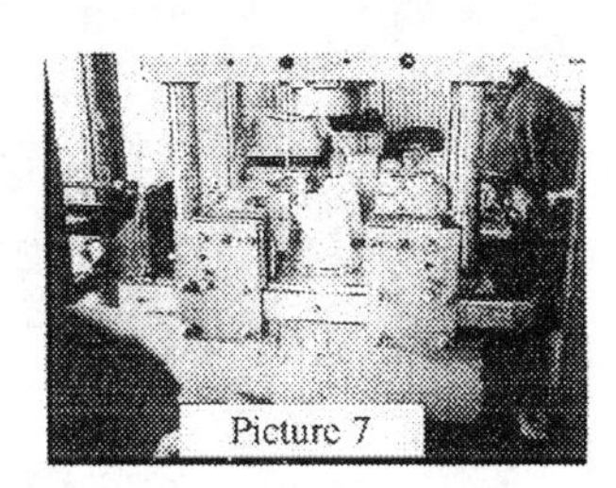
Picture 7

Picture 8

Actual productivity of the equipment is about 80.000 pieces per year.

References

L. Giamundo, A. Nanni , FRP ANCHORS: A BETTER WAY TO STABILIZE LOOSE SOILS OR ROCK - ICE '98 International Conference Exposition - Nashville (USA) February 1998.

Resin Flow Detection In RTM, Vacuum Assisted RTM And Other Resin Infusion Processes

David D. Shepard

Micromet Instruments, Inc., 7 Wells Avenue, Newton Centre, MA 02159-USA

Abstract

Monitoring of the resin flow during the part filling of resin infusion processes is critical because the flow is responsible for the final mechanical properties of the part. A new commercially available resin flow front analysis system is being introduced based on technology developed by the U. S. Army Research Laboratory. The system is used for the in-process monitoring of resin flow and cure in Resin Transfer Molding (RTM), SCRIMP, vacuum assisted RTM, and other resin infusion processes. The system consists of a sensor grid, an electronics package designed to rapidly interrogate the grid, and a Windows-based software program to control, record, and display the sensor data. The system measures the electrical properties at intersecting nodes of conductive wires or fibers that are manually laid out in the to form a grid pattern. When resin reaches each node, the electrical properties of that node will change and are recorded by the system. This provides a map of the part filling process and subsequent cure.

Introduction

Resin Transfer Molding (RTM) and other resin impregnation processes are widely used in the manufacturing of composite parts. RTM offers the advantages of relatively low tooling and processing equipment costs, short cycle times, and the ability to make complex shapes. In RTM and similar processes (vacuum assisted RTM, SCRIMP, etc.), a fiber preform is placed within a mold cavity or on a tool covered with a vacuum bag. Thermosetting resin is then injected into the mold where it saturates the preform and fills the mold. Once resin injection is complete, the resin system is designed to undergo a curing reaction to produce the finished part. The mold filling process, which can take a number of minutes, is critical to obtaining a good quality part. The resin must fully wet out the preform so that the part contains no voids or dry spots. Any voids present can result in a defect that can diminish the strength and quality of the cured part. Knowledge of how the resin fills the mold can aid in the optimization of the RTM process and result in faster process development and more consistent, high quality parts. The new **SMART*weave*™** sensing system can provide mold filling data by making in-process measurements at multiple locations within the mold to provide a map of the mold filling process.

The **SMART*weave*** system was developed and patented by the U.S. Army Research Lab for use in monitoring resin flow and cure in RTM and similar processes [1,2,3]. The U.S. Army has licensed the technology to Micromet Instruments, Inc. to develop a commercial **SMART*weave*** product.

Principles Of Measurement

The presence or absence of resin at any location within a mold can be detected by measurement of the dielectric properties of the medium between two electrodes at that location. If no resin is present between the electrodes, the conductance will be very low and thus the measurement can identify the absence of resin at that location. If resin is present, the higher conductance will indicate the presence of resin. If multiple sets of electrodes are located throughout the part and the conductance is repeatedly measured throughout the mold filling process, then a map can be constructed of the flow of the resin into the mold. During each cycle of measurements of the sets of electrodes, a "snapshot" is obtained of the progress of resin in filling the mold. Viewing the "snapshots" sequentially provides a "movie" of the mold filling process.

Dielectric measurements utilizing parallel plate or interdigitated comb electrodes are commonly used in the in-process cure monitoring of thermosetting resins. However, the use of these sensors to monitor resin flow fronts is impractical due to the number of sensors required to produce detailed information. Utilizing multiple conventional dielectric sensors involves a relatively high cost and requires connecting at least two electrical leads for each sensor. Furthermore, large numbers of conventional dielectric sensors placed in a part can affect the structural integrity of the part and could also disrupt the flow behavior. To avoid these problems, a low cost sensor grid compatible with the preform is used which provides multiple sensing locations with relatively few leads.

Sensing System

The **SMART*weave*** system consists of a sensing grid, an electronics package designed to rapidly interrogate the grid, and a computer running a LabVIEW software program (Figure 1). The sensor grid consists of electrically conductive filaments that cross on non-intersecting planes which results in a "sensing gap" between the filaments (Figure 2). The filaments are maintained on different planes by placing them on opposite sides of layers of the preform or by weaving the filaments into the preform. The system measures the conductance of the material within each sensing gap to determine the presence or absence of resin or the state of the resin. To enable rapid sampling of multiple sensing locations, a DC voltage measurement is made across each sensing gap.

Figure 3 shows a schematic of an 8 by 8 sensor grid. Each filament is connected to either the excitation (X) or response circuit (Y) of the electronics package. Each junction where an X and Y sensing filament cross constitutes one sensing node. Thus an 8 by 8 grid of X and Y filaments enables 64 locations within the mold to be measured with only 16 leads to be connected. A 16 by 16 grid enables 256 locations to be measured with only 32 connections. Figure 4 illustrates how, when the number of X filaments equals the number of Y filaments, the number of sensing locations increases as the product of the number of X and Y filaments. The geometry of the grid can be configured by the user to optimize the location of the sensing nodes.

The sensing filaments can be made out of any conductive filament and are generally low in cost. Commonly used filaments are a metal clad Aramid fiber and graphite tow fibers because they are similar in nature to preform materials. The Aramid fibers are convenient because they are very thin (about 0.15 mm in diameter). Other potential sensing filament materials include bare metal wires, fiberglass insulated thermocouple wire, or copper tape. When the preform is composed of graphite fibers, fiberglass cloth layers can be used to electrically insulate the sensor filaments from the graphite fibers. Alternatively, fiberglass insulated wires such as thermocouple wires, can be used with graphite preforms.

The sensing filaments are laid up so that one end of each filament extends out of the mold or vacuum bag. The filaments can easily be electrically insulated from metal mold or tool surfaces with tape or other insulating materials. The sensor filaments are connected to the electronics package via extension cables. Each extension cable contains eight individual wires which terminate in spring loaded clamps for quick connection to the sensor filaments. Up to 4 X and 4 Y extension cables can be connected to the electronics package for a maximum of 32 X and 32 Y filaments (1024 total sensing nodes).

Data Display

A LabVIEW software package is used to set up measurement parameters, define the configuration of the sensing grid, and display the data. During data collection, the progression of resin flow in the mold is displayed in real time through a depiction of the sensing grid. Different shades of a color can be displayed in proportion to the measured signal (voltage) level or a black and white display can be used to simply indicate the presence or absence of resin. Use of a shaded display enables the onset of resin gel to be monitored through the voltage decrease as the resin begins to cure. After the experiment is complete, the data can be replayed at varying speeds and paused at any time for detailed analysis and output to a printer.

A black and white scale is very effective is studying the mold filling process. Figures 5a, 5b, and 5c shows how the software depicts the filling of a vacuum-assisted RTM part at three discrete times during resin injection. An 8 x 8 sensor grid was laid up in the part and resin introduced in the lower right portion of the part. The white areas of the grid represent where the system has detected the presence of resin; the black areas represent regions where resin has yet to reach that location. The data shows that the resin tends to fill the lower portion of the part first before filling the upper portions.

Figures 6a, 6b, and 6c use shades of gray to show the gelling and onset of cure after the mold has been filled. In this experiment, a 6 x 6 sensor grid has been used and resin flow during mold filling was from left to right. The lighter areas represent higher voltages (and therefore lower degree of cure) while a totally black area means the resin has cured to a point where the system is no longer sensitive to additional cure. Figure 6a shows that gelling occurred first at the right side of the part (the last portion of the mold to fill). The resin at this location was injected into the mold first and required some time to travel to the far end of the part. Thus, this resin is the "oldest" and would start to react before "fresher" material that was injected at a later time. As time progresses (Figures 6b and 6c) gel and cure proceed throughout the remainder of the part. The sensitivity of the measurement to the cure state of the part will vary depending on the electrical characteristics of the resin, the diameter of the sensor grid filaments, the size of the sensing gap, and the sampling rate of the system. Future work will study the sensitivity of the system to the end of cure as compared to AC resistance measurements using interdigitated comb electrodes.

Other applications

In addition to resin flow front monitoring, dry spot detection, and resin gel and onset of cure information, the **SMART*weave*** technology has potential uses in other applications. Since the sensor grid is permanently cured into the structure, there is the opportunity for monitoring the health of the structure over its lifetime. Potential applications include gross damage detection and moisture absorption in composite structures.

Conclusion

SMART*weave* provides a new sensing technology for monitoring the resin flow and cure onset in RTM and other resin impregnation processes. The innovative use of the sensing grid allows multiple locations within a part to be conveniently monitored while maintaining low sensor costs. The information provided by the system can help RTM processes to be developed and optimized more efficiently and offers the promise of post molding inspection.

U4244 pl 2-2-98

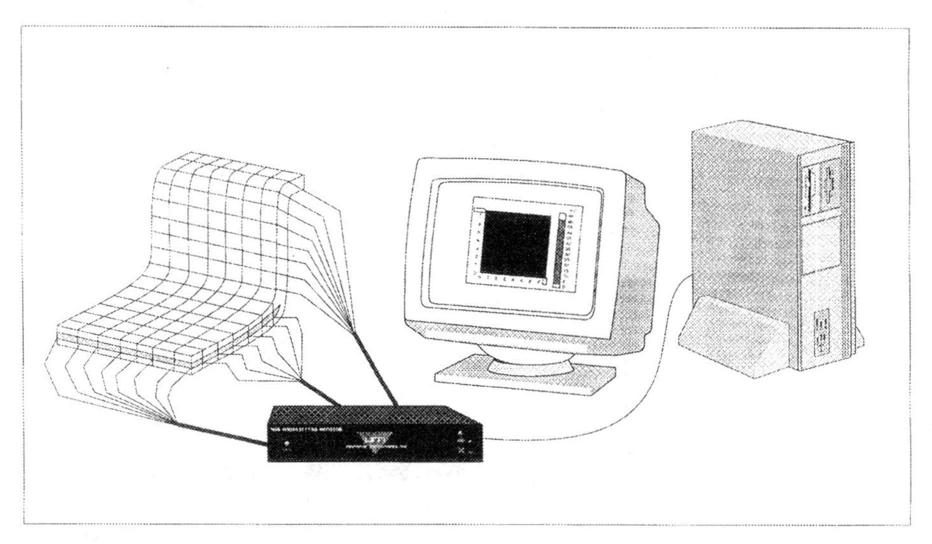

Figure 1. Schematic Diagram of **SMART*weave***™ System.

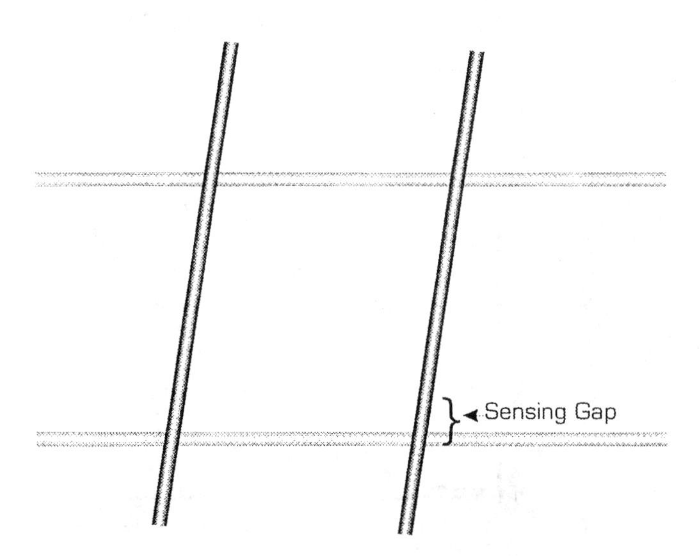

Figure 2. The sensing gap is the area between two intersecting filaments of the sensing grid. The system measures the conduction of the material within each sensing gap.

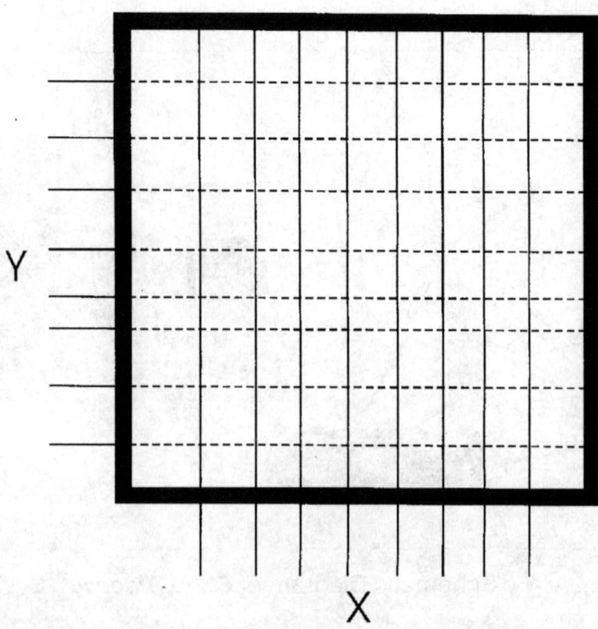

Figure 3. A **SMART*weave***™ grid; each intersection of an X and Y line represents a sensing node.

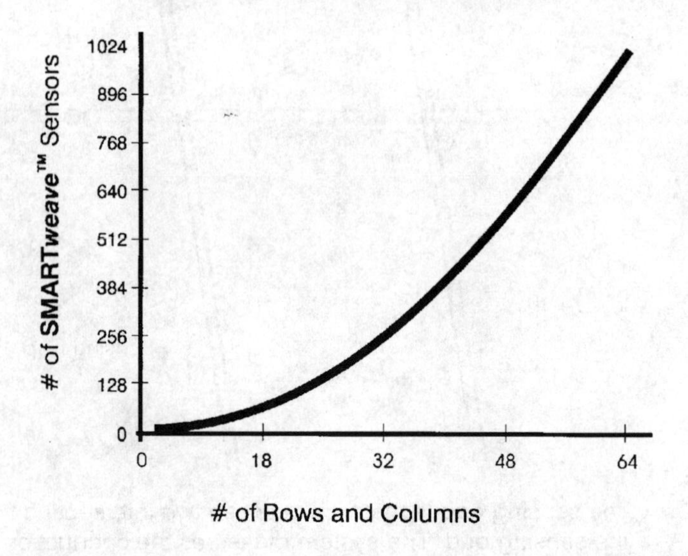

Figure 4. Calculation for number of sensing locations

04244 p3 2-2-98

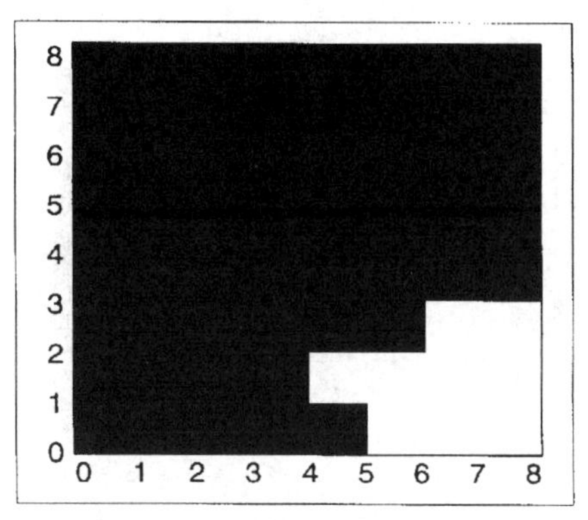

Figure 5a. Progression of resin flow into vacuum assisted RTM mold after 4 minutes

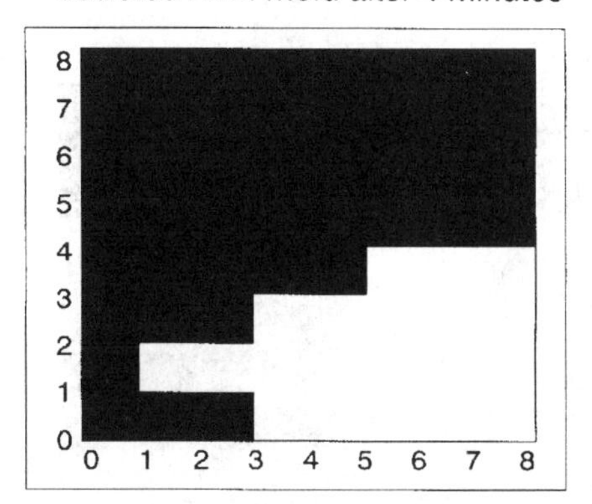

Figure 5b. Progression of resin flow into vacuum assisted RTM mold afte 13 minutes

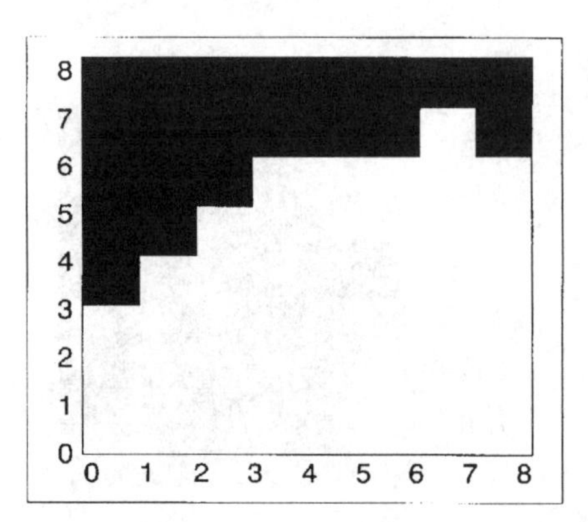

Figure 5c. Progression of resin flow into vacuum assisted RTM mold after 26 minutes

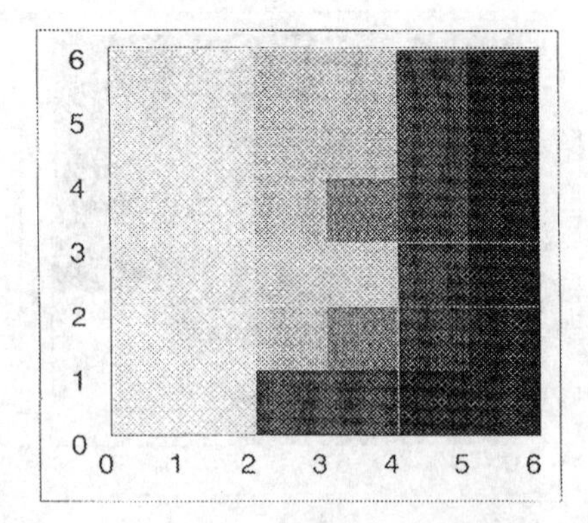

Figure 6a. Onset of resin gel and cure after 74 minutes

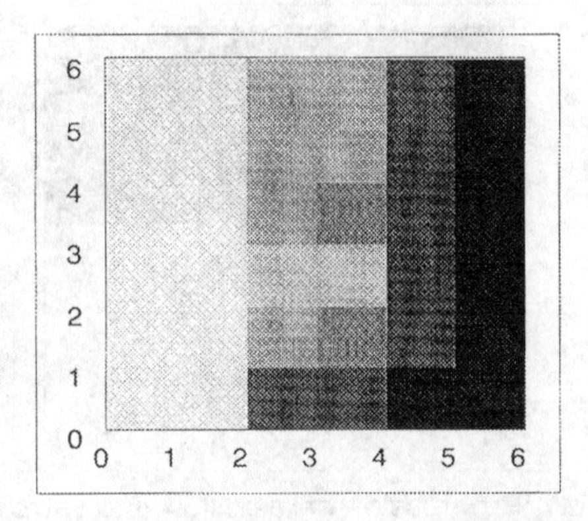

Figure 6b. Onset of resin gel and cure after 98 minutes

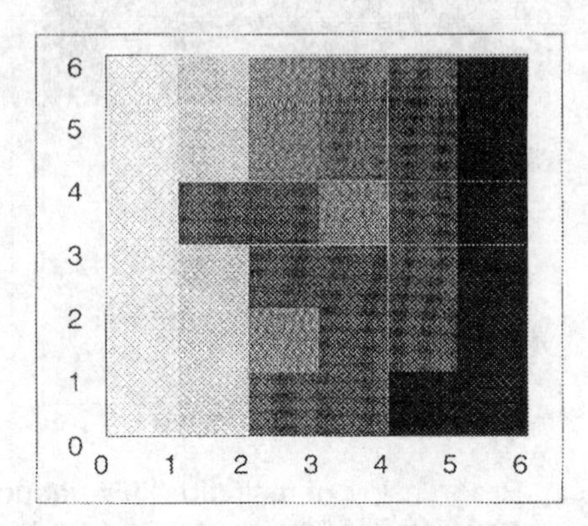

Figure 6c. Onset of resin gel and cure after 101 minutes

References

[1] S. *Walsh,* U.S. PATENT NUMBER 5,210,499 - Assignor to U.S. Government - May, 1993.

[2] S. *Walsh, C. Freese,* RESIN FLOW FRONT RECONSTRUCTION WITH APPLICATIONS TO RESIN TRANSFER MOLDING - 41st International SAMPE Symposium - Anaheim, CA USA -March, 1996.

[3] S. *Walsh, D. DeSchepper.* SMARTWEAVE: A SENSOR SYSTEM FOR RESIN TRANSFER MOLDING - Proceedings of the 41ST Sagamore Conference - 1994

Composite materials using RTM with ε-Caprolactam

D. Gittel*, B.Möginger*; P.Eyerer*

* Institut für Kunststoffprüfung und Kunststoffkunde, Universität Stuttgart, Pfaffenwaldring 32, D-70569 Stuttgart, Germany

Abstract

This presentation illustrates the most recent results obtained first from a study and later from projects with industries done at the Institute for Polymer Testing and Polymer Science in Stuttgart since more than two years. Objective our works are to investigate a new way to manufacture thermoplastic composite structure parts by utilization of anionic polymeri-zation by ε–caprolactam in presence of different kinds of fibers and fiber structures in a mould cavity. The chemical compatibility between fibers and the reactive monomer system was investigated and optimized. The technology must be adapted at the chemical reactivation and the kind of fiber in order to obtain finished structure parts with high mechanical properties and excellent surfaces. Furthermore it should be possible to combinate different fibers and fiber structures to achieve lightweight parts with partial reinforcements only in nessesary areas. Thereby is a cost-saving und a weight-reduction feasible.

1. Introduction

Castable polyamide (ε-caprolactam) is used in RIM technology to produce lightweight stiff parts. The textile structure can consist of knitted fabrics also with fiber alignment, fabric, non-woven or a combination of these structures (e.g. sandwich structure). If stitch bonding technique is applied this will result in a composite structure with excellent delamination resistance. The mentioned textile structures allow the combination respectively variation of different fibers (e.g. glass, polymer, carbon) in regions with different mechanical loading. The reinforcement can be partial in that regions and the number of fibers as well as the fiber direction can be adjusted depending on the external load. A very stiff and at the same time impact resistant thermoplastic composite can be realised using a sandwich structure and perhaps the Malimo stitch bonding technique.

Cut or preformed textile reinforcement structures are placed in a casting mould which then is closed. Modified RTM-technique is used to infiltrate the textile structure with the low viscous reactive monomer caprolactam. Due to the activator, catalyst and temperature the melt polymerizes to PA 6.

Tab. 1 is listing mechanical properties achieved with carbon and glass fiber reinforced composites.

properties	unit	results				
fibers		CF	CF	CF	CF	GF
fiber volume content	vol.-%	37,4	53,3	72,1	54,1	59,7
number of fiber layers		4	6	8	2	9
tensile test: modulus	GPa	34	54	68	25,8	26
tensile strength	MPa	430	530	740	380	390
elongation at break	%	1,3	1,1	1,1	3,1	1,8
flexural test: modulus	GPa	29	48	57	49	25
flexural strength	MPa	350	380	420	520	350
elongation at break	%	1,8	1,1	1,0	1,1	1,9
interlamin.strength	MPa	-	36	34	50	31
impact penetration test:	J	-	-	8	-	14

CF = carbon fibers; GF = glass fibers

Table 1: Mechanical properties of different composites

2. Procedure of processing

The final part is manufactured in one step using a modified RIM (reaction injection moulding) / RTM (resin transfer moulding) - compression moulding technique with a castable polyamide (caprolactam).

Costs will be low as an effective manufacturing technique with cheap base material is used. Using different reinforcement materials and techniques allow to adapt the structure of the composite to the applied loads.

From the point of ecology the material is a recyclable thermoplastic polymer. The manufacturing has low energy consumption and small amounts of waste. As light weight parts are manufactured there is an energy conservation during life cycle.

For vehicles in transportation and sometimes machine construction leigth weight construction is always a demand. High-performance supporting structures are more and more made of fiber reinforced polymers. These materials are not yet used for mass production (e.g. in transportation) due to the high material costs and long processing time resulting in high manufacturing costs. Today used technology is thermoforming of cut flat laminates which were during a former working cycle preimpregnated and consolidated. The cycle times of thermoforming are already quite short (< 2 min) but working- and investment-expenditure are high. The price of the semi-finished material is higher than industry can accept.

The objective of the current work is to develop a manufacturing technique for supporting structures made from thermoplastic filament reinforced material having partial reinforcement, short cycle times and high properties while costs are low which will make the process and material suitable for a mass production.

The costs for raw materials and processing could be half compared to thermo forming of preconsolidated prepregs. Reducing the production costs is possible by the minimizing the costly production steps.

Fig. 1 illustrate the procedure of producing composite materials using RTM.

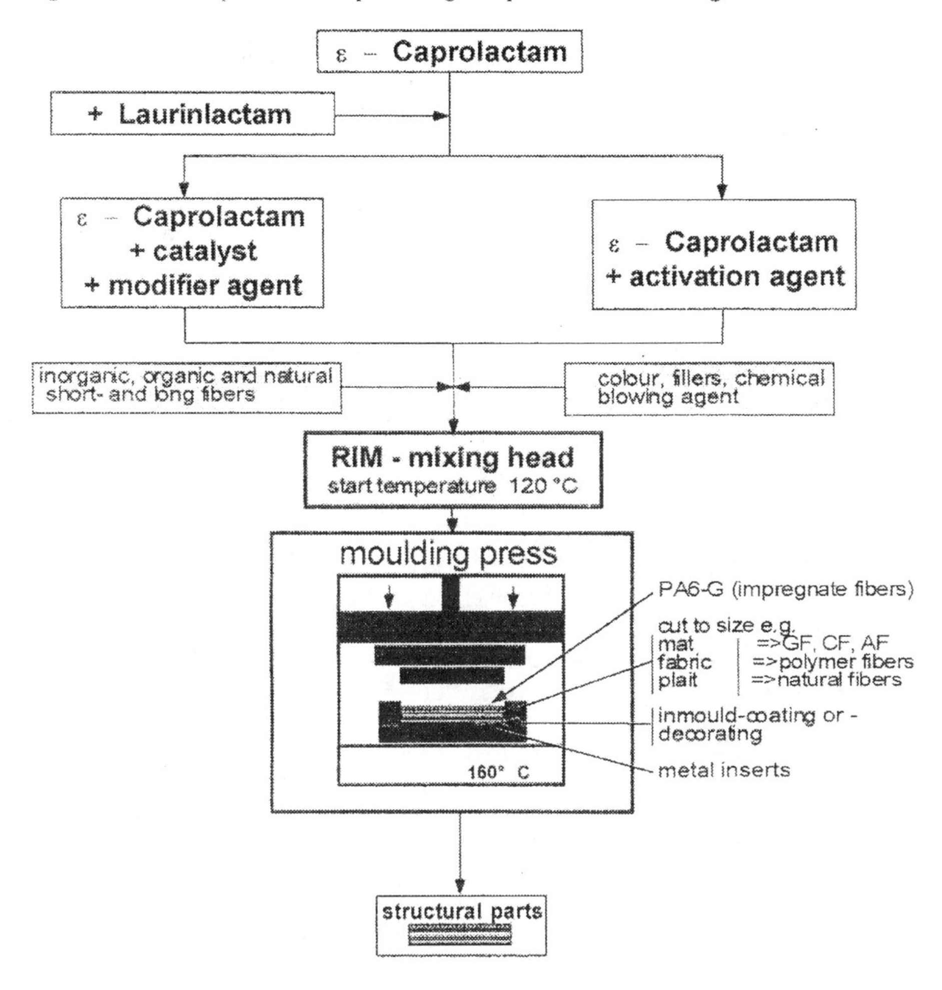

Fig. 1: Procedure of processing

A high fibre volume fraction is possible by compression in the mould cavity.

Different regions of the part are planned to be reinforced with reinforcement structures adapted to the occurring loading. Regions with less requirement concerning mechanical properties can be reinforced using cheap or even no reinforcement. The wall thickness of the part can be very different all over the part.

Mechanical properties and fibre fraction in the loaded regions of the produced parts should be like those of parts made by costly thermoforming while the weight should be lower due to partial reinforcement.

References

/1/ H. Mooij: „SRIM Nylon Composites"; Advanced Materials: Cost Effectiveness, Quality Control, Health and Environment; SAMPE / Elsvier Science Publishers B.V., 1991

/2/ J.D.Gabbert, R.M.Hedrick: „Fortschritte bei PA-Blockcopolymeren für die RIM-Verarbeitung"; Kunststoffe 75 (1985) 7, S. 416-420

/3/ J.E.Kresta J.Gabbert u.a: „Reaction Injection Molding-Nylon 6 RIM"; Polymer Chemistry and Engineering ACS, Symp. Series 270; Washington D.C.; 1985; S. 135-161

/4/ P.D.Coates u.a: „Organic fibre reinforced polymer matrix composites produced by SRIM"; ANTEC `94, Tagungsband, S. 1659 - 1663

/5/ Nelson(Dow): "Reinforcing structural foam RIM polymers with continuous fiber glass mat"; Proc. of SPI Ann.Techn. /Market. Conf., 29th.; SPI; NY, S. 146-153

/6/ Karger-Cocsis: „Fracture behaviour of glass-fibre mat-reinforced structure nylon techniques"; J. of Mat. Sci. (1993) 28 (9), S.2428 -2438

/7/ D.Gittel u.a.: „Fiber Reinforced Casting Polyamid", 15. Stuttgarter Kunststoff-Kolloquium; März 1997

/8/ D.Gittel: „Gußpolyamid mit Faserverstärkungen"; Poster Symposium-„Mit Kunststoffen zu neuen Produkten" Karlsruhe Mai 1997

Experimental investigation of the vacuum infusion moulding process

A. Hammami[1] and B.R.Gebart[2,†]

(1) Swedish Institute of Composites, P.O. Box 271, SE-941 26 Piteå, SWEDEN
(2) Associate professor, Division of Fluids Mechanics, Luleå University of Technology, SE- 971 87, SWEDEN

Abstract

The vacuum infusion moulding process is attracting a lot of attention due to its low emissions of volatile organic compounds and its potential for high repeatability. The process is similar to the RTM technique. However in the vacuum infusion technique, the stiff cover is replaced by a polymeric film, often called vacuum bag. The film is sealed and air is evacuated from the resulting mould cavity which will create a compaction of the reinforcement by the atmospheric pressure applied on the bag. Finally, resin is allowed to enter the mould cavity through a hole either in the stiff mould or in the vacuum bag.
In this paper, the results from a number of infusion experiments are presented. This makes it possible to identify, first, the main processing parameters, and then, to assess possible interactions that take place during the process. The parameters investigated in this paper are thought to be those of most interest for the process (compaction of the reinforcement, resin temperature dependency and infusion strategies).

1. Introduction

The composites industry is faced with some very difficult challenges today. Environmental concerns led to increased health and safety regulations to protect workers from exposure to volatile organic compounds (VOC). Furthermore and until recently, to produce large composite structures, hand lay-up and spray-up technologies were considered as the only cost effective available techniques [1]. However, these two techniques make extensive use of Unsaturated Polyester (UP), which contains typically between 30% and 45% of styrene as a reactive solvent. The latter has been identified by legislators as the main harmful substance to eliminate to reduce VOC emissions.
Among the existing alternatives, closed mould processes such as RTM attract considerable interest. However for large structures, the RTM technology is no longer cost effective since it involves high equipment costs, especially for building complex and heavy moulds to withstand the injection pressure. For manufacturing situations involving low volume production and large structures, injection strategies have been developed that uses the

† To whom correspondence should be addressed.

vacuum pressure as the only driving force to achieve impregnation of the reinforcement inside the mould cavity. As mentioned by Williams et al.[2], early interest for these techniques goes back to the 40's when the so-called Marco method was introduced. Over the past years, many developments have helped vacuum infusion processes become economically viable and various methods are available on the market today. Among them, one can mention the SCRIMP method, the Quick Draw VARTM method, the Resin Injection Recirculation Method (RIRM) and the Prestovac Method [1]. Compared to open mould technologies, vacuum infusion has obvious safety and health advantages through the reduction of styrene emission. Another advantage of the process is its low tooling costs and the improved mechanical properties of the produced parts [3].

Experimental observations have shown that the impregnation of the preform introduces a change in the pressure difference over the vacuum bag and hence the compressive force on the preform. As a result, the thickness of the cavity will change with time. Then, in order to derive the governing equations, we must understand the interaction between the compressibility of the preform and the resin inflow mechanisms during filling. A theoretical analysis of the process that agrees quite well with experiments was presented in [4].

In this paper, the main processing parameters involved in the vacuum infusion moulding process are identified. Then a number of infusion experiments were performed to investigate to what extent these parameters are affecting the process. The parameters investigated in this paper are thought to be those of most importance for the process (compaction of the reinforcement, resin viscosity and infusion strategies).

2. Experiments

2.1 Experimental set-up

The experimental set-up shown in Figure 1 is composed of a metallic plate on which 5 layers of Ahlström multi-axial +/-45°/0° were stacked. The reinforcement was 200 mm wide and 500 mm in length and oriented in the 0° direction. This reinforcement is a non-crimp stitch bonded fabric with a surface weight of 819.5 g/m^2. Near the inlet three layers of 450 g/m^2 Unifilo continuous random strand mat were placed to help distribute the flow laterally and to allow for a linear progression of the flow front inside the mould cavity. A transparent polymeric film (Polyamide 6) is sealed against the plate and air is evacuated from the formed cavity. The vacuum pressure acts as the only driving force for the resin to complete the filling of the mould cavity. The resin employed is an unsaturated polyester resin (S 910-4138, from NESTE Chemicals) especially developed for vacuum infusion moulding. Another reinforcement, Devold 400-E01, was also used. It consists of two layers oriented in the +45° and -45° directions respectively and stitched together with a polypropylene yarn, the fabric has a surface weight of 419.5 g/m^2.

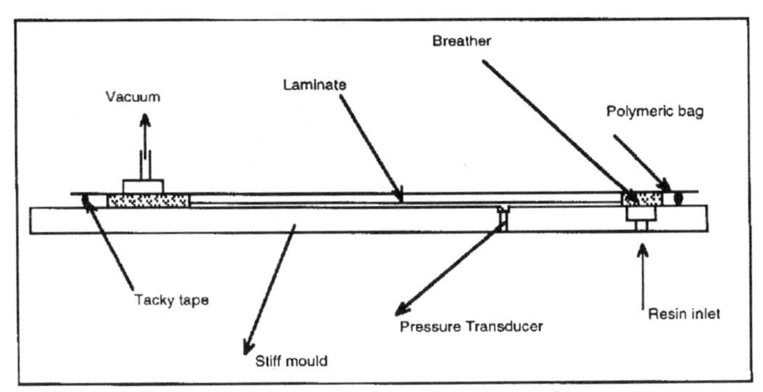

Figure 1. *Schematic of the experimental set-up.*

2.2 The reinforcement compaction

The compaction behaviour of a stack of continuous reinforcements, under compression forces applied normal to their planes has been extensively studied during the last decade. While some studies have been oriented toward the characterisation of a number of reinforcements under fixed conditions, in other cases the effect of a change in some selected parameters was investigated for a small number of reinforcements. A comprehensive review of the available compaction models is presented by Robitaille and Gauvin [5]. Toll [6] has recently suggested a compaction model that relates the compaction response of the reinforcement to the fibre bundle structure. The overall agreement with experiment is relatively good but the agreement for low compaction is not so good. This remark can be extended also to the model developed by Gutowski [7]. A plot of the different models is presented in Figure 2, one can see that for low applied compaction pressure, no acceptable agreement could be found between the models and the experimental data. The existing models over-estimate the fibre volume fraction for a given compaction pressure, even though, Gutowski's model is much closer to the experimental data. Actually, during vacuum infusion, the absolute pressure never exceeds 1 bar. Then according to the experimental results, in this region, even a small increase in the pressure introduces large changes in the fibre volume fraction. According to Gebart and Lidström [8], a change in the fibre volume fraction of 1% will result in a 10% variation in the permeability value of the preform. As a consequence, the pressure-dependency of the fibre volume fraction add to the complexity of deriving the governing equations, since the fibre volume fraction (V_f) can no longer be assumed constant.

For practical reasons, it is common that the mould cavity consists of a stack of 2 or 3 types of reinforcement. This is done either to enhance resin impregnation or to add some stiffness to the part. Such lay-up when compacted can exhibit different flow characteristics, even if each reinforcement has individually the same flow properties. Two fabrics were selected (Devold and Ahlström) that have the same permeabilities at 50% fibre volume fraction [9] and infusion experiment was performed using these two fabrics.

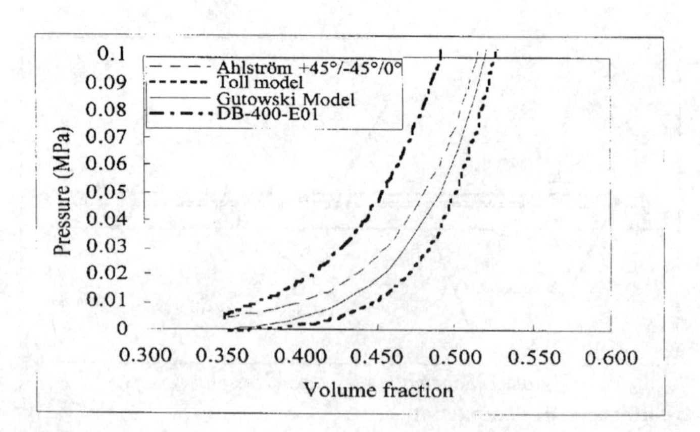

Figure 2. *Performance of analytical models at low compaction pressure.*

When the vacuum pressure is applied, one can expect a fibre volume fraction of 48.9 % for the Devold and slightly more than 51 % for the Ahlström according to the compaction data. The fabrics were laid side by side and were impregnated simultaneously with the same resin. Then, both fabrics are subject to the same operating conditions. A plot of the flow front profiles obtained with two fabrics is shown in Figure 3. Based on the compaction results, the impregnation should be faster with Devold, although preliminary experiment shows that resin impregnation is much faster inside the Alhström cavity. This delay between the two flow front profiles can be the result of a different compaction behaviour (see Fig. 2). Obviously Devold has a much higher slope than Ahlström. At this stage, it is not possible to make any final decision to explain this discrepancy. Additional experiments should be performed to eliminate any side effects such as a surplus of sizing agents that are dissolved by the resin and which may result in a slower impregnation.

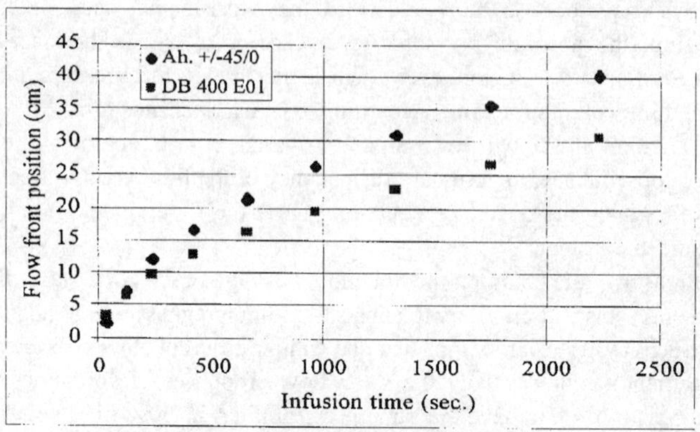

Figure 3. *Flow front positions recorded for two different fabrics.*

2.3 Infusion strategies

A major concern for vacuum infusion, especially when it comes to infuse large structures, is how to complete the infusion of the mould cavity within the shortest cycle time. Furthermore, depending on the infusion strategy involved, additional production costs can be expected. Two infusion strategies will be compared : (1) point infusion and (2) line infusion. Each of the aforementioned techniques is associated with some advantages and drawbacks. For instance, with point infusion, leakage occurrence and its effect on the filling time and the final part is minimised, since air is always driven from inside the mould cavity. In addition, no trimming operation is required, thus allowing the production of net shape parts. However, this method is always associated with longer infusion times. In case of line or edge infusion, one can expect to complete the filling stage in a shorter time. But avoidance of leakage is not an easy task, and prediction of the outlet is not straightforward. In addition, trimming of the finished part adds extra cost to the product.
Gebart *et* al. [10] have evaluated different injection strategies for resin transfer moulding (RTM). According to their results, line/edge injection was up to 10 times faster than point injection. The flow front position achieved through the two infusion methods are shown in Figure 4. Obviously line infusion is faster than point infusion, especially when using a coil wire holder as a distribution channel. The channel has an inner diameter of 4 mm and an outer diameter of 6 mm. As an example, to infuse a distance of 20 cm, only 1500 s is needed with line infusion compared to more than 3000s for point infusion. Actually, the area covered by line inlet for a given time depends on the length of the distribution channel (runner) but can be made much larger than for a point inlet. In a previous work [11], we define a new parameter called practical infusion length to achieve the infusion before resin gelation occurs. According to the results shown in Figure 4, this parameter can also be affected by the infusion strategy, i.e., one has to expect a smaller practical infusion length when using point infusion.

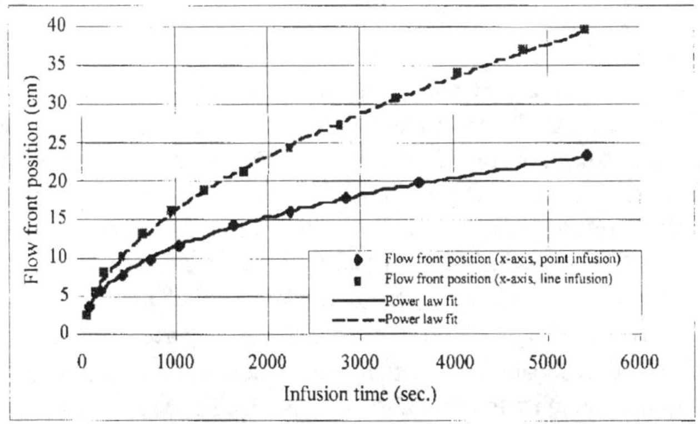

Figure 4. *Experimental Flow front position achieved through point infusion and edge infusion.*

2.4 Temperature effects

Typical parts produced with vacuum infusion are large structures such as boat hulls. These structures are normally infused at room temperature, and it is common that the infusion time exceeds 1 hour. Accordingly, one can expect that a temperature gradient may be established during infusion. The build-up of this gradient can be the consequence of a sudden change in the ventilation inside the production area. To elucidate the possible effect of this gradient on the infusion time, two infusion experiments were performed with different ambient temperature within the laboratory. It appears clearly that resin temperature affects significantly the filling of the mould cavity as shown in Figure 5 which depicts the flow front profile obtained in two successive infusion experiments. A difference of 2 degrees in the laboratory ambient temperature results in a delay of 15 min between the two flow fronts. This delay will increase for large structures and may result in an incomplete filling if resin gellation occurs. Then when preparing resin formulations, one has to take into consideration such perturbations.

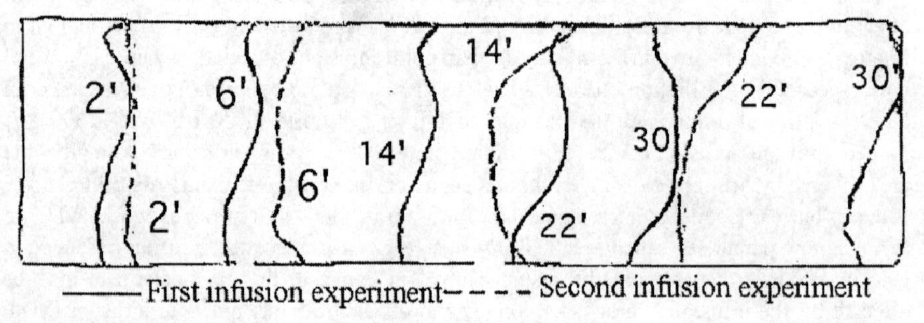

Figure 5. *Influence of resin temperature on preform impregnation ($\Delta T = 2°C$).*

2.5 Preform relaxation

In this experiment, a measurement equipment normally dedicated to compute the weaviness of craft paper was used (ASSI Tech Centre Laboratory). The measurement device consists of a laser transducer fixed to a special fixture. Two DC motors with four different speeds are used to allow for the movement of the laser transducer in the x and y directions. The lowest speed that corresponds to 25 mm/s, was used during the experiment. Scanning takes place always along the y direction. The surface to be scanned is considered as a grid having a number of lines ranging from 4 to 10. A square sample of 200 mm by 200 mm, consisting of 4 plies of Ahlström ± 45/0, was used. The infusion takes place in the 90° direction which represents the lowest permeability direction. Even though the lowest speed was selected, the scanning speed was higher than the infusion speed. To overcome this problem and be able to trace the effect of impregnation on thickness variation, surface scanning was repeated four times. Please keep in mind that the first scan as shown in Figure 6 is performed on the dry surface, thus it will act as a reference for the successive runs. Furthermore, one can consider the top line a snap shot of the dry fabric and the consecutive lines as snapshots taken at later times. It is clear, as illustrated in Figure 6 that preform impregnation causes a thickness change of the preform. As one may expect, the largest variation was recorded near the inlet. Then as the liquid impregnates the preform, this variation decreases until reaching a constant level. This level is certainly related to the critical infusion length and the permeability of the preform.

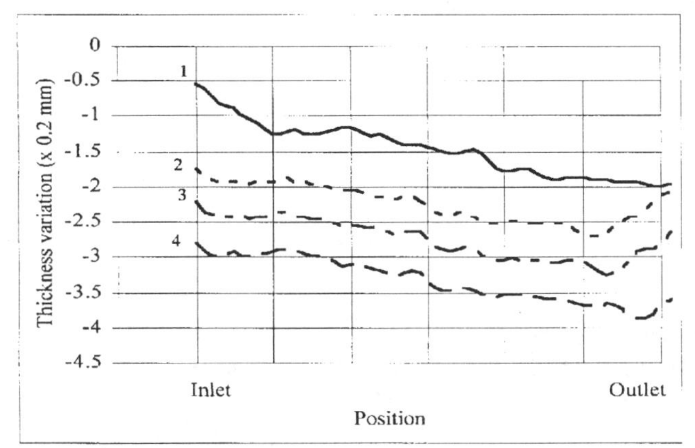

Figure 6. *Preform relaxation during resin impregnation. The lines correspond to $t = t_0 +$ (20,40 and 60s).*

3. Discussion and conclusions

Obviously, the parameters investigated are closely associated with the process and contribute to a large amount in achieving successful moulding situations. Since process knowledge has been developed mainly using trial and error, no real understanding exists among the users of the mechanisms involved in the process. Actually, the work presented in this paper is fundamental to gain a deep knowledge of the process in order to derive the proper flow model. Furthermore, it offers new learning for some practical moulding situations. For instance, infusion experiments performed have shown that existing compactions models are not successful in describing the compaction behaviour of the reinforcement. This is due mainly to the fact that these models under estimate the compressive response of the reinforcement at low pressures. According to preliminary results that are subject to further investigation, it appears that reinforcement stiffness also affects the time dependent compressive response of the reinforcement. A stiffer reinforcement is less sensitive to relaxation that takes place during resin impregnation. Practically, a judicious choice of a stacking sequence will contribute to reduce the thickness variation normally observed in infused parts. Another important area is to investigate how the choice of reinforcements and stacking sequence affects the surface the surface finish and

print through effects. Regarding infusion strategies, a comparison was made of point and edge infusion. Edge infusion is two times faster, especially when using a coil wire holder as a distribution channel. Nevertheless improvements can be made regarding two aspects: (1) leakage control, this is an important issue for large structures, and (2) production of net shape parts. Point infusion can be an alternative for small parts and using semi-rigid tools.
In conclusion, the work presented here is a first step toward a complete definition of the process. Further research is needed to fully understand all the mechanisms involved. For instance, an extensive research effort has to be dedicated to the compressive response of the reinforcements in saturated media. This can consolidate developed flow models [4] and contribute in defining appropriate simulations tools for the process.

References

[1] *P. Lazarus.* INFUSION- Professional Boat Builder, Oct./Nov. 1994, No. 31..

[2] *C D Williams, J Summerscales and S M Grove.* THE RESIN INFUSION UNDER FLEXIBLE TOOLING (RIFT): A REVIEW- Composites Manufacturing, 1996 Vol. 27 A, 517-524.

[3] *K F Karlsson and B T Åström.* MANUFACTURING AND APPLICATIONS OF STRUCTURAL SANDWICH COMPONENTS- Composites Manufacturing, 1997 28 A, 97-111.

[4] *A. Hammami and B.R. Gebart.* A MODEL FOR THE VACUUM INFUSION MOULDING PROCESS- Accepted for publication in FRC'98, New Castle, U.K. 1998.

[5] *F. Robitaille and R. Gauvin.* COMPACTION OF FIBER REINFORCEMENTS FOR COMPOSITE MANUFACTURING : A REVIEW OF ANALYTICAL RESULTS- Accepted for publication in Polymer Composites, 1997.

[6] *S Toll.* PACKING MECHANICS OF FIBRE REINFORCEMENTS- submitted to Polym. Eng. Sci, 1997.

[7] *T.G. Gutowski, T. Morigaki and Z. Cai.* THE CONSOLIDATION OF LAMINATE COMPOSITES- Journal of Composite Materials, 1987 Vol. 21, No. 2, 172-188.

[8] *B.R. Gebart and P. Lidström.* MEASUREMENT OF IN-PLANE PERMEABILITY OF ANISOTROPIC FIBER REINFORCEMENTS- Polymer Composites, 1996, Vol. 17, No.1, 43-51.

[9] *S.T. Lundström, B.R. Gebart and E. Sandlunnd.* IN-PLANE PERMEABILITY MEASUREMENTS ON FIBRE REINFORCEMENTS BY THE PARALLEL FLOW TECHNIQUE- Accepted for publication in FRC'98, New Castle, U.K. 1998.

[10] *B.R.Gebart, P.Gudmundson and C.Y.Lundemo.* AN EVALUATION OF ALTERNATIVE INJECTION STRATEGIES IN RTM- 47th Annual Conference, Composites Institute, Cincinnati, U.S.A 1992.

[11] *A. Hammami.* THE EFFECT OF PROCESS PARAMETERS IN THE VACUUM INFUSION MOULDING PROCESS- SICOMP Technical Report, 97-013.

Influences of material and process parameters on the properties of pultruded glass fiber / polypropylene composites

G. Bechtold*, R. Reinicke*, K. Friedrich*

* Institut für Verbundwerkstoffe (IVW), University of Kaiserslautern, 67663 Kaiserslautern, Germany

Abstract

In a series of experiments, optimum processing windows (preheating temperature, heated die temperature, pulling speed) for pultrusion of rectangular profiles out of commingled glass fiber/polypropylene (GF/PP) yarn with different glass fiber volume fractions and pre-impregnated GF/PP tapes were determined. Mechanical properties such as flexural modulus, flexural strength and shear strength were measured as a basis for comparison to studies of other investigators and as data for establishing process models. As a quality criterion, a shear strength of 80 % of the maximum achievable shear strength was chosen.
Furthermore, the influence of processing parameters on the void content of the pultruded beams was determined and compared to values predicted by a model approach. The maximum allowable void content was set in the model as 2 %. It resulted in a curve for the minimum processing time at a given heated die pressure. The results obtained were in good agreement with experimental data.

1. Introduction

Recent developments of impregnation technology for thermoplastic matrices have generated considerable interest in the possibility of thermoplastic pultrusion [1-3]. Successful works have however only been performed for simple cross-sections, and at pultrusion speeds not dramatically exceeding those known for commercial thermoset pultrusion (0.6 - 1.2 m/min). Major reasons for these deficits are the inherent difficulties associated with the thermoplastic matrices, such as high processing temperatures and high melt viscosities. An additional obstacle may be a lack of both fundamental understanding of the governing process mechanisms and adequate mathematical models for predicting the relationships between the various processing variables and the resulting structural/mechanical properties of the thermoplastic pultruded products [4].
Lee et. al [5] have presented a model which allows to predict the effects of pulling speed and die geometry on the required temperatures, the degree of consolidation, and the pulling force. Åström et. al. [4, 6] pultruded rectangular beams through a heating and cooling die system. The authors were interested in a comparison of experimental pultrusion results with model predictions of the temperature and pressure distribution in the pultrusion die. Mechanical properties of their pultruded carbon fiber/polyetheretherketon (CF/PEEK)-profiles were only slightly below those values presented in the literature for ideally compression molded standard samples. Additional works were also carried out with glass fiber (GF) / polypropylene (PP) tape material.
Gibson and co-workers [2] emphasized that the preheating section of their thermoplastic pultrusion process was of crucial importance to a successful operation.

Polybutyleneterephtalate (PBT) powder impregnated and sheath surrounded glass fiber bundles have been pultruded by Kerbiriou [7] into rectangular profiles. It was found that the temperature conditions in both the preheating zone and the heated die had a remarkable influence on the parts' quality. In addition, the pressure conditions in the heated die (independantly varied by the use of different taper angles and loads) and the pultrusion speed were of high importance. The results that were supported by a new model for predicting the remaining void content in the pultruded beams as a function of processing conditions, lead to the generation of an optimum processing window within which acceptable mechanical properties could be achieved. Additional results in Kerbiriou's previous works were also achieved with GF/PP-tapes, pultruded into circular rods and into hollow, rectangular profiles.
Tomlinson and Holland [1] reported on the pultrusion and properties of GF/PP matrix composites. The properties of their beams, as pultruded from tapes at a relatively low line speed of 4 mm/min, were slightly better than their hot pressed standard samples.
Additional studies on the pultrusion of unidirectional GF/PP-composites have been carried out by Michaeli and co-workers (e.g. [8]). By the use of various types of hybrid yarns the authors demonstrated the effects of cooling conditions, pulling speed and pressure profile in the heated die on the consolidation quality and the surface properties of flat rectangular profiles.
In the following, some new data on the pultrusion of glass fiber/PP fiber comingled yarn are presented and compared to results of previous investigators.

2. The Process

A schematic of a typical pultrusion line is shown in Figure 1. Fiber bundles are preheated in a 700 mm long hot air preheating zone and enter directly into the heated die. The temperature of the preheating zone can be varied, in order to find the optimum temperature range for achivement of maximum pulling speeds. The cavity of the heated die is tapered, and its angle can be varied without changing the final thickness of the beam. The surface of the cavity is polished and chrome plated in order to get smooth surfaces of the beams and to prevent rapid wear. Two pressure sensors incorporated in the die measure continuously the internal pressure profile. Static compression molding studies have shown that a compaction during cooling improves significantly the quality of laminates [9]. Therefore a water cooled die just behind the heated die is installed for further compaction and improvement of the pultrudate's surface quality. Heated and cooled die are fixed on a linear guide and are pulled against a load cell during the whole process, so that the pulling force is measured continuously (Figure 2). The beam is pulled by a Leimbach pulling mechanism, which can realize speeds between 0.01 and 30 m/min.
The temperature of the fiber bundles just before entering the heated die is varied between 20°C and 177°C, the heated die temperature between 177°C and 235°. Speeds from 0.03 to 0.6 m/min are realized. During pultrusion the temperature of the heated die is controlled by a thermocouple in the mold
The final cross section of the beams is 3.5 x 10 mm^2, i.e. sufficient to be able to make mechanical test specimens.

3. Processing Window and Properties

The goal of this study was to collect data about the influence of the parameters preheating temperature T_{ph}, heated die temperature T_{hd}, pulling speed v_p, glass fiber volume content V_f and types of material preforms (commingled yarn or preimpregnated tape) on the properties of the pultruded beams, such as flexural modulus E_F, flexural strength σ_F and shear strength τ. Based on these data, it is possible to define and to verify analytical impregnation models.

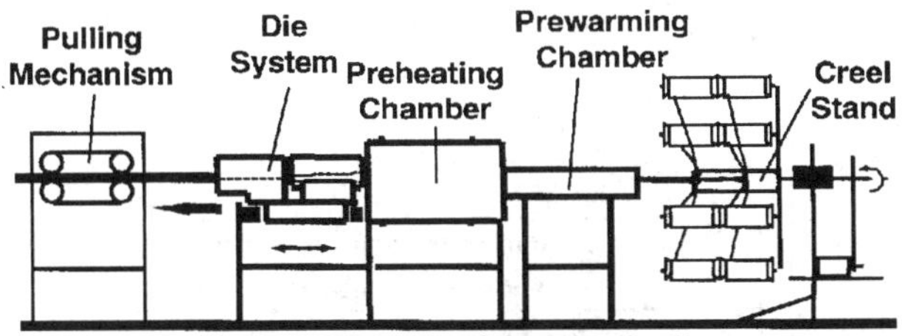

Figure 1. Schematic of the pultrusion line

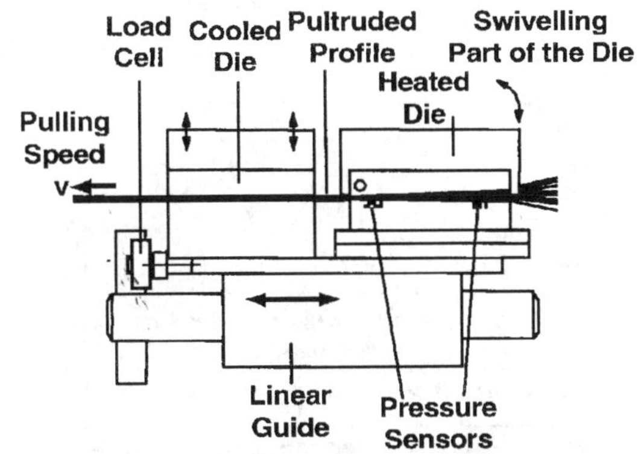

Figure 2. Schematic of the die system

Former works by Kerbiriou [9] exhibited that the shear strength, measured by the Lauke shear test [10], was most sensitive to variations in void content. In Figures 3 to 6, the influence of pulling speed on the shear strength of different materials is shown. With commingled yarns, the shear strength decreases slightly with increasing glass fiber volume content. This is due to the fact that more fiber - matrix interfaces exist, which decrease the shear strength. Up to a glass fiber volume content of 35%, the shear strength does not seem to be dependent on either the forming die temperature nor the applied pulling speeds. At 50 vol. %, the maximum shear strength is at about 0.1 m/min. Further, a temperature dependence is obvious: Forming die temperatures only slightly above the melting point (177 or 203°C) are not sufficient to ensure an acceptable quality of impregnation. This is the reason for the application of higher preheating temperatures at glass fiber volume fractions above 35%.
Preimpregnated tapes (Fig. 6) show a higher shear strength level, but the values decrease rapidly with increasing pulling speed.

Although Figures 3 and 4 rise the expectation that a further increase of pulling speed could be possible, a polymer build-up between forming die and cooled die at higher speeds foiled a continous process.

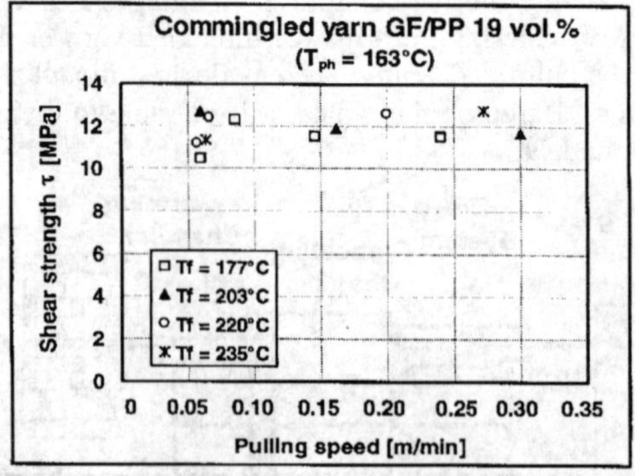

Figure 3. Shear strength of commingled yarn 19 vol. %

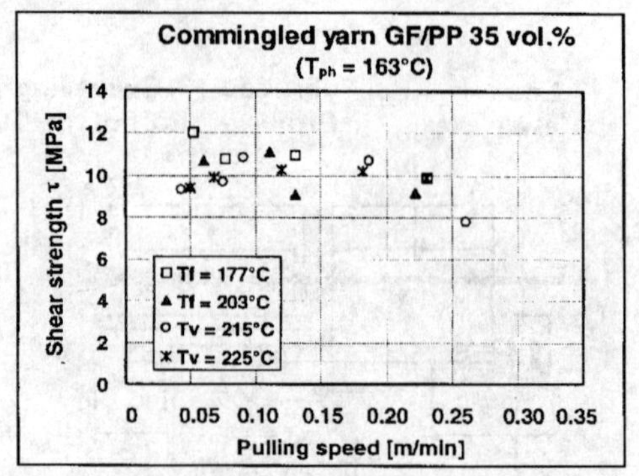

Figure 4. Shear strength of commingled yarn 35 vol. %

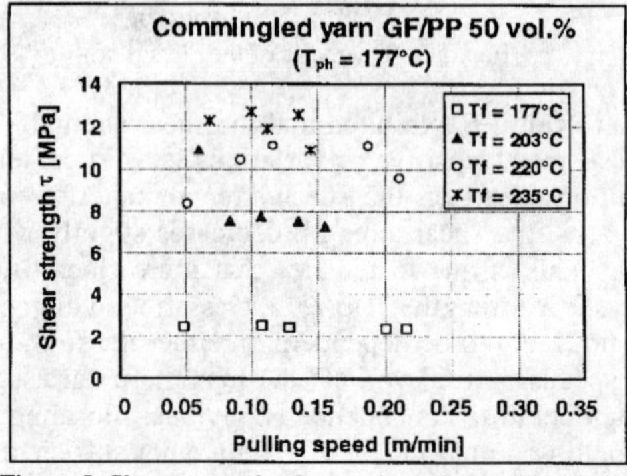

Figure 5. Shear strength of commingled yarn 50 vol. %

Further evaluations containing the results of commingled yarn with 19% glass fibers are presented in the following.

Figure 7 shows qualitatively the influence of processing speed and processing time (i.e. time in the heated die) on the shear strength and the void content (as measured by the Archimedes' bouyancy method).
The maximum properties at four different preheating temperatures are presented in Table 1. Columns 3 and 4 represent the corresponding values of T_{hd} and v_p, at which individual maxima of E_F, σ_F and τ were achieved. The table reads as follows: At T_{ph} = 20°C a maximum in E_F was achieved, when T_{hd} = 215°C and v_p = 0.225 m/min, whereas a maximum of σ_F was possible at T_{hd} = 203°C and v_p = 0.068 m/min.

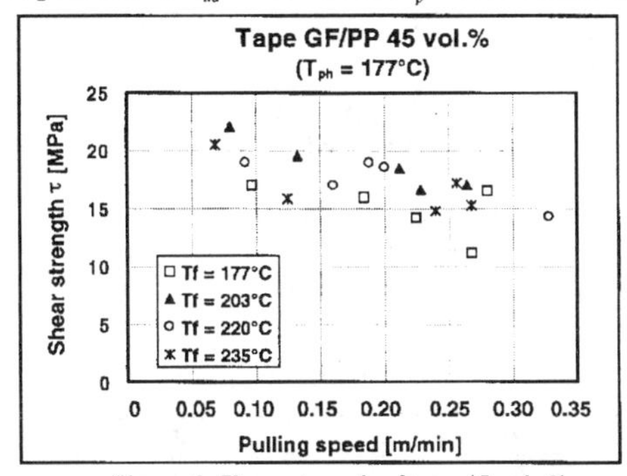

Figure 6. Shear strength of tape 45 vol. %

The results correspond quite well with those of Wilson et al. [10] and Tomlinson et al. [1], when compared on the basis of the same glass fiber volume fraction V_f = 19%. Only a comparison of the shear strength data is limited due to different testing methods. With the pultrusion line and materials used here, the following process parameters lead to at least 80 % of the maximum shear strength (τ_{max} = 14 MPa) at pulling speeds of at least 0.3 m/min:

Preheating temperature: T_{ph} = 143 - 163° C
Heated die temperature: T_{hd} = 185 - 215° C

It should be noted that the parameters partly influence each other or cannot be realized simultaneously.
A rectangular model based on Darcy's law for predicting the degree of impregnation (or the void content, respectively) during the pultrusion process was developed in [9]. It leads to the following equation for the void content X_v:

$$X_v = \frac{(h_0 - z) \cdot (1 - V_f)}{h_{1z}} \qquad (1)$$

where h_0 and h_{1z} represent geometrical dimensions of the rectangular model arrangement between PP matrix fibers and the GF agglomeration to be impregnated. V_f is the global glass fiber volume fraction, and z the flow path of the polymer melt transverse to the glass fiber agglomeration. Note that longitudinal flow was not considered because the PP fibers in their initial state are already continously, longitudinally arranged, i.e. parallel to the glass fibers, so that most probably transverse flow will take place under pressure once the PP fibers exceed their melting temperature.

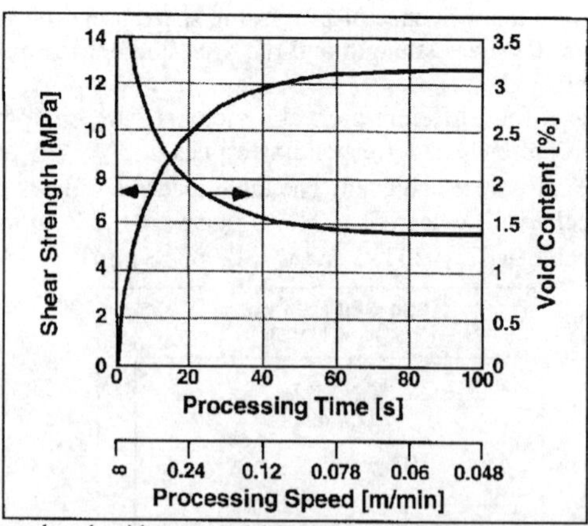

Figure 7. Shear strength and void content at different processing times, T_{ph} = 143° C, T_{hd} = 203° C, commingled yarn, V_f = 19%

T_{ph} [°C]	Properties	T_{hd} [°C]	v_p [m / min]
20	$E_{F\,max}$ = 12.5 GPa	215	0.225
	$\sigma_{F\,max}$ = 339.8 MPa	203	0.068
	τ_{max} = 13.03 MPa	203	0.068
114	$E_{F\,max}$ = 12.5 GPa	177	0.080
	$\sigma_{F\,max}$ = 339.8 MPa	185	0.077
	τ_{max} = 13.33 MPa	177	0.080
143	$E_{F\,max}$ = 15.2 GPa	203	0.167
	$\sigma_{F\,max}$ = 396.6 MPa	203	0.050
	τ_{max} = 14.11 MPa	185	0.050
163	$E_{F\,max}$ = 12.3 GPa	177	0.146
	$\sigma_{F\,max}$ = 349.5 MPa	177	0.060
	τ_{max} = 12.79 MPa	225	0.273

Table 1. Maximum properties at constant preheating temperatures

The impregnation time in order to reach a certain void content X_V can be calculated by inserting the corresponding value of z (according to equation (1)) into the following equation:

$$t = \frac{1}{n} \cdot \sum_{i=1}^{n} \left(\frac{z_{(x_i)}^2 \cdot 2 \cdot \eta \cdot k_{zz} \cdot \left(\frac{V_a}{V_{f(x_i)}} + 1 \right)}{r_{gf}^2 \cdot \left(\sqrt{\frac{V_a}{V_{f(x_i)}}} - 1 \right)^3 \cdot \frac{\Delta p}{\Delta L} \cdot x_i} \right) \quad (2)$$

where

k_{zz} : Permeability constant of the fibers (Gutowski) = 0.044
V_a : Initial fiber volume fraction
r_f : Fiber radius = 7μm
n : Number of sections in the heated die (in this case 80)

z	: Flow path through the agglomeration (calculated by equation (1))
$V_{f(xi)}$	: Local fiber volume fraction (calculated)
Δp	: Pressure difference (measured by the sensors)
ΔL	: Length of the heated die = 80 mm
η	: Viscosity of the matrix = 690 Pas (Temperature 185°, shear rate 10 s^{-1})
x_i	: Position of actual segment in pulling direction

Figure 8 contains the resulting information on reqired compaction time and pressure to reach a void content of less than 2%. Experimentally determined data enable a comparison to the calculated curves. The most significant sources of error in the latter are the viscosity and the permeability constant. The viscosity depends strongly on the temperature and the shear rate. Both are not constant at any point of the matrix. As the calculation estimated the processing time too short, it can be stated that the viscosity or the permeability constant have been chosen as too low. By a further development of the model, errors can be reduced, however, the former model will then lose its advantage of simplicity.

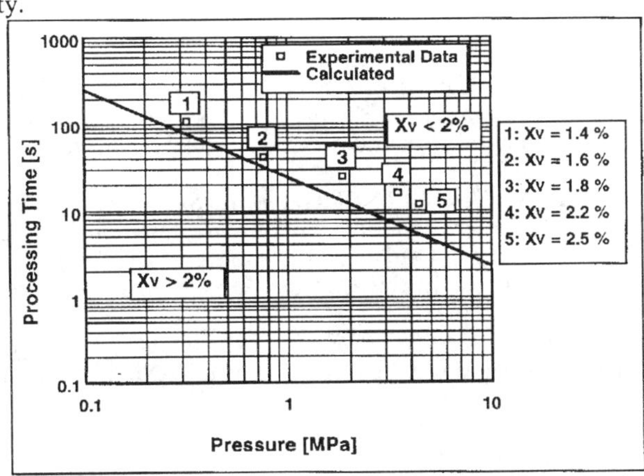

Figure 8. Required processing times at different pressures near die exit (almost maximum of the pressure profile), T_{ph} = 143° C, T_{hd} = 203° C

4. Conclusions

Pultrusion of comingled GF / PP yarn leads to profiles with acceptable mechanical properties. The processing window shows, that preheating temperatures marginally below the polymer melting point are ideal, whereas the heating die temperature should not exeed 215° C. Within the presented processing window, the pultrusion speed (as long as it stays in the range between 0.05 and 0.27 m/min) has not a strong influence on the quality of the parts. The main problem for the pultrusion line used is to overcome a polymer built-up between heated and cooled die occuring especially for materials with low glass fiber volume contents or at high process temperatures and speeds. This effect leads to unacceptable surface quality and fiber pile-up.

It is intended in a future step to improve the die system, e.g. by a direct combination of heated and cooled die and its extension to a longer length. Further, so called impregnation pins in the preheating zone are promising a considerable increase in process speed [11].

For the pultrusion process performed, a simple consolidation model can be applied to estimate the influence of processing speed and pressure on the void content of the final part quite well. However, it is very sensitive to measurement errors of viscosity and permeability.

Acknowledgements

Special thanks are due to Vetrotex, Chambery, France, who kindly supplied the material. Furthermore, the authors gratefully acknowledge the support of the Deutsche Forschungsgemeinschaft (DFG-Fr 675/20-1). In addition, the Max-Buchner Stiftung (MBFSt-Kennziffer 1857) funded a student for doing special studies in this field.

References

[1] *W. J. Tomlinson, J. R. Holland,* Pultrusion and properties of unidirectional glass fibre-polypropylene matrix composites - J. Mater. Sci. Letters 13 (1994), pp. 675-677.
[2] *B. J. Devlin, M. D. Williams, J. A. Quinn, A. G. Gibson,* Pultrusion of unidirectional composites with thermoplastic matrices - Composites Manufacturing 2 (3/4) (1991), pp. 203-207.
[3] *M. L. Wilson, J. D. Backley, G. E. Dickerson, G. S. Johnson, E. C. Taylor, E. W. Covington, III,* Pultrusion process development of a graphite reinforced polyetherimide thermoplastic composite - Proc. 44th Annual Conference, Society of Plastics Industry, Reinforced Plastics / Composite Institute, Dallas, Texas, USA, February 6-10 (1989), Paper Session 8-D.
[4] *B. T. Åström, P. H. Larsson, R. B. Pipes,* Experimental investigation of a thermoplastic pultrusion process - Proc. 36th International SAMPE Symposium, April 15-18 (1991).
[5] *W. I. Lee, G. S. Springer, F. N. Smith*: Pultrusion of thermoplastics - a model. - J. Composite Materials 25 (1991), pp. 1632-1652.
[6] *B. T. Åström, R. B. Pipes,* Modeling of a thermoplastic pultrusion process - Proc. 46th Annual Conference, Composites Institute, The Society of Plastics Industry, February 18-21, Session 4-A, 1-9 (1991).
[7] *V. Kerbiriou, K. Friedrich,* Pultrusion of thermoplastic composites - process optimization and mathematical modelling. – J. Thermoplastic Composite Materials (1998), submitted.
[8] *J. Blaurock, W. Michaeli,* Pultrusion of endless fiber-reinforced profiles with a thermoplastic matrix system - Engineering Plastics, Vol. 9 (4) (1996), 282-292
[9] *V. Kerbiriou,* Imprägnieren und Pultrudieren von thermoplastischen Verbundprofilen - Dissertation, Fachbereich Maschinenbau und Verfahrenstechnik, Universität Kaiserslautern, Germany (1996)
[10] *B. Lauke, K. Schneider, K. Friedrich,* Interlaminar Shear Strength Measurement of Thin Composite Rings Fabricated by Filament Winding - Proceeding 5th European Conference of Composite Materials, ECCM 5, 7 - 10 April 1992, Bordeaux, France, EACM-Publication, Bordeaux 1992, 423 - 428
[11] *A. H. Millner, N. Dodds, A. G. Gibson,* High speed pultrusion of thermoplastic matrix composites via the commingled and dry powder impregnation routes - Proceeding International Conference on Automated Composites, ICAC 97, 4 – 4 September 1997, Glasgow, UK, 169 - 192.

PUSH-PULL-PROCESSING OF UNREINFORCED AND SHORT FIBRE REINFORCED THERMOPLASTICS

A. Kech, H.-C. Ludwig, B. Möginger, U. Müller, U. Fritz, P. Eyerer
Institute for Polymer Testing and Polymer Science (IKP), University of Stuttgart
Pfaffenwaldring 32, D-70569 Stuttgart, Germany
Phone: +49 711 685 2676
Fax: +49 711 685 2066
email: armin@ikp.uni-stuttgart.de

Abstract

The content of this work is to show the influence of Push-Pull-Processing (PPP) on the mechanical properties of unreinforced and short glass fibre reinforced thermoplastics [1]. The fibre orientation of injection moulded fibre reinforced thermoplastics varies according to the dominance of either elongational or shear flow during mould filling [2]. The fibre orientation has tremendous effect on the mechanical properties of the composite parts. Push-Pull-Processing can be used to manipulate the fibre orientation distribution of fibre reinforced and even the molecule orientation of some unreinforced thermoplastics, which results in transversely isotropic mechanical properties. Within this paper results of tensile and bending tests are presented giving an overview about the potential of PPP for applications in order to reduce the thickness of parts. E.g., Young's modulus parallel to the flow direction of fibre reinforced nylon 6.6 can be increased by 40 to 50 % compared to conventionally injection moulded parts. Even long fibre reinforced VERTON RF 70010 shows no significant fibre breakage during PPP. Taking a look on unreinforced materials, like Polypropylene Young's modulus increases approximately 40 % parallel and 25 % perpendicular to the flow direction.

1. Introduction

PPP is an injection moulding technique developed by KLOECKNER FERROMATIK DESMA, Germany for multi-component injection moulding machines. A special software controls the injection units during filling, counter filling, holding and ejection, Fig. 1.

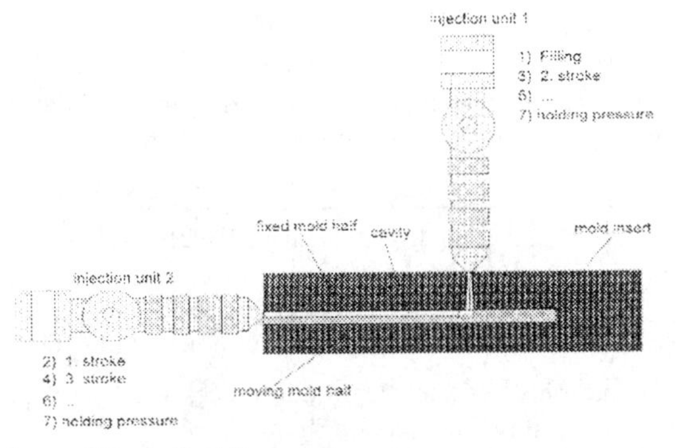

Figure 1: Procedure of Push-Pull-Processing

After complete filling of the cavity with the first injection unit, the screw in the second unit is driven forward and pushes the melt under high shear forces through the cavity.

This counter filling, called a stroke, can be repeated several times with both injection units. Fig. 2 shows recorded data of the pressure inside the cavity, the hydraulic pressure and the movement of the screws.

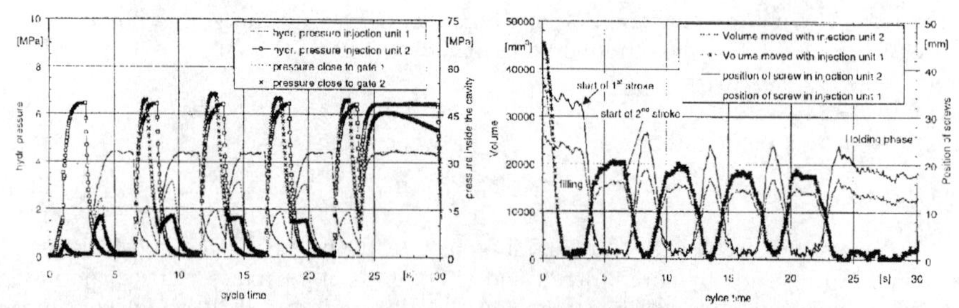

Figure 2: Recorded pressure data (left) and position of the screws (right) during PPP

2. Theory

PPP results in advanced mechanical properties of parts compared to conventional injection moulded ones as the fibre orientation distribution is more anisotropic. Fibres are aligned preferentially parallel to the flow direction and molecules in unreinforced materials like Polypropylene or LCP can be oriented as well. Therefore the mechanical properties of these materials differ parallel and perpendicular to the flow direction more than for conventional samples. To predict the Young's moduli parallel and perpendicular to the flow direction E_1 and E_2, respectively, one can use the Halpin-Tsai equations [3]. For conventionally injection moulded parts the fibre orientation is varying throughout the thickness of the part. A quasi-planar random fibre orientation is the result of the dominance of either shear or elongational flow.

$$E_{Random} = \frac{3}{8} \cdot E_1 + \frac{5}{8} \cdot E_2 \tag{1}$$

$$E_1 = E_M \cdot \frac{1+\xi_1 \cdot \eta_1 \cdot v_F}{1-\eta_1 \cdot v_F}; \eta_1 = \frac{(E_F / E_M) - 1}{(E_F / E_M) + \xi_1} \tag{2}$$

$$E_2 = E_M \cdot \frac{1+\xi_2 \cdot \eta_2 \cdot v_F}{1-\eta_2 \cdot v_F}; \eta_2 = \frac{(E_F / E_M) - 1}{(E_F / E_M) + \xi_2} \tag{3}$$

The fibre orientation using PPP results in an orthotropic material behaviour with elevated moduli parallel and lowered moduli perpendicular to the flow direction, Table 1.

Processing	$E_{1,Meas}$ [MPa]	$E_{2,Meas}$ [MPa]	E_1 [MPa]	E_2 [MPa]	E_{Rand} [MPa]
conventional	8797	7854	-	-	11067
PPP	13857	6060	18939	6344	-

Table 1: Values for Young's moduli of fibre reinforced nylon 66 (50 wt. %)

The difference in morphology between both processing techniques can easily be observed by investigating thin cuts under a light microscope with linear polarized light for Polypropylene, Fig. 3 and for fibre reinforced LCP, Fig. 4 and 5, respectively

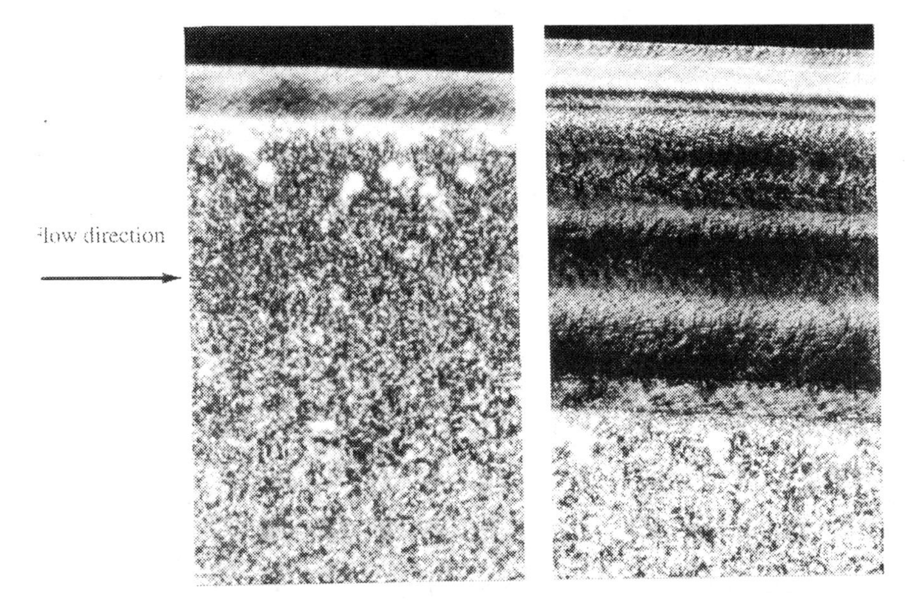

Figure 3: Morphology of Polypropylene: conventional (left) and PPP (right)

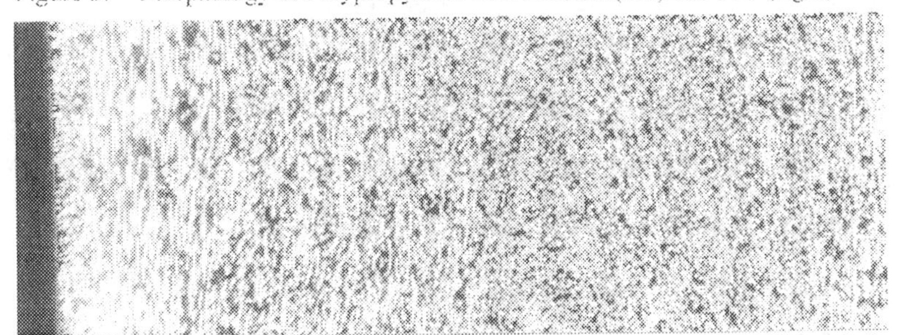

Figure 4: Polished cross section of conventionally injection moulded fibre reinforced LCP

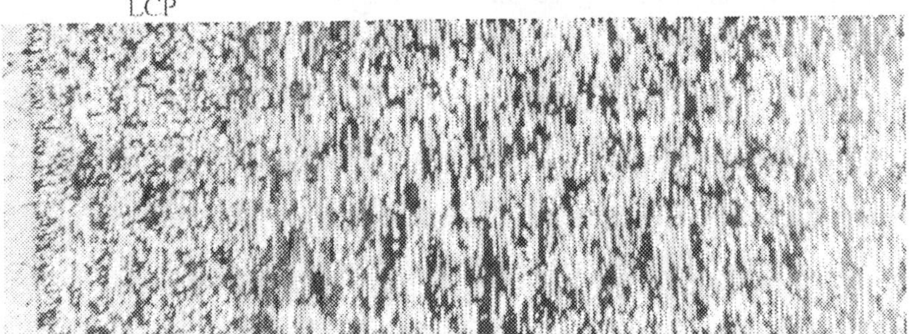

Figure 5: Polished cross section of fibre reinforced LCP using PPP

This visual impression can be quantified with fibre orientation measurements using an image analyzer system and the surface ellipse method. Conventionally and PPP injection moulded fibre reinforced nylon 66 (35 wt. % glass fibres) shows significant differences in the main orientation tensor components a_{11} and a_{22} representing orientation parallel and perpendicular to the flow direction, respectively. The geometry used for these samples is shown on the right side of Fig. 6.

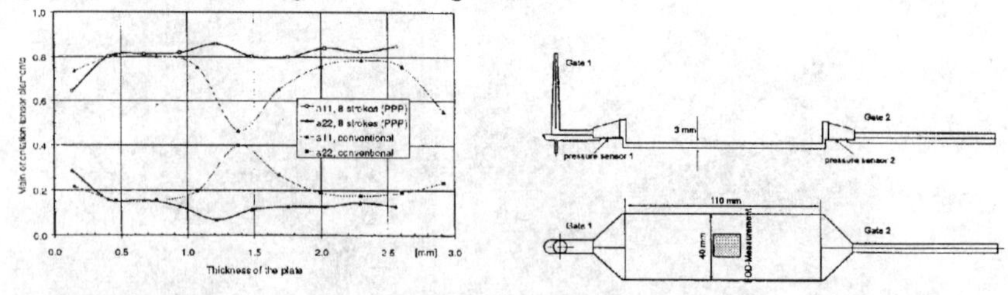

Figure 6: Fibre orientation measurements of fibre reinforced nylon 66 (left) and geometry used for PPP and conventional injection moulded parts

3. Experimental part

In order to determine Young's moduli parallel and perpendicular to the flow direction, tensile tests were performed according to DIN 53 455 using a universal tensile testing machine (ZWICK 1476). The draw ratio was 5 % l_0/min (gauge length, l_0 = 50 mm). Young's modulus was evaluated according to DIN 53 457 between 0.05 and 0.25 % strain. Additionally the tensile strength σ_{zR} and the strain at rupture ε_{zR} were determined in order to compare the mechanical properties of conventionally and PPP injection moulded specimens. The tensile bars were taken out of plates parallel and perpendicular to the flow direction.

Besides tensile tests three point bending tests of glass fibre reinforced thermoplastics have been performed according to DIN EN 63 using a universal tensile testing machine with gauge length 80 mm and test bars of the dimensions 5 mm thickness, 15 mm width and 100 mm length.

The crosshead speed was 2.5 mm/min (1,17 % flexural strain/min).

4. Results

The mechanical properties of 2 grades of nylon 66 with 50 wt. % glass fibres differ significantly for both processing techniques, Table 2.

Material	Direction	conventional	PPP	
ULTRAMID A3EG10	parallel	8797 MPa	13857 MPa	16 strokes
	perpendicular	7854 MPa	6060 MPa	16 strokes
VERTON RF70010	parallel	8312 MPa	13846 MPa	8 strokes
	perpendicular	9726 MPa	7308 MPa	8 strokes

Table 2: Young's moduli for short and long glass fibre reinforced nylon 66

Using PPP distinct stiffer parts parallel to the flow direction are obtained. The increase is in the range of 50 % for ULTRAMID and 60 % for VERTON. The stiffness perpendicular to the flow direction of conventionally processed VERTON is higher than parallel. The amount of long fibres inside the melt forms during conventional injection moulding a felt-like structure with a more or less planar random fibre orientation. As a result of PPP fibres align parallel to the flow direction and the stiffness perpendicular to the flow direction is decreasing (about 25 %) [4], [5].

Using PPP the tensile strength of VERTON is increased slightly, for ULTRAMID no significant change occurs. Regarding the strain at rupture PPP results in a decrease compared to the conventionally processed specimens.
Tensile tests on unreinforced Polypropylene show the very interesting effect that not only parallel to the flow direction Young's modulus increases but also perpendicular, Table 3

Direction	Processing	0 wt% GF	20 wt% GF	30 wt% GF
parallel	conventional	1600	2900	4400
perpendicular	conventional	1600	2100	2300
parallel	Push-Pull	2300	4000	6100
perpendicular	Push-Pull	2000	3500	2000

Table 3: Young's modulus of i-PP (VESTOLEN P7000, P7032, P7052)

The flexural bending strength σ_{bR} and the flexural modulus E_b parallel to the flow direction of the three point bending tests are summarized in Table 4.

Material	Processing	σ_{bR} [MPa]	E_b [MPa]
ULTRAMID	conventional	276	12064
A3EG10	PPP	320	14912
VERTON	conventional	267	11586
RF70010	PPP	268	12313
VESTOLEN	conventional	58 ($\sigma_{b3.5}$)	1572
P7000	PPP	61 ($\sigma_{b3.5}$)	2385

Table 4: Results of bending tests for fibre reinforced nylon 66 and unreinforced i-PP

The effect of PPP on the mechanical properties is different for flexural strength and stiffness. The flexural strength of short fibre reinforced PPP samples is slightly increased (about 15%) while there is no significant change for the long glass fibre reinforced grade. The flexural modulus E_b increases for both grades. The mechanical properties of the short fibre reinforced materials are higher parallel to the flow direction than the long glass fibre reinforced ones. The fibre orientation measurements prove that there is still a remarkable core-layer in the PPP VERTON samples, Fig. 7.

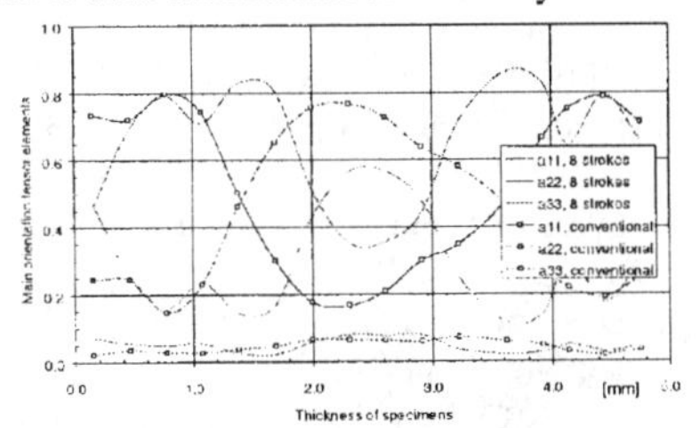

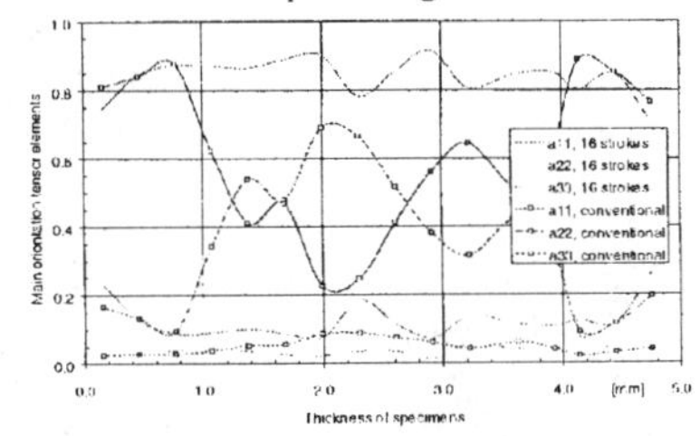

Figure 7: Fibre orientation of long (left) and short (right) discontinuous glass fibre reinforced nylon 66

As there is almost no core-layer in ULTRAMID for PPP, the mechanical properties parallel to the flow direction are increased (about 25% for flexural modulus). This situation changes for specimen taken perpendicular to the flow direction, as expected. Table 5 depicts the flexural strength and modulus for loads on specimens cut transversal to the flow direction.

Material	Processing	σ_{bR} [MPa]	E_b [MPa]
ULTRAMID	conventional	184	7270
A3EG10	PPP	109	5924
VERTON	conventional	196	7181
RF70010	PPP	123	5548

Table 5: Results of bending tests perpendicular to the flow direction

The flexural stress at rupture for VERTON is slightly higher than for ULTRAMID (5 to 10%). The alignment of fibres parallel to the flow direction with PPP results in reduced flexural strength in the range of 40%. Almost no difference is observed between the flexural stiffness of both materials, but an influence of the processing technique is present. The values for the short fibre reinforced nylon 66 grade vary for conventionally processed plates. This effect vanishes by using PPP, which on the other hand results in lower values for E_b compared to the conventionally processed ones (20 to 25 %).

5. Conclusion

The mechanical tests prove that the stiffness parallel to the flow direction is increased for fibre reinforced nylon 66 grades as well as for unreinforced Polypropylene using PPP. The properties of fibre reinforced materials perpendicular to the flow direction are decreased due to the alignment of the fibres parallel to the flow direction when using PPP. A very interesting effect is found for unreinforced i-PP. The special morphology of i-PP produced with PPP results in advanced mechanical properties both parallel and perpendicular to the flow direction. The amount of higher properties using PPP can be used to minimize the thickness of composite parts.

Acknowledgement

The authors thank ICI, NYLTECH, HOECHST, VESTOLEN and BASF for providing their granules and the European Communities for funding the work within a Brite-EuRam II-project, entitled Push-Pull Processing of Short Fibre Reinforced Thermoplastics and LCP's (Project No.: BE-8081).

References

[1] H.-C. Ludwig, G.Fischer, H. Becker, A Quantitative Comparison Of Morphology And Fibre Orientation In Push-Pull Processed And Conventional Injection-Moulded Parts, Composite Science and Technology **53** (1995), 235-239

[2] F. Folgar, C.L. Tucker III, Orientation Behavior of Fibers in Concentrated Suspensions, J. Reinf. Plast. Comp., 3, (1984), 98-119

[3] J.C. Halpin, J.L. Kardos, The Halpin-Tsai Equations: A Review, Polymer Engineering and Science, May 1976, Vol. 16, No. 5, 344 - 352

[4] A. Kech, H.-C. Ludwig, G. Fischer, P. Eyerer, N.C. Davidson, G. Archenhold, A.R. Clarke, T.E. Berland, E.L. Hinrichsen, M. Vincent, Push-Pull Processing Of Short Fiber Reinforced Thermoplastics, Preprints of the Topical Conference on Processing, Structure and Properties of Poymeric Materials, (1996) Chicago

[5] A. Kech, H.-C. Ludwig, G. Fischer, P. Eyerer, H. Becker, Push-Pull Processing of Unreinforced and Short Fiber Reinforced Thermoplastics and LCP: Processing and Mechanical Properties, Extended Abstracts of the PPS Europe/Africa Regional Meeting in Gothenburg, Sweden (1997), 6:9

Innovative filament winding technique with thermoplastics for the fabrication of hoop-wrapped composite cylinders

O. Christen*, M. Neitzel*, C. Rasche+

* Institut für Verbundwerkstoffe GmbH, D-67663 Kaiserslautern, Germany

+ Mannesmann Stahlflaschen GmbH, D-46535 Dinslaken, Germany

Abstract

In recent years, there has been an increasing use of fibre-reinforced polymers in the various industries. An example is the manufacturing of composite cylinders for the transportation of industrial gases or the on-board storage of compressed natural gas (CNG). The requirement for these kind of pressure cylinders is to decrease the weight but to reach properties comparable to monolithic steel or aluminium cylinders. A weight reduction of such cylinders can be achieved by replacing the metallic wall partly or fully with unidirectional fibre reinforced polymers of high specific strength and stiffness. Especially thermoplastic matrices offer conditions like high fracture toughness, unlimited shelf life, elimination of post curing cycles, recyclability and reduction of volatile solvent. The principle of the current thermoplastic filament winding technique is to apply heat and pressure to consolidate preimpregnated fibres (tapes). Nevertheless the material needed for this kind of tape winding is still very expensive. To reduce manufacturing costs but to incorporate from the advantages of the thermoplastic matrices a new filament winding technique has been developed. The innovation is to impregnate and consolidate the fibres and the thermoplastic matrix simultaneously during winding.

1 Introduction

In general, there are two main types of cylinder, namely the hoop-wrapped cylinder and the fully-wrapped cylinder. In the case of the hoop-wrapped cylinder (Figure 1,2), the metallic liner is generally designed for a service pressure of 100 bar, which is half the operating pressure for the composite cylinder. With respect to CEN- and ISO regulations the required minimum burst pressure of this metallic liner is more than 255 bar [1,2]. The overwrap, which is made of a fibre-reinforced plastic enables the completed cylinder to be operated safely at a pressure of 200 bar. An overwrap in the cylindrical portion of the liner alone is adequate in this case, because the stress on such a cylinder in circumferential direction is twice as high as those in the longitudinal direction and, as a rule, the cylinderends have an unnecessarily thicker wall than the cylindrical portion, as a result of liner forming operation.
The desired reduction in cylinder weight is achieved by the substitution of part of the liner section thickness with the fibre-reinforced composite material. The operating pressure or the burst pressure of a hoop-wrapped cylinder may be further increased and/or the thickness of its loadbearing liner may be reduced by using a full overwrap. For hoop-wrapped cylinders, only metallic liners made of a quenched and tempered steel or an aluminium-base alloy can be used to sustain the longitudinal stresses.
Fully-wrapped cylinders generally contain a metallic liner withstanding an operating pressure of approximatly 40 bar. The metallic liner may however be replaced with a plastic liner, in which case the plastic liner only serves as a barrier to prevent gas diffusion and as a core on which to make the overwrap.

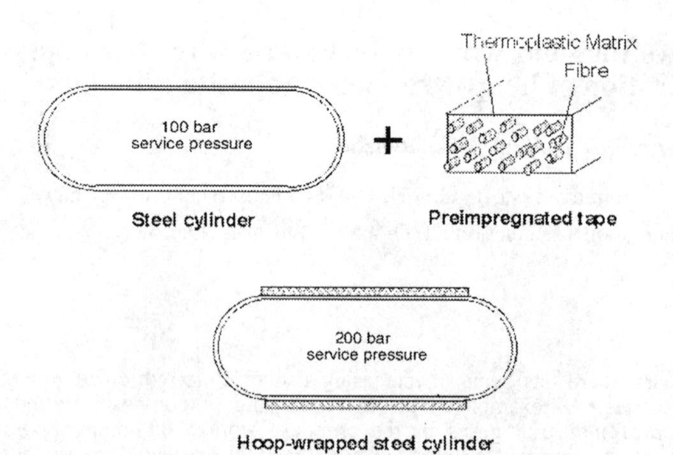

Figure 1: Hoop wrapped steel cylinder

In comparison with conventional steel and aluminium cylinders, weight reduction up to 75 % can be realised, dependent on the materials selected. In the case of monolithic cylinders, the use of conventional aluminium-base alloys in place of steel does not result in any significant reduction of cylinder weight because of their specific strength properties. In contrast, in the case of composite cylinders, the overwrap made of polymers reinforced with fibres of glass, aramid or carbon results in considerable weight reduction, depending upon the type of wrapping used. Each reduction in cylinder weight is associated with a corresponding increase in the manufacturing costs. In the case of a fully- wrapped cylinder, which has a thin liner, a large amount of reinforcing fibre has to be used, compared with a hoop-wrapped cylinder, which has a thick liner. Consequently, the total cost of fibre, which is very expensive anyway, accounts for the major part of the total cost of cylinder manufacture [3].

HYDROGEN-TRAILER (Mannesmann)

- 225 hoop-wrapped vessels
- service pressure of 200 bar
- 60 % higher storage capacity (37.125 l) compared to solid steel vessels

CNG-BUS

- 8 hoop-wrapped vessels
- service pressure of 200 bar
- storage capacity: 1.320 l

Figure 2: Applications for hoop-wrapped composite cylinders [4]

In general, fibre reinforced thermosets are used in the production of composite vessels by the wet winding method. Compared to conventional thermoset materials, a number of specific properties and fabrication-related advantages can be made by the application of fibre reinforced thermoplastics. These are for example higher fracture toughness, longer shelf life and good recyclability. Especially regarding filament winding thermoplastics possess further qualities like less emission of deleterious substances and continuous processing by elimination of long and costly curing cycles.

2 Thermoplastic filament winding with on-line impregnation

The innovation is to impregnate and consolidate the fibres and the thermoplastic matrix simultaneously during winding [5]. The most interesting advantages of this new filament winding technology compared to the tape winding method are:

- less energy and material wastage
- higher flexibility towards fibre-matrix-combination and possibility of a flexible and application-specific selection of the material
- less manufacturing costs due to less material costs

The key point of this new manufacturing process shown in Figure 3 is to achieve the same impregnation quality and fibre volume fraction as with the tape winding technique by comparable winding speeds.

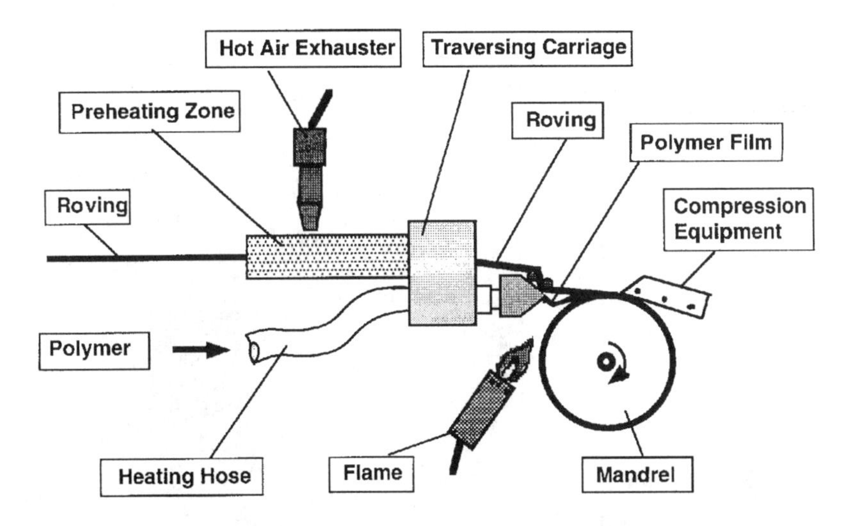

Figure 3: Scheme of on-line impregnation technique

3 Results and Discussion

For the winding trials the aramid fibre Twaron 2200 with 2520 dtex and the thermoplastic matrix polyamid 12 were used. Figure 8 shows a photo of an wounded ring at the beginning of the project and a photo of a ring with improved process parameters. For the optimisation of the process parameters a lot of winding trials were necessary. The characterisation of the impregnation quality of the filament wound rings was performed with different testing methods like the interlaminar shear test, the determination of the fibre volume fraction and the optical examination with a microscope. It has been found out that the most important winding parameters are the process temperatures, the filament guide and the matrix viscosity. But however these tests are not useful for analysing the performance of the rings. Due to this some more rings were wound for static and dynamic tensile tests.

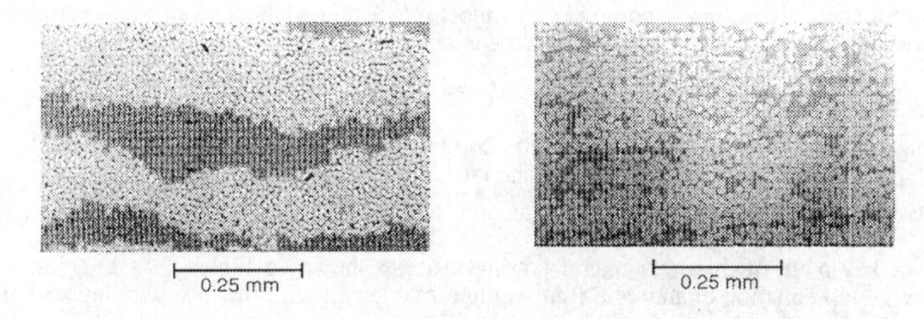

Figure 4: Fibre matrix distribution at the beginning of the project and with optimised parameters

3.1 Static tensile test

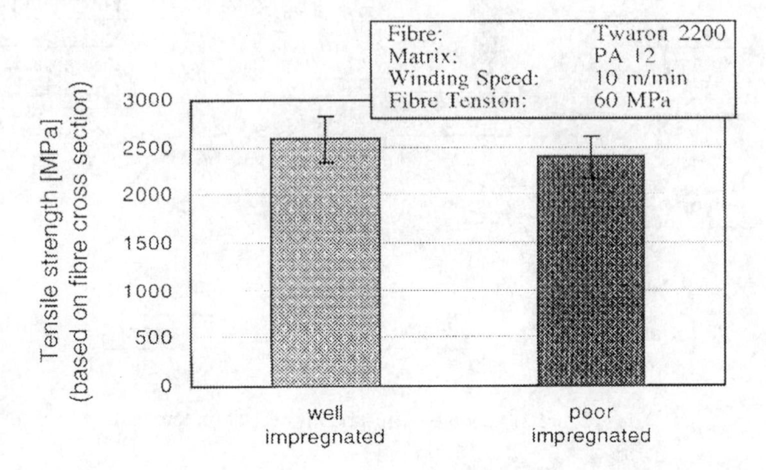

Figure 4: Tensile strength as a function of impregnation quality

The static tensile tests was carried out with a self made testing tool according to the NOL ring test designation ASTM D-2290. Figure 5 shows the results of the static tensile tests. As expected the impregnation quality has nearly no influence on the static tensile strength of the rings manufactured with the new on-line impregnation technique. The detected mean strength value is about 2400 MPa according to the fibre cross section.

3.2 Dynamic tensile test (cycle test)

To get information about the dynamic performance of the rings wound with the on-line-impregnation technique dynamic tensile test was carried out. To shorten the testing time the upper tension was adjusted at 1333 MPa based on the fibre cross section. This is twice as high as the tension of the hoop wrapped steel vessel during the cycle pressure test. The lower tension was adjusted at 380 MPa and the cycle frequency was 5 Hz.
Figure 6 shows that the average value of all tested rings wound with the optimised processing parameters reached round about 30000 cycles independent of the winding speed. For example the cycle pressure test for pressure vessels demands more than 12000 cycles by an upper tension of about 750 MPa.

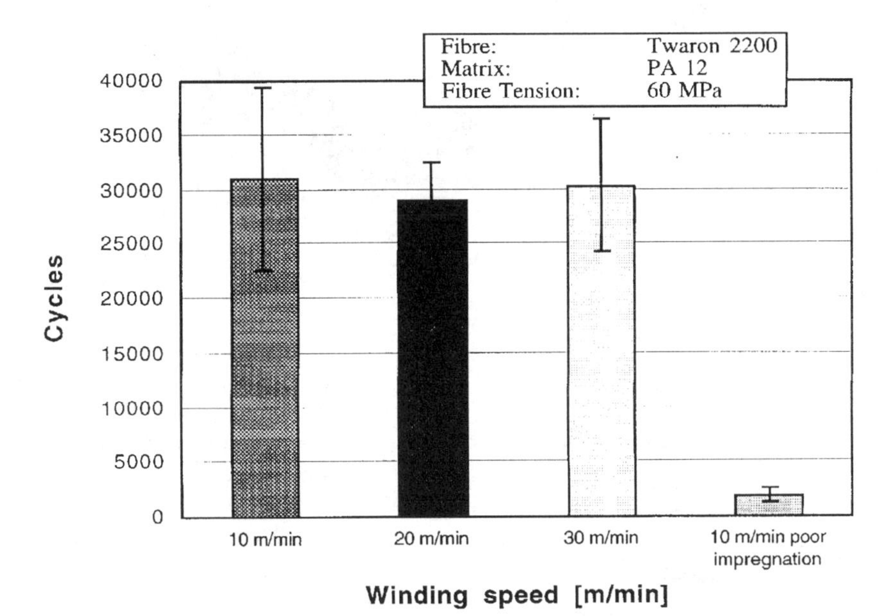

Figure 6: Fatigue as a function of the winding speed and impregnation quality

Furthermore poor impregnated rings (not impregnated fibre bundles with only a few matrix between the layers) were tested. In contrast to the static tensile test the impregnation quality is very important for the fatigue behaviour. The poor impregnated rings failed already after a few 1000 cycles.

5 Manufacturing and Testing of hoop-wrapped steel vessels

Because of the successful ring tests first hoop-wrapped steel cylinders were manufactured with the optimised winding parameters (Figure 7). Therefore 16 aramid rovings Twaron 2200 (2520 dtex) and polyamid 12 matrix has been processed with a winding speed of 16 m/min. Each pressure vessel was reinforced with 40 layers (laminate thickness: about 5mm). The continuous manufacturing time took about 90 minutes. The bonding of the reinforcement and the polyamid powder coated steel vessel was very good. The manufacturing time with the current tape winding is about 30 minutes. Due to this the winding speed will be optimised.

Figure 7: Manufacturing of a hoop-wrapped steel cylinder

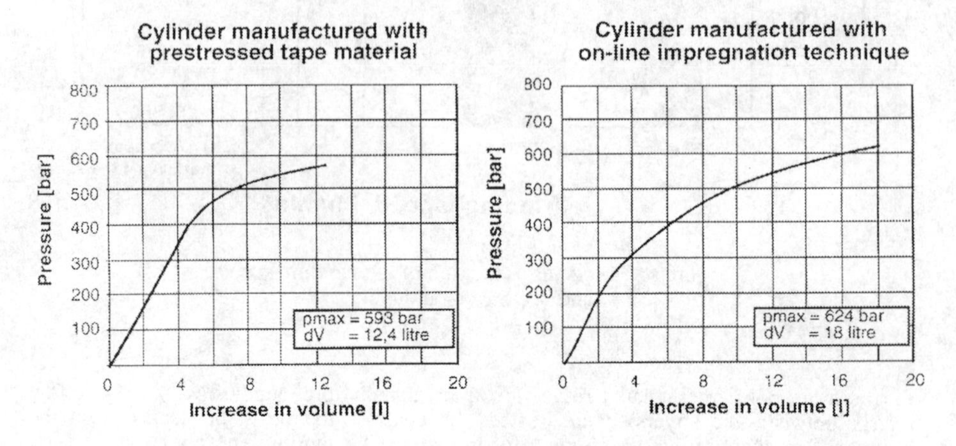

Figure 8: Testing of hoop-wrapped steel cylinders

The finished cylinders were tested under hydraulic pressure at ambient temperature. The burst test according to CEN (Specification for hoop-wrapped gas cylinders) demands a burst pressure of more than 501 bar, the initiation shall be by longitudinal failure in the cylindrical part and the liner shall remain in one piece. All tested cylinders initiated in that way in the cylindrical part and failed at burst pressures between 550 and 600 bar. The burst behaviour of the cylinder wound with the new impregnation method is the same compared to the current tape wound cylinders. But the performance during testing is different (Figure 8). The tape reinforced cylinder was wound with high tape tension that results in residual compressive stresses in the liner and tensile stresses in the reinforcement. Therefore the plastification of this cylinder starts later than the plastification of the cylinder manufactured with the new winding technique. Due to this the volume increase is much lower. To reach comparable fatigue life with the new cylinders they have to be pressurized before application (autofrettage).

Outlook

Further research work focuses on the manufacturing of hoop-wrapped composite cylinders in order to optimise the layer build up and to increase the processing speed. A lot of tests will be carried out like a pressure cycle tests and environmental tests etc.. The next main target is to manufacture fully wrapped composite vessels with thermoplastic matrix using the new on-line impregnation technique.

Acknowledgement

The authers would like to thank the „Bundesministerium für Bildung und Forschung" for funding this research project (OS 42180).

References

[1] Draft International standard ISO DIS 11439: High pressure cylinders for the on-board storage of natural gas as a fuel for automotive vehicles, 1997

[2] CEN draft prEN 12257: Transportable gas cylinders / Hoop-wrapped composite cylinders, Issue Feb. 1996

[3] M. E. Richards, C. Blazek, C. Webster et al.: Compressed natural gas storage optimization for natural gas vehicles, Report Gas Research Institut GRI-96/0364, Dec. 1996

[4] C. Düren, I. von Hagen, G. Junker, E. Lange, C. Rasche: Application of Thermoplastics Reinforced with Unidirectional Fibres in the Manufacture of Composite Vessels, The 3rd Biennial Int. Conf. & Exhibition on Natural Gas Vehicles, Conference Proceedings, p. 473-500, Göteborg (1992)

[5] Rasche, O. Christen, M. Neitzel: Filament winding thermoplastic composite cylinders with On Line Impregnation, SAMPE May 1997, Anaheim, Conf. Proceedings

Simulations of Pultrusion Process of GF/Polyester and GF/Vinyl-Ester Composite Rods

B. Suratno[1], L. Ye[1*], M. Hou[2], Y. Mai[1], and I. Crouch[2]

[1] Centre for Advanced Materials Technology (CAMT), Department of Mechanical and Mechatronic Engineering, University of Sydney, NSW 2006, Australia

[2] Cooperative Research Centre for Advanced Composite Structures Ltd (CRC-ACS), 361 Milperra Road, Bankstown, NSW 2200, Australia

Abstract

A commercial finite element package coupled with a kinetic curing model was used to investigate effects of pulling speed and die temperature setting on centreline temperature and degree-of-cure profiles in pultruded composite rods. The post-die analysis was also conducted since the curing reaction normally continues after the composite exits the die. Curing mechanisms of a thermoset resin for manufacturing pultruded composites play important roles on the final quality and mechanical performance of the products. A comparison study was conducted using commercial polyester and vinyl-ester resins to address some essential aspects for manufacturing pultruded composites.

1. Introduction

Composite structural components with a constant cross section are becoming important in many applications apart from the traditional aerospace industry [1]. There are several available techniques to produce composite components. Among those, pultrusion is one of the most cost-effective techniques, and it is applicable for both thermoset and thermoplastic matrix composites [1]. The process involves pulling fibres through a resin bath for impregnation, a heated die for consolidation and curing, a pulling mechanism for pulling at a constant speed, and a cut saw for cutting the material to the desired length.

To produce pultruded composite components with consistent and high quality, it is important to tailor and control the pultrusion processing parameters such as die temperature and pulling speed settings. Meanwhile, the on-line curing mechanisms of a thermoset matrix material are greatly dependent on the processing parameters. In the previous studies [1-5], simulations of pultrusion processes were mainly conducted using a self-developed finite difference or finite element program coupled with the curing kinetics. An iterative procedure has been developed by the authors to simulate the pultrusion process of thermosetting matrix composites, involving both heat transfer and curing sub-models using a commercial finite element software coupled with numerical approximation of the curing kinetics. Major pultrusion mechanisms can be simulated using such an approach [8].

* To whom all the correspondence should be addressed.

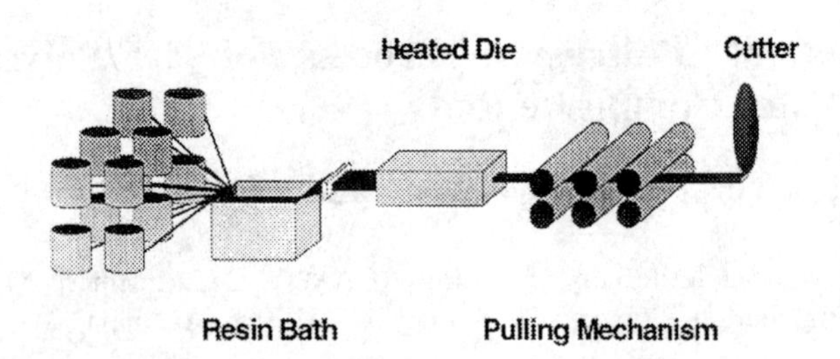

Figure 1. Diagram of a normal pultrusion process

From the previous studies, it was found that in cases where the die temperature setting is low or the pulling speed is high, a substantial amount of curing takes place outside the die. Therefore, it is necessary to incorporate a post die modelling in the simulation. As it is well known, the curing mechanisms of a particular resin play important roles in determining the quality of the final product. The present work presents a two-dimensional finite element model to simulate the temperature and curing profiles in both inside and post die sections. Some essential aspects for manufacturing pultruded composites were also addressed.

2. Process Modelling

The mechanics model to simulate pultrusion process basically includes a curing reaction sub-model and a heat transfer sub-model. In this study, a two-dimensional model was applied. For simplification, it is assumed that the process is in the steady state, the diffusion of resin during curing is negligible, the influence of pressure on the curing reaction is neglected, and the material is homogeneous. Considering those assumptions, the governing equation for the pultrusion process is [3]

$$\rho c u \frac{\partial T}{\partial x} - \frac{\partial}{\partial x}(k_x \frac{\partial T}{\partial x}) - \frac{\partial}{\partial y}(k_y \frac{\partial T}{\partial y}) - \Delta H \, \rho \, m_m \frac{\partial \alpha}{\partial t} = 0 \quad (1)$$

where T is the temperature, k the thermal conductivity, c the heat capacity, u the pultrusion line speed, ΔH the ultimate heat of reaction of the resin, m_m the mass fraction of matrix, α the degree-of-cure, ρ the density of the material.

The pultruded composite material inside the heating die mainly consists of resin, clay filler and fibres. Assuming the thermal properties are not temperature-dependent, the bulk thermal properties of the composite material can be approximated using the micromechanics analysis,

$$\rho = \rho_r v_r + \rho_c v_c + \rho_f v_f \quad (2a)$$

$$c = m_r c_r + m_c c_c + m_f c_f \quad (2b)$$

$$k = \left[\frac{m_r}{k_r} + \frac{m_c}{k_c} + \frac{m_f}{k_f} \right]^{-1} \quad (2c)$$

$$m_f = \frac{\rho_f v_f}{\rho}, \; m_c = \frac{\rho_c v_c}{\rho}, \; m_r = \frac{\rho_r v_r}{\rho} \quad \textbf{(2d)}$$

where the subscripts r, c and f refer to resin, clay filler and fibre, respectively, k the thermal conductivity, c the heat capacity, v the volume fraction, and m the mass fraction. The boundary conditions of the model can be defined as follows,

$$T_i = T_{ambient} \quad \textbf{(3a)}$$

$$T(x,0) = T_{die}(x) \quad \textbf{(3b)}$$

where T_i is the initial temperature and T_{die} is the die wall temperature.

In the post-die stage, heat loss from the surface of the pultruded composite to the ambient atmosphere occurs through convection. This condition is applied to the surface of the composite and expressed as a boundary condition,

$$q_c = h\,(T_{surface} - T_{sink}) \quad \textbf{(3c)}$$

where q_c is the convective heat flux from the face of the composite, h the convective heat transfer coefficient, and T_{sink} the ambient temperature.

Assuming the air flow around the pultrusion composite is laminar, the convective heat transfer coefficient, based on free convection for horizontal cylinders defined by Holman [7], was calculated for the surface of the composite using,

$$h = 0.27(\Delta T / d)^{1/4} \quad \textbf{(4)}$$

where d is the diameter of the material and $\Delta T = Tsurface - Tsink$

3. Curing Reaction of the Resin

The curing within the pultruded composite is a complex chemical reaction process that has not been clearly understood [3]. Differential scanning calorimetry (DSC) was normally applied to measure the heat reaction and the rate-of-cure. The relationship between the rate-of-cure ($d\alpha/dt$) and the degree-of-cure (α) can be approximated using a modified Arrhenius type equation [6],

$$\frac{d\alpha}{dt} = \frac{A\exp(-\Delta E / R.T)}{\alpha_{max}} \alpha^m (\alpha_{max} - \alpha)^n \quad \textbf{(5)}$$

where A is a pre-exponential constant, ΔE the activation energy, R the universal gas constant, T the absolute temperature, α_{max} the maximum degree-of-cure, m and n the order of reaction. Table 1 shows the kinetic parameters for the polyester and vinyl-ester resins used in experimental studies[6].

Table 1. Curing kinetic parameters of polyester and vinyl-ester resins

Material	A [1/s]	E [J/mol]	m	n	ΔH [J/g]
Vinyl-ester	1.015×10^5	47994	0.7	1.3	315
Polyester	6.606×10^6	63624	0.45	1.56	296

4. Numerical Simulation and Experimental Validation

From the governing differential equation (Equation 1), it can be seen that the temperature state and the degree-of-cure are coupled, and they have to be solved simultaneously. In this case, an iteration technique was applied [8]. Firstly, it was assumed that the degree-of-cure is zero everywhere in the composite. Then the finite element analysis was performed to obtain the initial state of the temperature for each element. Using the temperature profile, the rate-of-cure for each element was calculated using Equation 5. At the entrance of the die, the resin enters the die in the form of an uncured liquid such that starting from this entrance line, the degree-of-cure was set to be zero at any step of iteration. Since the process is assumed to be at steady state, the rate-of-cure can be defined by the derivation with respect to the x axis rather than to time.

$$\partial\alpha / \partial x = (1/u)\,\partial\alpha / \partial t \quad \textbf{(6)}$$

Then the degree-of-cure for the element further into the die is obtained by integrating the gradient [3]. This degree-of-cure profile is then applied to calculate the temperature profile for a new step of iteration using the finite element analysis based on ABAQUS finite element package.

Experiment studies were performed for pultruded composite rods with a diameter of 1.00 cm for both GF/Polyester and GF/Vinyl-ester resin systems. The length of the pultrusion die was 60 cm. To overcome the difficulty of measuring the die wall temperature during the process as experienced by Hackett [3], two parallel die channels of the same geometry were manufactured symmetrically between the heat platens. The die wall temperature profiles were measured using thermocouples along one die channel without material flow as the composite was pultruded in the other channel. This die wall temperature profile was then used as the boundary condition for the simulation. One can argue that the actual die wall temperature profile on the die wall will be somewhat different, because of heat convection between the die and the composite. However, the difference will be small due to the poor thermal conductivity and low heat capacity of the composite material, presented in Table 2.

Table 2. Thermal properties of constituents for GF/vinyl-ester and GF/polyester composite materials

Material	Density [g/cm^3]	Specific Heat [J/g.K]	Thermal Conductivity [W/m. K]
Glass fibre	2.56	0.67	11.1×10^{-3}
Vinyl-ester	1.04	1.72	1.22×10^{-3}
Polyester	1.15	1.883	1.92×10^{-3}
Clay filler	2.6	0.92	8.6×10^{-3}

The fibre volume fraction for all experiments was kept constant at 58 %. On-line temperature measurements were also performed by embedding a thermocouple inside the composite material as it passes through the die. Although attempts were made to locate the thermocouple at the centreline of the composite rods, the thermocouple missed the centreline in most cases.

5. Results and Discussion

Profiles of die wall temperature and the predicted centreline temperature distributions for GF/polyester and GF/vinyl-ester composites with the die temperature setting of 140°C at three different pulling speeds are presented in Figure 2. For both cases in the region near the die entrance, the die wall temperature is much higher than that of the centreline. This is due the facts that the die is releasing heat to the composite during the curing reaction and the thermal conductivity of the composite is very low. The length of this zone depends on the pulling speed; the higher the pulling speed is, the longer this zone [3]. The temperature of the composite beyond this zone is higher than that of the die wall due to accumulation of heat generated by the exothermic curing process. The predicted temperature profiles at the location of the thermocouple agree well with the on-line temperature measurements for both GF/polyester and GF/vinyl-ester composite materials. This indicates that the finite element modelling coupled with numerical approximation of the curing kinetic is in good agreement with the experiments.

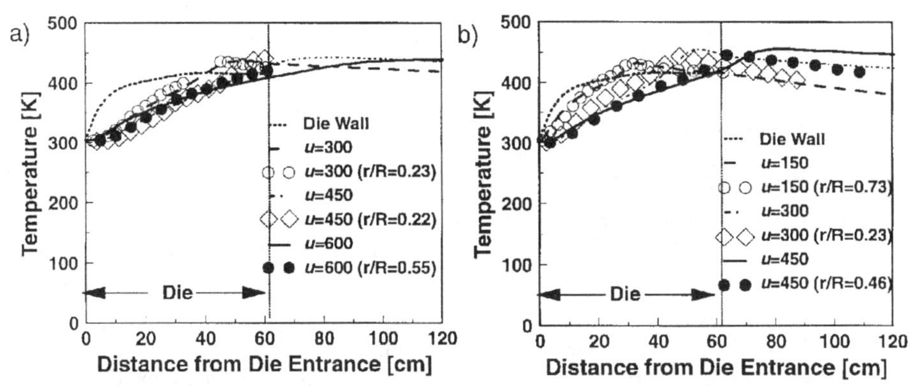

Figure 2. Effect of pulling speed on centreline temperature at die temperature setting of 140°C, a) GF/polyester composite and b) GF/vinyl-ester composite (symbols for experimental data and lines for calculations).

Figure 2 also illustrates the effect of pulling speeds on the centreline temperature profiles for GF/polyester and GF/vinyl-ester composites, respectively. The pulling speed was varied from 150 mm/min to 600 mm/min while the die temperature was 140°C. Evidently as the pulling speed is increased, the maximum exothermic temperature at the centreline is brought backward. This is because heat is conducted more slowly within the composite and the exothermic reaction takes longer to initiate for a high pulling speed. One can also notice that at a high pulling speed the pultruded composite attains a somewhat higher peak temperature at the centreline than that at a low speed. This is attributed to the fact that at a high speed a pultruded composite does not have enough time to release the exothermic heat due to the curing reaction through convection to the die wall.

The profiles of the surface and centreline degree-of-cure for pultruded GF/polyester and GF/vinyl-ester composites with the die temperature setting of 140°C at three different pulling speeds are presented in Figure 3. As shown, the curing reaction has not been completed at the die exit or even at the end the post die analysis (60 cm from the die exit), and the curing reaction still continues. These results propose that the die temperature setting is too low. Furthermore, it can be seen that as the pulling speed is increased, the centreline degree-of-cure decreases and the curing occurs slowly. Although the peak temperature achieved in the pultruded composite at high pulling speeds is increased, the degree-of-cure at the die exit decreases, indicating that the time of the composite passing through the die has an overriding role on the curing quality.

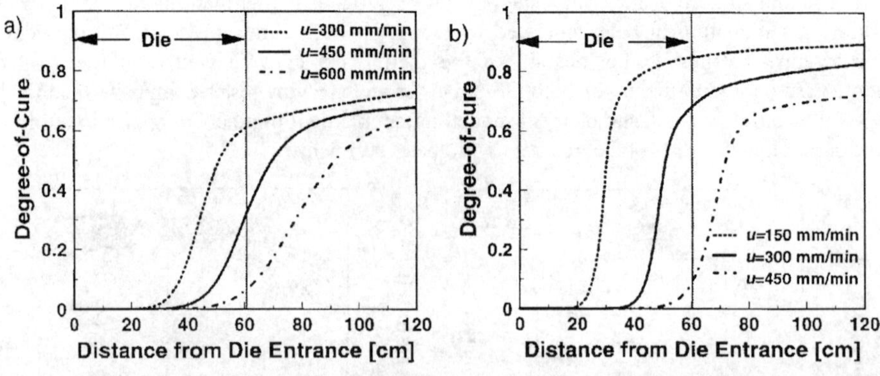

Figure 3. Effect of pulling speed on degree-of-cure profiles at die temperature setting of 140°C, a) GF/polyester composite and b) GF/vinyl-ester composite.

Figure 4 presents the centreline temperature profiles in GF/polyester and GF/vinyl-ester composites for various die temperature settings (140°C, 160°C, and 180°C) at a pulling speed of 600 mm/minute. As expected, the centreline temperature increases with the increase of die temperature setting; and at the low die temperature setting the peak temperature is brought backward, compared to that at the high temperature, due to the fact that at the high temperature setting the curing reaction proceeds faster than that at the low one. Meanwhile, at the high die temperature setting, the results show a high degree-of-cure at the die exit, shown in Figure 5a. However, for the GF/vinyl-ester composite the high die temperature setting does not promote a high degree-of-cure at the die exit, presented in

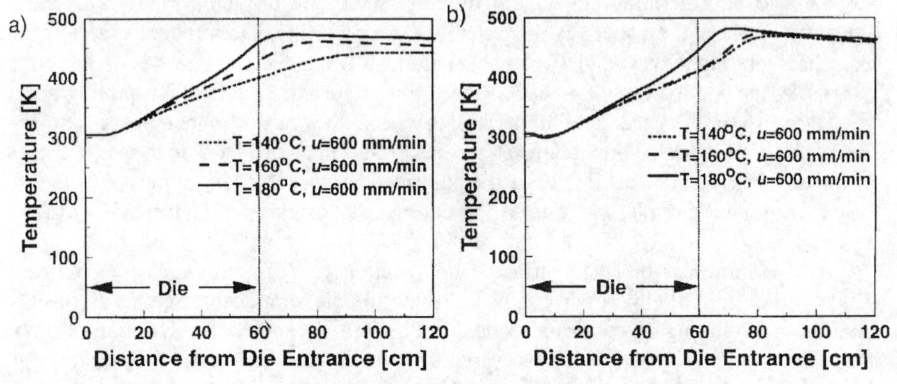

Figure 4. Effect of die temperature setting on centreline temperature profiles, a) GF/polyester composite and b) GF/vinyl-ester composite.

Figure 5b. This is because the vinyl-ester resin exhibits much more evident vitrification than the polyester resin [6]. Therefore, increasing die temperature setting does not necessarily increase the degree-of-cure for the GF/vinyl-ester composite.

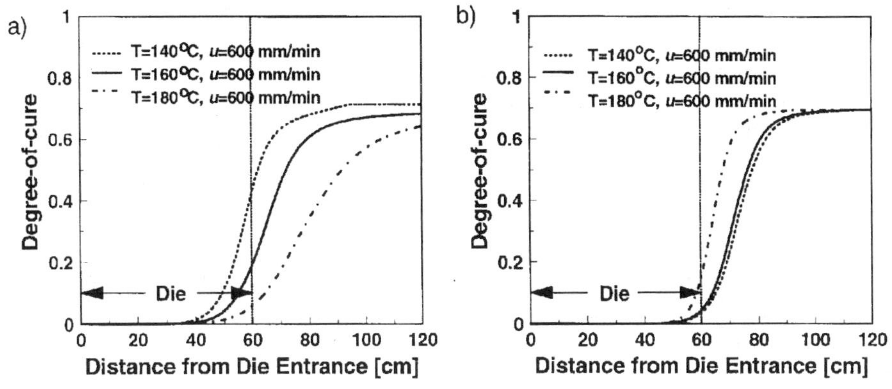

Figure 5. Effect of die temperature setting on degree-of-cure profiles, a) GF/polyester composite and b) GF/vinyl-ester composite.

Figure 6a shows centreline temperature profiles for both GF/polyester and GF/vinyl-ester composites at the die temperature setting of 140°C and a pulling speed of 300 mm/min. Through comparison, one can easily notice that in the GF/polyester composite the temperature profile is superseded by that in GF/vinyl-ester composite. This is because the ultimate heat of the vinyl-ester resin is higher than that of the polyester resin such that the GF/vinyl-ester composite experiences a stronger exothermic reaction than the GF/polyester composite. Figure 6b presents the degree-of-cure profiles at the centreline of the both pultruded composites. It can be seen that the GF/vinyl-ester composite has a higher degree-of-cure than the GF/polyester composite at the centreline. This may be due

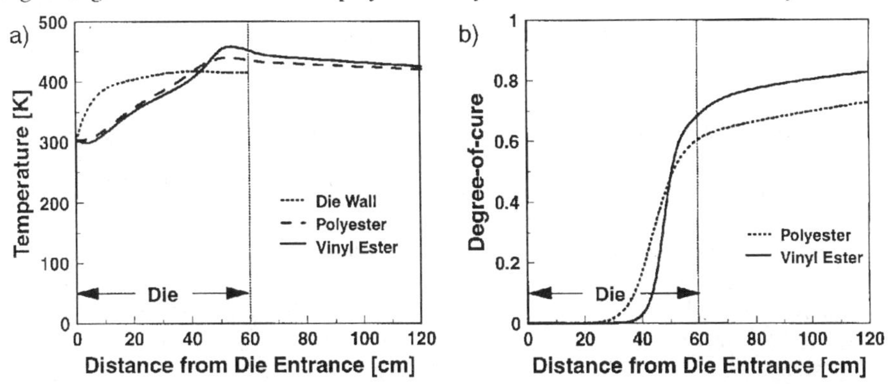

Figure 6. Centreline profiles for GF/polyester and GF/vinyl-ester at pulling speed of 300 mm/min and die temperature setting of 140°C, a) Temperature and b) Degree-of-cure.

to the facts that the peak temperature and curing rate in the GF/vinyl-ester composite are higher than those in the GF/polyester composite.

6. Conclusion

The simulations of pultrusion process of GF/polyester and GF/vinyl-ester composites at various experimental conditions, i.e. different die temperature setting and pulling speeds were performed. It was found that the simulated temperature profiles along the pultrusion die length were in good agreement with experimental data. At the same experimental conditions, the GF/vinyl-ester composite has a higher degree-of-cure than that of GF/polyester composite. Meanwhile, for the GF/vinyl-ester composite increasing die temperature setting does not increase the degree-of-cure of the final product. Using the methodology developed in this study, a procedure to simulate profiles of temperature and degree-of-cure of pultruded composites can be easily adopted using a commercial finite element package.

References

[1] *M. Valliappan, M., J. A. Roux, J. G. Vaughan, E. S. Arafat*, DIE AND POST-DIE TEMPERATURE AND CURE IN GRAPHITE/EPOXY COMPOSITES - Composites: Part B, 27B (1996) 1-9.

[2] *G. L. Batch, C. W. Macosko*, HEAT TRANSFER AND CURE IN PULTRUSION: MODEL AND EXPERIMENTAL VERIFICATION - AIChE Journal, 39 (1993) 1228-1241.

[3] *R. M. Hacket, S. Zhu*, TWO-DIMENSIONAL FINITE ELEMENT MODEL OF THE PULTRUSION PROCESS - Journal of Reinforced Plastics and Composites, 2 (1992) 1322-1351.

[4] *E. M. Ma, C. Chen*, THE DEVELOPMENT OF A MATHEMATICAL MODEL FOR THE PULTRUSION OF BLOCKED POLYURETHANE COMPOSITES - Journal of Applied Polymer Science, 50 (1993) 759-764.

[5] *Y. R. Chadchad, A. Roux, J. G. Vaughan*, EFFECT OF PULL SPEED ON DIE WALL TEMPERATURES FOR FLAT COMPOSITES OF VARIOUS SIZES - Journal of Reinforced Plastics and Composites, 15 (1996) 718-739.

[6] *B. R. Suratno, L. Ye, Y. W. Mai*, SIMULATION OF TEMPERATURE AND CURING PROFILES IN PULTRUDED GF/POLYESTER AND GF/VINYL-ESTER COMPOSITE RODS - (to be submitted to Composite, Part A. December 1997).

[7] *J. P. Holman*, HEAT TRANSFER - Third Edition, McGraw-Hill, New York, 1972

[8] *B. R. Suratno, L. Ye, Y. W. Mai*, SIMULATION OF TEMPERATURE AND CURING PROFILES IN PULTRUDED COMPOSITE RODS - Composites Science and Technology (in press, 1998).

CAD/CAM FOR ROBOTIC FILAMENT WINDING PROCESS DESIGN

L. Carrino*, G. Moroni*, S. Turchetta*

* Industrial Engineering Department - Università degli Studi di Cassino - via Di Biasio 43, 03043 Cassino, Italy - Fax +39 0776 310812, E-mail: moroni@ing.unicas.it

Abstract

Filament winding is a process to manufacture composite material axis-symmetric parts on numerically controlled machines. Not axis-symmetric parts are usually wound by hand. However, the need for high precision and cost reduction in manufacturing complex 3D parts is bringing to the development of robotic based deposition technologies. At the Industrial Engineering Department of the Cassino University a robotic filament winding work cell has been studied and built. This technology seems very promising for small batch production of parts, not only for the aeronautic and aerospace industrial sectors. The development of CAD/CAM software in order to optimize the process is expected to enhance its benefits. In this paper the structure and the implementation of a CAD/CAM software system for product design and process planning of complex three-dimensional parts to be manufactured by robotic filament winding is presented. Particularly, the filament path generation problem is deeply discussed.

1. Introduction

The need for high precision and cost reduction in manufacturing composite material parts is bringing to the development of robotic based deposition technologies [1-5]. In filament winding applications, a robot has to wind a composite filament on a 3D shaped support through complex trajectories, in a way similar to the human manufacturing for hand made objects. At the Industrial Engineering Department of the Cassino University a robotic work cell for filament winding manufacturing has been studied and built [6-8]. Its layout is shown on figure 1a. The basic elements are a 5 degree of freedom industrial robot, a feeding spool for the strand installed on the robot forearm, a suitable gripper device (deposition head), and an adequate shaping support, on which the robot has to wind the desired object avoiding collisions with obstacles.
This robot work cell has been tested on a real industrial part: a helicopter tension link (figure 1b). At the moment AGUSTA S.p.A. manufactures this component through a manual filament winding process by using composite material rowing, which is made by a tow of graphite fibers which are impregnated with thermosetting resin. It is worth to note that the tension link is composed by two symmetric components, with a real complex three-dimensional shape and each one separately wound.
The experiences have shown that the development of a robotic filament winding work cell is convenient to fulfill to the need of high precision, cost reduction and flexibility in manufacturing of three-dimensional composite material parts. However, the integration

with a CAD/CAM software in order to optimize the process is expected to enhance the benefits of this technology, which seems very promising for small batch production of parts, not only for the aeronautic and aerospace industrial sectors.
The state of the art on CAD/CAM for filament winding has been fully discussed in [9]. However, no existing tool is able to help the user in part and process design of complex 3D products, like the considered tension link. Therefore, at the Manufacturing Laboratory of the Cassino University a research, whose main goal is to develop such a system, is currently in advanced state of progress.
In this paper the structure and the implementation of a CAD/CAM software system for product design and process planning of complex three-dimensional parts to be manufactured by robotic filament winding is presented.

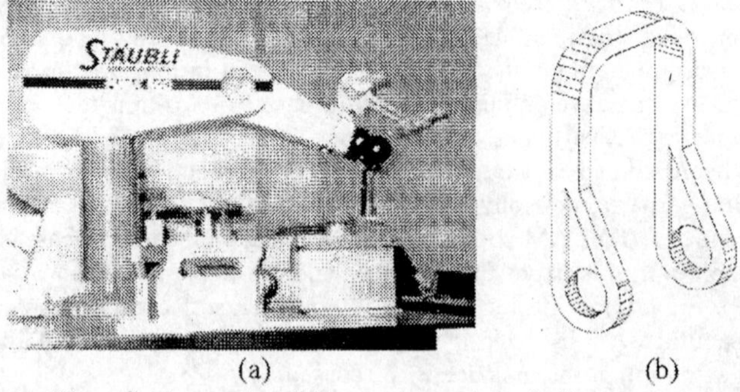

(a) (b)

Figure 1 - The robotic work cell layout (a) and the helicopter tension link (b).

2. CAD/CAM Architecture

The architecture of the CAD/CAM system under development is shown in figure 2a. A simplified version of the helicopter tension link (figure 2b) has been considered as benchmark to test the filament winding robotic work cell and the CAD/CAM system.

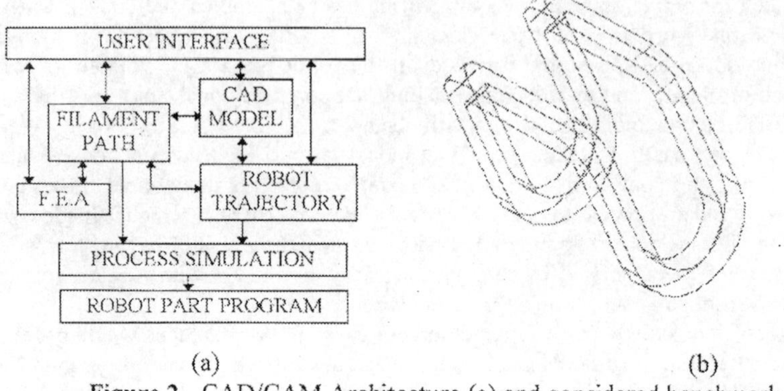

(a) (b)

Figure 2 - CAD/CAM Architecture (a) and considered benchmark (b).

An accurate analysis of the benchmark geometry has brought to a definition of a class of parts that the software should be able to design. From a geometric modeling point of view, it can be seen as a sweep of a rectangular section along a close not self-intersecting path. During the manual manufacturing of this part, the operator winds the rowing on supporting surfaces following a filament path close to the sweeping path, laying up the fibers until the whole section is obtained. Of course, these supporting surfaces are extremely important in order to obtain the desired object shape.
The CAD module of the CAD/CAM system under development has the main goal to help the user to design parts belonging to the described class [10]. In order to do this, interacting with the system the user has to define the shape of the section to be swept, the close path along which the sweep is performed, and the sides on which the part may be supported during the winding process (figure 3a). From this data, the software is able to generate automatically a draft model of the shaping support, as shown in figure 3b. Of course, the draft of the support may be saved and modified to obtain the final design of it.

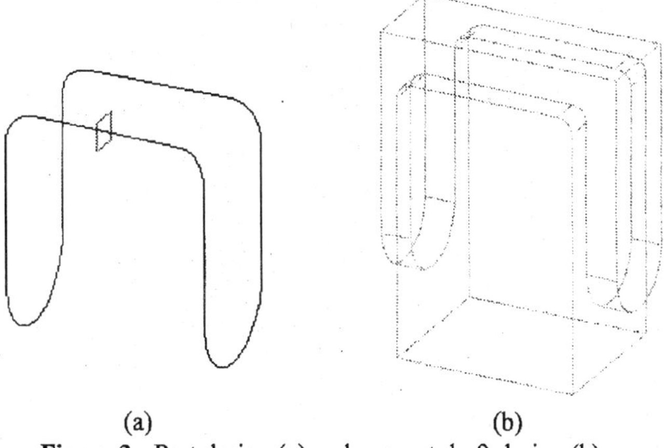

(a) (b)
Figure 3 - Part design (a) and support draft design (b).

Starting from these design data, the system should be able to generate different filament paths along which the fibers have to be wound, as shown in figure 4. Particularly, in the rest of the paper the filament path generation will be deeply discussed. Integration with a F.E.A. system completes the part design stage.
Given a filament path, the robot trajectory generator module has to define the trajectories to be followed by the deposition head to wind the part. In doing this, it is required that the filament path is the given one, the tension of the filament is as constant as possible, the deposition head and the robot arm move on collision free paths, and the free filament does not interfere with the support and the whole environment. Algorithms to perform this automatic planning have been presented in [9,10] and examples are shown in figure 5a. Finally, it results straightforward to generate the robot part program and to simulate the process, as in figure 5b.

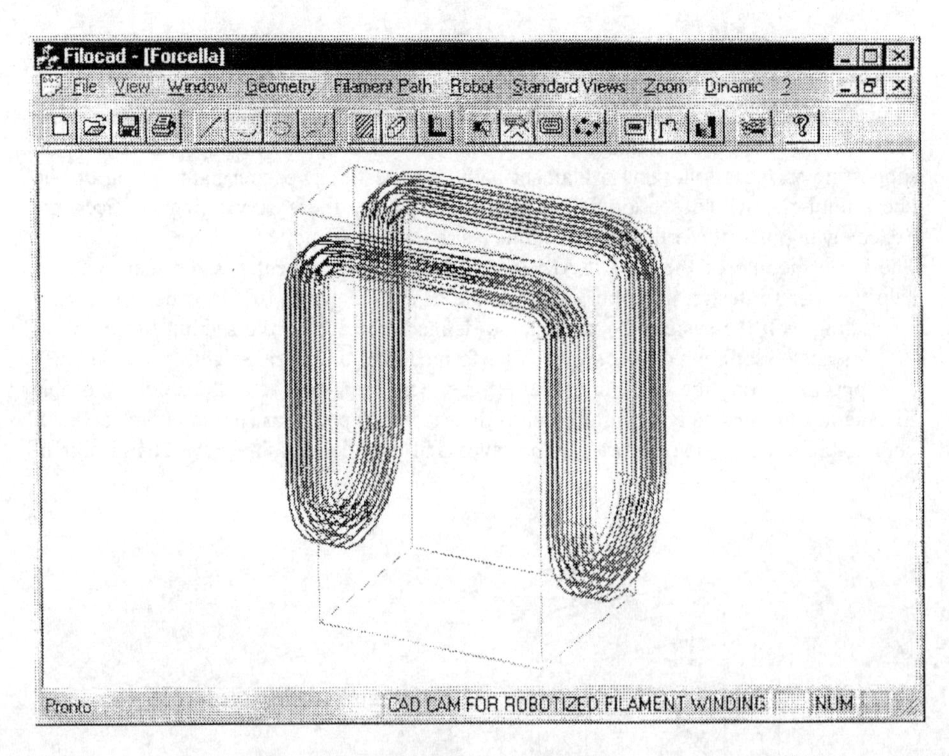

Figure 4 -Example of filament path generation for the benchmark object.

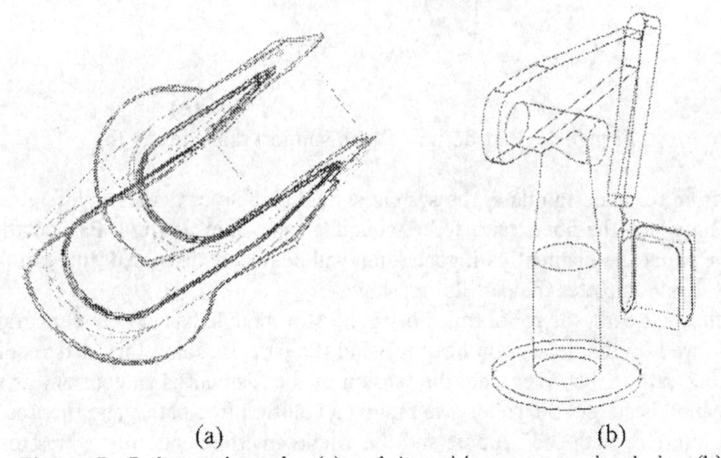

(a) (b)

Figure 5 - Robot trajectories (a) and deposition process simulation (b).

3. Filament Path Generation

The parts the software is able to deal with can be seen, from a geometric modeling point of view, as a sweep of a planar section along a close not self-intersecting 3D path. For each part of this kind the Filament Path Generation module should be able to determine how a unique continuos composite material filament has to be wound in order to fill its entire volume. In reaching this goal, two main effects are to be avoided: fiber bridges and fiber slippage.
In order to have a deposition without any bridge it is necessary to have a convex filament path. The filament path is closely related to the sweeping path of the part. Therefore, it is possible to reason on this sweeping path to evaluate the convexity of the part itself. To do so, the sweeping curve is discretized in a user definable number of points. For each point $\boldsymbol{P}_i$ the tangent unit vector $\boldsymbol{u}$ to the curve is determined and a point $\boldsymbol{P}_i^*$ on $\boldsymbol{u}$ direction and very close to $\boldsymbol{P}_i$ is chosen. Simultaneously, a particular support is generated in a way similar to that used to generate the shaping support and described in [10]. This time, however, the support is generated so that the sweeping curve lies on it. Then, if every points $\boldsymbol{P}_i^*$ of the discretization are either outside or on this particular support then the sweeping path is convex, otherwise it is concave.
Once the sweeping curve convexity is confirmed, the filament path could be generated. To avoid any fiber slippage during the deposition the generation algorithm will consider only those filament paths that correspond to geodetic curves for the supporting surfaces. For that curves the slippage tendency is null.
In the following the path generation algorithm will be examined in detail.

3.1. Path Generation Algorithm

As previously stated the software is able to deal with parts having rectangular planar sections. The composite material filament, called rowing, has itself a section that may be approximate with a rectangle. Therefore, in order to have the part volume completely filled by continuos rowing, it is possible to consider rowing sections displaced in the part section like in figure 6a. This rowing displacement may be obtained by sweeping each rowing section along the part sweeping-curve. However, doing so it is not possible to have continuos deposition, because a filament starting from position 1 in figure 6a will terminate in the same position, as shown in figure 6b. To be able to have a continuos deposition it is necessary that the filament starting in position 1 terminates, after a winding round, in a new position, like position 2, from which to start a new winding round. In figure 7a an example is given.

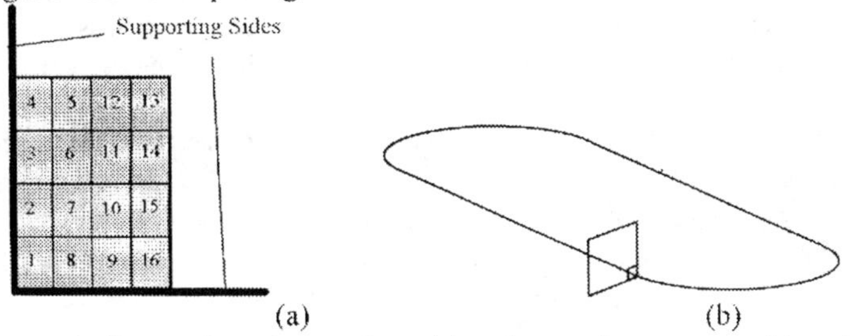

Figure 6 - Part and rowing sections (a) and sweeping curve problem (b).

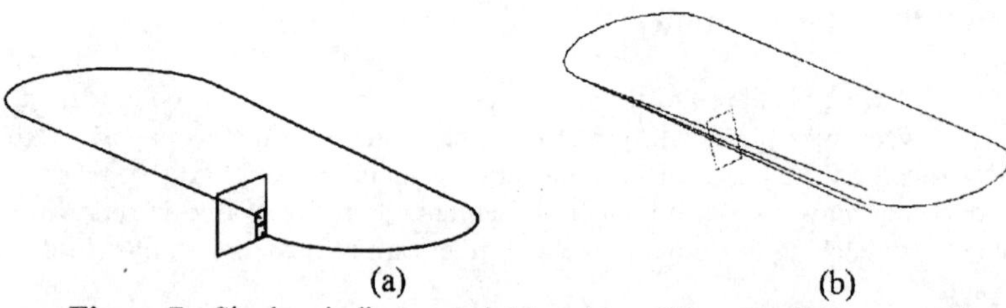

Figure 7 - Single winding round (a) and possible sweeping curves (b).

To be able to generate such a path by sweeping, it is necessary to modify the sweeping curve. The modified sweeping curve is an open curve, which is generated by substituting an element to the part sweeping-curve. Usually the substituted element is a straight line, because whatever is its direction a straight line on a planar surface is always a geodetic line, and the slippage is avoided.
In the previous example, referring to figure 6a, we consider a filament path going from position 1 to position 2. Following this method it is possible to lay up the rowing on the vertical supporting surface (in the plane of the section) going to position 3 and 4. Obviously, an alternative deposition path is the one that goes from position 1 to 8, 9 and 16. The choice of the main deposition direction is given to the user.
Once a layer is obtained, it is necessary to define how to obtain the other layers to complete the part volume. At the moment two different deposition strategies are possible: the continuos layering and the discontinuous layering.
The continuos layering is a strategy in which after obtaining a layer by aligning the rowing in one direction and versus, the following layer follow the same direction but the opposite versus. Considering figure 6a, if we consider a vertical deposition this situation corresponds to the deposition from position 1 to 2, 3, and 4 for the first layer, and from 5 to 6, 7, and 8 for the second layer, otherwise from position 1 to 8, 9, and 16 for the first layer, and from 15, to 10, 7, and 2 for the second layer. Of course the following layers are to be connected and a path going from position 4 to 5, in the first case, or from position 16 to 15, in the second one, are to be created.
In the discontinuous layering, instead, each layer will follow the same direction and versus. In the vertical deposition case, it means that after reaching position 4, end of the first layer, the second layer will start from position 8 and go up to position 7, 6, and 5. The connection between position 4 and 8 has to be generated.
Whatever is the chosen strategy, laying up the following layers it is possible to generate a continuo filament path, in which a rowing is able to fill the entire part volume. An example for the benchmark part is given in figure 4.
From a technological point of view, it is important to note that the rowing section is not exactly rectangular. This may bring to voids formation. To avoid this problem we consider the two dimensions of the rowing rectangular section as pitches in the two deposition directions. In this way, it is possible to control the overlapping degree of the rowing inside the generated part and avoid the void presence.

4. Conclusion

At the Industrial Engineering Department of the Cassino University a robotic filament winding work cell has been studied and built. This technology seems very promising for small batch production of parts, not only for the aeronautic and aerospace industrial sectors. The development of CAD/CAM software in order to optimize the process is expected to enhance its benefits. In the CAD/CAM system under development at the Manufacturing Laboratory a specific interface enables the user to design a filament-wound component. Starting from these data, the system is able to propose a draft design of the shaping support and to generate alternative filament paths, which represent parts with different fiber orientation. If no feasible path is able to respect the requirements, the user will be asked to change some design parameters or even the whole part design. Interacting with the system the designer should be able to perform a complete stress analysis on the composite part with a given fiber orientation. Simultaneously the manufacturability of the part may be evaluated by running a robot trajectory generator module. This module is able to generate collision free trajectories of the robot to wind correctly the fibers on the shaping support and give an estimation of the process cycle time and cost. Finally, a complete simulation of the process may be performed and the robot part program may be generated. The prototype is implemented according to the object-oriented paradigm using the Visual C++ compiler and the ACIS® geometric modeler.

Acknowledgements

The research presented in this paper has been partially supported by Centro Ricerche Compositi (Italy), whose management is gratefully acknowledged.

References

[1] Sholliers J. and van Brussel J., "Computer-integrated filament winding: computer-integrated design, robotic filament winding and robotic quality control". Composite Manufacturing, vol.5, n.1, pp.15-23, 1994.

[2] Castro E., Seereeram S., Singh J., Desrochers A.A., and Wen J.T., "A real-time computer controller for a robotic filament winding system". Journal of Intelligent and Robotic Systems, n.1, pp.73-91, 1993.

[3] Tornincasa S., Ippolito R., and Bellomo N., "New trends in robotics: the robotized lay up of carbon fiber tapes on surfaces with large curvature". Proceedings 21st Symposium on Industrial Robotics, pp.215-220, 1990.

[4] Hummler L., Lee S.K., and Steiner K.V., "Recent advances in thermoplastic robotic filament winding". CCM Report 91-06, University of Delaware.

[5] Brite/EuRam Project BREU0114 – "Design of structures in composite materials with CAD/CAM techniques - Achievement of a prototype of a fully automated equipment for filament winding".

[6] Ceccarelli M., Iacobone F., Carrino L. and Anamateros E., "On the Manipulation of

a Composite Material Rowing for a 3D Winding Manufacturing", 26th International Symposium on Industrial Robots, Singapore, pp. 597-602, 1995.

[7] Ceccarelli M., Volante G., Carrino L. and Anamateros E., "A Feeding Device for Composite Material Rowing in a Robotized 3D Winding Manufacturing", 27th International Symposium on Industrial Robots, pp. 987-992, Milano-Italy, 1996.

[8] Moroni G., Carrino L., Ceccarelli M., and Anamateros E., "Robotized Filament Winding Manufacturing: Some Experiences", RAAD'97, Cassino-Italy, June 26-28, 1997;

[9] Carrino L., Landolfi M., Moroni G. and Di Vita G., "CAM for Robotized Filament Winding", Advanced Manufacturing System and Technology, CISM Courses and lectures No.372, E. Kulianic (Ed.), Springer Verlag, pp.601-608,1996.

[10] Carrino L., Ceccarelli M. and Moroni G., "New software tool for robotized filament winding process design", Advancing with Composites'97, May 6-7, Milano, 1997.

Optimisation of the bending performance of GMT-produced low cost textile reinforced thermoplastic composites by an experimental design

Sofie Baeten *, Ignaas Verpoest (KULeuven, B), Karl Schulte (TUHamburg-Harburg, G), Walter Zäh (Karl Mayer, G), Edith Mäder (Institüt für Polymerforschung, G); Carl-Håkan Andersson (Lund University, S), Kjell Eng (Engtex, S)

*: to whom correspondence should be addressed. Department of Metallurgy and Materials Engineering, KULeuven, de Croylaan 2, B-3001 Leuven, Belgium

Abstract

A growth in the use of thermoplastic composites is currently driven by a reduction of the manufacturing cost and an optimisation of the part performance. To decrease the textile cost, weft-inserted warp knitting is chosen to produce a new split-warpknit textile, comprising both reinforcing glass fibres and PP matrix ribbons. Furthermore, it is shown that this new type of thermoplastic textile is processable with a fast GMT-like cold pressing method for the one-step manufacturing from textile preform into a final composite part. A conductive heating at low pressure is chosen for the external preheating of the textile layers, prior to forming in a cold mould. An experimental design method is applied to determine the optimal settings of the preheating temperature, preheating time, mould temperature, holding time and mould pressure for the flexural performance of the split-warpknit laminates. The optimal parameter array is then validated by performing tensile tests. The linear correlation between the flexural and tensile properties confirms the optimal process window.

1. Introduction

For minimising fuel consumption and reducing the weight in automotive applications, new lightweight, low cost materials and appropriate fast processing techniques are an interesting area of research.

As a thermoplastic matrix appears to be favourable for mass production with short cycle times, a cost-effective glass fibre reinforced thermoplastic textile preform has been developed, where the preform contains both the reinforcing fibres and the matrix material. The merit of such a structure is the good drapability combined with a comparable low production cost, whereas the composite part can eventually be made in only one step directly from the textile by simple pressing.

In all material developments for mass applications, the processability in an industrial environment, is an important issue. The three stage process for thermoplastic composites consists of preheating the matrix polymer above its melting point, followed by a compaction to wet out the reinforcement and to give the final shape to the part and a

cooling down [1]. The matrix must be heated sufficiently above its melting point to reduce the viscosity for full fibre impregnation.
The *matched-die moulding*, where a loose stack of textile layers is heated, formed and cooled down in the mould, involves high heating energy flow to and from the forming equipment, resulting in long cycle times and high operating costs. This hot pressing process is hence far too expensive for industrial use.
During an alternative *cold pressing* method, the fabric is preheated separately in an external oven prior transfer to the forming equipment, which is at a temperature below the glass transition temperature of the thermoplastic matrix. This technique is well established in the GMT (Glass Mat reinforced Thermoplastic) processing and eliminates the cycling of the mould temperature, avoiding large cycle times. However, in classical GMT-processes, pre-impregnated materials are used, and the GMT-processing is mainly chosen to shape the part.

Hence, the goal of the present work is the development of an adequate fast manufacturing technique to produce split-warpknit composites in a cost-effective manner. The feasibility of the GMT-based cold pressing technique, allowing for the use of industrially available equipment, for this type of material in non-pre-impregnated form, is evaluated.

2. Glass fibre reinforced SPLIT-WARPKNIT thermoplastic textile

To produce the split-warpknit textile, weft-inserted warp knitting is applied. The thermoplastic ribbon which will form later on the matrix is laid down parallel to the glass fibres (fig 1) and stitched together with a non-meltable PET binding yarn.

The advantages of this class of thermoplastic SPLIT-WARPKNITS are versatile [3][4]:
- the production cost of the textile is low because thermoplastic ribbons are used
- the structure shows a good drapability: complex parts can be made in a one step process from dry preform to final composite part
- the final composite part shows increased mechanical performance due to the non-crimp reinforced structure
- one textile structure is composed of different layers, with different orientations of the reinforcing fibres. It is hence a multi-layer, multi-axial preform which enhances the part cost efficiency
- the split-warpknit textile is processable with the GMT-based cold pressing technique, allowing for a fast production using commercially available moulding equipment. The composite plate is hence obtained by simple cold pressing the preheated textile preform, without any further addition of matrix or other kind of pre-treatment.
- as the hybrid yarn structure determines the homogeneity of the distribution of the reinforcing fibres in the polymer matrix and thus the mechanical performance, a good composite part can be made because the fibres and matrix are intensely mixed.

Special glass fibre sizings and a low viscosity PP matrix material are developed to get a good wetting out of the reinforcement and an easy, thus fast impregnation [2].

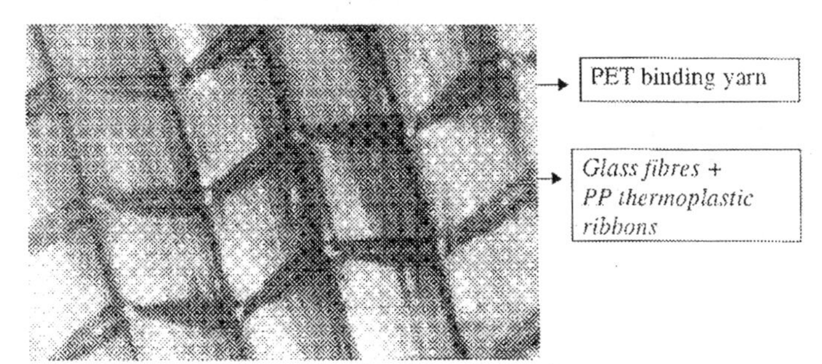

Figure 1: Split warpknit thermoplastic textile preform

3. GMT based processing technique

As the duration of the heating and the cooling phase is the bottleneck when optimising the cycle time in matched-die compression moulding, a GMT-like production process including an external preheating phase and a cold mould pressing, is used for this new split-warpknit material and optimised allowing for a fast production process using commercially available moulding equipment (see figure 2).

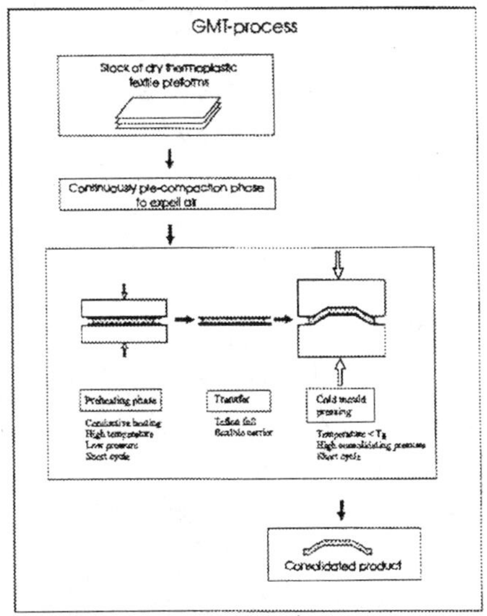

Figure 2: GMT-like cold pressing: one step process from textile to final composite part

A first aim of the study is to investigate whether an external preheating and a cold pressing is feasible for the one step manufacturing of the split-warpknit into a final composite part. Secondly, a systematic optimisation of the process parameters is performed.

4. Feasibility of an external preheating phase

An uncontrolled shrinkage of the stretched thermoplastic ribbons during free heating, causes wrinkling of the fibres and hence a distortion of the fabric. The shrinkage can be prevented by sideways clamping of the preform or by applying a flatwise pressure on the preform. This also means that the textile can not be heated by IR or hot air without a pre-consolidation to eliminate the shrinkage of the ribbons.

Only conductive heating of the textile layers between the platens of a hot press at very low pressure (< 2 bar) is feasible. After transferring the molten layers between two Teflon foils, a flat laminate is subsequently be pressed in a cold mould [5].

5. Systematic optimisation by an experimental design method

While during in-mould heating and cooling, the impregnation time is a programmable parameter, this is no longer the case for the GMT-like cold pressing technique. Hence, an experimental design method is applied for reaching the optimal processing window in an efficient way [5][6].

In a designed experiment, the series of tests are determined on beforehand where planned changes are made to the production parameter settings to observe and identify changes in the output properties [7][8][9]. A well known *Taguchi* study is performed to investigate the effects of processing parameters on the flexural behaviour of the split-warpknit laminates. The study encompassed variation in preheating temperature and time, mould pressure and mould temperature and holding time. An overview of the strategy is given in figure 3.

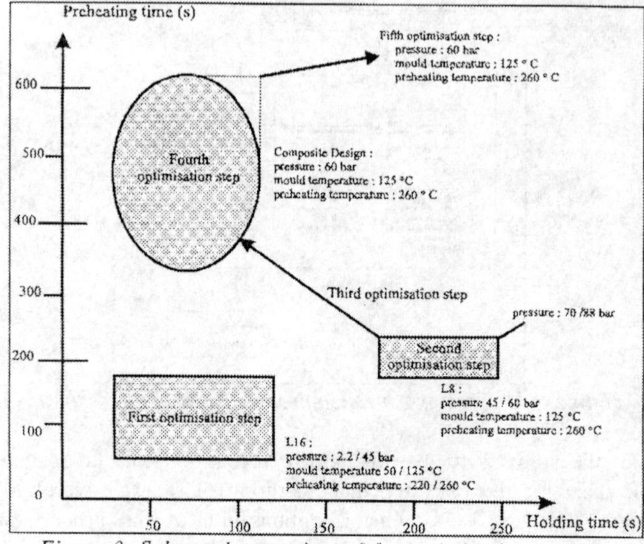

Figure 3: Schematic overview of the optimisation strategy

SCREENING
At first, the process parameters were investigated at two levels using an L16 array, performing 16 experiments (table 1).

	Low	*High*
Preheating temperature (°C)	220°C	260°C
Preheating time (s)	60 s	180 s
Mould pressure (bar)	2.2 bar	45 bar
Mould temperature (°C)	50°C	125°C
Holding time (s)	30 s	120 s

Table 1: Parameter settings of the screening

Cross ply laminates $(90,0,90)_{2s}$ of *the mono axial GF/PP with PET binding yarn* (see fig 1) were tested in transversal three point bending. Flexural tests are chosen because of the the combination of normal and shear stresses which allows a good characterisation of the impregnation quality [10]. A statistically based model is derived to predict the flexural properties of the split-warpknit composites at any processing level between the lower and upper limit of the parameter settings. The magnitude of the parameter effects is investigated by an analysis of means, ANOM, using a response table, while the significance of the effects is determined by performing an analysis of variance, ANOVA method [7].

Figure 4, shows the magnitude of the main effects on the flexural modulus. A similar behaviour is noticed for the bending strength. When ranking the processing parameters in order of importance, it is clearly seen that the preheating time, the holding time and the mould pressure are having the largest effect on the flexural stiffness and strength.

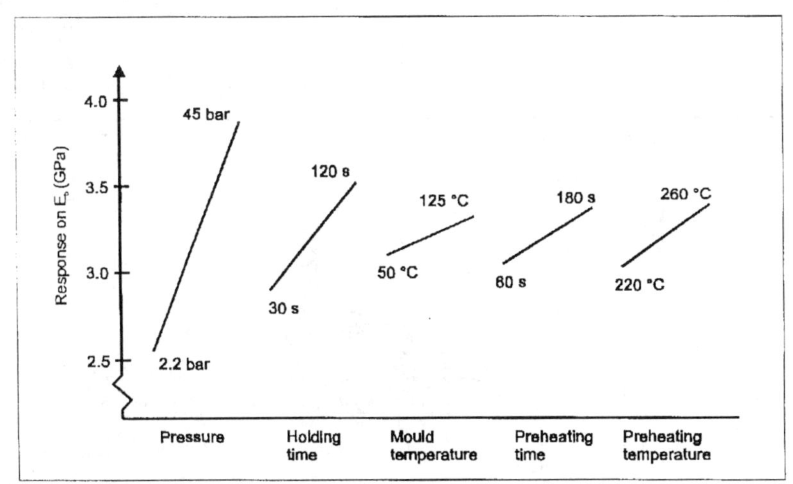

Figure 4: Effect of the parameter settings on the flexural modulus of split-warpknit laminates during a first screening

FIRST OPTIMISATION

As the preheating temperature is limited to the degradation temperature of the PP matrix, only the preheating time, mould pressure and holding time are further increased. The interaction between the preheating time and the holding time was further investigated (table 2). It could be seen that increasing the pressure further improves the bending performance of the laminate. At a low holding time, an increase in the preheating time is also beneficial.

	Low	***High***
Preheating time (s)	180 s	240 s
Mould pressure (bar)	45 bar	60 bar
Holding time (s)	180 s	240 s

Table 2: Parameter settings of the first optimisation step

SECOND OPTIMISATION

The pressure is further increased to the maximum allowable pressure (about 88 bar) and the interaction between preheating and holding time is studied towards a lower holding time (economically driven). The parameter values are calculated in the direction of the steepest slope of the plane of the measurements, obtained during the former optimisation step (table 3). The bending strength increases with a lower holding time and a higher preheating time (fig 5). It is not clear however from this set of experiments, which of both parameters plays the predominant role. This is examined in the third optimisation step. A higher pressure gave no further improvement. However, the applied pressure is much lower than the pressure of the classical GMT process (150-200 bars).

Holding time (s)	***Preheating time (s)***
160	276
140	312
120	348
100	384

Table 3: Settings for holding and preheating time during the second optimisation step

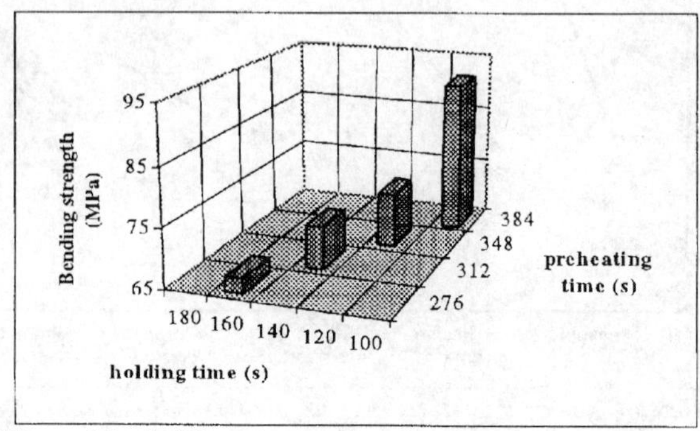

Figure 5: Effect of the holding time and preheating time on the flexural strength

THIRD OPTIMISATION

A composite design method of elliptical shape is applied to model the interaction of the preheating and holding time on the flexural behaviour. The settings, calculated to maintain an elliptical shape, are given in figure 6. The pressure is set at 60 bar, because a further increase did not show any improvement in the flexural behaviour. A quadratic regression model is computed to model the response surface for the flexural modulus (fig 7) and strength.

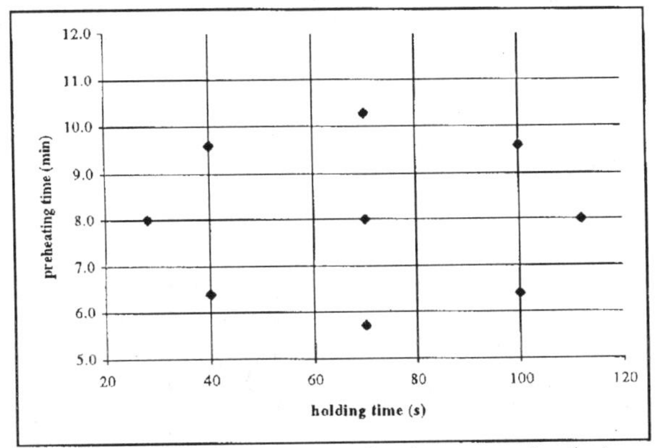

Figure 6: Elliptical parameter array for the composite design method

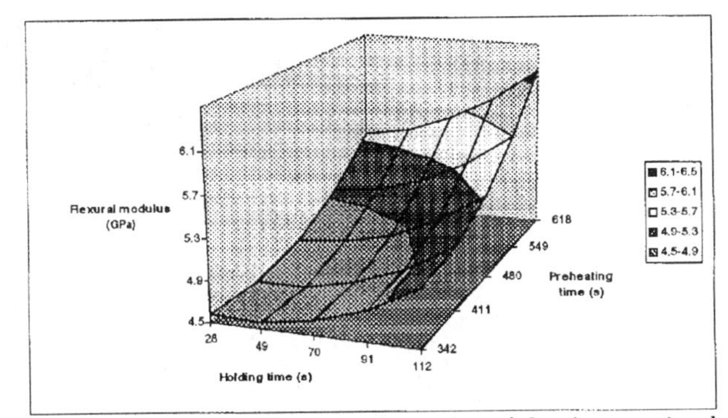

Figure 7: Modelled surface for the flexural modulus (composite design)

After evaluation of the effect of the different parameters, an optimal processing window for the GMT cold pressing of split-warpknit composite laminates is derived:

- *Preheating*
  - preheating time: 540 s
  - preheating temperature: 260°C
- *Cold pressing*
  - mould temperature: 125°C
  - mould pressure: 60 bar
  - holding time at pressure: 100 s

6. Validation of the optimal process

Tensile tests on cross ply laminates $(90,0,90)_{2s}$ were carried out to validate the optimal parameter set. The correlation between the tensile and bending modulus is shown in figure 8 as a function of processing parameters (see table 4). A linear behaviour can be seen for the modulus and strength.

Type	**preheating time (s)**	**preheating temperature (°C)**	**Mould pressure (bar)**	**Mould temperature (°C)**	**Holding time (s)**
1-GMT	180	260	2.2	50	30
2-GMT	180	220	45	50	30
3-GMT	180	220	45	125	120
4-GMT	240	260	60	125	180
5-GMT	180	260	45	125	120
6-GMT	180	260	45	125	120
7-GMT	380	260	60	125	100
8-GMT	610	260	60	125	70
9- Classical	-	-	20	240	300

Table 4: Parameter set of the laminates, also tested in tension

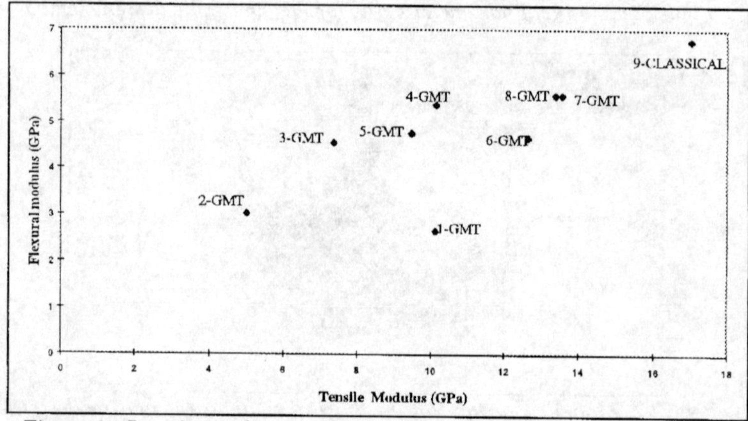

Figure 8: Correlation between the tensile and flexural modulus (table 4)

7. Conclusions

A new type of continuously glass fibre reinforced thermoplastic textile is developed where PP split-film ribbons are used as matrix material. Furthermore, it is proven that this textile is processable with a fast GMT-like cold pressing technique where the final composite part is manufactured in one step from the preform. No intermediate steps seem to be necessary. A Taguchi study is preformed to determine the optimal process window for the flexural and tensile properties of cross ply split-warpknit laminates.

Flow and Heat Transfer Analyses during Tape Lay-up Process of Thermoplastic Composite

H. J. Kim*, W. I. Lee*, S. W. Sihn, S. W. Tsai****

* Dept. of Mechanical Engineering, Seoul National University, Seoul 151-742, Korea
** Dept. of Mechanical Engineering, Stanford University, Stanford, CA 94305, USA

Abstract

Tape lay-up process of thermoplastic composites was analyzed. The heat transfer coefficient distributions along the surfaces of composite were evaluated via three-dimensional flow analysis. Surface temperatures of the composite were obtained through heat conduction analysis using two-dimensional FEM. To solve the conjugate heat transfer problem existing between the composite and the flow field around the consolidation region, the heat transfer coefficient and the surface temperature of the composite were obtained simultaneously in an iterative manner. Numerical simulations were done for APC-2. The major process variables considered in this study were nozzle exit temperature and gas flow rate of nitrogen. The heat flux and the heat transfer coefficient as well as the velocity distribution in the flow field showed that three-dimensional effect is apparent especially in the upstream region of the calculation domain..

1. Introduction

Figure 1 shows a typical configuration of the composite tape lay-up process. On a tool plate, the thermoplastic tape or tow is laid to fabricate composite laminates. Hot nitrogen gas flow exits from the nozzle of a torch, which is located between the incoming tow and the composite substrate, to provide heat for the consolidation. Once the incoming tow and the composite substrate are heated over the melting temperature, they are pressed by a compaction roller to achieve consolidation.

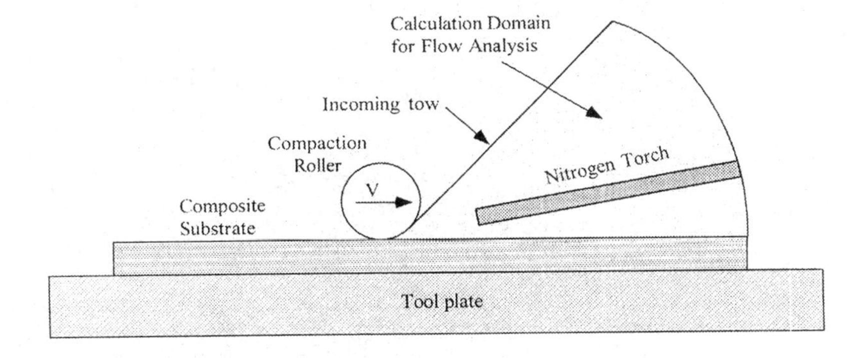

Figure 1. Schematic view of configuration of the thermoplastic tape lay-up process

Many studies had been don to analyze the temperature field inside the composite [1,2]. In their works, the heat transfer coefficient which is crucial in accurate prediction of temperature, was arbitrarily assumed. In this study, an effort was made to improve pre-existing models to predict the temperature profile more accurately and realistically, so that the model can be used as an engine for future process design.

2. Numerical Modeling

In the flow model, the domain between the incoming tow and the composite substrate where the hot nozzle flow has a significant influence was analyzed (see Figure 2). Refined meshes, as in Figure 2, for three-dimensional flow analysis in this domain were generated. A commercial package called *Star-CD* [3] was used for the flow analysis. Figure 3 shows boundary conditions for the flow analysis.

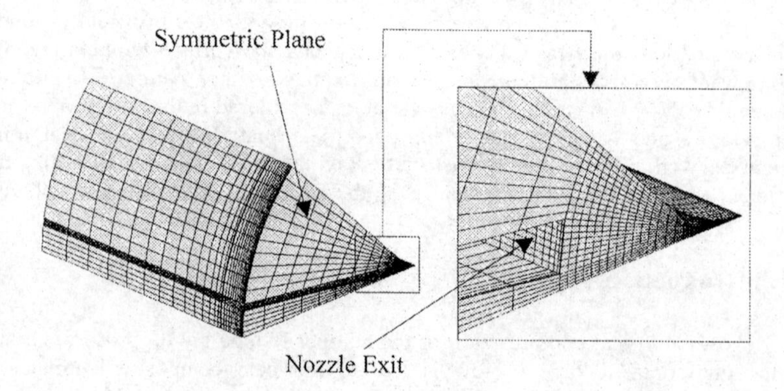

Figure 2. Refined mesh for three-dimensional flow analysis

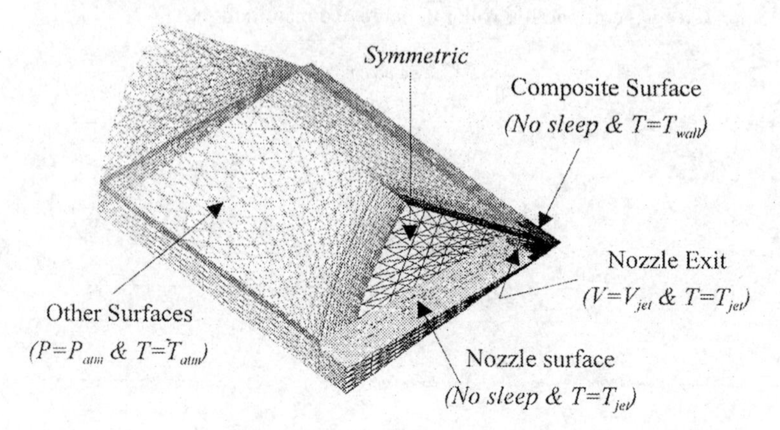

Figure 3. Boundary conditions for three-dimensional flow calculation

To consider turbulence in the flow analysis, some assumptions were included in the flow model. Turbulent intensity was taken as 5% of the nozzle exit velocity and the turbulent mixing length was set as 10% of the characteristic length of the nozzle. Also, standard k-ε model was adopted.

Two-dimensional finite element analysis was conducted to find temperature distributions inside the composite. The following governing equation [4] was solved for the boundary conditions shown in Figure 4.

$$\rho C_p(\vec{u} \cdot \nabla T) - \nabla(\tilde{k} \nabla T) = 0 \tag{1}$$

As shown in the figure, one end of the composite substrate is set as adiabatic, whereas the other end is set as constant room temperature. In the heat conduction analysis, volumetric heat capacity and anisotropic heat conductivity were considered as functions of temperature. Petrov-Galerkin shape function [5] was used in order to obtain stable solution. Along the upper surface of the laminate, heat transfer coefficient of the air ($h_{ambient}$) and room temperature ($T_{ambient}$) were given. Heat transfer coefficient along the upper and lower surfaces before the nip point (h_{upper} and h_{lower}, respectively) must be given in order to obtain the inside temperature distributions. These heat transfer coefficients were found by solving the *conjugate heat transfer* problem in the following iterative manner.

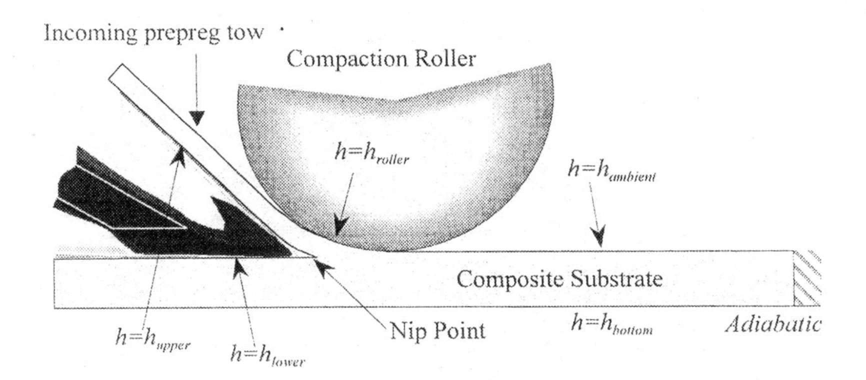

Figure 4. Boundary conditions for heat conduction analysis

First, in order to solve the conjugate heat transfer problem, surface temperature was given as the room temperature for the initial guess. Flow analysis was performed to obtain a heat transfer coefficient profile. Next, this calculated heat transfer coefficient profile on the composite surface was used as a new boundary condition to solve the heat conduction equation inside the composite (Eq. 1). New temperature profile on the composite surfaces was thus obtained. Using newly obtained surface temperature, next iteration of the flow analysis updates the heat transfer coefficient profile. This procedure is repeated until satisfactory convergence is reached. The calculation usually converged within a few iterations (5 or 6 iterations). These iterative steps are illustrated in Figure 5.

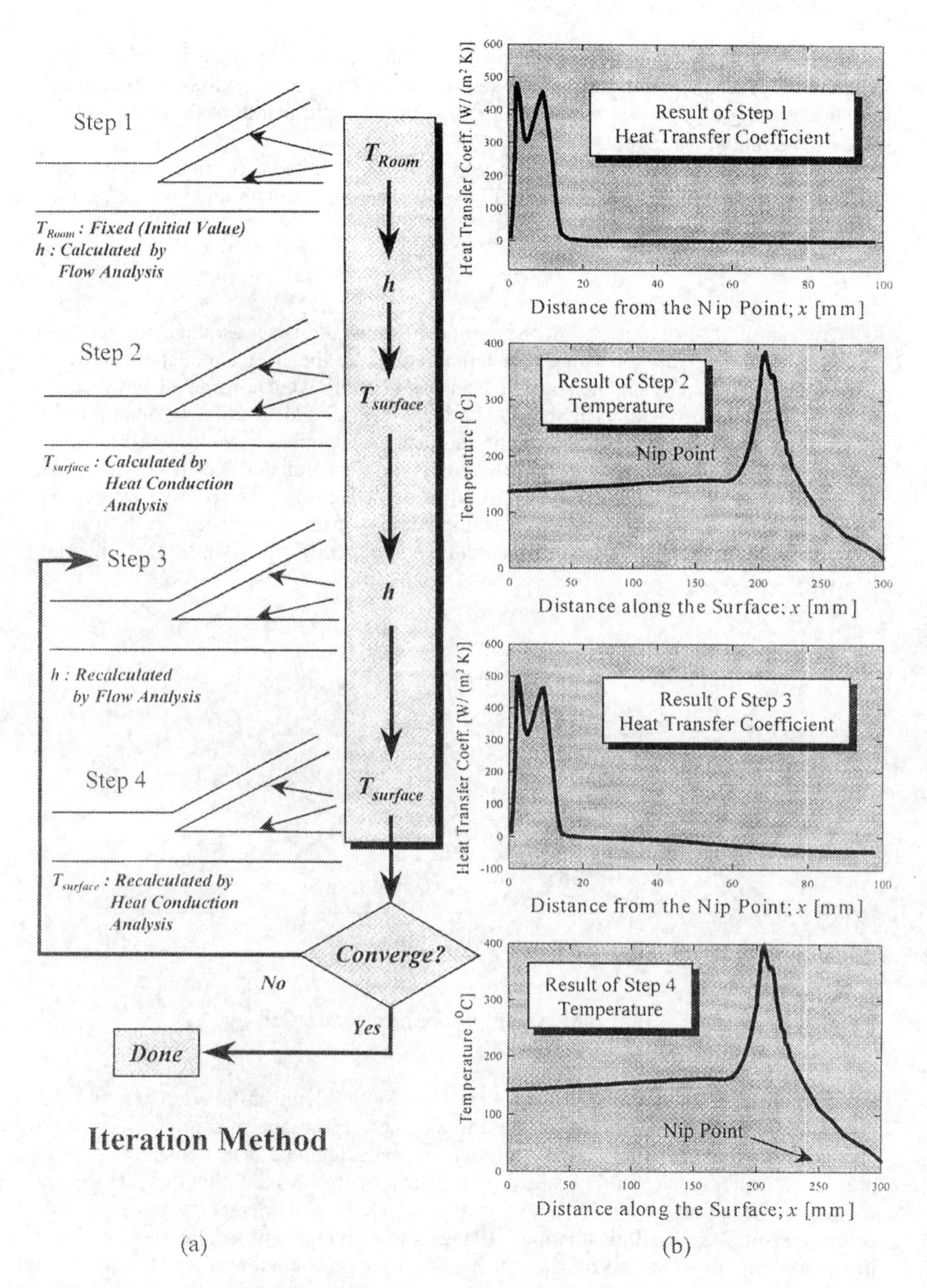

Figure 5. (a) Description of the iterative scheme for conjugate heat transfer problem
(b) Example plots using the iterative scheme

3. Results

Numerical simulations were performed to solve the problem described in the previous section. As mentioned earlier, the major process variables considered in this study were volumetric flow rate of and nozzle exit temperature hot nitrogen gas. Volumetric flow rate of nitrogen was varied from 50 to 150 liter/min and the nozzle exit temperature of nitrogen gas was from 850 to 920 *°C*. Figures 6 to 8 correspond to the case in which the volumetric flow rate was 100 liter/min and nozzle exit temperature was 850°C. In this case, the nozzle exit velocity was 74.5m/sec. The Reynolds number was about 30,000 which indicated that the flow was turbulent. The lay-up speed was given as 38.1 mm/sec (1.5 inch/sec).

After the flow analysis, the velocity and pressure distributions in the calculation domain as shown in Figure 6 were obtained. As can be seen from the calculated velocity distribution, three-dimensional flow analysis should be considered because the lateral flow, which escapes the composite surfaces, is not negligible in the adjacent region of the nip point. The pressure distribution on the symmetric plane indicates that most regions in the calculation domain remained at ambient pressure except the small region around the nip point.

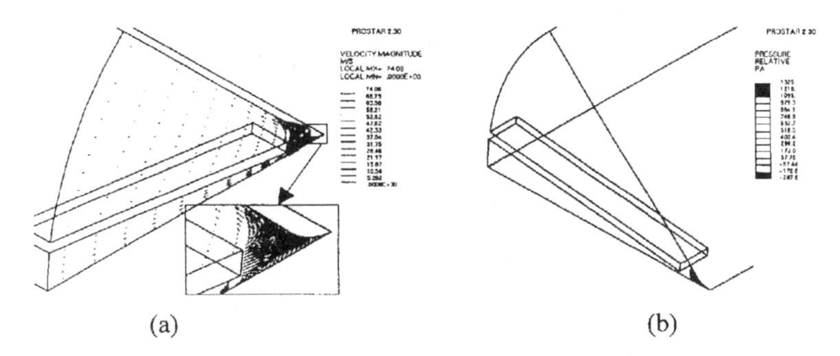

Figure 6. (a) Velocity distribution and (b) pressure distribution on the symmetric plane of calculation domain

Figure 7(a) shows the calculated heat transfer coefficient of the composite surfaces. Like the velocity distributions in Figure 6(a), it was found that three-dimensional effect is distinct due to the presence of lateral flow.

As shown in Figure 7(b), on the surface of the composite substrate, two peak points right before the nip point were observed. The maximum values are located right before the nip point (not at the nip point) due to the stagnation at the nip point. The rapid decrease after the short distance significantly lowers the nip temperature compared to the nozzle temperature.

It is noted that negative values of the heat transfer coefficient near the nip point were obtained. These negative values imply that the heat flows from the composite substrate to the ambient, which causes the nip temperature much lower than the torch temperature.

Figure 8 shows the result of heat conduction analysis after the fifth iteration. The temperature reaches the maximum value right before the nip point, then decreases to either the ambient temperature or the tool temperature. Since negative heat transfer coefficients appear as in Figure 7(b), the temperature along the composite substrate drops significantly at the nip point.

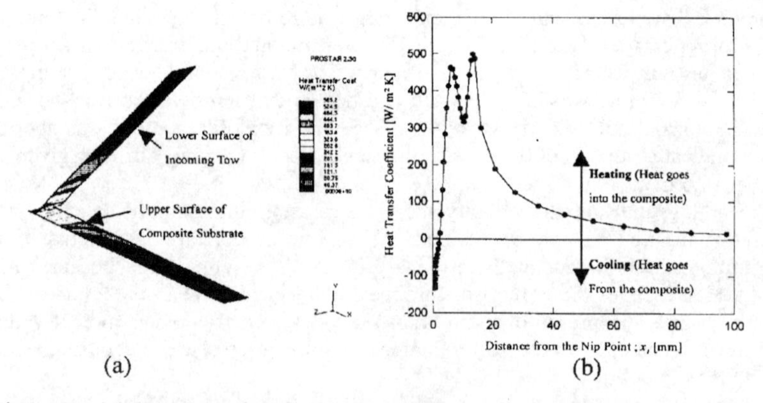

Figure 7. (a) Heat transfer coefficients along the composite surfaces
(b) Heat transfer coefficient profile along the surface of composite substrate averaged in the width direction

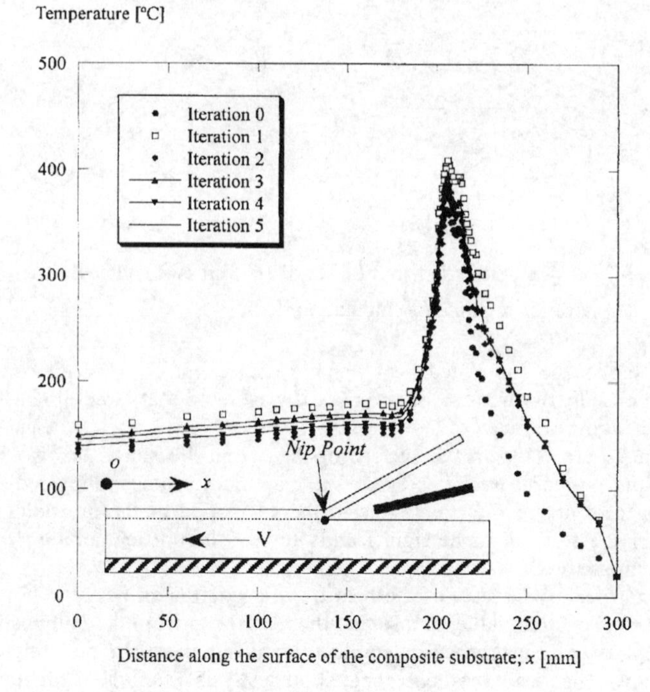

Figure 8. Temperature distribution of the surface of composite substrate

PRINCIPLES AND PRACTICE OF ULTRAFAST HEATING CYCLES FOR THERMOPLASTIC BASED COMPOSITES USING MAGNETIC FIELD TECHNOLOGY

C-H Andersson*, T Cedell*, A Schwartz+ and G Langstedt#

* Department of Production and Materials Engineering, Lund Institute of Technology, PO Box 118, SE-221 00 Lund, Sweden

\+ A.S.Separering, Wiesengrund 3, D-21217 Fleestedt, Germany

\# Linlan Composite AB, Forskningsbyn Ideon, SE-223 70 Lund, Sweden

Abstract

Magnetic field techniques ford heat generation are not new, but have undergone considerable development in recent years with respect to understanding of the field distribution and thus the control of temperature, heating rate and distribution of heat. The fundamental merit is the direct generation of heat in the tool and thus the limitations in heating rate are in practice only due to power input and control system. Balancing induction, magnetic hysteresis and anomalous loss heating, tools can be heated without warping. The magnetic field outside the equipment can be minimized using the Faraday cage principle for shielding.

1. Introduction

Magnetic field techniques for heat generation are not new, but have undergone considerable development in recent years with respect to understanding of the field distribution and thus the control of temperature, heating rate and distribution of heat. A fundamental merit is the direct generation of heat in the tool or work piece, and thus in principle no limitations in the heating rate due the heat transfer. The limitations of the heating rate are however given by the power input, the magnetic saturation in coils and pole pieces and by the response control systems. Heating rates of 10°C - 20°C per second are typical for moulding purposes using the equipment presented, with the practical limitations set in order to get even distribution of the emitted heat. This is however much faster than the heating rates obtained by any other commercially available process. Fast cycle processing with total cycle times of 1 – 2 minutes for 280°C has been performed for thermoplastic polyester pre-pregs, comingled and coknitted fabrics, with the directly oriented material loaded cold into the tools.
The three complementary mechanisms behind this kind of technique are electromagnetic induction, magnetic hysteresis and anomalous losses generating the heat in situ.

2. Electromagnetic rapid heating apparatus

Figure 1 illustrates the set-up of the electromagnetic rapid heating apparatus used here, with a flat top moulding tool, pole pieces, air gap, integrated thermal insulation and distributor controlling the field distribution.

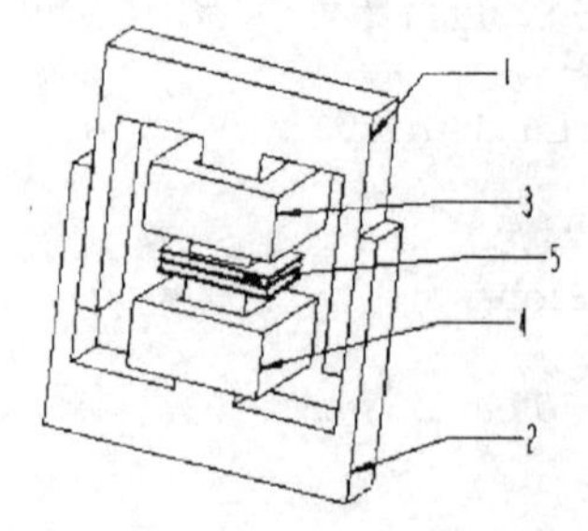

1,2 Magnetic core
3,4 Coils
5 Moulding tool with isolation and distributor

Figure 1: Schematical drawing of electromagnetic rapid heating apparatus

The magnetic field is created by coils charged by AC current. The frequency used is usually net frequency, but it is also possible to apply a higher frequency for heating thinner details. The moulding tool to be heated is placed symmetrically in the air gap and builds therewith a magnetic circuit with the pole pieces.
In order to avoid heat losses the moulding tool is thermally isolated from the pole pieces. Magnetic distributors are positioned on top and below the moulding tools and used to tune the field distribution in order to obtain uniform temperature in the tool. These distributors are adjusted to emit the tool geometry. Solid steel distributors as well as laminated pole pieces are used.

3. Physical basics of electromagnetic rapid heating

Figure 2 illustrates the three mechanisms of magnetic heat generation in a transformer core depending on frequency. Three different simultaneous mechanisms are generating the heat in the tool:

- Magnetic hysteresis losses
- Classical eddy current losses
- Anomalous magnetic losses

The magnetic hysteresis and the anomalous magnetic losses give in depth heating of the work piece. The eddy current losses give preferably surface heat generation demonstrated below. Warping of tools can thus be mastered.

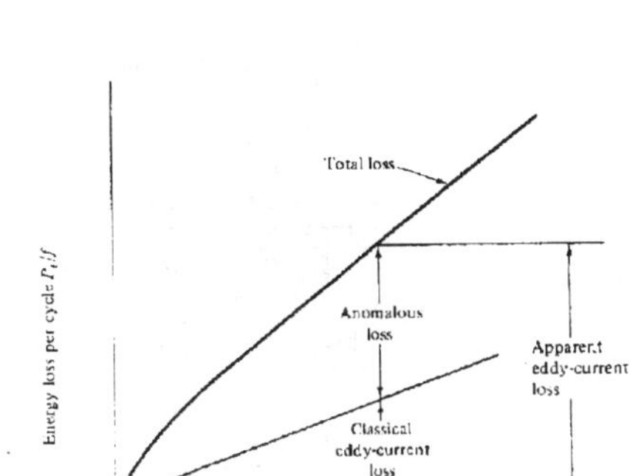

Figure 2: Mechanisms of magnetic heat generation in a transformer core

The distribution of eddy currents and the resulting heat losses can be calculated with the Maxwell field equations.
It is convenient to define a standard depth of penetration where the eddy current is 1/e (37%) of its surface value. This standard depth of penetration is given by the formula [8]:

$$\delta = \sqrt{\frac{2}{\omega \cdot \mu \cdot \sigma}} = 503 \cdot \sqrt{m/\Omega s} \cdot \frac{1}{\sqrt{f \cdot \mu_r \cdot \sigma}} \tag{3.1}$$

where σ is the conductivity in $S \cdot m / mm^2$, ω the angular frequency, f the frequency in s^{-1} and μ_r the relative permeability.
From this equation it can be seen that the standard depth of penetration:

- Decreases with an increase in frequency
- Decreases with an increase in conductivity
- Decreases with an increase in permeability

With the material properties for aluminum and steel (table 1) the standard depth of penetration is calculated according to equation 3.1 and shown in figure 3.

	Aluminum	Steel
Frequency [s^{-1}]	$10\text{-}10^6$	$10\text{-}10^6$
Relative permeability	1,000021	1500
Conductivity [Sm/mm^2]	35	12

Table 1: Assumed material properties for aluminum and steel

The depth of penetration for aluminum is approximately 10 times higher as for steel due to the difference in conductivity and relative permeability.

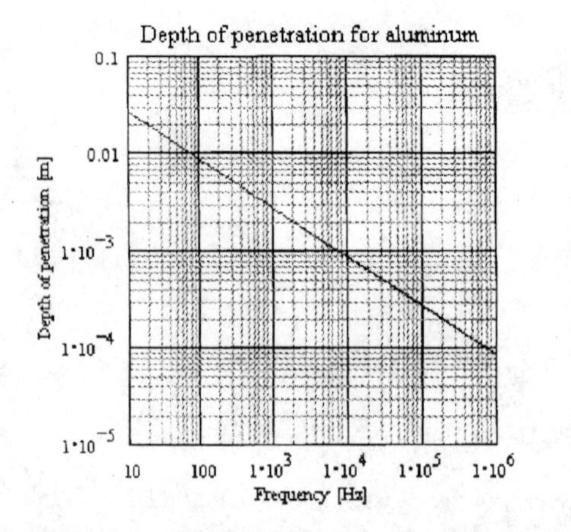

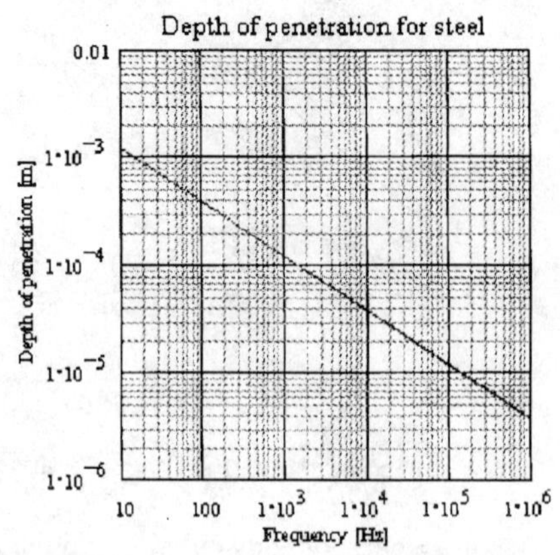

Figure 3: Standard depth of penetration for aluminum and steel

The machine is designed and tuned in order to minimize these eddy currents in the magnetic cores and optimize them in the moulding tool. The moulding tool has a comparatively low resistance hence producing high eddy current losses and thus efficient heat generation.

4. Heating of different moulding tools

4.1 Problems occurring during electromagnetic rapid heating

In thermoplastic composite materials processing the temperature uniformity in the moulding tool plays a vital role in determining the quality of the end product. The basic electromagnetic rapid heating apparatus as shown in figure 1 will in most cases not produce a uniform temperature distribution in a moulding tool.
First of all the pole pieces must be isolated from the tool that is heated. Otherwise the magnetic pole pieces will extract heat from the tool.
Secondly, accessory poles, called distributors in the continuing text, have to be used. These distributors are assembled on top and below the work tool to distribute the magnetic field. Generally steel distributors can be used.

4.2 Heating of the circular steel tool

The tool had an outer diameter of 125 mm and a total thickness of 20 mm. For the temperature measurements of the round steel tool thermocouple sensors were positioned at the two tool parts as illustrated in figure 4.

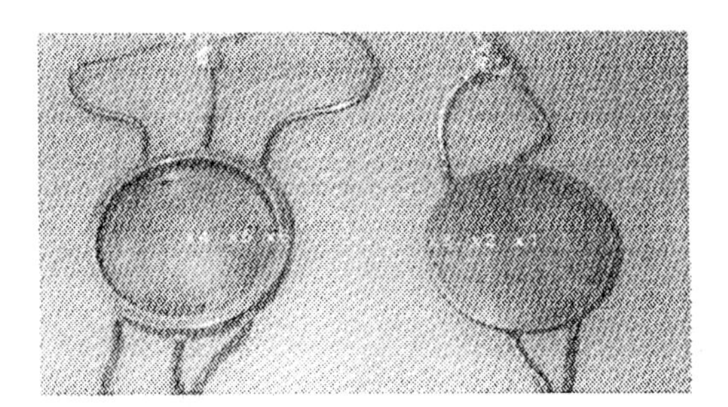

Figure 4: Set up of temperature sensors for round steel tool

The temperature distribution with thermal isolation between tool and pole pieces but without distributor can be seen in figure 5.

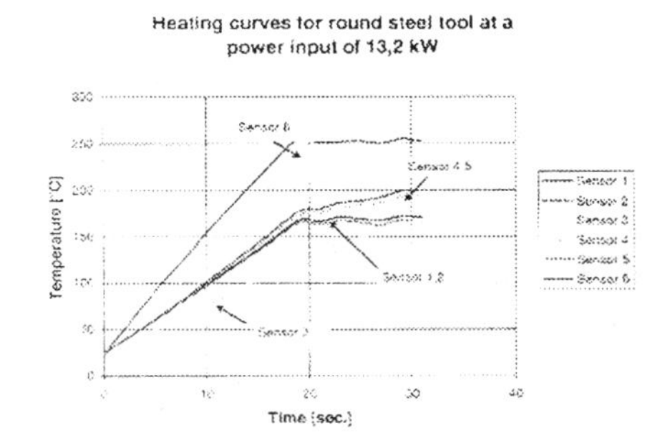

Figure 5: Temperature distribution with isolated tool but without magnetic distributors

In order to improve the temperature distribution in this application metal rings were sufficient. Using the distributors good temperature uniformity was obtained over the entire tool. The distributor set up of the tool however has to be individually adjusted to the fit the tool geometry.

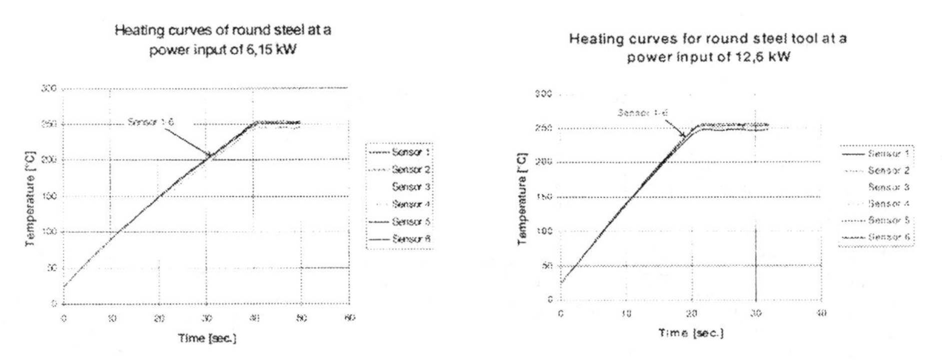

Figure 6: Temperature distribution with magnetic distributors and isolated tool at a power input of 6,15 kW and 12,6 kW

5. Efficiency of electromagnetic rapid heating

In industrial production processing, costs are of major interest. Therefore this chapter shows the efficiency of the electromagnetic rapid heating apparatus that was tested to make comparisons with other processes possible.
The voltage and current are not in phase and differ in the "so called" phase angle φ. Condensator batteries are used to minimize this phase angle deviation in order to optimize the efficiency of the machine.

5.1 Round steel moulding tool

The needed power input Q_{12} to heat a material from temperature T_1 to T_2 is given by:

$$Q_{12} = m \cdot c_p \cdot \Delta T \tag{5.1}$$

where c_p is the specific heat capacity of the material and m the mass. This equation does not consider any heat losses like heat radiation or convection.
If the given moulding tool with a weight of 2 kg is heated from 30 °C to 100 °C the specific heat capacity can be assumed constant $c_p = 0{,}465 \frac{kJ}{kg\,K}$. Hence, the needed power input is according to equation 5.1 $Q_{12} = 0{,}01810$. The measured values and the resulting efficiency are shown in the table below.

Heating rate	5 °C/sec.	10 °C/sec.
Total power input [W]	6,15	12,6
Total power input [kWh]	0,0239	0,0245
Efficiency [%]	75,7	73,9

Table 2: Efficiency for heating the round steel tool

6. Magnetic field outside the electromagnetic rapid heating machine

For many years the influence of magnetic fields on the human body have been discussed. Meanwhile, organizations like the The International Commission for Non-Ionising Radiation Protection and the European community have defined recommendation values for workplaces which should not be exceeded.

Organization	Frequency	Magnetic field strength
ICNIRP 1988/89 (The International Commission for Non-Ionising Radiation Protection)	50 Hz	500 µT
European Community 1993	50 Hz	400 µT

Table 3: Recommended limits for low frequency electromagnetic field [10]

6.1 Magnetic field around the circular steel tool

These values for the magnetic field strength were measured with and without shielding for 12,6 kW power input. The shielding was built with a 0,5 mm soft magnetic steel sheet material.

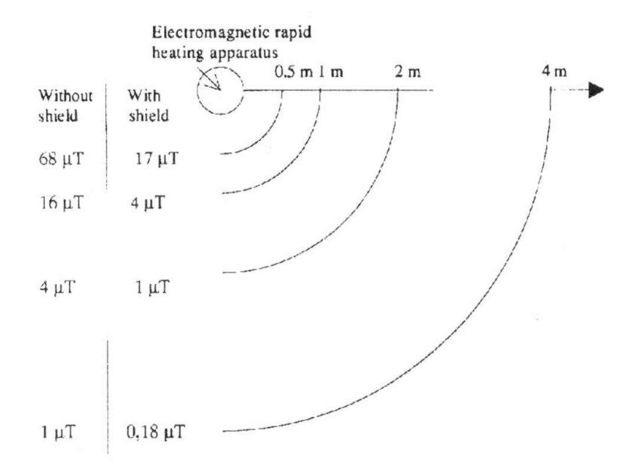

Figure 7: Magnetic field around circular steel tool at a power input of 12,6 kW

References

[1] Schwartz, A., *Electromagnetic rapid heating for thermoplastic composite material processing*, Technical University Lund, Lund, 1997
[2] US Patent No. 5.023.419, Göran Langstedt, 1991
[3] International patent applicationPCT/SE97/00060, Linlan Composite AB, 1997
[4] Andersson C.-H. and Eng K, *Split film based weft insert warp co-knitting and lost yarn preforming, two new routes for composite materials,* Proc. ECCM-6, Bordeaux 20-24 September 1993 pp 237-242
[5] Cedell, T., *Magnetostrictive Materials and Selected Applications*, Department of Production and Materials Engineering, Lund, 1995
[6] Boll, R., *Weichmagnetische Werkstoffe,* Vacuumschmelze GmbH, Hanau, 1990
[7] Cheng, D. K., *Field and Wave Electromagnetics*, 2nd ed., Addison-Wesley Publishing Company, Reading, Massachusetts, 1989
[8] Cullity, B. D., *Introduction to Magnetic Materials*, Addison-Wesley Publishing Company, Reading, Massachusetts, 1972
[9] T. Pisanikowski: "Production and properties of thermoplastic non crimed composites from split film co-knitted fabrics" Proc. Techcomp 7, Frankfurt am Main Germany 1997
[10] Eder H., *Grenzwerte für elektrische, magnetische und elektromagnetische Felder*, Bayerisches Landesamt für Arbeitsschutz, 1995

Advancements in processing fibre reinforced thermoplastic composites by Explosive Diaphragm Forming (EDF)

Frank Weiblen*, Markus Bucher*, Gerhard Ziegmann*

* Institute for Design Methods and Construction, Composites Laboratory, Swiss Federal Institute of Technology, Wagistrasse 13, CH-8952 Schlieren/Zürich, Switzerland

Abstract

The diaphragm technology with their inherent advantages concerning low tooling cost and high reproducibility of fibre orientation and fibre distribution for example, gets advanced processing capability when using it as the *Explosive Diaphragm Forming*. Investigations at the composites laboratory of the ETH Zürich showed processing advantages and cycle time reduction with an advanced machine concept incorporating fast heat-up of the laminate material preferably seperate from the deep-drawing unit, and processing and cooling within the deep-drawing unit both with forced air and a tool held at a considerable low temperature. The deep-drawing process itself is made faster in using high volume stream compressed air coming from a pressure tank integrated in the deep-drawing unit.

Introduction

Processing of fibre reinforced thermoplastics is known in principle and some techniques are available. The most popular one is pressforming of Glass Mat Thermoplastic material (GMT), which has become a standard industrial process for example in the automotive industry for secondary structures.
In order to allow more advanced design and extreme lightweight construction under the constraint of high production rates, economically feasible processing for engineering thermoplastics with reinforcement by continous fibres from glas, carbon, aramide and maybe polyester must be developed.
One contribution towards this may be the so called *Explosive Diaphragm Forming (EDF)*, a processing principle invented by the Composites Laboratory of ETH Zürich.

General

It is quite obvious that thermoplastic based composites are offering a lot of advantages in comparison to thermoset composites /1/. The production processes are not related to chemical reactions, but to physical processes, and thus much faster. The chemical/physical/mechanical characteristics are comparable or even better and the possibilities for recycling are superior to a high degree.

This invites the engineer to think about fast processing possibilities. Additionally for low material consumption and lightweight and stiff structures, continous fibre reinforcement will show up more and more.
To retain these advantages after the processing steps, the method used must fulfill some demands:
1) Fibre orientation must be controlled and retained to the design
2) Resonable deep-drawing rates must be achieved without generating wrinkles
3) Short cycle times

The diaphragm technology

For solid sheet material as well as sandwiches, a technology enclosing the composite lay-up inside two films, the so called diaphragms, sealing them by vacuum, and processing this lay-up into a rigid male or female mould by air pressure is established, and called Diaphragm Technology. It is a technology derived from the superplastic forming of advanced aluminium material and features some important advantages for composites over the pressforming technology, coinciding with the a.m. discussed demands:

1) During the process the diaphragms are stretched biaxially. The molten thermoplastic matrix creates a certain adhesion to these diaphragms, resulting in tension forces on the fibre fabric. This tension forces prevent fibre wrinkling, missalignment, missorientation and buckling. Therefore control of fibre orientation is possible, an important feature for design of high-performance composite structures. Additionally, because of the isostatic pressure distribution on the lay-up, homogenous consolidation is ensured, even when the geometry is very complex (i.e. double-curvatures, varied wall thickensses etc.).

2) The achievable deep-drawing rates are mainly restricted by the used diaphragm material, which is silicone rubber, polyamide-film or, for the high-end high-temperature thermoplastics like PEEK or PEI, is polyimide-film. These materials have even higher elasticity than the known superplastic aluminium brands, so resonable high deep-drawing rates as well as deep-drawing speeds are possible /2/, /3/.

3) Processing the diaphragm technique as a standard means using for example an autoclave to build up the necessary pressure and temperature for the deep drawing. This leads to slow processes /3/, /8/.

The Explosive Diaphragm Forming (EDF) Technology

The main goal when developing the *EDF* technology was to enhance productivity and thus shorten the process cycle time.

Various possibilities to achieve this were examined:

1) Heating of the material lay-up

The laminate material has to be heated together with the diaphragms. This can be done for example with heated air, infrared heaters or contact heating via convection from a heated surface. It could be shown, that heating should be executed separate from the processing /5/. This allows optimization of the heating process (outside the deep-drawing facility) as well as optimization of the cooling process (inside the deep-drawing facility).

The laminate and diaphragm lay-up is preferably heated via convection heating, with aluminium plates as contacting surfaces heated with infrared heaters. The heating rate is approx. 4 times higher than with infrared heaters alone /7/. The heaters are from quarz glass type using a medium grade wave length of 2.2 to 2.7µm /3/. Their surface temperature was chosen to 600°C to 700°C, being approx. 50°C to 150°C higher than used for infrared heating alone.

Additional shortening of the heating time is achieved when pressing the lay-up against the heating plates to asure proper contact. This may be done by vacuum, maybe assisted by pressure from the adjacent side of the lay-up.

This heating technique offers especially some benefits, if sandwich structures from foam core material are incorporated in the laminate lay-up. The heat transfer to the core is slower as when using infrared heating directly, because conduction applies only, in contrary to direct energy transmission. This avoids damage or degradation of the core material prior or during processing.

2) Temperature of the tooling and mould

Tools from castable ceramic material, from steel sheet material or solid aluminium have been examinated. Important for fast processing is the possibility to hold the tool on a relative low temperature (approx. 60°C to 100°C when processing PA 12 for example). This needs a cooling device incorporated in the tool.

Additionally an adopted surface treatment or surface material of the tool may be used when optimizing the temperature behaviour during the deep-drawing and cooling phase.

3) Rapid execution of the deep-drawing

The most important contribution for rapid processing with the diaphragm technology is a quick build-up of the process-pressure.

In order to reach high deep-drawing rates and deep-drawing speed it is necessary to realize a high pressure build-up to shape the part within a short time . This needs sufficient volume stream. Therefore a pressure tank was installed at the deep-drawing facility of the composites lab at the ETH. The tubes and hoses were from relative high diameter (approx. 20-25mm) and the distance from the pressure tank to the deep-drawing facility was kept as short as possible. A pressure rate of more than 2 bar/s was possible then, with a volume stream of more than 4000 cm^3/s, which is considerably more than with the standard compressed air supply (approx. 1000 cm^3/s), all related to the ETH sample process equipment.

The pressure used for deep-drawing an example sandwich laminate from glassfibre reinforced PA12 together with a PMI foam core was 4.5bar, with a pressure of approx. 1bar being sufficient as the clamping force of the diaphragm lay-up to the tool (mould).

A range of pressures used for deep-drawing from 3 to 10bar is suitable when processing with the *EDF*-technology.

4) Cooling of the material lay-up

The second quite important contribution to low cycle times is an effective cooling after the deep-drawing procedure.

Because of the two principally different contact surfaces of a diaphragm lay-up, there are two different possibilities of cooling these surfaces and thus the material.

The one surface of the material is covered by the diaphragm itself. Here cooling from convection, preferably forced convection is possible via cooled air. ETH experience showed good results using compressed air with a pressure of 3.5bar. Time for cooling the a.m. example sandwich laminate was 30s.

The other surface of the material is in contact with the tool (mould) until the formed part has dimensional stability. Therefore again convection is the technique to be used. Promising results have been achieved with the tool being held at a considerably low temperature of approx. 60°C to 100°C, or even room temperature. Then the cooling starts with the deep-drawing and thus allows fast processing and release from the tool. For continous processing the tool has to be equipped with a cooling device, otherwise it will heat up too much.

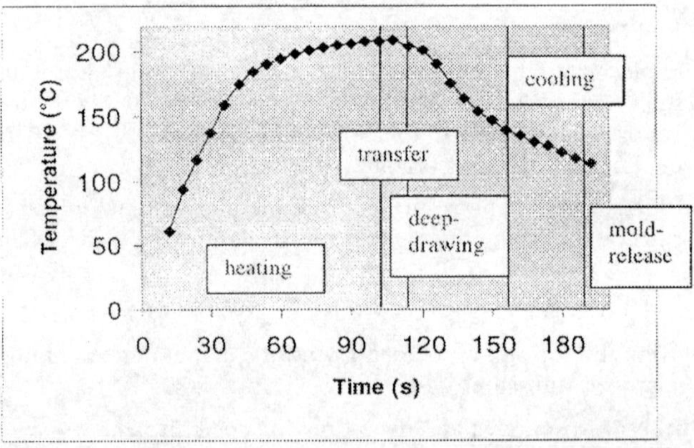

Figure 1: Example *EDF*-process with a Glassfiber/PA 12 sandwich, manual transfer from the preheating unit to the deep-drawing unit

The EDF-machine layout

When implementing the a.m. discussed possibilities to enhance productivity of the diaphragm technology into a machine concept, the following layout and design features may be envisaged:

1) heating unit seperate from deep-drawing unit, or integrated in a multifunctional plate, incorporating infrared heating, capable to be evacuated and pressurized

2) heating of the material via infrared heating or convection heating via plates heated through infrared heating

3) deep-drawing unit capable to press the diaphragm lay-up to the mould with approx. 1 to 3bar

4) deep-drawing unit with pressure ring above the diaphragm lay-up, capable to be evacuated, to be pressurized and to allow forced cooling via ventilation openings. This ring could be an element of the a.m. multifunctional plate.

5) pressure supply via pressure tank, thick tubes or hoses (diameter more than approx. 20mm) and in close vicinity to the deep-drawing-unit, capable to deliver 5-10bar pressure and a high volume stream. Pressure build-up should be possible at a rate of 2bar/s.

6) tool/mould capable to be evacuated (to assist consolidation and contact of the diaphragm lay-up to the tool/mould)

7) tool/mould capable to be held at a temperature well below the melting temperature of the matrix used in the composite laminate

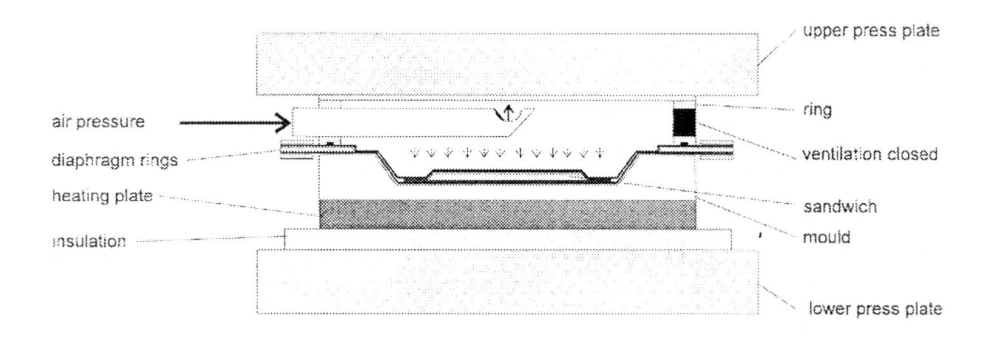

EDF-build-up with a sandwich-composite as an example

Conclusion

The process constraints discussed and the *EDF*-machine layout presented enables processing of continuous fiber reinforced thermoplastic composites in very short cycle times with reproducible high quality of the laminate. The limiting constraint is the time of the diaphragm lay-up being installed in the deep-drawing unit during processing. The heating is therefore preferably executed seperately, for example in a continuously operating feeder unit.

The process is not limited to special matrices. Various diaphragm materials are possible as well.

Experience has been gained with glassfiber and carbonfiber reinforcement in PA12, PET, PBT and PEI matrices at the composites laboratory at ETH Zürich, using diaphragms from PA66, silkon rubber, PEI or PI material.

Outlook

The limitations in deep-drawing ratio as well as the suitability of new diaphragm materials to enhance the spectrum of possibilities with this innovative processing technology have to be investigated further on.

The expected application range is thought to be high volume automotive part production with low-temperature thermoplastic matrices and reusable diaphragms in a highly integrated machine lay-out, as well as the processing of high-performance laminates from the known high-end matrices with complicated lay-up sequences, varied thicknesses, incorporating stiffeners for example, and future integrated functional design to be manufactured in one step.

References

/1/ *Ziegmann, Flemming, Roth*, FASERVERBUNDBAUWEISEN, BD. 1: FASERN UND MATRICES, BD. 2: HALBZEUGE UND BAUWEISEN - Springer Verlag, Berlin, 1995 und 1996

/2/ *M. E. Niedermeier*, ANALYSE DES DIAPHRAGMAFORMENS KONTINUIERLICH FASERVERSTÄRKTER HOCHLEISTUNGSTHERMOPLASTE - VDI Verlag, Düsseldorf, 1995

/3/ *S. Delaloye*, DIE DIAPHRAGMA-TECHNIK, EIN ANLAGEKONZEPT ZUR AUTOMATISIERTEN FERTIGUNG KONTINUIERLICH FASERVERSTÄRKTER THERMOPLASTE - Dissertation, ETH Nr. 11151, 1995

/4/ *M. Bucher, M. Hintermann, M. Renker, F. Weiblen, G. Ziegmann,* LATEST DEVELOPMENTS IN RTM- AND THERMOFORMING-PROCESSES FOR AUTOMOTIVE APPLICATIONS - 6th International EAEC Congress, 2-4 July 1997, Cernobbio, Italy

/5/ *K. Kilchenmann, R. Ruch,* HERSTELLUNG EINES FAHRRADES AUS FASERVERBUNDWERKSTOFFEN MIT DEM "EXPLOSIVE DIAPHRAGM FORMING" VERFAHREN UNTER UMSETZUNG VON NEU ZU ENTWICKELNDEN FAHRRADKONZEPTEN – Diploma Thesis Ingenieurschule Burgdorf in collaboration with ETH Zürich, March 1995

/6/ *M. Bucher,* EXPLOSIVE DIAPHRAGM FORMING - Semester Thesis Ingenieurschule Burgdorf in collaboration with ETH Zürich, September 1995

/7/ *A. Cerini,* OPTIMIERUNG DER PROZESSPARAMETER ZUR UMFORMUNG VON THERMOPLASTISCHEN HALBZEUGEN IM EDF-VERFAHREN - Semester Thesis ETH Zürich, June 1996

/8/ *P.J. Mallon, C.M. O'Bradaigh,* DEVELOPMENT OF A PILOT AUTOCLAVE FOR POLYMERIC DIAPHRAGM FORMING OF CONTINUOUS FIBRE REINFORCED THERMOPLASTICS – Composites, Vol. 19, 1988

Stiffness Analysis of Thermoformed Plain Weave Composites

J. Hofstee[a], F. van Keulen[b], H. de Boer[b] and K. Stellbrink[a]

[a] DLR - Institute of Structures and Design, Pfaffenwaldring 38-40, D-70569 Stuttgart, Germany.
[b] Laboratory for Engineering Mechanics, Delft University of Technology, P.O.Box 5033, NL-2600 GA Delft, the Netherlands

Abstract

An analytical method is presented for stiffness calculation of a balanced plain weave reinforced thermoplastic laminate after thermoforming. Thermoforming is assumed to cause in-plane shear and bending of the fabric. Stiffness calculation is carried out for a repeating element of the sheared fabric lamina, using two subsequent homogenization procedures. Initially, effective yarn stiffnesses are determined. Secondly, undulation is accounted for, using either stiffness averaging or compliance averaging. The undulated geometry is described by a model which is based on experimental cross section analyses. The secondary effect of shear on yarn undulation is schematized as well. Tensile and bending tests are carried out on macroscopically homogeneous specimen. A comparison between theory and experiment will be given.

1. Introduction

Continuous Fiber Reinforced ThermoPlastics (CFRTP) are increasingly being used in automotive and aeronautical industry thanks to their high stiffness and strength to weight ratios. Application of rapid forming techniques, such as rubber forming [10], further increases the advantages of CFRTP. Commonly used reinforcements are woven fabrics (WF). Woven fabric consists of two interlacing yarn directions, which are denoted 'warp' and 'fill' (Figure 1). Woven fabric reinforcements have the advantage of being easy to handle at the cost of a reduction in stiffness due to undulation of the yarns. Rapid forming techniques for woven fabric CFRTP have become even more interesting with the development of draping simulation methods [1,12] that give a good indication of the orientation of the fibers after thermoforming, without the requirement for expensive computer simulations. Combined with a model for the resulting laminate stiffnesses, these draping simulations provide the basis for numerical analysis of CFRTP products. Thus a proper model for the laminate stiffnesses is an indispensable component in advanced tools for the product designer, see, e.g., [9].

Stiffness analysis of WF reinforced laminates has received considerable attention in literature [3,7,11]. Several models have been developed based on Classical Laminate Theory (CLT). Of a WF lamina only a subcell is investigated, frequently referred to as Repeating Volume Element (RVE) [3]. The RVE is either regarded as a whole, or further divided into subelements. The (sub)elements are schematized as a stack of layers, representing the neat resin regions and the yarns. Most analytical models only investigate a weave in which the warp and fill yarns are mutually orthogonal. However, thermoforming of complex three dimensional structures is possible due to significant deformations of the fabric. The most important deformations are in-plane shearing and bending of the laminate, of which the latter is generally neglected on the length scale of an individual RVE. The influence of shearing on stiffness properties of deformed WF reinforced laminates is treated by Vu-Khanh and Liu [12]. The effect of undulation on stiffness is not included directly in their analytical model, but taken into account by means of a special experimental procedure. The present model combines the effect of yarn undulation and shear on WF laminate stiffnesses analytically. Furthermore, the interaction between in-plane shear and out-of-plane undulation is studied and implemented as well.

2. Geometric modeling

In thermoforming simulation methods, weave deformation is generally schematized by the simple shear or Trellis deformation mode [1,12]. During simple shearing, the fabric is schematized as a pin-joined net lying in the lamina plane (Figure 1). Warp and fill yarns are represented as rigid members. Pivot points represent the cross over points of the yarn centerlines. Arbitrarily curved three dimensional structures can be shaped due to a relative rotation of the members about the pivots. Thermoforming simulation methods generally predict the distribution of an inter yarn angle (θ), as shown in Figure 1. This angle will be used as input for the RVE stiffness calculation, and is assumed to be constant for each yarn throughout the RVE.

The influence of simple shearing on geometry is now schematized for a balanced plain weave. Orthogonal and sheared geometry are shown in Figures 2a and 2b, respectively. A translated yarn architecture is used, as defined by Chapman and Whitcomb [2], i.e. yarn volume is distributed continuously in in-plane direction along the undulated yarn centerline. The yarn centerline is schematized by arc segments for yarn sections that cross the other yarn direction, and straight lines for yarn sections crossing gaps. This approaches the goniometrical description often encountered in literature [3,7]. The present geometry description is somewhat easier to handle [11], especially when more complicated weave patterns such as twill and satin are to be described. It has been observed experimentally that cross section shape is determined by the yarn crossing it.

A number of significant consequences of simple shearing on balanced weave geometry can be derived from Figure 2. In-plane RVE dimensions (l) remain constant during simple shearing because the yarns are schematized as rigid members. Thus, the in-plane area of the RVE decreases with a factor ($\sin\theta$). Assuming constant volume, it follows that laminate thickness (t_{lam}) increases with a factor ($1/\sin\theta$). Furthermore, the distance between the centerlines of two parallel yarns decreases with the same factor ($1/\sin\theta$). Influence of shearing kinematics on undulation will be further analyzed based on experimental observations. To this effect, cross section analysis is carried out on specimens with varying amount of shearing. Bridged yarn width (a_b) and amplitude of the undulation (A) are measured. These parameters are normalized with respect to the in-plane RVE dimensions (l):

$$a_b^* = \frac{a_b}{l} \quad ; \quad A^* = \frac{A^0}{l} \quad . \tag{1}$$

Definions of these geometrical parameters are provided by Figure 2. Table 1 shows the normalized weave parameters measured from photomicrographs of a laminate consisting of 6 WF reinforced layers (carbon/PPS).

θ [deg]	a_b^*	A^*
90	0.468	0.017
82	0.455	0.018
71	0.468	0.021
58 [1)]	0.435	0.023

Table 1: Weave parameters as a function of shear, measured from photomicrographs of a thermoformed laminate, number of measurements n>70. [1)] Cross section obtained from a laminate that was deformed with a simple shear test [5] (n=30).

It is seen that the normalized bridged yarn width (a_b*) remains approximately constant, and even decreases for large shear. Because, in a sheared weave, the bridged yarn width (a_b) is a projection of the actual yarn width (a), the actual yarn width decreases with shear. The normalized amplitude (A*) increases. Because amplitude is a function of yarn thickness, these two observations imply that the yarn cross sections deform from a flat towards an annular shape [1]. This is explained by the observation made above, that the yarn centerlines shift towards one another. At a certain point, the gaps between the yarns will close, resulting in the described cross section deformation when shearing progresses. Because the measured bridged yarn width remains approximately constant, it is assumed that this point is already reached in the undeformed weave used in the current experiments.

The increase in undulation connected to the observations described above is a secondary effect relative to shear, and is therefore sufficiently accurate described by the following assumption. Bridged yarn width remains constant during shearing, whereas yarn thickness and yarn amplitude increase with the same rate as the actual yarn width decreases:

$$a_b^* = a_b^{0^*} \quad ; \quad A^* = \frac{A^{0^*}}{\sin\theta} \quad , \tag{2}$$

Parameters used to define the undulation for the stiffness calculation are the maximum crimp angles (Ω), the relative volume of the arcwise yarn segment with respect to the total yarn volume (φ_{arc}) and the WF lamina thickness (t_k). Direct determination of Ω from cross section photomicrographs is considered impractical because orientation of the fibers is often hard to distinguish. The volume fraction φ_{arc} simply follows from (Figure 2):

$$\varphi_{arc} = 2a_b^* \quad . \tag{3}$$

Maximum crimp angles are derived from geometric relations given in Figure 2. Assuming moderate crimp ($\Omega^2 \ll 1$), the maximum crimp angle can be derived from measured values as:

$$\Omega = \frac{4A^*}{1-a_b^*} \quad . \tag{4}$$

Combining (3) and (4) with (1) and (2), and deriving the lamina thickness from the number of layers (n) and the laminate thickness treated above, results in:

$$\varphi_{arc} = \varphi^0_{arc} \quad ; \quad \Omega = \frac{\Omega^0}{\sin\theta} \quad ; \quad t_k = \frac{t^0_{lam}}{n\sin\theta} \quad . \tag{5}$$

A comparison between theses relations and measurements of Table 1 is given in Figure 3. It is noted that the increase in maximum crimp angle is slightly underestimated with (5). For a balanced weave, cross section analysis in one yarn direction of an undeformed laminate would suffice for the definition of the undulation geometry as a function of θ. Even so, cross sections in both directions are recommended to establish a possible difference in warp and fill crimp (e.g. due to the weave process). The undeformed laminate used here shows a slight difference in crimp between warp and fill directions (about 1 deg), which further decreases as the laminate is sheared. Hence, an artificial balanced weave geometry with average weave parameters is assumed to be sufficiently accurate.

3. RVE stiffness determination

To calculate the laminate stiffnesses, the WF lamina RVE is divided into three representative regions. These are the fill and warp yarn regions and the neat resin region, referred to with subscripts 'f', 'w' and 'r', respectively. The yarn regions consist of impregnated fiber bundles, whereas the neat resin region fills up the remaining part of the RVE. Volume fractions of the representative regions (V_f , V_w and V_r) can be calculated from manufacturers or measured data [11]. The RVE for a single reinforcement layer is now replaced by a laminate consisting of three substitute layers that represent yarn and resin material. For the 6-layered laminate used here, the chosen number of substitute layers and layer stacking has negligible influence on bending stiffnesses calculated with CLT. The substitute layers have a constant thickness distribution over the RVE (Figure 4), with thicknesses:

$$t_f = V_f\, t_k \quad ; \quad t_w = V_w\, t_k \quad ; \quad t_r = V_r\, t_k \quad . \tag{6}$$

Effective stiffnesses of the individual substitute layers are obtained with a stiffness or compliance averaging which will be described in more detail below. Contribution of the substitute layers to the overall RVE stiffnesses is determined on the basis of straightforward application of CLT.

Constitutive relations of an infinitesimally small yarn element are calculated in the orthogonal coordinate system in which this element behaves transversely isotropic, denoted (1',2',3') in Figure 4. This is easily achieved using a micromechanics model usually applied to UD composites, see, e.g., [4]. Using the transformations for stresses and strains, as presented by Lekhnitskii [6], the corresponding constitutive relations are transformed to RVE coordinates *(x,y,z)*, which are shown in Figure 4 as well. The equivalent yarn element stiffnesses (**C**) and compliances (**S**) in RVE coordinates are thus a function of the constant transversely isotropic engineering properties, inter yarn angle (θ), and the local crimp angle (ω), which varies along the yarn.

A volume averaging procedure is applied to obtain either the effective stiffness ($\mathbf{C}_{eff}$) or the effective compliance ($\mathbf{S}_{eff}$) of one yarn undulation (as present in a single RVE). This implies that either a constant strain (ε) or a constant stress (σ) is assumed to act throughout the undulated yarn. For moderate crimp, and with θ constant, the stiffness averaging procedure (superscript $^{\varepsilon}$) leads to:

$$\mathbf{C}^{\varepsilon}_{eff} = \varphi_{arc}\left[\frac{1}{2\Omega}\int_{-\Omega}^{\Omega}\mathbf{C}(\omega)\,d\omega\right] + (1-\varphi_{arc})\left[\frac{\mathbf{C}(\Omega)+\mathbf{C}(-\Omega)}{2}\right] , \tag{7}$$

whereas the compliance averaging (superscript $^{\sigma}$) gives:

$$\left(\mathbf{C}^{\sigma}_{eff}\right)^{-1} = \mathbf{S}_{eff} = \varphi_{arc}\left[\frac{1}{2\Omega}\int_{-\Omega}^{\Omega}\mathbf{S}(\omega)\,d\omega\right] + (1-\varphi_{arc})\left[\frac{\mathbf{S}(\Omega)+\mathbf{S}(-\Omega)}{2}\right] . \tag{8}$$

Integration along one arc section suffices because the contribution of both arc sections is identical. The integrals in (7) and (8) can be solved analytically thanks to the selected geometry description.

4. Experimental validation

To validate the proposed stiffness model, tensile tests (ASTM D-3039) and three point bending tests (Dornier DON 128) were carried out. Specimens were taken from macroscopically homogeneous regions of the thermoformed product shown in Figure 1. Both tensile and bending tests were carried out in directions in which the balanced sheared laminate behaves macroscopically orthotropic, i.e. along the *x*-axis and *y*-axis in Figure 4, and along the yarn directions in an undeformed laminate. The laminate specimen is assumed to be a macroscopically homogeneous array of RVE's with identical geometry. Stiffness calculation of a single reinforced layer can then be limited to its RVE. Hence it is assumed that RVE dimensions and characteristic length of the macroscopic deformation pattern can be neglected with respect to specimen dimensions.

representative region	**$E_{1'}$ [GPa]**	**$E_{2'}$ [GPa]**	**$\nu_{1'2'}$**	**$\nu_{2'3'}$**	**$G_{1'2'}$ [GPa]**	**Ω^0 [deg]**	**φ_{arc}**	**V**
fill, warp	162.2	15.0	0.281	0.422	6.2.	7.3	0.47	0.357
resin	3.9	-	0.350	-	1.4	-	-	0.286

Table 2: Input parameters for analytical stiffness model, PPS matrix (Fortron 0214C1) reinforced with a balanced carbon plain weave (Ten Cate CD 0200 040 000 0000).

Input for the stiffness calculation is given in Table 2. Figure 5 shows the flexural E-modulus in x-direction as a function of θ for both compliance and stiffness averaging. Measured values of bending test specimen taken from the pyramid are indicated for a number of shear angles. Figure 6 presents the in-plane E-moduli of an undeformed and a 30 degrees sheared laminate for all in-plane orientations, again presenting stiffness and compliance averaging. Tensile test data are given for the two orthotropic directions in an undeformed laminate.

5. Discussion

A sufficient agreement is found between the proposed geometrical model and cross section analyses on a draped product. Crimp leads to significant differences in the stiffness predictions and should be accounted for. A difference of about 10 GPa is obtained for a moderate maximum crimp angle of 7.3 degrees. Compliance averaging along the yarns results in the lower prediction. Test results in yarn direction correspond with the predictions, but tests in yarn direction of a sheared laminate should be performed as well, taking into account resulting anisotropic behavior [8]. A good correlation exists between bending tests and predictions for E-moduli in orthotropic directions of a sheared laminate. Behavior in this direction is however highly non-linear due to the influence of the resin pockets.

Acknowledgement

Ten Cate Advanced Composites, Fokker Special Products B.V. and the Max Planck Gesellschaft are gratefully acknowledged for their contributions to the present work.

References

[1] **Bergsma, O.K.:** Three dimensional simulation of fabric draping, development and application, Ph.D.thesis, Delft University Press, Delft, 1996.

[2] **Chapman, C. and Whitcomb, J.:** Effect of assumed tow architecture on predicted moduli and stresses in plain weave composite, J. Composite Materials, Vol.29, No.16, 1995.

[3] **Hahn, H.T. and Pandey, R.:** A micromechanics model for thermoelastic properties of plain weave fabric composites, Engineering Materials and Technology, Vol.116, 1994.

[4] **Hashin, Z.:** Analysis of composite materials - a survey, Applied Mechanics, Vol.50, No.3, pp.481-505, 1983.

[5] **Keilig, Th. and Arendts, F.J.:** Ermittlung und Modellierung des Umformverhaltens von thermoplastischen Gewebeprepregs unter besonderer Berücksichtigung der Atlas 1/7 Bindung, Verbundwerkstoffe und Werkstoffverbunde, DGM-Tagung, Kaiserslautern, 1997.

[6] **Lekhnitskii, S.G.:** Theory of elasticity of an anisotropic elastic body, Holden-Day, San Francisco, 1963.

[7] **Naik, N.K.:** Woven fabric composites, Technomic, Lancaster, 1994.

[8] **Pagano, N.J. and Halpin, J.C:** Influence of end constraint in the testing of anisotropic bodies, Composite Materials, Vol.2, No.1, pp.18-31, 1968.

[9] **Polynkine, A.A., Van Keulen, F., De Boer, H., Bergsma, O., Beukers, A.:** Shape optimization of thermoformed continuous fibre reinforced thermoplastics, Structural Optimization, Vol.11, No.3/4, pp.228-234, 1996.

[10] **Robroek, L.M.J.:** The development of rubber forming as a rapid thermoforming technique for continuous fibre reinforced thermoplastic composites, Delft University Press, Delft, 1994.

[11] **Stellbrink, K.:** Micromechanics, Carl-Hanser Verlag, München, 1995.

[12] **Vu-Khanh, T. and Liu, B.:** Prediction of fibre rearrangement and thermal expansion behaviour of deformed woven-fabric laminates, Composites Science and Technology, Vol.53, pp.183-191, 1995.

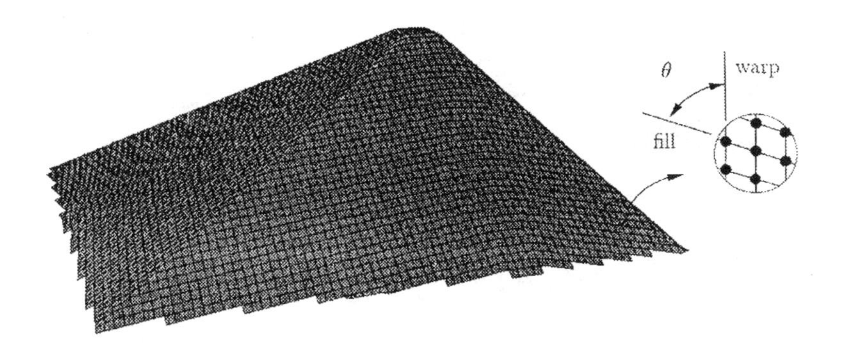

Figure 1: Drape simulation for a knotted pyramid. Warp and fill yarn directions are represented by rigid members, cross over points by pivots. Draping results in a simple shear deformation, characterized by the inter yarn angle (θ).

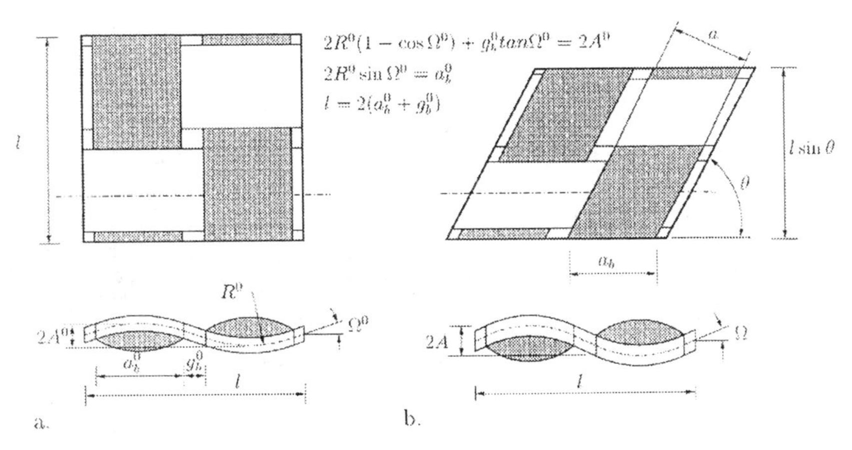

Figure 2: The effect of simple shear deformation on weave geometry, schematized for a balanced plain weave Repeating Volume Element, (a.) undeformed geometry, (b.) sheared geometry.

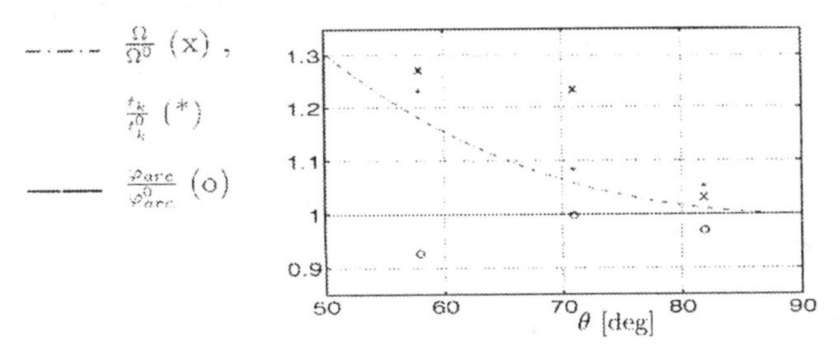

Figure 3: Maximum crimp angle (Ω), arcwise yarn segment volume fraction (φ_{arc}) and lamina thickness (t_k) as a function of θ with respect to values in undeformed condition. Theory versus experimental cross section analyses.

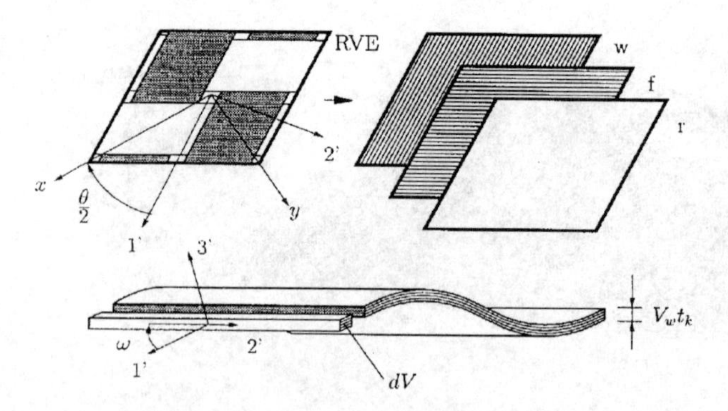

Figure 4: Further schematization of the RVE with three substitute layers for stiffness calculation. Warp yarn (w), fill yarn (f) and neat resin (r) layers have constant thicknesses.

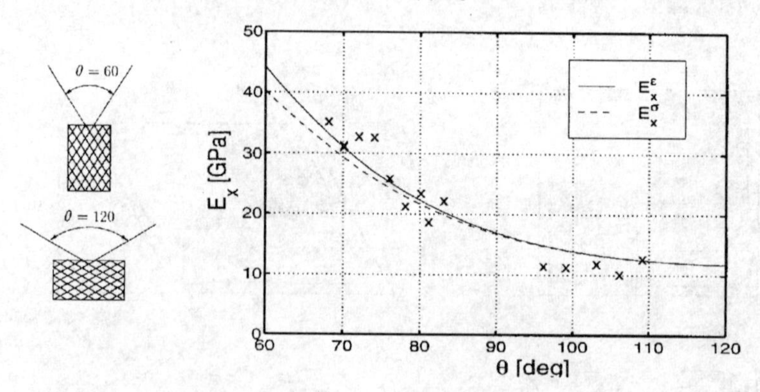

Figure 5: E-modulus in x-direction as a function of the inter yarn angle (θ). Prediction with compliance averaging (E_x^σ) and stiffness averaging (E_x^ϵ). Bending test results are marked (x).

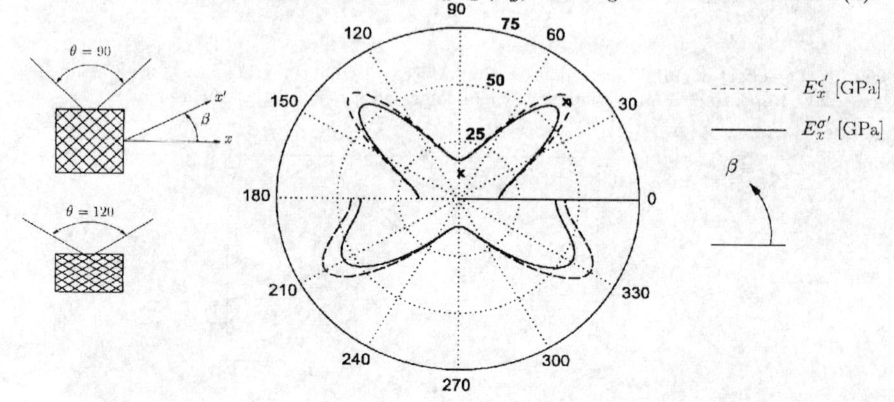

Figure 6: E-modulus as a function of β (E'_x) for an orthogonal laminate (upper half), and a laminate with θ = 120 [deg] (lower half). Compliance averaging ($E_x^{\sigma'}$) and stiffness averaging ($E_x^{\epsilon'}$) predictions are shown. Tensile test results in the orthotropic directions are marked (x).

Diaphragm Fusion Bonding Of Advanced Thermoplastic Composites

C.L. ONG* and M.F. SHEU*

Aeronautical Research Laboratory/CSIST Taichung, Taiwan, R.O.C.

Abstract

The joining of thermoplastic composite structures, Gr/PEEK using the fusion bonding technique was studied in this work. The PEI film was used as bonding agent between Gr/PEEK laminates. A serious of samples were prepared using diaphragm forming technology with different parameters: fusion bonding temperature (265-295℃), pressure (20-80 psi), time (15-60 min). Based on the 7702 psi single lap shear bond strength, the optimum fusion process (275℃ / 40 psi / 40 min) can be determined.

In addition, the environmental effect on fusion bonded Gr/PEEK has been studied. Results show that the bond strength:1) at 180℃ reduced to 45.4% of that at the room temperature, 2) increased to 105.6% for 30 days immersion in hot water, 3) reduced to 98.3% for 30 days immersion in JP-4, 4) increased to 102% for 500 hours salt spray. Furthermore, the fatigue tests of fusion bonded Gr/PEEK specimens reveal that the fatigue behavior are similar to those of metals. Therefore, the Gr/PEEK composite structures are better in the application for low cycle fatigue with higher fatigue stress.

1. Introduction

The fastening of composite structures could be made by-rivetting, adhesive bonding, and/or combination of two. Besides the above methods, the thermoplastic composites can also be joined by yet another method. This method is called fusion bonding or welding and involves the melting of the surfaces to be bonded. Several techniques [1,2,3] are available to introduce the energy required to melt the surfaces-the design of fusion bonding are generally classified by these energy input methods. The optimum one is the healing of amorphous thermoplastic fusion bonding from the evaluation so far.

In the past years, several publications has reported the study of the joint of thermoplastic composite-Gr/PEEK by amorphous thermoplastic fusion bonding. They used hot press to carry out the fusion bonding. The advantage of hot press is easy to operate; but it can't obtain uniform pressure for complex configuration structures during the forming. At present, the diaphragm forming is the advanced method for manufacturing thermoplastic composite structures. The advantages of the diaphragm forming method are as follows: 1)uniform pressure 2)ease in forming complex configuration structure 3)less defect in structures.

Recently, a serious of studies has been carried out for diaphragm forming technology by ARL research group, such as the evaluation of performance of the thermoplastic composite, repair of Gr/PEEK and the manufacture of thermoplastic composite structure of fighter aircrafts [4, 5, 6, 7]. In this report, we study the fusion bonding of the thermoplastic composite-Gr/PEEK using diaphragm forming process and evaluate the affecting parameters on the bond strength including 1)environmental effect 2)solvent effect 3)salt spray effect, and 4)fatigue behavior.

2. Experiment

Raw materials:
Gr/PEEK: ICI Fiberite product; unidirectional prepreg.
PEI (Polyetherimide): GE company product; 5 mil thickness, 12″ width.
PI (Polyimide)film (R-type): UBE company product, U.S.A.
Equipment:
The equipment for diaphragm forming is shown in Figure 1.

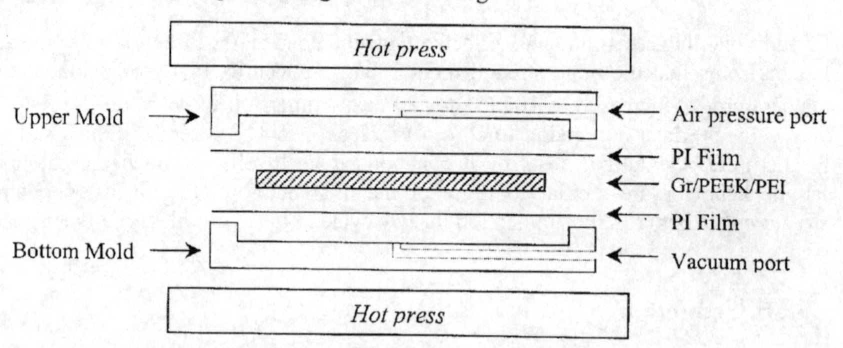

Fig. 1 The equipment of diaphragm forming process.

Gr/PEEK/PEI laminate preparation: Cut 8-ply of 12"x12" Gr/PEEK prepreg and fix each layer by hot air gun, add 1-2 layers PEI film above the top layer(Figure 2). Put the stacked laminate into the diaphragm forming machine and consolidate laminate according to the forming cycle shown in Figure 3.

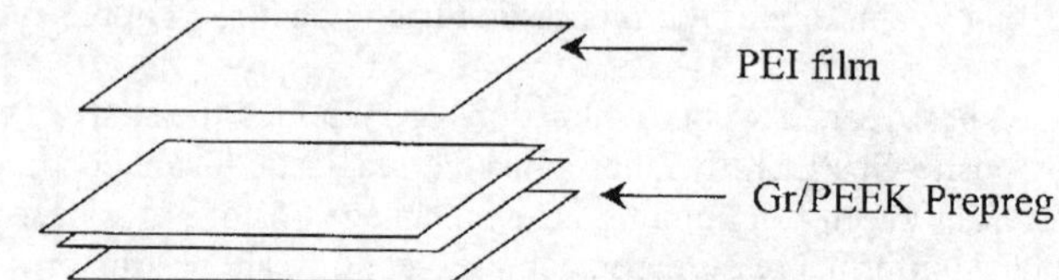

Fig. 2 The stacking sequence of laminate-Gr/PEEK/PEI

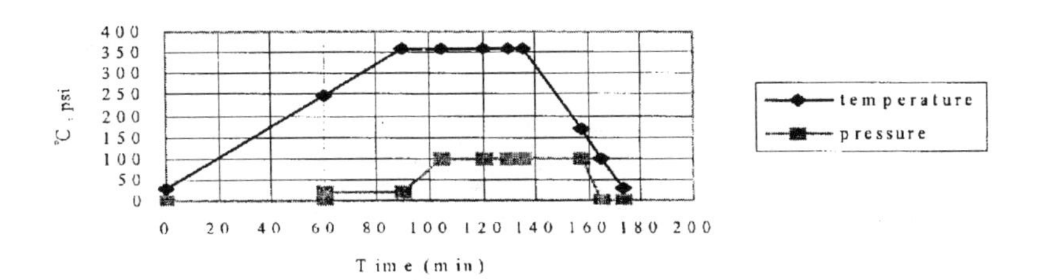

Fig. 3 The consolidation cycle of Gr/PEEK/PEI

The specimen preparation and testing: Cut Gr/PEEK/PEI composite panels into 12″ x 6″ size of standard specimens. On the pre-joining PEI surface of specimens, all specimens were roughened a 0.5″ width of overlap region(with 120 grit sandpaper) and wash with pure water to remove the residual release agent of PEI ,and finally, put the treated specimens into oven to dry at 100℃ for 1 hour. To overlap two treated specimens and the fusion bonding process was carried at diaphragm forming equipment.

ENVIRONMENTAL TEST

Temperature effect: The fusion bonded specimens were put into temperature controlled chamber, and the single lap shear strength test was carried out when the setting temperature was reached. The temperature range was from ambient to 180℃.
Humidity effect: The bonded specimens were put into 70℃ hot water for 15 days and 30 days, and the specimens were tested after removing from the water.
Solvent effect: The bonded specimens were put into JP-4 solvent for 15 days and 30 days, and the bond strength were tested after removing the solvent.
Salt spray test: Firstly, spray 5 % NaCl sloution onto the bonded specimens and stay for 500 hours, then, the bond strength were tested after removing the salt water from specimens surface.
Fatigue test: The MTS-810 testing machine was used to evaluate the fatigue property of bonded specimens, and the whole testing were carried out at r (σmin/σmax) of 0.1 and frequency of 30 Hz.

RESULTS AND DISCUSSION

THE STUDY OF FUSION BONDING CONDITION

The evaluation of fusion bonded design: The fusion bonding of thermoplastic composite (Gr/PEEK/PEI) depends upon the PEI layer. Three designs of fusion bonding were fabricated (Figure 4). The tested results were shown in Table 1 and the preferable bond strength were B and C types at 285℃. The lower bond strength of A design was due to incomplete fusion bonding of PEI layers of two specimens at 265℃. For B and C bonded types, the PEI layers of two specimens could fuse together completely at 285℃ and result in higher bond strength.

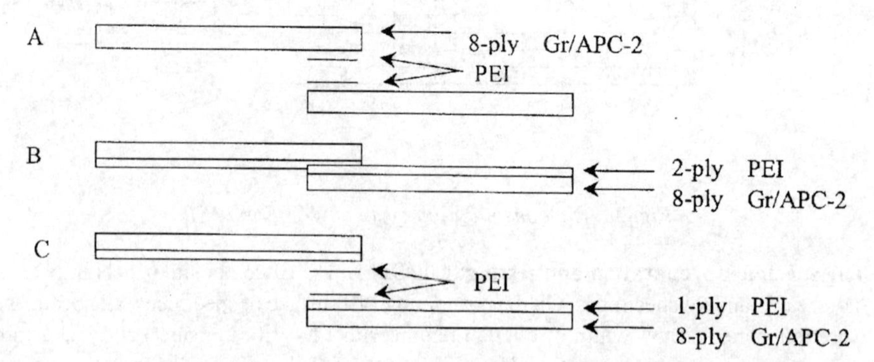

Fig.4 Three designs of fusion bonding configuration

Table1.Fusion bonding temperature versus bond strength

Bond strength / Temperature / Bond type	Lap Shear Strength (psi)	
	265℃	285℃
A	1824	3119
B	2874	6366
C	4780	6929

Fusion bonding temperature effect on the bond strength: The glass transition temperature (Tg) of amorphous thermoplastic PEI is 215℃. Theoretically, the fusion bonding process could be performed if the bonding temperature was over 215℃. The higher bond strength is obtained under higher bonding temperature. Table 2 shows the results of B and C designs at various temperature under 40psi for bonding time 40 min. In Table 2, the bond strength of Gr/PEEK/PEI was over 6000 psi in the temperature range 275℃~285℃. At high temperature, the viscosity of PEI is low and leads to complete fusion of the PEI; but at higher temperature, the fusion bonded PEI layer was extruded by pressure and led to decreasing bond strength.

Table 2.The relationship between bonding temperature and the bond strength

Lap shear strength / Bonding temperature(℃)	B-type (psi)	C-type (psi)
265	2874	4780
275	6586	7702
285	6366	6929

The effect of number of PEI layers on the bond strength: The thickness of adhesive affects the bond strength. In bonded joint of thermoset composite, the optimum bond thickness is a range of 10-20 mil. With lower or higher range of thickness, the bond strength would decrease. The similar phenomena were found for thermoplastic composites. In this article, we found that the number of PEI layers affect the bond strength with C-type bonded joint. For the fusion bonded condition:275℃/40psi/40min,the four PEI layers provided best bond strength (see Table 3).

Table 3.The effect of PEI layers on the bonded strength of thermoplastic composites

Number of PEI layers	3	4	5
Single Lap shear Strength, (psi)	6690	7702	6197

The effect of bonding pressure on the bond strength: The forming pressure could affect the miscibility and extrusion of PEI during the fusion bonding process. For B and C bonded designs at bonding condition 275℃/40min the bond strength at the forming pressure from 20psi to 80 psi were shown in Table 4. The highest bond strength was obtained at 40 psi. At the pressure, the miscibility of PEI layers was complete and the extruded PEI resin was the least. The miscibility of PEI layers was incomplete at lower pressure and PEI resin extruded at higher pressure, and lower or higher forming resulted in lower bond strength of thermoplastic composite.

Table 4. Bonding pressure effect on the bond strength

Pressure(psi) / Bonded type	20	40	60	80
B	4451 psi	6586 psi	6420 psi	6100 psi
C	5813 psi	7702 psi	6508 psi	6078 psi

Note: Fusion bonded condition: 275℃/40min

The effect of duration time on the bond strength: Besides the temperature and pressure, the duration also affect the bond strength of Gr/PEEK/PEI. Generally speaking, the longer the forming time, the more complete the degree of miscibility of PEI is. In this experiment, the fusion bonded condition was 275C/40psi and the forming time were 15, 30, 40 and 60 min with B-type bonding. The results were shown in Figure 5. The bond strength was dependent on time and the optimum forming time was 40 min.

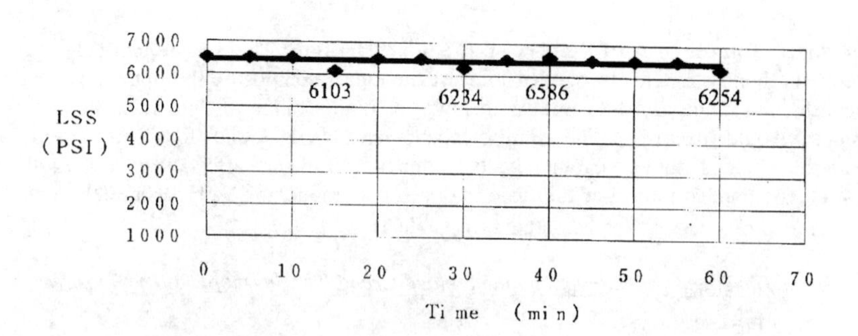

Fig. 5 The relationship between forming time and bond strength

ENVIRONMENTAL EFFECT ON BOND STRENGTH

Temperature effect: The bond strength of the thermoplastic composite-Gr/PEEK/PEI was tested at various temperatures and the results were shown in Figure 6. The bond strength at 180℃ reduced to 45.4% of that at the ambient temperature.

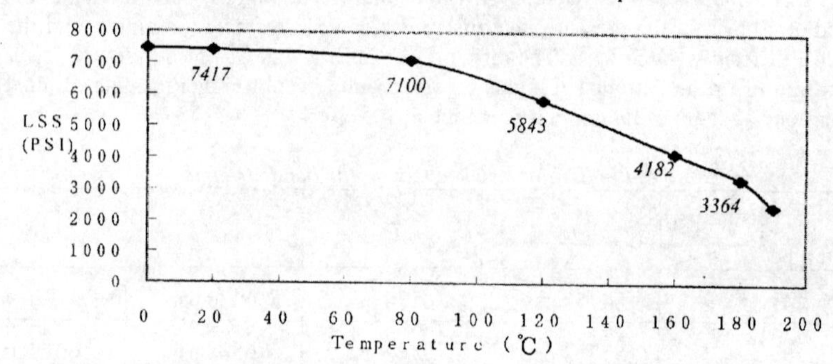

Fig. 6 The relationship between bonded strength and tested temperature of Gr/PEEK/PEI composite

Humidity effect: Because the PEEK and PEI molecules have hydrophobic functional groups, the PEEK and PEI have low humid absorption. When the specimens of thermoplastic composite (Gr/PEEK/PEI) were put into 70℃ hot water, the test results of immersion specimens were shown in Table 5. The bond strength increased after immersion. The reason is that the water molecules entered the interior of Gr/PEEK/PEI composite and formed hydrogen bonding with the molecules of PEEK and PEI leading to higher shear strength.

Table 5. Humidity effect on the bond strength

Immersion time, day	Single lap shear strength, psi
0	7417
15	8015
30	7832

Solvent effect: The PEEK and PEI have resistance to solvent attack from JP-4. The bonded specimens of Gr/PEEK/PEI composite were put into JP-4 for 15 days and 30 days. The bond strength reduced to 98.5% and 98.3% for 15 days and 30 days immersed in JP-4 (See Table 6), however the reduction is not significant.

Table 6. Solvent effect on the bond strength of Gr/PEEK/PEI composite in JP-4

Immersion time, day	Single lap shear strength, psi
0	7417
15	7304
30	7291

Salt spray effect: The bonded specimens of thermoplastic composite were sprayed with 5% NaCl solution for 500 hours. The bond strength increased to 7563psi(See Table 7). The result shows that the Gr/PEEK/PEI composite was a resistant salt spray material.

Table 7. Salt spray effect on bonded strength of Gr/PEEK/PEI composite

5% NaCl salt spray time, hour	Single lap shear strength
0	7417
500	7563

Fatigue behavior: The fatigue properties of Gr/PEEK/PEI composite were tested and the results were shown in Figure 7. The S-N curve of thermoplastic composite (Gr/PEEK/PEI) was similar to that of the metal(curve A). On the other hand, the S-N curve of thermoplastic composite (curve C) bonded with EA9346 was similar to that of the thermoset composite-Gr/Epoxy (curve B). So the bonded joint of structural parts of Gr/PEEK composite with PEI material would be better in the application for low cycle fatigue with higher fatigue stress, and the thermoplastic composite structural parts bonded with thermoset adhesive EA9346 were better in the application for high cycle fatigue with higher fatigue stress.

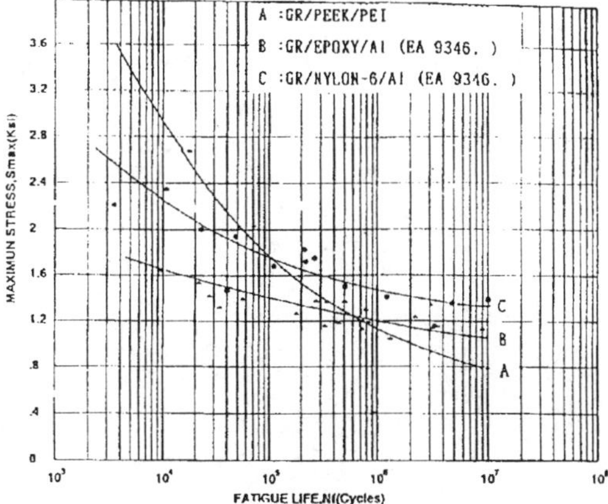

Fig. 7 The S-N curves of various composite

CONCLUSIONS

From the above experimental results, we could conclude as follows,

1. The optimum fusion bonding condition of thermoplastic composite (Gr/PEEK/PEI) is 275°C/40psi/40min.
2. The best bond strength of Gr/PEEK composite is the addition of two layers of PEI for C-design joint.
3. The bond strength of thermoplastic composite (Gr/PEEK/PEI) reduces as the temperature increases.
4. The bond strength of Gr/PEEK increases to 105.6% for 30 days immersion in hot water.
5. The bond strength of Gr/PEEK reduces to 98.3% for 30 days immersion in JP-4.
6. The bond strength of Gr/PEEK increases to 102% for 500 hours salt spray.
7. The fatigue behavior (S-N curves) of Gr/PEEK composite bonded with PEI is more similar to that of metal.

REFERENCES

[1]. *Avraham Benatar, Timothy G. Guyowdki,* METHODS FOR FUSION BONDING THERMOPLASTIC COMPOSITES - SAMPE Quaterly October 1988 P.35.

[2]. *Daniel M. Maguire,* JOINING THERMOPLASTIC COMPOSITE SAMPE J. Vol.25 No.1 Jan/Feb 1989 P.11.

[3]. *E.M. Silverman, R.A. Griese,* JOINING METHODS FOR GRAPHITE/PEEK THERMOPLASTIC COMPOSITES - SAMPE J.Vol.25 No.5 Sept/Oct 1989 P.34.

[4]. *C.L. Ong, M.F. Sheu, Y.Y. Liou,* THE REPAIR OF THERMOPLASTIC COMPOSITE AFTER IMPAC – The 34th SAMPE symposium; Vol.34 No.1 May 8-11 1989.

[5]. *C.L. Ong and H. Chin,* DESIGN FOR STRENGTH AND RIGIDITY OF A THERMOPLASTIC COMPOSITE SPEED BRAKE - J. of Theoretical and Applied Fracture Mechanics; Vol.14, No.1 1990 P.1-12.

[6]. *C.L. Ong and Y.Y. Liou,* CHARACTERIZATION OF MECHANICAL BEHAVIORS OF ADVANCED THERMOPLASTIC COMPOSITE AFTER IMPACT - Acta Astronautica; Vol.29 No.1 1993 P.99-108.

[7]. *C.L. Ong, Y.C. Chang, M.F. Sheu,* THE DEVELOPMENT OF THE THERMOPLASTIC LANDING GEAR DOOR OF A FIGHTER AIRCRAFT - The 42the SAMPE Symposium, Vol42, No.2 1997.

Heat Transfer and Consolidation Modelling in Resistance Welding of Thermoplastic Composite Coupons

C. Ageorges[1], L. Ye[1*], M. Hou[2] and Y.-W. Mai[1]

[1]Centre for Advanced Materials Technology
Department of Mechanical & Mechatronic Engineering
University of Sydney, NSW 2006, Australia
[2]Cooperative Research Centre for Aerospace Structures Ltd.
361 Milperra Road, Bankstown, NSW 2200, Australia

Abstract

A transient three-dimensional heat transfer-consolidation finite element model for resistance welding of thermoplastic matrix composite lap shear specimens is established. Different welding configurations of lap shear specimens are evaluated, namely APC-2 laminate/PEEK film, APC-2 laminate/PEI film and CF-PEI laminate/PEI film. The heat transfer model assumes orthotropic heat conduction in the composite parts and accounts for heat losses by radiation as well as natural convection. "Time to melt" and "time to cause thermal degradation" are predicted. The consolidation process is described in terms of intimate contact and autohesion concepts. The effects of power level on the time to achieve full intimate contact is determined. A processing window for CF-PEI/PEI configuration is established, which closely agrees with experimental data.

1. Introduction

Continuous fibre reinforced thermoplastic composites have been introduced as structural materials for high performance aerospace and industrial applications. High strain to failure, increased fracture toughness, better impact resistance, short processing cycle time, infinite shelf life of prepreg, environmental resistance, recyclability and reparability are some of the reasons cited for their growing popularity. However, due to deformation restrictions of continuous fibre reinforcement only relatively simple geometry can effectively be produced from these materials [11. To make large and complex structures, several of these simple components must be joined together. Resistance welding is a fusion bonding technique available for joining thermoplastic matrix composite materials, which offers many advantages over adhesive bonding and mechanical fastening, such as little or no surface treatment required, low cost, short process time, re-processability, and potential for on-line control [2, 31. Several models of the resistance welding process have been developed; namely, Holmes and Gillespie

* To whom all the correspondence should be addressed

[4], Don *et al.* [5], Bastien and Gillespie [6], Xiao *et al.* [7] and Maffezzoli *et al.* [8]. These models were either one-dimensional or two-dimensional. Bastien and Gillespie [7] model included non-isothermal healing theory while Maffezoli *et al.* [8] predicted crystallinity.

2. Modelling

The resistance welding process for lap shear coupons is depicted in Figure 1. The heating element made from a unidirectional single ply of prepreg is subjected to an electric current providing heat at the interface between the two laminates to be welded together. Neat resin films are placed on each side of the heating element in order to provide resin rich layers for fusion bonding and to prevent current leaking in the laminates. The system is assumed to be symmetric about the X-Z plane and the X-Y plane (Figure 1), accordingly only one eighth of the complete geometry is modelled as shown in dotted lines.

The laminate considered was a 16-ply unidirectional configuration with a thickness of 2.032 mm [5]. The thickness of the neat resin film was 0.076 mm [5]. A single composite prepreg layer was taken as the heating element, being 0.127 mm in thickness. The thickness of the insulator was set to 5 mm. For the simulation of the APC-2/PEI configuration, i.e. Thermabond process [9], two neat PEI films were introduced between the laminate and the heating element, corresponding to the PEI films co-moulded to the laminate and to the heating element. For the CF-PEI/PEI welding configuration [1], the laminate considered was a 10-ply $[90/0]_5$ configuration (i.e. fibres in first layer are parallel to the axis of the heating element, second layer perpendicular to the heating element and so on), 2.6 mm in thickness.

In the simulation, three modes of heat transfer are present, i.e. conduction, convection and radiation. For heat conduction, the composite material, i.e. the laminate and heating element, was assumed to be macroscopically homogeneous and orthotropic. The orthotropic thermal heat conduction coefficients were calculated from those of the constitutive materials using the rule of mixtures. Meanwhile, the neat polymer films and the insulator were considered as isotropic media. For the natural heat convection between the surfaces of the welding stack and the surrounding media, the heat transfer coefficients were assumed to be temperature dependent. For heat radiation, a constant emissivity of 95% [4] was selected for every free surface. The top surface of the insulator was assumed to be a heat sink, having a constant temperature equal to the ambient temperature of 20°C.

The consolidation process was modelled in terms of intimate contact [10] and autohesion [6]. The diagram in Figure 2 describes the process. In the intimate contact stage, two rough surfaces are brought into contact though a laminar flow mechanism, the initial roughness of the surface is modelled by a series of similar rectangles defined by the geometric parameter g*, Figure 2. Once intimate contact is achieved, region 'b' in Figure 2, molecular chains are free to move across the interface in the following stage of autohesion, region 'c' in Figure 2. Bonding refers to the areas where both intimate contact and autohesion are completed. The definitions of the degree of intimate contact, D_{ic}, the degree of autohesion D_{au} and the degree of bonding, D_b, are described in Figure 2.

3. Results and Discussion

For the APC-2 /PEEK welding configuration, density, specific heat and thermal conduction of APC-2 and PEEK materials were assumed to be dependent on temperature. The geometric parameter g* was determined to be 0.147, fitting the time required to achieved intimate contact on the data point corresponding to a power level of 52 kW/m^2 and an electrified time of 192.5 seconds [5]. In Figure 3, the predicted time to melt, corresponding to the time at which every node of the welding interface has a temperature above the melting temperature of PEEK, i.e. T_m=343°C, is plotted against the power level. The time to melt is in close agreement with that determined experimentally [5]. In Figure 3, the time required to achieve full intimate contact, t_{ic}, is plotted as a function of the power level, and matches the experimental electrified time. It can be seen that it is essential to keep the interface in the molten state for a certain period of time in order to allow for the full intimate contact to develop in the joint interface. The degree of degradation, D_d, in Figure 3, corresponds to the percentage of the welding interface at which thermal degradation has occurred, i.e. local temperature greater than the degradation temperature of PEEK (550°C [11]). For power levels in excess of 90 kW/m^2, the whole welding interface is subjected to temperatures greater than the degradation temperature of PEEK, Figures 3 and 4. Meanwhile, the average degree of intimate contact and the average degree of bonding are plotted against time for input power levels of 52 kW/m^2 and 120 kW/m^2 in Figure 4 (a) and (b), respectively. At a specific time of 50 seconds, segment 'l' in Figure 4 (a) corresponds to the region of intimate contact where autohesion has developed, physically it represents the region 'c' in Figure 2 (b). Segment 'm' refers to the region where intimate contact has been reached but autohesion has not yet occurred (region 'b' in Figure 2 (b)), while segment 'n' corresponds to the zone where intimate contact is not yet achieved, region '(b_0+w_0)-b' in Figure 2 (b). As time increases, the curves for intimate contact and bonding join each other because the time required for autohesion drops dramatically with increasing temperatures. In other words, achievement of intimate contact controls the consolidation process.

In Figure 5, the results concerning the APC-2/PEI lap shear coupon are presented. The time at which the interface temperature exceeds 260°C (Minimum recommended processing temperature [9]) is comparable with the time to melt. A value of 0.26 for g* was determined by fitting on the data point corresponding a power level of 52 kW/m^2 [5]. Similar results are presented for the CF-PEI/PEI welding configuration in Figure 6, where g*=0.33. A processing window for CF-PEI/PEI lap shear coupons generated from the FEM simulations is superimposed with that constructed experimentally [1] (Figure 7). The lower bound of the predicted processing window corresponds to a criterion of acceptable consolidation where 80% bonding is achieved. The upper bound of overheating was determined by a fully degraded interface, i.e. D_d=1, with a degradation temperature of 550°C determined by TGA.

The effect of the welding pressure on the development of intimate contact was investigated by varying the applied pressure which was assumed to remain constant during each simulation. In Figure 8, the time required to achieve intimate contact is plotted as a function of the applied pressure for the APC-2/PEEK welding configuration under a constant power level of 52 kW/m^2. As expected the greater the consolidation pressure, the faster intimate contact is achieved.

4. Conclusions

The resistance welding process of lap shear coupons was studied using a transient three-dimensional heat transfer-consolidation finite element model. The heat transfer model accounts for the orthotropic nature of the composite materials and features conduction, natural convection and radiation. The time to melt predicted by the model for different welding configurations; namely, APC-2 laminate/PEEK film, APC-2 laminate/PEI film (Thermabond process) and CF-PEI laminate/PEI film configurations, is in close agreement with experimental data. Thermal degradation in the heating element for high power levels was exhibited and quantified. The degradation of the heating element was shown to be reduced in the APC-2/PEI configuration as compared with the APC-2/PEEK system. The bonding time was predicted by the consolidation model for the three configurations, being comparable to the experimental electrified time. The processing window generated by the finite element model for the CF-PEI/PEI welding configuration was in a close agreement with that mapped experimentally.

Acknowledgements

This study is supported by the Australian Research Council (ARC) under a large research grant. C. Ageorges is supported by an Overseas Postgraduate Research Scholarship (OPRS) and a postgraduate scholarship from the Department of Mechanical & Mechatronic Engineering, the University of Sydney.

References

1. Hou M., L. Ye, and Y.-W. Mai, , *ICCM-11*, (1997) VI 36-45, *Australia, 1997, Australian Composite Structure Society, Ed. M.L. Scott.*
2. Silverman E. M., R. A. Griese, *SAMPE Journal*, **25** (5) (1989) 34-38.
3. Maguire D. M., *SAMPE journal*, **25** (1) (1989) 11-14.
4. Holmes S. T. and J. W. Gillespie Jr., *J. Reinf. Plas. And Comp.*, **12** (1993) 723-736.
5. Don R. C., L. Bastien, T. B. Jakobsen, J. W. Gillespsie, Jr., *SAMPE Journal*, **26** (1) (1990) 59-66.
6. Bastien L.J., J.W. Gillespsie Jr., , *Polym. Eng. and Sci.*, **31** (1991) 1721-1730.
7. Xiao X. R., S. V. Hoa and K. N. Street, *J. Comp. Mat.*, **26** (1992) 1031-1049.
8. Maffezzoli A. M., J. M. Kenny, L. Nicolais, , *SAMPE Journal*, **25** (1) (1989) 35-39.

9. Smiley A. J., A. Halbritter, F. N. Cogswell, P. J. Meakin, , *Polym. Eng. Sci.*, **31** (1991) 526-532.
10. Mantell S.C., G.S. Springer, *J. Comp. Mat.*, **26** (16) (1992) 2348-2378.
11. Cogswell F.N., "Thermoplastic Aromatic Polymer Composites", Butterworth-Heinemann Ltd, 1992

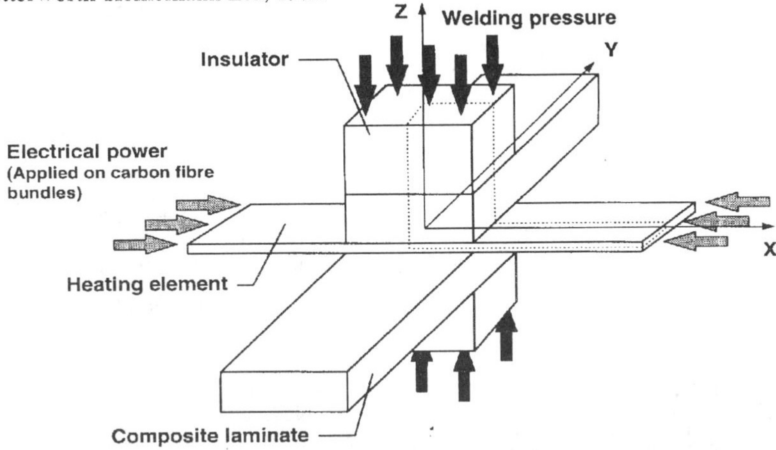

Fig. 1 Resistance welding configuration for lap shear specimens.

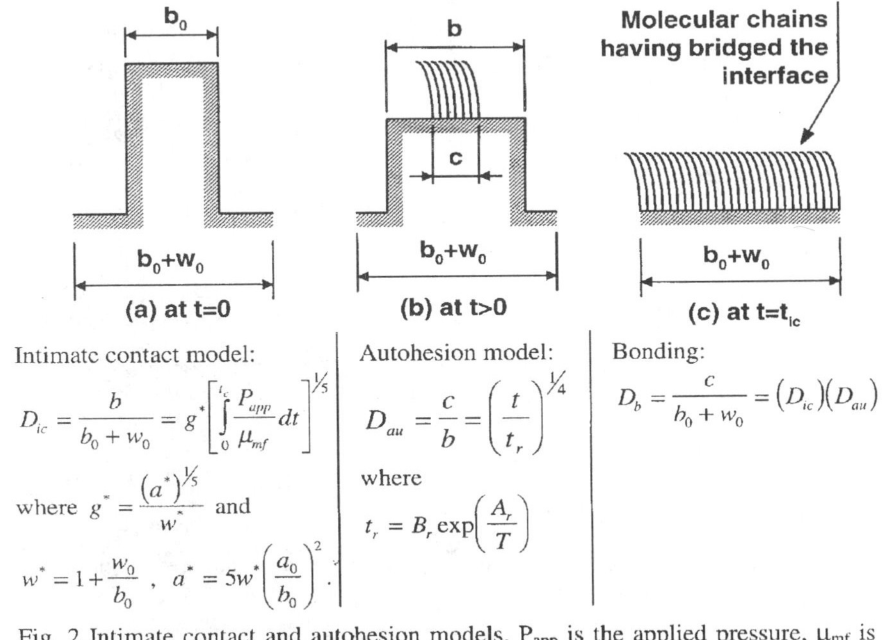

Intimate contact model:

$$D_{ic} = \frac{b}{b_0 + w_0} = g^* \left[\int_0^{t_c} \frac{P_{app}}{\mu_{mf}} dt \right]^{1/5}$$

where $g^* = \frac{(a^*)^{1/5}}{w^*}$ and

$$w^* = 1 + \frac{w_0}{b_0}, \quad a^* = 5w^* \left(\frac{a_0}{b_0} \right)^2.$$

Autohesion model:

$$D_{au} = \frac{c}{b} = \left(\frac{t}{t_r} \right)^{1/4}$$

where

$$t_r = B_r \exp\left(\frac{A_r}{T} \right)$$

Bonding:

$$D_b = \frac{c}{b_0 + w_0} = (D_{ic})(D_{au})$$

Fig. 2 Intimate contact and autohesion models. P_{app} is the applied pressure, μ_{mf} is the viscosity of the matrix-fiber system and A_r and B_r are experimental parameters.

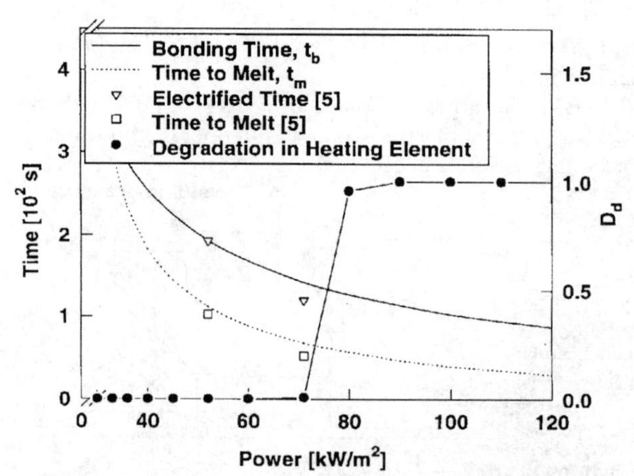

Fig. 3 Simulations of APC-2/PEEK welding configuration.

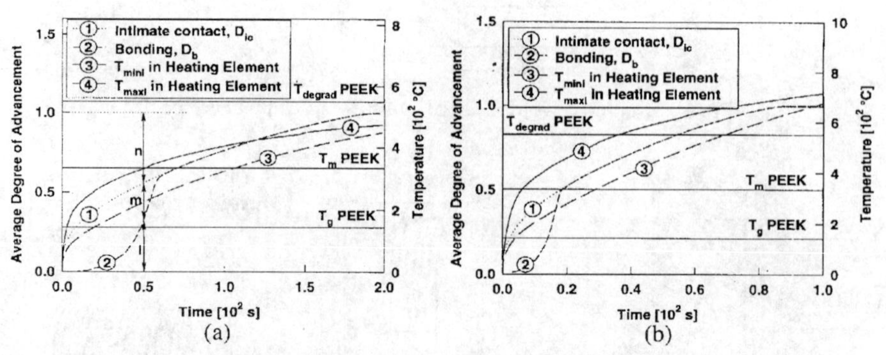

Fig. 4 Simulations of APC-2/PEEK, Power level 52 kW/m^2 (a) and 120 kW/m^2 (b).

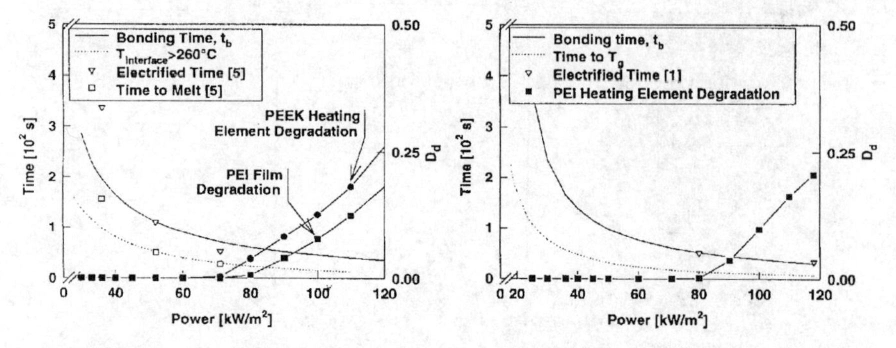

Fig. 5 Simulation of APC-2/PEI welding configuration.

Fig. 6 Simulations of CF-PEI/PEI welding configuration.

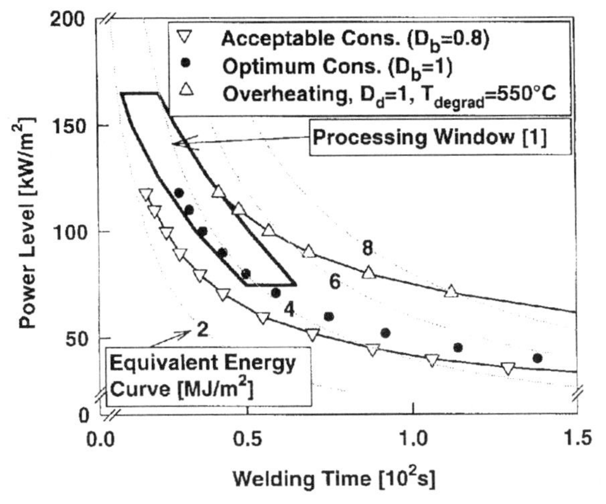

Fig. 7 Simulated and experimental processing windows for CF-PEI/PEI welding configuration.

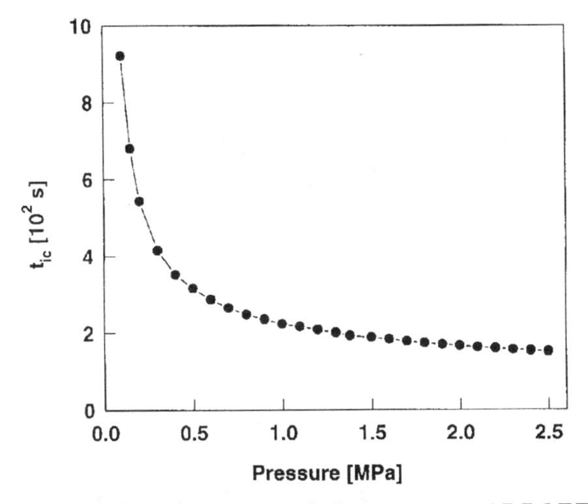

Fig. 8 Effect of pressure on intimate contact, APC-2/PEEK system, P=52 kW/m^2.

Considerations in infiltration studies and measurement of permeability in the processing of polymer composites

C.Lekakou and M.G.Bader

School of Mechanical and Materials Engineering
University of Surrey
Guildford, Surrey GU2 5XH, UK

Abstract

A mathematical model is proposed to describe the macro- and micro-infiltration through reinforcements of bimodal porosity distribution in Resin Transfer Moulding. The model is based on Darcy's law incorporating mechanical, capillary and vacuum pressures and covers three modes of infiltration. Permeabilities and capillary pressures are calculated at macro- and micro-level. A numerical analysis is presented where at each numerical location the flowrate is split between three possible types of flow on the basis of mass balance and a combination of permeability magnitude and local flow potential. Parametric computational studies are carried out to study the flow of a model Newtonian fluid through woven cloths where the following parameters are varied: fibre volume fraction, fibre tow diameter and injection pressure. Predicted variables include micro- and macro-infiltration times and apparent global permeability. A study of inertia effects on infiltration at high injection pressures relates their presence to the pore Reynolds number.

1. Introduction

Impregnation of the fibre reinforcement by the matrix is a key factor in the processing of composites since it concerns the establishment of the two-phase system, the consolidation of plies, the elimination of voids and the interfacial properties between matrix and fibre bundles or yarns. Parameters affecting impregnation are resin rheology, applied mechanical pressure and/or vacuum, flow path length in the mould, permeability of reinforcement and geometrical parameters of the mould design and reinforcement architecture.

With regards to the modelling of impregnation, Darcy's law of flow through a homogeneous porous medium has been used to describe the global infiltration of viscous resin through the fibre reinforcement. On the basis of Darcy's law, experimental techniques have been developed for the measurement of the global in-plane permeability in radial or rectilinear flow under constant pressure or constant flowrate. However, several experimental studies illustrated that the flow behaviour of Newtonian fluids in both saturated and unsaturated reinforcements deviated from Darcy's law with the permeability being a function of porosity, superficial velocity of permeating fluid and pressure drop [1-4]. Attempts have been made to explain some of these effects by using generalised flow models including non-Newtonian and inertia effects [5-7] or by modelling the deformation of preform [8]. However, the dependence of the permeability on the superficial velocity was observed even at low superficial velocities and injection pressures [1-2] where inertia effects and preform deformation were negligible. Chan et al [1] speculated that the observed increase in permeability with flowrate could be due to a preferential channelling of the liquid through the large pores (macro-flow) rather than though the micro-pores within fibre tows (micro-flow).

Another group of published papers focus on microscopic studies of void formation in composites manufacturing [9,10 amongst others]. Void formation has been related to resin impregnation and more specifically to viscous flow and wetting properties of resin related to the reinforcement. It has also been observed that the measured global permeability may vary, depending on the infiltrating fluid and its wetting properties. In flow visualisation studies it has been observed that at a low flowrate micro-flow was leading whereas at a high flowrate macro-flow was leading. This illustrates the need for developing a mathematical model for the macro- and micro-infiltration [11,12].

The scope of the current study is to carry out a theoretical analysis of infiltration of reinforcement in RTM, including wetting effects for low and intermediate injection perssures, and inertia effects for high injection perssures.

2. Mathematical model of macro- and micro-infiltration

Darcy's law has been employed to describe the relation between flowrate and pressure drop, at low flowrates and injection pressures, in both macro- and micro-impregnation where the impregnating fluid was considered to be Newtonian. Since only viscous effects are considered in Darcy's law, this analysis is applicable to very low pore Reynolds numbers, Re_p, where

$$Re_p = \frac{\rho U D_e}{\mu} \tag{1}$$

and D_e is the equivalent pore diameter.

The two aspects of macro- and micro-infiltration are described by Darcy's law according to the following general relations:

$$U_i^{sup} = \frac{K_i}{\mu} \nabla (P^{mech} + P^{v} + P_i^{c}) \tag{2}$$

where U^{sup} is the superficial velocity, K is the permeability tensor and subscript i refers to different types of permeability and capillary pressures due to differences in pore size and flow direction with respect to fibre direction. The pressure gradient includes the terms of mechanical injection pressure, P^{mech}, vacuum pressure, P^v, and capillary pressure, P^c. The present study considers viscous and capillary flow in Darcy's law for both macro- and micro-infiltration.

The study focuses on in-plane global infiltration of fibrous reinforcements, as is common in RTM. Regarding the modelling of the combination of in-plane macro- and micro-infiltration of an assembly of woven cloths, three modes of flow have been identified. In mode I the fibre tows have not been fully impregnated in the radial direction; the flow then is considered as channelled mainly into the macro-pores and from there the fluid simultaneously impregnates the fibre tows in the radial direction. In mode II the fibre tows are considered as fully impregnated in the radial direction; the flow then is considered as being split between macro-impregnation and micro-impregnation along the fibre direction. In mode III the fibre tows are considered as fully impregnated in the radial direction and the flow front is ahead inside the tows; the flow is then modelled as a combination of micro-impregnation along the fibre direction and simultaneous transverse flow from the tows into the macropores if sufficient driving pressure exists.

In each mode of impregnation, the flow at each point in the mould is split between the different types of impregnation according to the relations

$$\textit{Darcy's law:} \quad Q_i = \frac{K_i A_i}{\mu} \nabla (P^{mech} + P^v + P_i^c) \tag{3}$$

$$\textit{Mass balance:} \quad Q_{tot} = \Sigma Q_i \tag{4}$$

where i refers to the following types of flow: (i) macro-flow, (ii) micro-flow transverse to the fibres and (iii) micro-flow parallel to the fibres. This approach is equivalent to network models for Poisseuille flow of Newtonian viscous fluids where common mechanical pressures are considered at junctions, pressure drops are associated with network bonds and there is a balance of volumetric flowrates at junctions. A linear relationship exists between pressure difference (flow potential) and flowrate of Newtonian fluids which also features a bond conductance or resistance. Provided, then, that the network branches are associated with common inlet and outlet pressures, the volumetric flowrate in each branch is proportional to the branch conductance. However in equation (3), where individual capillary pressures are considered at the flow front in each branch of the network, the total pressure differences in each branch differ and the flowrate Q_i in each branch is proportional to the product of branch conductance and pressure difference in this branch.

The calculation of flowrates, total filling times and filling lengths allowed the calculation of a global apparent permeability under constant injection pressure on the basis of a filled length and corresponding filling time which are associated with the progress of macro-flow front when macro-flow is ahead of micro-flow, and with the progress of micro-flow front when micro-flow proceeds ahead of macro-flow. The evaluated apparent global permeabilities are equivalent to the global permeabilities measured in rectilinear or radial flow experiments in which the flow front is monitored for example by taking photographs at regular intervals. In this case, the monitored flow front normally corresponds to the type of flow which advances first.

3. Parametric studies of macro- and micro-infiltration

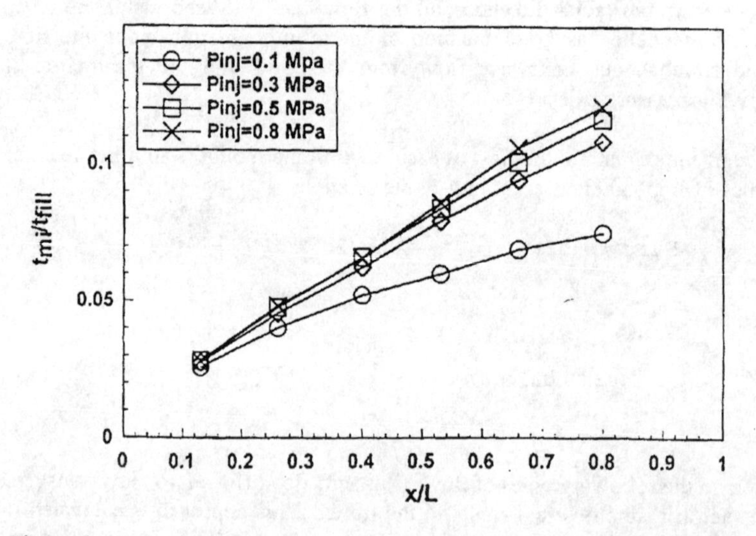

Figure 1 Micro-impregnation times normalised by the corresponding filling times in rectilinear infiltration studies under constant injection pressure

A series of computational parametric studies were carried out concerning the in-plane infiltration of five layers of a woven glass cloth of an areal density of 0.546 kg/m^2, thickness 0.48 mm, containing 650 ends/m and 650 picks/m. The employed infiltrating Newtonian fluid was silicone oil of a viscosity of 0.142 Pas. The capillary pressure is calculated according to Young-Laplace relation [13]. The surface tension and contact angle between silicone oil and glass fibres were taken from Patel et al's [14] experimental study as σ=21x10^{-3} N/m and θ=0°. The first set of studies focused on the evaluation of local macro- and micro-infiltration times under different injection pressures. A case of rectilinear macro-flow under constant injection pressure was considered with a filling length of L=0.15 m. Figure 1 presents the micro-infiltration times, normalised by the total filling time, as a function of distance from the edge gate. Low injection pressures were associated with long filling times and as a result, micro-impregnation could be considered to occur instantaneously: for example $t_{mi}/t_{fill} \sim 0.05$. At high injection pressures, filling occurs in short times which are more comparable to micro-impregnation times as Figure 1 illustrates. The driving factors for micro-impregnation in the suggested model are the capillary and local mechanical pressures. Since the local mechanical pressure decreases as the fluid moves away from the gate, the micro-impregnation time increases.

In the next set of parametric studies where the layers of cloth were compressed to different fibre volume fractions, changes were taken into account at macro-level, i.e. macro-porosity and average size of macro-pores, while bundle dimensions and micro-porosity were assumed to remain the same. Permeabilities were evaluated according to Carman-Kozeny relationship. The following micro-properties were calculated:

$\varepsilon_{mi} = 0.42$ $K_{mi,II} = 2.6 \times 10^{-12}$ m^2 $K_{mi,\perp} = 1.3 \times 10^{-13}$ m^2
$P^c_{mi,II} = 11720$ Pa $P^c_{mi,\perp} = 5860$ Pa

Low injection pressures were generally employed so that the assumption of neglecting inertia effects would be justified. In the low flowrate or injection pressure regime it has been reported in the literature [1,2] that an increase of Q or P_{inj} may cause an increase in the measured global permeability of assemblies of plain woven cloths. Figure 2 presents the effects of the injection pressure on the apparent permeability for different fibre volume fractions. At low V_f macro-flow is important and as a result flow modes I and II are encountered. At low P_{inj} the micro-capillary pressures are important and produce micro-flowrates which cannot be neglected. As a result the macro-flowrate is reduced and this reduces the overall permeability which is calculated on the basis of macro-flow in the combination of flow modes I and II. As P_{inj} increases, capillary effects become less important and the flowrate is channelled towards the largest permeability, i.e. the macro-pores. The macro-flow front then moves faster and this raises the apparent permeability. As Q is raised further in the parametric studies the capillary effects become negligible and the flow is channelled primarily as macro-flow lleading to constant K. On the other hand, as V_f is increased the macro-permeability is reduced. So, the effects of flow channelling into the macro-pores are less important and K changes less with P_{inj}. At V_f = 0.58 macro- and micro-impregnation become almost identical and no change of permeability is observed when P_{inj} is varied.

Figure 3 presents the effects of changing injection pressure on the evaluated apparent

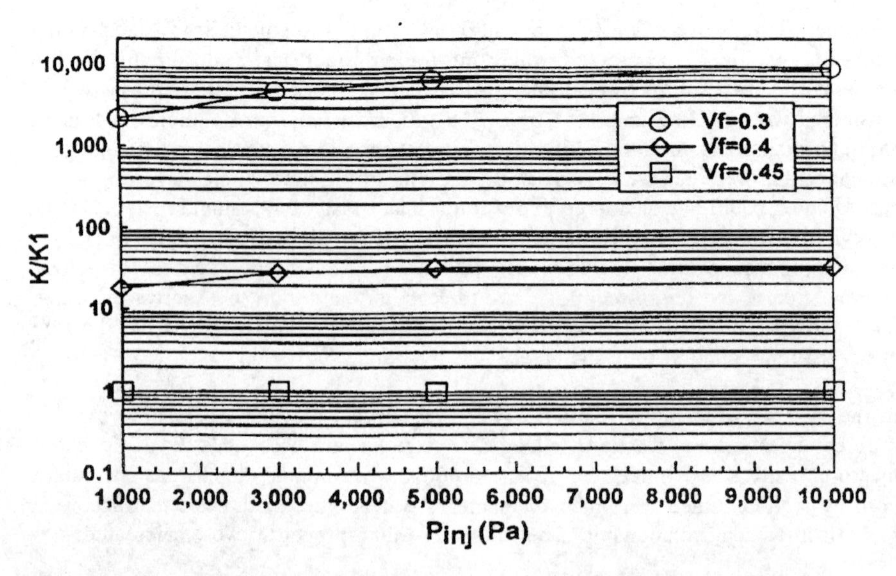

Figure 2 Normalised global apparent permeability as a function of injection pressure for three values of global fibre volume fraction

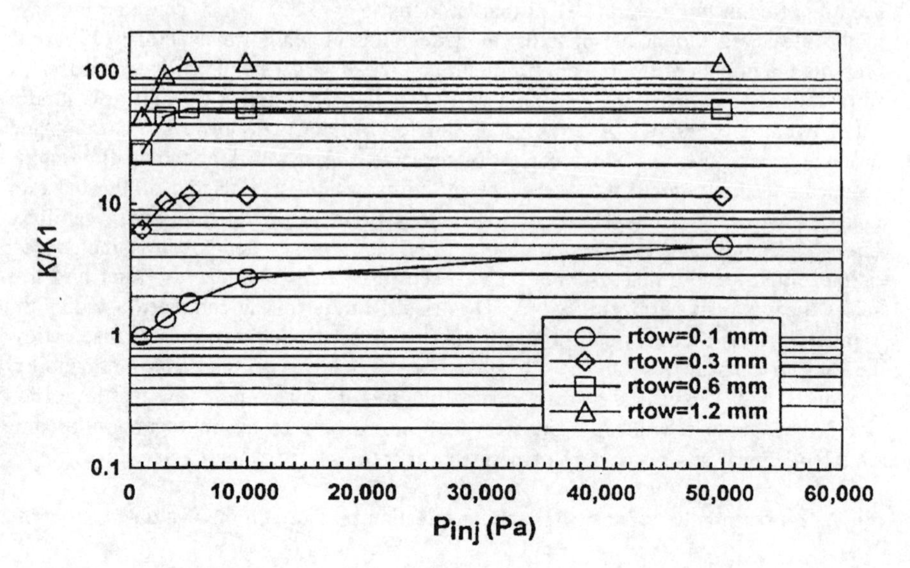

Figure 3 Normalised apparent permeability as a function of injection pressure for four values of fibre tow radius

permeabilities for different tow sizes. Larger fibre tows correspond to larger equivalent macro-pores and as a result the flow channelling effect is more apparent at very low P_{inj}. On the other hand, smaller fibre tows correspond to small macro-pores but generate higher macro-capillary pressures. Therefore, at small injection pressures the effects of changing P_{inj} on the apparent permeability are not so important but once P_{inj} becomes comparable to P^c_{ma} these effects grow significant. So it was found that whereas for r_{tow}=0.3 - 1.2 mm the apparent permeability stabilises in the pressure range of 5 - 10 kPa, for r_{tow}=0.1 mm the apparent permeability stabilises around a P_{inj} of 50 kPa. This has important implications in the measurement of permeability where a suitable range of injection pressures needs to be identified, depending on the type and structure of fabric and the fluid wetting properties at the fluid/fibre interface.

4. Study of the inertia effects

Inertia effects are present at relatively high injection pressures which are not usually employed in permeability measurements. Momentum transport for the general case of flow though a porous medium, including both inertia and viscous terms, is described by the following relation in rectilinear coordinates:.

$$\frac{Ah\Delta P}{\mu Q\Delta x} = \frac{1}{K} + a\frac{\rho Q}{\mu Ah} \tag{5}$$

Relation (5) can be easily considered as the outcome of a combination of Ergun's and Darcy's equations. K denotes a global permeability and a denotes an inertia coefficient to be determined from experimental data. Relation (5) can be used at high injection pressures where capillary effects may be ignored.

A search of appropriate data in the literature yielded Trevino et al's study [4] in which a non-linear relation between pressure and superficial velocity was reported and injection pressures reached $8x10^5$ Pa. In that study there was no data of fibre or pore size and, hence, it was not possible to estimate the macro-flow Reynolds number. However, when relation (5) was applied to Trevino et al's data, a had a negative value which is not acceptable. Therefore, other effects such as deformation of the fibre reinforcement under the injection pressure at the inlet is suspected for Trevino et al's study. According to our experimental data inertia effects become present at $Re_p > 0.1$.

5. Conclusions

A mathematical model has been suggested for macro- and micro-impregnation in the processing of polymer composites, including Darcy's and capillary effects. Separate permeabilities and capillary pressures were included for macro- and micro-pores. On the basis of the proposed model in which micro-impregnation is affected by the sum of capillary, mechanical and vacuum pressures, it was demonstrated that micro-impregnation times increase as the resin front moves away from the inlet port due to the corresponding decrease of local mechanical pressure. An interesting parameter that it was possible to predict in the parametric infiltration studies under constant injection pressure was the apparent global permeability, which is equivalent to the global permeabilities obtained from in-plane infiltration experiments where the progress of flow front is monitored. Inertia effects are usually negligible under RTM conditions. The value of pore Reynolds number should be a good indication for the presence of inertia effects, and, if present, numerical procedures have been suggested for the determination of global permeability and inertia coefficient from experimental data.

Acknowledgements

The case-studies of this publication are related to the EPSRC grant GR/J76972 the support of which the authors gratefully acknowledge.

References

1. Chan A.W., Larive D.E. and Morgan R.J., J. of Composite Materials, 27(10), 1993,pp.996-1008.
2. C.Lekakou, M.A.K.Johari, D.Norman and M.G.Bader, Composites: Part A, 27A, 1996, pp.401-408.
3. Martin G.Q. and Son J.S., Proc. ASM/ESD 2nd Conf. on Advanced Composites, ASM International, Materials Park OH, 1986, pp.149-157.
4. Trevino L., Rupel K., Young W.B., Liou M.J. and Lee L.J., Polymer Composites 12(1), 1991, pp.20-29.
5. Skartsis L., Khomami B. and Kardos J.L., J. of Advanced Materials, 25(3), 1994, pp.38-44.
6. Cai Z., J. of Advanced Materials, 25(1), 1993, pp.58-63.
7. Cai Z., J. of Composite Materials, 29(2), 1995, pp.257-278.
8. Parnas R.S., Schultheisz C.R. and Ranganathan S.,Polymer Composites, 17(1), 1996, pp.4-10.
9. Mahale A.D., Prud'homme R.K. and Rebenfeld L., Polym.Eng.Sci., 32(5), 1992, pp.319-326.
10. Nowak T. and Chun J.H., Composites Manufacturing, 3(4), 1992, pp.259-271.
11. Chan A.W. and Morgan R.J., SAMPE Quart., October 1992, pp.45-49.
12. Chan A.W. and Morgan R.J., Polymer Composites, 12(3), 1991, pp.146-152.
14. Patel N., Rohatgi V. and Lee L.J., Polym.Eng.Sci., 35(10), 1995, pp.837-851.

Mathematical and experimental studies of the draping of woven fabrics in resin transfer moulding (RTM)

U.Mohammed, C.Lekakou and M.G.Bader

School of Mechanical and Materials Engineering
University of Surrey
Guildford, Surrey GU2 5XH, UK

Abstract

This study includes experimental and finite element analyses of the draping of dry fabrics. Draping is considered over the hemispherical surface of a tranparent mould for four different types of woven cloths, namely a loose plain weave (basket weave), a tight plain weave, a twill and an 8 harness satin weave. A 20x20 points grid has been drawn on the fabric to be used for each experimental study, where the grid deformation is observed after draping. The various fabrics are compared in terms of maximum shear deformation and wrinkling. Further experiments involve shear tests for each fabric where the emphasis is on the "locking shear angle" attained, i.e. the maximum shear angle before wrinkles appear. The theoretical analysis of draping is based on a finite element, solid mechanics approach, where elastic properties are assigned to both fabric and the interfacial volume between fabric and mould.

1. Introduction

Woven fabrics have a great potential for use by many types of industry ranging from automobile to aerospace manufacturing, due to the variety of woven patterns available which could suit different types of applications and specifications. In many cases, resin transfer moulding (RTM) is employed in the fabrication of geometrically simple or complex components of large or medium size. RTM is a versatile, cheap and well controlled technique. In the process of producing components using this technique, woven fabrics are draped onto surfaces of varying geometric complexity (mould) before being infiltrated with liquid resin. The draping process results in changes in the fibre orientation, local volume fibre fraction and porosity. Consequently, there is a change of permeability over different parts of the fabric, as well as local variation in the mechanical properties of the component. This is an important reason for devoting research effort towards producing a methodology that explains and predicts the deformation of fabrics during draping. The ultimate aim is to improve processability, rate of production and mechanical properties of the final component

In the mathematical modelling of draping, many authors [1-3] have adopted an idealised kinematic mapping approach to predict the deformation of woven fabrics. Their model considers the fabric as a kind of inextensible, pin-jointed network, in which the fibres are pinned at the crossover points so that fabric deformation is determined by the rotation of the fibres at the intersections only. The model is found to be mostly applicable to woven fabrics and to simple geometries where the fabric deformation does not approach the so-called "locking angle". Another approach such as the visco-elastic model has been traditionally adopted to model the forming of thermoplastic composite sheets[4] containing continuous unidirectional fibres, bi-directional stitched reinforcement and continuous random mat. A third approach is based on the continuum, solid mechanics model in which the fabric is considered as a solid continuum with certain mechanical properties and under the influence of external factors such as loads or friction. This type of model can be incorporated in a finite element analysis (FEM) [5] and offers great future potential.

Various theories have been put forward to explain the way fibres or fabrics deform when draped over surface geometries. One of the theories is that reported by Porter [6] in which he proposed that two mechanisms are responsible for the deformation of the fibres in conformity with the surface geometry: fibre slip as fibres move over each other in response to the draping forces and rotation of the fibres (or inter-fibre shear) at the tow intersection. The first mechanism could not be adequately quantified and so little attention was given to it. The second mechanism is thought to be the most important and has drawn much interest by many authors and researchers. A third plausible mechanism that takes place when fibres are sheared beyond the so-called "locking angle" is that of fibre buckling. Fibre buckling is said to be associated with the local in-plane compression and subsequently, leads to the formation of certain undesirable features such as wrinkles and folds. This problem has great repercussions in high volume production.

In the light of this, several shear test investigations were performed quite recently using various techniques such as the "strain automated analysis system" developed by Long et al [7] for example, and used to measure the maximum shear angle for non-crimp, stitch-bonded, glass fabrics. A simple hand stretching shear experiment was performed by Andrews et al. [8] to determine the locking angle of woven fabrics.

The scope of this paper is to (a) carry out experiments in order to study the draping of various types of weaves over a hemi-spherical surface, (b) perform shear tests for the same weaves and relate the obtained data to the conclusions of the draping experiments and (c) carry out a finite element analysis based on the continuum solid mechanics approach.

2. Materials

The following types of woven E-glass fabrics were tested: a loose plain weave (LPW), a tight plain weave (TPW), a twill weave (TWILL) and an 8 harness satin weave (8HSW). Their specifications are given in Table 1.

Table1: Specifications of the fabrics tested.

Fabric	Fabric	Thickness (mm)	Areal density (kg/m^2)
LPW	Basket weave	0.58	0.529
TPW	Y0212*	0.48	0.546
TWILL	Y0185*	0.28	0.331
8HSW	Y0227*	0.23	0.297

* supplied by Fothergill Engineering Fabrics.

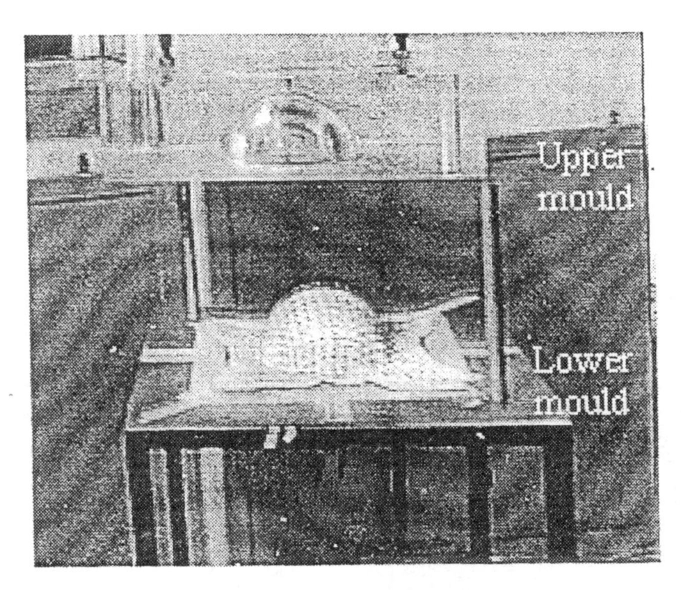

Fig 1. Hemispherical drape mould.

3. Draping experiments

This section focuses on experiments of draping of each type of fabric onto a hemispherical mould as displayed in Fig.1. The mould consists of two parts: a lower aluminium platen with a hemispherical solid dome of a radius of 97 mm and an upper transparent block, from poly(methyl methacrylate), with a hemi-spherical cavity. Draping experiments were carried out for all fabrics presented in Table 1 in order to compare them in terms of ease of shearing, maximum degree of shear attained and occurrence of wrinkling. The first stage was to cut a layer of fabric of 360x360 mm and inscribe on it a regular, square grid of 20x20 divisions. The fabric was then placed on the mould so that its centre coincided with the pole of the hemispherical part. The fabric was gently smothered onto the mould shape manually to ensure good contact and best fit between the fabric and the mould surface. The upper block was then lowered down until it housed the fabric in the hemispherical mould.

Fig.2 presents photographs of the four draped fabrics where the deformation of the inscribed grid can be observed as well as the occurrence of wrinkles. The basket weave has a very loose pattern which allows easy deformation by shear without wrinkling. The 8-harness satin is also easy to deform up to a maximum shear angle. After this angle, locking is expected which leads to wrinkling that is particularly emphasised on the particular fabric due to its small thickness (see Table 1). Wrinkling is worse for the twill due to its lower shearing ability. The tight weave has a lower ability for shear than the satin and the loose basket weave. However, it has not as many wrinkles as the twill because this particular plain woven fabric is rather thick.

4. Shear test experiments

The second part of experimental work focused on the determination of maximum shear angle before the appearance of wrinkles in each of the fabrics presented in Table 1. Two methods were used for the determination of maximum shear angle. In the first method a sample of fabric of dimensions 185x185 mm was placed in an orthorhombic shear frame which was sheared in a tensile Instron machnie. The deformation of the frame was monitored and photographs were taken at regular intervals, in order to identify the time when the first wrinkles would appear. The second method involved shearing of the fabric sample manually until the "locking angle" beyond which wrinkles would appear. Table 2 displays the "locking shear angle" detected by each method for all four tested fabrics. The basket weave displays the highest locking shear angle due to its loose structure. This is consistent with the results of the draping experiment (Fig.2c) in which no wrinkles appeared when this fabric was draped onto the mould of Fig.1. The 8 harness satin weave is the next best fabric in terms of locking shear angle due to its loose weave pattern. The tight plain weave Y0212 displays the smallest locking shear angle, as expected.

Table 2: "Locking shear angle" for a loose plain weave (basket weave), tight plain weave, twill and 8-harness satin weave.

Fabric	Frame shear test	Manual shear test
LPW	56	54
TPW	45	43
TWILL	49	50
8HSW	51	49

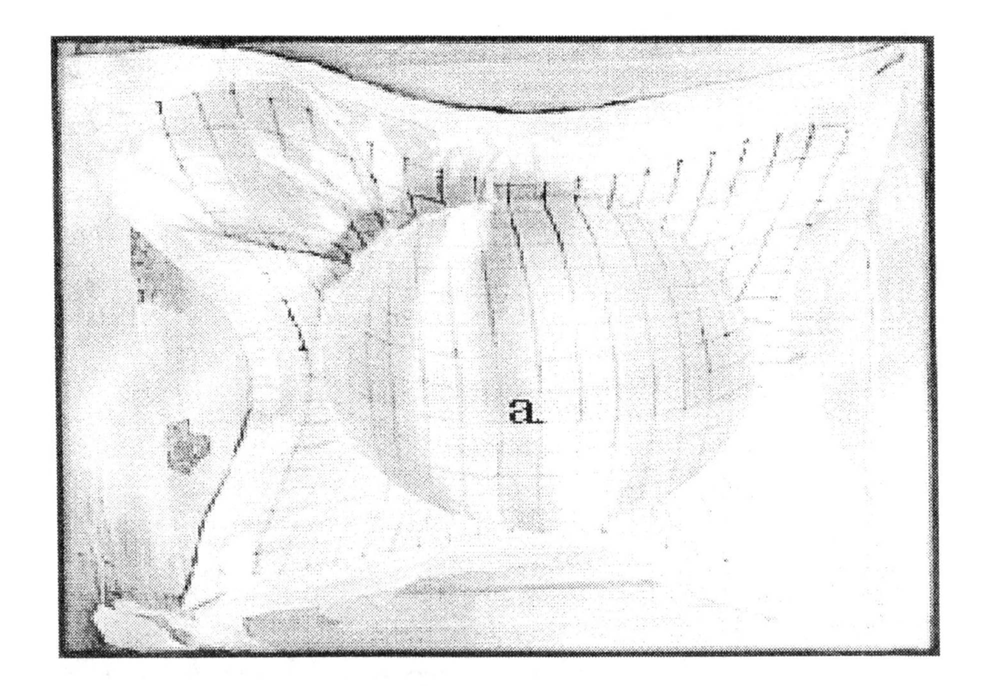

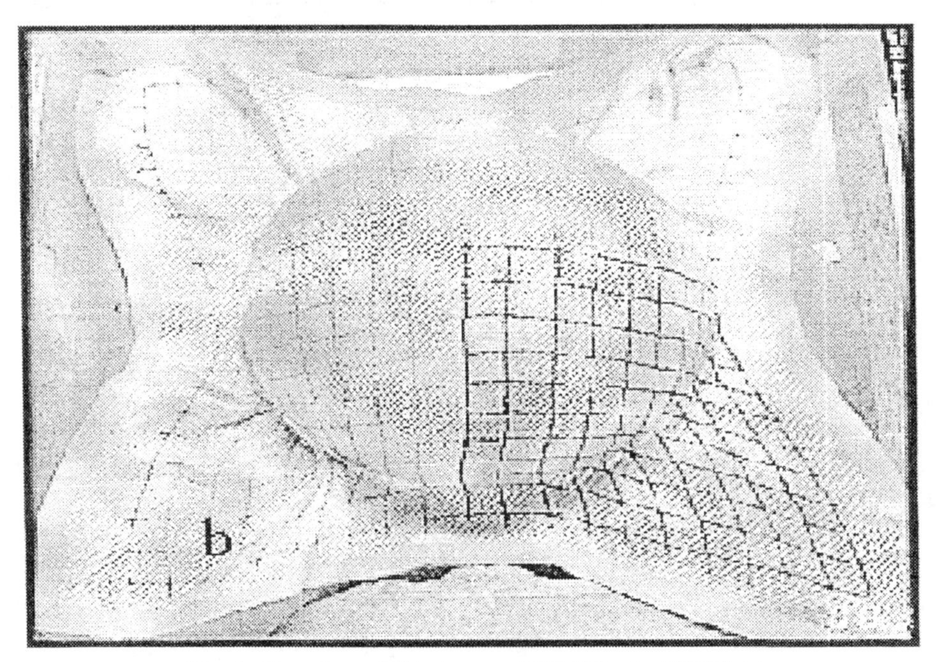

Fig 2: Draping of fabrics: a. (8HSW), b. (TWILL).

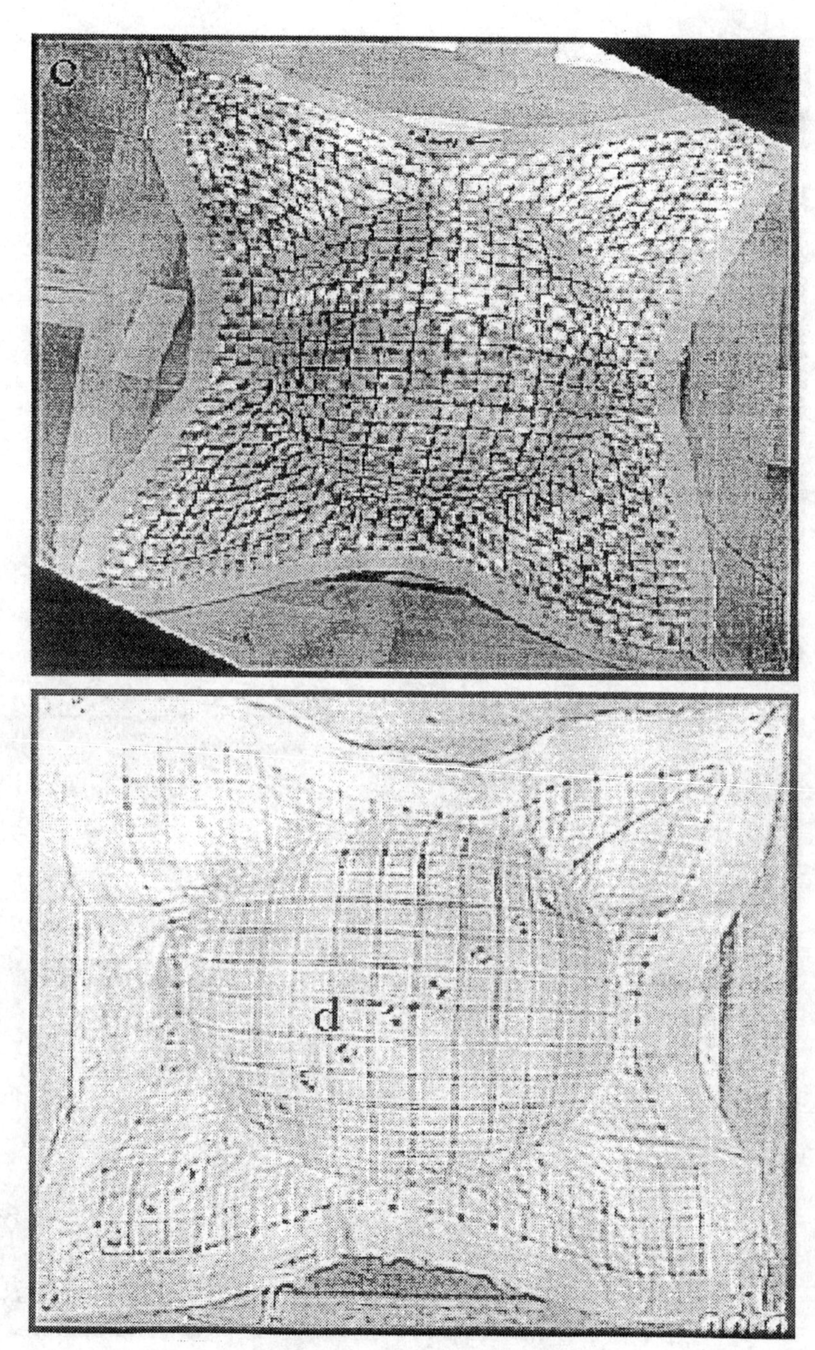

Fig.2. Draping of fabrics: c.(basket weave), d.(tight weave).

5. Finite element continuum solid mechanics of draping

This section includes the mathematical modelling and computer simulations of draping by using the continuum solid mechanics approach. In this approach the fabric is represented by a solid surface or a thin solid body with a numerical finite element mesh. The mesh should be sufficiently fine to model accurately local features of draping, such as wrinkles, but may not reach the fabric micro-scale. The finite element analysis for static or quasi-static cases is based on the principle of conservation of virtual work which includes body forces, concentrated loads and surface forces. The displacement in each element, $u^{(e)}$, is derived from the interpolation of the nodal element displacements, $a^{(e)}$, by using the shape function matrix, $N^{(e)}$.

$$u^{(e)} = N^{(e)} a^{(e)} \tag{1}$$

The strains in each element, $\varepsilon^{(e)}$, are related to $a^{(e)}$ via the strain-displacement matrix

$$\epsilon^{(e)} = B^{(e)} a^{(e)} \tag{2}$$

Constitutive relationships are used between stress and strain where the mechanical/viscous material properties feature. By using the virtual displacement theorem, the equilibrium equations of the assembly of elements are derived:

$$R = Ka \tag{3}$$

where K is the structure stiffness matrix and R is the structure force vector including element body forces, element surface forces, initial stresses and strains and concentrated loads.

Fif.3 illustrates the numerical draping of a fabric over a hemi-spherical mould. The fabric is modelled as a thin, flat, solid sheet which is meshed numerically using a regular grid of quadrilateral, orthotropic, solid elements. As a first approximation, orthotropic, elastic properties were given to the sheet. The volume between the sheet and the solid surface is modelled by flexible solid elements of a modulus of 0.1 MPa. The finite element simulation was performed using LUSAS 11, (FEA Ltd.) software package.

6. Conclusions

This paper focuses on the drapability of four types of fabrics, namely a loose plain weave, a tight plain weave, a twill and an 8 harness satin weave. Two types of experiments were used to assess drapability: (a) draping over a hemispherical mould and (b) shear tests where the locking shear angle was determined, beyond which wrinkling would appear. Loose weave structures like the 8 harness weave or the basket weave seem to favour large locking shear angles and no or low wrinkling after draping. Finite element simulations of draping on the basis of solid mechanics approach are feasible and offer advantages in terms of a variety of possible results to be obtained such as coordinates of the deformed mesh, local fibre orientation, local fibre volume fraction and stresses required for draping.

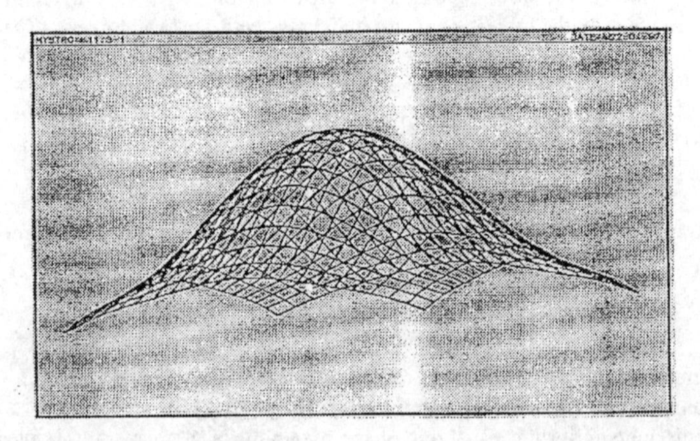

Fig.3. Predicted deformed shape of woven cloth after draping over a hemispherical surface.

7. References

1. R.E.Robertson, E.S.Hsiue, E.N.Sickafus and G.S.Y.Yeh 'Fibre arrangements during the moulding of continuous fibre composites', Polymer Composites, 2(3), 1981, pp.126-131.

2. S.Albert and G.Gutowski 'The kinematics for forming ideal aligned fibre composites into complex shapes', Composites Manufacturing, 1(4), 1990, pp.219-228.

3. A.C.Long, C.D.Rudd, M.Blagdon and P.Smith 'Characterising the processing and performance of aligned reinforcements during preform manufacture', Composites, 27A, 1996, pp.247-253.

4. T.A.Martin, D.Bhattacharyya and I.F.Collins 'Bending of fibre-reinforced thermoplastic sheets', Composites Manufacturing, 6(3-4), 1995, pp.177-187.

5. H.Sabhi, J.C.Gelin, P.Boisse and A.Cherouat 'Modelling the constitutive behaviour of glass fabrics', 2nd Int.Symp. TEXCOMP2, Leuven, 1994.

6. K.D.Potter 'The influence of accurate stretch data for reinforcements on the production of complex structural mouldings', Composites, 10, 1979, pp.161-167.

7. A.C.Long, C.D.Rudd, M.Blagdon and M.S.Johnson Experimental analysis of fabric deformation mechanisms during preform manufacture' Proc. ICCM-11, Gold Coast Australia, 1997.

8. P.Andrews, R.Paton and J.Wang 'The drape forming of a rudder tip preform', Proc. ICCM-11, Gold Coast Australia, 1997.

Experimental comparison of rigid joints for pipes in composite materials

G. La Rosa, G. Mirone

Istituto di Macchine, Facoltà di Ingegneria, Università di Catania
Viale Andrea Doria 6, 95125 Catania-Italy

Abstract

In the present paper a new type of butt joints for fibreglass pipes was analysed. The joints are realised by bevelling the tubes with low angles (4 to 20 degrees) with respect to their axis and connecting them with a sleeve of reinforced layers along all the thickness extending from the vane obtained between bevellings to the external side of the sleeve itself. The behaviour of this kind of joints has been compared to that of the traditional joints having higher bevelling angles and no reinforcement in the vane, it being filled only with a special glue. The results obtained clearly showed the improvement in the mechanical characteristics of this kind of joints with respect to the traditional one. The bevelling angle has not large influence in the range tested. Bi-and tri-dimensional FEM analysis also confirmed the experimental results.

1. Introduction

It is not possible to design a system of fibreglass pipelines without information regarding the performance of the pipes at joints and of the resistance characteristics of the different types of joint used. The traditional method of making a fixed joint uses a resin adhesive to unite the two pieces of pipe, but the technique which may be called «glue-less» is becoming increasingly common. In glue-less joints, fibreglass reinforcing is also used to reconstitute the thickness of the pipe which was removed during preparation. The present paper aims to analyse the behaviour of glue-less butt joints uniting fibreglass reinforced polyester resin pipes, manufactured using the filament winding technique.
The authors determined what effect the geometry of the surfaces to be united has on the resistance of the component. Further, the resistance characteristics obtained with this jointing technique were compared with those found for traditional, «glued» joints.

2. Description of joint typologies

To create a joint with resin adhesive, the surfaces of the two extremities of the pipes are dressed by means of lapping, using marble abrasive disks to give the surface a high degree of roughness. The area of the surfaces to be treated is given in the relevant specifications. The edges are then bevelled, in the case of butt joints, at an angle of 40 degrees to the axis. The extremities of the pipes are then aligned using a centering tool to leave the correct distance between the two pieces of pipe, into which an elastic ring in steel or an inflatable balloon is inserted to prevent the resin dripping into the pipe. The

centering and sealing tools are removed only when the joint has been completed. The tapered area is first coated with a layer of non-resistant liner, consisting of two coats of C-glass. The successive layers of adhesive resin are laid down over the liner only after it has completely hardened, so that the absence of gaps and bubbles can be verified. Each layer of adhesive is applied with a thickness of no more than a few millimetres to avoid delamination while it is shrinking. Having reconstituted the thickness of the pipe, layers of fibreglass reinforcing are then added in the form of mats or bi-directional woven roving (composed of two mono-directional layers of roving superimposed orthogonally) according to the lamination sequence reported in Table 1. The adhesive, without catalyst, has the following composition:
Vinylester resin: 10 kg, Accelerator: 50 g, FlakE-glass : 2 kg Aerosil: 1 kg

For the so-called glue-less joint, the operations are analogous for the preparation of the extremities of the pipes. In this phase, in fact, the only difference is that, to obtain a certain continuity between the reinforcing present in the pipes and that used for the joint, the angles of bevelling are much lower than that those used in the joint with adhesive. The stratification of the joint begins with the layer of liner, followed by alternating layers of mat and woven roving according to the lamination sequence reported in Table 1. By convention, the mat and woven roving layers are indicated by M and W, respectively. The bi-directional woven roving is always applied with one of the two directions of the fibre orientated along the axis of the pipe, while for the mat, which has short, randomly orientated fibres, all the directions of the lamination plane are equivalent from the mechanical point of view.

Joint with adhesive	M-W-M-M
Joint without adhesive	M-M-W-M-W-M-W-M-W-M-M

Table 1

The different types of glue-less joints differ in terms of the angle of bevelling. Reducing the angle is understood to give a better coupling between the fibres of the various layers of the pipe and those of the reinforcing closest to the internal diameter of the pipe itself. Conversely, using high bevel angles implies a poor pipe-joint continuity since the reinforcing of the internal layers of the pipe tend to be bonded to the internal layers of the joint over a much reduced surface area. At the limiting value of 90 degrees, the contact between corresponding layers in the joint and in the pipe is one of direct butt jointing. The bevel angles of the test specimens examined are 4, 7, 12 and 20 degrees with respect to the axis of the pipe, while for the joint with adhesive the angle is, as mentioned above, about 40 degrees.

3. Experimental trials

3.1 Determination of the laminae characteristics

Three different types of fibreglass reinforcing are used for the pipe and joint and, therefore, these types of lamina are present in the finished article, together with the isotropic lamina equivalent corresponding to the resin adhesive alone. The characteristics of the transversely isotropic (roving, woven roving) and the isotropic (mat, adhesive) materials making up the different laminae are described in Table 2.

Lamina (resin coupled with various reinforcements) :	Roving	Woven roving	Mat	Adhesive
Lamination angle (degrees)	55	0	0	0
Nominal thickness (mm)	0.6	0.47	0.74	------------
E_1 (MPa)	30350	16300	8420	3600
E_2 (MPa)	6350	16300	8420	3600
G_{12} (MPa)	2690	2900	3800	1295
ν_{12}	0.316	0.32	0.355	0.39
R_1(MPa)	320	263	140	60
R_2(MPa)	90	240	140	60

Table 2

The mechanical characteristics of the laminae with reinforcing were calculated by means of strain gauge tests on laminae or laminates consisting of a single sample of each of the types of reinforcing used. Some of the characteristics (all those regarding mono-directional material and the normal moduli of mat and woven roving) were also calculated analytically (mixture theory, stiffness averaging), referring to E-glass reinforcing (isotropic) with characteristics E = 69,000 MPa and ν = 0.22. A more than satisfactory agreement between the two series of results was found.
The effective dimensions of the test samples, shown schematically in Figure 1, are reported in Table 3 as a mean of about ten samples for each family of joint.

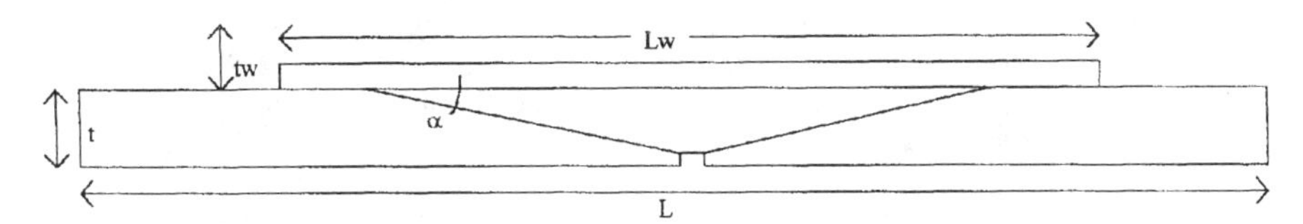

Figure 1

Joint	4 deg.	7 deg.	12 deg.	20 deg.	40 deg. Glue
L (mm)	400	400	400	400	400
Lw (mm)	180	160	160	160	160
tw (mm)	6.2	6.3	6.2	6.1	6.1
t (mm)	6.3	6.3	6.4	6.2	6.4

Table 3

As can be seen from the comparison between the observed thickness values and those calculated for the single laminae, there is an excess of resin on the finished article which increases the thickness of the pipe from 4.2 to 6 mm, and the total thickness of the joint from 7.06 to 11 mm. Clearly, the excess adhesive in the joint is also greater because, near the vertex of the bevel, the surface area is reduced in the axial direction and so is not occupied by the reinforcing in an intensive manner.

3.2 Experimental trials on joints

Having acquired the necessary preliminary information regarding the materials used, the joints were tested in a series of experimental trials. The specimens tested were obtained by cutting pieces of jointed pipe along two generatrices to produce prisms with a cross-section which was a sector of circular crown with a mean radius of 125 mm (the radius of the pipe), subtending a chord of 25 mm. The trials performed involve tensile and bending tests on the specimens described above, while lengths of uncut jointed pipe were subjected to burst and fatigue (rotational bending) tests.
Five groups (four types of glue-less joints and one group of glued joint) of ten specimens were tested to failure by traction and bending (three point bending), using a Galdabini test machine. The failure load value indicates the complete breakage of the piece and was considered, for each group, to be the mean value of those obtained on all ten specimens of that group. Table 4 reports these values, expressed as nominal stresses referring to the overall cross-section of the pipe, that is where the thickness is that measured, including the excess resin present.

Joint	4 deg.	7 deg.	12 deg.	20 deg.	40 deg. Glue
σ_R tensile (MPa)	66.7	68.2	68.2	54.76	50.8
σ_R flexural (MPa)	18	13.75	9.72	8.05	6.18

Table 4

Another significant finding in these tests is the position where failure began: in the tensile trials, the onset of failure was located at the extremity of the joint for all the glue-less joints, while for the glued joint it was located on the mid-line of the joint, on the internal diameter. In the bending tests, instead, the failure began in all pipes at the extremity of the joint, corresponding to the apex of the cavity produced by the bevelling of the pipes.
The burst trials were performed by pumping water under pressure into pieces of jointed pipe with an overall length of 1120 mm, an internal diameter of 300 mm and thickness of 8 mm. The pipes were sealed by means of a metal chuck with six segments. Each trial was monitored using a pair of strain gauges placed on two circumferences, one corresponding to the mid-line of the joint and the other in a part of the pipe as far as possible from both the joint and the sealed extremity. The longitudinal deformation was not determined since the sealing chuck blocked, or limited, axial lengthening of the pipe, behaving as a substantial rigidity applied in parallel to the pipe. The samples tested in these trials were glued joints and glue-less joints with bevel angles of 7 and 20 degrees.
Even though the angles of bevel are significantly different, both types of glue-less joint exhibited a practically identical behaviour, showing a maximum difference of about 6% in the measured deformation, at equal internal pressure, at 90 MPa, the highest pressure used. Both of the glue-less joint configurations supported this pressure without failing, and it was considered unnecessary to increase the loading given that it was already well over the projected working pressure. The performance of the joint with adhesive was substantially different, with localised cracks, and a consequent loss of pressure, already forming at internal pressures in the order of 40 MPa, however beyond the design requests. The results of the burst trials are reported in Table 5 in terms of deformation

on the two circumferences monitored, for the internal pressure values and the nominal circumferential stresses (hoop stress) at which the strain gauge measurements were made. The values in brackets refer to the data obtained on the glued joint. These data show a slight reduction in the modulus of proportionality, probably due to effects of viscosity and plasticity which are quite likely given the excess of resin present in both the joint and the pipe.

Internal Pressure (MPa)	Hoop stress (MPa)	Hoop ε tube μm/m	Hoop ε joint μm/m
10	18.75	837 (860)	570 (645)
40	75	3567	2619
60	112.5	5675	4083
80	150	7880	5338
90	168.75	9185	6377

Table 5

The final series of trials consisted of applying a rotational bending load to pieces of pipe (internal diameter 75 mm and thickness 6 mm) jointed without adhesive (bevel angles of 7 and 20 degrees) and with adhesive. During these trials, performed at constant load, the surface temperature of the piece was recorded using an infrared thermal scanner. Tests were performed on samples of the three series using applied loads of 100, 200 and 300 N. The configuration used was that of a clamped beam loaded on the free end with the joint immediately in front of the clamp constraint. The pipe, integral with the rotating chuck which acts as clamp, rotates at a speed of 150 rpm. The temperature maps were recorded for the specimen without load, and after 500 and 1000 revolutions under load. The digitised temperature maps at null load were subtracted from those of the 500th and 1000th cycle, providing a map of the variation in temperature which is indicative of the stress state of the surface of the specimen and of its fatigue behaviour. As far as possible, the curve N-T was obtained for different loads $\Delta\sigma$ to determine the initial gradient dT/dN. Reconstructing the trend of $\Delta\sigma$ as a function of dT/dN, it is possible to extrapolate the value of $\Delta\sigma$ corresponding to dT/dN=0, that is the fatigue limit sought [2-5]. After 1000 cycles, the maximum load, corresponding to a fixed end moment of about 270 Nm, produced a mean temperature variation of 0.5°C over the whole surface of the two specimens of pipe jointed with adhesive and bevel angles of 7 and 20 degrees, while the pipe jointed with adhesive showed a mean temperature variation of 1.5°C. The corresponding maximum (local) variations were 0.9, 0.9 and 1.5°C, respectively, for the three joint configurations analysed. No appreciable temperature differences were found at lower loads, indicating that the corresponding stress values could be under the fatigue limit.

4. Discussion of results of the experimental trials

In brief, the experimental trials show how, except in the static bending tests on specimens cut from jointed pipes, the behaviour of the joints examined was more or less homogeneous for all the models of the series without adhesive, regardless of the angle of bevelling which varied quite considerably (from 4 to 20 degrees). The joint with adhesive exhibited significantly different behaviour, however, with a generally inferior performance in all the trials. Within the group of glue-less joints, only the static bending test evidenced any connection between the performance of the joint and the angle of bevelling, with the nominal failure stress (stress evaluated at breaking for an equivalent isotropic material having the same tensile Young's modulus in the main direction) more than doubling over the step from 20 to 4 degrees. In order to evaluate the possible mechanisms of failure and the differences in behaviour of the joints under different loads, a finite element analysis was performed for representative configurations of the specimens taken from the jointed pipe.

5. Finite element analysis

The models reproduced, with two different approaches, the four configurations of the specimens cut from the pipes jointed without adhesive and bevel angles of 4, 12 and 20 degrees, and the pipe jointed with adhesive and a bevel of 40 degrees. The MSC/Nastran programme was used with the analysis limited to the linear elastic field. Two models were developed, one two-dimensional with laminate type elements and orthotropic material and the other three-dimensional with brick type elements and anisotropic material (transversely isotropic). The need to construct a 3-D model was the result of the consideration that the laminate analysis programme used is based on the classic theory of lamination. This theory does not allow a correct evaluation of the stresses and tangential deformations which could have a determinate role in the case, as here, where the mechanism of failure is a result of unsticking. With regard to tensile loading only, both types of analysis confirmed the uniformity of the behaviour found experimentally, but they also provided information helpful for an understanding of the clear dependence of the failure load on the angle of bevelling in the bending tests. In the case of the glued joints, in fact, it was seen that the internal layers of reinforcing on the two pieces of pipe, near to it but separated by non-rigid adhesive, cannot transmit the load with sufficient uniformity, leading to a concentration of stresses. Although these stresses are not particularly elevated, they act on the adhesive, which is the weak link in the chain.

Instead, when the fibreglass reinforcing of the joint is already present from the internal radius on, regardless of the angle of bevelling, the load is transmitted between the internal layers of the reinforcing of the two pieces of pipe without excessive discontinuity. The peak of critical stress is therefore moved from the vertex of the bevelled area on the mid-line of the joint on the internal diameter, to the area at the extremity of the joint itself on the external diameter. This behaviour coherently reflects the experimental findings since failure is effectively triggered in those areas which the numerical models indicate are the most stressed. The analysis of the two-dimensional models gave, among various outputs, the failure index according to Hill, defined following the hypothesis that the tensile and compressive ultimate stresses are equal for each material. The tangential failure load values, which are one of the material characteristics required by the programme, were extrapolated from failure loads obtained experimentally and the stiffness values found in the literature for materials analogous to those used in the manufacture of the pieces under examination. Some images are given below, showing the trend of the failure index of the two-dimensional laminate and of the Von Mises stresses on the 3-D model, regarding the glued joint and the glue-less joint with 4degrees bevel (it should be noted that the behaviour of all the models of glue-less joints was practically identical). All the models were subjected to the load which leads to failure of the glued joint under traction. In the model corresponding to the joint with adhesive, in fact, the failure index is very close to 1, while in the model of the joint with a 4 degrees bevel there is a narrow safety margin (about 10%) before the conditions of failure are reached. The trend of the stresses in the 3-D models shows that the lack of continuity in the reinforcing at the internal radius of the pipes jointed with adhesive is responsible for its diminished performance. The stress distributions in the 3-D models of the pipes jointed without adhesive were found to be essentially the same on varying the bevel angle, confirming the experimental findings.

6. Conclusions

An experimental study was made of the behaviour of joints produced using resin adhesive and bevelled at 40 degrees and joints in which also the cavity created between the bevelling of the pipe (with angles varying between 4 and 20 degrees) is occupied, at least in part, by the fibreglass reinforcing. The characterisation consisted of a series of static tensile and bending tests on specimens cut from the jointed pipes, of burst and bending tests on lengths of jointed pipe, and finite element analysis of the specimens under tensile loading. Most of the experimental trials showed that a distinct difference in terms of behaviour can only be made between joints with adhesive and those without, given that no significant difference in performance was observed among the "glue-less" joints with different angles of bevelling. Only the static three point bending tests showed a clear dependence of the failure load on the bevel angle, as well as on the presence of resin adhesive or fibreglass reinforcing in the cavity formed by the bevel. In the light of the results of the finite element analysis, it was possible to conclude that the determining factor for the resistance of the joint is the presence of fibreglass reinforcing in the area of the cavity between the bevels closest to the apex. Clearly, the manual placement of layers of reinforcing in this area requires that the bevel has an angle well below 40°, above which, given the thickness of these laminae (about 6 mm), there would be an extremely reduced axial extension of the reinforcing. In practice, the key to the mechanism of joint resistance is to reconstitute the continuity of the internal reinforcing layers of the pipe, without which the structure of the specimen is highly asymmetrical and unbalanced with respect to the mean plane. Then, when this area is subjected to increasing tensile stresses towards the intrados (bending stresses on the mid-line section of the specimen), the presence of a "bridge" of reinforcing, localised at the intrados alone, is no longer sufficient. Under these conditions, the determining factor is the quantity of reinforcing in the more internal and external layers, a quantity which increases on decreasing the angle of bevelling. However, under the working conditions of the pipe, not even bending loads result in variations in performance (with the angle of bevel) which were found in the specimens. This is because the neutral axis is so distanced from the mean plane of the laminate that the stresses are practically constant along the thickness, so that the functioning of the pipe is the same as the tensile-compressive behaviour found in the specimens. From the discussion, it is clearly superfluous to strive for very low angles of bevelling, since the behaviour becomes sufficiently uniform as soon as the bevel allows a contact, between the fibreglass reinforcing of the pipe and that of the joint, which can be considered one of overlapping rather than of butt jointing. An angle of 20 degrees already results in behaviour which is close enough to that found for lower angles, indicating that this value may be adopted as one providing a performance close to the upper limit, while reducing both the time spent in preparing the pipes and the quantity of reinforcing material required when compared to joints with bevel angles of 4, 7 and 12 degrees.

Acknowledgements

The authors would like to thank Engineer Fichera of VED for his kind collaboration in the provision of the components used in this study, and the Istituto Tecnico Industriale "Archimede" in Catania in the person of Engineers Vazzano and La Rosa for his invaluable help in all the experimental endeavour.

References

[1] *A. Bogdanovich,* MECHANICS OF TEXTILE AND LAMINATED COMPOSITE, Chapman et Hall
[2] *Curti G., La Rosa G., Orlando M., Risitano A.,* ANALISI TRAMITE INFRAROSSO TERMICO DELLA TEMPERATURA LIMITE IN PROVE DI FATICA, XIV Convegno Nazionale AIAS, Catania 1986
[3] *Geraci A., La Rosa G., Risitano A,.* ON THE NEW METHODOLOGY FOR THE DETERMINATION OF THE FATIGUE LIMIT OF MATERIALS USING INFRARED THERMAL TECHNIQUES, VDI IMEKO-GESA Symposium, Dusseldorf (Germany), 1992
[4] *Geraci A., La Rosa G., Risitano A.,* INFLUENCE OF FREQUENCY AND CUMULATIVE DAMAGE ON THE DETERMINATION OF FATIGUE LIMIT OF MATERIALS USING THE THERMAL INFRARED METHODOLOGY, XV Polish National Symposyum on Experimental Mechanics of Solids, Warsaw (Poland), 1992
[5] *Guglielmino E., Guglielmino I.D., La Rosa G., Pasta A.,* ANALISYS OF A LAMINATE COMPOSITE UNDER CYCLIC TESTING USING THERMAL INFRARED IMAGERY, International Conference on Material Engineering, Gallipoli (Italy) 1996
[6] *Tsai S. W. , Hahn H. T.,* INTRODUCTION TO COMPOSITE MATERIALS, Technomic Publishing, 1980
[7] *Hull P., Cline T. W.,* AN INTRODUCTION TO COMPOSITE MATERIALS, Cambridge Solid State Science Series
[8] *Capponi T., Gargiulo C., Marchetti M., Philippidis T.P.,* STABILITY OF LAMINATED CYLINDRICAL TUBES UNDER UNI- AND MULTIAXIAL STRESSES, AGARD Program, Project G-80, Rome (Italy)-Patras (Greece), 1995
[9] *Morohoshi T., Sawa T.,* ON THE CHARACTERISTICS OF RECTANGULAR BOLTED FLANGED CONNECTIONS WITH GASKETS SUBJECTED TO EXTERNAL TENSILE LOADS AND BENDING MOMENTS, Journal of Pressure Vessel Technology, Vol 116, pag 207
[10] *Ochoa O. O., Reddy J. N.,* FINITE ELEMENT ANALISYS OF COMPOSITE LAMINATES, Solid Mechanics and its Application, Vol. 7, Kluwer Academic Publishers, 1992

General Rules for Scaling in Curing of Thermosetting Flat Plate Composites

Mehdi Hojjati

Department of Mechanical Engineering , Tarbiat Modarres University
Tehran , Iran

ABSTRACT

General rules for scaling in curing of flat plate composites is developed. The transient heat conduction equation coupled with the kinetic equation and the initial and boundary conditions are non-dimensionalized. Dimensionless parameters are extracted from the non-dimensional governing equations and initial and boundary conditions. Scaling rules are constructed based on these dimensionless parameters. Application of scaling rules to design a model is explained. By application of the transformation law for thermal conductivity, it is shown that the fiber orientation is one of the important parameters involved in determination of model specifications. An equation is derived to calculate the model fiber orientation based on the material properties and prototype fiber orientation. By numerical examples based on the available material physical properties, it is shown that this equation does not have the solution most of the time. Therefore, the model can be designed and constructed just in some special cases.

KEY WORDS: Thermoset Resins/Polymer, Similarity, Model Laws, Curing.

1 INTRODUCTION

Application of thick thermosetting composites is growing in the aerospace industry and construction and military applications. To fabricate a high quality thick thermosetting composite part, the proper choice of the processing parameters during curing process is very important. The pertinent processing parameters in curing are time, temperature, and pressure.

Performing experiments to obtain optimum processing parameters for curing of thick composite requires the expensive prepreg material and labor work. Using simulation to obtain optimum processing parameters [1-4] is economical. However, an accurate expression for kinetic equation is required for simulation. Kinetic equation is not derived for most of the resins. Even the available kinetic equations do not predict the rate of reaction accurately [3,4]. Similarity and modeling (scaling) are developed for curing of flat plates which consist of just unidirectional layers in the principal directions [4]. In model work, one tries to predict the performance of a final system called the *prototype* from tests on the performance of another actual system called the *model.* To predict the performance of the prototype from that of the model, the similarity rules and model laws should be established.

Establishment of the *general rules for scaling* for curing of thermosetting composite parts is the subject of this study. Model laws can be used to design a model based on the prototype (i.e. thick composite) in such a way that the performance of the prototype (i.e. temperature and degree of cure distribution inside composite) can be predicted from the performance of the model. The transient heat conduction coupled with kinetic equation and initial and boundary conditions are non-dimensionalized and dimensionless parameters are extracted from them. Any relationship among dimensionless groups of model and prototype can be expressed in the form of a model law. It is shown how these model laws can be used to design a model based on the prototype. By application of the transformation law for thermal conductivity, it is shown that the fiber orientation is one of the important parameters involved in determination of model specifications. An equation is derived to calculate the model fiber orientation based on the material properties and prototype fiber orientation. It is shown that this equation does not have the solution most of the time. Therefore, the model can be designed and constructed just in some special cases.

2 ANALYSIS

Consider the manufacture of a flat-plate composite (as prototype) composed of unidirectional fiber reinforced thermosetting resin matrix prepreg materials. As shown in Figure (1), the composite has a rectangular cross-section of width M, thickness N and length L. The fiber orientation is θ. It is assumed that the fiber volume fraction ν_f throughout the composite is the same. This composite is subjected to a specified cure cycle inside autoclave and temperature and degree of cure are monitored inside the composite at different points and times. The objective is to design a model for that prototype using the same resin but different fibers. The model dimensions L_m, M_m, and N_m, fiber orientation θ_m, and fiber volume fraction ν_{fm} as well as model cure cycle should be determined.

2.1 Mathematical Model

The transient heat conduction equation with an internal heat generation source term is used to obtain the temperature distribution inside the composite. This equation can be expressed as[6]

$$k_{11}\frac{\partial^2 T}{\partial x_1^2} + 2k_{12}\frac{\partial^2 T}{\partial x_1 \partial x_2} + k_{22}\frac{\partial^2 T}{\partial x_2^2} + k_{33}\frac{\partial^2 T}{\partial x_3^2} + \dot{g} = \rho c_p \frac{\partial T}{\partial t} \tag{1}$$

where k_{11},k_{22},k_{12} and k_{33} are the thermal conductivity tensor components. The term $\dot{g}$ represents internal heat generation. ρ is the composite density and c_p is the specific heat of the composite. T and t are absolute temperature and time, respectively. The thermal conductivity tensor components can be obtained from the composite thermal conductivities in the principal directions (K_x, k_y, k_z) and transformation law as [9]

$$\left\{ \begin{array}{c} k_{11} \\ k_{22} \\ k_{12} \end{array} \right\} = \left[\begin{array}{cc} m^2 & n^2 \\ n^2 & m^2 \\ mn & -mn \end{array} \right] \left\{ \begin{array}{c} k_x \\ k_y \end{array} \right\} \tag{2}$$

where $m = cos\theta$, $n = sin\theta$, and $k_{33} = k_z$.

The heat generation term is defined as

$$\dot{g} = \rho_r(1 - \nu_f)\frac{d\alpha}{dt}H_R \tag{3}$$

where α is the degree of cure and H_R is the total heat of reaction of the resin. ρ_r is the resin density and ν_f is the fiber volume fraction. $\frac{d\alpha}{dt}$ is the rate of cure and should be calculated from kinetic equation. The kinetic model chosen for this study is a simplified Arhenius type first order rate equation expressed as [7]

$$\frac{d\alpha}{dt} = A\exp(\frac{-E_o}{RT})(1 - \alpha) \tag{4}$$

where A is the pre-exponential factor and E_o is the activation energy. R is the universal gas constant. It should be mentioned that the first order kinetic equation is not representative of many of the new resins where the conversion reactions are not well understood. We have chosen this equation only for simplicity.

Initial conditions are

$$T(x, y, z, 0) = T_o \quad , \quad \alpha(x, y, z, 0) = 0 \tag{5}$$

where T_o is the initial temperature inside the composite. Boundary conditions are

$$\begin{array}{lll} T(0, y, z, t) = T_a & , & T(L, y, z, t) = T_a \\ T(x, 0, z, t) = T_a & , & T(x, M, z, t) = T_a \\ T(x, y, 0, t) = T_a & , & T(x, y, N, t) = T_a \end{array} \tag{6}$$

where T_a is the autoclave temperature. T_a can be either a constant or a function of time.

2.2 Non-dimensional Governing Equations

The model laws can be extracted from non-dimensional set of governing equations and initial and boundary conditions[5]. The non-dimensional approach reduces the large number of variables involved in the process to a few groups which each group has a physical meaning. Each dimensionless parameter is called as a π. In terms of non-dimensional parameters, the heat conduction equation[6] and kinetic equation[7] can be expressed as

$$\pi_1\frac{\partial^2\overline{T}}{\partial\overline{x}_1^2} + \pi_2\frac{\partial^2\overline{T}}{\partial\overline{x}_1\partial\overline{x}_2} + \pi_3\frac{\partial^2\overline{T}}{\partial\overline{x}_2^2} + \frac{\partial^2\overline{T}}{\partial\overline{x}_3^2} + \pi_4\frac{d\alpha}{d\tau} = \frac{\partial\overline{T}}{\partial\tau} \tag{7}$$

$$\frac{d\alpha}{d\tau} = \pi_5 \exp(-\frac{\pi_6}{\overline{T}})(1-\alpha) \tag{8}$$

where

$$\begin{aligned} \pi_1 &= \frac{k_{11}}{k_{33}}(\frac{N}{L})^2 \\ \pi_2 &= \frac{2k_{12}}{k_{33}}\frac{N^2}{LM} \\ \pi_3 &= \frac{k_{22}}{k_{33}}(\frac{N}{M})^2 \\ \pi_4 &= \frac{\rho_r(1-\nu_f)H_R}{\rho c_p T_\circ} \\ \pi_5 &= \frac{\rho c_p N^2 A}{k_z} \\ \pi_6 &= \frac{E_\circ}{RT_\circ} \end{aligned} \tag{9}$$

$\overline{x_1}$, $\overline{x_2}$ and $\overline{x_3}$ are dimensionless independent variables in the new system of coordinates. $\overline{T}$ and τ are dimensionless temperature and time, respectively.

The initial and boundary conditions in non-dimensional form can be written as

$$\overline{T}(\overline{x_1},\overline{x_2},\overline{x_3},0) = 1 \quad , \quad \alpha(\overline{x_1},\overline{x_2},\overline{x_3},0) = 0 \tag{10}$$

and

$$\begin{aligned} \overline{T}(0,\overline{x_2},\overline{x_3},\tau) = \pi_7 \quad &, \quad \overline{T}(1,\overline{x_2},\overline{x_3},\tau) = \pi_7 \\ \overline{T}(\overline{x_1},0,\overline{x_3},\tau) = \pi_7 \quad &, \quad \overline{T}(\overline{x_1},1,\overline{x_3},\tau) = \pi_7 \\ \overline{T}(\overline{x_1},\overline{x_2},0,\tau) = \pi_7 \quad &, \quad \overline{T}(\overline{x_1},\overline{x_2},1,\tau) = \pi_7 \end{aligned} \tag{11}$$

where

$$\pi_7 = \overline{T}_a \tag{12}$$

Equations (7)-(12) employ the following dimensionless parameters

$$\begin{aligned} \overline{T} = \frac{T}{T_\circ}, \quad \overline{T}_a = \frac{T_a}{T_\circ}, \quad \tau = \frac{k_z t}{\rho c_p N^2} \\ \overline{x_1} = \frac{x_1}{L}, \quad \overline{x_2} = \frac{x_2}{M}, \quad \overline{x_3} = \frac{x_3}{N} \end{aligned} \tag{13}$$

T and t are temperature and time, respectively. T_a is autoclave temperature.

2.3 Model Laws

The similarity rules and model laws for curing of thermosetting composite can be expressed as [5]

$$\pi_{im} = \pi_{ip} \qquad i = 1, \cdots, 7 \tag{14}$$

Subscripts m and p refer to model and prototype, respectively. If one composite system(model) has the same non-dimensional equations and boundary conditions with the same value of the π's which appear in these equations with another composite system(prototype), then similarity of behavior between two systems is completely achieved.

3 RESULTS

The use of similarity rules and model laws to determine the composite model specifications from composite prototype specifications is presented. Suppose there is a composite flat-plate (as prototype) with dimensions L_p, M_p, N_p, fiber orientation θ_p and fiber volume fraction ν_{fp} which is subjected to a specified cure cycle. The physical properties of model constituent as well as resin kinetic equation are given for both model and prototype. The task is to determine the model dimensions L_m, M_m, N_m and fiber orientation θ_m and fiber volume fraction ν_m and the new cure cycle during which model should be subjected to.

The only feasible way to design a model is to use the same resin, but different fibers [4]. By this fact and π_6 and π_7, one can show that the initial temperature of model should be the same as prototype. It also shows that the model should be subjected to the same cure cycle as the prototype.

The most important step in obtaining the model specifications is to calculate model fiber volume fraction ν_{fm}. Application of π_4 for model and prototyp ($\pi_{4m} = \pi_{4p}$) will give us an equation to calculate the model fiber volume fraction [4].

The next step is to calculate model physical properties. The relation proposed by Tsai and Springer[8] is used to calculate thermal conductivities in the y and z directions. The model thermal conductivity in the x direction,density, and specific heat are obtained by applying the rule of mixture.

The last step is to calculate the model dimensions L_m, M_m, and N_m as well as the model fiber orientation θ_m. First of all, θ_m should be determined. Combination of π_1, π_2, and π_3 results in the following equation

$$(\frac{k_{12}^2}{k_{11}k_{22}})_m = (\frac{k_{12}^2}{k_{11}k_{22}})_p \tag{15}$$

The right hand side of this equation is given and we call it C. Combination of equations (2) and (15), give us the final relation for θ_m as

$$sin^2\, 2\theta_m = \frac{4C k_{xm} k_{ym}}{(1-C)(k_{xm}-k_{ym})^2} \tag{16}$$

where θ_m is the model fiber orientation. k_{xm} and k_{ym} are calculated from the previous step. After calculation of θ_m the model dimensions can be obtained by application of π_1, π_3, and π_4.

Resin density	ρ_r	$1.26 \times 10^3 \frac{kg}{m^3}$
Specific heat of resin	c_{p_r}	$1.26 \times 10^3 \frac{J}{kg\,^\circ K}$
Thermal conductivity of resin	k_r	$0.167 \frac{W}{m\,^\circ K}$
Fiber density	ρ_{fp}	$1.79 \times 10^3 \frac{kg}{m^3}$
Specific heat of fiber	$c_{p_{fp}}$	$7.12 \times 10^2 \frac{J}{kg\,^\circ K}$
Thermal conductivity of fiber	k_{fp}	$26 \frac{W}{m\,^\circ K}$

Table 1: Material properties of Hercules AS4/3501-6 (prototype)

Fiber density	ρ_{fm}	$2.60 \times 10^3 \frac{kg}{m^3}$
Specific heat of fiber	$c_{p_{fm}}$	$9.50 \times 10^2 \frac{J}{kg\,^\circ K}$
Thermal conductivity of fiber	k_{fm}	$1 \frac{W}{m\,^\circ K}$

Table 2: Typical physical properties of Glass fiber (model)

As an illustration of applying the model design procedure for this case, consider the manufacture of the Hercules graphite epoxy prepreg tape, commercially called as AS4/3501-6. The prepreg specifications are shown in Table 1. The kinetic equation for this resin is given in reference [7]. The objective is to replace the graphite fiber with glass fiber and design and fabricate a model using Glass/3501-6 material for doing experiments and monitoring temperature and degree of cure. The typical physical properties of glass fiber are given in Table 2.

The procedure is described in Ref. [4]. The only difference is to calculate θ_m. this numerical example and many other examples show that equation (16) does not have solution most of the time. Just for small angles (θ_p less than 5), there is a solution. Otherwise, the model can not be designed.

4 CONCLUSIONS

For curing of thermosetting composites, the similarity rules and model laws were developed. The transient heat conduction coupled with kinetic equation and initial and boundary conditions were non-dimensionalized. Similarity rules and model laws were constructed based on dimensionless parameters which were extracted from non-dimensional equations. Application of model laws to design a model was shown. Generally a composite model made of less expensive materials can be used to determine the optimum processing parameters for a composite prototype made of more expensive materials. The relations between different variables and processing parameters for model and prototype are determined through the model laws.

By application of the transformation law for thermal conductivity, it was shown that the fiber orientation is one of the important parameters involved in determination of model specifications. An equation is derived to calculate the model fiber orientation based on the material properties and prototype fiber orientation. By numerical examples based on the available material physical properties, it is shown that this equation does not have the solution most of the time. Therefore, the model can be designed and constructed just in some special cases.

References

[1] A.C. Loss and G.S. Springer,"Curing of Epoxy Matrix Composites " *Journal of Composite Materials*, 17,135-169 (1983).

[2] T.E. Twardowski, S.E. Lin and P.H. Geil,"Curing in Thick Composite Laminates: Experiment and Simulation" *Journal of Composite Materials*, 27(3),216 (1993).

[3] M. Hojjati and S.V. Hoa,"Some Observations in Curing of Thick Thermosetting Laminated Composites" *Science and Engineering of Composite Materials* , 4(2),89-107 (1995).

[4] M. Hojjati and S.V. Hoa,"Model Laws for Curing of Thermosetting Composites" *Journal of Composite Materials*, 29(13),1741 (1995).

[5] S.J. Kline, *Similitude and Approximation Theory*, McGraw-Hill Book Company, (1965).

[6] M.N. Ozisik, *Heat Conduction* , John Wiely and Sons, (1980).

[7] Woo Il Lee,A.C. Loos and G.S. Springer,"Heat of Reaction, Degree of cure, and Viscosity of Hercules 3501-6 Resin *Journal of Composite Materials*, 16,510 (1982).

[8] G.S. Springer and S.W. Tsai," Thermal Conductivities of Unidirectional Materials" *Journal of Composite Materials*, 1,166 (1967).

[9] Tsai S.W. and T.H. Hahn,"Introduction to Composite Materials " Technomic Publishing co. (1980)

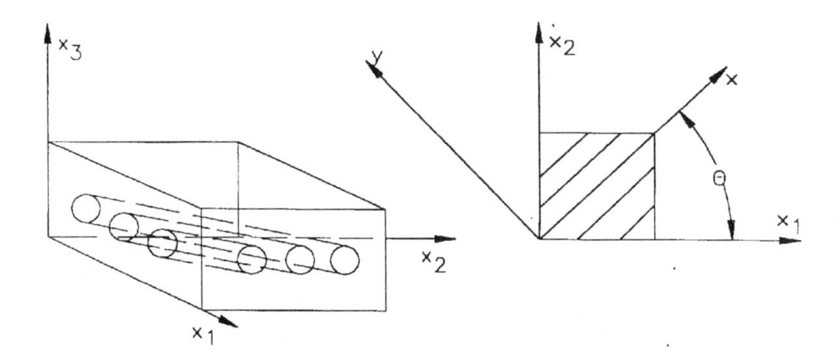

Figure 1: Composite Geometry

NEW METHODS IN TECHNOLOGY OF POLYMER COMPOSITS, REINFORCED BY FIBRES

V.N.Stoudentsov*, I.V.Karpova*, A.S.Sergeienko*

Department of Chemical Technology - Technological
Institute of Saratov State Technical University.-
Svoboda sg.,17 ,413100 Engels - Russia

Abstract

In this work, we report main directions of study doing by authors at the department of Chemical Technology of Technological Institute of SSTU. Purposes of this directions are : economizing technology and produced materials, improving technological properties of prepregs, regulation characteristics of produced materials by the help nontraditional physical influences.

Theoretical basing of observing phenomena is conducting except decision practical problems. Some of new methods may be used as supplemental stages in traditional technology. Main positive effect consists in the increasing on the several tens percents of strength characteristics of materials in comparison with materials, produced by traditional technology. Comparatively high durability of improvement reinforced by fibres materials permits to use this materials as in things of domestic use as in more important constructions.

1.Introduction

In the traditional way of wares production from materials,r einforced by fibres (PC) on the basis of cross - linced polymers (Fig.1) roving from bobbin 1 is given in the bath 2,where resin solution is saturating fibres. Roving, saturated by binding, runs to reception mechanism 4 over warming - up pipe 3,which is necessary for removal the solvent. Different versions of overwork produced prepregs exist. It is possible to wares produce by means of wind of continuous fibres on the reception mechanism 4,with following thermal treating. If it is necessary, we cut prepreg and then overwok it, for example, by straight pressing.

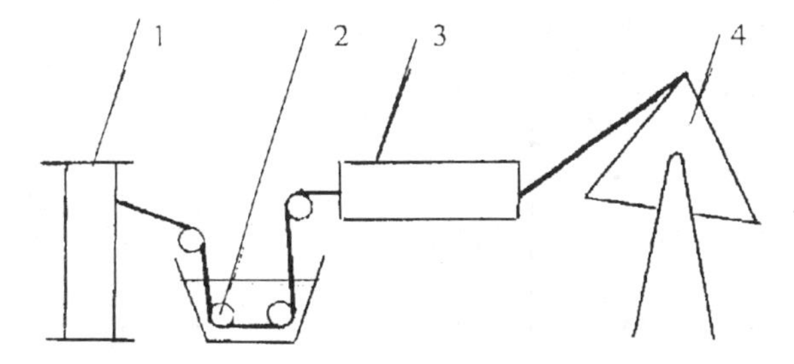

Fig.1.,Scheme of traditional method of wares production from PC, reinforced by fibres:1-bobbin with fibres ;

2 -bath for saturating roving ;3 -warming-up pipe ;4 -reception mechanism.

Conditional markings :

σi - bending static stress-at-break ;
σ_r - blow stress-at-break ;
α - specific shock viscosity ;
p - density ;
W - water-absorption ;
VF - viscose fabres ;
Kapron - polykaproamide fabres ;
Nitron - poliacrilonitrile fabres ;
Fenilon- poiyaramide fabres ;
CF - carbon fibres ;
DAF- cellulose diacetate fabres ;
TAF-cellulose triacetate fabres ;
TETRA - triethylentriamine ;
TEA- triethanolamine ;
APFR - anilinophenoloformaldehyde resin ;
ER - epoxy resin ;
PCF -polycondensational filling ;
CSP - components separate putting ;
CLP - components layer putting ;
MT - magnetic treating .

(Table 1),that may aggravate PC characteristics in comparison with material,

Table 1

Characteristics of reinforced by short fibres materials, produced by policondensational filling of epoxy resin

Filler	Method of PC production	σi, MPa	σr, MPa	α,kJ/m^2	P,kg/m^3	W,%
Kapron	Traditional	9	4	46	1060	26
Kapron	PCF	5	4	41	980	23
VF	PCF	19	14	8	1100	16
VF	PCF+MT+CLP	8	28	40	950	31

produced by traditional method .This deficiency is compensated

partially by combination PCF with other methods ,which are given further.

3.Methods of increasing permissible storage of prepregs

The problem of adjusting permissible storage of prepregs ,when we use bindings, for which application of harden is necessary ,successfully has decided by selection harden system now .Two methods of deciding this problem by technological way mainly ,when we use hardens ,which may work ,at the normal temperature ,are suggested in this work : in the method of components separate putting (CSP) some elements of filler is saturated by binding with plenty of resin function groups and thus type 1 prepreg is produced, others elements of filler is saturated by binding with plenty of harden function groups [2] comparatively to steheometria correlation of resin and harden and type 2 prepreg is produced so. Prepregs of both types may be preserved separately about two months in the normal conditions .Prepreg ,which is produced by traditional method and contains epoxy resin and one among examined hardens ,for example ,polyethylenpolyamine (PEPA),must be treated several hours past production of it.

In ware formation prepregs of both types is mixed and plenty of resin in the prepreg type 1 is hardened by plenty of harden in the prepreg type 2.Dffusion difficulties increase in this method ,that let us possibility to get porous materials ,which may be employed for example, in the quality of heat-isolation materials (table 2),with more hagh strength than usual heat-isolation materials.

Table 2

Characteristics of materials on the basis of epoxy resin, filled by short fibres

Method of PC production	Harden	λ,W/mK	σ_i,MPa	σ_r,MPa	α,kJ/m^2	W,%
Traditional	TETRA	0,096	81	28	80	36
CSP	TETRA	0,057	5	3	10	68
CSP	TEA	0,032	8	2	8	61

Low coefficient of heat conduction is index of high heat isolation properties of material ,water-absorption is in proportion of porousness.

In the method of components layer putting (CLP) every element of filler is saturated by resin at first, then by harden system (fig.2).

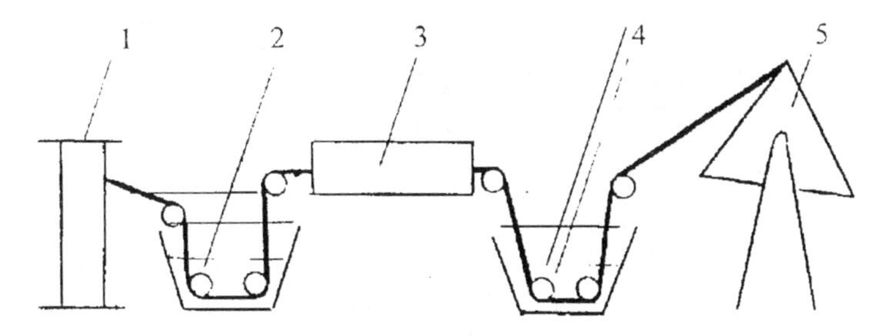

Figure 2.Sheme of method of components layer putting (CLP) :

1-bobbin with roving ;2-bath for solution of resin ;3-warming-up pipe ;4-bath for harden system ;5-reception mechanism.

CLP increases permissible storage of prepregs to two weeks and more and improves all the PC durability characteristics comparatively with materials, produced by traditional method. Concentrations of solutions, composition of harden system, linear speed of roving, temperature of between-baths thermotreating in warming-up pipe 3 essential influence on the properties of PC (table 3).

Different types of CLP depend on combination of those parameters.

Table 3

Characteristics of materials on the basis of epoxy resin, filled by long rovings

Method, type of method of PC production	Filler	σi, MPa	α,kJ/m^2	P ,kg/m^2	W,%
1	2	3	4	5	6
Traditional	nitron	67	68	1150	4
CLP,1	nitron	62	74	1150	6
CLP,2	nitron	47	81	1180	5
CLP,3	nitron	74	60	1120	6
CLP,4	nitron	90	52	1180	6
CLP,5	nitron	79	103	1010	1
Traditional	VF	45	144	1200	3
CLP,5	VF	71	152	1100	3
Traditional	capron	77	99	1200	4
CLP,4	capron	78	143	1200	7

4.Magnetec treating

Magnetic treating (MT) and another non traditional physical influences may be employed on the different stages of traditional technology and of new technologies. In result of interaction between external magnetic field (MF) and magnetic moments of oligomer molecules or segments of macromolecules probability of definite orientation of binding particles increases [3],that leads to formation of anisotropy structure of polymer in the process of hardening. This may positive influence on the PC durability (table4).External MF increases adhesion between binding and filler, that positive influences on all durability characteristics of PC.

Type of MT depends on temperature and duration of treating, from her place in the technological process, from direction, polarity, and stress of MF.

Table 4

Influence MT on the durability of PC, filling by long rovings

Method, type of method of PC production	Resin	Filler	σi, MPa	α,kJ/m^2
1	2	3	4	5
Traditional	APFS	fenilon	107	17
MT,1	APFS	fenilon	110	39
MT,2	APFS	fenilon	204	25
MT,3	APFS	fenilon	169	40
Traditional	ES	capron	94	157
MT,1	ES	capron	221	283
Traditional	ES	CF	187	82
MT,1	ES	CF	364	189
Traditional	ES	DF	90	308
MT,3	ES	DF	151	377
Traditional	ES	TAF	73	165
MT,3	ES	TAF	148	235
Traditional	ES	CF	170	89
MT,4	ES	CF	188	120
Traditional	ES	CF	532	104
MT,5	ES	CF	595	120

5.Conclusion

Technically simple and effective methods is offered, that permits to decide many of problems and to increase all durability characteristics of PC, reinforced by fibres :

Method	Increase of σi,%	Increase of α,%
CLP	10-58	6-51
MT	3-135	15-135

Prof. V.N. Stoudentsov is author : on PCF -one invention, on CSP-one invention, on CLP-patent and two inventions, on MT-patent and two inventions.

6.Acknowlegment

Authors express deep gratitude to ECCM-8 Organizing Committee and International Science Foundation (ISF) for organization of publication of this work.

References

[1] Дьячковский Ф.Е., Новокшснова Л.А. Синтез и свойства полимеризационно -наполненных полиолефинов//Успехи химии.-1984.-Т.53,№2.-с.200-222.

[2] Pat.3390037 USA,Cl.156-148/Process for preparing preimpregnated strands of fibres and use of resulting products in making reinforced composites/Samuel H. Christie, Warren Township.

[3] Родин Ю.П., Молчанов Ю.М. Влиянис конформационных изменений, вызванных воздействием однородного постоянного магнитного поля, на процессы отверждения эпоксидной смолы//Механика композитных материалов.-1988.-№3-с.497-502.

Mechanics of the Squeeze Flow of Planar Fibre Suspension Systems such as SMC and GMT

G. Kotsikos, A.G. Gibson

Centre for Composite Materials Engineering - University of Newcastle upon Tyne, Herschel Building, Newcastle upon Tyne, NE1 7RU, UK

Abstract

The squeeze flow mechanics of planar fibre suspension systems has been investigated by isothermal axisymmetric squeeze flow testing of GMT and SMC between parallel circular plates. Tests have been performed over a range of squeeze rates and the analysis is based on simple power law constitutive equations. A key part of the work is determining the relative contributions of shear and extensional flow to the overall squeezing force response. By developing a variational flow model the behaviour has been found to be predominantly extensional, with shear effects becoming important at very small plate separations. Also, a predictive model for the radial pressure distribution during squeezing has been developed, based on fibre suspension rheology. Since fibres are generally long or continuous, treating the material as a statistically homogeneous fluid does not yield accurate results in this respect. The paper will describe the squeeze flow technique and discuss the derivation and application of these models to experimental results on both GMT and SMC materials.

1. Introduction

Press mouldable polymer composites like sheet moulding compounds (SMCs) and glass mat thermoplastics (GMTs), with a structure similar to planar glass fibre suspensions, have achieved considerable commercial success. Both materials are processed by non isothermal compression moulding, although with different boundary conditions. Sheets of SMC, initially at room temperature, are shaped in a hot press at around 150°C (at this temperature the thermosetting resin curing reaction is triggered). GMT is preheated in an oven to melt the thermoplastic matrix and rapidly introduced in a cold press (approx. 60°C) for processing.
Despite their differences the flow characteristics of SMC and GMT are rather similar and both materials lack a reliable method for rheological characterisation [1,2,3,4,5]. In this work an attempt has been made understand the mechanics of squeeze flow of GMT and SMC by utilising squeeze flow between parallel plates method

2. Experimental

The materials investigated in this work were commercial GMT, chopped (50mm fibre length) and continuous glass fibre mat, supplied by Symalit AG and commercial SMC, 30% glass content by weight and 25-30mm fibre length supplied by Autopress Mouldings – UK. Both materials were supplied in the form of sheets (GMT: 3.8mm thick and SMC: 3mm thick). The apparatus used in this work consisted of two parallel circular plates, 150mm in diameter, mounted on a computer controlled universal testing machine. One of the plates incorporated pressure transducers at three radial positions to monitor the pressure distribution during squeezing Fig 1. The plates were heated by a set of 250W cartridge heaters inserted at various positions in the plate. The temperature of the plates was monitored via 5 thermocouples inserted at various points (and depths) around the plate circumference.

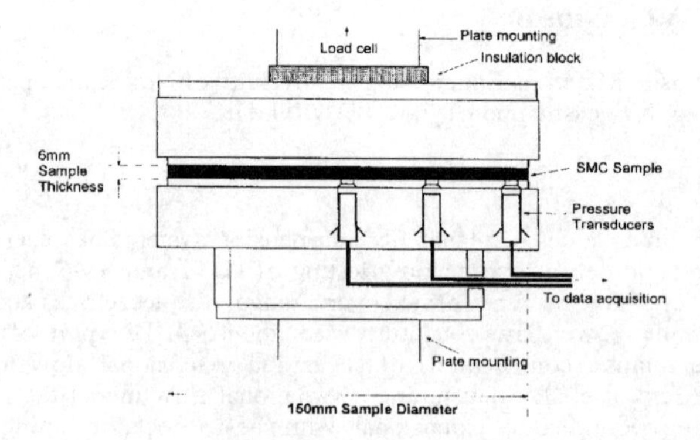

Fig 1. Squeeze flow die

A sample of GMT (or SMC) was placed between the plates and when temperature equilibrium was reached the plates were brought together at constant speed to 25% of the original plate separation (sample thickness). Although the commercial process is non-isothermal, for the sake of simplicity, the isothermal squeeze flow behaviour is considered in this work. Tests for GMT were carried out at 180°C and for SMC at 30 and 70°C. The pressure transducer readings, squeeze load and plate position for each test speed were fed to a data acquisition unit for subsequent analysis.

3. Theory

Squeeze flow has been used for many years as a way of studying high viscosity materials on account of its relative simplicity. In this work the analysis and experimental work was carried out on constant area squeeze flow. The analysis is based on simple power law expressions for shear and extensional type flows, which have been used in the past to model short fibre thermoplastic injection moulded materials [6,7] and the solution for the squeezing force due to pure biaxial extension is given by:

$$F_{ext} = \pi R^2 A_e(\varepsilon)\left(\frac{\dot{h}}{h}\right)^n \tag{1}$$

and for pure shear flow [8] by:

$$F_{shear} = \left(\frac{2m+1}{m}\right)^m \left(\frac{2\pi A_s R^{m+3}}{m+3}\right)\left(\frac{\dot{h}^m}{h^{2m+1}}\right) \tag{2}$$

The velocity and pressure distribution profiles associated with these types of flow are shown in Fig 2. For pure shear, the pressure varies across the plate with a maximum value at the centre dropping of to zero at the edges. For pure extension, the pressure is constant over the whole plate area. In most cases the flows are a mixture of shear and extension and have been successfully modelled in the past by pure additive relationships.

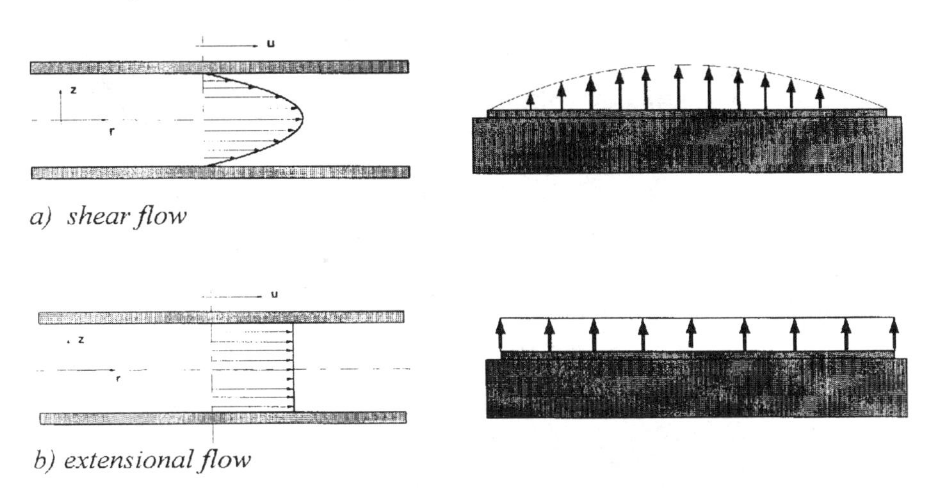

Fig 2. Velocity and pressure distribution profiles assiciated with a) shear flow and b) pure biaxial extension

Variational flow model

This is a more detailed approach to the summation model in which the shape of the velocity profiles during squeezing is allowed to vary continuously in order to determine the relative importance of shear and extensional dissipation. This approach has been used in the treatment of plasticity in metals and briefly states that any form of velocity field which satisfies continuity for a particular case, will cause energy dissipation greater than or equal to the actual dissipation. So the field which minimises the energy dissipation will be the actual field.

Although there have been some doubts on the applicability of this approach to metals, it is generally deemed suitable in the absence of inertia and memory effects. Applying this to the squeeze flow experiment gives a solution of the form shown below.

$$P_{squeeze} = \Phi_1 \cdot A_e \cdot \dot{\varepsilon}^n + \Phi_2 \cdot A_e \dot{\varepsilon}^n \cdot X \cdot \left(\frac{2}{n+3}\right) \cdot \left(\frac{2n+1}{n}\right)^n \quad (3)$$

Where

$$X = \frac{A_s}{A_e} \cdot \left(\frac{R}{h}\right)^{n+1}$$

and n is the power law index, which is determined from experiment. Here $\Phi 1$ and $\Phi 2$ are functions of X normally being close to unity and change with plate separation. The functions $\Phi 1$ and $\Phi 2$ are numerical integrals which are evaluated for a number of valid and kinematically admissible profiles, the actual solution for the overall pressure being the one that corresponds to the velocity field which minimises overall pressure. The actual solution can then be identified by an optimisation process.

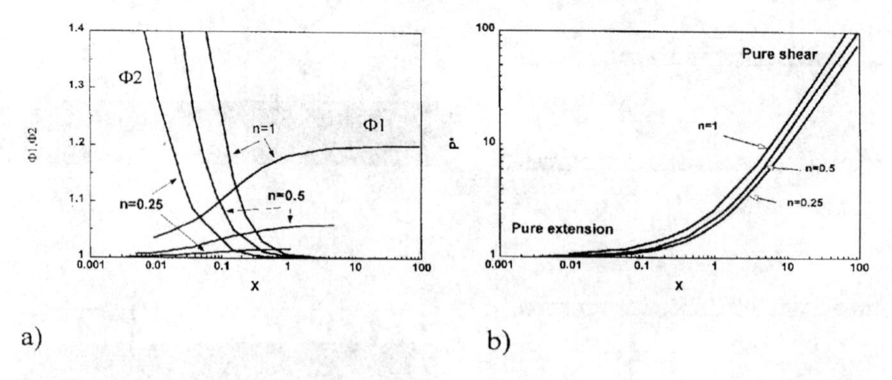

a) b)

Fig 3 a) variation of *Φ1* and *Φ2* with parameter *X* and b) variation of pressure *P** with parameter *X*

Equation (3) has two asymptotes one for a predicted behaviour of pure biaxial extension at low values of *X* and one for pure shear behaviour for large values of *X*, Fig 3. It can also be seen from Fig 3b, that the power law index *n* makes a very small difference to the overall solution.

4. Results

The flow behaviour at strain rates within the region of commercial interest can generally be characterised by a power law relationship between applied stress and strain rate. It can be seen that for both GMT and SMC the relationship between applied stress and strain rate in a log-log plot is linear, Fig 4a,b. with the presence of strain hardening [9].

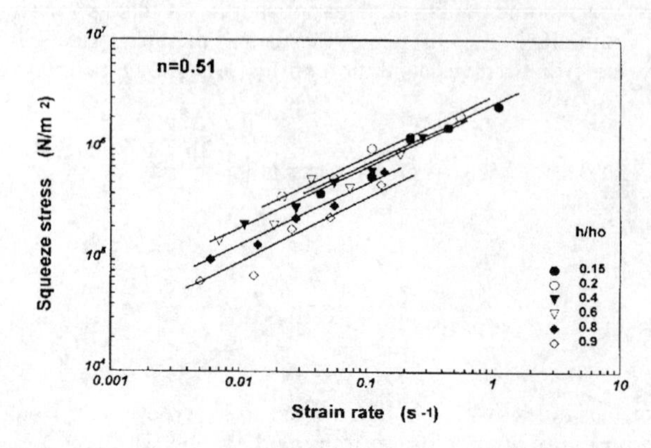

Fig 4. Plot of stress vs atrain rate during squeeze flow testing of a) commercial continuous fibre 40% glass content by weight GMT.

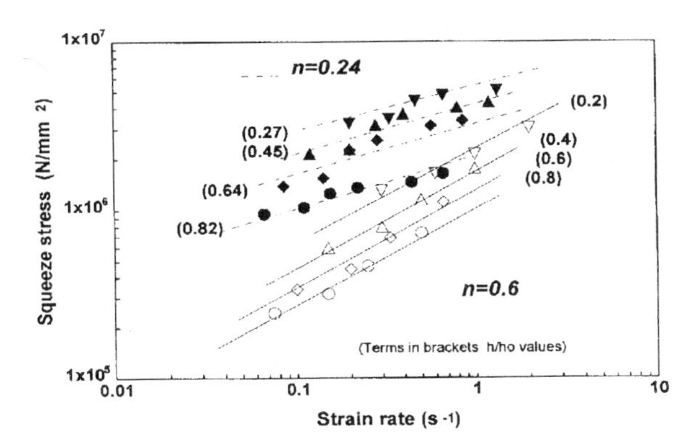

Fig 5. Plot of stress vs strain rate during squeeze flow testing of commercial 30% glass content by weight SMC.

Pressure transducer measurements, at all speeds tested, are shown in Figs 6 & 7. These show a parabolic profile with a maximum pressure in the centre dropping to near zero at the edges. This type of profile resembles that of simple shear flow of isotropic fluids.The pressure distribution for the chopped fibre GMT, shown on the same figure for comparison, has a similar distribution profile but the pressure levels are lower than the continuous fibre GMT, Fig 6. Calculations show that unreasonably high viscosities would be needed to produce such a strong pressure variation by shear flow suggesting the existence of another mechanism responsible for such a behaviour. The pressure distribution measurements for SMC do not show such a pronounced pressure variation but a flatter pressure distribution profile associated to an extensional flow regime Fig 7. The variational flow form analysis was carried out for for a squeeze rate of 1mm/s both for GMT and SMC and the results are shown in Fig 8a,b.

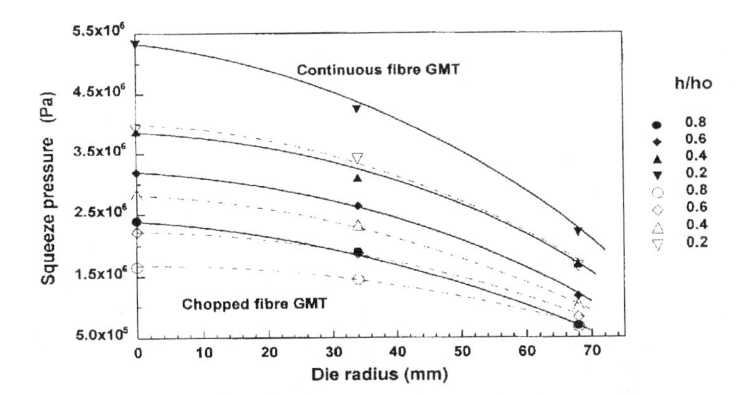

Fig 6. Pressure distribution profiles during squeeze flow testing of continuous (solid line) and chopped (dotted line) fibre GMT.

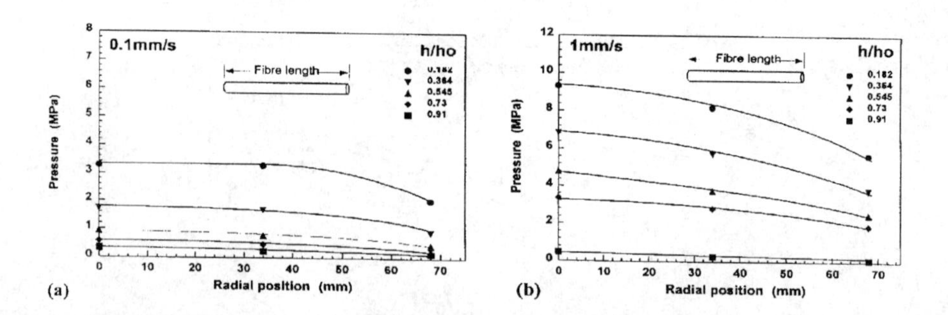

Fig. 7. Pressure distribution along plate radius during squeeze flow of SMC at 30°C.
a) 0.1 mm/s squeeze rate, b) 1 mm/s squeeze rate

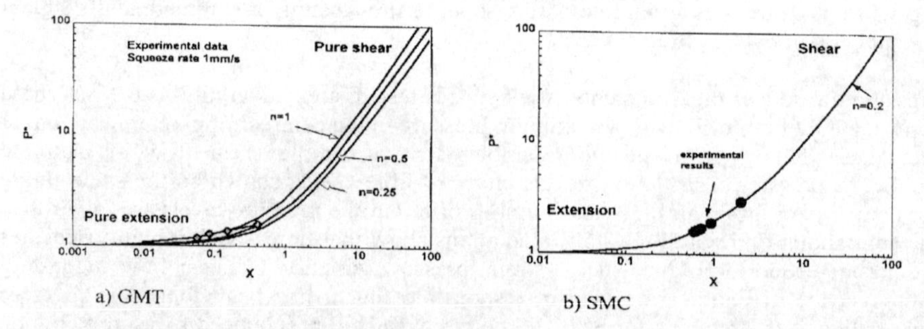

a) GMT b) SMC

Fig. 8. Variation of pressure P^* with parameter X as obtained from the upper bound theory analysis (solid lines) for a squeeze speed of 1mm/s. The experimental results for the 1mm/s squeeze speed tests are also shown (symbols).

5. Discussion

GMT tests

Modeling of the squeeze force results suggests that biaxial extension is the dominant flow form. However the data from the pressure distribution profiles during testing do not correspond to the flat pressure profile expected for biaxial extension. This would suggest that either shear flow is present or some other factor affects the pressure distribution profile.

Adopting the variational flow form model in these results it was found that the contribution from the shear component was very small during squeeze flow to explain the pressure distribution levels observed. Any shear flow effects were becoming significant only at very small plate separations, which anyway are irrelevant in the commercial processing of either GMT or SMC.

Looking closely at the structure of GMT it can be seen that in a circular disc sample, continuous random in plane fibres would vary in length according to their position and orientation in the sample. Thus fibres near the centre of the disc will be longer than samples close to the edges. This variation in fibre length and the motion of the fibres (or fibre bundles) in an extensional flow field may be connected to the pressure distribution profile observed. In fact Batchelor (10) has developed a model for the variation of the extending stress along a fibre, for fibre suspensions in Newtonian fluids which has also been extended to power law fluids (11).

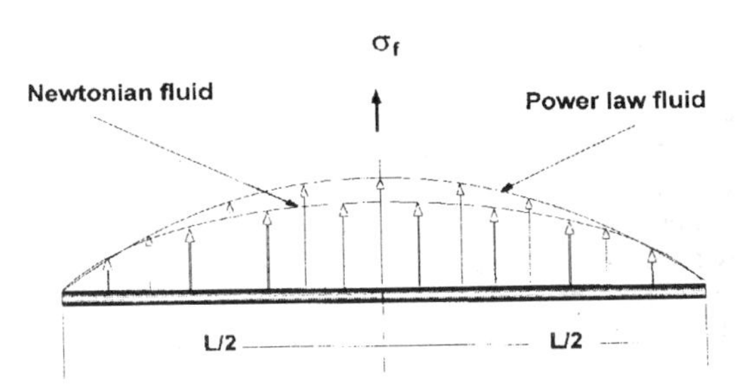

Fig 9. Variation of the extending stress along a fibre in an extensional flow field.

The variation of the extending stress along the length of the fibre, as predicted by these models, is as shown in Figure 9. The extending stress is a maximum at the mid-point of the fibre falling to zero at the fibre end. For the case of a circular sample, the stress contribution from each individual fibre at a given point will be directly related to the length of the fibre concerned and the distance of the considered point from the midpoint of the fibre. The total stress at that point may then be found by summing the separate contributions from all the fibres passing through that point. Integrating this expression over the disc radius gives the total stress variation over the disc. The resulting pressure distribution is similar to the one observed experimentally, Fig 10.

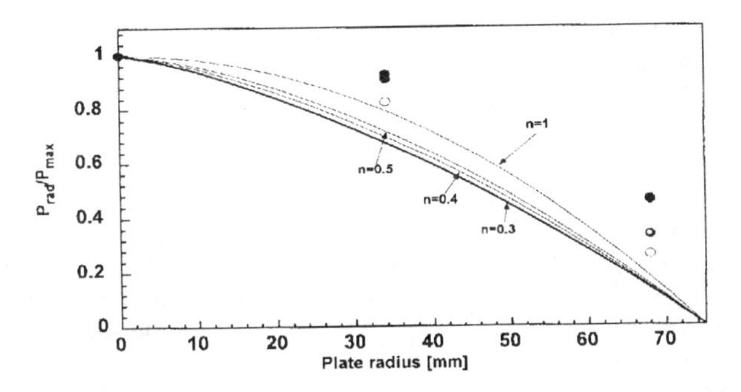

Fig 10. Pressure distribution profile as derived by the fibre stress model. The power law index n does not affect greatly the overall solution. Symbols indicate the experimental results for comparison.

This result is supported by the tests on chopped fibre GMT. As the fibres in this material are of a smaller length (50mm) there would be less variation in the pressure distribution as well as lower magnitude. The fibre length in the chopped fibre GMT tested is comparable to the die radius so as to show a dramatic difference in the resulting profiles. Nevertheless, the result can be treated with confidence as the reduced magnitude in the pressure profiles has been observed for all squeeze velocities.

SMC tests

The mechanisms involved in the moulding flow behaviour of SMC have been discussed in the literature. Barone & Caulk [3] have thus described the flow to be a combination of slip and uniform extension. This effect was seen to diminish with reduced charge thickness and disappeared for charges three layers thick. Lee, Marker and Griffith [2] carrying out similar tests described the behaviour of SMC to be a combination of biaxial extension and transverse shearing. In the present work the squeeze flow behaviour of SMC was expected to be different from the one described above. The material used in the aforementioned studies was model SMC which did not include the initiator so as not to allow the curing reaction to take place. It is not clear what rheological effect the presence of the initiator might introduce even in the moderate temperatures tested in this work. It can be seen that the relationship between squeeze stress and strain rate is linear with the presence of a strain hardening effect which has also been observed in GMT. Also the slope is 0.24 for the 30°C tests and 0.6 for the 70°C indicating the strong temperature dependence of the flow behaviour of SMC. The variational flow model results show that the relationship is predominantly extensional with shear becoming significant near the end of squeezing. The pressure distribution measurements support this finding as the profile is rather flat indicating biaxial flow. The variation near the plate edges is due to fibre stress edge effects which was described above for GMT. The fact that the fibre length in SMC is comparable to the plate radius allows for the profile to appear flatter than in the case of chopped fibre GMT.

6. Conclusions

Squeeze flow testing between parallel plates can be successfully used to characterise the moulding flow behaviour of planar fibre suspension systems such as fibre polypropylene matrix based commercial GMT and polyesteric commercial SMC. For both materials biaxial extension has been found to be the dominant flow form. Pressure distribution measurements during squeeze flow testing to investigate the presence of shear flow effects have shown a parabolic velocity profile which has been attributed to 'edge effects' arising from the fact that the fibres near the centre of the sample are longer than the ones near the edges. The longer fibres will contribute more to the overall stress than the shorter ones at the edges. This statement is supported by the pressure distribution measurements on SMC This effect has been modeled in continuous fibre GMT by using the variation of extending stress along the length of a single fibre in an extensional flow field. Summing the individual contributions from each fibre across the sample diameter gives a pressure distribution similar to the one measured experimentally in this work.

Squeeze flow testing of SMC. The squeeze flow behaviour of GMT is strongly influenced by temperature given that a modest increase of the test temperature from room temperature to 30°C can dramatically alter the behaviour of the material. The flow of SMC is found to be predominantly extensional. The upper bound theory can be successfully used to characterise the squeeze flow of SMC at 30 and 70°C.

References

[1]. *Silva-Nieto, R.J., Fisher, B.C. and Birley A.W.,* RHEOLOGICAL CHARACTERISATION OF UNSATURATED POLYESTER RESIN SMC, Polymer Engineering and Science, Vol. 21, No. 8, 1981, pp. 499-506.

Effects of Sizings and Reinforcement Levels on Processing and Performance of Composites

V. M. Karbhari* and L. Kabalnova*

* Division of Structural Engineering, University of California, San Diego, MC-0085
La Jolla, Ca 92093-0085, USA

Abstract

The increasing use of resin infusion type processes for the fabrication of large scale primary structural components in the marine and civil infrastructure areas using lower cost processes such as pultrusion, RTM, and resin infusion has generated a critical need for a fundamental understanding of mechanisms such micro- and macro-flow during infusion, and details of cure kinetics as related to the choice of sizing and finish. Although fiber sizings and loading levels have been studied at length to date, the primary emphasis has been on the investigation of effects related to bond and adhesion and performance of composites, rather than effects related to processing. The paper discusses advances in the understanding of phenomena occurring at the interface level related to cure kinetics, emphasizing the important role of sizings and loading levels in the use of such processes.

1. Introduction

Composites are increasingly being considered for large, low-cost applications such as in the marine and civil infrastructure area. Needs of these areas are increasing the use of low-cost processes such as SCRIMP, resin infusion, VA-RTM and pultrusion, which are capable of processing large structural parts with tailored preforms, without the use of autoclaves. However, the efficacy of such processing schemes lies in the ability to either continuously fabricate modular sections at high rates (such as in pultrusion) or infuse large sections in batch processes such as in resin infusion, without problems associated with incomplete wetting, viscosity changes and fabric impregnation/infusion, and of uncontrolled or non-uniform cure. Since cost-effectiveness is a major driver in such applications, the use of resin systems such as Vinylesters and low cost epoxies rather than the aerospace grade epoxies becomes a necessity. Further, because of stiffness driven applications, use will be made of carbon fibers either in traditional form (small tow size) or increasingly in the form of larger tow sizes (48K and above), or as part of glass-carbon hybrids. It is quickly becoming clear that materials and quality levels used in thin skin, sandwich type constructions used in the pleasure craft area if used for these applications, are unsuitable due to effects that can be related to the interphase formed between the fiber and the matrix through interactions between the fiber sizing and/or finish and the infusing resin system.

It is becoming increasingly clear that reinforcement-resin interactions affect two aspects of composite performance: (1) composite behavioral characteristics and (2) processing performance. In both cases, the interactions that occur on a microscopic level affect macroscopic behavior. The relationships among microscopic effects, macroscopic effects, composite behavioral characteristics, and processing performance are shown schematically in Figure 1. On the microscopic level, the behavior of a formed part is affected by the formation of interphase zones that can lead to variations in local material properties such as modulus and glass transition temperature. Such compositional and material property gradients affect microcracking and interfacial strength, which influence macroscopic characteristics such as strength and fracture toughness. Similarly, on a microscopic level, processing has been shown to be affected by reinforcement-resin interactions through the development of stoichiometric imbalances and compatibility effects, which influence wet-out and local flow behavior. Such microscopic effects in turn influence process performance by affecting cure kinetics (i.e., exotherms) and macroscopic flow (i.e., mold filling).

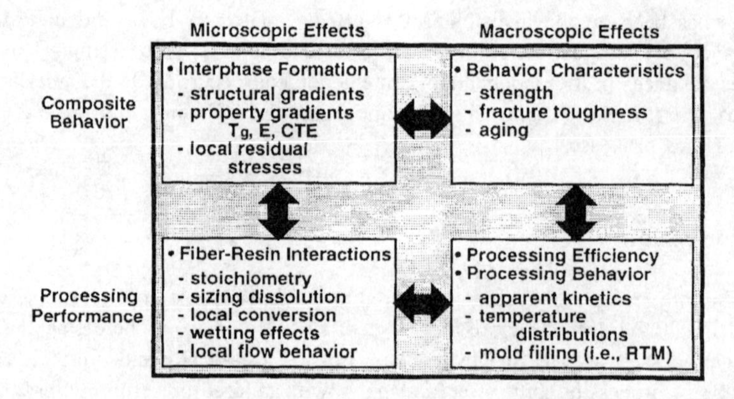

Figure 1: An Integrated Approach to Interphasial Effects

Although the interphase has been studied at length in the past, previous research by other groups has concentrated primarily on investigating performance issues in an isolated fashion, neglecting issues of processing that could bear on critical aspects related to both materials design and final part performance. This has resulted in the continuation of the field of sizings/surface interactions as a black art, rather than based on processing science. The effects associated with low strengths and moduli of carbon/vinylester systems especially in compression, and the nonuniform gelation in large resin-infused parts, amongst other problems, can be traced to this deficiency. In this paper we investigate the effect of carbon fibers, loading levels, and amount of sizing on the cure kinetics of a vinylester resin system. Comparisons are also made with two epoxy systems to emphasize the differences between the carbon-vinylester system and the carbon-epoxy systems.

2. Materials and Test Procedure

For the purposes of this investigation, two different carbon fibers, T300 and AS4 which have roughly equivalent mechanical performance levels were used. The T300 carbon was milled so as to enable good mixing at high loading levels for use in DSC investigations, whereas the AS4 carbon fiber was used with four different levels of sizing, i.e. unsized, 0.8%, 1.1-1.2%, and 1.3%. Three different resin systems were used so as to cover the range of available systems, i.e. a Dow Derakane 8084 Vinylester, an ambient temperature cure Shell 862 epoxy and an elevated temperature cure Shell 826 epoxy. The Derakane 8084 resin is an elastomer-modified vinylester in which the elastomer is reacted onto the backbone to enhance toughness. The Epon 826 resin is a low viscosity liquid bisphenol-A based epoxy, whereas the Epon 862 is a liquid bisphenol-F and epichlorohydrin based epoxy. All specimens were tested using dynamic scanning calorimetry. In order to investigate the effect of sizing content on rheological behavior including flow, a parallel investigation is underway using the AS4 fibers.

3. Results and Discussion

Results of the isothermal thermograms of the unreinforced vinylester system tested over a range of temperatures are shown in Figure 2. It can be seen that increasing temperatures result in a shift to the left, i.e. a shorter time period to peak. A comparison of results achieved through the addition of milled carbon fibers at 70°C and 90°C can be seen from Figures 3 and 4 respectively. The addition of fibers results in an increase in degree of cure with the degree of cure increasing with loading level as is listed in Table 1, and as shown in Figure 5.

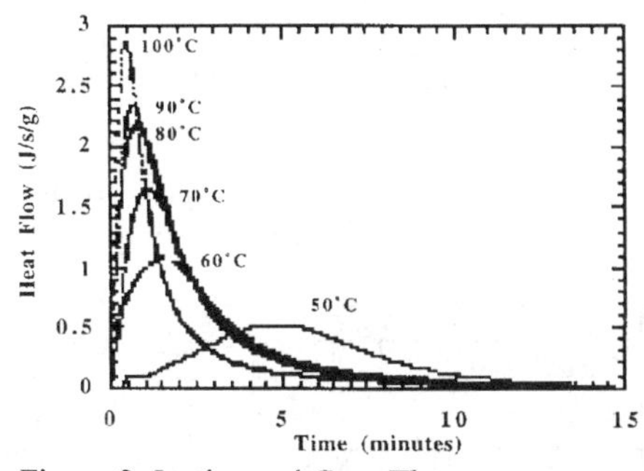

Figure 2: Isothermal Cure Thermograms for Unreinforced 8084 Dow Derakane Vinylester Resin

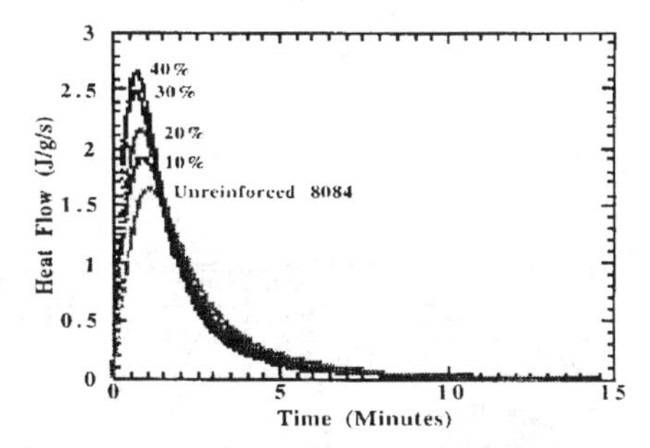

Figure 3: Heat flow as a Function of Loading Level (wt %) of Milled Carbon Fiber to 8084 Vinylester at 70°C

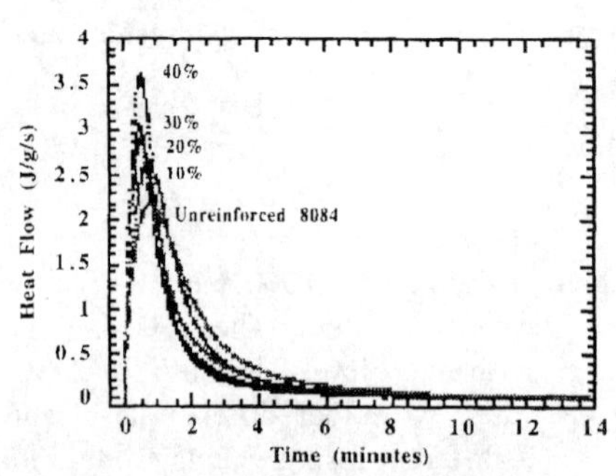

Figure 4: Heat flow as a Function of Loading Level (wt %) of Milled Carbon Fiber to 8084 Vinylester at 90°C

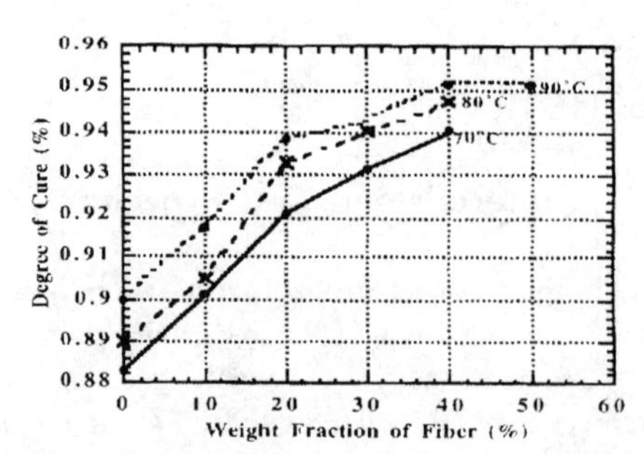

Figure 5: Degree of Cure as a Function of Loading Level (wt %) of Milled Carbon and Temperature

Further the peak in the isothermal thermograms is also seen to increase, as well as occur earlier with increase in loading level. As seen in Figure 5, the addition of reinforcement can not only result in a higher degree of cure, but also at lower temperatures. Reinforcement content and local variations can thus be expected to be very important in the curing of large and thick section parts wherein as seen above, the presence of fibers in increasing percentages acts similar to an accelerator. The use of neat resin cure kinetics for the cure modeling and prediction of such composites should hence be done with significant care, since effects can be critical as shown through changes in values of m (an exponent in the second order auto-catalytic kinetic equation, wherein $m+n = 2$), and in the rate constant, k. It can also be seen that the value of the rate constant is not only dependent on the temperature, but also on the loading level, with increases due to loading levels being significant and larger than those due to changes in temperature. A change in temperature level from 70° C to 90° C results in a 55% increase in the rate constant for the neat resin, whereas the addition of 40 wt. % of milled carbon fibers results in increases of 128%, 126% and 106% over the levels for the unreinforced resin at 70°, 80° and 90°C, respectively.

Table 1: Comparison of Cure Characteristics for the Milled T300 Carbon Fiber - Dow Derakane Vinylester System

Loading Wt. (%)	70°C			80°C			90°C		
	α (%)	m	k (1/min)	α (%)	m	k (1/min)	α (%)	m	k (1/min)
0	0.88	0.4243	0.843	0.89	0.3840	1.023	0.90	0.3249	1.306
10	0.90	0.5590	1.220	0.91	0.4330	1.475	0.92	0.4517	1.912
20	0.92	0.6080	1.260	0.93	0.4837	1.586	0.94	0.4321	1.956
30	0.93	0.5610	1.480	0.94	0.5126	1.744	0.94	0.4817	2.413
40	0.94	0.6150	1.920	0.95	0.5905	2.317	0.95	0.4658	2.687

These effects are especially important in the fabrication of large length or thick section parts using processes such as resin infusion which are largely uncontrolled, and have significant potential for the encapsulation of heat sinks through the use of cores, or areas with metal inserts, wherein local hot spots would be expected.

It is important to realize that the effects are not only generated by the presence of the fiber, but also through the type and amount of sizing that is present. The reaction products formed through chemical interactions between the sizing and the resin has been noted through past studies to alter the nature of the polymeric resin material in the vicinity of the fiber surface and more significantly affect cure and flow response [1,2]. The effect of the use of different levels of sizing on a Dow Derakane 8084 Vinylester system loaded with 20 wt % of AS4 carbon fiber is shown in Figure 6. It can be seen that the peak level of heat generation as well as the total amount of heat increase with increase in percentage of sizing. It is noted that the effect of increase in the percentage of sizing using the same fiber loading can be significant, with the peak rate of heat evolution being as much as 84% greater for the 1.3% sizing sample than the base line level for the unreinforced resin system.

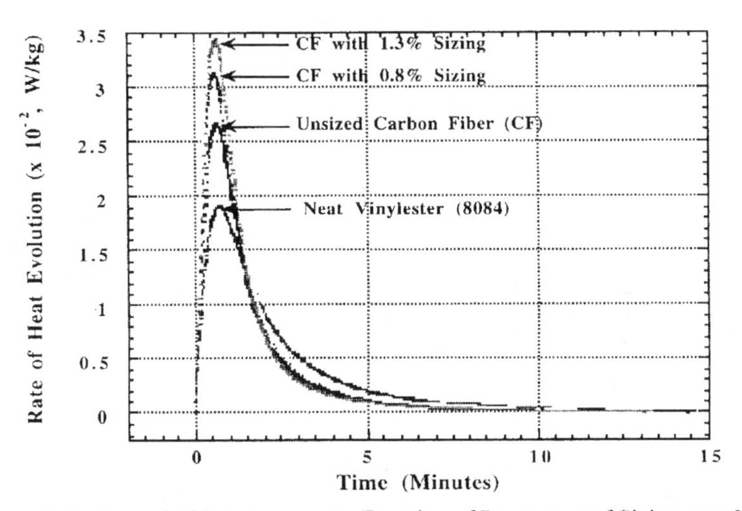

Figure 6: Isothermal Thermograms as a Function of Percentage of Sizing on a 20% Weight Fraction AS4/Dow Derakane Vinylester System at 80°C

Figures 7 and 8 show the effect of the sizing level on heat flow for two different epoxy systems, Shell Epon 826 and Shell Epon 828 at 120°C, reinforced with 20 wt. % of AS4 fibers with varying levels of sizing. It can be seen that in the former the effect of level of sizing is clear, whereas in the latter there appears to be only a marginal difference in peak levels of heat flow with changes in sizing level. In fact in the case considered in Figure 8, there is insignificant change in response between levels of 0 and 0.8% of sizing. This effect could be related to the fact that between the two resin systems, the 862 is rated for ambient cure, whereas the 826 is an elevated cure resin. However, a comparison with the results for the ambient cure Vinylester system described earlier suggest that the relationship is not as simple, but could in fact be related to surface interaction effects.

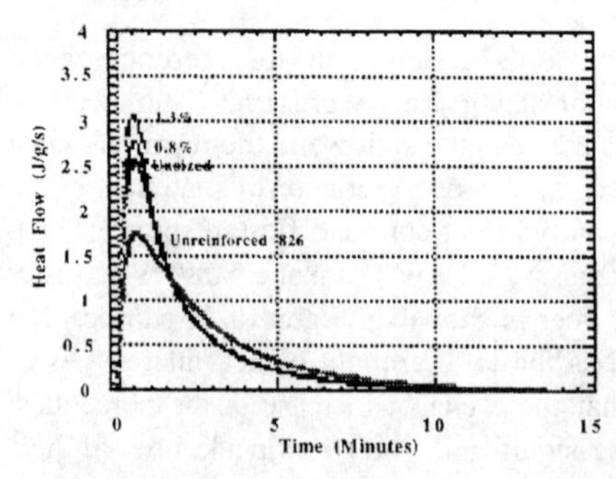

Figure 7: Effect of Sizing level for Epon 826 Resin System at 120 °C

Figure 8: Effect of Sizing level for Epon 862 Resin System at 120 °C

Table 2: Comparison of Rate Constant, k (1/min) as a Function of Resin System and Level of Sizing at 20 wt % Fiber Loading at 120°C

Resin System	Unreinforced	Reinforced (20 wt %)			
		Unsized	0.8%	1.1-1.2%	1.3%
862	1.366	1.441	1.450	1.439	1.433
826	0.869	1.189	1.194	1.202	1.205

The comparison shown in Table 2 shows that the rate constant is more sensitive to change at 120°C for the 826 system, than for the 862 system, wherein the change in rate constant between the unreinforced system and the 20 wt. % reinforced system at the 1.3% sizing level is only about 5% which is well within scatter bounds expectable in experimentation. It is however interesting to note that there is a strong dependence on temperature as shown in Table 3.

Table 3: Effect of Temperature and Sizing Level on Rate Constant, k (1/min)

Resin System	Temperature °C	Unreinforced	Reinforced (20 wt. %)			
			Unsized	0.8%	1.1-1.2%	1.3%
862	80	0.237	0.254	0.260	0.265	0.272
	100	0.661	0.710	0.734	0.732	0.744
	120	1.366	1.441	1.450	1.439	1.433
826	110	0.640	0.769	0.850	0.857	0.875
	120	0.869	1.189	1.194	1.202	1.205

4. Summary

It is shown that the presence of carbon fibers can substantially increase the rate of cure with the rate of cure being faster both with an increase in loading level, and an increase in the percentage (i.e. thickness) of sizing. Whereas this is not surprising for some systems, it is of considerable significance in the development of a processing science base for the use of low-cost processes such as resin infusion for the fabrication of large (and/or thick) components such as being developed for use in the civil infrastructure and marine areas. These effects are seen to not only affect part performance (as has been noted previously) but also affect processing through cure kinetics and rheological aspects. The effects need to be carefully considered in the use of process models as well as in the design of these parts. Further it is envisaged that advantage could be taken of these effects in the tailoring of preform structures for optimum processing through sizings and surface finishes, not just at the level of type of chemistry used, but also through selective use of levels of sizing/finish through the preform depending on position as a function of flow path and part thickness or cure regime.

Acknowledgment

The support of the National Science Foundation CAREER program through award # 9702560 is gratefully acknowledged.

References

[1] V.M. Karbhari and G.R. Palmese. Sizing Related Kinetic and Flow Considerations in the Resin Infusion of composites, Journal of materials Science, 32, 5761-5774 (1997).

[2] G.R. Palmese and V.M. Karbhari, Effects of Sizing on Microscopic Flow in Resin Transfer Molding, Polymer composites, 16[4], 313-318 (1995).

Comparison of Techniques for the Machining of Thermoplastic Fibre Reinforced Plastics

F. Klocke, C. Würtz
Fraunhofer Institute of Production Technology IPT
Steinbachstr. 17, 52074 Aachen, Germany

in cooperation with

J. van Lindert, J. Bauder
Bond-Laminates GmbH
Weigheimerstr. 11, 78647 Trossingen, Germany

Abstract

The fact that thermoplastic fibre reinforced plastics can be near-net shaped and that they can be recycled make them an important alternative to thermoset composite materials. The manufactured components will, however, usually require a certain degree of post-processing. The low thermal stability of the thermoplastic matrix materials reduces the cutting velocity which can be used during milling: the milling strategy must therefore be modified to suit the individual material. Alternatively, water jet or laser beam cutting processes can be used instead of milling. These techniques are very flexible in terms of the geometries which can be produced due to their narrow beam/jet geometry but certain material requirements must be fulfilled in order for them to be used for cutting. The aim is to select the suitable machining process by considering material-specific parameters. Further evaluation is carried out in terms of economic and qualitative aspects.

1. Introduction

Transforming the developments in the field of materials technology into industrial practice is highly dependent on the availability and the economics of the necessary manufacturing processes. New materials such as fibre reinforced plastics (FRPs) offer designers superior properties which enable specific component requirements to be met. The implementation of concepts such as these requires that one is very aware of the limitations of the manufacturing technologies when planning the implementation of new materials.

Despite the large number of papers on the subjects of FRP machining and the production of the end contours of FRP components, the current state of materials development, particularly in the case of fibre reinforced thermoplastics, can generally only offer even more expensive and complicated manufacturing solutions. In the past, this has often led to a lower than expected turnover [1]. Applications of fibre reinforced plastics are therefore mainly found in those fields in which their advantageous properties, e.g. high weight-related strength and/or stiffness or temperature-related behaviour, compensate for the high manufacturing costs.

One reason for the problems mentioned above lies in the lack of knowledge available on this group of materials: experience from metal machining processes cannot be directly applied to FRPs. This in turn is partly based on the fact that the term *composite material* covers a multitude of possible fibre and matrix combinations which each need to be machined in a different way, depending on the particular application. The fundamental knowledge about cutting mechanisms or the selection of suitable machine tools as well as the process control, the quality criteria or health and safety aspects must all be considered.

The machining process itself is frequently quoted as an obstacle to using these materials in functional components as it often represents a significant cost factor in the production chain. The main machining process used during manufacturing is milling; less frequently used operations are drilling and turning. New technologies such as laser beam, water jet and abrasive water jet cutting processes are being brought into use, particularly for the purpose of contour machining.

The aim of this study is to optimise the utility of these milling, laser beam and water jet technologies and the quality of the achieved cut when machining glass fibre reinforced thermoplastics. The material in question is produced by the Bond-Laminates GmbH in Germany.

2. Milling of Glass Fibre Reinforced Thermoplastics

Optimising the FRP milling process is largely influenced by (a) the fibres which must be cut in and near to the top layers and (b) the thermal stress on the contact surface. This second factor, which is caused by the substantial abrasiveness and the friction between material and tool, directly effects the quality of the cut surface. The quality indicators in milling thermoplastic composites are influenced by thermal stress which, in turn, are determined by the plastic matrix. The abrasive nature of the glass fibres, which is inevitably reflected in the mechanical properties of the laminate, means that fraying is not as problematic here as it is in the case of carbon fibre reinforced composites. High cutting velocities combined with low feeds lead to the matrix melting. Molten matrix can then end up stuck to the surface of the laminate or to the chips of GFRP which have already been cut off. If thicker materials or larger cutting depths cause this stress to increase, the matrix could even burn, (Fig. 1). A significant reduction in cutting velocity does remedy this problem to a certain extent. An increase in feed reduces the period of contact between the cutting edge and the laminate thereby lowering the thermal stress. At this point, the quality reaches the lowest permissible level with inadequately cut fibres (tufts) remaining attached to the surface layers.

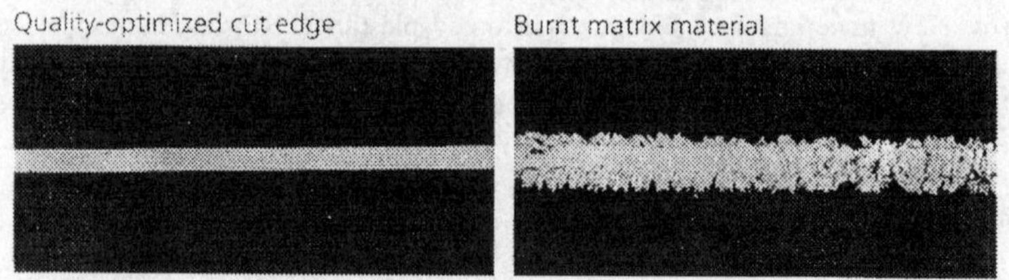

Figure 1: Quality achieved when machining GFRP laminates

The machining quality is evident, not only from the visible fraying on the cut edge or the molten/burnt surface, but also from the quantitative measurement of the surface roughness R_z. The values for GFRP also reflect the ability to endure thermal stress. The roughness of the GFRP cut edge is approx. 16 μm. One can expect that an increase in cutting velocity, cutting depth and material thickness will lead to poorer surface quality, regardless of the matrix material. The roughness increases form 5 to approx. 20 μm with increases in cutting velocity or cutting depth (Fig. 2). When thick laminates are machined, the process dynamics caused by system vibrations become weaker: the roughness values deteriorate by only 2μm, despite the fact that thicker laminates cause significantly higher mechanical stress.

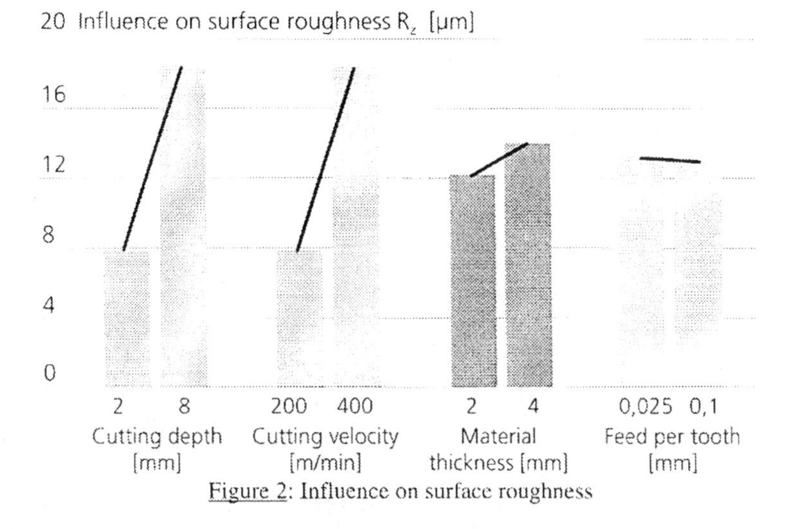

Figure 2: Influence on surface roughness

In comparison to carbide cutting materials, PCD tools do not in general alter the resulting machined quality. The roughness measurements also range between 5 and 6μm. The machining of polyamide with PCD tools has a slightly positive effect on surface quality: the roughness values also lie between 5 and 6 μm. The fact that these values remain more or less constant over longer milled lengths can be traced back to the longer PCD tool life compared to carbide. The cut edge shows signs of fraying at the surface layer at an earlier stage than with carbide tools. This can be attributed to initial rounding of the cutting edge, a typical characteristic of PCD tools, which occurs relatively early on in the tool life. Further milling, however, results in a constant level of quality. This phenomenon arises later in the case of carbide tools and immediately signifies the end of the useful tool life.

The end of the tool life is marked by a slight worsening in surface roughness (approximately 4 - 6 μm, see Fig. 3). Fraying in the top layers is a much clearer indicator of the end of the tool life as, at this point, it is no longer possible to remove it by wiping the edge by hand.

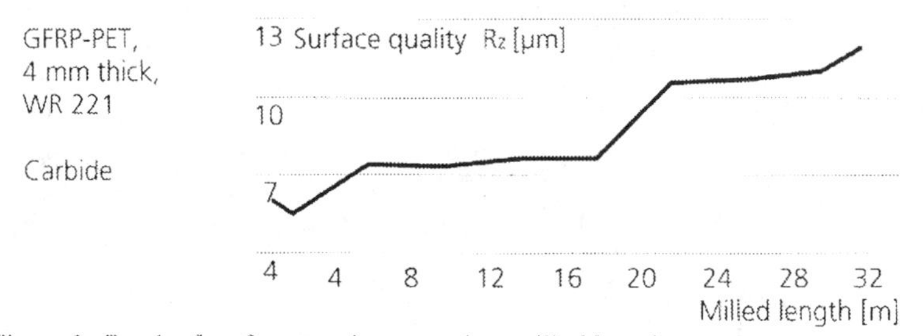

Figure 3: Graph of surface roughness against milled length

3. Water Jet Cutting of Glass Fibre Reinforced Plastics

The seam angle of the cut edge and the surface roughness are both used as evaluation criteria for the quality of the machining process. The quality criteria in turn are mainly affected by the feed rate - the most significant parameter for the process [2]. An increase in feed rate leads to a drastic deterioration in surface roughness and an increase in the seam angle for all laminate thicknesses. This is due to the jet energy introduced at various points into the workpiece. The absorbed energy reduces as the feed rate increases resulting in poorer surface quality. In the worst case, the energy is just sufficient to separate the laminate but not enough to produce a *quality cut*. The best results in terms of the seam geometry and the quality of the cut surface of glass fibre reinforced composites are achieved at low feed rates, regardless of the laminate thickness. This relationship is based on the amount of energy introduced into the workpiece material. The seam angle values when machining using an abrasive water jet are around 5° for thin laminates and only around 2° for thicker laminates.

The amount of abrasive material which is added to the water jet in order to increase the cutting capacity, particularly in the case of thicker laminates, also has a relatively important effect on the quality of the cut. (Fig. 4). Pure water jet cutting is marked by a significant loss in the technically feasible feed rate and, because of this a poor cutting capacity, and can only be used for laminates thinner than 2 mm.

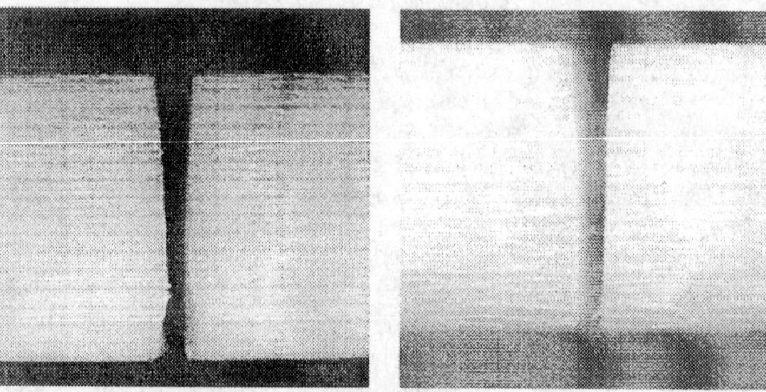

Figure 4: Qualities of the cut edge with 150 and 300 g/min use of abrasives

An assessment of the surface roughness confirms the conclusions regarding the influence of feed rate: the R_z values deteriorate with an increase in feed rate . The values worsen by an average of 15 µm, even with the use of abrasives (Fig. 5). Experience shows that this deterioration is most significant on the exit side of thicker laminates where the R_z value can change by 20 µm. The typical scoring inclined away from the feed direction can also have roughness peaks of up to 70 µm; on the jet entry side of the laminate, they are around 32 µm. In terms of the thinner laminates, changes of up to only 2 µm over the laminate cross-section have been measured: the influence of the feed rate on the results are therefore almost negligible.

The influence of jet pressure is stronger for thicker laminates than for thinner laminates. The machined result is largely independent of the pressure: these laminates should therefore be machined using low water pressure from a qualitative point of view.

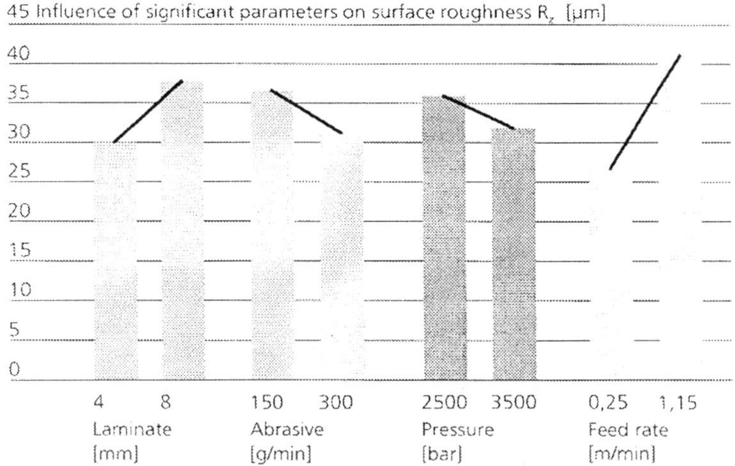

Figure 5: Effect of machining parameters on surface roughness

The selection of the nozzle diameter has a deciding influence on the results produced when machining GFRPs. A diameter of 0.15 mm allows for the best machined results in terms of the shape of the seam angle and the surface roughness. The use of a smaller diameter can largely compensate for a loss in quality at higher feed rates. This is, however, not possible for thicker laminates: abrasives must always be used and a smaller nozzle diameter can therefore not be utilised.

4. Laser Beam Cutting of Glass Fibre Reinforced Thermoplastics

Laser beams can be used on fibre reinforced plastic components for final processing. It is possible to trim, deburr and cut contours into components with large surface areas using laser beams, and the technique has certain benefits over conventional metal cutting machining processes such as milling. By focusing the laser beam, very high power densities and temperatures are generated at the processing point so that, in comparison to other machining techniques, very narrow/thin cutting seams and high feed rates can be achieved [3].

A very high energy density (> 10^5 W/cm^2) is achieved within the focused laser beam and the optical penetration depth is low for plastics as is the ability to dissipate heat into the surrounding material: relatively high temperatures and extremely high heating rates are therefore achieved in the cutting seam when processing plastics. As a result, a fusible phase is not usually generated during processing. Even when machining thermoplastics, the material only melts very slightly at the edge of the cutting seam and the material within the cutting seam almost entirely decomposes thermally.

For the processing of fibre reinforced plastics, one must note that the workpiece material is not homogeneous but composed of fibres and matrix materials with different thermophysical properties. This has a significant influence on processing and the results thereof. In this context, the width of the thermal damage at the cutting surface margin represents an important evaluation criteria for the purpose of assessing cutting quality.

Using light microscopy, one can divide the cutting surface margin into two basic zones: a very thin carbonised margin at the surface of the cut edge largely covered with evapourated material and, connected to it, a much wider heat-affected zone where most of the matrix material has evaporated, i.e. the matrix has reacted chemically, leaving only the fibres.

Using a laser beam to cut FRPs does not result in a homogenous cut surface margin. As the laminates are built up with the fibres aligned in two directions, the situation will arise where only matrix material or only fibres will be cut. The generally lower decomposition temperatures of the matrix materials compared to those of the fibres means that the fibres will protrude from the cut surface margin, i.e. the matrix material is forced back. This behaviour is particularly evident in the case of glass fibre reinforced composites. Figure 6 shows the cross-sectional photograph of GFRP cut using a laser beam.

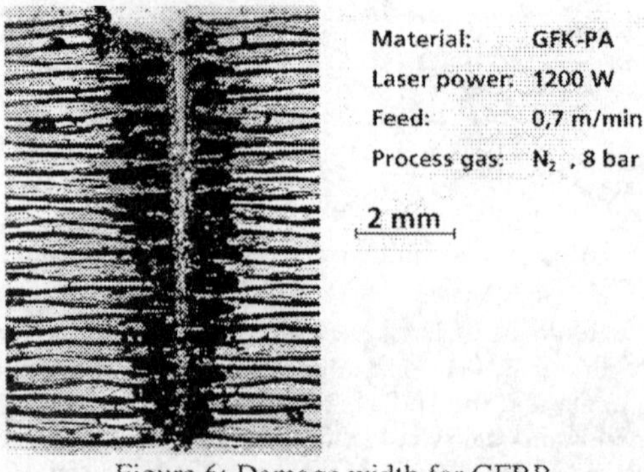

Figure 6: Damage width for GFRP

Figure 7 shows the threshold feed rates for glass fibre reinforced polyamide with varying laser beam power. For a thinner laminates, the material can already be cut at a relatively low laser power and very high feed rates of up to 8 m/min. With increasing material thickness, the threshold feed rate is significantly lower even with a large increase in laser power.

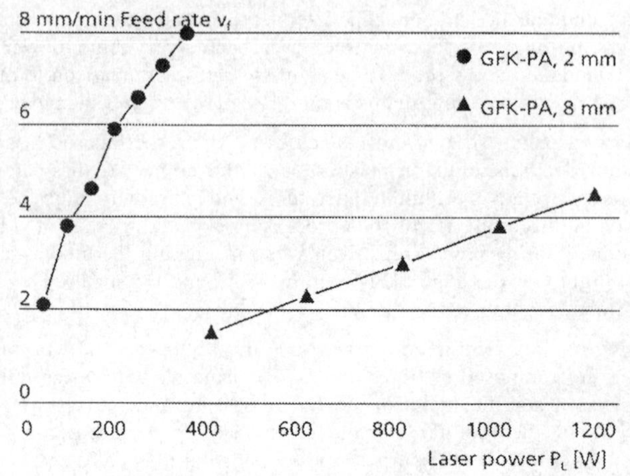

Figure 7: Dependence of threshold feed rate on material thickness and laser power for glass fibre reinforced polyamid

5. Comparision of the Machining Techniques

A comparison of these FRP machining techniques shows that the beam/jet cutting processes are more flexible than milling processes in terms of the geometries which can be produced because of the narrow beam/jet. The use of small tool diameters is limited because of the thermal instability of the matrix and the high process heat generated during milling, particularly in the case of the fibre reinforced thermoplastics investigated here. Milling equipment is, however, cheaper. Both water jet and laser beam cutting processes are associated with high plant costs which affect the specific processing costs. The high fixed costs can be compensated for by the high flexibility of these techniques as already mentioned as well as through higher processing speeds (up to 10 m/min depending on the laminate thickness). Such high speeds can not be used during milling processes due to the thermal instability of the matrix material. Good quality cuts are, however, associated with higher costs for the beam and jet techniques.

All the techniques discussed here present various environmental problems. Despite the GFRP matrix being made of a ductile material, one must always assume that the milling process will result in the emission of fine dust particles which can be inhaled. The laser beam cutting process is also associated with the emission of very fine soot particles and solid emissions are an important factor to consider when designing the plant. The emission of gaseous substances, which is also linked to laser beam cutting, must be controlled (extracted) by supplementary equipment. Emissions are less significant in the case of water jet cutting: disposal of the particles which have been removed from the workpiece during the machining process is more of a financial problem, as they are suspended or mixed in with the water [4]. This waste material as well as the original abrasive material must be separated out using several filters and then disposed of appropriately.

The use of any one of these techniques cannot simply be considered in terms of the quality of the resultant cut: as already mentioned, the machining process is affected by various factors which in turn affect both the quality and the cost effectiveness of the technique. Both material and component-related parameters must be taken into consideration. It is therefore necessary to adapt the machining technique to suit the individual component geometry.

References

[1] Rummenhöller, S., Werkstofforientierte Prozeßauslegung für das Fräsen kohlenstoff-faserverstärkter Kunststoffe, Doctoral Thesis at the RWTH Aachen, 1996, Germany

[2] Schmelzer, M. P., Mechanismen der Strahlerzeugung beim Wasser-Abrasiv-strahlschneiden, Doctoral Thesis at the RWTH Aachen, 1994, Germany

[3] Trasser, F.-J.; Laserstrahlschneiden von Verbundkunststoffen, Doctoral Thesis at the RWTH Aachen, 1992, Germany

[4] N.N.; Demonstrationszentrum für Faserverbundkunststoffe, Ergebnisse der Begleitforschung, Gutachten des DIW, Deutsches Institut für Wirtschaftsforschung, Berlin, 1994

The influence of injection pressure on the permeability of unidirectional fibre preform in RTM

I. Crivelli Visconti, A. Langella and M. Durante

Department of Materials and Production Engineering-University of Naples "FedericoII"
P.le Tecchio, 80-80125 Naples-Italy

Abstract

Resin transfer moulding is a process to produce fibre reinforced polymer structures having high performance characteristics with a low cost of equipment. Resin transfer moulding is a closed mould process in which a dry reinforcement preform is impregnated with a reactive polymer.
The permeability of the "bed of fibres" is a very important parameter which indicates all the interactions between the flow of the fluid and the fibre preform architecture.
This parameter is described by Kozeny-Carman relationship, but generally the permeability results in dependence of the particular conditions of the process.
In this work the permeability of the unidirectional fibres preform is valued in relationship to the different injection pressures and to the different fibre volumetric fraction.

1. Introduction

Resin transfer moulding (RTM) is a process to produce composite materials structures ranging from small to large articles and from simples to complex shapes. The advantages this process presents are high performance characteristics of the products and low costs of equipment.
Resin transfer moulding is a closed mould process in which a preplaced, dry reinforcement preform is impregnated with a reactive polymer in a transfer process. After the mould is filled, time is allowed for polymerization and when the part has gained sufficient strength it is demolded.
Reinforcement content can be varied in function of reinforcement type and orientation. The lower limit of reinforcement content should be maintained at or above a minimum level that will fill the entire cross section of the part in order to avoid resin rich surfaces or corners that may crack because of excessive resin shrinkage if unreinforced. Maximum reinforcement contents up to approximately 80% in weight are possible but the injection time will be long. Levels of random reinforcement content will be lower than those for oriented reinforcement because of the lower packing efficiency of the fibres.

When the polymer, generally a two component resin system mixed in a static mixer, is metered into the mould through a runner system, the air inside the closed mould cavity is displaced by the advancing resin front and escapes through vents located at the high or last areas of the mould to fill. When the mould has filled the vents and the resin inlet are closed and the resin within the mould cures.
In RTM the process conditions are determined empirically with moulding experiments, that make process development expensive and processing parameters difficult to optimize. Some of these parameters are associated with fluid flow in porous media.
The one dimensional flow through a porous bed was modelled by Darcy. Darcy's law relates the linear volumetric flow rate q to a coefficient depending by the permeability k, the pressure drop Δp, the bed length L, the area σ of the section of the bed and the viscosity of the fluid μ through the equation

$$q=k\Delta P\sigma/\mu L$$

Kozeny [1] extended this equation introducing the hydraulic radius concept. He suggested that the relationship between the flow rate q and the average velocity u in a capillary should be :

$$q=(1-V_f)\sigma u$$

where Vf is the reinforcement volume fraction.
Carman [2] introduced a correction in this relationship including the tortuosity term k_1:

$$q\,k_1=(1-V_f)\,\sigma u$$

Then the permeability in the direction along to the fibres of an unidirectional reinforcement with periodic arrangement of the fibres can be written as

$$K=\frac{R^2}{c}\frac{(1-V_f)^3}{V_f^{\,2}}$$

where R is the radius of the single fibre , Vf is the volumetric fraction of fibres and c is a constant depending by fibre arrangement.
But this relationship may only be applied if fibres have either uniform or truly random packing. In reality placing the dry preform of reinforcement into the mould, the space fibre distribution is uneven and furthermore the distribution can be partially distorted by the movement of flow front at elevated values of injection pressures. These values of the pressure are necessary to obtain complex products and to reduce the mould fill time from which the cost of the process depends.
In this work the permeability of unidirectional fibre preform was valued in function of the different volumetric fibres fraction and of the injection pressure. The injection pressures used are included into a range from 1 Bar to 5 Bar. In this condition of flow it has to be considered the deformation of preform and the channelling of the flow

between the clusters of fibre due to the relation between macro- and micro-flow at low volume fibre fraction.
The macroflow [3] develops between fibre bundles or yarns and usually controls mould filling and elimination of large-scale dry spots. The microflow develops between individual fibres inside fibre bundles or yarns and controls the elimination of microporisity and quality of reinforcement-resin interface. At low flowrate and high volumetric fibre fraction the microflow was leading whereas at high flowrate and low volumetric fibre fraction the macroflow was leading.
Some researchers[4] used a specific hydraulic radius to model the effect of variations in the reinforcement architecture on the flow rate. The flow rate in a clustered array of fibres was predicted to be significantly greater than for uniform distribution of fibres at the same volume fraction. Thirion et al. [5] have shown that the linear flow rate was more rapid in commercial fabrics when clustered flow enhancing tows were present at same volumetric fibre fraction.
But the presence [6] of uneven fibre distribution may lead to degradation of the mechanical properties of continuous fibre reinforced laminates.
Some studies [7] have shown that voids form inside the tows of a preform at the moving front of the fluid during the injection, and that it occurs because the length scale that governs the microflow impregnating the tows is smaller than the scale governing the movement of flow front.
In this work a large range of values of volume percentage of fibres has been considered starting from a value of 0.4 to as high value as 0.6, analysing the permeability in different conditions in which the macroflow or microflow prevails respectively.

2. Experimental materials and apparatus

The tests have been conducted using a mould made assembling three plates. A rectangular window was obtained on the central plate.
The other two plates were placed on and down respect to the central plate to close the mould. These plates have a big thickness (2 cm) and are blocked along the perimeter of the mould with a metallic flange to detect and prevent problems due to deflection of the mould during the injection of the fluid. The plates material was Lexan to permit observation of the flow front.
The seal between the plates was assured by a cloth rubber packing with 0.1mm of thickness.
Calibrated transducers were mounted in three positions on the down plate of the mould, located at the extremes and in the middle point of the bed of reinforcement.
The transducers were connected to an acquisition system to measure and to storage the pressures of the flow in porous media with time. The flowrate was measured with weighing method.
On the down plate two hole were made at the extremes of the mould to allow the way in and way out of the fluid into the mould. The pressure of the fluid coming in the mould was measured by a manometer.

In this work the reinforcement was unidirectional fibre glass type MEE 250, the fluid injected in the mould was glycol.
The tests have been conducted with different volumetric fraction of fibre : 40%, 45%, 50%, 55%, 60%; injecting the glycol at different constant pressures: 1,2,3,4,5 bar. The tests at 60% of volumetric fibre fraction were conducted only with injection pressure of 3,4,5 bar.
Having measured the flowrate Q crossing the media and the pressure gradient ΔP of the flow in three point where the transducers were placed, the permeability of the bed of reinforcement was calculated using Darcy's law.

3. Tests and results

For every type of test characterized by injection pressure and volumetric fibre fraction, three specimen were used.
In fig.1 the data of permeability, as an average of three results, are reported in function of volumetric fibre fraction. It is possible to note the permeability decreases remarkably as the volumetric fibre fraction increases.
The differences in permeability are very strong at the 40% in volume of fibre for different values of injection pressures.

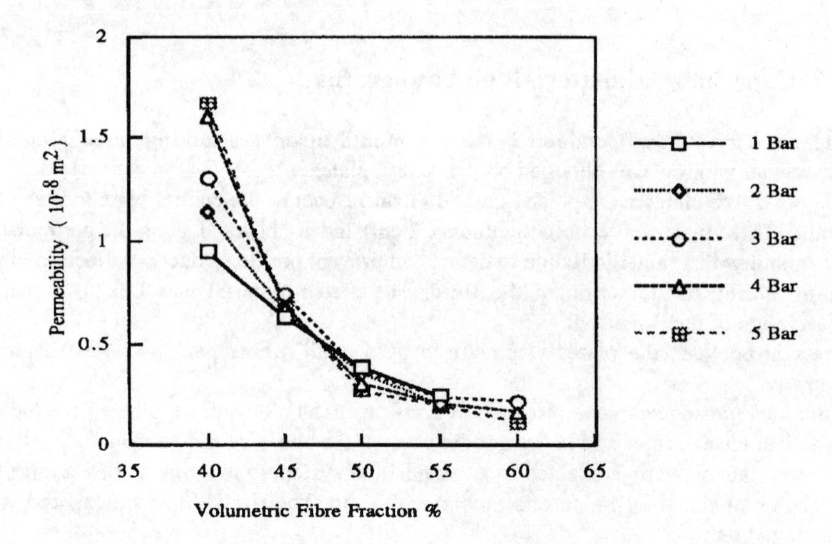

Fig.1 Permeability in function of volumetric fibre fraction for different injection pressures.

These differences decrease as the volumetric fibre fraction increases; it is possible to see easily this in fig.2 where the values of permeability in function of the injection pressure are reported for different volumetric fraction.
The fig.2 shows also that the permeability is constant for different pressures at 45% in volume of fibre, but for high values of volume fibre fraction the permeability decreases lightly as the injection pressure increases. To explain this it is necessary to consider the relationship between the values of permeability respect to the direction of the flow impregnating the tow.

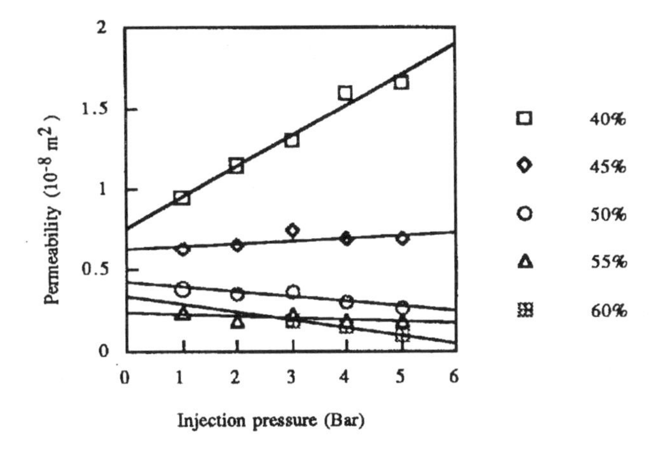

Fig.2 Permeability in function of injection pressures for different volumetric fibre pressures.

When the flow impregnates the tows in radial direction the permeability K_2 is smaller than the permeability K_3 due to a flow parallel to the fibre direction.
Then at low volumetric fibre fraction and high injection pressure the impregnation of the tows is in radial direction but the flowrate related to the macroflow in macroporosities is more important than the flowrate of radial impregnation, while at high volumetric fibre fraction and low pressure the impregnation is in direction parallel to the fibres, the microflow leads and the permeability tends to the value of K_3. But at high volume fibre fraction and high injection pressure the flowrate relative to macroflow decreases because the diameter of macroporosities decreases and the radial permeability makes the total permeability to decrease at values smaller than the permeability in direction parallel to the fibres relative to the microflow.
It is possible to confirm these effects in fig.3 where the values of pressure inside the reinforcement bed measured by the central transducer are reported in function of the time.

After having achieved the maximum value the pressure decreased suddenly for a radial impregnation of the tows, successively it increased softly perhaps because a microflow parallel to the fibres arises, and it became stable to intermediate value. The phenomenon is more strong as the injection pressure increased.

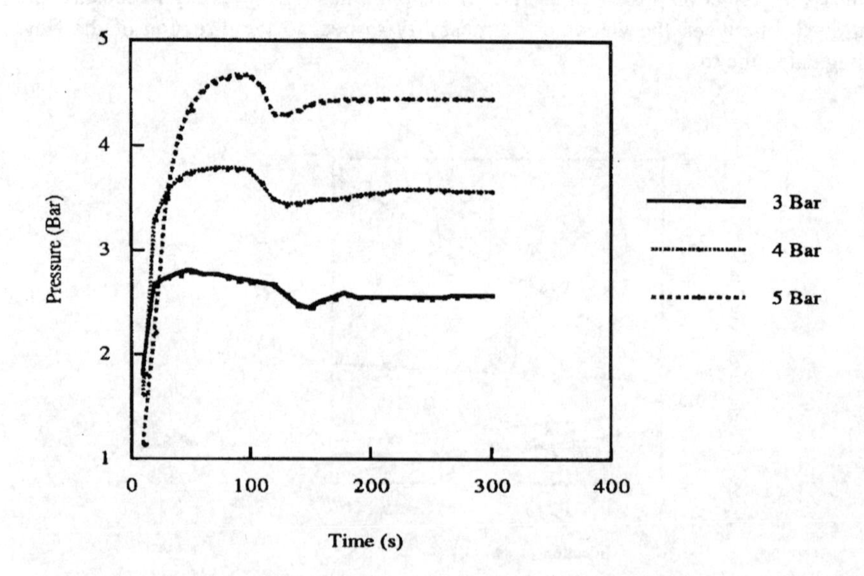

Fig. 3 Pressures measured by central transducer in function of the time at 50% in volume of fibres.

At the end as the volumetric fibre fraction increases, the section of macroporosities between the tows could vary along the mould axis for the imperfect arrangement of the fibres and furthermore the macroporosities could be interrupted by some clusters of fibres displaced in direction not parallel to the axis of the preform. In these conditions the macroflow could be pushed to cross the fibres transversely. It could be the cause of the strong decrement of permeability for high injection pressures at 60% in volume of fibre.

4. Conclusions

The tests conducted in this work show an effect of channelling for low value of volumetric fibre fraction with an increase in permeability, exalted increasing the injection pressure. For high values of volumetric fibre fraction, this effect is not important because the diameters of the macroporosities are small, but in this conditions

another effect becomes interesting: a soft decrease of the permeability as the injection pressure increases.
It could be explained with the presence of a transversal flow due to a macroflow generated by high pressure impregnating the tow in radial direction. In this condition the ratio between flowrate crossing the fibres in transversal direction and flowrate crossing the macroporosities is relevant respect with the same ratio in condition of flow at low volumetric fibre fraction and high pressure.
The small value of permeability in the case of transversal flow involves a lower value of the total permeability than the value due to inner microflow to the tow.
As the injection pressure decreases microflow becomes important then the flow is split in micro and macroflow impregnating the fibres in parallel direction.

REFERENCES

[1] *J. Kozeny* 1927 Sitzungsberichte Akademie der Wissenshaft, Wien, Abt. IIa, 136: 271-306

[2] *P.C Carman.* 1938 J.Soc. Ind. Trans., 57: 225-234.

[3] *G.Lekakou, M.G. Bader* Composites 29(1) 1998: 29-37

[4] *J. Summerscales* Composites Manufacturing 1993 4(1) 413-419

[5] *J.M Thirion., H Girardy. And U. Waldvogel* Composites (Paris) 1988, 28(3), 81-84

[6] *D. M Basford., P. R Griffin., S. M Grove. and T .Summerscales,* Composites 26 1995 675-679

[7] *R.S. Parnas, A.J Salem., T. Sadiq, Hsin Peng Wang, G. Suresh Advani,* 1994, Composites Structures 27: 93-107

ECCM-8

EUROPEAN CONFERENCE ON COMPOSITE MATERIALS

SCIENCE, TECHNOLOGIES AND APPLICATIONS

AUTHOR'S INDEX

Finito di stampare
nel mese di maggio 1998
nella Tipolitografia
R. ESPOSITO
Via Diocleziano, 154 - Napoli - Italia

Woodhead Publishing Ltd
Abington Hall
Abington
Cambridge CB1 6AH
England

ISBN 1 85573 408 7 (Vol 2)
ISBN 1 85573 377 3 (Four volume set)

WOODHEAD PUBLISHING LIMITED